Power:

$$1\ \textbf{hp} = 550\ \text{ft}\cdot\text{lbf/s} = 33\,000\ \text{ft}\cdot\text{lbf/min} = 2545\ \text{Btu/h} = 0.746\ \text{kW}$$

$$1\ \textbf{W} = \text{J/s} = \text{N}\cdot\text{m/s} = \text{volt}\cdot\text{ampere} = 1.34\times 10^{-3}\ \text{hp} = 0.239\ \text{cal/s} = 9.49\times 10^{-4}\ \text{Btu/s}$$

Pressure:

$$1\ \textbf{atm} = 101.3\ \text{kPa} = 1.013\ \text{bar} = 14.696\ \text{lbf/in.}^2 = 33.89\ \text{ft of water} = 29.92\ \text{inches of mercury} = 1.033\ \text{kgf/cm}^2 = 10.33\ \text{m of water} = 760\ \text{mm of mercury} = 760\ \text{torr}$$

$$1\ \textbf{psi} = \text{atm}/14.696 = 6.89\ \text{kPa} = 0.0689\ \text{bar} = 27.7\ \text{in. H}_2\text{O} = 51.7\ \text{torr}$$

$$1\ \textbf{Pa} = \text{N/m}^2 = \text{kg/m}\cdot\text{s}^2 = 10^{-5}\ \text{bar} = 1.450\times 10^{-4}\ \text{lbf/in.}^2 = 0.0075\ \text{torr} = 0.0040\ \text{in. H}_2\text{O}$$

$$1\ \textbf{bar} = 10^5\ \text{Pa} = 0.987\ \text{atm} = 14.5\ \text{psia}$$

Psia, psig:

Psia means pounds per square inch, absolute. Psig means pounds per square inch, gauge, i.e., above or below the local atmospheric pressure.

Viscosity:

$$1\ \textbf{cp} = 0.01\ \text{poise} = 0.01\ \text{g/cm}\cdot\text{s} = 0.001\ \text{kg/m}\cdot\text{s} = 0.001\ \text{Pa}\cdot\text{s} = 6.72\times 10^{-4}\ \text{lbm/ft}\cdot\text{s} = 2.42\ \text{lbm/ft}\cdot\text{h} = 2.09\times 10^{-5}\ \text{lbf}\cdot\text{s/ft}^2 = 0.01\ \text{dyne}\cdot\text{s/cm}^2$$

Kinematic viscosity:

$$1\ \textbf{cs} = 0.01\ \text{stoke} = 0.01\ \text{cm}^2\text{/s} = 10^{-6}\ \text{m}^2\text{/s} = 1\ \text{cp/(g/cm}^3) = 1.08\times 10^{-5}\ \text{ft}^2\text{/s} = \text{cp/(62.4 lbm/ft}^3)$$

Temperature:

$$\textbf{K} = {}^\circ\text{C} + 273.15 = {}^\circ\text{R}/1.8 \approx {}^\circ\text{C} + 273 \qquad {}^\circ\textbf{C} = ({}^\circ\text{F} - 32)/1.8$$

$${}^\circ\textbf{R} = {}^\circ\text{F} + 459.67 = 1.8\ \text{K} \approx {}^\circ\text{F} + 460 \qquad {}^\circ\textbf{F} = 1.8{}^\circ\text{C} + 32$$

Concentration (ppm):

In the air pollution literature and in this book, ppm applied to a gas always means parts per million by volume or by mol. These are identical for an ideal gas, and practically identical for most gases of air pollution interest at 1 atm pressure. Ppm applied to a liquid or solid means parts per million by mass.

For perfect gases at 1 atm and 25°C, 1 ppm = (40.87 · molecular weight) μg/m^3

Common Units and Values for Problems and Examples:

See inside back cover.

AIR POLLUTION CONTROL ENGINEERING

McGraw-Hill Chemical Engineering Series

Bailey and Ollis: *Biochemical Engineering Fundamentals*
Bennett and Myers: *Momentum, Heat and Mass Transfer*
Carberry: *Chemical and Catalytic Reaction Engineering*
Coughanowr: *Process Systems Analysis and Control*
de Nevers: *Air Pollution Control Engineering*
de Nevers: *Fluid Mechanics for Chemical Engineers*
Douglas: *Conceptual Design of Chemical Processes*
Edgar and Himmelblau: *Optimization of Chemical Processes*
Gates, Katzer, and Schuit: *Chemistry of Catalytic Processes*
Gupta and Kumar: *Fundamentals of Polymers*
King: *Separation Processes*
Luyben: *Essentials of Process Control*
Luyben: *Process Modeling, Simulation, and Control for Chemical Engineers*
Marlin: *Process Control: Designing Processes and Control Systems for Dynamic Performance*
McCabe, Smith, and Harriott: *Unit Operations of Chemical Engineering*
Middleman and Hochberg: *Process Engineering Analysis in Semiconductor Device Fabrication*
Perry and Green: *Perry's Chemical Engineers' Handbook*
Peters and Timmerhaus: *Plant Design and Economics for Chemical Engineers*
Reid, Prausnitz, and Poling: *Properties of Gases and Liquids*
Smith, Van Ness, and Abbott: *Introduction to Chemical Engineering Thermodynamics*
Treybal: *Mass Transfer Operations*
Wentz: *Hazardous Waste Management*

McGraw-Hill Series in Water Resources and Environmental Engineering

Consulting Editor

George Tchobanoglous, *Universtiy of California, Davis*

Bailey and Ollis: *Biochemical Engineering Fundamentals*
Bishop: *Pollution Prevention: Fundamentals and Practice*
Canter: *Environmental Impact Assessment*
Chanlett: *Environmental Protection*
Chapra: *Surface Water-Quality Modeling*
Chow, Maidment, and Mays: *Applied Hydrology*
Crites and Tchobanoglous: *Small and Decentralized Wastewater Management Systems*
Davis and Cornwell: *Introduction of Environmental Engineering*
de Nevers: *Air Pollution Control Engineering*
Eckenfelder: *Industrial Water Pollution Control*
Eweis, Ergas, Chang, and Schroeder: *Bioremediation Principles*
LaGrega, Buckingham, and Evans: *Hazardous Waste Management*
Linsley, Franzini, Freyberg, and Tchobanoglous: *Water Resources and Engineering*
McGhee: *Water Supply and Sewage*
Mays and Tung: *Hydrosystems Engineering and Management*
Metcalf & Eddy, Inc.: *Wastewater Engineering: Collection and Pumping of Wastewater*
Metcalf & Eddy, Inc.: *Wastewater Engineering: Treatment, Disposal, Reuse*
Peavy, Rowe, and Tchobanoglous: *Environmental Engineering*
Sawyer, McCarty, and Parkin: *Chemistry for Environmental Engineering*
Tchobanoglous, Theisen, and Vigil: *Integrated Solid Waste Management: Engineering Principles and Management Issues*
Wentz: *Hazardous Waste Management*
Wentz: *Safety, Health, and Environmental Protection*

AIR POLLUTION CONTROL ENGINEERING

Second Edition

Noel de Nevers
University of Utah

Boston, Massachusetts Burr Ridge, Illinois
Dubuque, Iowa Madison, Wisconsin New York, New York
San Francisco, California St. Louis, Missouri

McGraw-Hill Higher Education

*A Division of The **McGraw-Hill** Companies*

AIR POLLUTION CONTROL ENGINEERING

This book is printed on acid-free paper.

4 5 6 7 8 9 BKM BKM 0 9 8 7 6 5 4

ISBN 0-07-039367-2

Publisher: *Thomas Casson*
Executive editor: *Eric M. Munson*
Editorial coordinator: *Michael Jones*
Senior marketing manager: *John T. Wannemacher*
Senior project manager: *Beth Cigler*
Production supervisor: *Kari Geltemeyer*
Freelance design coordinator: *Laurie J. Entringer*
Cover illustration: *PhotoDisc*
Supplement coordinator: *Rose M. Range*
Compositor: *Techsetters, Inc.*
Typeface: *10/12 Times Roman*

Library of Congress Cataloging-in-Publication Data

de Nevers, Noel (date)
Air pollution control engineering / Noel de Nevers. – 2nd ed.
p. cm.—(McGraw-Hill series in water resources and environmental engineering) (McGraw-Hill chemical engineering series)
Includes index.
ISBN 0-07-039367-2
1. Air—Pollution. 2. Air—Purification—Equipment and supplies. I. Title. II. Series. III. Series: McGraw-Hill chemical engineering series.
TD883.D42 2000
628.5′3—dc21 99-29945

http://www.mhhe.com

ABOUT THE AUTHOR

Noel de Nevers received a B.S. from Stanford University in 1954, and M.S. and Ph.D. degrees from the University of Michigan in 1956 and 1959, all in chemical engineering.

He worked for the research arms of what is now called the Chevron Oil Company from 1958 to 1963 in the areas of chemical process development, chemical and refinery process design, and secondary recovery of petroleum. He has been on the faculty of the University of Utah from 1963 to the present in the Department of Chemical and Fuels Engineering.

He has worked for the National Reactor Testing Site, Idaho Falls, Idaho, on nuclear problems; for the U.S. Army Harry Diamond Laboratory, Washington, DC, on weapons; and for the Office of Air Programs of the U.S. EPA in Durham, NC, on air pollution.

He was a Fulbright student of Chemical Engineering at the Technical University of Karlsruhe, West Germany, in 1954–1955; a Fulbright lecturer on Air Pollution at the Universidad del Valle, in Cali, Colombia, in the summer of 1974; and then at the Universidad de la República, Montevideo, Uruguay, and at the Universidad Nacional Mar del Plata, Argentina, in autumn 1996.

He was a member of the Utah Air Conservation Committee (the state's air pollution control board) from 1972 to 1984 and its chair in 1983–1984. He served on the Utah Governor's Citizen Advisory Task Force on the Protection of Visibility in 1986, the Utah Legislature's Hazardous Waste Task Force in 1988, the Utah Governor's Clean Air Commission in 1989–1990, and the Western Governor's Conference Grand Canyon Visibility Transport Commission Citizen's Advisory Board, 1992–1996.

His areas of research and publication are in fluid mechanics, thermodynamics, air pollution, technology and society, energy and energy policy, and explosions and fires. He regularly consults on air pollution problems, explosions, and fires.

In 1991, his textbook, *Fluid Mechanics for Chemical Engineers*, Second Edition, was issued by McGraw-Hill.

In 1993 he received the Corcoran Award from the Chemical Engineering Division of the American Society for Engineering Education for the best paper (" 'Product in the Way' Processes") that year in *Chemical Engineering Education.*

In addition to his serious work he has three "de Nevers's Laws" in the latest "Murphy's Laws" compilation, and won the title "Poet Laureate of Jell-O Salad" at the Last Annual Jell-O Salad Festival in Salt Lake City in 1983.

CONTENTS

PREFACE

This book is intended for university seniors and graduate students who would like an overview of air pollution control engineering. It may be of value as a reference work to engineers who are professionally active in air pollution control, but they will probably find the treatment somewhat simpler and less detailed than their own personal experience. They may, however, find use for the treatment of areas in which they are not personally experienced.

About half of the book is devoted to control devices, their theory and practice. The other half is devoted to topics that form some of the background for the selection of such devices, e.g., air pollution effects, the structure of U.S. air pollution law, atmospheric models, etc. These topics interact strongly with the device selection and design, which is the reason for their inclusion.

I have tried to make the book direct and clear enough that an experienced engineer can read and understand any part of it without help. I have also tried to base it as completely as possible on the basic chemical engineering disciplines of stoichiometry, thermodynamics, fluid mechanics, heat transfer, mass transfer, and reaction kinetics so that senior students in chemical engineering will see that this is a field in which they can use all that they have previously learned. I have also tried to select the level of treatment so that any interested chemical engineering faculty member can teach a senior level course using the book (and the solutions manual) without requiring that the faculty member have a personal background in air pollution control engineering. The chemistry in this book is presented at a level corresponding to a background of one year of university chemistry because when I teach our course there are mechanical and civil engineering students present, who have that chemistry background.

I have been guided by two pedagogical maxims: "The three rules of teaching are, from the known to the unknown, from the simple to the complex, one step at a time," and "If you don't understand something at least two ways, you don't understand it." I have devoted more space and effort to determining numerical values of pertinent quantities than do most authors. I believe students need to develop a feel for how big? how fast? how hot? and how much?

In many areas of the book the treatment in the text is simple, with a more complex treatment outlined or discussed in one of the problems. Students are encouraged at least to read through all the problems, to see where more complex and complete treatments are either described or referred to. In many places in the book there are digressions not directly applicable to air pollution and problems not directly related to air pollution. Some of these are there because they show interesting related technical issues that do not apply directly to air pollution control. I include these because I think they help students build mental bridges to other parts of their personal experiences. The more the students are able to integrate the new information in this book into their existing knowledge base by such connections, the more likely they are to retain it and be able to use it.

I will be very grateful to readers who point out to me typographic errors, incorrect equation numbers, incorrect figure numbers, or simply errors of any kind. Such errors will be corrected in subsequent editions or printings. In the second edition I have tried to update those parts that change with time (e.g., regulations, atmospheric trends, control technology). I have added a few more examples and problems. There is some reorganization of topics, in response to reader suggestions. I thank all the students, faculty, and others who have pointed out errors or poor explanations in the first edition. Those who criticize you in a soft voice in private are your friends.

Noel de Nevers

NOTATION

Symbol	Brief description	Units	
		English	SI
A	coal ash content	wt %	wt %
A	area	ft^2	m^2
A	area of city = LW	ft^2	m^2
A	constant in Antoine equation	—	—
A	constant in Arrhenius equation (sometimes called "frequency factor")	1/s	1/s
A	constant in Cunningham correction factor	—	—
a	acceleration	ft/s^2	m/s^2
a	length parameter	1/ft	1/m
a	mass transfer area per unit volume	ft^2/ft^3	m^2/m^3
A, B, C	chemical species in reaction rate equations	—	—
A/F	air-fuel ratio	lbm/lbm	kg/kg
A, B, C, K	arbitrary constants	various	various
a, b	characteristic dimensions	—	—
a, b	polynomial coefficients	various	various
b	background concentration	not used	$\mu g/m^3$
b	time parameter	1/h	1/s
B, C	constants in Antoine equation	°R	°C or K
C	carbon content of fuel	wt %	wt %
C	Cunningham correction factor	—	—
C_d	drag coefficient	—	—
C_P	heat capacity at constant pressure	Btu/(lbm or lbmol) · °F	J/(kg or mol) · °C
C_V	heat capacity at constant volume	Btu/(lbmol or lbm) · °F	J/(kg or mol) · °C
c	concentration	(lbm or lbmol)/ft^3	(kg or mol)/m^3
$\mathcal{D}$	diffusivity	ft^2/s	m^2/s
D	diameter or particle diameter	ft	m
D_a or D_{pa}	aerodynamic diameter, or aerodynamic diameter of a drop or particle	not used	$\mu(g/cm^3)^{0.5}$

D_b	diameter of barrier	ft	m
D_{cut}	"cut diameter," the diameter at which the efficiency = 50%	ft	m
D_D	droplet diameter	ft	m
D_{mean}	mean particle diameter (arithmetic or logarithmic)	ft	m
D_o	outside diameter of a cyclone separator	ft	m
D_p	particle diameter	ft	m
E	electric field strength	V/ft	V/cm
E	excess air	lbmol/lbmol	mol/mol
E_A	activation energy	Btu/lbmol	kcal/mol
E_o	electric field strength where particles are charged	V/ft	V/cm
E_p	electric field strength where particles are collected	V/ft	V/cm
EF	emission factor	various	various
F	force	lbf	N
F	packing factor in flooding equation, or packing factor for absorbers	—	—
F_d	drag force	lbf	N
F_g	gravity force	lbf	N
f	fugacity (for ideal gases = partial pressure)	psia	Pa
f_s	saturated fugacity at this T ($\approx$ vapor pressure)	psia	Pa
G	Gibbs free energy	Btu/lbmol	J/mol
G	molar flow of nontransferred component in gas phase	lbmol/s	mol/s
G_m	gas molar mass velocity	$\text{lbmol/ft}^2 \cdot \text{s}$	$\text{mol/m}^2 \cdot \text{s}$
G'	gas mass velocity	$\text{lb/ft}^2 \cdot \text{s}$	$\text{kg/m}^2 \cdot \text{s}$
g	acceleration of gravity	ft/s^2	m/s^2
H	effective stack height	ft	m
H	height in the vertical direction, or the direction in which particles are collected	ft	m
H	Henry's law constant	atmospheres	Pa
H	humidity, lbm water/lbm dry air	—	—
H	hydrogen content of fuel	wt %	wt %
H	mixing height	ft	m
h	enthalpy or molar enthalpy	Btu/(lbm or lbmol)	J/(kg or mol)
h	height above floor in a gravity settler	ft	m
h	height of slit	ft	m
h	physical stack height	ft	m
Δh	plume rise	ft	m
j_m	mass transfer factor	—	—
K	coefficient in pressure drop equations	—	—
K	constant in Langmuir equation	1/atm	1/Pa
K	equilibrium constant	various	various
K	turbulent dispersion coefficient	$\text{ft}^2\text{/s}$	$\text{m}^2\text{/s}$
K	mass transfer coefficient	$\text{lbmol/ft}^2 \cdot \text{s}$	$\text{mol/m}^2 \cdot \text{s}$
K_p	equilibrium constant with activities in atm	various	various

k	Boltzmann constant = R/Avogadro's number	not used	1.38×10^{-23} kg · m^2/K · s^2
k	coefficient in modified Deutsch-Anderson equation	—	—
k	kinetic rate constant	various	various
k	permeability	ft^2	m^2
k	ratio of specific heats (C_P/C_V)	—	—
k	reaction velocity constant	1/s	1/s
k_f, k_b	forward and backward reaction rate constants	various	various
k'_g	mass transfer coefficient	lbmol/ft^2 · s	mol/m^2 · s
L	length	ft	m
L	length of city in downwind direction (in box models)	ft	m
L	length of collector in flow direction	ft	m
L	length of piston stroke	in.	m
L	mixing height (Fig. 6.9 only)	not used	m
L	molar flow of nontransferred component in liquid phase	lbmol/s	mol/s
L'	liquid mass velocity	lb/ft^2 · s	kg/m^2 · s
L_V	visual range constant	not used	km · µg/m^3
M	molecular weight	lbm/lbmol	g/mol
m	mass	lbm	kg
$\dot{m}$	mass flow rate	lbm/s	kg/s
N	nitrogen content of fuel	wt %	wt %
N	number of particles, or of people, or of turns in a cyclone separator	—	—
N	number, number of transfer units	—	—
N_D	rate of droplet flow	number/s	number/s
N_s	separation number = characteristic dimension/Stokes stopping number [see Sec. 8.2]	—	—
n	exponent in rate equation and Freundlich equation	—	—
n	age	year	year
n	distance in direction of interest in Gaussian plume derivation (Chapter 6)	ft	m
n	exponent in series expansion	—	—
n	number of mols	lbmol	mol
$\dot{n}$	molar flow rate	lbmol/s	mol/s
O	oxygen content of fuel	wt %	wt %
pH	negative $\log_{10}$ of the H^+ activity ($\approx$ concentration) expressed in mol/liter	—	—
P	gas pressure	psia or atmospheres	Pa or mb
Po	power	ft · lbf/s or hp	kW
p	penetration = 1 − collection efficiency	—	—
p	vapor pressure	psia	Pa
p_{water}	vapor pressure of liquid water	psia	Pa or mb
Q	emission rate	lbm/s	g/s
Q	volumetric flow rate = $V \cdot A$	ft^3/s	m^3/s
Q_G	gas volumetric flow rate	ft^3/s	m^3/s

Q_L	liquid volumetric flow rate	ft^3/s	m^3/s
q	charge on a particle	C	C
q	emission rate per unit area	$lbm/hr \cdot mi^2$	$g/s \cdot m^2$
$\mathcal{R}$	Reynolds number	—	—
$\mathcal{R}_p$	Reynolds number for particles	—	—
RH	Relative humidity $= \dfrac{\text{Humidity}}{\text{Saturation humidity}}$	—	—
R	universal gas constant (see Appendix A)	$psi \cdot ft^3/lbmol \cdot °R$	$N \cdot m/mol \cdot K$
r	radius	ft	m
r	reaction rate	various	various
S	sulfur content of fuel	wt %	wt %
Sc	Schmidt number	—	—
s	standard deviation	various	various
T	absolute temperature	°R	K
t	quench zone thickness	in.	m
t	thickness	ft	m
t	time	s	s
$t_{1/2}$	half-life	s	s
U	overall heat transfer coefficient	$Btu/h \cdot °F \cdot ft^2$	$W/m^2 \cdot K$
u	wind speed	ft/s	m/s
u	internal energy or molar internal energy	Btu/(lbm or lbmol)	J/(kg or mol)
V	voltage (or potential)	V	V
V	volume	ft^3	m^3
V	velocity	ft/s	m/s
V_{avg}	average gas velocity	ft/s	m/s
V_c	particle or gas velocity on a circular path	ft/s	m/s
V_D	drop velocity	ft/s	m/s
$V_{D\text{-fixed}}$	drop velocity relative to fixed coordinates	ft/s	m/s
V_G	gas velocity	ft/s	m/s
V_{rel}	relative velocity	ft/s	m/s
V_S	stack gas velocity	ft/s	m/s
V_S	superficial velocity	ft/s	m/s
V_t	terminal velocity	ft/s	m/s
W	mass of solids/(volume of gas × cake density)	—	—
W	width of a collecting device	ft	m
W	width of city	ft	m
w	drift velocity (in electrostatic precipitators)	ft/s	m/s
w	weight fraction	—	—
w	weight of a particle sample	lbm	kg
w^*	equilibrium amount adsorbed	lbm/lbm	kg/kg
[X]	activity or concentration of compound X	not used	atm, or mol/cm^3
X	molar humidity of air, mol water/mol dry air	—	—
X	amount emitted in Lagrangian Gaussian plume equations	lb	kg

X	liquid content of transferred component	lbmol/lbmol	mol/mol
x	distance	ft	m
x	independent variable	various	various
x	mol fraction in the liquid phase	—	—
x	mol number of carbon in hydrocarbon fuel	—	—
x	small quantity in series expansion	—	—
x_{mean}	mean value of independent variable	various	various
x, y	distance in x and y directions	ft	m
x, y	indices in hydrocarbon formulae, C_xH_y	—	—
x, y, z	coordinate directions or lengths	ft	m
Y	gas content of transferred component	lbmol/lbmol	mol/mol
y	mol fraction in gas or vapor	—	—
y	mol number of hydrogen in hydrocarbon fuel	—	—
y^*	equilibrium mol fraction	—	—
z	elevation or vertical distance	ft	m
z	number of standard deviations from mean, $= (x - x_{\text{mean}})/\sigma$ in the normal distribution, $= [\ln(D/D_{\text{mean}})]/\sigma$ in the log normal distribution.	—	—
α	constant defined by Eq. (12.17)	1/s	1/s
α	constant in Freundlich equation	mixed	mixed
α	filter medium resistance	ft	m
α	dummy variable in flooding equation	—	—
β	dummy variable in flooding equation	—	—
ε	dielectric constant	—	—
ε	porosity	—	—
ε_O	permittivity of free space	not used	8.85×10^{-12} C/V · m or 8.85×10^{-12} F/m
Φ	cumulative distribution function	—	—
ϕ	equivalence ratio	—	—
ϕ	latitude	deg	deg
η	efficiency	—	—
λ	latent heat of vaporization	Btu/lbm	J/kg
λ	mean free path	ft	m
λ	normalized A/F ratio	—	—
λ	wavelength of maximum emission	(never used)	μm
μ	micron or micrometer	not used	$= 10^{-6}$ m
μ	viscosity	cP	Pa · s
ν	kinematic viscosity $= \mu/\rho$	ft^2/s	m^2/s
ρ	density or molar density	(lbm or lbmol)/ft^3	(kg or mol)/m^3
ρ'_L	liquid density at normal boiling point	lbm/ft^3	kg/m^3
σ	$(\text{variance})^{0.5}$	various	various
σ	constant in Gaussian, or normal, distribution function	various	various
σ	Stefan-Boltzmann constant	Btu/hr · ft^2 · °R^4	W/m^2 · K^4
σ_y	horizontal dispersion coefficient	not used	m
σ_z	vertical dispersion coefficient	not used	m
Ψ	specific gravity in flooding equation	—	—
ω	angular velocity	radians/s	radians/s

CHAPTER 1

INTRODUCTION TO AIR POLLUTION CONTROL

Air pollution is the presence of undesirable material in air, in quantities large enough to produce harmful effects. This definition does not restrict air pollution to human causes, although we normally only talk about these. The undesirable materials may damage human health, vegetation, human property, or the global environment as well as create aesthetic insults in the form of brown or hazy air or unpleasant smells. Pollutants are known that may do all of these things. Many of these harmful materials enter the atmosphere from sources currently beyond human control. However, in the most densely inhabited parts of the globe, particularly in the industrialized countries, the principal sources of these pollutants are human activities. These activities are closely associated with our material standard of living. To *eliminate* these activities would cause such a drastic decrease in the standard of living that this action is seldom considered. The remedy proposed in most industrial countries is to continue the activities and control the air pollutant emissions from them.

1.1 SOME OF THE HISTORY OF AIR POLLUTION CONTROL IN THE UNITED STATES OF AMERICA

Although air pollution control actions go back at least as far as the thirteenth century [1],* most of the major effort in the world has taken place since 1945. Before then, other matters were higher on society's priority list (and are still higher in developing

*Numbers in brackets refer to items listed in the reference section at the end of each chapter.

countries). In the 1930s and 1940s, a factory smokestack issuing a thick plume of smoke was considered a sign of prosperity, and some government agencies included it in their official symbols.

Before 1945, industrial air pollution control efforts were directed at controlling large-factory emissions of pollutants that had led to conflict with neighbors of the factories. Much of this did not involve governmental action, but rather was a response to nuisance damage suits or the threat of such suits.

Between 1945 and 1969, as awareness of air pollution problems gradually increased, some worthwhile local efforts to control air pollution were initiated, notably in Pittsburgh, Los Angeles, and St. Louis. Between 1963 and 1967 the federal government began to oversee and coordinate local and state air pollution control efforts.

In 1969 and 1970, the United States experienced a great environmental awakening. Today's students may not realize how rapid or drastic a change that was. Compare some major newspapers from 1968 with the same papers from 1970. Environmental matters were scarcely mentioned in newspapers in 1968, but the same newspapers had an environmental story every day in 1970. This period saw the passage of the National Environmental Policy Act and the Clean Air Act of 1970, both of which have had sweeping effects and have greatly changed our way of dealing with air pollution. Similar changes took place throughout the industrial world at about the same time, with similar effects.

The sudden and sweeping change in air pollution law brought about by the Clean Air Act of 1970 came as a great surprise to most major American industries. At first the leaders of the older "smokestack" industries (steel, copper, some electric power) fought the new regulations, in the courts, in the press, and in Congress. Twenty-five years later their successors mostly have decided that the air pollution regulations are here to stay and that their goals should be to influence the regulatory process to make the regulations as clear and practical as possible and then to comply with the regulations in as efficient and economical a way as possible. The best of the industry leaders are always looking at the next generation of regulations so that when those regulations appear, they will be prepared for them and will not have to change what they did for today's generations of regulations. Most major industries try to be *at least* as well informed (and if possible *better* informed) on air pollution technical matters as any of the other participants in the regulatory process.

In the late 1980s, a new theme entered the air pollution arena: global air pollution. Until 1980, most air pollution problems were perceived as local problems. The pollutants of interest had short lifetimes in the atmosphere, or were emitted in such small quantities that they were not perceived as a problem far beyond the place from which they were emitted. Thus, it seemed logical to let local or state governments deal with them. (If a stinky factory provides jobs, the conflict between those who enjoy the economic benefits of the factory and those offended by its smell can be settled in a local election.) In the 1980s, three problems emerged involving longer-lived pollutants and pollutants that are transported a long way before they do their damage: acid rain, destruction of the ozone layer by chlorofluorocarbons, and the buildup of carbon dioxide in the atmosphere. The legal and administrative structure developed in the 1970s to deal with local air pollution problems seems useless to deal with these international or global problems. We shall return to these three problems in Chapter 14. The history in other countries has been similar to that in the United States; the industrial countries responded at about the same time and in about the same way as the United States. The developing countries responded later than the United States, and used a mixture of the ideas combined from the United States and the World Health Organization, which seek similar goals by somewhat different means.

1.2 WHY THE SUDDEN RISE IN INTEREST IN 1969–1970?

Why did air pollution awareness increase in 1969–1970? This is a subject for historical debate, but some of the reasons are obvious. A great deal of the anti-Vietnam war activism was diverted into the environmental arena quite suddenly. The communications media jumped on the bandwagon vigorously at about the same time that the Santa Barbara oil spill provided a visible example of pollution problems and attracted wide attention. There are certainly other causes.

Environmental concern is often considered a luxury only wealthy nations can afford, and the United States had become very wealthy. To people who are worried about their next meal or whether they will have a home or be able to pay for medical care, air pollution does not seem very important. To a person whose basic physical needs are satisfied, air pollution can be a much greater cause for concern. Certainly the people who participated in the environmental awakening were mostly upper middle class, including many college students. There were not many poor people involved, or many people who had lived through the Great Depression of the 1930s.*

Furthermore, when the principal cause of death was infectious disease such as influenza, tuberculosis, and typhoid fever, the effects of air pollution on health, which are slow and cumulative, were seldom observed. As we have learned to prevent or treat these diseases, we have doubled our average life span, surviving long enough to die of long-term diseases such as arteriosclerosis, heart malfunctions, stroke, emphysema, and cancer, all of which are related to environmental factors, including air pollution. The same observation can be made about cigarette smoking; before

*Although the environmental movement was mostly an activity of the upper middle classes, the poor are most often exposed to more severe air pollution (and other environmental insults) than are the rich. The highest concentrations of air pollutants are found in the central cities, where poor people live, not in suburbs where wealthier people live. The price of homes in Los Angeles is related to local air pollutant concentrations; those near the beaches or high on the foothills, where the air pollutant concentrations are lowest, normally command the highest prices. The same is true of industrial exposure; only poor people work in jobs with severe exposure to potentially harmful materials. This is also true of the location of unpleasant facilities; the slaughterhouse, landfill, and municipal incinerator are rarely located in rich neighborhoods.

we learned to treat these contagious diseases, smoking probably had little effect on overall life expectancy. Now that these other causes of death are practically gone, we live long enough that smoking has a real effect on life expectancy. So also with air pollution.

It is useful to contrast the air pollution situation, for which we have taken action so recently, with water pollution, for which we have had active programs for over a century. The worst water problems were caused by contamination of drinking water with human sewage. This quickly spreads cholera, typhoid, and amoebic dysentery. These diseases are sudden and dramatic in onset and often swiftly fatal. Their connection with polluted water is easily demonstrated. Thus, we responded to the water pollution problem much sooner and more vigorously than we have to the air pollution problem.

Evidence of the effects of air pollution on health (see Chapter 2) is much less dramatic than that for water pollution. One can seldom point to a pile of corpses and say, "They died of air pollution," as one can after a cholera outbreak due to polluted water. The effects are more like those of smoking; we seldom say, "He died of smoking," but we know that smoking has been shown to decrease the life expectancy of the smoker and to increase the incidence of certain well-defined illnesses in smokers and in those who breathe secondhand smoke. The fact that so many people—including educated people—smoke demonstrates that this type of argument is not as persuasive as the sight of the corpses after an epidemic spread by water pollution. Many people do not take very seriously the loss of life and health due to air pollution, like that due to smoking, because they believe it is "only statistical."

The effects of air pollution and of smoking are also analogous in that many people who have lived in badly air-polluted environments all of their lives have excellent lungs and hearts. Similarly, everyone knows someone who lived to be a vigorous 95 and smoked cigarettes or cigars every day. Those examples exist; the counterexamples died younger, of diseases caused or aggravated by air pollution or smoking.

Public awareness of air pollution developed at a period when the problem was less severe in many respects than it had been previously. Before the introduction of natural gas as the principal fuel in most U.S. cities, winter air was much dirtier with coal soot than it is now. Likewise, early in this century, the emissions of sulfur dioxide from copper smelting in cities such as Tacoma, Salt Lake City, El Paso, and Anaconda were much greater than they are now. At those times, there must have been dissatisfaction about these sources of pollution, but presumably not at the level we have had in the past few years.

This increase in awareness is partly explained by the increased wealth of the country, as mentioned before. We once thought these pollutants were necessary concomitants of a prosperous economy; we now know otherwise. Similarly, we once believed that nothing could be done about such problems. Now that we have learned to read the genetic code and put people on the moon, it is harder to argue that we cannot control air pollution. We can; this book explains the technical bases and some of the details of how to do it.

1.3 DIRTY AIR REMOVAL OR EMISSION CONTROL?

Example 1.1. The area of the Los Angeles basin is 4083 square miles. The heavily polluted air layer is assumed to be 2000 ft thick on average. One solution to Los Angeles' problems would be to pump this contaminated air away. Suppose that we wish to pump out the Los Angeles basin every day and that the air must be pumped 50 miles to the desert near Palm Springs. (We assume the residents of Palm Springs won't complain.) Assume also that the average velocity in the pipe is 40 ft/s. Estimate the required pipe diameter.*

The flow rate required is

$$Q = \frac{AH}{\Delta t} = \frac{4083 \text{ mi}^2 \cdot 2000 \text{ ft}}{24 \text{ h}} \cdot \frac{(5280 \text{ ft/mi})^2}{3600 \text{ s/h}} = 2.63 \times 10^9 \frac{\text{ft}^3}{\text{s}}$$

$$= 7.47 \times 10^7 \frac{\text{m}^3}{\text{s}}$$

and the required pipe diameter is

$$D = \sqrt{\frac{4Q}{\pi V}} = \sqrt{\frac{4 \times 2.63 \times 10^9 \text{ ft}^3/\text{s}}{\pi \times 40 \text{ ft/s}}} = 9158 \text{ ft} = 2791 \text{ m}$$ ■

This is about six times the height of the tallest man-made structure, and far beyond our current structural engineering capabilities. Similar calculations (Problem 1.1) show that the power required to drive the flow exceeds the amount of electrical power generated in the Los Angeles basin. We are unlikely to solve our air pollution problems by pumping away the polluted air, although this solution is still frequently proposed. Instead, we must deal with those problems by reducing emissions, the principal subject of the rest of this book.

1.4 ONE PROBLEM OR A FAMILY OF PROBLEMS?

In Table 1.1 we see emissions estimates for the major man-made pollutants for the United States in 1997. From this table, we see the following:

1. There are six individual pollutants listed, which are the major regulated pollutants in the United States. There is a much longer list of other pollutants, emitted in much lesser quantities and regulated in a different way in the United States (see Chapters 3 and 15).
2. Some of the pollutants come mostly from transportation (motor vehicles) and others come mostly from industrial sources.
3. There is no entry for "General air pollution." The public thinks in terms of "general air pollution" and wonders if the problem is mostly industry (them) or autos (us).

**Note:* The symbol ■ indicates the end of an example.

TABLE 1.1
National emissions estimates for 1997 (Values in millions of short tons/yr)

Source category	PM_{10}	SO_2	CO	NO_x	VOC	Pb
Transportation	0.7	1.4	67.0	11.6	7.7	0.00052
Fuel combustion	1.1	17.3	4.8	10.7	0.9	0.00050
Industrial processes	1.3	1.7	6.1	0.9	9.8	0.0029
Miscellaneous	—	0.0	9.6	0.3	0.8	—
Total	3.1	20.4	87.5	23.5	19.2	0.0039
Percentage of 1970 total	—	65%	78%	116%	70%	1.7%

PM_{10} = particulate matter, 10 μ or smaller; see Chapter 8. SO_2 = all sulfur oxides, mostly SO_2; see Chapter 11. CO = carbon monoxide; see Chapter 15. NO_x = all nitrogen oxides, mostly NO and NO_2. The mass shown is based on all NO being converted to NO_2; this is referred to as "NO_x expressed as NO_2"; see Chapter 12. VOC = volatile organic compounds; see Chapter 10. Pb = lead; see Chapter 15.

No value is shown for PM_{10} emissions as a fraction of 1970 emissions because no reliable estimate is available for PM_{10} emissions in 1970. Forest fires are the most important of the "Miscellaneous" sources, for most pollutants. This table contains no entry for O_3, which is a major pollutant, but which is almost entirely a secondary pollutant for which there are no major primary emission sources. VOC are listed not because they are directly harmful to human health, but because they are a major primary precursor of secondary O_3.

Source: Ref. 2.

Engineers recognize that there is not *one* air pollution problem but a group of related problems, and that some of the problems are mostly caused by industry and others are mostly caused by motor vehicles. The public and many politicians hope to find a simple, one-step, inexpensive solution to "the air pollution problem." Engineers recognize that we are unlikely to find such a solution, and must continue to apply limited solutions to parts of the family of air pollution problems.

4. From 1970 to 1997, the United States has made significant progress in reducing emissions of lead (mostly by taking lead out of gasoline) and modest progress in reducing emissions of the other major pollutants. The air pollutant emission situation can be roughly approximated by

$$\begin{pmatrix}\text{Air pollutant}\\ \text{emissions}\end{pmatrix} = \text{population} \cdot \begin{pmatrix}\text{economic activity}\\ \text{per person}\end{pmatrix} \cdot \begin{pmatrix}\text{pollutant emissions}\\ \text{per unit of economic}\\ \text{activity}\end{pmatrix} \tag{1.1}$$

Since the environmental awakening of 1969–1970, the population of the United States has increased by about 30%, our economic activity per person by about 80%, and our motor vehicle usage by about a factor of 4. But the pollutant emissions per unit of economic activity have declined steadily because of stringent programs of emission control. Thus, in most of the United States, the emissions and hence the measured concentrations of most pollutants in the atmosphere declined steadily between 1970 and 1997. The decline has not been as rapid as many have wished, or as rapid as many predicted, and there are exceptions to this decline (e.g., increases in acid rain in the northeastern United States). In general, however, the installation of ever-more-effective pollution control equipment has allowed us to increase our population and increase our level of economic activity per person while decreasing most measured air pollutant concentrations. Unfortunately, the law of diminishing returns applies to air pollution control: the pollution control steps taken to date have been easier and cheaper than the ones we will have to take in the future.

1.5 EMISSIONS, TRANSPORT, RECEPTORS

Figure 1.1 is a schematic of the air pollution process. Some source emits pollutants to the atmosphere. The pollutants are transported, diluted, and modified chemically or physically in the atmosphere; and finally they reach some receptor, where they damage health, property, or some other part of the environment. Some of the pollutants are removed from the atmosphere by natural processes, so that they never find a receptor.

In this book, in any discussion of air pollution, or any study of the regulatory structure of air pollution control, one finds myriad details. One also finds that what is done for one kind of source or one particular pollutant is different from what is done for another source or pollutant. Some of these differences result from historic accidents and some result from the very different sources and control technologies for the various major pollutants. Faced with this diversity of details, one would do well to look occasionally at Fig. 1.1 to see how that particular detail fits into the overall air pollution schematic shown here.

In Fig. 1.1 we also see a major reason why air pollution is different from water pollution or industrial hygiene. If the same figure were drawn for water pollution, the atmospheric transport box would be replaced by a box for groundwater or stream transport. Those mechanisms are indeed complex, but not nearly as complex as atmospheric transport. We would also see that the chemical or biological form in which most water pollutants are emitted is the one that causes harmful effects. The same is not true of air pollution: many of the major pollutants are formed in the atmosphere and are called *secondary pollutants* to distinguish them from their precursors, the *primary pollutants.* The industrial hygienist, who is responsible for protecting

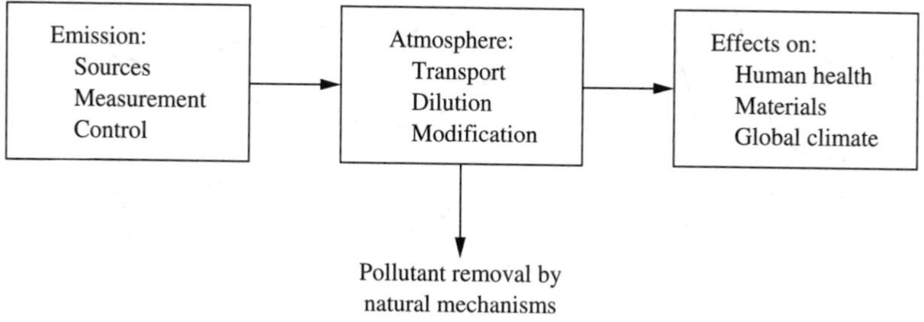

FIGURE 1.1
Air pollution schematic, showing the interrelations among emissions, transport, dilution, modification, and effects.

workers in factories and other workplaces, is often concerned with the same emissions as is the air pollution control engineer, but the industrial hygienist normally has a more easily defined transport path between emission and those affected, and rarely deals with secondary pollutants.

Several of these ideas are illustrated in Fig. 1.2, where we see smoothed average concentrations of four air pollutants for one day in Los Angeles. CO and NO are primary pollutants, emitted mostly by automobiles (Chapter 13), as is hydrocarbon (HC), not shown on this figure. The peak concentrations of CO and NO occur during the morning commute period. NO_2 and O_3 are secondary pollutants formed in the atmosphere by a complex set of reactions, summarized (see Appendix D) as

$$NO + HC + O_2 + \text{sunlight} \rightarrow NO_2 + O_3 \tag{1.2}$$

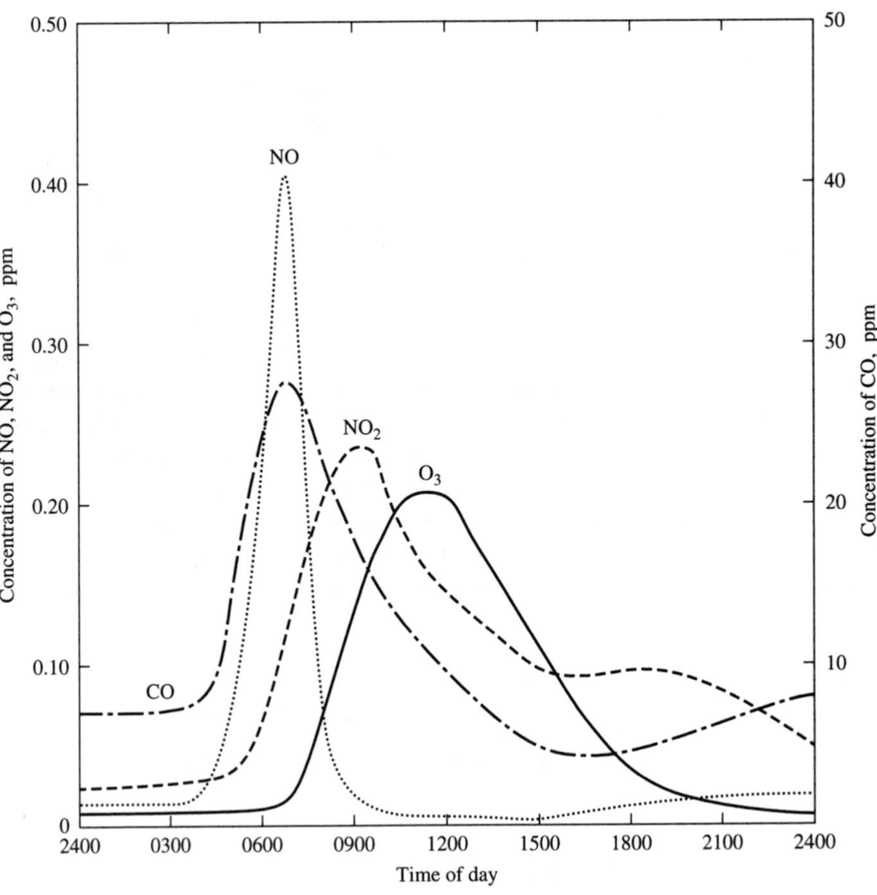

FIGURE 1.2
Smoothed average daily concentrations of selected pollutants in Los Angeles, California, July 19, 1965 [3, 4]. Observe the progression NO → NO_2 → O_3 and the different behavior of CO, which does not undergo rapid chemical reactions in the atmosphere.

The peak concentration of NO_2 occurs before the peak for O_3 because the reaction sequence, which is much more complex than the simplification in Eq. (1.2), forms NO_2 first, then O_3. The CO concentration peak, which is shown on the right-hand scale as being $\approx$70 times the peak concentration of NO, does not decline as rapidly as the NO peak because the CO concentration is reduced only by atmospheric mixing and dilution (Chapter 6) whereas the NO concentration is reduced by dilution and mixing and by the chemical reaction in Eq. (1.2). The afternoon commute also produces increases in NO and CO, but the measured concentrations are not as large as the morning peaks because the average wind speed is higher and the atmospheric mixing is stronger in the afternoon than in the morning (Chapter 5), thus causing more rapid dilution. It has also been observed that the highest peak O_3 concentration normally occurs about 30 to 60 miles downwind of the place that had the maximum morning emission of NO and HC because the polluted air mass can ride the wind that far in a day. Thus, any regulatory scheme for these pollutants (Chapters 3, 10, and 12) must account for the fact that the worst pollutant exposure may occur in a different city, state, or country from the major emission source.

The two pollutants of greatest current (late 1990s) health concern are both secondary: ozone, as described above, and fine particles. The very small particles that enter most deeply into our lungs and that are believed to be most harmful are largely formed in the atmosphere by reactions that can be summarized (in very simplified form) as

$$\text{Hydrocarbons} + \text{sulfur oxides} + \text{nitrogen oxides} \rightarrow \text{fine particles} \tag{1.3}$$

See Chapter 8.

1.6 UNITS AND STANDARDS

In this book, both English and SI units are used. As much as possible, we use the units most commonly used in the United States in that particular part of the air pollution control field. Historically, scientists have used metric or SI (often the cgs version of metric) whereas engineers have used the English engineering system. The regulators have used mixed systems. The permitted emissions from automobiles (Chapter 13) are stated in g/mile, a mixed metric–English unit! This seems like an illogical unit, but it is not. The emission data are used in mathematical models (Chapter 6) that express emissions in g/s. The available data on automobile usage are all in vehicle miles driven/hour, and the federal automobile fuel efficiency standards, which are tested by the air pollution branch of the U.S. EPA, are in miles/gallon. The prudent engineer will accept the units in use, clearly state the units on any quantity, and always check the units in every calculation.

Most "practical" air pollution books present formulae that are unit-specific, whereas most "scientific" or "theoretical" books present equations that are independent of units. For example, the power requirement of a low-pressure fan or blower (Chapter 7) is

$$\text{Po} = \frac{Q \Delta P}{\eta} \tag{1.4}$$

where Po is the power requirement, Q the volumetric flow rate, ΔP the pressure increase across the blower, and η the efficiency of the blower or of the motor-blower combination. This equation is correct in any set of units. One regularly sees it written as

$$\text{Po} = \frac{Q \Delta P}{\eta} \cdot \frac{1}{33{,}000} \tag{1.5}$$

which is only correct if the power is expressed in horsepower, the flow rate in cubic feet per minute (cfm), and the pressure in lbf/ft^2. That is an uncommon unit for pressure, so one is quite likely to misuse this equation. If we use the more common lbf/in^2 (psi), then this becomes

$$\text{Po} = \frac{Q \Delta P}{\eta} \cdot \frac{144}{33{,}000} = 0.00436 \frac{Q \Delta P}{\eta} \tag{1.6}$$

which is only correct for horsepower, cfm, and psi.

In this book all equations are of the type of Eq. (1.4), correct in any consistent set of units, except if there is an explicit statement to the contrary. Some of the problems ask the reader to convert from the universal form to "practical" forms like Eqs. (1.5) and (1.6).

In the United States, a concentration expressed in parts per million (ppm) is almost always ppm **by volume or by mol** if it is concentration in a **gas**, and ppm by **mass or weight** if it is concentration in a **liquid or solid**. (For a liquid or a solid with a specific gravity of 1.0, such as water or dilute solutions in water, ppm is the same as mg/kg, which is also widely used.) This mixed meaning for ppm continues to be a source of confusion when both liquid or solid and gas concentrations appear in the same problem. One often sees this concentration written as ppmv, to remind the reader that for gases it is most often ppm by volume. (The same is true of parts per billion; ppb = μg/kg for a solid or liquid material with specific gravity of 1.0.)

When standard conditions for a gas are referred to, there seems to be only one choice for pressure, the *standard atmosphere,* whose values in a variety of systems of units are shown inside the back cover. Unfortunately, there is no comparable agreement as to which temperature should be used. Values of 0°C, 18°C, 20°C, and 25°C are used. Throughout this book, unless stated otherwise, air and process gases are assumed to be at 1 standard atmosphere and 20°C (= 68°F). The properties of air and water at this temperature and several others are shown inside the back cover as well. (Unfortunately, many EPA regulations are based on a standard temperature of 25°C = 77°F.)

1.7 THE PLAN OF THIS BOOK

There are many possible ways to arrange an Air Pollution book, no one of which seems to please all readers. The plan of this book is first to discuss topics that are common to all pollutants, and then to discuss individual pollutants. For each pollutant, the control technology is adapted to the sources and the physical and chemical nature of that pollutant. Chapters 1–7 cover general topics in air pollution. Chapters 8–12 cover the four major air pollutants that have been and continue to be the focus of most of society's air pollution control efforts. Chapter 13 covers motor vehicles, which play a unique role in air pollution and contribute significantly to urban air pollution problems. Chapter 14 discusses larger-scale problems, including global ones. Chapter 15 treats five additional specific air pollution topics briefly.

1.8 SUMMARY

1. Air pollution is the presence of man-made harmful materials in the air, in quantities large enough to produce harmful effects.
2. Public interest in air pollution was low before 1969. About that time, it increased dramatically, and has remained high.
3. We are unlikely to solve our air pollution problems by blowing the polluted air away; we will have to solve them by reducing pollutant emissions.
4. There is not one "air pollution problem" but rather a family of related problems. We are unlikely to find a cheap, easy way to solve these problems. Instead, we will have to make many small steps to reach our air quality goals, and these will probably be more expensive than the steps we have taken so far.
5. The overall air pollution problem takes the following form: emissions → transport, dilution, and modification in the atmosphere → effects on people, property, and the environment. Although the details may differ from pollutant to pollutant, all fit this pattern.
6. Some of the most important air pollutants are secondary pollutants, formed in the atmosphere from primary pollutant precursors.
7. Ppm means ppm by volume or mol when applied to gases and ppm by mass or weight when applied to liquids and solids.
8. For all problems and examples in this text, unless stated otherwise, the pressure is 1 atm and the temperature is 20°C = 68°F (see inside the back cover).

PROBLEMS

See Common Units and Values for Problems and Examples, inside the back cover.

1.1. In Example 1.1:
(*a*) Estimate the pressure drop required.
(*b*) Estimate the pumping power required.
See any fluid mechanics textbook for methods of making these estimates.

1.2. (*a*) In Table 1.1 we see that 57 wt % of the listed pollutants are CO. Does it follow from that table that 57 percent of the air pollution problem in the United States is a CO problem?
(*b*) The same table shows that 57 wt % of all the listed pollutants come from transportation (mostly automobiles). Does it follow that 57 percent of our national air pollution problem is an automotive problem?
(*c*) If the answer to these questions is no, explain your answer.

1.3. In Table 1.1 we see that most of the nitrogen oxides are emitted by transportation and power generation, with much smaller emissions from other sources. Why are these other sources less important?

1.4. On May 18, 1980, Mount Saint Helens in Washington state ejected into the atmosphere an estimated 540 million tons of ash [5].
(*a*) How does that compare with the emissions of PM_{10} from human activities for 1997 shown in Table 1.1?
(*b*) Is it reasonable to make this comparison? Why or why not?

1.5. On November 4, 1996, José Angel Conchello, the secretary of the second-largest political party in Mexico (PAN), wrote to the mayor of Mexico City, proposing that four helicopters be flown over the city to disperse the air pollutants. He said, "Extraordinary situations require extraordinary solutions....I refer to the use of the helicopters of the Federal District, as if they were huge ventilators to cause turbulence and vertical columns of contaminated air to diminish the poisoning in the streets." [6] Comment on the practicality of this proposal. Sketch the air flow generated by hovering helicopters.

1.6. The "law of diminishing returns" is widely discussed in economonics texts. The author's favorite example is that the first hour of cleaning a messy house produces a very visible improvement in its appearance, but that the next hour of cleaning effort produces less visible effect, and subsequent ones even less. Suggest other examples from daily life of the law of diminishing returns. Suggest how it applies to air pollution control.

REFERENCES

1. Halliday, E. C.: "A Historical Review of Atmospheric Pollution," in *World Health Organization Monograph Series,* No. 46, Geneva, 1961.
2. *National Air Quality and Emissions Trends Report, 1997*, EPA-454/R-98-016, and *National Air Pollution Emissions Estimates, 1940–1990*, EPA-450/4-91-026.
3. "Comprehensive Technical Report on All Atmospheric Contaminants Associated with Photochemical Air Pollution," TM-(L)-4411/002/01, System Development Corporation, Santa Monica, Calif., June 1970.
4. "Air Quality Criteria for Nitrogen Oxides," AP-84, U.S. EPA, 1971.
5. Tilling, R. I.: *Eruptions of Mt. St. Helens: Past, Present and Future,* U.S. Department of the Interior/USGS (No date), p. 17.
6. Article, *La Capital, Mar del Plata, Argentina*, Nov. 5, 1996, p. 6.

CHAPTER 2

AIR POLLUTION EFFECTS

This is a book about air pollution control. But any competent engineer begins any engineering task by asking, among other things, "Why are we doing this at all?" We control air pollution because it causes harmful effects on human health, property, aesthetics, and the global climate. This brief chapter reviews what we know about these effects on human health and property and on visibility. Chapter 14 considers global effects. Because the air pollution laws in the United States and other industrialized countries are mostly concerned with protecting human health, we will consider the effects on human health first.

2.1 EFFECTS OF AIR POLLUTION ON HUMAN HEALTH

In Bhopal, India, in December 1984, a release of methyl isocyanate from a pesticide plant killed about 2500 people. Similar leakages of hydrogen sulfide from natural gas processing plants have killed hundreds of people. These tragic events attract wide attention. Normally, they are not considered air pollution events, but rather industrial accidents. The damages to human health caused by air pollution are of a very different type. The materials involved are rarely as toxic as methyl isocyanate or hydrogen sulfide. They are generally not released in concentrations nearly as high as those that cause such disasters. Their effects normally do not result from a single exposure (methyl isocyanate and hydrogen sulfide can kill in a minute or two), but from repeated exposure to low concentrations for long periods.

Table 2.1 lists the air pollutants that are regulated in the United States in 1998 because exposure to them is harmful to human health. The majority of the air pollution efforts in the United States (and most of this book) is devoted to the control

TABLE 2.1
Air pollutants believed dangerous to human health and currently regulated in the United States

Pollutants regulated by National Ambient Air Quality Standards (NAAQS) as described in *40CFR50* (as of July 1, 1998). These are called *criteria pollutants* because before the standards were issued, documents called Air Quality Criteria were issued.

- Sulfur oxides
- Fine particulate matter
- Carbon monoxide
- Ozone
- Nitrogen dioxide
- Lead

Pollutants regulated by National Emission Standards for Hazardous Air Pollutants (NESHAP) as described in *40CFR61* (as of July 1, 1998). These are called *hazardous air pollutants* or *air toxics.*

- Asbestos
- Benzene
- Beryllium
- Coke oven emissions
- Inorganic arsenic
- Mercury
- Radionuclides
- Vinyl chloride

The Clean Air Amendments of 1990 expanded this list to 189 chemicals. The regulations for those in addition to the above 8 are currently in the regulatory pipeline (see Chapter 15).

of the pollutants on this list. Extensive, detailed reviews of the health effects of air pollutants are regularly published [1–5]. The rest of this section presents some basic ideas about the health effects of these pollutants.

At least since the time of Paracelsus (1493–1541), people have known that it is meaningless to speak of any substance as harmful unless we specify how much of the substance is administered. He said, "There is poison in everything and no thing is without poison. It is the dose that makes it harmful or not." The same is true of air pollution. To make any meaningful statements about air pollution effects on human health, we must consider the dosages people receive, that is,

$$\text{Dosage} = \int (\text{concentration in air breathed})\, d(\text{time}) \tag{2.1}$$

Current interest in air pollution and health is mostly directed at long-term, low-concentration exposures (which lead to *chronic effects*). Short-term, high-concentration exposures (which lead to *acute effects*) occur only in industrial accidents (such as the Bhopal tragedy) or *air pollution emergency episodes;* the latter occurred occasionally in the past [6], but are now very rare in countries with modern pollution control regulations.

To determine what dosage is harmful, we wish to construct a *dose-response curve.* Such a curve can be plotted only for individual pollutants, not for "air pollution in general." (Synergism, the effect of two pollutants together being greater than the sum of the separate effects of the two, may occur; that is believed to be the case with sulfur oxides and fine particles, and perhaps some other pollutant combinations as well.) Figure 2.1 is a dose-response curve for a hypothetical homogeneous population exposed to a single hypothetical pollutant for a specific time period. We know most about dose-response curves from pharmacology, where experimental subjects are regularly given carefully measured doses of experimental pharmaceuticals and their responses are measured. From theory and experiment, we know that for pharmaceuticals, the most common dose-response curve is the no-threshold curve, which passes through the origin [7].

However, in industrial hygiene it has been observed that there is some concentration of pollutants called the *threshold value* that "represents conditions under which it is believed that nearly all workers may be repeatedly exposed day after day, without adverse effect" [8]. These values, called *threshold limit values* (TLVs), are

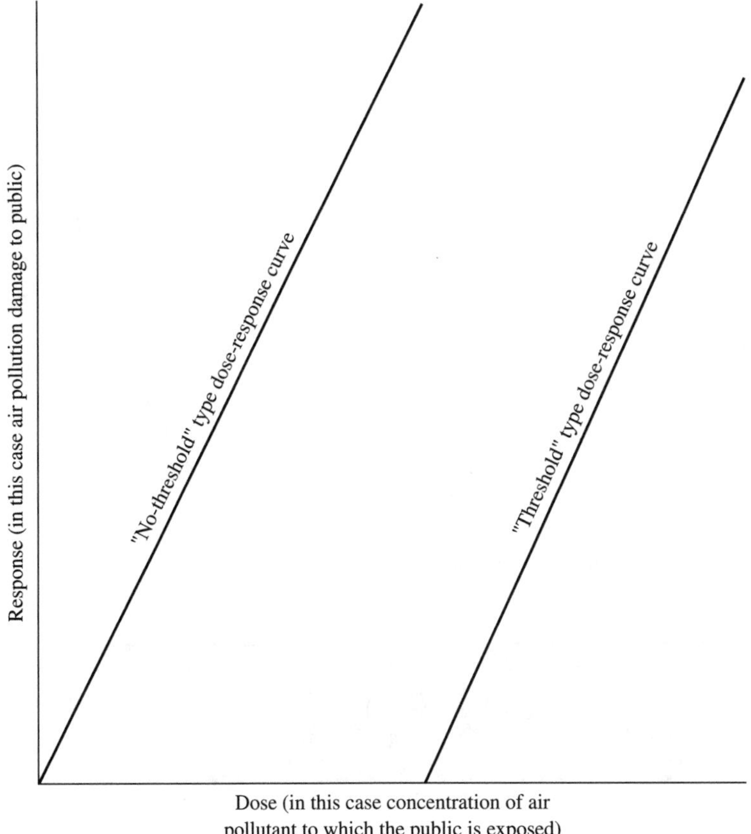

FIGURE 2.1
Threshold and no-threshold dose-response curves. The straight lines are an admission of ignorance; we generally do not know the true shapes of these curves.

established by industrial hygiene boards [9]; industrial plants are expected to prevent the exposure of workers to concentrations higher than the TLVs. These TLVs do not represent true *no-effect concentrations;* rather, they represent concentrations at which the health effects (if they exist) are less than the variation in health of the general populace; hence the "signal" (health effect) is lost in the "noise" of the general health variation of the population. This idea is sketched in Fig. 2.2. If the idea of threshold values were literally true, then the true dose-response curves would be like the threshold curve in Fig. 2.1. Ghering et al. have presented theoretical grounds for believing that such true thresholds exist [10]. Their theory is illustrated by hydraulic analogy in Fig. 2.3. If a first elimination mechanism can handle the entire pollutant input into our bodies, then the second elimination mechanism will not come into play. However, if the first mechanism is saturated, then the second will come into play. If the first mechanism is harmless but the second mechanism creates harmful degradation products within the body or harms some bodily organ, there will be no damage to our bodies as long as the first elimination mechanism can handle the entire input, but harm will result if the input exceeds the capacity of the first elimination

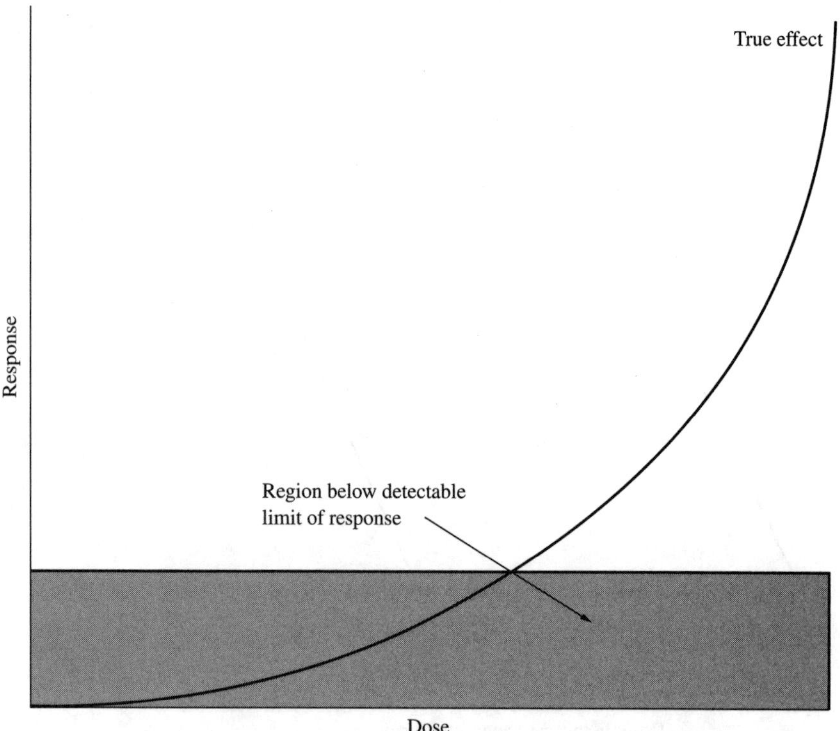

FIGURE 2.2
The true dose-response situation may be that at low doses the effect is not truly zero but instead is too small for us to detect.

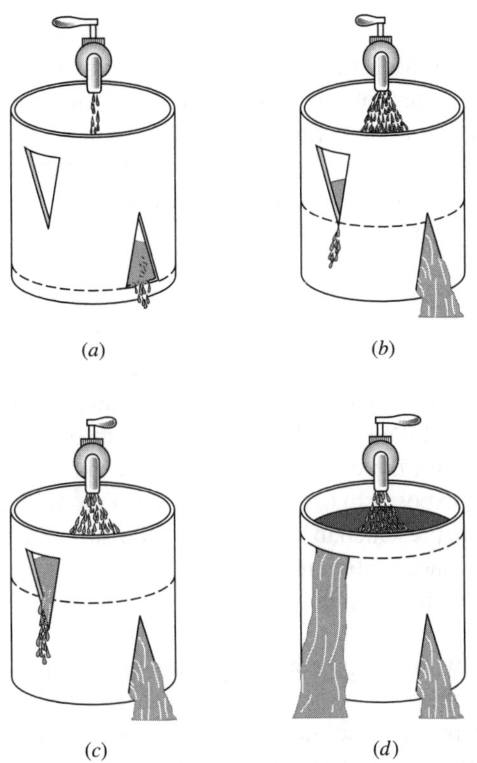

(a) (b) (c) (d)

FIGURE 2.3
A fluid-mechanical analog of the biological mechanism that could result in a true threshold value for a toxic substance. For flow rates less than (b), no flow exits by the higher opening. If the degradation products by the lower route are harmless, and those by the higher are harmful, then the true threshold would correspond to an intake rate equal to that shown in (b). (After Ghering et al. [10].)

mechanism. Under this theory, we would have to modify Eq. (2.1) to

$$\text{Harmful dosage} = \int [(\text{intake rate due to breathing}) - (\text{removal rate by first mechanism})]\, d(\text{time}) \quad (2.2)$$

There are known thresholds for some substances in our diet, such as selenium [11]. Selenium is an essential nutrient; a zero-selenium diet is fatal. Large doses, however, are poisonous; a high-selenium diet is also fatal. Therefore, there must be two thresholds, a lower and a higher one, between which there is a selenium dietary intake level that is harmless (or at least not fatal). Fortunately, the range between the two fatal conditions is fairly wide.

There are theoretical (and some experimental) grounds for believing that there are some substances for which there is no threshold; for such substances, any input is harmful. (Such an input need not cause harm to every person exposed. Instead, it may raise the statistical probability of contracting some disease, e.g., cancer.) In terms of the hydraulic analogy, there is no harmless elimination mechanism (or its effects are so small as to escape experimental detection). Most of the substances believed to have no thresholds are either carcinogens or emitters of ionizing radiation. Establishing the existence or nonexistence of such thresholds experimentally is difficult.

If we wish to establish the dose-response curve for a pollutant, we have three possible approaches: animal experiments, laboratory experiments with humans, and epidemiological studies of human populations.

2.1.1 Animal Experiments

A good example of an animal experiment is given in [12]. Two groups of mice (the ozone group and the control group) were simultaneously exposed to an aerosol containing *Streptococcus* C bacteria, which killed up to 80 percent of the mice. The ozone group had previously been exposed for three hours to various concentrations of ozone; the control group had not been exposed to ozone. The observed mortality values for the ozone and control groups are presented in Table 2.2, and the difference in mortality is plotted against the ozone concentration in Fig. 2.4.

From this experiment we observe the following:

1. It is hard to perform any experiment with living beings and get as good reproducibility as one can with inanimate objects. The control groups in all 10 trials were exposed to what was intended to be the same concentration of bacteria each time. The observed mortality varied from 0 to 15 percent. The data on differences in mortality have significant scatter as well. The negative mortality difference is almost certainly the result of scatter in the experimental data. It is hard to imagine a mechanism by which exposure to 0.07 ppm of ozone would protect mice from subsequent bacterial infection.

TABLE 2.2
Experimental results from exposure of mice to ozone and then *Streptococcus* C bacteria

	Percent mortality		
Ozone concentration, ppm	**Control group**	**Ozone group**	**Difference**
0.52	13	80	67
0.35	0	60	60
0.30	3	40	37
0.20	8	50	42
0.18	0	63	63
0.17	8	45	37
0.10	8	35	37
0.08	15	38	23
0.07	15	35	20
0.07	8	5	−3

Source: Ref. 12.

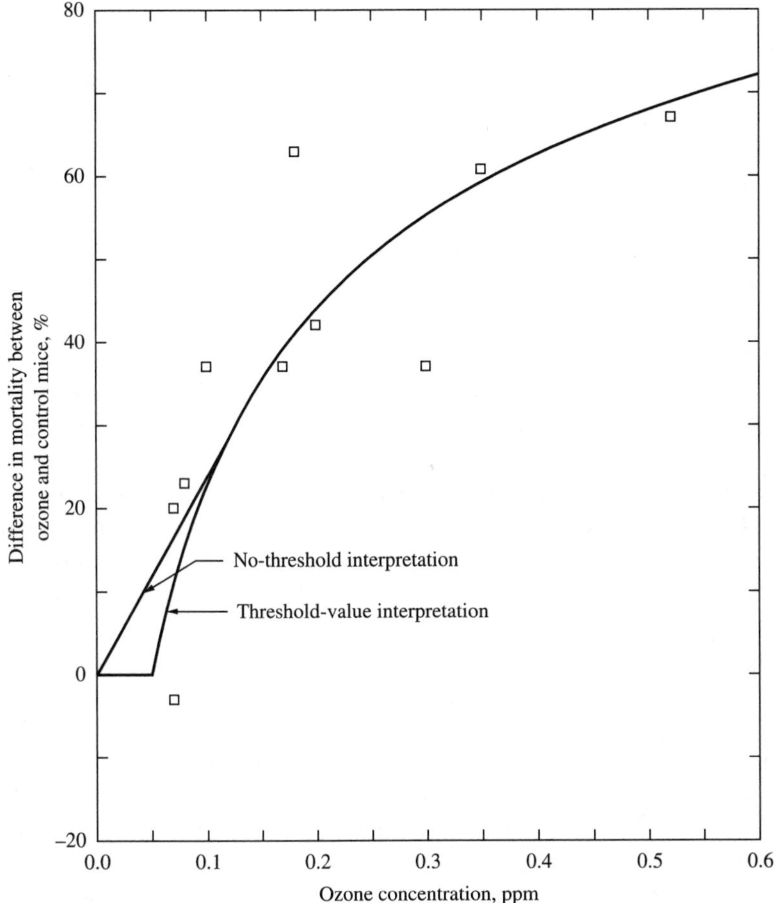

FIGURE 2.4
Experimental data from Table 2.2 on the difference in mortality between mice exposed to ozone and an unexposed control group, both subsequently exposed to *Streptococcus* bacteria [12], with two possible interpretations.

2. Ozone exposure produces a significant effect on mortality at concentrations above about 0.10 ppm, and the effect increases with increasing ozone concentration.
3. Here, the air pollution effect was indirect. No mice died as a result of ozone exposure alone. Rather, the ozone, which is a respiratory irritant, presumably irritated the lungs of the exposed mice, making it easier for lethal numbers of bacteria to enter the bloodstream. The authors of the study concluded that the ozone damaged some of the white cells that defend the body against bacterial invasion. If we didn't know the history of the test, we could conclude that exposure to high concentrations of ozone led to increased mortality, but we would probably not know the mechanism of that mortality. This uncertainty about mechanism is

common in the epidemiological air pollution studies described in Section 2.1.3. Conversely, if we merely looked at autopsy reports, we would have no way to know which mice had been exposed to ozone and which not, or that ozone exposure had played any role in their deaths. The autopsy reports would simply say, "Died of bacterial infection."

4. Although the data scatter is annoying, it is not nearly as great as it would have been if we had used human subjects. The mice in this kind of study are highly inbred so that the genetic variation among them is thousands or millions of times less than that in human populations. Their environment from birth is controlled to make them as similar as possible; the same is not true for humans. Even so, their response shows considerable variation.

5. From this kind of test, we can *estimate* the effects on humans of similar exposures. For new drugs not yet in public use, animal experiments are the only way we have of making such estimates. However, what is harmful to one animal may not be harmful to another. For example, before thalidomide (a sedative) was approved for human use, it was extensively tested on mice (including pregnant mice), and showed no harmful effects. In humans, it produced very severe birth defects. Thus, animal tests only *suggest* what the human health effects of such exposures will be.

6. These tests measured only acute effects, those seen in a few hours. They give us some guidance about human short-term exposures. Because we are the longest-lived of all mammals, we are concerned with lifetime exposures. Most laboratory animals do not live very long, so it is hard to expose a laboratory animal to some pollutant for more than a year or two. Such short-term tests tell us little about lifetime exposures of humans to the same concentrations of the same pollutants.

7. This experiment was quick, simple, and cheap. Only small numbers of mice were involved, and the effect considered, death, is easy to detect. To do a similar test for carcinogenicity, one would have to expose mice for much longer and then do an autopsy on each mouse. If one did not know which organ was likely to develop the cancer, one would have to examine every organ of every mouse.

8. Two interpretations of the data appear in Fig. 2.4: a threshold-value interpretation and a no-threshold interpretation. Based on these data alone, one cannot say which of these interpretations (if either) is correct. This flaw is typical of all such animal tests; at high concentrations, the results are rather clear, but at low concentrations the uncertainty and scatter introduced by the variability of even highly inbred mice make it impossible to determine the true shape of the curve. It is estimated that if one wished to settle completely the threshold or no-threshold question for one substance suspected of being a carcinogen using mice as the experimental animal (which does not necessarily settle the questions for humans), then an experimental program involving at least a million mice would be needed (the "megamouse experiment").

9. The concentration at which significant effects are seen is near the currently permitted value (NAAQS) of 0.08 ppm in the United States. However, the pathogen exposure that followed the ozone exposure was much more severe than humans normally encounter. It quickly killed up to 15 percent of the control mice.

2.1.2 Short-Term Exposure of Human Volunteers

Ample published data show that short-term laboratory exposures of healthy young adults to air pollutant concentrations much higher than those ever measured in the ambient air produce no *measurable, irreversible* short-term or long-term effects [13, 14]. (Such tests show *reversible* changes in lung function and other physiological parameters; these changes disappear a few hours after the tests.) However, because we are interested in the effects of long-term exposure, and because we are interested in the health effects not only on healthy young adults but also on the most sensitive members of our society (young children, asthmatics, and the very old), it seems clear that short-term laboratory tests on healthy young adults will not provide the data we need. Such tests are useful for looking for the detailed physiological mechanisms of air pollution damage, but the only way we can ultimately settle health-effect questions is through sophisticated epidemiology.

2.1.3 Epidemiology

Several attempts have been made to do the required epidemiological studies. Perhaps the most interesting is the Community Health and Environmental Surveillance System (CHESS) study [15]. It has received vigorous technical criticism [16] and has been vigorously defended [17]. However, in spite of its technical shortcomings, the general approach of this study is ultimately the one most likely to allow us to construct accurate dose-response curves for air pollutants. In one part of the CHESS study, four cities were selected at various distances from a large copper smelter in the Salt Lake Valley. The cities had a demonstrable gradient of sulfur dioxide concentration because of the prevailing wind patterns, and their different distances from the smelter. The study team attempted to select neighborhoods in each of these cities in which they could match sociological characteristics. For each neighborhood, the study team attempted to measure the health of the populace, with specific emphasis on health problems believed to be influenced by sulfur dioxide (asthma, chronic bronchitis, and lower respiratory disease in children). They then sought a relationship between SO_2 exposure and such diseases. They claim to have demonstrated such a relationship, a conclusion their critics have vigorously denied.

If we assume, for the sake of argument, that their data are valid, we can examine those data to see if they lead to an unambiguous definition of the dose-response relationship for exposure to one specific air pollutant. Figure 2.5 is a plot of the incidence of lower respiratory disease among children as a function of annual average concentration of SO_2 in the four cities. It reveals the following:

1. The health effect considered is not zero for zero pollutant exposure; even in the cleanest environment, a significant fraction of children will have lower respiratory

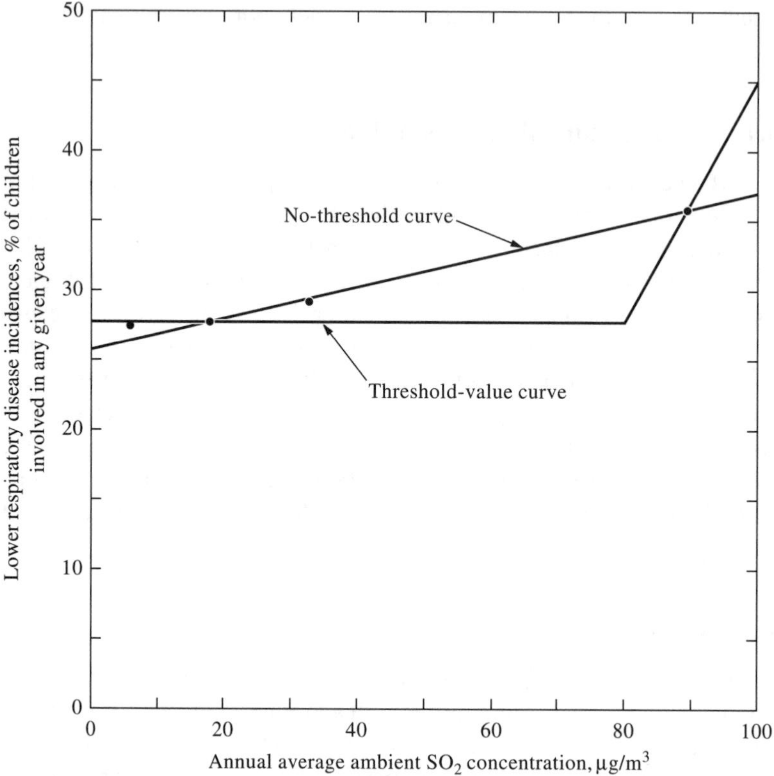

FIGURE 2.5
Some data from the CHESS study [15]. The points represent study areas in (left to right) Ogden, Salt Lake City, Kearns, and Magna. The SO_2 concentrations are influenced by the distances from a large copper smelter, and by prevailing wind patterns.

disease in any year. If the curve is of the threshold-value type, then it must proceed horizontally from the zero-exposure value, as shown, until the threshold value is reached, where it will turn upward. The threshold-value curve shown in Fig. 2.5 turns upward at 80 $\mu g/m^3$, which is the annual average SO_2 NAAQS in the United States.

2. The data do not unambiguously support either the threshold-value or the no-threshold interpretation. Given this data set alone, one would be hard pressed to select the better interpretation.
3. The health effects are plotted versus concentration of SO_2, the most easily measured sulfur oxide. It is far from clear that this is the biologically active agent; it may be serving as a proxy for all sulfur oxides. There is evidence suggesting that the biologically active agent is acid aerosol, created by the deposition of sulfuric acid on fine particles [18]. The CHESS study contains many more data than are shown in Fig. 2.5. This particular data set was chosen because it is not

complicated by the effect of smoking, which severely complicates all of the adult data.

4. This location was chosen for study because a 1960s copper smelter emitted large amounts of SO_2 but only small amounts of particulates, producing a sharp gradient in SO_2 concentration without a corresponding gradient in particulate concentration. In most industrial areas SO_2 and particulate concentrations are more or less proportional, so that it is hard to study the effect of one without the other. Since the date of this study, the smelter has reduced its emissions enough that in 1996 the highest annual average ambient SO_2 concentration at any of the four cities shown in Fig. 2.5 was about one-tenth of the highest value shown in Fig. 2.5. Since 1970 air pollutant emissions from major sources in industrial countries have been reduced enough that the measurements shown in Fig. 2.5 are unlikely to be repeated in the United States or other industrial countries.

An alternative epidemiological approach has been to correlate deaths or hospital admissions with measured air pollution concentrations. These can be carried out by looking at historical records (a *retrospective* study) or by choosing one or more suitable populations and following their health or longevity over time, together with the air pollutant concentrations to which they are exposed (a *prospective* study). Figure 2.6 shows the results of a retrospective study of the December 1952 London pollution episode. An unusual meteorological situation caused five consecutive days of very low wind speeds over London, England. The concentration of pollutants, mostly derived from coal combustion, increased to values rarely encountered in large cities. Schwartz [19] reported,

> ...There was a 2.6-fold increase in deaths in the second week (of December). Increases were seen in all age groups, but the largest relative increases were in ages 65–74 (2.8-fold) and ages 75 and over (2.7-fold)....The largest relative increases were seen for bronchitis and emphysema (9.5-fold), tuberculosis (5.5-fold), pneumonia and influenza (4.1-fold), and myocardial degeneration (2.8-fold).

From this report we see:

1. The observed particle concentrations are very high. Such concentrations have rarely if ever been observed since 1952 in technologically advanced countries.

2. The increase in deaths followed the increase in particle concentration by a day. It is commonly found in such studies that the concentration the previous day or the average over the previous several days is the best predictor of the daily death rate.

3. Other pollutants were present, but statistical analysis of the data shows a better correlation with particulate concentrations than with other pollutant concentrations or combinations of concentrations.

4. Most of the deaths were not of healthy young persons. Rather they were of susceptible persons, mostly older persons with pre-existing respiratory or circulatory

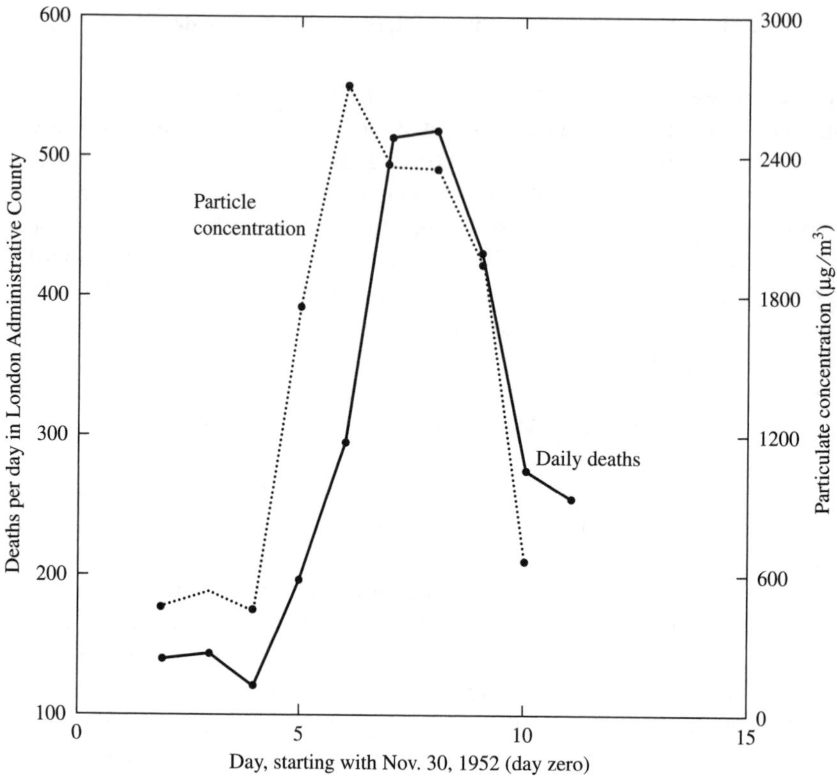

FIGURE 2.6
Daily death rates and particle concentrations for the December 1952 London pollution episode, after Schwartz [20].

problems. The air pollution episode did not kill them, but rather hastened their deaths or shortened their lives.

Figure 2.7 shows the results of a prospective study of mortality [21]. Large groups (1200 to 1600) or participants were selected in six cities. For 14 to 16 years their health and survival were measured, along with concentrations of pollutants in the six cities. The survival rate (fraction of the original study population still living) was highest in the least polluted cities. Figure 2.7 plots the ratio of the annual death rate in each of the cities to that in the cleanest city (Portage, WI). This ratio is obviously 1.00 for Portage, increasing to 1.26 for Steubenville. This study, by the highly respected air pollution group at the Harvard School of Public Health, was one of the major bases for the change in U.S. particulate standards in 1997 [22]. From the study we see:

1. The death rate, adjusted for smoking and some other factors, seems to be linearly proportional to the fine particle concentration (particles with diameters $< 2.5\ \mu$).

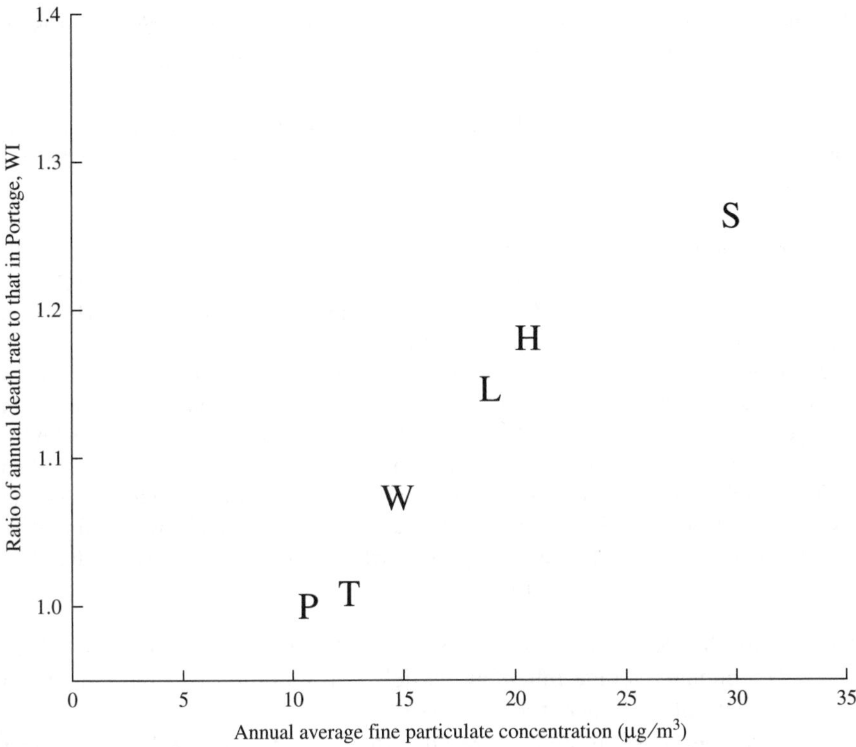

FIGURE 2.7
Ratio of death rates to that in Portage, WI, as a function of fine particle concentration. Here P = Portage, WI; T = Topeka, KA; W = Watertown, MA; L = St. Louis, MO; H = Harriman, TN; and S = Steubenville, OH. After Dockery et al., "An Association between Air Pollution and Mortality in Six U.S. Cities," *New England J. of Medicine*, Vol. 329, pp. 1753–1759, 1993. *Copyright ©1993 Massachusetts Medical Society, All rights reserved.* [21]

Other air pollutant concentrations or combinations of them did not correlate the mortality data as well.

2. There does not appear to be any threshold.
3. The concentrations are quite low. The values are not directly comparable to those in Fig. 2.6 because of different measuring methods, but using the best estimates of the correspondence of those methods [23], one concludes the peak value of $\approx 2500\ \mu g/m^3$ on Fig. 2.6 would correspond to about $1500\ \mu g/m^3$ on Fig. 2.7. However, the value on Fig. 2.7 is an annual average, and those are typically about one-third of the highest-day value, so the proper ratio between the highest values on the two figures is roughly $[1500/(30 \cdot 3)] \approx 17$.
4. Figure 2.7 is a comparison of annual death rates as a function of particle concentration whereas Fig. 2.6 is of daily death rates. Various statistical studies have shown that the effect is similar over most studies, both prospective and retrospective.

The finding is that an increase in particle concentration of 100 $\mu g/m^3$ causes an increase of about 6% in both the annual and the daily death rates.

The two previously reported studies were of death rates (*mortality*). Other studies concern sicknesses (*morbidity*). The results are similar; for example, Schwartz [20] reports the results of a

> ... fortuitous natural experiment. Pope ... examined hospitalization for respiratory illnesses in children in three adjoining counties in Utah—Utah County, Cache County and Salt Lake County. All had similar housing and demographic patterns. However, in the mid 1980s, the rate of hospitalization for respiratory illness in children in Utah County was approximately twice as great as that in the two adjoining counties. Utah County had an integrated steel mill in a valley subject to temperature inversions. In August 1986, the steel mill shut down due to a strike. It remained closed for 13 months. In that period the rate of hospitalization of children for respiratory conditions in Utah County fell dramatically and was indistinguishable from the rate in the neighboring counties. When the steel mill reopened, the rate of childhood hospitalization for respiratory conditions grew in Utah County and reached a level about twice as high as that in the adjoining counties once more.

(The steel mill changed owners and has significantly reduced its emissions since 1986–87. The annual average inhalable particle concentration in Utah County in 1995–96 was 64% of the concentration in 1988–89.)

These epidemiological studies are all difficult, and their results are subject to challenge. Most require analyzing the data statistically and adjusting the data to account for extraneous variables like smoking, accidental deaths, epidemics, and the like. Often the results are plausible, but of only modest statistical significance. They almost never lead to results as unambiguous as those in Figs. 2.4 and 2.6. Nonetheless, they appear to be the best measures we have of the effects of air pollutants on human health, at concentrations to which human populations are regularly exposed.

2.1.4 Regulations to Protect Human Health

Given the difficulty of obtaining unambiguous dose-response curves, we should not be surprised that there is controversy over how clean people want the air to be (or how much people are willing to spend for clean air). Faced with this problem, the U.S. Environmental Protection Agency (EPA), acting under the Clean Air Act, has commissioned studies by outstanding medical scientists and concluded that the first six pollutants listed in Table 2.1 have thresholds, while the last eight do not have demonstrable thresholds. The Clean Air Act requires the EPA to establish National Ambient Air Quality Standards (NAAQS) (which are maximum allowable levels of contamination) for threshold-value pollutants. The values are to be set to "protect the public health, with an adequate margin of safety." For pollutants for which there does not appear to be a demonstrable safe threshold value, such a standard cannot be set. The Clean Air Act regulates the eight no-threshold pollutants via the National

Emission Standard for Hazardous Air Pollutants (NESHAP) regulations, discussed briefly in Chapter 3 and in more detail in Chapter 15. In 1990, Congress listed 189 chemicals as hazardous air pollutants.

Industrial exposures to pollutants in the United States are regulated by the Occupational Safety and Health Administration (OSHA) and the ACGIH (American Conference of Governmental Industrial Hygienists) [9]. They have determined permitted occupational exposure concentrations for some of the pollutants listed in Table 2.1 (as well as many others to which the public is not exposed). The two sets of values are presented in Table 2.3.

From Table 2.3, we see that the permitted industrial concentrations are generally much higher than the permitted ambient air concentrations. This difference reflects two facts: we are exposed to ambient air 168 hours a week but are on the job only 40 hours a week; and the working population does not contain the most susceptible members of the population (infants, asthmatics, and very old people). In addition, people who are especially susceptible to irritation by a certain pollutant will quit a job where the concentration of that pollutant is insufficient to bother average people; unlike ambient air quality standards, industrial standards are not intended to protect everyone.

In Chapter 15, we consider the health effects of CO and lead and the special problems of indoor air pollution and radon, which are not normally considered air pollution.

2.2 AIR POLLUTION EFFECTS ON PROPERTY

In the early history of air pollution control, a great deal of attention was paid to air pollution damage to property. Today we pay little attention to it. The reason for this change is that 50 years ago, there were pollutants that caused visible damage to plants and animals. The owners of these plants and animals sued the emitters for damages and thus contributed to the early development of air pollution science and engineering. Today there are few such sources because we have imposed strict controls on them to protect human health.

A few examples of this kind of damage remain. Metals corrode faster in the polluted environments of our cities than they do in cleaner environments. Paints do not last as long in polluted environments as in clean ones; tires and other rubber goods fail due to ozone cracking, caused by atmospheric ozone, if they are not made with antioxidant additives (which most now have); and some green plants are harmed by air pollutants. Figure 2.8 is a summary of the effects of nitrogen dioxide on plants [24]. As expected, the damage depends on the concentration and the duration of exposure. Like humans, plants can survive short-term exposures to high concentrations of NO_2 without measurable ill effect; the longer the exposure time, the lower the concentration needed to produce damage.

In the case of crop damage caused by a single, well-identified emitter, historically it has been cheaper for the emitting facility to pay the neighboring farmers a

TABLE 2.3
Comparison of air quality standards and industrial exposure standards

Substance	Permitted ambient concentrations (NAAQS)[a]	Permitted industrial concentrations (TWA and STEL)[a]
Sulfur dioxide	80 μg/m^3 (0.03 ppm),[b] annual average, 365 μg/m^3 (0.14 ppm), 24-h average.[c]	2 ppm, 8-h average. 5 ppm, 15-min peak.
Ozone	0.08 ppm (157 μg/m^3), 8-h average.	0.1 ppm, 8-h average.
Nitrogen dioxide (NO_2)	0.053 ppm (100 μg/m^3), annual average.	3 ppm, 8-h average. 5 ppm, 15-min peak.
Carbon monoxide	9 ppm (10 mg/m^3), 8-h average. 35 ppm (40 mg/m^3), 1-h average.	25 ppm, 8-h average.
Inhalable particles (PM_{10})[d]	50 μg/m^3, annual average. 150 μg/m^3 24-h average.	Standards exist for specific kinds of particle, but not for PM_{10}.
Fine particles ($PM_{2.5}$)	25 μg/m^3, annual average. 65 μg/m^3, 24-h average.	Standards exist for specific kinds of particle, but not for $PM_{2.5}$.
Lead	1.5 μg/m^3, quarterly average.	50 μg/m^3, 8-h average.
Asbestos	No NAAQS.	A special standard, in number of fibers per cc, exists.
Benzene	No NAAQS.	10 ppm, 8-h average.
Beryllium	No NAAQS.	2 μg/m^3, 8-h average. 10 μg/m^3, 15 min peak.
Coke oven emissions	No NAAQS.	No standard for these as a group, standards for individual components.
Inorganic arsenic	No NAAQS.	10 μg/m^3, 8-h average.
Mercury	No NAAQS.	25 μg/m^3, 8-h average plus a lower standard for alkyl mercury compounds.
Radionuclides	No NAAQS.	No comparable standard.
Vinyl chloride	No NAAQS.	5 ppm, 8-h average.

[a]The NAAQS (National Ambient Air Quality Standards) are current EPA values. The TWA (time-weighted average) and STEL (short-term exposure limit) values are current ACGIH (American Conference of Governmental Industrial Hygienists) values.

[b]For gases, the standards can be expressed as ppm or μg/m^3; most tabulations show them both ways (for standard temperature and pressure of the gas). For solids such as PM_{10}, $PM_{2.5}$, or asbestos, the molecular weight is generally not known, so representation as ppm by volume or by mol is generally not possible; the standards are expressed as μg/m^3 or its equivalent (again assuming the gas is at standard temperature and pressure).

At 1 atm and 25°C, one m^3 of any perfect gas contains 40.87 moles. One ppm is 40.87 micromoles. The weight concentration (see above) of any gaseous pollutant is

$$\text{Concentration}\left(\frac{\mu g}{m^3}\right) = \text{ppm} \cdot 40.87 \cdot (\text{molecular weight, g/mol})$$

[c]For SO_2, NO_2, CO, and lead the short-term NAAQS (8 or 24 h) are not to be exceeded more than once per year, and the annual average standards are not to be exceeded in any year. For O_3, PM_{10}, and $PM_{2.5}$, the standards are statistical, requiring that some percentile of the annual distribution not exceed the standard.

[d]The standard for Total Suspended Particulates (TSP) was revoked and replaced by the PM_{10} standard in 1987. The $PM_{2.5}$ standard, promulgated in 1997, operates in parallel with the PM_{10} standard.

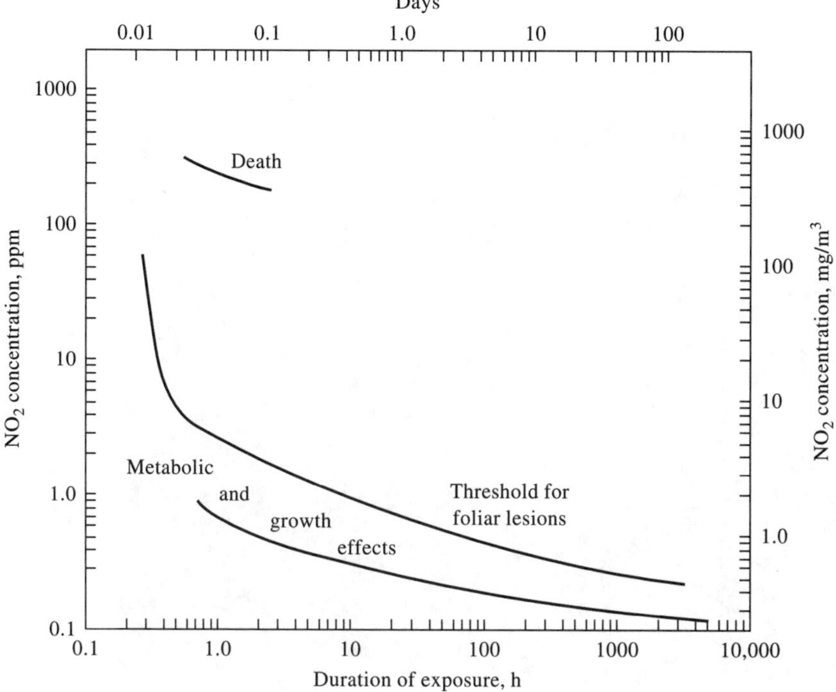

FIGURE 2.8
Threshold curves for the death of plants, foliar lesions, and metabolic or growth effects as related to the nitrogen dioxide concentration and the duration of the exposure [24]. The concentrations shown are much higher than the NAAQS for NO_2, 0.053 ppm annual average. (Reprinted with permission from Springer-Verlag and Professor D. C. MacLean.)

small damage settlement than to reduce their emissions to zero. This practice has created amusing situations like the one in the Salt Lake Valley, in which farmers near the Kennecott Copper smelter regularly planted alfalfa, which is particularly sensitive to the SO_2 emissions from the smelter, and then claimed damages from Kennecott for the demonstrable SO_2 damage to their crops. When the smelter emissions were greatly reduced in the 1970s to protect human health, those farmers stopped growing alfalfa because, without the annual damage payments, it was not economical to grow alfalfa in that location.

We have internalized many property damage costs: city dwellers expect to paint their houses more often than country dwellers, and we are used to paying a bit more for tires that contain antioxidants. Occasional studies have estimated the increased costs of such damages, and the calculated amounts are substantial. However, our concern with them is not comparable to our concern with human health.

One type of property damage of great concern is the damage to historical monuments. If alfalfa production is reduced, the petunias in our garden wilt, or our tires wear out a bit faster because of ozone damage, we can mitigate the damage for

FIGURE 2.9
An example of acid precipitation damage to an outdoor statue. The statue, made of porous sandstone, was created in 1702 as part of the gable of the entrance of the Castle at Herten, near Recklinghausen, Germany. The left photo, taken in 1908, shows some stains and the loss of the left hand, but most of the face and right hand were intact after 206 years of exposure. The right photo, taken in 1969, shows the loss of most of the detail of the statue over 61 years [25]. (Reprinted with permission from the Westfälisches Amt für Denkmalpflege.)

small amounts of money. Unfortunately, air pollution (chiefly acidic precipitation) is damaging the sandstone and marble statues and monuments of Europe and the northeastern United States. Those are not easily replaced. Figure 2.9 shows an example of damage to a European statue, caused by acidic deposition. The most famous statues at the Parthenon—the caryatids—have been moved into an air-conditioned museum; fiberglass and epoxy replicas now stand outdoors in their place.

2.3 AIR POLLUTION EFFECTS ON VISIBILITY

Most gaseous air pollutants are totally transparent. The only common exception is NO_2, which is brown. (Fluorine, chlorine, bromine, and iodine are also colored, as are some organic vapors, but these are rarely emitted to the atmosphere in significant quantities.) Some urban smogs appear brown because of the NO_2 they contain. Most visible effects of air pollution are caused by the interaction of light with suspended particles. Figure 2.10 shows the possible interactions of a light photon with atmospheric particles.

In Fig. 2.10*a*, we see the light from an auto headlight coming to an eye. Some of the photons (1) come directly to the eye; the eye sees those. Some (2) are scattered

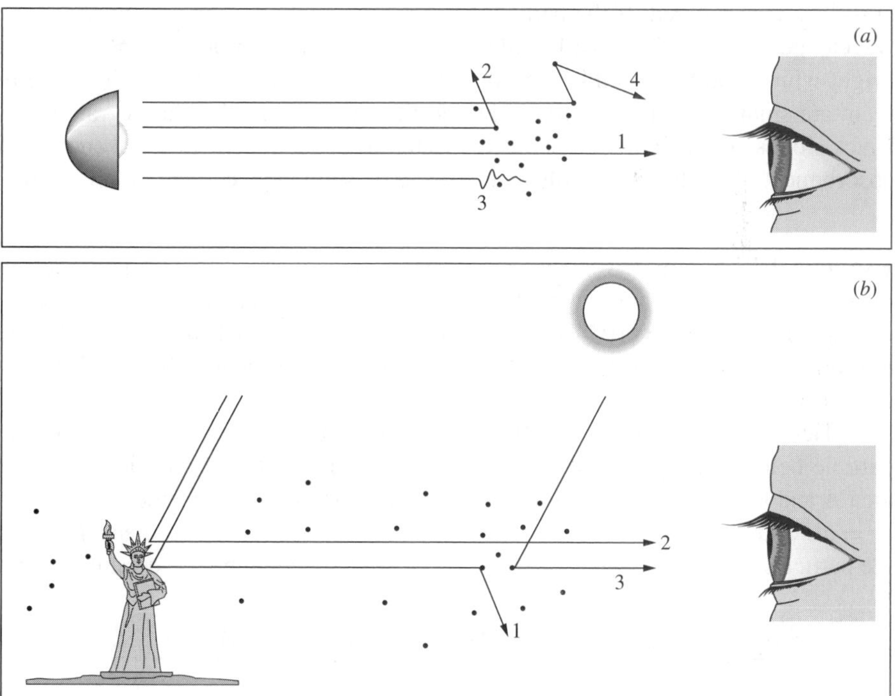

FIGURE 2.10
Possible interactions of light photons with particles: (*a*) light from an auto headlight coming to an eye, (*b*) view of the Statue of Liberty coming to an eye.

away by particles in the air between the lamp and the eye; the eye does not see those. Some (3) are absorbed by the particles; the eye does not see these photons either. Some (4) are scattered by particles more than once and come to the eye from a different direction than from the headlight. You have probably observed that on a foggy night, at first you do not see an oncoming car, then you see a diffuse glow, and finally as the car approaches, you see the shape of the car's headlights. You do not see the car at first because the light from the headlights is either scattered away or absorbed by the fog particles (water droplets) before it reaches your eyes. When you see the diffuse glow, some of the light photons from the headlights have been scattered out of the direct line of sight and then scattered again by a second collision with a droplet so that they come to your eyes from a direction other than the direct line of sight from the headlight. Finally, as the car approaches, most of the photons coming to your eyes come directly, without being scattered, so you see a clear image of the headlight.

Whether a photon is absorbed or scattered by a particle is mostly determined by the ratio of the diameter of the particle to the wavelength of the light. If the particle diameter is much larger than the wavelength, the photon will be absorbed (or reflected back if the particle is highly reflective). If the particle diameter is much smaller than the wavelength of the light, the photon will pass right by it, neither absorbed nor scattered. If the particle has approximately the same diameter as the wavelength of the light, it will scatter the light. You have probably observed that bright white clouds are unlikely to produce rain; the particles (water droplets) in them are small enough to scatter light and thus are too small to fall as raindrops. Black clouds are likely to rain on us; the water droplets in them are large enough to absorb most of the light that falls on them and thus are large enough to fall as rain.

Example 2.1. Figure 2.10 shows the possible fates of a photon of light passing through an air mass containing particles. If such a mass contains 50 $\mu g/m^3$ of particles, all of diameter $0.3 \cdot 10^{-6}$ m, and the distance between the headlight and the eye is 1 km, what fraction of the light photons would be expected to collide with one of the particles?

Here we ignore the possibility of a photon hitting two particles, or of one particle being directly behind another. We consider a prism of air with projected area A and length L, whose volume is $V = AL$. The mass of contained particles is $m = cV$ and the number of particles is $N = m/[(\pi/6)D^3\rho]$. The projected area of the particles is $A_{\text{projected,particles}} = N(\pi/4)D^2$. We may combine these values to find

$$\begin{aligned}\frac{A_{\text{projected,particles}}}{A} &= \frac{N(\pi/4)D^2}{A} \cdot \frac{m}{N(\pi/6)D^3\rho} \cdot \frac{cLA}{m} = \frac{1.5cL}{D\rho} \\ &= \frac{1.5 \cdot 50 \cdot 10^{-6}\ \text{g/m}^3 \cdot 1000\ \text{m}}{0.3 \cdot 10^{-6}\ \text{m} \cdot 2 \cdot 10^6\ \text{g/m}^3} = 0.125\end{aligned} \tag{2.3}$$

■

This simplified calculation suggests that 12.5% of the photons would be expected to contact a particle. (See Problems 2.8 and 2.9 for more detailed estimates.)

In Fig. 2.10*b*, we see how we view the Statue of Liberty from a distance on a sunny day. The statue itself is not emitting light; we see it by the sunlight reflected off of it. These reflected photons can be either absorbed or scattered by particles between it and us (1) or can come to our eyes (2). In addition, particles in the air between us and the statue can scatter sunlight to us. When we speak of air being hazy, we normally mean that it contains particles that scatter sunlight (or moonlight or streetlight) toward us, which prevents us from seeing distant scenes clearly. In Fig. 2.10*b*, if a cloud were to shade the air between us and the statue while the statue itself remained in the sun, then we would see the statue more clearly than we do when the sunlight is scattered from the particles in the air between us and the statue. The same effect is produced by a dirty windshield; if a cloud covers the sun, the visibility improves dramatically.

Gas molecules are, in effect, very small particles (diameter $\approx 0.0005\ \mu = 0.5$ nm). They also scatter light (a phenomenon called *Rayleigh scattering* [26]) but not nearly as efficiently as particles with diameters close to the wavelength of visible light (≈ 0.3 to $0.6\ \mu$). Scattering by molecules or particles changes the color of the light. Since the wavelength of blue light is shorter than that of red light, the wavelength-to-particle-size ratio is smaller for blue than for red, making blue easier to scatter than red. That is why the sky appears blue: when we look away from the sun, we see the blue part of sunlight scattered toward us, mostly by oxygen molecules. That is also why sunsets are orange or red: at sunset and sunrise, we see the sun through a longer column of air than at noon, so more of the blue light is scattered away. Normally, sunsets are redder than sunrises. Solar heating of the ground during the day and the resulting atmospheric turbulence produce a higher concentration of particles in the air at sunset than at sunrise. These particles scatter all light, so that the solar intensity is lower at sunset than at sunrise; and they scatter blue light more efficiently than red, so more of the light that reaches us is shifted toward the red end of the spectrum.

Figure 2.11 shows the visible haze caused by a layer of urban smog, trapped close to the ground and containing many fine particles. That haze is visible both because it scatters the image of the buildings and streets below it and because it scatters sunlight to our eyes. The fine particles in the second photo are mostly secondary particles, produced in the atmosphere by chemical reactions among primary pollutants emitted by human activities. There are nonhuman sources of visibility-impairing pollutants as well, e.g., secondary particles formed from hydrocarbons emitted by vegetation, wind-blown dust, and fine salt particles emitted from ocean sprays. In most major cities, particularly during periods of low winds, these secondary particles, caused by human activity, can cause a very perceptible haze.

Visibility is normally much better in dry climates than in moist ones, mostly because fine particles absorb moisture from the atmosphere and thus grow to a size at which they are more efficient light scatterers.

FIGURE 2.11
Two views of downtown Los Angeles: (*a*) a day with strong winds from a nonpolluted area brings clean air into the region; (*b*) during the morning of a day with a strong inversion (Chapter 5) with clean air above the inversion and smoggy air below. Visibility is obscured by numerous fine particles (about 0.1 to 1 μ in diameter), mostly formed in the air from hydrocarbons, sulfur dioxide, and nitrogen oxides. (Reprinted with permission from South Coast Air Quality Management District.)

The light-scattering and -absorbing properties of particles are used as a way of estimating the emissions of particles in plumes from chimneys and other sources. This phenomenon is discussed as *plume opacity* in Chapter 4.

In cities, these hazes may be beneficial because they alert the public to the fact that invisible pollutants are probably also present. These visible hazes have encouraged citizens to pay the cost of controlling air pollution, including control of invisible pollutants that may be more dangerous to their health than are the visible haze particles. In remote scenic areas, the hazes are annoying because they obscure the view. The 1977 Amendments to the Clean Air Act established a policy to protect visibility in scenic areas and to restore the visibility in scenic areas where it has been degraded by human activities. So far, the scenic areas referred to in the act have included large national parks and wilderness areas.

2.4 SUMMARY

1. Before 1960, our principal concern about air pollution effects was with property damage. Since 1960, we have been concerned primarily with human health.
2. Quantifying the health effects of short-term exposure to high concentrations of the common air pollutants is easy, but those high concentrations occur only in laboratory tests. It is much harder to quantify the health effects of the real situation we face: long-term exposure to low concentrations of these pollutants.
3. The visibility effects of air pollutants are often the effects most obvious to the public. They are now regulated in national parks and some other scenic areas.

This chapter shows that the quality of our experimental basis for deciding on the proper concentration standards for air pollutant concentrations to which the public is to be exposed is poor. At high concentrations (e.g., Figs. 2.4 and 2.6), the effects are clear and frightening. But at the concentrations to which the people of industrial countries are regularly exposed, our knowledge is much less complete and is largely based on extrapolations of the higher concentration data. We have only limited confidence in these extrapolations. Nonetheless, we must make important public health (and economic) decisions based on these inadequate data.

There is no way to escape this dilemma. If we decide to wait for more data, we are, in effect, deciding to continue doing what we are currently doing, which may be a serious mistake. We must make the best decisions we can based on the inadequate data now available. Engineers generally wish to make conservative decisions, but in this case it is not clear what is a conservative decision. If we decide to err on the side of public health by spending a large sum of the public's money on air pollution control, is that conservative? If we decide to err in the other direction, risking the public's health to save their money, is that conservative? There is no widely accepted answer to these questions.

PROBLEMS

See Common Units and Values for Problems and Examples, inside the back cover.

2.1. Industrial representatives claim that the NAAQS for SO_2 is so low that one exceeds it if one strikes a simple wooden match in a modest-sized room. Is this true?

(*a*) Calculate the concentration expected for striking such a match in a room that is 15 ft by 15 ft by 8 ft. A typical 2-inch wooden match contains $\approx$ 2.5 mg of sulfur.

(*b*) Compare the resulting concentration to the annual average SO_2 ambient air quality standard.

2.2. The NAAQS for fine particles (PM_{10}, annual average) is 25 $\mu g/m^3$. Every time you breathe, you take in about 1 liter of air.

(*a*) Assuming that the air contains 25 $\mu g/m^3$ of fine particles, how many grams of fine particles do you take in with every breath?

(*b*) Assuming that all the particles are spheres with a diameter of 1.0 μ, how many particles do you take in with every breath?

(*c*) If you are an industry representative, which of these numbers will you cite? If you represent an environmental organization, which will you cite?

2.3. The NAAQS for sulfur dioxide (annual average) is 80 $\mu g/m^3$. Every time you breathe, you take in about 1 liter of air. Assume the air is exactly at the NAAQS for SO_2.

(*a*) With every breath, how many grams of SO_2 do you take in?

(*b*) How many molecules of SO_2 do you take in? A gram of $SO_2 = 9.4 \times 10^{21}$ molecules.

2.4. Using the description in the text, draw a dose-response curve for selenium in the diet. Plot percent fatalities vs. dietary selenium input, g/day.

2.5. Suggest reasons for the following observations [27] about daily mortality:

(*a*) It is higher on Christmas and New Year's Day than on other days.

(*b*) It is higher on Monday than on Wednesday or Thursday.

2.6. Epidemiological studies are all correlations of observed health effects with measured air pollutant concentrations. In all such studies there is the hazard that some important variable has been overlooked, and that it is the true cause of the observed health effects. Careful investigators work **very hard** to avoid this error. It is widely reported that the monthly rate of death by drowning is well correlated with the monthly consumption of watermelons. Would it be safe to conclude that one of these caused the other? Or is there another variable to consider?

2.7. Figure 2.8 shows a summary of experimental data for the effect of NO_2 exposure on plants. If we consider only exposure for the time interval 0 to 10 h (i.e., up to one day but not the associated night) and if we assume that the plant breathes in and out at a rate of 1 L/h, and that such a plant will be killed if it breathes in a total of 0.001 g of NO_2 during any period less than 24 h, what would the Death curve on that figure look like? Show only a rough sketch, with no numerical values. For exposures less than 10 h, assume that the plant is exposed for Δt hours to concentration c and then spends the rest of the day in an unpolluted environment with $c = 0$.

2.8. The most widely used equation for estimating visibility is the Koschmeider equation:

$$L_V = \frac{1200 \text{ km} \cdot \mu\text{g/m}^3}{\text{Particle concentration}} \tag{2.4}$$

where L_V is the *visual range*, the distance at which an average person can barely distinguish a dark object (such as a mountain or skyscraper) against the sky. This equation is an approximation, based on an average set of atmospheric particles.

(*a*) Use Eq. (2.4) to estimate the visual range when the particulate concentration is equal to the annual average and to the maximum 24-hour NAAQS for PM_{10}.

(*b*) In the Grand Canyon and the surrounding area, on clear days, one can easily see mountains 100 miles away. What is the probable concentration of particles in the atmosphere when one can see that far?

(*c*) If the particle concentration in the atmosphere is increased by 1 μg/m^3, what is the percentage decrease in the visual range if the initial visual range is 20 km? If the initial visual range is 200 km, what is the percentage decrease?

2.9. The most general approach to visibility is

$$dE = -Eb_{\text{ext}}\, dx \tag{2.5}$$

where E is the light intensity of a collimated beam of light, b_{ext} is the *extinction coefficient*, which is the sum of the four separate b values for light scattering and absorption by gas molecules and by particles, and x is the distance [28]. If the air mass through which one views a distant object is uniform, so that b_{ext} is constant, this may be integrated to

$$\frac{E}{E_0} = \exp(-b_{\text{ext}}\, \Delta x) \tag{2.6}$$

The Koschmeider equation [Eq. (2.4)] is based on the visual range corresponding to $E/E_0 = 0.02$ when $L_V = \Delta x$.

(*a*) What is the relation between b_{ext} and L_V?

(*b*) What is the assumed relation between particle concentration c and b_{ext} in Eq. (2.4)? Reference 28, page 134, shows values for various kinds of particles, with ranges of b_{ext}/c from 0.4 to 5 (m^2/g). How does the value you compute here, which is intended to be an average over all conditions, compare with those values?

(*c*) Another measure of visibility is the *number of deciviews* [29], defined as

$$\text{Number of deciviews} = 10 \ln \frac{b_{\text{ext}}}{10^{-5}/\text{m}} \tag{2.7}$$

What is the relation between number of deciviews and visual range? What advantage might this measure have over b_{ext} and L_V as a measure of visual range? Here, 10^{-5}/m is the value of b_{ext} for air containing zero particles at an elevation of about 5500 ft.

(*d*) Show the relation between Eq. (2.3) in Example 2.1 and Eqs. (2.5) and (2.6).

(*e*) What is the value of b_{ex}/c for Example 2.1?

2.10. Figure 2.12 on page 38 shows the effect of sulfur dioxide on the corrosion of mild steel [30]. Can the data on this figure be represented by a simple equation?

2.11. By what chemical or physical mechanism does acid rain cause the destruction of statues shown in Fig. 2.9?

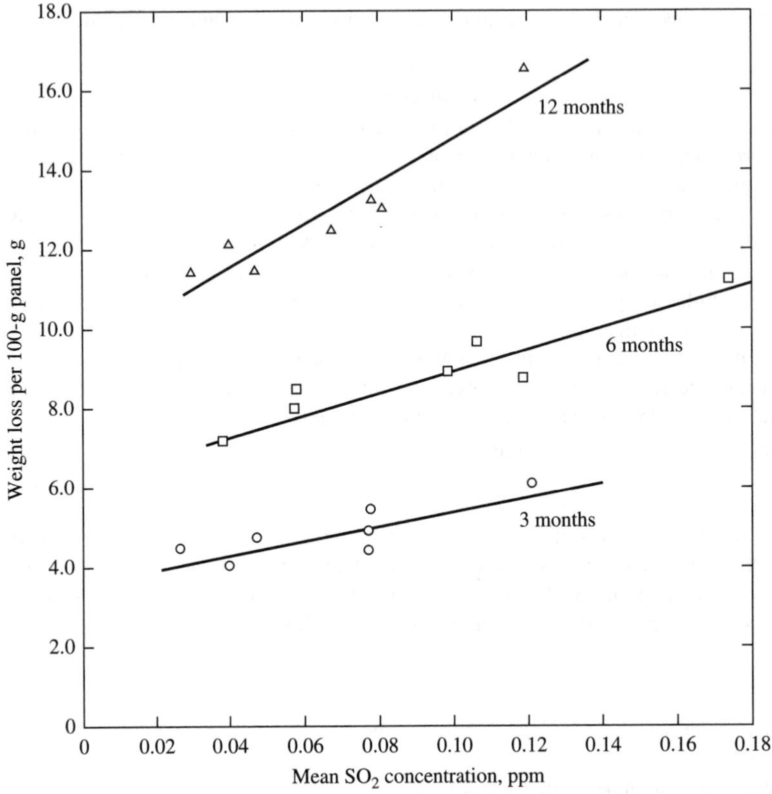

FIGURE 2.12
Relationship between corrosion of mild steel and corresponding mean sulfur dioxide concentrations for various exposure times at seven sites in Chicago (September 1963–1964) [30]. Reprinted with permission from the Air and Waste Management Association.

REFERENCES

1. Simmons, W. S., S. P. Rinne, N. S. Tesche, and B. R. Weir: "Toxicology of Fossil Fuel Combustion Products, Vols. 1 and 2," *EPRI EA-3920*, EPRI, Palo Alto, CA, 1985.
2. Ferris, B. J. J.: "Health Effects of Exposure to Low Levels of Regulated Air Pollutants: A Critical Review," *J. Air Pollut. Control Assoc.*, Vol. 28, p. 482, 1978.
3. Shy, C. M., J. R. Goldsmith, J. D. Hackney, M. D. Lebowitz, and D. B. Menzel: *Health Effects of Air Pollution*, American Lung Association, New York, 1978.
4. Watson, A. Y., R. R. Bates, and D. Kennedy: *Air Pollution, the Automobile and Public Health*, National Academy Press, Washington, DC, 1988.
5. Bascomb, R., et al.: "Health Effects of Outdoor Air Pollution," *Am. J. Resp. Crit. Care Med.*, Vol. 153, pp. 3–50, 1996.
6. Halliday, E. C.: "A Historical Review of Atmospheric Pollution," *World Health Organization Monograph Series*, No. 46, Geneva, Switzerland, pp. 9–37, 1961.
7. Goldstein, Q., L. Aronow, and S. M. Kalman, *Principles of Drug Action*, 2d ed., John Wiley & Sons, New York, p. 89, 1974.
8. Stokinger, H. E.: "Concepts of Thresholds in Standard Setting," *Arch. Environ. Health*, Vol. 25, p. 153, 1972.

9. *Threshold Limit Values and Biological Exposure Indices*, American Conference of Governmental and Industrial Hygienists, Cincinnati, OH, annual editions.
10. Ghering, P. J., P. G. Watanabe, J. D. Young, and J. E. Lebeau: "Metabolic Thresholds in Assessing Carcinogenic Hazard," in E. C. Blair (ed.), *Human Health and the Environment: A Collection of Dow Scientific Papers*, Vol. 2, Dow Chemical Company, Midland, MI, 1977.
11. Shamberger, R. J.: *Biochemistry of Selenium*, Plenum, New York, p. 32, 1983.
12. Coffin, D. L., E. J. Blommer, D. E. Gardner, and R. S. Holzman: "Effects of Air Pollution on Alteration of Susceptibility to Pulmonary Infection," *Proceedings of the Third Annual Conference on Atmospheric Contaminants in Confined Spaces*, Aerospace Medical Research Laboratory, Wright Patterson AF Base, Dayton, OH, pp. 71–80, 1968.
13. Amdur, M. O.: "Aerosols Formed by Oxidation of Sulfur Dioxide: Review of Their Toxicology," *Arch. Environ. Health*, Vol. 23, p. 459, 1971.
14. Higgins, I. T. T.: "Effects of Sulfur Oxides and Particulates on Health," *Arch. Environ. Health*, Vol. 22, p. 584, 1971.
15. "Health Consequences of Sulfur Oxides: A Report from CHESS 1970–1971," *EPA-650/1-74-004*, United States EPA, 1974.
16. Subcommittee on Special Studies, Investigation and Oversight and the Subcommittee on Environment and Atmosphere of the Committee on Science and Technology: "The Environmental Protection Agency's Research Program with Primary Emphasis on the Community Health and Environmental Surveillance System (CHESS): An Investigative Report," *USGPO NO 77-590*, U.S. House of Representatives, 1976.
17. Knelson, J. H.: "Evidence for the Influence of Sulfur Oxides and Particulates on Morbidity," *Bull. N.Y. Acad. Med.*, Vol. 54, pp. 1137–1154, 1978.
18. Amdur, M. O.: "Sulfuric Acid: The Animals Tried to Tell Us," *Appl. Ind. Hyg.*, Vol. 4, pp. 189–197, 1989.
19. Schwartz, J.: "What Are People Dying of on High Air Pollution Days?" *Env. Res.*, Vol. 64, pp. 26–35, 1994.
20. Schwartz, J.: "Air Pollution and Daily Mortality: A Review and Meta Analysis," *Env. Res.*, Vol. 64, pp. 36–52, 1994.
21. Dockery, D. W., Arden Pope III, Xiping Xu, J. D. Spengler, J. H. Ware, M. E. Fay, B. G. Ferris, Jr., and F. E. Speizer: "An Association between Air Pollution and Mortality in Six U.S. Cities," *New England J. of Medicine*, Vol. 329, pp. 1753–1759, 1993.
22. U.S. Environmental Protection Agency: "National Ambient Air Quality Standards for Particulate Matter: Final Rule," *Code of Federal Regulations, Chapter 40, Part 50, Federal Register*, July 18, 1997.
23. Dockery, D. W., and C. A. Pope III: "Acute Respiratory Effects of Particulate Air Pollution," *Anna. Rev. Public Health*, Vol. 15, pp. 107–132, 1994.
24. MacLean, D. C.: "Stickstoffoxide als phytotoxische Luftverunreinungen (Nitrogen Oxides as Phytotoxic Air Contaminants)," *Staub Reinhalt. Luft*, Vol. 35, pp. 205–210, 1975.
25. Schmidt-Thomsen, K.: "Steinzerstoerung und-konservierung in Westfalen-Lippe (Stone Destruction and Conservation in Lippe-Westfalia)," *Third International Clean Air Congress, Dusseldorf*, VDI Verlag, pp. A93–A97, 1972.
26. Modest, M. F.: *Radiative Heat Transfer*, McGraw-Hill, New York, p. 398, 1993.
27. Buechley, R. W., W. B. Riggan, and V. Hasselblad: "SO_2 Levels and Perturbations in Mortality: A Study in the New York–New Jersey Metropolis," *Arch. Environ. Health*, Vol. 27, p. 134, 1973.
28. National Research Council: *Protecting Visibility in National Parks and Wilderness Areas*, National Academy Press, Washington, DC, 1993.
29. Pitchford, M. L., and W. C. Malm: "Development and Applications of a Standard Visual Index, " *Atmospheric Environment*, Vol. 28, pp. 1049–1054, 1994.
30. Upham, J. B.: "Atmospheric Corrosion Studies in Two Metropolitan Areas," *J. Air Pollut. Control Assoc.*, Vol. 17, p. 398, 1967.

CHAPTER 3

AIR POLLUTION CONTROL LAWS AND REGULATIONS, AIR POLLUTION CONTROL PHILOSOPHIES*

Most air pollution control activities in the United States take place in response to or in anticipation of air pollution laws and regulations. These laws and regulations change with time. The details of the laws and regulations presented in this book are current as of the date of publication, 2000, but the laws and regulations are sure to change soon after the book is published. This chapter discusses the basic structure and underlying philosophies of U.S. air pollution law and regulations, which have not changed substantially in the past 30 years. Understanding that structure and philosophy will help the reader to understand the current laws and the changes that will occur in the future.

3.1 U.S. AIR POLLUTION LAWS AND REGULATIONS

Most air pollution control engineers work with permits. Major facilities (such as steel mills, copper smelters, and chemical plants) must have a permit in order to operate in the United States. These permits are authorizations by local, state, or federal authorities, normally expressed as "The emissions of pollutant *X* from the main stack at factory *Y* shall not exceed *Z* pounds per hour," for each stack in the plant,

*Much of this chapter is adapted from Refs. 1 and 2.

together with information about monitoring, reporting emissions to the regulatory agency, test procedures, and so forth. The legal authority for these permits is derived as shown in Fig. 3.1.

The permits ultimately are based on the U.S. Constitution, our basic legal and court system, and common law. The Clean Air Act of 1963, as amended in 1970, 1977, and 1990, passed by Congress and signed by the president, provides the legal basis of air pollution laws in the United States. The U.S. Environmental Protection Agency (EPA) prepares and publishes detailed regulations showing how those laws shall be applied. These regulations are the subject of public hearings, approval by the Office of Management and Budget (OMB), and litigation. When they have survived those tests, they have the force of law. Some of these regulations

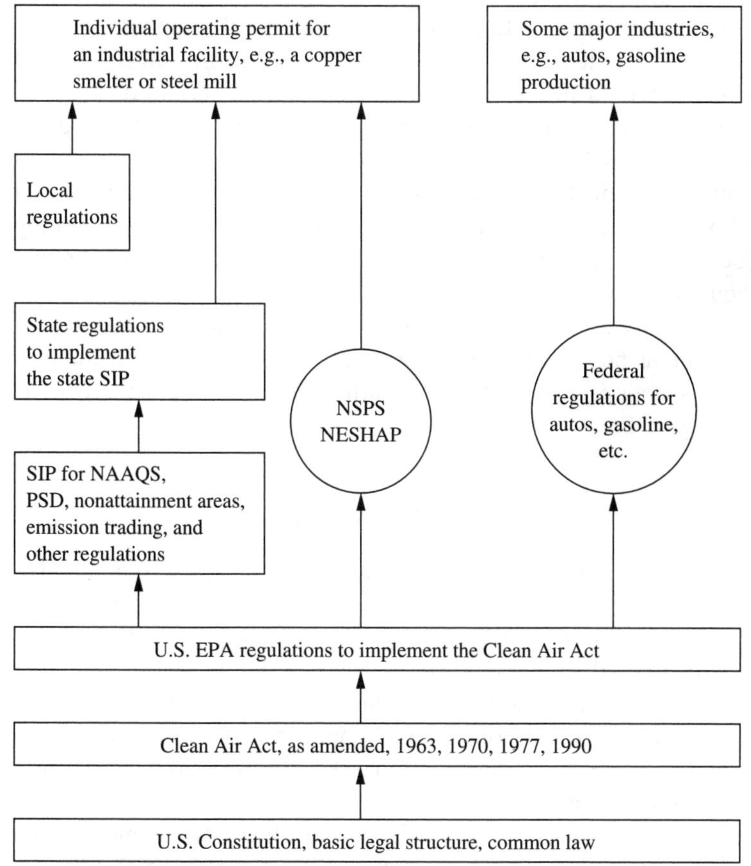

FIGURE 3.1
The flow of legal authority leading to air pollution operating permits in the United States. SIP (State Implementation Plan); NAAQS (National Ambient Air Quality Standards); PSD (Prevention of Significant Deterioration); NSPS (New Source Performance Standards); NESHAP (National Emission Standard for Hazardous Air Pollutants).

(NSPS and NESHAPS, discussed below) apply nationwide. Their corresponding EPA regulations govern a local facility's operating permits directly.

The EPA regulations are published in the *Federal Register* and compiled in the *Code of Federal Regulations, Chapter 40.* In the July 1, 1997, compilation the air pollution regulations occupy 7261 pages. These regulations include detailed instructions to the states on how to prepare *State Implementation Plans* (SIPs) for controlling air pollution in their states. These plans must undergo public review and approval in the states and then be reviewed and accepted (or modified) by the EPA. Based on their SIPs, the states prepare regulations that include the detailed operating permits for facilities in their state. The state permit often includes the direct federal regulations (NSPS, NESHAPS), and may include local regulations as well.

Federal regulations direct the states to require a permit for each facility that has the potential to emit 100 tons/yr of criteria pollutants or 25 tons/yr of hazardous pollutants. Here "has the potential to emit" means that if the facility were to operate with all its pollution control devices turned off, it would emit that amount. Thus, a facility with a potential to emit 100 tons/yr, but which has 99% efficient control and actually emits 1 ton/yr, would still be required to obtain such a permit. In addition, if the facility is located in a region with severe pollution problems, the above values at which permits are required can be smaller.

Some nationwide products, such as automobiles and gasoline, are regulated directly by the EPA. An automobile manufacturing plant must have a state operating permit for the local air pollutant emissions it creates and an EPA certification that its autos meet the federal emission standards.

Individuals generally do not need such permits. They are directly affected by local regulations (such as no open burning of garbage) and state traffic regulations and auto emission inspections, and are indirectly affected by the federal regulations on automobiles and gasoline.

3.2 AIR POLLUTION CONTROL PHILOSOPHIES

The wish of all concerned with air pollution is to have a completely unpolluted environment at no cost to anyone. That appears to be impossible, so our logical goal is to have an appropriately clean environment, obtained at an appropriate cost, with this cost appropriately distributed among industry, car owners, homeowners, and other sources of pollutants. An air pollution control philosophy is a fundamental set of ideas about how one determines what constitutes an appropriately clean environment, appropriate cost, and appropriate distribution of that cost. These ideas form the basis of the laws and regulations shown in Fig. 3.1.

The detailed regulations can be set in a strict way or a lax way (one may choose to err on the side of strict control or on the side of minimum control cost). Whether one should be strict or lax in applying any philosophy is independent of the choice of air pollution control philosophy.

A perfect air pollution philosophy and its implementing regulations are cost-effective, simple, enforceable, flexible, and evolutionary. A cost-effective philosophy gains the maximum possible benefits (reduced damages or discomforts) for the resources expended on pollution control. A simple philosophy and its implementing regulations are understandable to all involved in the pollution control effort and do not require legal interpretation of every word of the laws and regulations. An enforceable philosophy clarifies the responsibilities of all parties involved in a way that courts of law will enforce. A flexible philosophy can deal with special difficulties (such as control equipment breakdown and delays in control equipment delivery). An evolutionary philosophy enables us to utilize new information on the effects of pollution and new developments in control technology without major overhauls of our legal structure or major revisions of existing industrial plants.

3.3 THE FOUR PHILOSOPHIES

Actual regulations are often based on mixtures of the philosophies shown below. Which of these philosophies is the basis for some regulation is frequently not obvious. Nonetheless almost all air pollution regulations are based directly or indirectly on these philosophies. The four philosophies discussed here are *emission standards, air quality standards, emission taxes,* and *cost-benefit standards.* The first two are in current use in the United States and other industrial countries and are fairly well understood. The latter two have mostly been the subject of academic publications and have not had much practical testing. They are related to and interact with the first two.

3.3.1 The Emission Standard Philosophy

The basic idea of the emission standard philosophy is that there is some maximum possible (or practical) degree of emission control. This degree varies between various classes of emitters (e.g., autos, cement plants) but presumably can be determined for each class. If this degree of control is determined for each class, and every member of that class is required to limit emissions to this maximum degree possible, then the pollutant emission rate will be the lowest possible. Because emission rate and air cleanliness are inversely related (see Chapter 6), it follows that if this philosophy is carried out rigorously we will have the cleanest possible air. Thus this might be called a *cleanest possible air* philosophy.

Apparently the first large-scale application of this philosophy was the Alkali Acts in England starting in 1863 [3]. These followed the introduction of the Leblanc process for manufacturing an alkali, soda ash, Na_2CO_3. In the original form of the process, the hydrochloric acid (HCl) byproduct was emitted from the plant's smokestack as a vapor or mist. This emission devastated vegetation downwind and led to controversy and legislation. The legislation created a corps of "alkali inspectors" whose duty was to regularly inspect all alkali plants and to find the best techniques for minimizing the emission of harmful air pollutants. Once a technique had been shown to be effective in one plant, the inspectors forced all of the other plants to adopt it. Thus the emission limitations were steadily made more stringent as the control technology improved, and each member of the class was obliged to meet the same emission limitation as the cleanest member of the class. This application is called the *best technology* type of emission standard because all members of a class are required to employ the best technology currently available for controlling emissions and to keep the control equipment in good operating condition. In this type of regulation there is generally no specified emission rate or emission test; the operator who installs and operates properly the "best technology" is deemed to be complying with the regulation. (The British equivalent phrase for best technology is *best practical means.* The British air pollution control agency was called "The Alkali Inspectorate" until 1974.)

The best technology approach is still widely used in cases where determining the emission rate in pounds per hour would be difficult. For example, federal regulations for large gasoline storage tanks require that such tanks have floating roofs with well-designed and well-maintained seals (see Chapter 10). Similarly, most states require gasoline stations to use "Stage I Vapor Recovery" (see Chapter 10), which requires the station's underground tanks and the trucks that fill them to be connected in a way that minimizes emissions due to fuel transfer. The regulation consists of the technical description of the equipment and its operation and maintenance.

The prohibition against open burning of garbage and agricultural wastes is a kind of emission standard, because open burning generates more air pollutants per unit of waste than land fill, closed incineration, recycling, or composting. By forbidding open burning, we force waste disposers to use better technology.

Visible emissions from stacks and vents, particularly from the chimneys of coal-burning furnaces, are indicative of emissions of air pollutant particles. (The relation between emissions visibility and mass emission rate is far from linear, see Section 4.9). Regulations limiting these visible emissions are a form of emission standard. The common test for visible emissions, introduced by Ringleman [4], is a cheap, rapid, widely applied tool for emission regulation and enforcement.

Fuel sulfur content and gasoline olefin content maxima and gasoline oxygen content minima are also emission standards because most of the sulfur in fuels enters the atmosphere as sulfur dioxide, because olefins are more effective in causing photochemical smog than equivalent amounts of other hydrocarbons, and because autos using oxygen-containing gasolines emit less CO than those using other gasolines.

A final kind of emission standard is a numerical one. For example, under current EPA regulations, a coal-fired electric power plant whose construction commenced after September 1978 may not emit to the atmosphere more than 0.03 pound of particulates per 10^6 Btu of fuel burned, as determined by stack test, nor more than 1 percent of the ash in the fuel, whichever is less. Similarly, automobiles made in 1993 and later may not emit more than 0.25 gram/mile of hydrocarbons in a well-defined test procedure [4] (see Chapter 13).

All of these kinds of emission standards have the same general idea: there is some degree of emission control that it is practical to impose upon all members of a well-defined class of emitters, and that degree of control is required of all members of that class. This philosophy was the basis of most of the air pollution control activities in the industrial world from 1863 to 1970. In current U.S. air pollution law, two sections are "pure" emission standards. These are the Standards of Performance for New Stationary Sources (commonly called *new source performance standards* [NSPS]) and the National Emission Standards for Hazardous Air Pollutants (NESHAP).

The NSPS (see Table 3.1) prevent a firm that plans to construct a new facility from "pollution shopping" among states and localities to find the one with the least stringent air pollution control standards. No state or locality in the United States can become a pollution haven by offering a standard less stringent than the NSPS. Before the enactment of the Clean Air Act of 1970, which instituted the NSPS, some states and localities in the United States regularly invited industry to locate there and enjoy the lax pollution regulations, and some industries encouraged this policy. (No such rules exist between nations, and facilities have been moved from nations with strict standards to those without them, simply to reduce air pollution control costs.) NESHAP regulations cover pollutants that are believed to have no threshold (see Fig. 2.1). For them, any exposure is likely to produce some harm. Thus we wish to reduce emissions as much as possible, by applying best technology emission standards to all emitters of this category of pollutants (Chapter 15).

TABLE 3.1
Federal standards of performance for new stationary sources (commonly called *new source performance standards* [NSPS])

This list is an excerpt from the 1998 version of *40CFR60*. Standards are listed there for 68 industrial categories. New categories are regularly added, and existing ones modified. This excerpt shows the kind of regulations that are contained in that much larger compilation.

1. *Coal-fired power plants* whose construction started after September 18, 1978, may not emit the following to the atmosphere:
   a. Particulate matter more than 0.03 lb/10^6 Btu, or 1% of the ash solids in the fuel, whichever is less.
   b. Sulfur dioxide more than 1.2 lb/10^6 Btu, or more than 30% of the SO_2 that would be formed if all the sulfur in the coal were converted to SO_2, whichever is less.
   c. Nitrogen oxides more than 0.6 lb/10^6 Btu for most coals, or 0.5 lb/10^6 Btu for sub-bituminous coal.
2. Large *incinerators* shall not emit to the atmosphere gases that contain more than 27 mg/dry standard cubic meter of particulates. There are also limits of opacity, cadmium, lead, mercury, and acid gases.
3. *Portland cement plants* shall not emit to the atmosphere the following:
   a. Gases from the kiln containing more than 0.30 lb/ton of kiln feed (dry basis).
   b. Gases from the clinker cooler containing more than 0.10 lb/ton of feed to the kiln (dry basis).
4. *Nitric acid plants* shall not emit gases containing more than 3.0 lb of NO_2 per ton of nitric acid produced.
5. *Sulfuric acid plants* shall not emit gases containing more than 4 lb of SO_2 and/or 0.15 lb of sulfuric acid mist/ton of acid produced (100% basis).

The above regulations also limit the opacity of the plumes from these plants, mostly as a control measure, and have very detailed discriptions of testing and monitoring requirements.

These two parts of current U.S. law are "pure" emission standards in the sense that the emission rates permitted were determined strictly on the basis of best technology. On the other hand, the emission standards for motor vehicles [5] were determined not by inquiring what was the best available technology but rather by deciding on the basis of the ambient air quality standard philosophy (discussed later) what emission level was permissible and then making that emission level the standard. Because the emission standards computed in this way were more stringent than could be met by the then-current (1971) best technology, they are referred to as *technology-forcing* emission standards.

Permits of many state and local air pollution control agencies for individual facilities are based partly on their assessment of what is best technology and partly on an overriding application of the air quality standards philosophy discussed later in this chapter.

3.3.1.1 The advantages and disadvantages of emission standards. Table 3.2 compares emission standards (and the three other philosophies to be discussed later) with the list of qualities previously given. The cost effectiveness of the emission standard philosophy is very bad. If we uniformly apply the same emission standards to an entire class of emitters, including both those at remote locations and those in industrial, densely populated areas, then for a stringent standard, the remote plants will make a large expenditure to produce a small reduction in damage to receivers and hence a small benefit. If the standard is lax, then plants in industrial areas will not be controlled to the degree that minimizes damage to the surrounding population. This consequence follows naturally from application of a common standard ("cleanest possible air") to both densely and sparsely populated areas.

The simplicity of the emission standard philosophy is excellent. The entire set of regulations consists of the permitted emission rates and the description of the test method to be used to determine whether the emission standards are being met.

The problem of the trade-off between cost effectiveness and administrative simplicity of the emission standard strategy is exemplified by the history of emission standards for automobiles. In 1967 automobile manufacturers petitioned the United States Congress to write uniform motor vehicle emission standards for the whole

TABLE 3.2
Comparison of air pollution control philosophies

Desirable quality	Emission standard	Air quality standard	Emission taxes	Cost-benefit analysis
Cost effectiveness	Very bad	Good	Fair	Excellent
Simplicity	Excellent	Poor	Excellent	Terrible
Enforceability	Excellent	Fair	Excellent	Unknown
Flexibility	Poor	Fair	Unnecessary	Unknown
Evolutionary ability	Fair	Fair	Good	Good

United States and to forbid the states from individually writing their own. They did so because they feared the complexity of having to produce a multiplicity of different vehicles to meet different state standards [5]. In 1973 they petitioned Congress to do the reverse and allow them to use a "two-car strategy" in which vehicles that met stringent emission standards would be sold in areas with severe air pollution problems and vehicles that satisfied somewhat less strict standards would be sold in areas without severe air pollution problems [6]. They did this because they believed that the extra expenditure to produce and operate cars meeting stringent standards was not cost-effective in areas without severe air pollution problems. Congress refused their request.

The enforceability of the emission standard philosophy is excellent. Once standards are set and test methods defined, one knows whom to monitor and for what. Violation criteria can easily be written and penalty schedules formulated.

The flexibility of this philosophy is poor. If a plant orders pollution control equipment in good faith and the equipment fails to meet the manufacturer's predicted performance criteria (and hence the emission standards), it may take years to replace it. How should the air pollution control authorities deal with this plant? Under this philosophy they can close the plant, fine its operators, or give it a *variance* to operate until the equipment is fixed. Experience shows that plant closing is politically impossible, serious fines are politically very difficult, and the variance is an invitation to infinite delays; but under this philosophy there are no other obvious alternatives.

The evolutionary ability of this philosophy is fair. If a new technology makes it possible to set a lower standard, it can be implemented for all sources built after a certain date. This method works fairly well for autos, whose lifetime in the economy averages 10 years, but poorly for industrial plants whose lifetime is 30 to 50 years. Mandating a lower emission standard for plants built after a certain date will help the air quality in areas undergoing growth after that date but not those without growth.

Most of the progress in air pollution control between 1863 and 1970 was made by application of this philosophy. The best technology approach made sense for the Leblanc soda ash plants because their pollutant could be collected and sold at a profit. It made sense for coal-burning furnaces because their black soot emissions were wasted fuel. But most of the air pollutant emissions that can be recovered and sold at a profit are now being so collected and sold. Further progress in control of air pollutants (either to achieve cleaner air or to maintain current air cleanliness as the population grows) will be made by applying more stringent controls than those now in use, both to new and to existing sources. The emission standard philosophy is useless as a guide to deciding how stringent those controls should be.

This uselessness is illustrated by the question of the design efficiency of electrostatic precipitators for large emitters of particles, e.g., coal-fired electric power plants. The typical particle collection efficiencies by new installations have risen steadily over the past several decades, from 90 percent to 99+ percent. There appears to be no reason that precipitators cannot be built with recovery efficiencies of 99.9 percent or 99.99 percent, or better. The most general simple design equation for these precipitators is the Deutsch-Anderson equation (Chapter 9),

$$\text{Control efficiency} = 1 - \exp\left(\frac{-wA}{Q}\right) \tag{3.1}$$

where w = "drift velocity," an appropriate average particle movement velocity toward the collecting plates

A = area of the collecting surface

Q = volumetric flow rate of gas being cleaned

The cost of an electrostatic precipitator is roughly proportional to the area of the collecting surface, A, so that for a given installation (and hence a constant Q and w) we can say approximately

$$\text{Control efficiency} = 1 - \exp(-\text{some constant} \times \text{cost}) \tag{3.2}$$

Thus, according to Eq. (3.2), if it costs N dollars to install a 90 percent-efficient precipitator, it will cost $2N$ for 99 percent, $3N$ for 99.9 percent, $4N$ for 99.99 percent, etc. (This calculation is only approximate because precipitators collect big particles preferentially. As efficiency goes up, the average value of w goes down. See Chapter 9 for detail.)

Given this approximate cost/efficiency relation, what is the best technology or cleanest possible air value for this kind of installation? Clearly, we can mandate any degree of control efficiency we wish, and precipitators can be built to meet it. If the level of best technology is deemed to be 99.5 percent (a typical current value) and some plant installs a precipitator that is 99.95 percent efficient, shall we then mandate that all future plants should install precipitators that efficient? We could design and build even more stringent control devices without limit if we wished. Should we?

If society had infinite resources and were willing to commit them to the control of this one air pollutant, this question would not be difficult. But society has finite resources and will probably only commit some fraction of them to air pollution control. It would seem folly to commit all of them to this particular kind of pollutant. But the best technology philosophy or cleanest possible air philosophy, if carried to its logical conclusion, would lead inevitably to that. For this reason, those who apply this philosophy have generally tempered it with some qualifier like "taking costs into account." In current federal regulations there are defined values of *best available control technology* (BACT), *reasonably available control technology* (RACT), *maximum available control technology* (MACT), and *lowest achievable emission rate* (LAER). These all represent some kind of "best technology" that is believed suitable for some class of emitters, even though the requirements can be quite different from each other [7]. These all reflect the fact that this philosophy, if pursued to its logical conclusion, leads to impossible results. Although the emission standard philosophy has been useful in the past, it provides little guidance for the future.

3.3.2 The Air Quality Standard Philosophy

If the emission standard philosophy is logically a "cleanest possible air" philosophy, the air quality standard philosophy is logically a "zero-damage" philosophy. In Chapter 2 we discussed the idea of threshold values below which no air pollution damage would occur. The air quality standard philosophy is based on the *assumption* that the true situation for most major air pollutants is the threshold value situation sketched in Fig. 2.1. If that assumption is true, and if we can determine the pollutant concentration values (including time of exposure) that correspond to such threshold values, and if we can regulate the time, place, and amount of pollutant emissions to guarantee that these threshold values are never exceeded, then there can be *no air pollution damage, ever, anywhere.* The U.S. air pollution community is trying to do precisely that, by carrying out the basic air quality standard philosophy of the Clean Air Act.

To implement this philosophy, someone must study the available dose-response data and determine the threshold values. In U.S. air pollution law, these are to be set "with an adequate margin of safety…to protect human health" [8] and are called National Ambient Air Quality Standards (NAAQS). (Note the upbeat wording: this really means "permitted levels of contamination.") The EPA has established such standards for six major pollutants, shown in Table 2.3 [9]. (The process of setting these standards calls for issuing documents called "Air Quality Criteria," for which reason the pollutants on this list are called *criteria pollutants.*) The states are now attempting to manage air quality to ensure that those standards will not be exceeded, ever, anywhere. The procedure is illustrated in flowchart form in Fig. 3.2 on page 50.

The process for a specific pollutant at a specific locality begins with a measurement of the ambient air quality. If the measured pollutant concentration is acceptable (i.e., less than the NAAQS), then the air quality at some time in the future is predicted. If this is acceptable, no action is needed. If the future concentrations (taking into account population and industrial growth) exceed the standards, then emission regulations must be devised to prevent this predicted violation.

If the current pollutant concentrations are greater than the permitted values, then emissions must be reduced to bring the current values into compliance with the standards. Determining which emissions to reduce and how much to reduce them requires some way of estimating the relation between emissions and ambient air quality, normally an air quality model (Chapter 6).

Using these models, one computes the needed emission reductions and enacts the regulations to compel the emitters to reduce their emissions. (These are usually a set of emission standards, based not necessarily on best available technology, but rather on a computation of the emission reductions needed to meet the NAAQS.) Once this set has been enforced and the emissions have been reduced, one again measures the ambient air. If the standards are not met (and the emissions have indeed been reduced as required by the model), then the modeling exercise has produced incorrect results and the entire cycle must be repeated until the standards are met. This process was initiated in the United States in April 1971 [10], with all

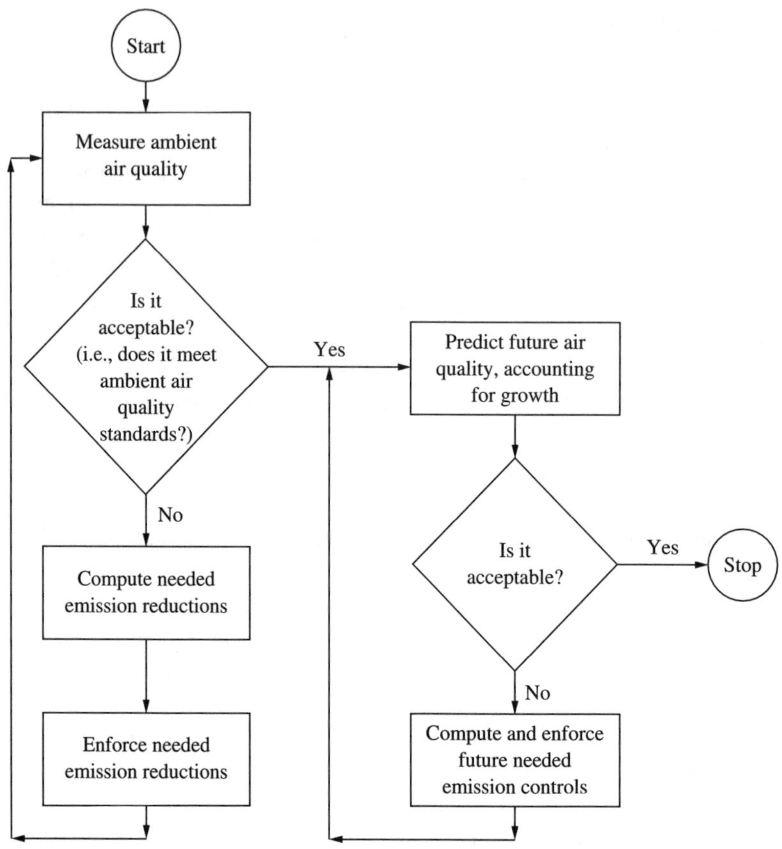

FIGURE 3.2
Flow diagram representation of the National Ambient Air Quality Standards (also called Air Quality Management) process.

of the standards to be met by 1975. The states were required to prepare SIPs and regulations to implement them for each of the major pollutants. Delays in meeting the standards were granted in some cases as provided by law. Meeting the standards has been more difficult than was imagined in 1971. In 2000, 25 years after the original deadlines, many of the standards have not been met in the regions with the most difficult problems. For the most part, we failed to meet the standards because we underestimated the total emissions, overestimated the efficacy of control measures, and used optimistic models to predict future air quality. The states and the EPA are now on their third or fourth time around the loop shown in Fig. 3.2, trying to bring the ambient air pollutant concentrations down to the NAAQS.

The process shown in Fig. 3.2 took place simultaneously for each of the six pollutants for which we have NAAQS and in each *Air Quality Control Region* (AQCR) in each state. (Some AQCRs are multistate, e.g., New York City and the adjacent part of New Jersey.) For states and pollutants where the standards have not been met (in EPA language, they have not attained the standards and hence are *nonattainment areas*), the process continues until such time as the air becomes as clean as the standards require.

3.3.2.1 The advantages and disadvantages of the air quality standard philosophy. Comparing the air quality standard philosophy with the list of desirable qualities previously described in Table 3.2, we see that its cost effectiveness is good but not excellent. It has the virtues of concentrating pollution control expenditures in the areas with the worst pollution problems and, in principle, of allowing higher emission rates (and lower pollution control expenditures) in areas with less serious problems. However, once a set of NAAQS is in place, they must be met everywhere, even in areas people seldom or never visit. Thus, this philosophy requires some control expenditures for which the damage reduction benefits are small.

No one has found a way to write a simple set of regulations based on this philosophy. The EPA's best efforts to write a simple set of regulations to enforce the NAAQS part of the Clean Air Act, which is based on the air quality standard philosophy, have been the subject of a seemingly unending set of legal challenges. Some critics have referred to the Clean Air Act as the "Lawyers' Full Employment Act." The reason for this complexity is that we are attempting to control the concentration of pollutants *in the ambient air.* Those concentrations are influenced by a wide variety of emitters, some nearby, some far away. The connection between the emissions and air quality at a given location depends on the meteorological transport and dispersion of the pollutants and on atmospheric reactions of the pollutants (see Chapter 6). None of these subjects is well enough understood to allow exact and unequivocal calculations of the contributions of individual emitters to specific local concentrations in urban areas. Given this uncertainty, regulations attempting to deal with local and long-distance polluters have been promulgated, and contested in court, with resulting modifications and complexities.

The enforcement difficulty of this philosophy results from the same cause as its complexity, namely, that one is trying to enforce air quality. When the air quality standard is not met, the culprit is not generally obvious. If the pollutant has only one major source in the region, then assigning responsibility is easy. If the pollutant is a secondary pollutant like ozone, formed in the atmosphere by the interaction of several other pollutants (volatile organic compounds—VOCs—and nitrogen oxides—NO_x) emitted by a variety of sources, then assigning responsibility is much more difficult.

The flexibility of the air quality standard philosophy is fair. Because of the multiple ways by which air quality standards can be met, those managing the air quality have some flexibility, and each state or local agency can write those detailed regulations it considers best, within limits. Special cases and emergencies can be handled locally.

The evolutionary ability of the air quality standard philosophy is fair. As new data appear, standards can be changed; but such changes require completely new emission regulations, which are expensive and time-consuming. When the EPA added the $PM_{2.5}$ standard to the existing PM_{10} standard (1997, see Table 2.3), each of the states had to write a new SIP section and the appropriate regulations for $PM_{2.5}$. The state air quality management plans (SIPs) are regularly updated, taking new data and information on control technology improvements into account.

One clear difficulty with the air quality standard philosophy, which led to court action in the United States, concerns *nondegradation* or *nondeterioration.* If it were absolutely true that there was no damage *at all, of any kind* at concentrations below the threshold values, then there could be no logical objections to polluting up to those concentrations. In effect, the EPA guidelines to the states for developing their SIPs took this view [10]. In contesting those regulations, a consortium of environmental groups showed that this interpretation was not apparently the intent of Congress nor was it even completely consistent with the EPA's own regulations issuing the standards [11]. Aside from these purely legal questions, the logical bases for opposing this view are (1) that the setting of threshold values is bound to be based on limited data, so that we cannot be absolutely certain that we will not cause harm in pure-air areas by polluting them up to the levels of the standards, and (2) that visibility (see Section 2.3) is not a threshold-value property. Hence, if we were to pollute up to the NAAQS, most of the scenic areas of the Southwest and Rocky Mountain states would experience a marked and significant degradation of their traditionally high visibility and clear skies.

This controversy was litigated for five years before Congress settled it for the United States by writing the *Prevention of Significant Deterioration* (PSD) section into the 1977 amendments to the Clean Air Act. Under these regulations the pollutant concentration in clean air areas is allowed to increase, but only by small, regulated amounts [12]. This problem and its legislative solution clearly reveal the most basic difficulty with the air quality standard or zero-damage philosophy: it is completely dependent on the assumption that there are threshold values below which there is zero damage. For visibility this assumption is demonstrably false (see Problems 2.8 and 2.9). Thus the strongest intellectual basis for the PSD doctrine is this attack on the basic premise of the air quality standard philosophy, which is certainly false in the case of visibility. However, as more and more data accumulate on air pollution effects on humans, it becomes harder to believe that the threshold-value idea applies to human populations [13]. If it becomes clear that threshold values do not apply, then it will be equally clear that the ambient air quality standard or zero-damage philosophy is without intellectual foundation. If that is the case, we can still use ambient air quality standards if we wish; but we will have to choose the values on some philosophical basis other than threshold values and zero damage.

3.3.3 Emission Tax Philosophy

Most of the current U.S. air pollution laws and regulations are based on the two preceding philosophies. We know a great deal about their advantages and drawbacks. The two philosophies discussed next are not in use to any significant extent anywhere in the world, but rather are ideas that have had theoretical discussion in academic journals. They represent possible future alternatives.

Laws based on an emission tax philosophy would tax each emitter of major pollutants according to its emission rate; e.g., X cents per pound of pollutant Y for all emitters. This tax rate would be set so that most major polluters would find it more economical to install pollution control equipment than pay the taxes. In its pure form these laws would exert no legal or moral sanction against an emitter who elected to pay the tax and not control emissions at all. In the pure form, the emission tax philosophy is clearly quite different from the air quality standard or emission standard philosophy. Emission taxes have also been proposed in combination with the air quality standard philosophy; in this combination, emission taxes would act as an added incentive to reduce emissions to lower levels than those required to meet air quality standards [14]. In this case the two philosophies would work in parallel.

Emission taxes can be considered as one member of a larger class of philosophies called *economic incentives.* The other members of this class are tax rebates, low-interest-rate loans from the government for the installation of air pollution control equipment, and direct public subsidies for pollution control. These rebates, loans, and subsidies have not been proposed as separate and complete philosophies (i.e., they have no pure form) but rather have been proposed and applied mostly as ways of distributing the costs of implementing the air quality standard or emission standard philosophies.

The emission tax philosophy assumes that the environment has natural removal mechanisms for pollutants (with chlorofluorocarbons—CFCs—as a possible exception, see Chapter 14) and that at any particular contaminant level the environment has a finite, renewable absorptive or dispersive capability. If this is so, and if that capability is seen as public property, then it should logically be rented to private users to return maximum revenue to the public treasury, and it should not be overloaded; the analogy with publicly owned forest or grazing land seems obvious. For this reason one might think of the emission tax philosophy as a *market allocation of public resources* philosophy, as compared with the cleanest possible air and zero-damage bases of the two previous philosophies.

If we take that view and apply the pure form of emission taxes, then we accomplish two desirable results. First, the degree of pollution control by the individual firm becomes an internal economic decision. In the two previously discussed philosophies, if the individual firm can persuade (or litigate) the control authorities into a less restrictive regulation, that firm saves money and possibly gains an advantage over its competitors who are not able to do so. In the emission tax philosophy, each firm chooses the degree of control efficiency that will minimize the sum of control costs and taxes for it. Industry is good at such economic choices.

Second, the emission tax philosophy should minimize the misallocation of pollution control resources. If we use it, small emitters will presumably find it economical to pay the taxes rather than put economically wasteful control devices on their plants. Large emitters will find the taxes on their emissions prohibitive and will install high-quality control equipment. Overall, this should allocate pollution control resources well.

Many versions of emission taxes have been proposed and discussed, but none has reached the state of legislation. Many states charge permit fees that are proportional to emissions, so these are a form of emission tax. The amounts involved are generally much smaller than the taxes that would be charged in a pure emission tax regulatory scheme. Comparison of the emission tax philosophy with the list of desirable qualities must be based on impressions of how the legislation would work.

The cost effectiveness should be fair because an emission tax philosophy would allow each emitter the choice of controlling emissions or paying the taxes (or controlling to some economic degree and paying for the rest). Making the decisions—whether or not to control and what the degree of control should be—a matter of the internal economics of major emitters would probably result in a better overall cost effectiveness than is possible with uniform emission standards. However, uniform national emission taxes may result in some remote plants installing control equipment at large cost to minimize taxes without a corresponding reduction in damages.

Most schemes proposed so far only envisage taxes on large sources. For these, the tax rates and emission test methods constitute the whole of the regulations. If an attempt was made to extend the tax to all emitters of a particular pollutant, then the problem would become much more complex [15]. For sulfur oxides, for example, one could tax motor vehicle and home-heating fuels, based on sulfur content, at a rate comparable to that for sulfur emissions from large industrial sources. This would be simple. But there seems to be no comparably simple scheme for particulate or NO_x emissions from home-heating sources, autos, etc.

If tax schemes are limited to large sources, then enforceability should be excellent. The emission-testing industry would have to be expanded, and certification of emission test firms instituted; but once a certified body of independent emission testers was available, their test values would be readily accepted as the basis for tax payments. Recording emission meters in exhaust stacks would also be most useful.

Flexibility to deal with the kinds of problems previously discussed would be unnecessary. Other philosophies need flexibility to deal with the problem of an emitter who cannot economically meet an area-wide standard, or who cannot meet it by a statutory deadline, or who has a control equipment breakdown. In an emission tax system, the emitter simply pays the tax. (With the control equipment out of service, the emission tax meter will run very fast, providing a strong incentive to get the equipment back in service quickly!)

The evolutionary ability should be good because the tax rate could be changed as necessary. Caution would be required, because industry has complained about their difficulties with changing standards. (They speak of the difficulty of "shooting at a moving target," which apparently adds greatly to the pleasure of duck hunting, but is not as much fun in industry.) However, raising a tax rate for existing plants causes much less economic disruption than lowering an emission standard. In the case of the tax rate increase, the existing plant would probably elect to pay the higher tax, whereas for the lowered emission standard, it would probably have to replace its existing pollution control equipment with more effective equipment.

Although the emission tax philosophy is widely favored by economists in pure or mixed form and was proposed in one of President Nixon's messages to Congress [14], it has generally been anathema to American industry. One industrial group stated [16], "As a matter of principle, the right to pollute the environment should never be for 'for sale.' " If we compare this view to that of holders of grazing rights on the public domain, we see that it is the same. Those who enjoy free or subsidized use of the public domain are reluctant to pay the fair market price for that use.

In a pure emission tax philosophy we need some way to set the emission tax rates. Generally the suggestion is that tax rates will be raised on a previously announced schedule, continuing until the air is "clean enough." If we decide on the basis of assumed threshold values, then emission taxes become merely a novel way (possibly a good one) of implementing the air quality standard philosophy, and not a freestanding philosophy at all. If the basic assumption of the air quality standard philosophy proves incorrect, then using it as a basis for determining "clean enough" in the emission tax philosophy has the same drawbacks as discussed previously.

We could choose not to consider air quality at all in deciding on our tax rates and use some purely economic criteria, e.g., maximum tax revenue or marginal cost of pollution control equal to some current best technology value. Such an approach would presumably include no consideration of air pollution damage to the public.

3.3.4 Cost-Benefit Philosophy

The cost-benefit approach assumes that either there are no thresholds or, if there are, they are low enough that we cannot afford to have air that clean. If so, then we must accept some amount of air pollution damage to someone, somewhere. This philosophy suggests that we attempt to decide, in as rational a manner as possible, how much damage we should accept and correspondingly how much we should be willing to spend to reduce damages to this level.

The idea is illustrated in Fig. 3.3 on page 56. At the right, a high ambient air concentration of pollutant corresponds to zero pollution control cost. The ambient air concentration can be reduced by air pollution control expenditures. The control cost goes up steeply as the ambient air concentration becomes small. At zero concentration we have zero damage costs; the damage cost rises slowly at first and then more rapidly at high concentrations. The sum of the two costs has a minimum value at some intermediate concentration. This minimum corresponds to the optimum pollution control expenditure; expenditures above or below it are economically wasteful.

Figure 3.3 is an example of the classic "minimization of the sum of two costs" problem that appears in economics and engineering texts. The minimum occurs when the slopes of the two cost curves are equal and opposite, or

$$\frac{d(\text{pollution control costs} + \text{pollution damage costs})}{d(\text{ambient air pollutant concentration})} = 0 \tag{3.3}$$

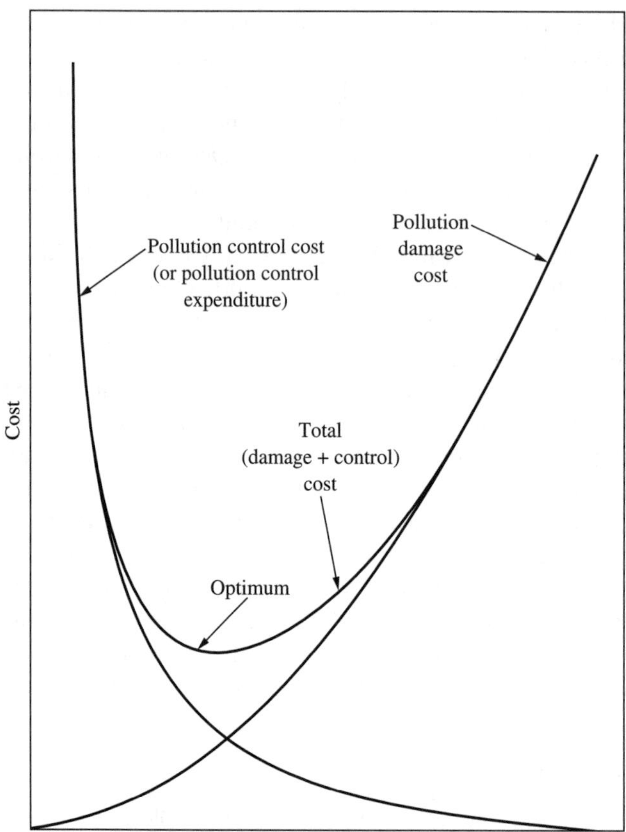

FIGURE 3.3
Schematic representation of the relation between damage, control, and total costs, for one pollutant at one location.

Figure 3.3 is a great simplification because it shows one control cost curve, one damage cost curve, and one atmospheric concentration. In reality there is a damage curve for each individual exposed to air pollution, a control curve for each emitter (including autos, household space-heating plants, etc.), and a concentration dimension for each pollutant at each location. Thus instead of a one-dimensional optimization, we have a multidimensional optimization with the number of dimensions being at least as large as the number of people in the world.

The simple application shown in Fig. 3.3 and Eq. (3.3) does not consider the questions of "Whose costs, whose benefits?" If the pollutant is emitted by our autos, then the cost of controlling their emissions will probably be distributed over the population in the same way as the damages are. But for a pollutant emitted by one factory, and injuring the rest of the community, that distribution is quite unequal; and questions of justice and equity must be included with the questions of economics.

This approach is frequently criticized by those who say, "You can't reduce *X* to monetary terms," where *X* may be human health, human life, or the quality of a clear sky, or air pollution damage to the cathedrals of Europe, or something else. Such values are hard to assign, but society obviously does. The value we place on health is indicated by how much each of us spends to safeguard or improve our own personal health and by how much society spends to improve community health. The value society places on human life is indicated by how much society will spend to prevent one accidental death [17]. Juries set financial values on loss of life and health every day. The value we place on clear skies is indicated by how much people will give up to live in areas with such clear skies. Frequently, the person making this criticism disagrees with society's evaluation. The author disagrees with some of society's evaluations, but that is not grounds for saying that society cannot and does not evaluate these things.

Comparing the cost-benefit philosophy with the list of desirable properties, we see that its cost effectiveness is excellent. Since the goal of this philosophy is to solve the cost-benefit minimization problem, if that minimization is done properly the result must have the best possible cost effectiveness. (*Cost-benefit* means the process sketched in Fig. 3.3, in which costs and benefits are explicitly computed. *Cost effectiveness* means trying to find the minimum-cost way to reach some goal or objective whose benefits are not explicitly computed, e.g., meet the NAAQS for ozone (O_3) in Los Angeles, or conduct a manned mission to Mars.)

The cost-benefit philosophy is not simple. The problem of solving Eq. (3.3), with as many variables as there are people and enacting regulations to enforce it, is far beyond our current capabilities. Because of this complexity, we will likely never have air pollution regulations based directly on cost-benefit analysis or on the direct application of Eq. (3.3). More likely, greatly simplified approaches will be used, for example, the use of cost-benefit analysis to set emission standards or to set air quality standards. For many years, laws based on the emission standard philosophy have included words suggesting that standards be set "taking into account the cost..." [18] or analogous words about reasonableness or practicality. In deciding what is reasonable or practical, those writing the regulations have consciously or unconsciously attempted to decide what the benefits of a given control measure would be and balanced these benefits against the cost. If it should become clear that one or more of the major air pollutants are no-threshold pollutants, air quality standards will probably be set on a cost-benefit basis.

Because Table 3.2 is written for "pure" philosophies, the enforceability and flexibility of the cost-benefit philosophy are listed as "unknown." No one has published any clear idea of how a set of regulations based on pure cost-benefit analysis would be written. The evolutionary ability should be good. As new air pollution damage data or new control technology appears, we can introduce these into our cost-benefit equation and modify the regulations to take them into account.

If this philosophy will most likely not be used as a "pure" philosophy, but rather as a guide for setting emission or air quality standards, why list this as a philosophy? The major purpose of this chapter is to elucidate the true philosophical bases on which we are currently acting. If we have been applying this philosophy informally, then we ought to admit it. If emission standards and/or air quality standards are not really our basic philosophy, then we ought to devote the thought and effort necessary to putting our cost-benefit decisions on a sound basis. To do so will require public exposure of the assumptions and value judgments that are needed to do cost-benefit calculations involving human health damage, aesthetic damage, etc. Exposing such value judgments will be painful and controversial, but the alternative will be making air pollution control decisions in a less-informed way.

3.4 MARKET CONTROL AND EMISSION RIGHTS

In carrying out the mandates of the Clean Air Act of 1970, most of the states enacted regulations that placed a numerical emission limit on each stack of each plant. Industries concluded that they could often meet the overall plant emission limitation more economically by controlling large sources more stringently than their permits required, and not controlling smaller ones. This practice led to legal controversy over how much flexibility individual emitters had in meeting overall emission goals. Industry calls the detailed, stack-by-stack regulation approach taken by the EPA and most of the states the *command and control* approach. They dislike it.

Industry also suggested that if two factories each emitted *X* pounds/year of some pollutant, and the applicable SIP required a 20 percent reduction in the emission of that pollutant, then it could be cheaper for one factory to reduce emissions by 40 percent than for both to reduce by 20 percent. Presumably both factories would share the cost of 40 percent control at one factory.

If a factory were to reduce emissions more than the minimum required by the applicable regulations, industry asked the state and federal regulators to permit them to bank, sell, or trade the credit for that extra emission reduction. This request was accepted in some EPA regulations in the 1980s and strengthened in the 1990 revision of the Clean Air Act. This provision is particularly important in areas that do not meet the NAAQS (nonattainment areas). By current EPA rules a new facility that wishes to locate in a nonattainment area must produce somewhere in the area an emission reduction that is larger than the new facility's permitted emissions. Banking, trading, or selling emissions credits with other facilities allows this to be done. That raises the philosophical question of whether someone who has always emitted *X* pounds/year of pollutant *Y* has a marketable property right to do that in the future. In current U.S. law the answer is yes. Consequently, for some old, dirty factory, the most valuable asset is the marketable right to emit air pollutants. In Los Angeles the oil refineries have concluded that they can reduce regional hydrocarbon emissions more cheaply by buying up and junking old cars (thus taking them off the road) than they can by improving the already efficient emission controls in their refineries. The U.S. government from 1980 to 1992 was strongly market-oriented and supported many schemes to bring market forces into the area of air pollution regulation. It is too early to say whether that has had long-term beneficial effects.

3.5 PRINCIPAL U.S. AIR POLLUTION LAWS

The body of U.S. air pollution law is contained mostly in the Clean Air Act and the regulations (local, state, and federal) that implement it. The principal parts of this complex law are listed in Table 3.3. The law contains many other provisions that are procedural, legal, and budgetary and that are generally of less interest to pollution control engineers than those listed here. Air pollution laws interact with water pollution and solid waste laws as well.

TABLE 3.3
The most important sections of the Clean Air Act of 1970, as amended in 1977 and 1990

Section	Title	Principal provisions
107	Air Quality Control Regions (AQCR)	Divides the country into regions. States must administer air quality in each such region, under federal supervision
109	NAAQS	Establishes National Ambient Air Quality Standards
110	Implementation plans (SIP)	Requires states to prepare and enforce State Implementation Plans. Gives details on how it is to be done
111	NSPS	Establishes the Standards of Performance for New Stationary Sources, commonly called the *new source performance standards*
112 and 301–306	NESHAP	Establishes national emission standards for hazardous air pollutants, also called air toxics
160–169	PSD	Lays out rules and regulations for regions with air cleaner than the NAAQS and for the protection of visibility, principally in large national parks and wilderness areas
171–192	Nonattainment areas	Gives detailed descriptions of what must be done in areas where NAAQS are not currently met
202–235	Mobile sources	Places control of motor vehicle emissions mostly in the hands of the federal government; sets motor vehicle and fuel composition standards
401–416	Acid deposition control	Establishes a federal acid deposition control program
601–618	Stratospheric ozone protection	Establishes programs for protection of the stratospheric ozone layer

3.6 SUMMARY

1. All major sources of air pollutants in the United States of America are required to have permits that regulate their emissions. These are mostly issued by states, as directed by the federal Clean Air Act and its implementing regulations.
2. From 1863 to 1970 air pollution control efforts were largely based on the emission standard or cleanest possible air philosophy. Since then, the air quality standard or zero-damage philosophy has been dominant in U.S. air pollution law. The emission tax, or market allocation of public resources, philosophy has been proposed and discussed as an alternative to these two philosophies.
3. Air quality standards, which are based on the assumptions of true threshold values, answer the question, "How clean should the air be?" Emission standards and emission taxes do not answer that question at all. If the basic assumption of the air quality standard philosophy proves incorrect, then none of these three philosophies will answer that question.
4. Informally, or unconsciously, pollution control agencies have answered that question by some kind of estimate of costs and benefits.
5. Market methods of allocating emission rights are favored in current U.S. air pollution laws and regulations.

PROBLEMS

See Common Units and Values for Problems and Examples, inside the back cover.

3.1. If an automobile uses 1 gallon of fuel for each 15 miles traveled, if the fuel density is 6 lb/gal, and if the hydrocarbon (unburned gasoline) emission standard for autos is 0.25 g/mile (in the exhaust gas) (Federal Standard for 1993 and later automobiles) and the emissions equal this standard, what fraction of the fuel fed to the car is emitted (unburned) in the exhaust gas?

√ **3.2.** (*a*) What percent efficiency must an ash collector for a coal-fired power plant have to meet the NSPS for coal-fired power plants (Table 3.1)? See inside the back cover for the properties of typical coal.

(*b*) Is the 0.03 $\text{lb}/10^6$ Btu rule more or less restrictive than the 1 percent emission rule?

(*c*) At what percent ash in the coal would the two rules be equally restrictive?

(*d*) Why are there two separate restrictions? (To answer this part of the question, you must know some of the history of these regulations beyond that presented in this book.)

3.3. Do the regulations for coal-fired power plants and for cement plants (kiln) in Table 3.1 lead to the same required control efficiency for particles? The uncontrolled particle emissions from a typical cement kiln [19] are about 180 lb/ton of kiln feed.

√ **3.4.** Many industrial countries are relocating factories with high air pollution (and other environmental) control costs to developing countries, and operating them with air and other pollutant emissions much larger than would be tolerated in any industrial country. (Lower wages are also a factor.) The industries and the less-developed countries argue that, although no one likes the pollutants emitted, those less-developed countries have much worse environmental problems than those caused by the industry (e.g., most of the population has no safe drinking water, and children regularly die of waterborne diseases that do not occur in industrial countries). The taxes and payroll from the industries will help develop safe drinking water, schools, hospitals, etc. The developing countries consider the air and other pollution from these factories a small price to pay for these economic and environmental benefits.

Environmental groups say this amounts to exporting pollution and to exploitation of the poor in other countries to support the wasteful lifestyles in industrial countries.

Who is right? Are both right? If you as the prime minister of a developing country were offered a plant that would be a severe air polluter but would generate enough taxes to pay for the installation of a safe drinking water system for a community that has none, would you accept?

REFERENCES

1. de Nevers, N., R. E. Neligan, and H. H. Slater: "Air Quality Management, Pollution Control Strategies, Modeling and Evaluation," in A. C. Stern (ed.), *Air Pollution,* Vol. 5, 3d ed., pp. 3–40, Academic, New York, 1977.
2. de Nevers, N.: "Air Pollution Control Philosophies," *J. Air Pollut. Control Assoc.,* Vol. 27, pp. 197–205, 1977.
3. Halliday, E. C.: "A Historical Review of Atmospheric Pollution," *Air Pollution,* World Health Organization Monograph Series No. 46, pp. 9–37, Geneva, Switzerland, 1961.
4. "Control of Air Pollution from New Motor Vehicles and New Motor Vehicle Engines; Certification and Test Procedures," *40CFR86.*
5. Mann, T.: "Statement on Behalf of Automobile Manufacturers Association," *Air Pollution—1967 (Automotive Air Pollution),* Hearings before the U.S. Senate Subcommittee on Public Works, 90th Congress, U.S. Government Printing Office, Washington, DC, Part I, p. 395, 1967.
6. Iacocca, L. A.: "Testimony at Hearings Before the Subcommittee on Air and Water Pollution of the Committee on Public Works, United States Senate," U.S. Government Printing Office, Washington, DC, No. 5270-01940, p. 1103, 1973.
7. *National Emission Standards Study:* Report of the Secretary of Health, Education, and Welfare to the Congress in Compliance with the Clean Air Act of 1967, U.S. Government Printing Office, Washington, DC, p. 106, April 1970. See also J. E. Bonine, letter in *J. Air Pollut. Control Assoc.,* Vol. 25, p. 1099, 1975.
8. *The Clean Air Act of 1970,* PL 91-604, Sec. 109(b)-1.
9. *40CFR50,* National Primary and Secondary Ambient Air Quality Standards.
10. Environmental Protection Agency: "Requirements for Preparation, Adoption, and Submittal of Implementation Plans," *Federal Register,* Vol. 36, p. 8186, April 1971.
11. The air quality standard document (Ref. 10) (Sec. 410.2(c)) says: "The promulgation of national primary and secondary ambient air quality standards shall not be considered in any manner to allow significant deterioration of existing air quality in any portion of any state." Furthermore, the Clean Air Act (Sec. 101b(1)) says: "The purposes of this title are (1) to protect and enhance the quality of the Nation's air resources. . . ." These two quotations plus the legislative history of the act formed the basis for the conservation group's successful suit to prevent pollution of the air up to the level of the standards.
12. de Nevers, N.: "Some Alternative PSD Policies," *J. Air Pollut. Control Assoc.,* Vol. 29, pp. 1139–1144, 1979.
13. Knelson, J. H.: "Developing Health Standards from Epidemiological and Clinical Data," presented at the 3rd Life Sciences Symposium, Los Alamos, New Mexico, October 1975.
14. Nixon, R. M.: "The President's 1972 Environmental Program," *Weekly Compilation of Presidential Documents,* Vol. 8, Issue 7, p. 220, 1972.
15. Lees, L., et al.: "Smog—A Report to the People," California Institute of Technology Environmental Quality Laboratory EQL Report No. 4, Pasadena, CA, January 1972.
16. Anon.: "Council Report, National Industrial Pollution Control Council," U.S. Dept. of Commerce, p. 13, February 1971.

17. "Air Quality and Automobile Emission Control," Vol. 4: "The Costs and Benefits of Automobile Emission Control," *A Report by the Coordinating Committee on Air Quality Studies, National Academy of Sciences, National Academy of Engineering,* prepared for the Committee on Public Works, U.S. Senate, Serial 93-24, p. 258 et seq., September 1974.
18. "The Clean Air Act" as amended, (42 USC 1857-18571), as amended by PL 91-604 and PL 101-549. (The first version of this act was passed in 1963, but the major, sweeping revision was in 1970. There were also major revisions in 1977 and 1990.)
19. "Compilation of Air Pollutant Emission Factors, Volume I: Stationary Point and Area Sources," 4th ed., AP-42, (U.S. EPA, Office of Air Quality Planning and Standards, Research Triangle Park, NC 27711), 1985, with updates through 9/91. pp. 8.6–8.

CHAPTER 4

AIR POLLUTION MEASUREMENTS, EMISSION ESTIMATES

There are two kinds of air pollution measurements: ambient measurements (concentrations of pollutants in the air the public breathes, or *ambient monitoring*) and source measurements (concentrations and/or emission rates from air pollution sources, or *source testing*). Both are required in the ambient Air Quality Standard philosophy (Chapter 3), the principal basis of air pollution law in the United States. Concentrations in the ambient air must be measured to determine whether that air is indeed safe to breathe (i.e., it meets the NAAQS). To control pollutant concentrations, we must regulate the time, place, and amount of their emissions. Thus emission rates of various sources of air pollutants (e.g., factories, power plants, automobiles) must be measured.

Even if we did not have legal requirements for these tests, we would need them to evaluate the performance of air pollution control devices, which normally are sold with performance guarantees. The buyer will usually not pay for the control device until tests demonstrate that the device meets these performance guarantees in actual plant operation.

In most air pollution control agencies the monitoring and source testing are done by different people, who use different terminologies to discuss their work.

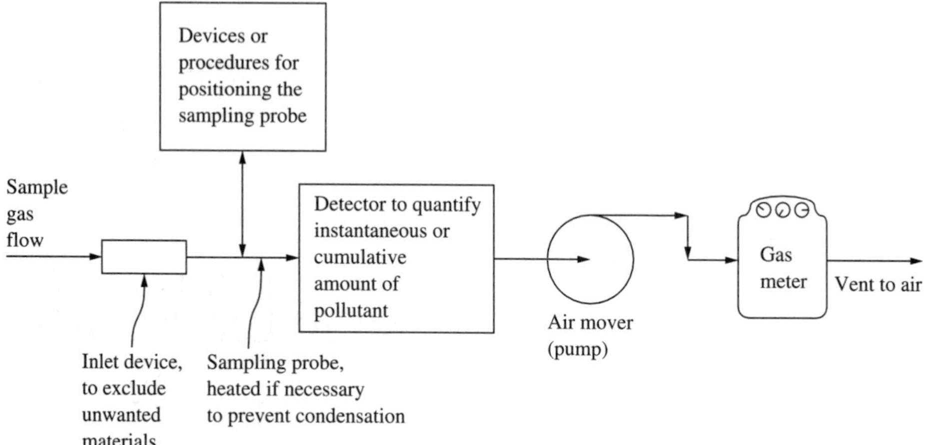

FIGURE 4.1
The components of any ambient-monitoring or source-sampling device. If the detector functions in real time (not cumulative), then the gas meter is not needed, but some kind of signal integrator or recorder is.

This chapter tries to treat them both as one, because they have so much in common. Where there is a significant difference between what the two groups do, that is noted.

Almost all air pollution measuring devices (ambient monitoring or source sampling) have some or all of the various parts shown in Fig. 4.1. As we discuss the details in subsequent sections, look back at this figure to see how each piece fits into this overall view.

4.1 A REPRESENTATIVE SAMPLE

Any air pollution measurement involves two problems. The first is to obtain a suitable, representative sample; the second is to determine the concentration of the pollutant of interest in it correctly. Generally the first is harder.

What constitutes a representative ambient air sample has been the topic of prolonged legal and technical controversy. Some of the problems are illustrated in Fig. 4.2. The air inside the parking structure normally contains much more CO than the NAAQS allows for ambient air. So if one takes a sample inside such a structure one finds a violation of the NAAQS. If one takes a sample directly across the street from such a structure, in most cases the concentration will be an order of magnitude less than inside the structure. A block away, the concentration will be even less. On the sidewalk directly adjacent to the structure the concentration will be perhaps twice as high as on the opposite side of the street. Which, if any, of these locations is suitable for obtaining a sample of ambient air? Generally the ambient air sampler should be located at the place *to which the public has free access* where the pollutant

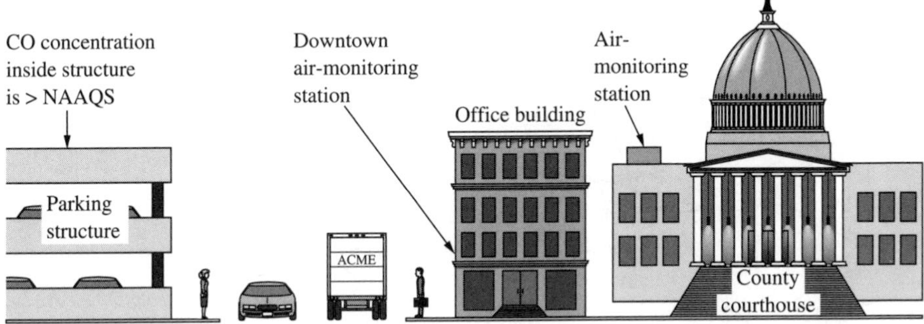

FIGURE 4.2
Illustration of some of the problems of choosing a sampler site to measure ambient CO in a city.

concentration is *highest*. This excludes all indoor spaces and plant sites to which the public has no access.

An ambient monitor must be placed where it has power, shelter from rain and snow, perhaps a constant temperature environment, easy access for monitoring personnel, protection from vandalism, and, if possible, free rent. The traditional place has been the roof of the county health building or of the county courthouse. Unfortunately the concentration of auto-related pollutants measured there is often much lower than at street level at the busiest intersection downtown.

The U.S. EPA has very detailed guidelines for the proper placement of intakes for air samplers that are meant to represent ambient air [1]. Carbon monoxide measurements must be made at street level, downtown. In Fig. 4.2 the air pollution control agency could meet this requirement by renting an office on the second floor of a downtown building and hanging its sampling probe out of the window, about 10 feet above the sidewalk. Other choices may be equally plausible, but those of the EPA are probably as good as any and have the merit of being uniform across the country.

In source testing, the representative sample problem is equally difficult. Gas flow in a large industrial flue or smokestack may be steady and well-mixed across the diameter of the stack, in which case any sample taken any time and any place in the stack will be representative. But for most such stacks the velocity and concentration in the stack vary from point to point and from time to time, so that many separate measurements must be made and averaged. Figure 4.3 on page 66 shows the measured local velocities and concentrations in a duct carrying a particle-bearing gas stream. The differences in velocity and particle concentration from place to place in the duct are substantial. Clearly if one had measured the velocities and concentrations at only one point, for example, near the bend on the inside, one would have computed a much lower overall gas flow rate and emission rate than the true values. Even 36 feet farther downstream, the velocity and concentration data do not indicate that the flow has become uniform, although the nonuniformity is much less than it was close to the bend.

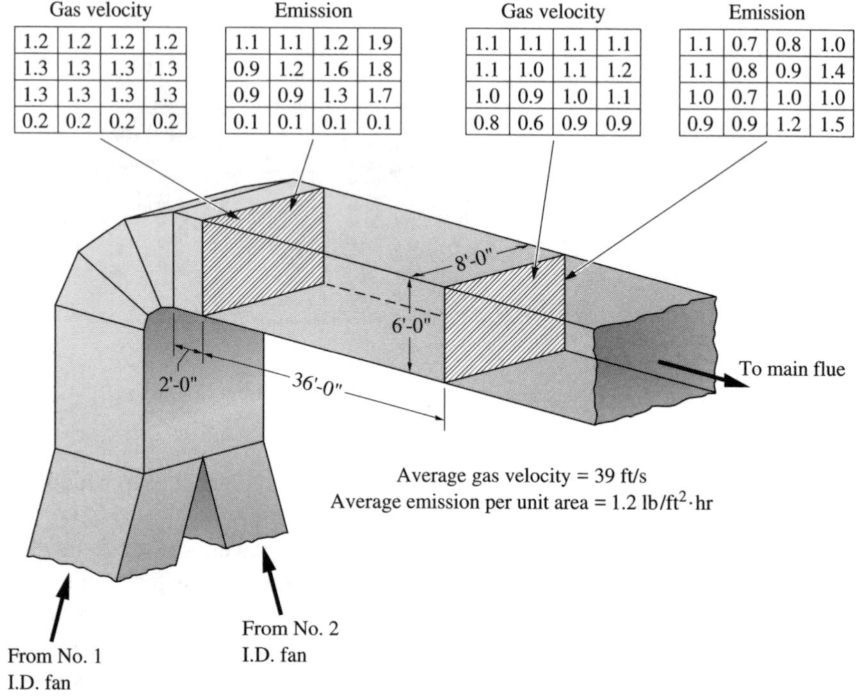

FIGURE 4.3
Measured velocities and particle mass flow rates (velocity × concentration) in a complex duct. The values shown are the ratio of the observed value to the average value for the whole duct. For example, near the bend, where the velocity is shown as 1.2, the measured velocity was 1.2 × 39 ft/s = 47 ft/s = 14.3 m/s [2].

Example 4.1. In a source test, the stack was divided into four sectors, each of which had the same cross-sectional area. The following velocities and pollutant concentrations were measured in these sectors:

Sector number	Velocity V, m/s	Concentration c, mg/m^3
1	10	500
2	12	600
3	14	650
4	15	675

What is the average concentration in the gas flowing in this stack?

The average concentration is

$$c_{\text{avg}} = \frac{\text{Total mass}}{\text{Total volume}} = \frac{\sum VAc}{\sum VA} = \left[\frac{\sum Vc}{\sum V}\right]_{\text{for equal areas}}$$

$$= \frac{10 \times 500 + 12 \times 600 + 14 \times 650 + 15 \times 675}{10 + 12 + 14 + 15} = 616\ \frac{\text{mg}}{\text{m}^3}$$

If the sampled sectors did not have equal areas this calculation would be more complex. ■

Presumably if one went far enough downstream in this duct one would find that the flow velocities and particle concentrations had become uniform. If one can do that, one should. But a more typical sampling situation is shown in Fig. 4.4. In those ducts, one cannot find a place "far downstream from any change of direction or other flow disturbance." In newer plants designers have sometimes considered the problems of obtaining a uniform gas flow and have provided access and a suitable

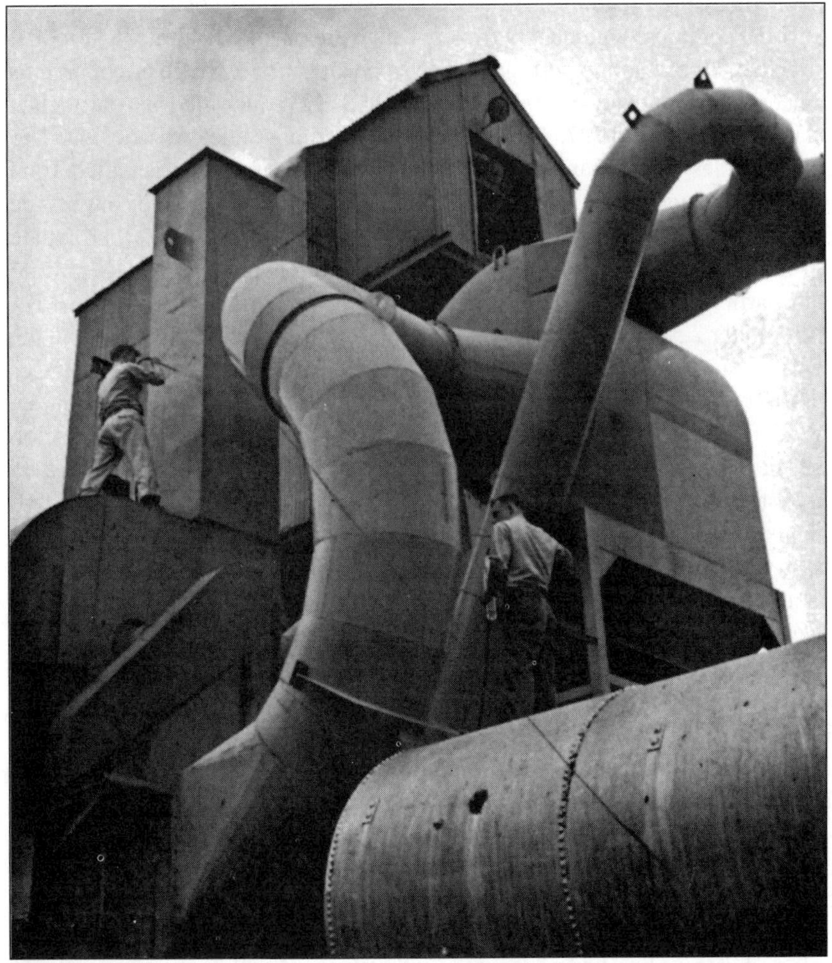

FIGURE 4.4
The long straight duct and the comfortable place to stand and rest one's instruments, with rain cover and power supplied, are not the norm for source testing [3]. (Reproduced with permission of Academic Press and the University of Minnesota.)

location for the source sampler to place the required instruments. In older plants such convenience is rarely the case, and in many newer plants it is more economical to make the plant efficient, and let the source tester work harder and under more adverse conditions. Procedures adopted by the EPA attempt to standardize the number and location of the samples on a technically sound basis [4].

4.2 GETTING THE REPRESENTATIVE SAMPLE TO THE DETECTOR

Many sampling instruments have some kind of device on their inlets to exclude unwanted materials. For example, insects sucked into a particulate sampler nozzle lead to erroneously high readings. A bug screen will exclude all but the smallest of these. Air often contains large dust particles, which are of little health concern, that weigh more than all of the fine particles in the same air sample, which are of serious health concern. As indicated in Table 2.3, the U.S. EPA modified its particulate samplers in 1987 and in 1997. Before 1987, the sampler inlet was designed to exclude all particles larger than 50 microns; the quantity sampled was called Total Suspended Particulate (TSP). The 1987 modification changed the inlet to exclude all particles larger than 10 microns; the quantity sampled is called PM_{10} (particulate matter 10 microns or smaller). The 1997 modification changed the inlet again to exclude all particles larger than 2.5 microns; the quantity sampled is called $PM_{2.5}$ (particulate matter 2.5 microns or smaller). The 1997 modification also changed the flow rate.

In all sampling devices there is the possibility that gases may condense in the sampling device or react with the solids they encounter there. Many combustion stack gases have a high water content and will condense on the walls of an unheated sampling probe; probes are normally heated to prevent this. Acid gases like SO_2 will react with alkaline solids on a filter, thus increasing the weight of solids on the filter.

If a *grab sample* taken in the field is brought to a laboratory for analysis, the sample container must not react with or modify its contents in transit. This problem is real; even apparently inert materials like glass react with some air pollutants.

4.3 CONCENTRATION DETERMINATION

Once a representative sample has been obtained, the concentration of the pollutant in it must be determined. For some pollutants this measurement can be done fairly easily by real-time instruments. Most of these operate optically; the sample passes through a cell in which a light beam of a suitable wavelength is absorbed by the pollutant of interest, or the pollutant is allowed to enter a fast, light-producing chemical reaction with some reagent and the resulting light emission is measured. If a wavelength can be selected that is absorbed or emitted by that specific pollutant and not by any other pollutant then this determination may be quick, simple, accurate, and cheap. Generally, however, some other pollutant called an *interference* also absorbs or emits at the wavelength of interest.

The problem of interferences is not trivial. Measuring SO_2 in nitrogen gas is easy. One passes the gas through a dilute solution of NaOH, in which the reaction is

$$SO_2 + 2NaOH \rightarrow Na_2SO_3 + H_2O \tag{4.1}$$

and measures the change in NaOH concentration by simple acid-base titration. However, if the problem is to measure SO_2 in air, CO_2 in the air will cause an interference by this reaction,

$$CO_2 + 2NaOH \rightarrow Na_2CO_3 + H_2O \tag{4.2}$$

and the measured change in NaOH concentration will be due partly to the SO_2 and partly to the CO_2. In this case, the CO_2 concentration in the air is known, so a correction could be made in principle (but not in practice, see Problem 4.6); but in the other cases the concentration of the interfering component may not be known, so that the uncertainty in the resulting measurement is also unknown. This type of problem can trap even very competent researchers. The U.S. EPA adopted a new method for measuring NO_2 and then had to withdraw it when it became clear that the method was not adequately protected from such interferences [5].

4.4 AVERAGING

If we are measuring ambient air quality with real-time instruments, we generally want to know the average concentration over some period of time so that we can compare it with the applicable ambient standards, which all have some measuring period. This is found by

$$\text{Average concentration} = c_{\text{avg}} = \frac{1}{\Delta t}\int c\,dt \tag{4.3}$$

where c = the instantaneous concentration indicated by the instrument

t = the time of measurement

Most of the real-time instruments present their results as an electronic signal that can be easily averaged by built-in electronics for any suitably chosen averaging time.

The older instruments for gases as well as the current instruments for particulates are not real-time instruments but rather are averaging instruments. For example, one EPA-required method for ambient particulate sampling is the $PM_{2.5}$ sampler. It consists of a special inlet that excludes particles larger than 2.5 μ in diameter, a filter, a fan, a flow-measuring device, and a suitable housing. A preweighed filter is placed in the filter holder, and air is sucked through it for 24 hours at a measured rate. The concentration of particles is computed from

$$\text{Average concentration} = c_{\text{avg}} = \frac{\text{Increase in filter weight}}{\text{Air flow rate} \times \Delta\text{time}} \tag{4.4}$$

Example 4.2. A $PM_{2.5}$ sampler ran for 24 hours at an average flow rate of 16.7 L/min. The tare weight of the fresh filter was 0.1400 g, and the gross weight of the filter, dried to the same humidity as the fresh filter, was 0.1405 g. What was the average $PM_{2.5}$ concentration in the air drawn through the sampler?

From Eq. (4.4), we calculate that

$$c_{\text{avg}} = \frac{(0.1405 - 0.1400)\text{ g}}{16.7\text{ L/min} \times 24\text{ h}} \times \frac{\text{h}}{60\text{ min}} \times \frac{1000\text{ L}}{\text{m}^3} = 2.08 \times 10^{-5}\,\frac{\text{g}}{\text{m}^3} = 20.8\,\frac{\mu\text{g}}{\text{m}^3}$$

This answer would normally be reported as 21 $\mu g/m^3$. ■

In this example we see that the sample size for a conventional $PM_{2.5}$ sampler is quite small. The value of 21 $\mu g/m^3$, which is 84% of the annual average NAAQS, is typical; and the measured change in filter weight is only 0.5 mg. High-quality weighing and sample humidity control are needed. If the weighings are only reliable to ± 100 μg, then our confidence in the difference of the two measurements is ± 200 μg, and our confidence in the ambient concentration is only about 2 parts in 5 or ± 40 percent. If the weighing uncertainty were ± 10 μg, then our uncertainty in the concentration would be 2 parts in 50 or ± 4 percent. The example also shows that the resulting measurement is the average over the past 24 hours. This type of instrument is not nearly as suitable for determining hourly variations or trends as are the real-time instruments. But efforts to develop a suitable real-time instrument for $PM_{2.5}$ have so far not been completely successful, mostly because of this small sample size problem (see Problems 4.10 and 4.11). (The 16.7 L/min flow rate for $PM_{2.5}$ samplers is close to the average human breathing rate, so that the sample weight increase, 500 μg, is close to the mass of fine particles breathed in per day by an average person when the $PM_{2.5}$ concentration is 20.8 $\mu g/m^3$.)

Some older measuring schemes for gaseous pollutants operated somewhat like this one, passing a measured volume of gas through a bubbler that contained a solution that reacted specifically (no interferences!) with the gas to be measured and then titrating the solution to determine the concentration of the pollutant in the gas or measuring the color of the solution. The calculation is the same as in Eq. (4.4) except that the weight change of the filter is replaced by the change in number of equivalents of reagent reacted times the molecular weight ratio or the color of the solution times some suitable weight conversion factor.

Similarly, many source sampling devices use these cumulative measuring schemes. For example, Fig. 4.5 shows the sampling train recommended by the EPA for measuring the concentration of SO_2 in a stack. The Pitot tube is used to measure the gas velocity in the stack. The sample of gas is pulled by the pump through the sampling probe, a midget bubbler, and three midget impingers (i.e., glass bubblers in which the gas contacts a suitable reagent). Then the gas passes through a needle valve and rotameter, which are used to ensure the flow rate of gas is in the right range, and into a dry gas meter, which is the ultimate measure of the amount of gas that has flowed through the system.

In practice, a sampling train like the one in Fig. 4.5 would be used to take samples at various points in the stack, with the flow rate at each point adjusted by the needle valve and rotameter so that sampling rates are proportional to local velocities as indicated by the Pitot tube. In this way the concentration determined by a single

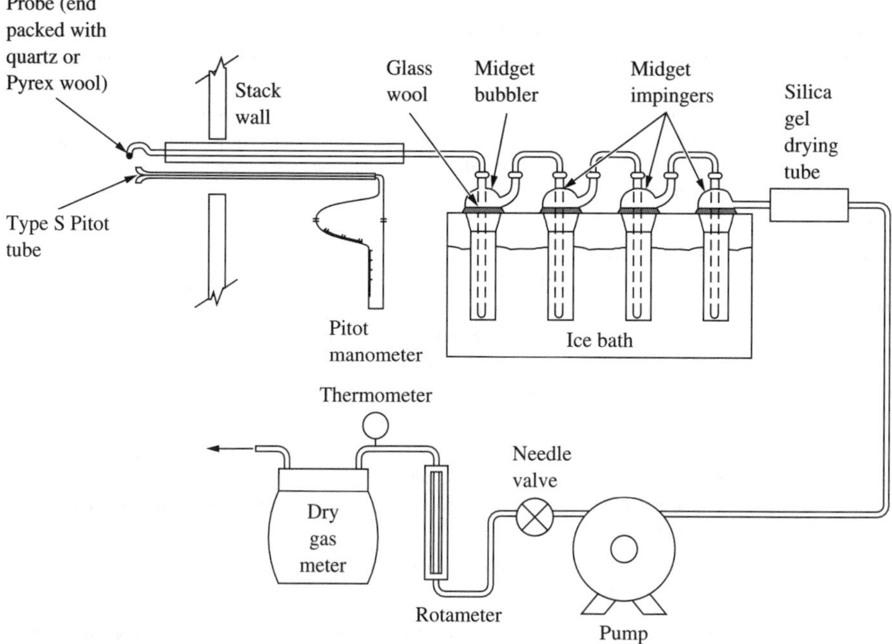

FIGURE 4.5
U.S. EPA "Method 6" sampling train for SO_2. Glass wool excludes particulate matter from the rest of the sampling train. The midget bubbler contains an aqueous isopropanol solution, which removes SO_3 but not SO_2; its contents are discarded after the sampling is completed. The first two midget impingers contain an aqueous solution of hydrogen peroxide; the third impinger is empty and traps carryover liquid from the second. At the end of the test, the contents of the three midget impingers, plus the water used to rinse them, are combined and titrated with barium perchlorate, using a thorin indicator. The silica gel drying tube protects the pump, rotameter, and dry gas meter from moisture carried over from the impingers [4].

chemical analysis of the reagents in the impingers and by the cumulative reading of the dry gas meter would be as representative as possible of the average concentration in the stack. This procedure averages the readings, just as was done in Example 4.1, and finds the measured value with one set of concentration measurements, rather than by numerically averaging a group of many individual measurements.

4.5 STANDARD ANALYTICAL METHODS

EPA has *standard sampling methods* for various pollutants. Often these are different for ambient monitoring than for source sampling. Table 4.1 on page 72 shows standard methods for monitoring *ambient* air [1]. In effect, these methods *define* the pollutants. Thus, for legal and regulatory purposes, in ambient air SO_2 is defined as that material which is detected by the SO_2 method shown in Table 4.1, the West-Gaeke method. However, in a powerplant stack SO_2 is defined as that material that is detected by "Method 6" (Fig. 4.5), which is chemically quite different from the West Gaeke method. This difference does not seem to have caused much trouble.

TABLE 4.1
Test methods for major air pollutants in ambient air

In EPA terminology, for each major air pollutant there is a *reference method*, which is the test method that is considered the standard against which other methods can be tested, and there are *equivalent methods*, which have been checked against the reference method and found to give similar results. State and local ambient monitoring agencies mostly use the equivalent methods, which are generally simpler, cheaper, and easier to use than the reference methods. This table lists only the reference methods. All of the material in this table is described in much more detail in Ref. 1.

Particulate Matter, TSP, PM_{10}, and $PM_{2.5}$. There are three standard methods. In all three a sample is drawn through an inlet designed to exclude particles larger than a certain size (50 μ, 10 μ, and 2.5 μ, respectively), and then collected on a filter for 24 hours. The filter's gain in weight is divided by the measured cumulative air flow through the filter to determine the particle concentration (see Example 4.2). The filter size and air flow are much larger for the TSP and PM_{10} devices than for the $PM_{2.5}$ device. The TSP (total suspended particulate) filter is used only for the lead measurement, described below. Both PM_{10} and $PM_{2.5}$ are used to test compliance with the applicable NAAQS.

Sulfur Dioxide (SO_2). In the West-Gaeke method a known volume of air is bubbled through a solution of sodium tetrachloromercurate, which forms a complex with SO_2. After several intermediate reactions, the solution is treated with pararosaniline to form the intensely colored pararosaniline methyl sulfonic acid, whose concentration is determined in a colorimeter.

Ozone (O_3). The air is mixed with ethylene, which reacts with ozone in a light-emitting (chemiluminescent) reaction. The light is measured with a photomultiplier tube.

Carbon Monoxide (CO). The concentration is measured by nondispersive infrared (NDIR) absorption. Here *nondispersive* means that the infrared radiation is not dispersed by a prism or grating into specific wavelengths; rather, filters are used to obtain a wavelength band at which CO strongly absorbs.

Hydrocarbons (Nonmethane). The test is for hydrocarbons excluding methane. The gas is passed through a flame ionization detector (FID), where the hydrocarbons burn in a hydrogen flame. Hydrocarbons cause more ionization than hydrogen; this ionization is detected electronically. Part of the sample is diverted to a gas chromatograph, where methane is separated from the other gases and then quantified. Its concentration is subtracted from the total hydrocarbon value from the FID. Although there is no NAAQS for hydrocarbons, its measurement in ambient air is required as part of the control program for O_3, for which it is a precursor.

Nitrogen Dioxide (NO_2). NO_2 is converted to NO, which is then reacted with ozone. The light from this chemiluminescent reaction is measured. Because ambient air contains NO (often more than NO_2), a parallel sample is run without conversion of the NO_2 to NO, and the resulting NO reading is subtracted from the combined NO and NO_2 reading to give the NO_2 value. The instrument normally reports the NO concentration as well.

Lead. A TSP filter is extracted with nitric and hydrochloric acids to dissolve the lead. Atomic absorption spectroscopy is then used to determine the amount of lead in the extract.

4.6 DETERMINING POLLUTANT FLOW RATES

The mass flow rate of pollutant is the product of the concentration in the gas and the molar or mass flow rate of the gas, e.g.,

$$\text{Pollutant molar flow rate} = (\text{molar flow rate of gas}) \times (\text{pollutant molar concentration in gas}) \tag{4.5}$$

Example 4.3. The sampling train shown in Fig. 4.5 indicates that the concentration of SO_2 in a stack is 600 ppm. The Pitot tube and manometer in the same figure indicate that the flow velocity is 40 ft/s. The stack diameter is 5 ft. The stack gas temperature and pressure are 450° F and 1 atm. What is the SO_2 flow rate?

The molar flow rate of the gas is

$$\text{Molar gas flow rate} = VA\rho$$

$$= 40\,\frac{\text{ft}}{\text{s}} \times \frac{\pi}{4}(5\ \text{ft})^2 \times 2.59 \times 10^{-3}\,\frac{\text{lbmol}}{\text{std ft}^3} \times \frac{528°\text{R}}{910°\text{R}}$$

$$= 1.18\,\frac{\text{lbmol}}{\text{s}} = 536\,\frac{\text{mol}}{\text{s}}$$

The molar flow rate of SO_2 is

$$1.18\ \text{lbmol/s} \times 600 \times 10^{-6} = 7.08 \times 10^{-4}\ \text{lbmol/s} = 0.32\ \text{mol/s}$$

Multiplying by the molecular weight of SO_2, we have

$$7.08 \times 10^{-4} \times 64 = 4.53 \times 10^{-2}\ \text{lb/s} = 163\ \text{lb/h} = 20.6\ \text{g/s} = 74.1\ \text{kg/h}$$ ■

This simple calculation would be suitable for a stack whose velocity and concentration are the same at every point and time in the stack. Otherwise, averaging would be needed.

4.7 ISOKINETIC SAMPLING

In stack sampling for particulates—but not in any sampling for gases—one must maintain *isokinetic* flow into the sampling probe. The problem is illustrated in Fig. 4.6. If the gas velocity inside the sampling probe is the same as the gas velocity in the stack from which the sample is being taken, the sampling condition (bottom sketch, Fig. 4.6) is isokinetic ($V_n = V_s$), and the measured concentration (c_m) will

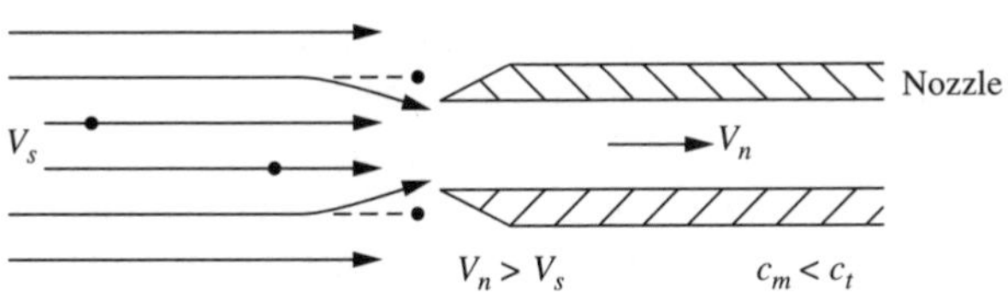

Sample *is not* representative.

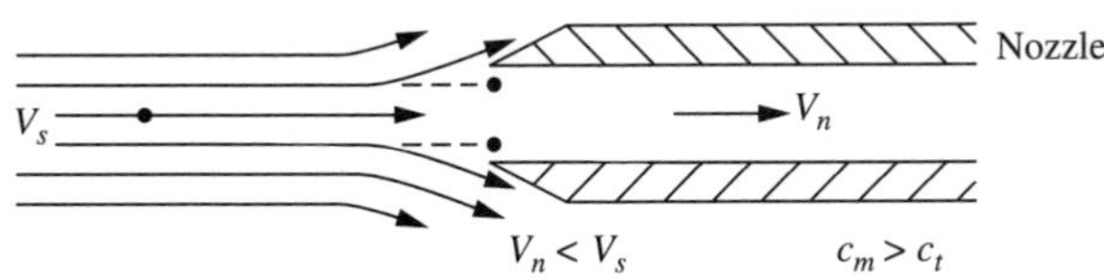

Sample *is not* representative.

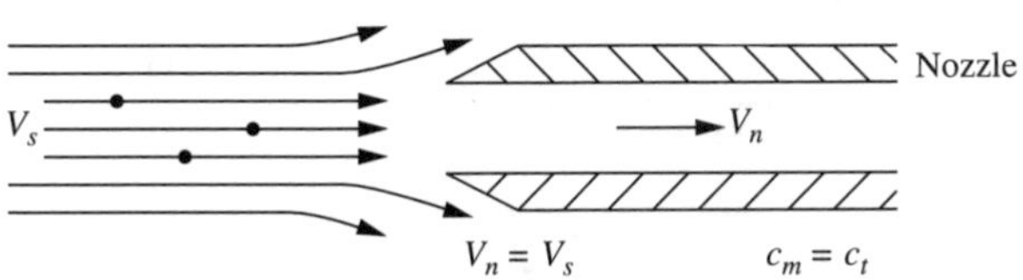

Sample *is* representative.

FIGURE 4.6
The isokinetic sampling for particulates.

equal the true concentration (c_t). However, if the velocity in the nozzle exceeds that in the stack (top sketch, Fig. 4.6), then the gas streamlines will bend into the nozzle, and the inertia of the particles will carry some of them past the nozzle even though the gas they were in will be collected, and thus the measured concentration will be less than the true concentration. Conversely, if the velocity in the nozzle is less than the velocity in the stack (middle sketch, Fig. 4.6), then the gas streamlines will bend away from the nozzle and their inertia will carry some particles into the nozzle, away from the gas that accompanied them. The measured concentration will then be larger than the true one. Various ingenious schemes have been devised to adjust the flow in the nozzle to keep its velocity the same as the flow in the stack and thus preserve isokinetic conditions. More details on this topic appear in Refs. 3 and 6.

4.8 EMISSION FACTORS

Emission testing is expensive. For simple, well-defined sources (e.g., a powerplant stack), it can be tedious but it is not difficult. For a poorly defined source (e.g., road dust from an unpaved road or CO from a forest fire), reliable test results are difficult to get. Furthermore, such testing is only possible after the facility is in place; often we want to know what the emissions from a new facility will be before it is built.

To meet these needs, the EPA has produced a very useful set of *emission factor* documents [7]. These are commonly referred to by their original publication number, AP-42. These are summaries of the results of past emission tests, organized to make them easy to apply.

Example 4.4. Table 4.2 shows part of two tables from one section of the EPA emission factors library. It shows the estimated emissions from the combustion of bituminous coal if no control devices are used. These are the emissions *going into* the control devices. Comparing them to the permitted emissions *coming out* of the plant (see Table 3.1), one can estimate the degree of control required.

Using this table, estimate the emissions from a 500-MW power plant at full load, burning a typical Pittsburgh seam coal (see inside the back cover and Appendix C). The thermal efficiency is 35 percent (this is a high-efficiency, modern plant; the electric power industry would refer to it as having a "heat rate" of (3413 Btu/kwh)/0.35 = 9751 Btu/kwh). The power plant's boiler is assumed to be of the PC, wall-fired, dry bottom type.

All of Table 4.2 is in terms of tons of coal burned. We can compute the coal consumption rate by

$$\begin{pmatrix}\text{Coal consumption}\\ \text{rate}\end{pmatrix} = \frac{\text{Power output}}{\text{Efficiency}\cdot\text{coal heating value}}$$

$$= \frac{500\ \text{MW}}{0.35 \times 13{,}600\ \text{Btu/lb}} \times \frac{3413\ \text{Btu}}{\text{kWh}} \times \frac{1000\ \text{kW}}{\text{MW}}$$

$$= 3.585 \times 10^5\ \frac{\text{lb}}{\text{h}} = 179\ \frac{\text{ton}}{\text{h}} = 163\ \frac{\text{tonne}}{\text{h}}$$

TABLE 4.2
Emission factors for bituminous and subbituminous coal combustion without control equipment

	Emission factor, lb/ton of coal burned[a]				
Furnace type[b]	**All particles[c]**	**PM^c_{10}**	**$SO_x^{d,e}$**	**NO_x^f**	**CO**
PC, wall-fired, dry bottom	10A	2.3A	38S	21.7	0.5
PC, wall-fired, wet bottom	7A	2.6A	38S	34	0.5
PC, tangential fired, dry bottom	10A	2.3A	38S	14.4	0.5
Cyclone	2A	0.26A	38S	33.8	0.5
Spreader stoker	66	13.2	38S	13.7	5
Hand-fired	15	6.2	31S	9.1	275

Source: Tables 1.1-3 and 1.1-4 of EPA Emission Factors Book [7]. Section 1.1 of that document (Bituminous and Subbituminous Coal Combustion) is 46 pages long and has 19 tables, 6 figures, and 77 literature citations.

[a] To obtain emission factors in kg/MT, divide table values by 2.

[b] The various furnace types are described in [7] and in combustion books. PC means pulverized coal.

[c] The letter A on some particulate and PM_{10} values indicates that the weight percentage of ash in the coal should be multiplied by the value given. Example: If the factor is 10A and the ash content is 8%, the particulate emissions before the control equipment would be 10 · 8 or 80 lb of particulate per ton of coal.

[d] S = the sulfur content, which plays the same role as A in the preceding footnote.

[e] SO_x is expressed as SO_2. It includes SO_2, SO_3, and gaseous sulfates.

[f] NO_x is expressed as NO_2. It includes NO and NO_2.

The particulate emission rate is

$$\begin{pmatrix}\text{Particulate}\\ \text{emission rate}\end{pmatrix} = \begin{pmatrix}\text{emission}\\ \text{factor}\end{pmatrix}\begin{pmatrix}\text{coal flow}\\ \text{rate}\end{pmatrix} = 10\text{A}\,\frac{\text{lb}}{\text{ton}}\begin{pmatrix}\text{coal flow}\\ \text{rate}\end{pmatrix}$$

$$= (10 \times 8.7)\,\frac{\text{lb}}{\text{ton}} \times 179\,\frac{\text{ton}}{\text{h}} = 1.56 \times 10^4\,\frac{\text{lb}}{\text{h}}$$

$$= 7.8\,\frac{\text{ton}}{\text{h}} = 7.1\,\frac{\text{tonne}}{\text{h}}$$

(See Notes *c* and *d* to Table 4.2 for the meanings of A and S.) Of these 7.8 ton/h of particulates, (2.3/10) = 23% are PM_{10} (smaller than 10 microns). In the same way, we can calculate the SO_2 emission rate, or

$$\begin{pmatrix}\text{SO}_2\\ \text{emission rate}\end{pmatrix} = \begin{pmatrix}\text{emission}\\ \text{factor}\end{pmatrix}\begin{pmatrix}\text{coal flow}\\ \text{rate}\end{pmatrix} = 38\text{S}\,\frac{\text{lb}}{\text{ton}}\begin{pmatrix}\text{coal flow}\\ \text{rate}\end{pmatrix}$$

$$= (38 \times 1.6)\,\frac{\text{lb}}{\text{ton}} \times 179\,\frac{\text{ton}}{\text{h}} = 1.09 \times 10^4\,\frac{\text{lb}}{\text{h}}$$

$$= 5.4\,\frac{\text{ton}}{\text{h}} = 4.9\,\frac{\text{tonne}}{\text{h}}$$

For carbon monoxide we can see that the emission rate is 0.5 lb/ton · 179 ton/h = 89.5 lb/h, and for nitrogen oxides, 3884 lb/h. ■

Table 4.2 contains two types of emission factors, those that can be calculated approximately and those that rest entirely on measurement.

Example 4.5. If all of the ash in the coal in Example 4.4 were emitted with the gas stream, what would the emission factor for particulates be?

The factor is 10A lb/ton, where A is the ash percentage in the coal. If the coal is A percent ash, then the coal contains

$$\text{Ash content} = \frac{\text{A}}{100} \times \frac{2000\text{ lb}}{\text{ton}} = 20\text{A}\ \frac{\text{lb}}{\text{ton}}$$

and thus, if all of it were emitted, the factor would be 20A instead of 10A. ■

If we knew nothing about the behavior of pulverized coal furnaces, we could estimate the particulate emission factor by simple stoichiometry; as shown in Example 4.5, we would find an answer that is twice the value in Table 4.2. Data in that table reflect the observation that, on the average, 25 percent of the ash in this type of furnace falls to the bottom *(bottom ash)*, 25 percent falls to the bottom of the economizer, and is also called bottom ash, while the remaining 50 percent is carried along with the gas stream *(fly ash)*. This is an example of a semicalculable emission factor. The emission factor for SO_2 is also semicalculable (see Problem 4.13). In Table 4.2 there is no simple way to calculate the emission factors for carbon monoxide and aldehydes, which are products of incomplete combustion. (Advanced combustion models can estimate them.) Instead, their values in Table 4.2 are simply the average of many test results.

We can prepare an estimate of the emissions from a new or existing facility quickly and cheaply using emission factors. The Emission Factors book gives a quality grade to each of the individual emission factors presented. Those shown in Table 4.2 range in quality from A (best) to E (worst), indicating that some of these values are highly reliable, based on ample, high-quality test data, whereas others are based on limited and/or questionable data. The original test data sources are usually cited. A prudent engineer always consults the emission factor library before deciding to do an emission test. Perhaps the test is not needed. If it is, the library will provide a good estimate of what to expect in the test and lead to literature on how previous investigators tested this type of emission source and what difficulties they encountered.

In principle, there is no difference between emission tests for stationary sources (e.g., factories) and for mobile sources (e.g., autos). In practice, they are different, because the normal function of a mobile source is to move about, whereas for testing purposes it must normally sit still because the measuring instruments are stationary. States and local agencies often use the emissions at idle to find and correct maladjusted or malfunctioning cars. For federal certification of new auto models the tests are done on a chassis dynamometer, in which the auto's drive wheels are placed on rollers that allow the vehicle to operate at significant engine speeds while it is standing still and, by changing the resistance of the rollers, to appear to accelerate or to go uphill or down. See Chapter 13 for more about mobile source emissions.

4.9 VISIBLE EMISSIONS

Section 2.3 showed that airborne particles can scatter and absorb light. If the particle concentration in the plume from some source (a smokestack, a diesel truck, an oil-burning auto) is high enough, the plume will be visible. As the plume flows downstream and mixes with the surrounding air the particle concentration is diluted, and the plume's *opacity* or *optical density* decreases (see Section 6.3). In principle one should be able to look at a plume and estimate the particle emission rate from what one sees. In practice it is relatively easy to measure the opacity, using either trained observers or electro-optical devices. But relating that opacity to the mass emission rate of the particles in the plume, which is usually the regulated quantity, is not easy.

The reason for this difficulty is evident in the following equation:

$$\begin{pmatrix}\text{Plume}\\ \text{opacity}\end{pmatrix} = f\left[\begin{pmatrix}\int c\,dx\\ \text{across the}\\ \text{plume}\end{pmatrix}, \begin{pmatrix}\text{particle}\\ \text{size}\\ \text{distribution}\end{pmatrix}, \begin{pmatrix}\text{particle}\\ \text{optical}\\ \text{properties}\end{pmatrix}, \begin{pmatrix}\text{solar}\\ \text{illumination}\\ \text{angle}\end{pmatrix}, \begin{pmatrix}\text{moisture}\\ \text{content of}\\ \text{plume}\end{pmatrix}\right] \tag{4.6}$$

where $c =$ particle concentration

$x =$ distance in the viewing direction

If all of the terms except $\int c\,dx$ were constant and known, then we could easily relate plume opacity to mass emission rate (if we knew the wind speed and dispersion parameters; see Chapter 6). But the remaining terms in Eq. (4.6) vary from one stack to another and from one time of day to another, so that relating opacity to mass emission rate is difficult.

In spite of this difficulty, visual measurements of plume opacity have played a major role in air pollution control. Professor Maximilian Ringleman devised a system for making such measurements in about the year 1890 [8]. He marked five grids of various densities on a piece of white cardboard. An observer viewed the plume and then the cardboard, comparing the plume's optical density to that of the various grids. From that visual comparison the observer could determine if the plume was lighter or darker than the five grids (called Ringleman #1, . . . ,#5, corresponding to 20 percent, . . . , 100 percent opacity). With practice, trained observers could make this determination without Ringleman's grids. State air pollution control agencies regularly conducted (and some still do conduct) *smoke schools* at which the students learned to determine opacity to ±10 percent by simply looking at the plume, thus acquiring "calibrated eyeballs."

Before 1970 there were no numerical emission limits (of the form "no more than X lb/h of pollutant Y") for most emission sources. Most pollution control agencies relied heavily on opacity measurement for regulating particulate emission sources. The regulations normally forbade a plume opacity greater than 20 percent, or in some cases 40 percent, for more than a few minutes at a time. The control agency's enforcement officers could make a measurement from the road outside the offending plant, without notifying the plant owner. The cost of the test was negligible, and it could be made on short notice. (This is the only kind of test discussed in this text that does not require a device or set of devices as shown in Fig. 4.1). The courts generally upheld the validity of the tests and the rules they were intended to enforce. In that period many major sources regularly exceeded these opacity values; the control agency's inspectors, using this cheap and simple method, helped bring them under much better control.

Since the passage of the Clean Air Act of 1970, most major sources of airborne particulates in the United States must submit to regular stack tests to demonstrate compliance with numerical emission limits. Although the visible emissions regulations became less important, the regulations still often contain a limit on plume opacity, and they require large sources to install in-stack electro-optical detectors to record the plume opacity. This serves to detect occasional periods of high emission rates and also responds to the public's belief that no one should be allowed to emit a thick plume of smoke. Alas, that belief also causes many industrial sources to expend money and resources to eliminate steam plumes, which are simply condensed water, but which the public assumes are harmful.

There is continued interest in using optical methods to estimate pollutant concentrations. For example, devices are now being developed that pass an infrared beam through the exhaust of individual automobiles as they pass the monitoring site on a street or freeway ramp. Those autos that are gross CO emitters are easily detected. This method, or others like it, may play a significant role in future emission testing and control.

4.10 SUMMARY

1. The basic problems in ambient monitoring and source testing are the collection of a representative sample and the correct analysis of that sample. Generally, the collection of a representative sample is the harder part.
2. A substantial literature exists on the details of ambient monitoring and source testing [3, 9–14].
3. The Emission Factors compilations produced by the EPA allow us to estimate emissions from existing or new sources quickly and cheaply. Often these compilations provide a practical substitute for emission measurements and a guide to the literature on testing of many kinds of sources.
4. Visible emissions regulations played a major role in air pollution control enforcement before 1970. They now play a minor role.

PROBLEMS

See Common Units and Values for Problems and Examples, inside the back cover.

4.1. In Fig. 4.3, estimate the particle concentration (lb/ft^3) in the section near the bend, where the emission rate is shown as 0.1 times the mean emission rate.

4.2. Estimate the total volumetric flow rate of gas and the total mass emission rate of the particles in Fig. 4.3.

4.3. We need to sample the gas flow in a circular duct with axisymmetric flow. We want to take three samples that represent equal areas perpendicular to the flow. If the radius of the duct is r_0, what are the three radial distances from the center (r_1/r_0, r_2/r_0, and r_3/r_0) at which to locate the sample probe to give equal-area sampling?

4.4. We have tested a plant's emissions of particulates and found the following data:

Sample number	Stack velocity, ft/s	Particle concentration, g/m^3
1	50	0.50
2	60	0.30
3	40	0.70

Each sample represents the average over a 20-min period. What was the average particle concentration in this stack over the whole 60 min of this test?

4.5. In Example 4.1, if each of the sectors represents a cross-sectional area perpendicular to the flow of 1 m^2, what is the total flow rate of pollutant (g/s or g/min or equivalent) in this duct?

4.6. Section 4.3 suggests that one could sample for SO_2 in the presence of CO_2 using a simple collection in an NaOH solution, and measure the amount of reagent consumed. The normal concentration of CO_2 in the atmosphere is about 360 ppm. In ambient air the SO_2 concentration is rarely more than 0.1 ppm. If we measured the NaOH consumption for a sample of gas, and if we knew the CO_2 concentration *exactly* (which we rarely do), how many significant figures would we need in our measured consumption in order to know the SO_2 concentration to $\pm$ 1%?

4.7. Figure 4.6 shows that, if the flow velocity into the probe is not the same as the flow velocity in the free stream being sampled for particles, the measured concentration will not be the same as the true concentration in the stream being sampled. The computation of how big an error is made by failing to match these velocities is fairly complex [3]. Some of the tools needed for making this calculation are presented in Chapter 9. However, if you are clever and think about the problem physically, you can make a reasonable estimate of how large an error is created for the following case.

The true particle concentration in the stream being sampled is 0.1 g/m^3. The particles are spheres, 10 μ in diameter. The probe has an inside diameter of 0.5 cm, and negligible wall thickness. The velocity in the free stream is 10 m/s, and the velocity inside the probe is 1 m/s. No particles stick to the inside wall of the probe.

What is the approximate particle concentration in the gas flowing in the probe?

4.8. In Example 4.1 we have now discovered that the velocity measuring device was improperly calibrated. We have recalibrated it and find that the velocities, in order, should have been 8, 12, 16, and 20 m/s. The concentration measurements are unchanged. What was the average pollutant concentration of the gas flowing in the stack?

4.9. In Example 4.1 we have now discovered that the four sectors in which we measured the velocity did not have equal areas, as we had previously assumed. Based on careful measurements, we find that sectors 1 and 2 had areas of 1 m^2 and sectors 3 and 4 had areas of 2 m^2. Making these corrections, estimate the average pollutant concentration of the gas flowing in this stack.

4.10. All natural fibers (paper, cotton, wool) absorb measurable amounts of moisture from a humid atmosphere. Their weights can increase a few percent in going from a dry to a moist atmosphere. Particulate samplers use filters made from glass or synthetic fibers, which absorb much less moisture. But even they absorb some. In Example 4.2, how much would the reported concentration change if the analyst failed to dry the filter to exactly the same moisture content before and after sampling, so that the filter itself increased in weight by 0.1 percent between its initial and its final weighing?

4.11. We wish to design a rapid-response, portable PM_{10} sampler. The specifications call for an air flow rate of 1 L/min, and a 10-min sampling time.
(*a*) If the ambient concentration of PM_{10} is 25 $\mu g/m^3$, how large a sample will we collect?
(*b*) Can we detect that on the ordinary balances in your laboratory?

✓ **4.12.** Before the early 1990s, the EPA regulators called for converting measured ambient concentrations to standard temperature and pressure (25°C, 1 atm).
(*a*) In Example 4.2, if the sampler were at Salt Lake City, where the atmospheric pressure is typically 0.85 atm, and the temperature was 25°C, what would the reported concentration be?
(*b*) If a person breathes 16.7 L/min at sea level, and if a person breathes the same **mass** of air at all elevations, would the correction shown in part (*a*) result in making the measurement correspond to the true mass of $PM_{2.5}$ inhaled?
(*c*) Physiological measurements [15] indicate that up to 10 000 ft of elevation, humans breathe in the same **volume** of air at all elevations. If that is correct, does the correction to sea level underpredict, correctly predict, or overpredict the mass of $PM_{2.5}$ that a person breathes? (Current EPA regulations do not require this conversion to STP.)

4.13. (*a*) Repeat Example 4.5 for SO_2. Remember that the oxidation of 1 kg of S produces 2 kg of SO_2.
(*b*) Compare your result to the 38S in Table 4.2. What does your result mean physically?

4.14. Table 4.2 shows that, on a weight of pollutant per weight of fuel burned basis, the nitrogen oxide emissions from hand-fired furnaces are about one-third as large as the emissions from cyclone furnaces. Explain why. Your answer should only require a sentence or two. The reasons are discussed in detail in Chapter 12.

4.15. The emission factor for CO for coal combustion in large furnaces is 0.5 lb/ton (see Table 4.2). What is the corresponding emission factor for automobiles, in lb CO/ton of fuel burned? The permitted CO emission for new cars (Chapter 13) is 3.4 g/mi. Assume that the fuel economy is 25 mi/gal and the gasoline density is 0.72 kg/L.

4.16. A typical person at rest produces 0.007 mL (stp)/min of CO, which is exhaled (Chapter 15). The typical human diet in wealthy countries like the United States is about 600 g/day of food (dry basis).
(*a*) What is the emission factor (g CO/g food) for a human?
(*b*) What is the ratio of that emission factor to the emission factor for a 1990s auto (g CO/g fuel)? (The latter factor is calculated in Problem 4.15.)

✓ **4.17.** Example 4.4 estimated the uncontrolled emissions for a typical, modern coal-fired power plant with no emission control. The permitted particulate emission rate for such a plant is shown in Table 3.1. Estimate the required percentage particulate control efficiency if this plant is to meet the standards in Table 3.1.

4.18. Based on Table 4.2, what fraction of the ash in the coal is expected to be emitted from the following uncontrolled furnaces:
(*a*) PC, wall-fired, wet bottom furnace?
(*b*) Cyclone-type furnace?

Explain the differences in terms of the different physical arrangements of these furnaces.

✓**4.19.** In the production of coke for steelmaking, coal is heated to decompose it. The hot coke is pushed into railroad cars. Then enough water is poured onto the hot coke to quench it so that it will not burn in the storage pile. (If you watch a steel mill you will see that every few minutes a large white cloud issues from a short tower next to the coke ovens; it comes out for only about a minute. This is the steam-water mixture from the coke-quenching operation.)

Normally the water used for quenching contains dissolved solids, so that evaporation of this water will cause those solids to form dry particulate air pollutants.

(*a*) Estimate the emission factor for the formation of these particles (in pounds of particles/ton of coke quenched or the equivalent) from the following information:
The concentration of dissolved solids in the quench water is c (ppm by weight).
The coke leaves the oven at 1700°F and is quenched to 212°F.
The average heat capacity of coke in this range is 0.325 Btu/lb°F.
The heat of vaporization of water, starting from the liquid at 70°F and ending as steam at 212°F, is 1112 Btu/lb.

Assume that only enough water is put on the coke to cool it to 212°F and that all of that water evaporates. (Operating steel mills use about five times this much water. Some of the particles are captured by that excess water, so your estimate should be higher than what is actually observed.)

(*b*) Compare your estimate with the value of 5.24 lb of particulates per ton of coal charged, for quenching with water with $\geq$ 5000 mg/L of dissolved solids, given in Section 7.2-16 of Ref. 7.

4.20. Stack tests are expensive. For that reason typically only about three tests will be run, and their values averaged. The permitted emission rate of pollutant X for some new facility is 100 lb/h. Three tests were run, with measured values of 95, 98, and 102 lb/h.

(*a*) What is the average measured emission rate?

(*b*) What is the statistical probability that the true average emission rate is $\geq$100 lb/h? To answer this question use Student's t statistics, as described in any statistics book.

REFERENCES

1. The EPA's regulations for locating ambient air monitors are in *40CFR58* App. E. The description of the analytical methods to use for measuring pollutants in ambient air are in *40CFR50* App. A–H.
2. Hawskley, P. G., S. Badzioch, and J. H. Blackett: *Measurement of Solids in Flue Gases*, British Coal Utilization Association, Leatherhead, Surrey, England, 1961.
3. Paulus, H. J., and R. W. Thron: "Stack Sampling." In A. C. Stern (ed.), *Air Pollution*, Vol. 3, 3d ed., Academic Press, New York, pp. 525–587, 1976.
4. The EPA's regulations for source testing are found mostly in *40CFR60* App. A.
5. "Air Quality Criteria for Nitrogen Oxides," EPA-600/8-82-026, Chapter 4, pp. 7–9, U.S. Government Printing Office, Washington, DC, 1982.
6. Suggs, H. J.: "Systems for Sampling of Ducts and Stacks." In S. V. Hering (ed.), *Air Sampling Instruments*, 7th ed., American Conference of Governmental Industrial Hygienists, Cincinnati, OH, p. 275, 1989.
7. "Compilation of Air Pollutant Emission Factors Volume I: Stationary Point and Area Sources," 4th ed., AP-42, U.S. EPA, Office of Air Quality Planning and Standards, 1985, with updates through September 1996. There is also Volume II, "Mobile Sources," for autos and other vehicles (also regularly updated). These emission factors are available and searchable at <http://www.epa.gov/ttn/chief/ap42supp.html>.
8. Purdom, P. W.: "Source Monitoring." In A. C. Stern (ed.), *Air Pollution*, Vol. 2, 2d ed., Academic Press, New York, pp. 537–560, 1968, p. 541.
9. Hering, S. V., ed.: *Air Sampling Instruments*, 7th ed., American Conference of Governmental Industrial Hygienists, Cincinnati, OH, 1989.
10. Cooper, H. B. B., Jr., and A. T. Rossano, Jr.: *Source Testing for Air Pollution Control*, Environmental Research and Applications, Inc., Wilton, CT, 1971.
11. Katz, M., ed.: *Methods of Air Sampling and Analysis*, 2d ed., American Public Health Association, Washington, DC, 1977.

12. Brenchley, D. L., C. D. Turley, and R. F. Yarmac: *Industrial Source Sampling*, Ann Arbor Science Publishers, Ann Arbor, MI, 1973.
13. Stevens, R. K., and W. F. Herget: *Analytical Methods Applied to Air Pollution Measurements*, Ann Arbor Science Publishers, Ann Arbor, MI, 1974.
14. Leithe, W.: *The Analysis of Air Pollutants*, Ann Arbor Science Publishers, Ann Arbor, MI, 1971.
15. Lillquist, D. R., J. S. Lee, and D. O. Wallace: "Pressure Correction Is Not Required in Particulate Matter Sampling," *J. Air Waste Mgmt. Assoc.*, Vol. 46, pp. 172–173, 1996.

CHAPTER 5

METEOROLOGY FOR AIR POLLUTION CONTROL ENGINEERS

Most of us are interested in meteorology because we want to know if it will rain on our picnic, freeze our tomato plants, or be suitable weather for lying on the beach. Air pollution control engineers have picnics, grow tomatoes, and lie on the beach, but their professional interest in meteorology is mostly with *wind speed and direction* and with *atmospheric stability*. Chapter 6 shows how these two atmospheric variables enter into the most commonly used air pollutant concentration models. Most often they are the only meteorological variables used in those models. This chapter gives some background on how our global weather system works and then considers these two topics of special interest to air pollution control engineers.

5.1 THE ATMOSPHERE

The global atmosphere is roughly 78 percent nitrogen, 21 percent oxygen, 1 percent argon, and other trace gases (see inside the back cover). Those ratios change very little with place or time in most of the atmosphere. The moisture content of the atmosphere, either as water vapor or as liquid drops or ice crystals, changes significantly with place and time and is responsible for many of the exciting, beautiful, and destructive things the atmosphere does. A typical water content (20°C, 50 percent RH) is 1.15 mol (or volume) percent. [*Relative humidity,* RH, is the ratio of the water content of the air to the saturation water content; see Eq. (5.10).] The water content of the atmosphere, at saturation, increases rapidly with increasing temperature. At any temperature the absolute amount of water in the air, expressed either as mol percent or lb water per lb air, increases with increasing RH.

The atmosphere has a perfectly well-defined—but quite uneven—lower boundary, the surface of the land and the oceans. Its upper boundary is not as well-defined; the atmosphere simply becomes thinner and thinner with increasing height until it is as thin as outer space. One-half of the mass of the atmosphere is within 3.4 miles of the surface; 99 percent is within 20 miles of the surface. If the atmosphere were peeled off the earth and had its edges stitched together to make a pancake shape, it would have an approximate thickness of 20 miles and a diameter of 16,000 miles (see Problem 5.1).

This large width and small depth mean that most of the motions in the atmosphere must be horizontal. Except for very vigorous storms, the vertical motions in the atmosphere are one or two orders of magnitude smaller than the horizontal ones (the vertical component of the wind velocity is one or two orders of magnitude less than the horizontal component of the wind velocity). Similarly, atmospheric storms and systems are thin. A tropical storm is typically 200 miles or more across and 10 miles from top to bottom.

5.2 HORIZONTAL ATMOSPHERIC MOTION

The horizontal movement of the atmosphere (the horizontal component of winds) is driven mostly by uneven heating of the earth's surface and modified by the effect of the earth's rotation (Coriolis force) and the influence of the ground and the sea.

5.2.1 Equatorial Heating, Polar Cooling

Averaged over the year, the solar heat flow to the earth's surface at the equator is 2.4 times that at the poles [1]. The atmosphere moves in response to this difference in heating, and in so doing transports heat from the tropics to the Poles, partly evening out the temperature difference from equator to poles, just as air movement in a room distributes the heat from an electric heater in one corner of the room to the whole room. In both cases the distribution of heat results from warm air rising at the heat source (electric heater or solar-heated equator) and cold air sinking where the surroundings are coldest (the part of the room away from the heater or the Poles).

In the room, heating can be accomplished by one simple circulatory cell, illustrated in Fig. 5.1. For the earth one might logically assume that the same would occur, with air rising at the equator and sinking at the Poles so that there would be equator-to-Poles flow at high altitudes and Poles-to-equator flow at the surface. However, because the atmosphere is quite thin relative to its width, that flow is mechanically unstable and breaks up into subcells. Any odd number (but not an even number! see Problem 5.2) of such cells could exist in each hemisphere; on the earth there are normally three cells in each, as shown in Fig. 5.2.

In the Northern Hemisphere we see from the circulation cells sketched at the edges of the figure a south-to-north flow at high altitude and a north-to-south

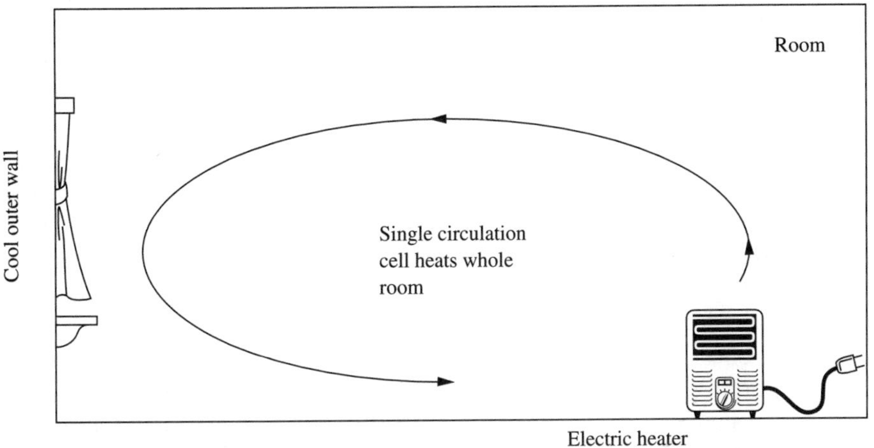

FIGURE 5.1
Air circulation in a room heated by an electric heater. There is a single circulation cell, rising at the heater and descending at the cold wall.

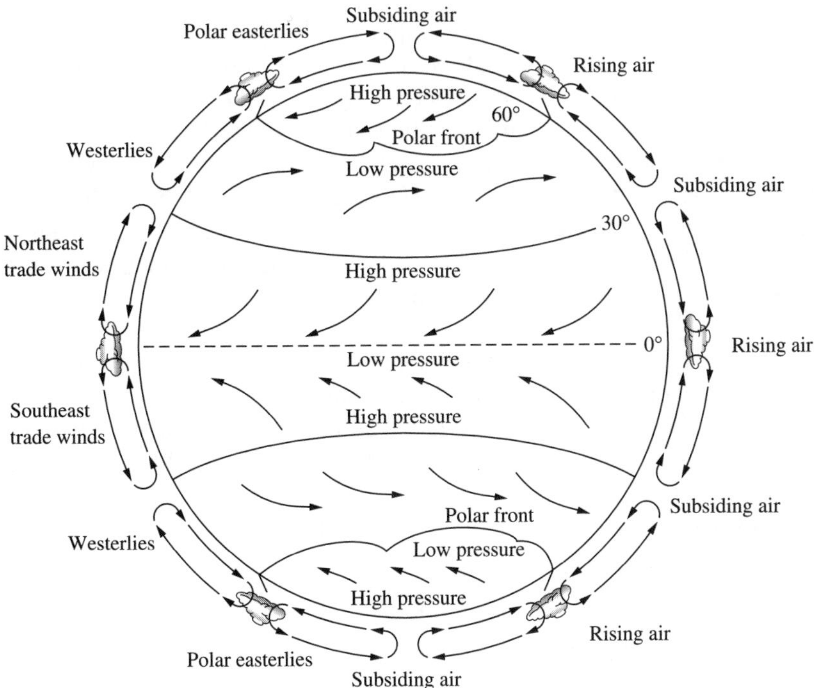

FIGURE 5.2
Schematic representation of the general circulation of the atmosphere. (Frederick K. Lutgens/Edward J. Tarbuck, *The Atmosphere,* 5e, ©1992, p. 170. Reprinted by permission of Prentice Hall, Englewood Cliffs, New Jersey.)

flow at the surface in the tropical and polar cells with oppositely directed flows in the temperate cell. There are seven boundaries between cells on the globe, one at the equator and two in each hemisphere and two at the Poles. At the boundary at the equator and the two between the temperate and polar cells the air is rising; at the boundaries between tropical and temperate cells and at the Poles (which are "hole in the donut" boundaries) the air is sinking. We shall see later that rising air is cooled and produces rain, while sinking air is heated and becomes relatively dry. The rising boundaries (equator and temperate-polar boundaries) are regions of higher than average rainfall; most of the world's rain forests are located near the equatorial rising zone, and the great temperate forests are near the temperate-polar rising zones. The sinking boundaries (the Poles and the tropical-temperate boundaries) are regions of lower than average rainfall; most of the world's great deserts are located near the temperate-tropical sinking zones. The Poles also have little precipitation; they are cold deserts, where the small amount of precipitation remains as ice and snow because there is negligible evaporation or melting. We will return to the wind directions shown in Fig. 5.2 shortly.

5.2.2 The Effect of the Earth's Rotation

The preceding simple picture is greatly complicated by the rotation of the earth. Figure 5.3 shows what would happen if two people threw a ball back and forth at the North Pole. We know the earth rotates once a day. While the ball is in flight from pitcher to catcher, an observer in a nonrotating spaceship hovering above the earth would see the ball go straight; the observer would also see the catcher move to the left of the ball, riding on the rotating earth. The catcher, riding on the earth, would see the ball curve away to the left. Thus, from the viewpoint of any observer riding with the earth, the ball appears to curve to its right. If the same experiment were conducted at the South Pole, the ball would appear to curve to its left.

Example 5.1. In Fig. 5.3, at the North Pole, the pitcher throws a standard baseball (0.32 lbm) at a speed of 90 mi/h (132 ft/s). The distance thrown (pitcher's mound to home plate) is 60 ft. The ball is thrown directly at the catcher. From the viewpoint of an observer on a nonrotating space station, how far does the catcher move to the left while the ball is in flight?

The earth completes one revolution per day so that

$$\omega = \frac{2\pi \text{ radians}}{\text{day}} \times \frac{\text{day}}{24 \times 3600 \text{ s}} = 7.27 \times 10^{-5} \text{ s}^{-1}$$

Here the distance traveled by the catcher is

$$\text{Distance traveled by catcher} = r\omega\,\Delta t = r\omega\frac{\Delta x}{V}$$

$$= 60 \text{ ft} \times 7.27 \times 10^{-5} \text{ s}^{-1} \times \frac{60 \text{ ft}}{132 \text{ ft/s}}$$

$$= 0.00198 \text{ ft} = 0.024 \text{ inch} = 0.60 \text{ mm}$$ ■

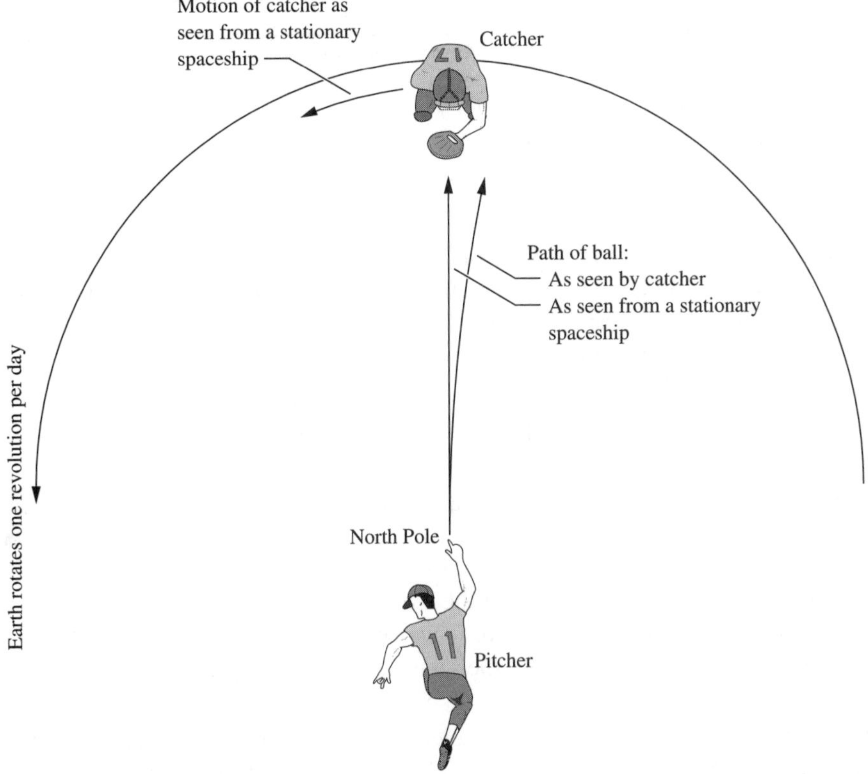

FIGURE 5.3
Path of a ball thrown from the North Pole, as seen by an observer on a stationary space platform, and as seen by people riding on the rotating earth.

A remarkably sharp-eyed catcher would notice that, although the ball started coming straight at her, it actually arrived 0.6 mm to the left of its original target. It would appear to her and to all earth-bound observers that it curved to its right by that amount while in flight. However, the observer on the nonrotating spacecraft would see it fly perfectly straight and would recognize that the earth rotated while the ball was in flight.

One could solve all engineering problems from the viewpoint of a person in a hovering spacecraft, but most of us prefer to solve them from our own viewpoint, that of a person riding on the earth. The most common way of adjusting for the observed curvature shown in Fig. 5.3 is to introduce an adjustment for the switch of frames of reference, called the *Coriolis force,* which, when added to the other forces in Newton's law of motion, correctly predicts the observed behavior. Unlike gravitational and centrifugal forces, which are independent of the motion of the body being acted upon, the Coriolis force (or Coriolis acceleration) acts at right angles to the motion of the body, is proportional to the velocity of the moving body, and is given by

$$\text{Coriolis acceleration} = \frac{\text{Coriolis force on a body}}{\text{Mass of the body}} = 2V\omega \sin\phi \tag{5.1}$$

where V = velocity of the moving body
ω = angular velocity of the earth
ϕ = latitude

Example 5.2. Repeat Example 5.1 from the viewpoint of someone riding on the earth, using the Coriolis force. The Coriolis acceleration is given by Eq. (5.1). From any physics book one may find that the horizontal deflection due to any constant acceleration acting over a short time period is

$$\text{Deflection} = 0.5a(\Delta t)^2 = 0.5 \cdot 2V\omega \sin\phi \left(\frac{\Delta x}{V}\right)^2$$

At the North Pole $\sin\phi = 1$, and a little algebra shows that the right side is $r\omega\,\Delta t$, the same as in Example 5.1. ■

This example shows that by including the Coriolis force we find the same deflection (which we would actually observe if we conducted the experiment) from the viewpoint of an observer riding with the rotating earth (see Problem 5.3).

Example 5.3. Estimate the Coriolis acceleration for a body moving 10 ft/s at 40° North latitude. Using the earth's angular velocity from Example 5.1, we find

$$\text{Coriolis acceleration} = 2 \times 10\ \frac{\text{ft}}{\text{s}} \times 7.27 \times 10^{-5} \frac{1}{\text{s}} \times \sin 40°$$

$$= 9.35 \times 10^{-4}\ \frac{\text{ft}}{\text{s}^2} = 2.85 \times 10^{-4}\ \frac{\text{m}}{\text{s}^2}$$ ■

The Coriolis acceleration for this velocity, a typical wind velocity, is 2.9×10^{-5} as large as the acceleration of gravity. This is small enough that most computational models in meteorology completely omit it for *vertical* motions, where gravity dominates. However, for *horizontal* motions, at right angles to the direction of gravity, the other accelerations are all small, so it plays a much more significant role. We do not notice the Coriolis acceleration as we walk or run; it is so small compared to gravity or wind resistance that we cannot perceive it.

The principal accelerating forces causing (or retarding) horizontal flow in the atmosphere are the Coriolis force, pressure gradient forces, and frictional resistance at the surface of the earth.

Example 5.4. Estimate the acceleration of the air caused by a pressure gradient of 1 mb/100 km. (Meteorologists always use the bar [= 10^5 Pa = 0.9872 atm] as a unit of pressure, and [mb = millibar = 10^{-3} bar = 100 Pa]. Meteorologists almost always state pressure gradients in mb/100 km; engineers would state this as 0.01 mb/km = 1 Pa/km.)

Here we apply Newton's law to 1 cubic km of air (a cube with edge length = x = 1 km) and use the standard sea-level air density, finding

$$\begin{aligned}\begin{pmatrix}\text{Pressure}\\ \text{acceleration}\end{pmatrix} &= \frac{F}{m}\\ &= \frac{A\,\Delta P}{V\rho} = \frac{x^2\,\Delta P}{x^3\rho} = \frac{\Delta P}{x\rho}\\ &= \left(\frac{1\text{ Pa}}{1\text{ km}\times 1.21\text{ kg/m}^3}\right)\left(\frac{\text{km}}{1000\text{ m}}\right)\left(\frac{\text{kg}}{\text{s}^2\times\text{m}\times\text{Pa}}\right)\\ &= 8.3\times 10^{-4}\,\frac{\text{m}}{\text{s}^2} = 2.7\times 10^{-3}\,\frac{\text{ft}}{\text{s}^2}\end{aligned}$$ ■

This is a typical atmospheric pressure gradient. The computed acceleration is about three times the Coriolis acceleration computed in Example 5.1, but only about 0.00008 times the acceleration of gravity. Away from the earth's surface (e.g., at high altitude) the Coriolis and pressure gradient forces together determine the wind velocity and direction; close to the surface, friction between the moving air and the ground or ocean makes the picture more complicated. Pressure-gradient acceleration is inversely proportional to air density, which means that if the horizontal pressure gradient is the same at high altitude as at low (which it practically is) then (1) the acceleration will be greater at higher altitudes than at low because air density declines with altitude and (2) high-altitude winds will be faster than low-altitude winds, which they generally are.

So far, we have not said in which direction the Coriolis acceleration (or force) operates. As we have defined it, it always operates at right angles to the velocity of the moving body; other definitions are possible. In the Northern Hemisphere it turns the body to the right (as seen in Fig. 5.3) and in the Southern Hemisphere to the left. Returning to Fig. 5.2, we can now see the reason for the direction of the wind arrows. If there were no Coriolis force, we would expect them all to point either north or south. However, in the Northern Hemisphere they curve to the right as shown, and in the Southern Hemisphere to the left. The result is that near the equator the surface wind is from the east (trade winds), in the midlatitudes the surface wind is from the west (prevailing westerlies), and in the polar regions the surface wind is from the east (polar easterlies).

5.2.3 The Influence of the Ground and the Sea

Figure 5.2 would be a better predictor of the world's horizontal atmospheric flows if all the land were flat, and if land and sea had the same response to solar heating. Neither of these is the case. The highest mountains rise above most of the atmosphere (the top of Mt. Everest, at 29,028 ft, is above 70 percent of the atmosphere). Major mountain ranges like the Himalayas, Rockies, Alps, and Andes are major barriers to horizontal winds, and regularly have very different climates on one side than on the other. Even smaller mountains and valleys can strongly influence wind direction, albeit on a smaller scale.

The surface of the ground heats and cools rapidly from day to night and from summer to winter because solid ground is a poor conductor of heat. The surface of oceans and lakes heats and cools slowly, mostly because their surface layers are stirred by the winds and by natural convection currents, thus mixing heat up and down. Solid ground is not stirred by the wind or convection currents, so heat cannot mix up and down; its surface temperature changes more rapidly than that of bodies of water. Thus the heating (or cooling) of the air layer adjacent to solid ground is much faster than that of air over bodies of water. Probably the most spectacular example of this phenomenon is the monsoon weather of India and parts of East Africa. The summer sun warms the air above India more than the air over the surrounding oceans, which causes strong upward motion of the air over India. Moist air from over the surrounding warm oceans flows inward to fill the low-pressure region caused by this rising air. This moist air rises, cools, and forms the monsoon rains on which Indian agriculture depends. The same phenomenon leads to summer thunderstorms over the southwestern United States, but they are not as strong as those of the Indian monsoon.

The simple picture of the *general circulation of the atmosphere* presented in Fig. 5.2 is obviously a great simplification of what nature does. Its predictions are better near the equator than near the Poles. As Fig. 5.2 shows, the boundaries between the polar cells and the temperate cells are irregular. Those boundaries move enough that their motion plays a significant role in the climate of North America and Europe. In spite of this simplification, this picture explains many of the observed facts about the movement of air and the location of deserts and forests. Most of the interesting and exciting weather phenomena are superimposed upon and ride with the general circulation shown in Fig. 5.2.

5.3 VERTICAL MOTION IN THE ATMOSPHERE

Vertical and horizontal motions in the atmosphere interact; the horizontal flows in Fig. 5.2 are driven by rising air at the equator and sinking air at the Poles. In the atmosphere any parcel of air that is less dense than the air that surrounds it will rise by buoyancy, and any parcel more dense than the surrounding air will sink by negative buoyancy. Most vertical motions in the atmosphere are caused by changes in air density.

5.3.1 Air Density Change with Temperature and Humidity

The density of any part of the atmosphere is given almost exactly by the perfect gas law,

$$\rho = \frac{MP}{RT} \tag{5.2}$$

At one particular altitude (and hence one particular P) the density is determined by M and T.

Example 5.5. Estimate the change in air density due to a 1°C increase in temperature (for dry air), and a 1 percent increase in relative humidity, both at 20°C.

Differentiating the natural log of Eq. (5.2), we find

$$\frac{d\rho}{\rho} = \frac{dM}{M} + \frac{dP}{P} - \frac{dT}{T} \tag{5.3}$$

(no term appears for R, because $dR = 0$). At constant M and P we have

$$\frac{d\rho}{\rho} = -\frac{dT}{T} = -\frac{1°\text{C}}{(20 + 273.15)\text{K}} = -0.0034$$

or

$$\frac{d\rho/\rho}{dT} = -0.0034/°\text{C}, \quad \text{at } 20°\text{C}$$

The average molecular weight of air is given by the equation

$$M_{\text{avg}} = y_{\text{water}} M_{\text{water}} + (1 - y_{\text{water}}) M_{\text{air}} = M_{\text{air}} + y_{\text{water}}(M_{\text{water}} - M_{\text{air}}) \tag{5.4}$$

where y_{water} is the mol fraction of water vapor. At 20°C

$$y_{\text{water}} \approx 0.023 \text{ RH}$$

so that

$$M_{\text{avg}} \approx 29 - 0.023 \text{ RH}(29 - 18) = 29 - 0.253 \text{ RH}$$

and

$$\frac{d\rho}{\rho} = \frac{dM_{\text{avg}}}{M_{\text{avg}}} = \frac{-0.253\, d\text{RH}}{29 - 0.253 \text{ RH}} \approx \frac{-0.253\, d\text{RH}}{29}$$

$$\frac{d\rho/\rho}{d\text{RH}} = \frac{-0.253(0.01/\%\text{RH})}{29} = -8.7 \times 10^{-5}/\%\text{RH}$$ ■

We see that about a 40 percent increase in relative humidity is required to produce the same effect as a 1°C increase in temperature. This explains why most of the vertical motion of the atmosphere is driven by changes in temperature rather than by changes in humidity.

5.3.2 Air Density Change with Pressure

The *basic equation of fluid statics,* also called the *barometric equation,* states that

$$\frac{dP}{dz} = -\rho g \tag{5.5}$$

where z = vertical distance

g = acceleration due to gravity

It is correct for solids, liquids, or gases. The pressure at any point in the atmosphere or in the oceans or inside the ground is that pressure needed to support the weight of everything above that point. If we substitute Eq. (5.2) for density in Eq. (5.5), we have

$$\frac{dP}{dz} = -\frac{gMP}{RT} \tag{5.6}$$

or

$$\frac{dP}{P} = -\frac{gM}{RT}\,dz \tag{5.7}$$

If T and M did not change with elevation, we could integrate this to find the relation between pressure and elevation. The changes in M are not important, as we saw in the previous example, but those of temperature are. To see why, consider a parcel of air in a flexible balloon that is moving upward in the atmosphere (see Fig. 5.4). According to the assumptions shown in that figure, the parcel of air does work on its surroundings as it expands, and it exchanges negligible amounts of heat with its surroundings. Thus it undergoes a *reversible, adiabatic* process. Here the balloon serves only to isolate a parcel of air for our consideration. If the air in the balloon

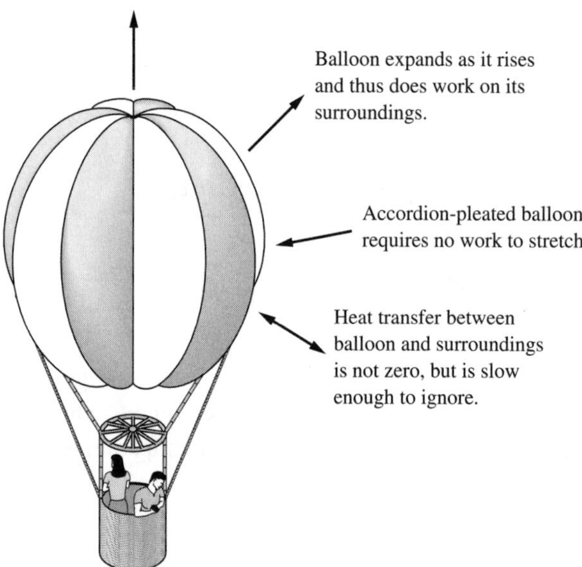
Balloon expands as it rises and thus does work on its surroundings.
Accordion-pleated balloon requires no work to stretch.
Heat transfer between balloon and surroundings is not zero, but is slow enough to ignore.

FIGURE 5.4
A parcel of air contained in a hypothetical, flexible, rising balloon: As the balloon rises, it does work on the atmosphere by expanding. If it also has negligible heat transfer with the surroundings, then the behavior of the air in the balloon is practically reversible and adiabatic.

were in the open but did not mix significantly with its surrounding air, it would behave the same way. Thus, for any parcel of air, moving upward without significant mixing with the air around it, reversible adiabatic behavior would be observed. From any thermodynamics book one finds that, for a perfect gas undergoing a reversible, adiabatic process (also called an *isentropic* process),

$$\frac{dP}{P} = \frac{C_P}{R}\frac{dT}{T} \qquad \text{reversible, adiabatic, perfect gas} \tag{5.8}$$

where C_P is the heat capacity of the gas at constant pressure. Eliminating dP/P from Eqs. (5.7) and (5.8) and rearranging, we find

$$\left(\frac{dT}{dz}\right)_{\substack{\text{adiabatic,}\\ \text{perfect gas}}} = -\frac{gM}{C_P} \tag{5.9}$$

which is true for a reversible adiabatic atmosphere of any perfect gas. It is also correct on Mars or Venus, which have very different atmospheres from the earth. (The adiabatic assumption is probably a poor one for permanently cloudy Venus.) For air at the gravity of the earth, we calculate that

$$\left(\frac{dT}{dz}\right)_{\substack{\text{adiabatic,}\\ \text{perfect gas}}} = -\frac{gM}{C_P} = -\frac{9.81\ \text{m/s}^2 \times 29\ \text{g/mol}}{3.5 \times 8.314\ \text{m}^3 \cdot \text{Pa/mol} \cdot \text{K}} \times \frac{\text{kg}}{1000\ \text{g}} \times \frac{\text{Pa} \cdot \text{m} \cdot \text{s}^2}{\text{kg}}$$

$$= -0.00978\ \frac{\text{K}}{\text{m}} = -9.78\ \frac{°\text{C}}{\text{km}} = -5.37\ \frac{°\text{F}}{1000\ \text{ft}} \approx -10\ \frac{°\text{C}}{\text{km}}$$

(Here we have taken the C_P of air as $3.5R$.)

This temperature gradient is called the *lapse rate,* and it is normally stated as a positive number. In the preceding calculation, the *adiabatic lapse rate* would be reported as 5.4°F per 1000 ft or 10°C per km. If the numerical value of the lapse rate is greater than this (e.g., 12°C per km), it is called a *superadiabatic lapse rate,* and if it is less than this (e.g., 8°C per km), it is called a *subadiabatic lapse rate.* For the purposes of calculation, meteorologists and aeronautical engineers have defined a "standard atmosphere" that represents the approximate average of all observations, day and night, summer and winter, over the whole United States [2]. This average of observed temperatures is compared in Fig. 5.5 on page 94 with the adiabatic lapse rate just calculated. The lapse rate in the standard atmosphere is 6.49°C/km = 3.56°F/1000 ft, about 66 percent of the adiabatic lapse rate.

On Fig. 5.5 one also sees that in the "standard atmosphere" the temperature declines linearly up to 36,150 ft and then remains constant up to 65,800 ft. The lower region, which contains about 75 percent of the mass of the atmosphere, is called the *troposphere.* In the region just above it, called the *stratosphere,* the temperature does not continue to decrease with increasing height, because at that elevation some chemical reactions occur that absorb energy (or heat) from the sun, so that the adiabatic assumption is not followed.

The fact that the "standard atmosphere" has a lapse rate only about 66 percent of the adiabatic lapse rate simply indicates that the adiabatic assumption is not appropriate for all circumstances, or even the average of all circumstances. The

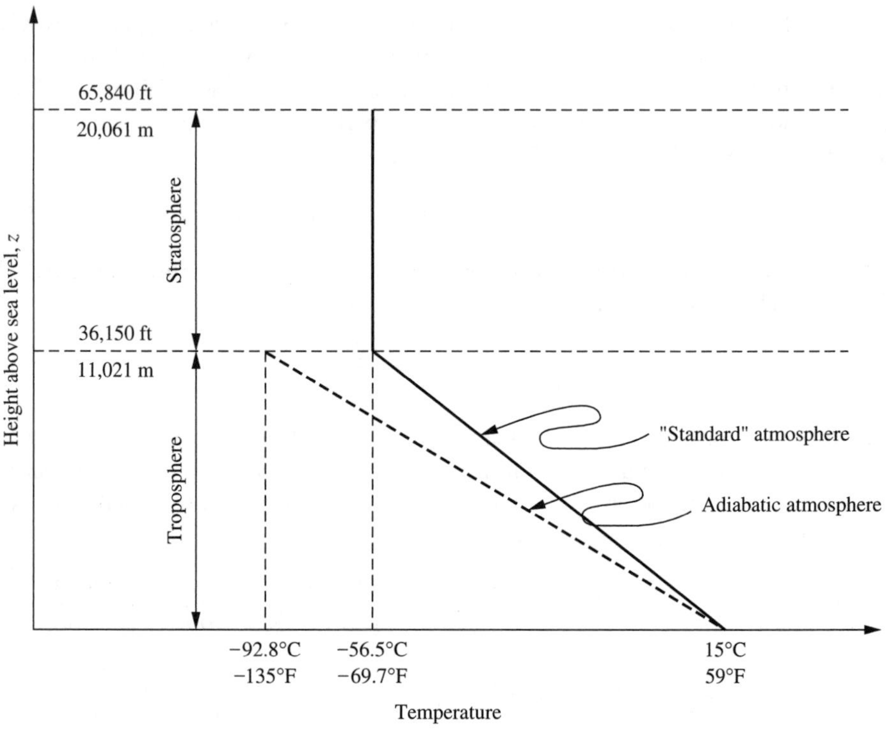

FIGURE 5.5
Comparison of the temperature-elevation relations in the adiabatic atmosphere and the standard atmosphere.

atmosphere is practically transparent to visible light, but it absorbs and emits heat significantly at infrared wavelengths mostly because of the water in the atmosphere. This absorption causes the air to be heated from below, by infrared radiation from the earth, thus making the upper troposphere warmer than it would be without it. Many readers have observed that on a clear summer night in the desert the temperature drops much more than on a comparable clear night in a moist climate; in the latter, more of the outgoing heat is absorbed and then emitted back to the earth than in a dry climate. This topic is discussed in more detail in Chapter 14. In addition, water vapor, rising from the ground and condensing to form clouds, transports heat upward from the surface, making the upper troposphere warmer than it would be under adiabatic conditions.

The adiabatic lapse rate just computed does not include this possibility of condensation of moisture; for that reason it is called the *dry adiabatic lapse rate.* In a few pages we will consider the *moist adiabatic lapse rate.* For any lapse rate (adiabatic, standard, some other) one may calculate pressure-height relations (Problem 5.6) and density using the preceding equations.

5.3.3 Atmospheric Stability

The temperature-elevation relationship sketched in Fig. 5.5 is the principal determinant of atmospheric stability. The reason is sketched in Fig. 5.6. Four cases are shown: for each there is a sketch of elevation-temperature and of the mechanical analog of the stability. On each of the temperature sketches the adiabatic lapse rate, $dT/dz = -5.4°\text{F}/1000$ ft, is shown as the dashed line, whereas the actual lapse rate is shown as a solid line.

In part (a) the actual lapse rate is greater than the adiabatic lapse rate $|dT/dz| > |-5.4°\text{F}/1000 \text{ ft}|$; this is a *superadiabatic* situation. If some parcel of air is moved up or down quickly, as by a bird flying by or by having the parcel pass over a hill or a low spot, there will not be enough time for much heat to transfer to or from the surrounding air. So the air parcel will follow the adiabatic curve in Fig. 5.6, not only in part (a) but also in parts (b, c, and d). In part (a) this means that if the air parcel starts at some point where its temperature is the same as that of the surrounding parcels and it moves upward along the adiabatic curve, it will be at a higher temperature than the surrounding parcels in its new location, so buoyancy

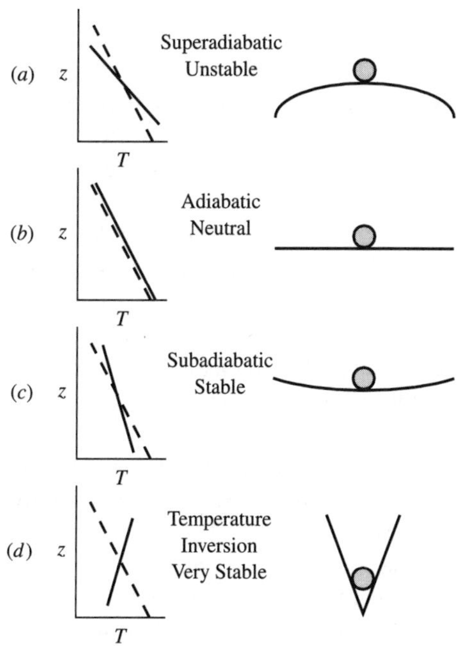

Note: In all of these plots the dashed line represents the adiabatic lapse rate, $dT/dz = -5.4°\text{F}/1000$ ft ($dz/dT = -185$ ft/°F).

FIGURE 5.6
Relation between actual lapse rate, adiabatic lapse rate, and atmospheric stability (see Problem 5.17).

will force it to continue to move upward. If, instead of moving upward, it is forced to move downward, then, following the adiabatic path, its temperature will be lower than that of the surrounding parcels and negative buoyancy will cause it to move downward. This situation is like that of the ball sketched at the right of part (*a*): it is *unstable*. Any disturbance will cause it to continue to move in the direction of the disturbance. In this situation vertical movements in the atmosphere occur spontaneously.

In part (*b*) the actual lapse rate is the same as the adiabatic lapse rate. If the parcel is moved up or down and it follows the adiabatic lapse rate, its temperature will be the same as that of the surrounding parcels in the new location, and buoyancy will move it neither up nor down. This is the *neutral stability* situation, sketched at the right of part (*b*).

In part (*c*) the actual lapse rate is less than the adiabatic lapse rate (*subadiabatic*). If a parcel of air is moved upward, it will follow the adiabatic lapse rate and be colder than the surrounding air. Negative buoyancy will force it back toward its starting spot. If the air parcel is moved downward, it will be warmer than the surrounding air, and buoyancy will force it back toward its starting spot. This is a *stable* situation, as shown in the mechanical analog at the right. Any vertical motions in the atmosphere are damped.

In part (*d*) the actual lapse rate has the opposite sign from the adiabatic lapse rate; temperature increases with elevation. This is a *temperature inversion* (or simply, *inversion*). By the same arguments as shown for part (*c*), in this situation vertical atmospheric movement is damped. In the case of inversion damping is very strong and vertical movement is possible only if there is a strong external driver, e.g., a big forest fire.

Where and when might we encounter stable, neutral, and unstable atmospheres? We would expect all three at the same place, at different times of day, on any clear, dry, sunny day with low or average winds, anywhere on land. Figure 5.7 shows how this happens. We assume that (1) this is a typical spring day in the Mojave Desert of California, which is dry enough that moisture plays no role; (2) there are no clouds; and (3) winds are light or moderate. The curve at the left shows the situation at dawn.

All night the ground surface has been cooling, and at dawn its temperature is perhaps 50°F. At infrared wavelengths the ground is an almost perfect blackbody radiator, so it is quite efficient at radiating heat to outer space. The ground surface has also been cooling the layer of air above it. The cooled air layer nearest the ground cools the layer of air above it, so that there is a steady flow of heat downward from the air to the ground by conduction, slight convection, and radiation. (Dry air, which is practically transparent to visible light, is not transparent to infrared radiation and does transfer some heat by infrared radiation.) At dawn, temperature increases with elevation up to perhaps 1000 ft. At that point the "cooling wave" from the ground runs into the lapse rate left over from the previous day, and the temperature continues along up the standard atmosphere curve.

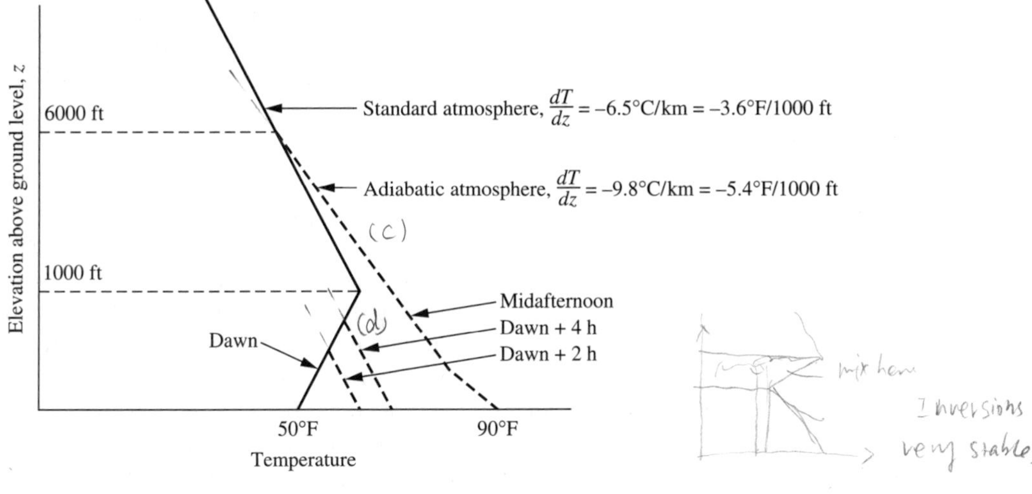

FIGURE 5.7
Vertical temperature distribution at various times on a cloudless day with low or average winds in a dry climate.

Below 1000 feet the temperature increases with height. This pattern is called an *inversion;* such inversions occur every clear or slightly cloudy night, with low or average winds, on most of the world's land surface. Inside the inversion the situation is extremely stable; vertical disturbances are strongly damped out. Above the inversion, in the region with the standard lapse rate, the situation is mildly stable, vertical disturbances are damped, but not nearly as strongly as inside the inversion. This kind of inversion is the most common one and is called a *radiation inversion.* Other types of inversions are discussed in Section 5.5.

When the sun comes up, it heats the ground surface, which heats the layer of air above it, by conduction, convection, and radiation. That layer heats the next layer above it, and so on. Two hours after dawn the ground temperature will be perhaps 70°F. There will be a layer of warmed air near the ground, in which the lapse rate is practically the adiabatic lapse rate. At its top, this layer encounters the remainder of the previous night's inversion. Rising air from below cannot penetrate that inversion, for stability reasons. But at the boundary it mixes with the inversion, slowly destroying it, so that by four hours after sunrise the warmed air layer has grown and almost eliminated the inversion.

By midafternoon, enough heat has been transferred from the warmed ground surface to the adjacent air that the inversion is gone. The heated air, which now has an adiabatic lapse rate, extends to perhaps 6000 ft, where it encounters the more stable air above, with a lapse rate at or near that of the standard atmosphere. In the few hundred feet closest to the ground the lapse rate is even greater than the adiabatic lapse rate (i.e., superadiabatic), and the air is unstable.

Shortly before sunset, the ground surface begins to cool by radiation and to cool the air layer nearest it. By sunset there will be a weak inversion close to the ground. All night this inversion will grow in strength and size, until by dawn of the next day the temperature profile will be practically the same as that shown for dawn in Fig. 5.7.

This picture of the daily behavior of the atmosphere is very important for understanding air pollution meteorology; it is worth our while to spend some time thinking about it. Most readers of this text have had the opportunity to observe soaring birds—eagles, hawks, vultures, perhaps even condors if they are lucky. These birds stay aloft for long periods without moving their wings, riding on the vertical updrafts caused by ground heating and described in detail below. Soaring is not the same as flying, which implies flapping the wings. Birds cannot soar at dawn because these vertical updrafts do not form until the ground has been heated. Likewise, on a strongly overcast day these big birds cannot soar because the kind of solar heating that produces these updrafts does not occur. Human glider pilots also fly by finding these rising air currents (which they call *thermals*) and riding them up; human glider pilots do not try to soar at dawn or on a totally overcast day. These kinds of updrafts and their associated downdrafts are dangerous for hot air balloons; they mostly fly at dawn. Even without this thermal instability one can find updrafts where the horizontal wind is forced up over a cliff or bluff or hill; soaring birds and human glider and hang-glider pilots take advantage of those terrain-induced updrafts as well.

In the situation sketched as midafternoon in Fig. 5.7, if there are updrafts, there must be downdrafts too, because the overall motion of the atmosphere has only a small vertical component. One might imagine that there would be equal areas of ascending and descending air, but nature seems to prefer to have a few small columns of rapidly rising, fairly hot air surrounded by a large area of slowly falling, slightly cooled air. One can see this phenomenon in the form of a dust devil, commonly seen in all desert areas on sunny afternoons, as sketched in Fig. 5.8. There we see that when the ground is much hotter than the air above it, a layer of hot air forms next to it. When a disturbance allows some of this air to rise, it does so as a column. This air is hotter, thus less dense, than the air around it, so the pressure at the ground at the point just below the rising column is less than the pressure a few feet away in all directions. Air flows into this low-pressure spot and then up the rising column. Meanwhile, to make up for this upward flow there is a general downward flow around it, replenishing the hot air layer at the ground.

Often a dust devil will break away from the surface to form a "bubble" of hot (superadiabatic) air. These bubbles rise and mix with the surrounding cooler air, eventually disappearing. One can often see them by watching for groups of soaring birds circling inside them as they rise. Champion glider pilots become champions by being better than their competitors at finding these rising bubbles.

This rising column would not be visible or stay together very well if it were not for the Coriolis acceleration. As the air flows in along the ground toward the rising column, the Coriolis force makes each parcel turn to the right so that, as seen from

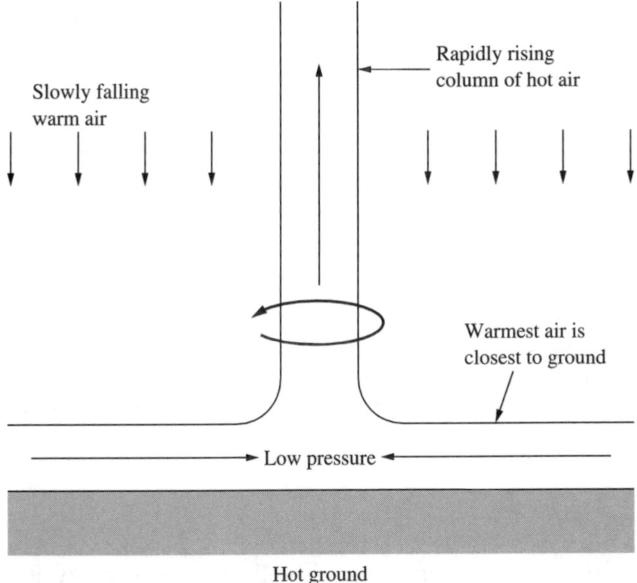

FIGURE 5.8
Flow in and around a dust devil.

above, the incoming flow is rotating counterclockwise in the Northern Hemisphere (see Fig. 5.2 and Problem 5.15). As it flows in, conservation of angular momentum requires its velocity to increase to make up for the decreased radius of the column. The rotational speed is small far from the center, but quite large at the center. If the ground is dry, the high-velocity wind at the center will pick up dust and carry it up, forming a visible dust devil. The rotation also stabilizes the upward flow, holding it together better than it would if it were not rotating.*

Dust devils are fun to watch and unlikely to produce more harm than dropping dust in the eyes of the unwary. They do not become strong enough to do much damage because the rate of energy input from solar heating per unit of ground surface is not very large. The tornadoes (also called *cyclones,* or *twisters*) that regularly cause fatalities, injuries, and property damage are described equally well by Fig. 5.8, except that their scale is much larger, their temperature gradients are steeper, and their wind

*If you doubt the description of increasing speed due to decreasing radius, recall how figure skaters do spins. They begin their rotation with their arms out and then pull them in, to increase their velocity by reducing their angular moment of inertia. Less obvious cases are (1) divers who can spring in the air, perform several spins, and then stop their rotation in flight and enter the water with no apparent rotation and (2) cats that, when dropped upside down from a height, manage to land on their feet. Both accomplish these feats by starting with some angular momentum and then changing their angular moment of inertia to control the rate of spin. An even less obvious example is "pumping" a playground swing to gain elevation; that also is a simple change of angular moment of inertia.

speeds are higher, making their destructive power much greater. To create a strongly unstable atmosphere over a large area one cannot simply rely on the sun shining on the ground. Instead, one must have bulk air movement in which a fast-moving cold air mass rides **over** a moist, warm air mass 2 to 3 km thick. As a result the lapse rate at the boundary between the two air masses is much larger than the adiabatic lapse rate, leading to the very strong upward-moving columns of air called tornadoes. These occur most often in spring and summer in the southeastern and midwestern United States. The rising air columns are normally produced simultaneously with strong thunderstorms [3]. (The overall cause is solar heating, but for tornadoes it is the stored solar heat accumulated over several days in the warm air layer, rather than the currently supplied amount that drives a dust devil.) Tornadoes do not begin at the ground, as does a dust devil, but rather begin at the hot-cold air interface and grow downward toward the ground from there. They become visible when they suck in dust and/or raindrops.

5.3.4 Mixing Height

Figure 5.7 is also an illustration of a key concept in air pollution meteorology, the *mixing height.* In that figure, for the midafternoon condition, there will be vigorous vertical mixing from the ground to about 6000 ft and then negligible vertical mixing above that height. The rising air columns that provide good vertical mixing induce large-scale turbulence in the atmosphere. This turbulence is three-dimensional, so it also provides good horizontal mixing. Pollutants released at ground level will be mixed almost uniformly up to the mixing height, but not above it. Thus the mixing height sets the upper limit to dispersion of atmospheric pollutants.

In the same figure we can see that in the morning the mixing height must be much lower and that it grows during the day. Similarly, we would expect that the mixing height would be larger in the summer than the winter (see Table 5.1 [4]).

Students have seen these mixing heights, although they may not have recognized them.* Figure 5.9 shows the situation in which there are many clouds, generally small, with spaces between. The tops of the clouds are not perfectly uniform, but they are all at practically the same height, which corresponds to the mixing height. Up to the mixing height rising, unstable air brings moisture up from below to form the clouds. Above the mixing height there is no corresponding upward flow.

A stronger form of this mixing-height phenomenon exists at the troposphere-stratosphere boundary (see Fig. 5.5). The stratosphere is practically isothermal, and very stable against mixing from below. Commercial airliners often fly above this boundary [whose height fluctuates up and down around the average value of 11 021 m (36 150 ft) shown in Fig. 5.5]. When the airliner is above the boundary the sky is clear and blue; as the airliner descends through the boundary the sky becomes brown or gray. The boundary is often very sharp and clearly visible.

---

*In Fig. 5.7 we see that at dawn there is an inversion, below which there is negligible mixing; the mixing height reported here is the value after solar heating has removed that inversion.

TABLE 5.1
Typical values of the mixing height for the contiguous United States

	Mixing height, m	
	Range	**Average**
Summer morning	200–1100	450
Summer afternoon	600–4000	2100
Winter morning	200–900	470
Winter afternoon	600–1400	970

Source: Ref. 4.

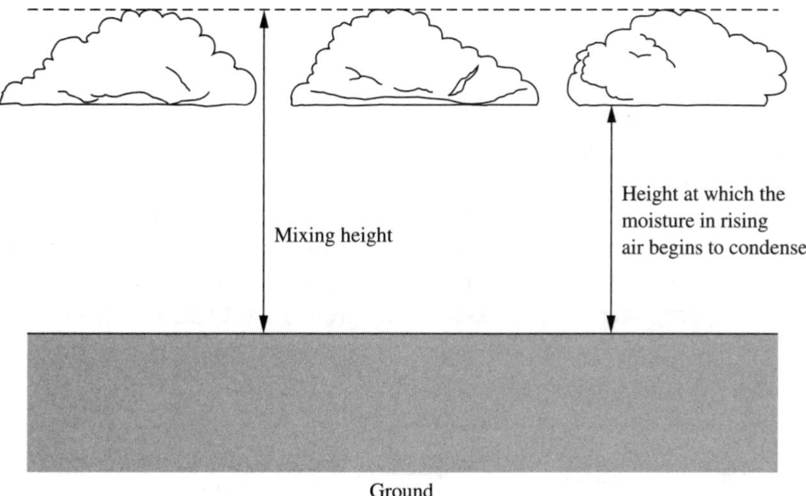

Clouds have irregular tops and sides but flat bottoms. The cloud tops are all at close to the same elevation, which is the mixing height.
Mixing height
Height at which the moisture in rising air begins to condense
Ground

FIGURE 5.9
During the day many small clouds with a common top elevation show the height of the mixing layer. The flat bottoms show the elevation at which condensation begins (see Sec. 5.3.5).

Figure 5.7 shows a strong radiation inversion at dawn, as would be caused by radiation to the night sky in a dry region like the Mojave Desert. In more humid areas, atmospheric moisture partly blocks the loss of heat from the ground to outer space, and a thin or partial cloud cover also reduces this heat loss. The resulting weaker inversions can serve as quasi mixing layers within which there may be substantial mixing, but whose tops retard further upward mixing. Other types of inversions caused by sea breezes or the drainage of cold air from hills often act the same way as mixing layers. The smoggy layer in Fig. 2.11 is an example of an inversion serving to determine the height of the mixing layer.

5.3.5 Moisture

So far, the discussion has concerned the behavior of dry air, or air that did not contain enough moisture to condense at ambient temperature. Moisture greatly complicates all of this. Most of the moisture in the atmosphere is evaporated from tropical oceans; the rain that falls on your picnic in Chicago likely evaporated from the Gulf of Mexico. Figure 5.10 shows the overall water balance for the world oceans and the land. From it one might infer that the raindrop in Chicago had more chance of forming from moisture evaporated over the land; that would ignore the fact that the majority of land evaporation takes place in tropical rain forests. The average residence time of a water molecule in the world atmosphere is about nine days [5].

When a parcel of moist air is displaced upward by solar heating or by some mechanical disturbance, its temperature behavior is almost the same as that of a parcel of dry air. In Eq. (5.9) the M and C_P are slightly perturbed by the moisture content (see Problem 5.9) but the effect is small. However, as the parcel is raised, its relative humidity, described by Eq. (5.10), increases.

$$\text{Relative humidity} = \frac{\text{Humidity}}{\text{Saturation humidity}} = \frac{y_{\text{water}} P}{p_{\text{water}}} \tag{5.10}$$

As a mass of air rises, the total pressure P decreases. However, the vapor pressure of pure water, p_{water}, also decreases because it depends only on the temperature, which also decreases as the elevation increases. The combined effect of these two opposing factors is shown in Fig. 5.11, which is based on the "standard atmosphere." We see that as the parcel of air rises, the ratio of its temperature to the surface temperature declines, but the ratio of the pressure to the surface pressure declines more rapidly. The ratio of the pure water vapor pressure to its ground level value declines more

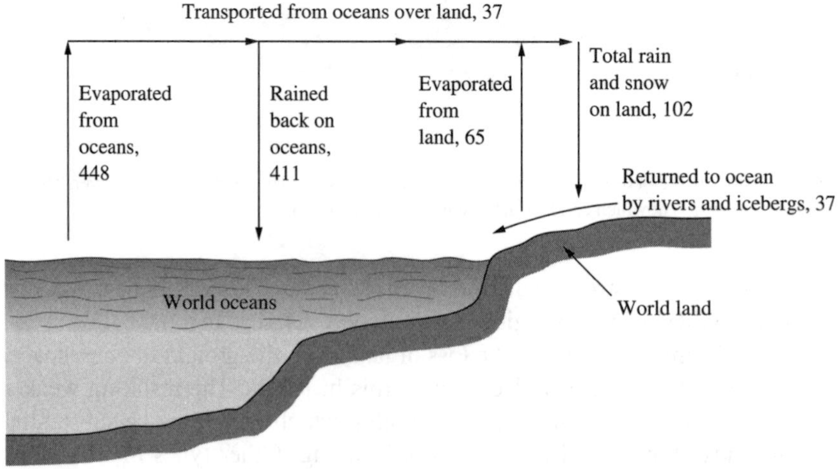

FIGURE 5.10
The annual evaporation, rainfall, and runoff of the whole world, in 10^3 km^3/year.

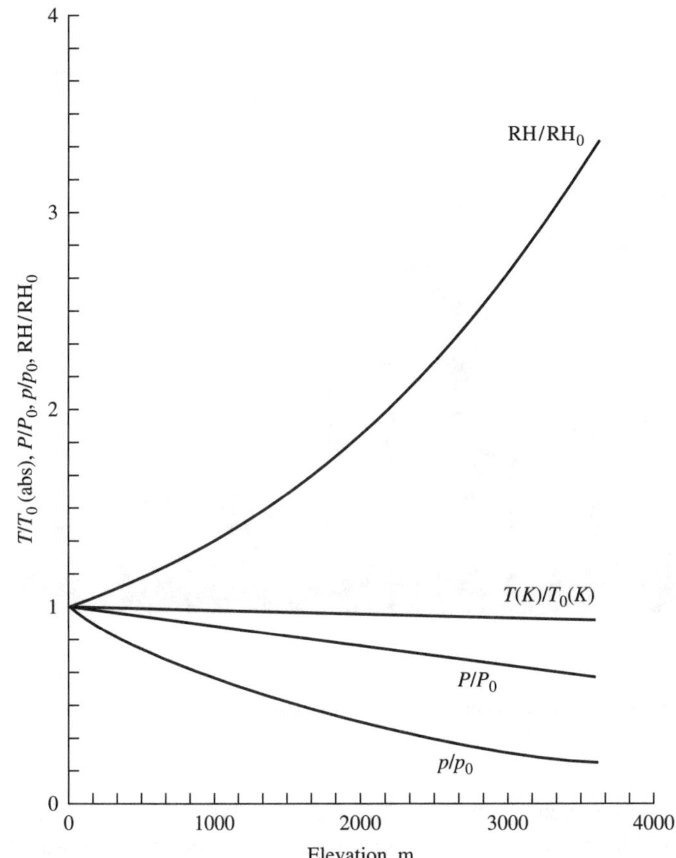

FIGURE 5.11
Changes with elevation of temperature, total pressure, vapor pressure of pure water, and relative humidity, for the standard atmosphere. The zero subscript indicates conditions at $z = 0$.

rapidly than either of these, because it is roughly proportional to $\exp(-1/T)$. We also see that the ratio of the relative humidity to the surface relative humidity increases. Thus, if the relative humidity at the surface were 50 percent, it would reach 100 percent (and moisture would just begin to condense if enough condensation nuclei were present) at 2150 m (7052 ft). This is slightly higher than the highest mountain in the United States east of the Mississippi River; air of 50 percent RH must be lifted a long way to condense. From the same plot we can compute that if the air is initially at 90 percent RH, it would only have to increase its relative humidity by a factor of 1.11, corresponding to about 450 m (1480 ft), to condense.

Figure 5.12 on page 104 shows an air mass flowing up over a mountain and down the other side. If this air is at 20°C and 50 percent RH at sea level, then at about 2150 m its moisture will begin to condense and it will form a cloud over the mountain. As the air flows down the other side it will warm, and the cloud will

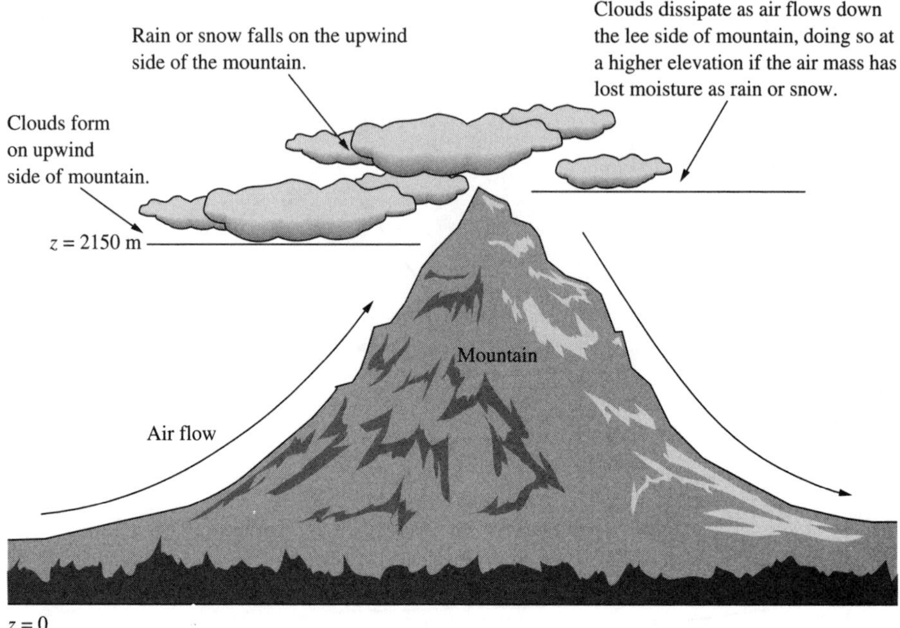
Rain or snow falls on the upwind side of the mountain.
Clouds dissipate as air flows down the lee side of mountain, doing so at a higher elevation if the air mass has lost moisture as rain or snow.
Clouds form on upwind side of mountain.
z = 2150 m
Mountain
Air flow
z = 0

FIGURE 5.12
Cloud formation by flow of initially unsaturated air up over a mountain. Here, for air at 20°C and 50 percent RH at $z = 0$, condensation begins at $z = 2150$ m.

evaporate. Anyone who lives near mountains has seen this type of cloud. If the mountain is high enough that the cloud produces rain or snow (thus removing water from the air mass), then the air mass that descends its lee side will be drier than the air mass that went up its windward side. On the lee side the cloud will evaporate at a higher elevation than the one at which it formed on the windward side. The driest deserts in the United States are formed that way, in the *rain shadow* of the Sierra Nevada mountains. In the Hawaiian Islands, the windward sides of the islands are the wet sides, and the lee sides the dry sides.

When the temperature is lowered enough that water begins to condense, the heat released by condensation becomes significant. One may show that if condensation is occurring with increasing elevation, then (see Problem 5.11)

$$\frac{dT}{dz} \approx \left(\frac{dT}{dz}\right)_{\text{adiabatic, dry}} - \frac{\Delta h_{\text{condensation, water}}}{C_{P,\text{air}}} \frac{dX}{dz} \tag{5.11}$$

where X is the molar humidity, expressed as mols of water vapor/mol of dry air. Condensation makes (dX/dz) negative, so the rightmost term in Eq. (5.11) is always positive, and thus the *moist adiabatic lapse rate* is always less than the dry adiabatic lapse rate. The numerical value of the rightmost term in Eq. (5.11) is not a constant (or even nearly a constant), as is the dry adiabatic lapse rate, because (dX/dz) has very different values in different parts of the atmosphere. It is a relatively large number in the tropics, where X is large and condensation occurs at low elevations. It is much smaller near the Poles, where X is small and condensation occurs mostly at high elevations.

A typical value of the moist adiabatic lapse rate is about 6.5°C/km. This is close to the lapse rate in the "standard atmosphere." The difference between the dry adiabatic lapse rate and the lapse rate in the "standard atmosphere" (the average of most observations) is due to several factors, of which the upward transport of the heat of condensation by rising moist air is one of the most important.

In the discussion of Fig. 5.6 we saw that the stability of an atmosphere depends on its lapse rate. In that figure we considered only air that did not have the possibility of water condensing. This need not be dry air, only air whose relative humidity is less than 100 percent during the whole process. If a parcel of air is at or near its saturation point (RH = 100%), then if it is moved up, condensation will occur, and it will follow the moist adiabatic lapse rate rather than the dry one. If it contains droplets (a cloud or fog) and it is moved down, some of the droplets it contains will evaporate and its temperature will follow the moist adiabatic lapse rate rather than the dry one. If the surrounding air is dry and has the dry adiabatic lapse rate, then the surrounding air would be neutral for an intruding parcel of dry air but quite unstable for an intruding parcel of moist air. This is the reason for the growth of clouds and thunderstorms. If an air parcel rises into a region where the water in it can condense but where the surrounding air has a lapse rate greater than the moist adiabatic lapse rate, then that parcel will rise and condense, and continue to do so until most of its moisture is condensed or until it reaches a place where the lapse rate is less than the moist adiabatic lapse rate (see Problem 5.10). If it does not find such a place, it can grow explosively upward to form a large thunderstorm. This calculation assumes that the rising air mass must take the condensed water with it. If it loses that condensed water as rain, snow, or hail, then its density decreases even faster than we would calculate (see Problem 5.10), so it becomes even more unstable for upward motion.

The inverse of this scenario occurs below a thundercloud. If the thundercloud releases large water drops into relatively dry air below its base, they will evaporate as they fall. That cools the air, making it more dense than the surrounding air. It descends rapidly, causing a strong downdraft. At the surface it spreads radially outward in all horizontal directions. If such a horizontal wind should suddenly overtake an aircraft from the rear, it will reduce its airspeed and hence its lift, causing it to fall rapidly. If it is close to the ground on takeoff or landing, it will crash; that has happened on several occasions. This meteorological event is called a *wind shear* [6], although *downdraft* or *downburst* is more descriptive of what happens.

In Fig. 5.5 it is clear that the stratosphere is very stable, because its lapse rate = 0. Thus, while the troposphere is fairly well mixed, the stratosphere is stratified. There is relatively little mixing between the two. Pollutants injected into the stratosphere (e.g., by major volcanic eruptions) remain aloft much longer than they do in the much better mixed troposphere.

5.4 WINDS

The general circulation pattern of our world's winds is shown in Fig. 5.2. There are numerous local variations, both in space and in time.

5.4.1 Velocities

The highest ground-level wind velocities are those in tornadoes, up to 200 mi/h (89.54 m/s). The average ground-level wind velocity in most of North America (day and night, summer and winter) is about 10 mi/h (4.5 m/s). The wind rarely blows less than about 2 mi/h (1 m/s). If you are standing outdoors in a 2 mi/h wind, you cannot feel it; the only way you can tell which way the wind is blowing is to observe the behavior of leaves, flags, smoke, or steam plumes, which will show that there is a wind even if you cannot feel it. Most weather services report any wind less than about 2 mi/h as "calm" because their wind-measuring instruments, called *anemometers*, become unreliable for such low velocities.

Wind speed increases with elevation, most of the time, in most of the troposphere. The reason is that ground friction slows the wind. Typically the wind will reach its frictionless velocity (called the *geostrophic* or *gradient* velocity) at about 500 m (1640 ft) above the ground. The region below this elevation, where ground friction plays a significant role, is the *planetary boundary layer.* The ground-level wind velocity is largely determined by how well this layer is coupled to the fast-moving geostrophic layer above it. When the atmosphere is stable or has an inversion, there is little vertical movement; and the coupling between the planetary boundary layer and the geostrophic wind is weak. Thus, inversions and stable atmospheres are normally associated with low ground-level wind velocities.

When the planetary boundary layer is unstable (midafternoon in Fig. 5.7) there is a great deal of vertical motion in the lower atmosphere and thus a great deal of momentum transfer between the planetary boundary layer and the geostrophic wind. Thus, unstable atmospheres have higher ground-level wind velocities than stable ones. From Fig. 5.7 we would expect higher ground-level winds in the early afternoon than in the morning or the late afternoon or night. Sailboat races are always scheduled for the early afternoon.

The increase in ground-level wind caused by instability is self-limiting; these winds tend to destroy the atmospheric instability that caused them. Strong winds provide good mixing, both horizontal and vertical, which makes the temperature gradient approach the dry adiabatic gradient. When the wind speeds are greater than about 6 m/s, or 13 mi/h, the observed stability is almost always neutral. High winds improve the mixing of hot air near the ground with the cooler air above it so that an extremely hot layer of air does not form near the ground, and thus no strong rising air columns can be formed. On a hot, dry summer day the atmosphere "chooses" the wind speed that balances this stability-producing trend with the atmospheric instability that helps cause the wind.

5.4.2 Wind Direction

We are often as concerned with wind direction as with wind speed. (Does the wind blow from my house toward a smelly feedlot or the other way?)

Superimposed on the general circulation shown in Fig. 5.2 are a series of disturbances called high-pressure zones (*highs,* or *anticyclones*) and low-pressure zones (*lows,* or *cyclones*). They are formed from large-scale instabilities, often involving the boundaries between the three circulation zones in each hemisphere. Their properties are compared in Table 5.2.

On a typical day there will be at least one of each of these over the contiguous United States. Major storms are normally associated with low-pressure systems. Near the center of low-pressure areas the winds associated with them can be strong enough to overwhelm all of the local effects described in this section. Thus, the following discussion considers only what happens during periods between the passages of these storms.

Mountains, valleys, and shorelines all influence wind direction and magnitude as well as other meteorological parameters. On a clear night the ground is cooled by radiation to outer space, and a layer of air forms adjacent to it that is colder and hence more dense than the air above it. If the ground were perfectly flat, this layer would be perfectly flat, and gravity would not tend to move it. But if the ground is not flat, then this more dense layer will tend to flow downhill. The steeper the hill, the faster it flows. In any valley cold air flows down to the bottom, and then the collected cold air flows down the valley the same direction that the stream or river in the valley flows.

During the day the opposite occurs. The sun-heated ground heats the air adjacent to it, which then rises by buoyancy. Normally one side of the valley will be more strongly heated by the sun than the other, so the air will begin to rise on that side,

TABLE 5.2
Behavior of high- and low-pressure areas

	High	Low
Pressure, compared with average atmospheric pressure	High, typically 1020 to 1030 mb	Low, typically 980 to 990 mb
Average vertical air motion near the ground	Sinking	Rising
Behavior of moisture in the air they contain	Evaporates, causing clear skies	Condenses, causing clouds and precipitation
Winds they generate	Out from the center, clockwise in the Northern Hemisphere, weak	In toward the center, counterclockwise in the Northern Hemisphere, strong

causing a rotating flow with its axis along the axis of the valley. This is superimposed on a net upward motion of the heated air, causing an upslope and upvalley daytime flow that is generally not as strong as the night flow.

Mountains can act as barriers to low-level winds. The Los Angeles Basin has high mountains on its north, east, and southeast. These impede the wind. They also trap air masses in the basin, preventing dilution of the emissions. Los Angeles's problem is compounded by the effect of the nearby ocean. Figure 5.13 shows this effect. In the early afternoon the ground has heated to perhaps 90°F, while the ocean is at perhaps 75°F. The heated air over the land rises, and cold air flows in from the ocean. This is the cool *sea breeze* that has drawn people to the beach on hot summer days throughout history. At night the ground cools, to perhaps 60°F, while the ocean does not cool much, because its upper layer is well-mixed. So the air over the ocean rises, and cooled air flows from the land back out over the ocean. This is the *land breeze* that we would go to the shore to enjoy if we liked cool breezes from the land at 4 A.M. as well as we like cool breezes from the ocean on hot summer afternoons.

This cool sea breeze makes the lower-level air mass in Los Angeles cooler, and hence more stable, than it would be if the ocean were agricultural land. The sea breeze is one of the contributors to the meteorological situation that makes air

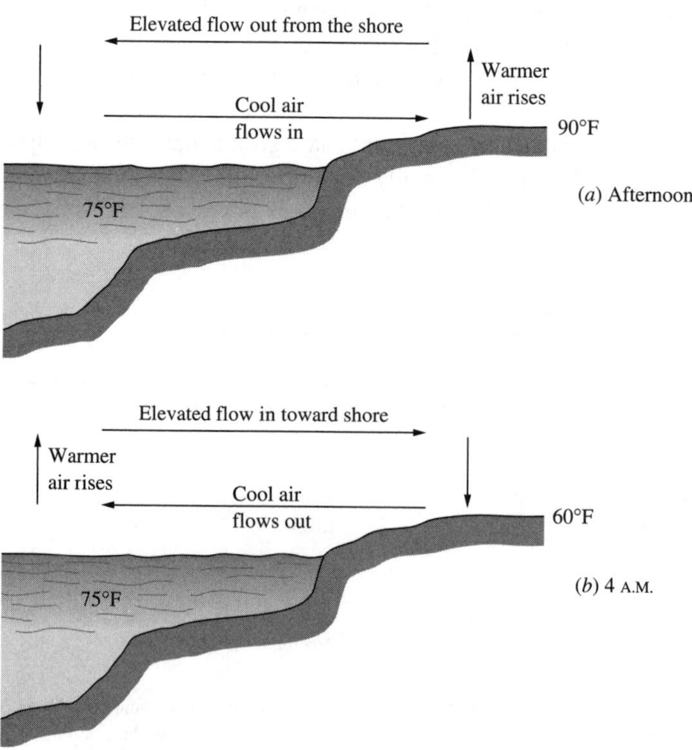

FIGURE 5.13
Onshore sea breezes in the day and offshore land breezes at night.

pollution control particularly difficult in Los Angeles. The same situation occurs for any city at the edge of the ocean or a large lake. In Los Angeles the sea breeze situation interacts with the mountains on the other side of the city to trap the air. One might think that the cool land breeze at night would take polluted air out to sea, only to have it return the next day. That must occur to some extent, but most measurements suggest that there is enough dispersal over the ocean at night that the sea breeze the next day is much less polluted than the land breeze was the night before.

In estimating the wind direction at any time and any location, one can use the following rules of thumb:

1. Major, rapidly moving storms and fronts overwhelm all local influences; local ground-level winds blow the way that the major storms dictate.
2. In deep valleys the daily alternation—wind up the valley in the daytime, down at night—overcomes most other influences and determines most of the local flow when no major storm or frontal passage dominates. The valley effect is greater in deep valleys than in shallow, in steep valleys than in gentle ones, at night than in the daytime, and under conditions of light wind and clear sky than of strong wind or cloudiness.
3. Onshore and offshore breezes dominate when there is no major storm. They are more likely to control the wind direction in light wind, clear sky conditions than in the opposite conditions, and more likely to control in the daytime than at night.
4. Absent all of the preceding or any other effects of local topography, the wind direction is more likely to be that shown in Fig. 5.2 than any other. Figure 5.2 is a better predictor near the equator than near the Poles.

Meteorological services regularly prepare *wind roses* like that shown in Fig. 5.14 on page 110. These summarize the frequency of winds of varying velocities and directions at one location. Normally one speaks of and plots a wind in terms of the direction *from which it comes.* A west wind blows from west to east. The wind rose in Fig. 5.14 is for a valley in the western desert of Utah. The surrounding mountains run practically north-south. The most common winds, governed by local topographic effects, are the up- and downvalley winds, north and south. The strongest winds come from the northwest and are associated with the passage of winter storms.

The same wind rose format is used to show many other properties, e.g., concentration of some atmospheric contaminant as a function of wind direction. Detailed meteorological data for various locations in the United States are regularly published by the National Climatic Data Center, Federal Building, Asheville, NC 28801. Some of their data files are specifically prepared for easy matching with the EPA air quality models discussed in Chapter 6.

Table 5.3 on page 111 shows the average wind speed (average of all measurements, independent of wind direction) and prevailing (most common) wind direction for a selection of U.S. cities [7].

From this table we see the following:

1. The average speeds range from 6.2 to 12.6 mph, with the overall average near 10 mph.
2. Most of the values are from airports where the National Weather Service (NWS) or the Federal Aviation Administration (FAA) maintains weather observation facilities. For the three cases where we have comparable measurements from both downtown stations and airports, the downtown wind velocities are less than those at the airport. High buildings in the downtown increase the coefficient of friction between the wind and the ground, thus lowering the wind speed there. Airports have no such high buildings.

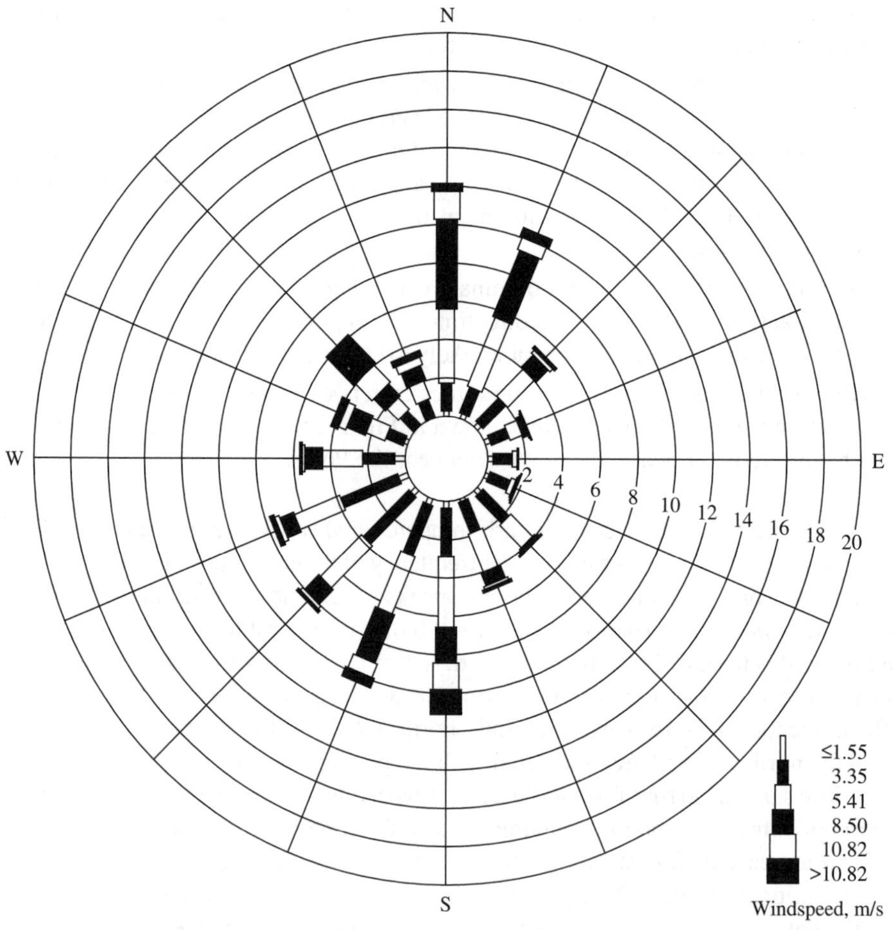

FIGURE 5.14
Wind rose for 1988 at Newfoundland, Utah. The concentric circles represent frequencies. For example, the wind blew from the south 11 percent of the time. It blew from the south with wind velocity > 10.82 m/s about 1 percent of the time; about $3\frac{1}{2}$ percent of the time it blew from the south with velocities between 3.35 and 5.41 m/s.

TABLE 5.3
Wind speed and direction for some U.S. cities

City	Mean wind speed over the entire year, mi/h	Prevailing wind direction
Seattle–Airport	9.2	SSW
San Francisco–Downtown	8.7	SE
–Airport	10.5	WNW
Los Angeles–Downtown	6.2	W
–Airport	7.4	W
Salt Lake City–Airport	8.7	SSE
Phoenix–Airport	6.2	E
Denver–Airport	9.0	S
Chicago–Midway Airport	10.3	W
St. Louis–Airport	11.2	SSE
Dallas–Airport	10.9	S
Boston–Airport	12.6	SW
New York–Central Park	9.4	NE
–La Guardia Airport	12.2	WNW
Charleston, SC–Airport	8.8	NNE
Miami–Airport	9.1	ESE
Anchorage–Airport	6.7	N
Honolulu–Airport	11.8	ENE

3. The prevailing wind direction is the direction from which the wind blows most frequently, not necessarily the direction from which the strongest wind comes. These wind directions are mostly governed by local topography, e.g., onshore or offshore winds, or local mountains and valleys.
4. These values are averages for a year. The source used data from different years at different sites, so some are from 1975, some 1976, and some 1977.
5. Chicago is unjustly called "The Windy City;" others have more wind [8].

5.5 TEMPERATURE INVERSIONS

Temperature inversions play a significant role in air pollution meteorology. Within an inversion the air is stable against buoyant vertical motion. That stability also lessens the exchange of wind energy between the air layer near the ground and high altitude winds, so that both horizontal and vertical dispersions of pollutants are hindered.

There are four ways to produce an inversion: Cool a layer of air from below, heat a layer of air from above, flow a layer of warm air over a layer of cold air, or flow a layer of cold air under a layer of warm air [9]. All of these occur. The first, cooling from below, is the very common radiation inversion discussed in Section 5.3.4 and Fig. 5.7; the other three are discussed here.

Heating an air layer from above can occur if a cloud layer absorbs incoming solar energy, but it most often occurs when there is a high-pressure region (common in summer between storms) in which there is a slow net downward flow of air and light winds. The sinking air mass will increase in temperature at the adiabatic lapse rate and often become warmer than the air below it. The result is an *elevated inversion,* also called *subsidence inversion* or *inversion aloft.* These normally form 1500 to 15 000 ft above the ground, and they inhibit atmospheric mixing. These inversions are common in sunny, low-wind situations, e.g., Los Angeles in summer.

Nighttime flow of cold air down valleys often leads to inversions at the bottom of the valley, with cold air flowing in under warmer air. In the winter this nighttime flow of cold air causes *drainage inversions.* In effect the valley collects all the ground-cooled air from the whole watershed above it. If condensation results, forming a fog, then the sun cannot get to the ground during the day, and the inversion will persist for days until a major storm clears it out. The presence of snow on the uphill ground makes these inversions stronger because snow is a good reflector of sunlight and a good emitter at infrared wavelengths. Thus the daily average net heat input is less for snow-covered surfaces than for bare ground or vegetation. This type of inversion can fill large valleys like the Central Valley of California with a cold fog for several days at a time in winter. Sea or lake breezes also bring cold air in under warm air, and can cause inversions or add to existing inversions.

Air flowing down the lee side of a mountain range is warmed by adiabatic compression. Air flowing down the east side of the Rocky Mountains is often warmed to temperatures higher than that of the air at the foot of the mountains. The warm air rides over the cold air, thus forming a strong inversion that can be very persistent.

All inversions, either at ground level or at higher elevations, inhibit atmospheric mixing and thus lead to the accumulation of pollutants. In summer, with clear skies, heating of the ground by the sun will normally eliminate an inversion every day, as shown in Fig. 5.7. However, local effects, like cool onshore breezes, may be powerful enough to maintain inversions. In winter the sun is often not strong enough to eliminate such inversions, and they may persist until a major storm brings winds strong enough to overcome the local topographic effects and sweep them away. If the inversion is strong enough to form a fog in a valley, it will reflect away sunlight, making the inversion persist longer than it would without the fog. Persistent drainage inversions in closed or semiclosed basins often lead to maximal pollutant concentrations.

5.6 FUMIGATIONS, STAGNATIONS

If a pollutant source is located in a region that has a strong, ground-based inversion, then its plume of pollutants will be trapped in the inversion and will travel with the local wind with very little dilution, as sketched on the right of Fig. 5.15. In the left side of that figure we see the lower atmospheric temperature as a function of time. At 6 A.M. there is a strong ground-based radiation inversion, caused by nocturnal cooling of the ground. As soon as the sun hits the ground, its temperature rises, and an unstable layer is formed that eats away at the bottom of the inversion. By 9 A.M.

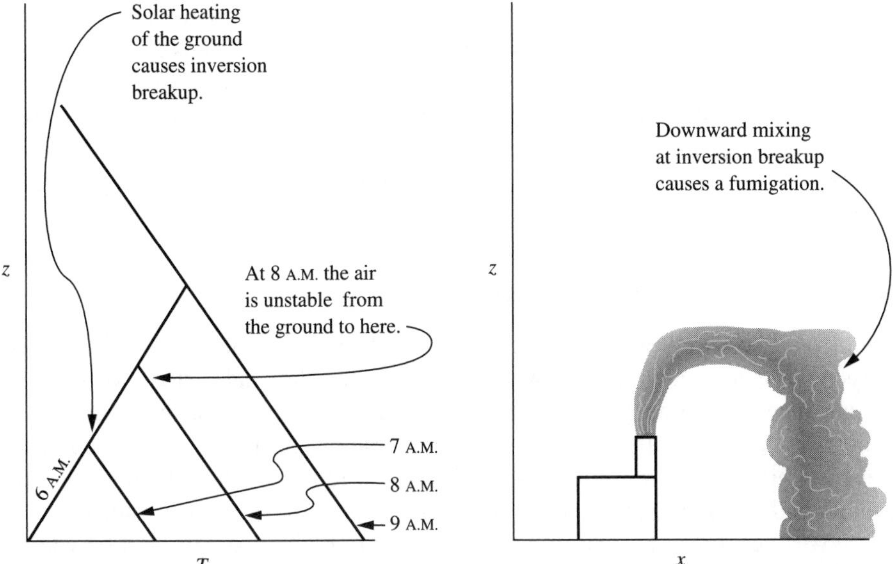

FIGURE 5.15
The development of a *fumigation.* The figure at the left shows the vertical temperature profile at various times. At 6 A.M. there is an inversion that is slowly destroyed by solar heating. On the right, the plume from a factory is shown inside an inversion. It flows horizontally with little mixing or dispersion due to the strong stability of the inversion. When the unstable mass of air from the heated ground reaches the plume, it mixes it to the ground, often at a high concentration, producing a short-term fumigation. This is most likely to occur with clear skies and light winds.

the inversion is gone. Returning to the right side of this figure, we see that when the unstable layer reaches the plume, at perhaps 8:30 A.M., the plume will mix down to the ground. In this instance the plume will not have been diluted much from its initial concentration, so that the ground level concentration at that point and that time will be surprisingly high. The high concentration will not last long, but such short, intense exposures can damage crops, etc. This kind of event is called a *fumigation.* A famous historical example of that type is discussed in Chapter 7.

Fumigations can also occur if the plume from a shoreline source is carried inland by a stable onshore breeze. As the breeze passes inland it encounters warmed air from the solar-heated soil, which mixes in from below; if the stability of the shore breeze is strong enough, it will prevent the plume from mixing upward. When the ground-based heating reaches the plume, that mixing from below may pull the plume to ground, producing a fumigation. Comparing this type of fumigation with that in Fig. 5.15, we see that the temperature pattern is the same except that instead of showing changes with time at one location, the left part of the figure shows changes with travel distance from the shoreline at one time.

In most of the eastern United States there is a more or less regular alternation of air masses from the Gulf of Mexico (warm, humid) and from central Canada (cold, dry). In the autumn one of these air masses will sometimes remain in place for four or more days. When it does, atmospheric pollutant concentrations rise, sometimes to harmful values. These events are called *stagnations.* With improved air pollution control in the United States in the past 30 years these events no longer lead to very high pollutant concentrations, but they still can cause or contribute to the highest pollutant concentrations normally encountered in this region. They occur in other parts of the world as well.

5.7 METEOROLOGICAL FORECASTS

People who develop forecasts for local TV rely heavily on data, analyses, and forecasts developed by the National Weather Service, the National Meteorological Center, and other centers operated by the National Oceanic and Atmospheric Administration (NOAA). Some rely on forecasts and graphic images provided by commercial weather services. We would all like to know the future weather, both to plan our picnics and to forecast the air pollution consequences of various emissions.

Weather forecasting is difficult. We can all predict that July will be warmer than December in most of the Northern Hemisphere, or that next year's precipitation in our city will be within one or two standard deviations of the mean for the past few years. Often that kind of projection of past history is all that is needed for air pollution analysis. The question really being asked by regulators and plant engineers is, "If we had installed this new facility or control device last year, what would the effects have been, given the weather we had last year?" The assumption is that next year, and the future years for which this new plant will be in operation, will be like those past years for which we have historic weather data.

That assumption is not always safe. The past few years may have been drier or wetter or windier than the long-term average, and the global climate may be changing. Until we have something better we will operate on the unstated assumption that the meteorological future will be like the meteorological past, and that the more past data we have, the more confident we are of our predictions of the future.

Probably the largest collection of supercomputers outside the U.S. military is the group at the National Center for Atmospheric Research in Boulder, Colorado. It is devoted to devising, testing, and using advanced numerical models of the atmosphere. Our knowledge is still imperfect; even with acres of supercomputers we cannot predict the future behavior of the atmosphere with much certainty.

5.8 SUMMARY

1. This chapter has only introduced the topic of air pollution meteorology. Much more complete and detailed books are available [10–13]. The application of meteorological information to practical and regulatory problems is shown in the next two chapters.
2. The two meteorological parameters of greatest interest to air pollution engineers are the atmospheric stability and the wind speed. In general, stable atmospheres

(low lapse rates or inversions) and low wind speeds lead to the highest ground-level pollutant concentrations (from ground-level or low-level sources). Unstable atmospheres and high wind speeds lead to the lowest ground-level pollutant concentrations.

3. The topic of global climate change is postponed to Chapter 14.

PROBLEMS

See Common Units and Values for Problems and Examples, inside the back cover.

5.1. If a pancake is $\frac{1}{4}$ in. thick (average for fluffy pancakes) and has the same ratio for thickness to diameter as the atmosphere, what would its diameter be? If a pancake has a diameter of 6 in. (a typical value) and has the same thickness-to-diameter ratio as the atmosphere, what would its thickness be?

5.2. Show why the number of circulatory cells in any one hemisphere must be odd, not even. A simple sketch and a few sentences should suffice.

5.3. Example 5.1 and Fig. 5.3 show the simplest instance of how to use the Coriolis force (or acceleration) to reconcile the views of the observer on a fixed space station and the observer on the rotating earth. It is the simplest because (1) only one of the two players moves (i.e., from both viewpoints the pitcher does not move) and (2) the pitcher does not impart any velocity component at right angles to the throw, due to the motion of the earth. For all other locations of the pitcher and catcher, the problem is more complex. The most important step in solving such problems is to draw the right figure, correctly showing both pitcher and catcher, both from their viewpoint and from that of an observer on a fixed space station. Repeat that example for the following situations:

(*a*) Both pitcher and catcher are 30 ft from the North Pole on opposite sides of the Pole, and the pitcher throws the ball directly toward the catcher over the Pole. Are the results the same? Should they be? How does the answer change if the pitcher throws the ball slower?

(*b*) The pitcher is at the equator and throws the ball due north to the catcher, who is 60 ft away. (First work the simpler problem of two people riding on railroad trains, traveling on parallel tracks at constant, equal speed. They throw the ball back and forth at right angles to the tracks, and air resistance is negligible. Show the view from above, both from their viewpoint and from that of an observer in a space station. Then draw the corresponding diagram for this case.)

(*c*) The pitcher and catcher are both 1000 ft south of the North Pole, with the pitcher 60 ft east of the catcher. Then repeat this exercise with the pitcher 60 ft west of the catcher.

5.4. In Examples 5.1 through 5.3 we estimated the angular velocity of the earth using its rotation period of 24 hours. That corresponds to the *solar* day. If one bases the rate of rotation on the stars rather than the sun, one finds a rotation period of 23.93 hours, which corresponds to the *sidereal* day.

(*a*) Which day should have been used in these calculations?

(*b*) How much difference does it make?

(*c*) Why are solar and sidereal days different?

5.5. Starting with Eqs. (5.8) and (5.9), work out the following relations for the adiabatic atmosphere:

$$\frac{T_2}{T_1} = \left(1 - \frac{R}{C_P} \cdot \frac{gM\,\Delta z}{RT_1}\right); \qquad \frac{P_2}{P_1} = \left(1 - \frac{R}{C_P} \cdot \frac{gM\,\Delta z}{RT_1}\right)^{C_P/R}$$

where the subscripts 1 and 2 can stand for any two locations in the atmosphere. One may show that

$$\frac{R}{C_P} = \frac{k-1}{k}$$

where k is the ratio of specific heats; one often sees these equations written with that substitution. Some texts use γ where we use k.

5.6. (*a*) At what height does the equation for an adiabatic atmosphere (Problem 5.5) indicate that the temperature in the air would be 0 K? Assume a surface temperature of 15°C = 59°F.

(*b*) What is the physical significance of this prediction?

(*c*) What is the predicted pressure at this altitude?

5.7. For the "standard atmosphere" shown in Fig. 5.5, perform the following calculations:

(*a*) Derive the pressure-height relation for the troposphere.

(*b*) Calculate the pressure at the troposphere-stratosphere boundary.

(*c*) Derive the pressure-height relation for the stratosphere.

5.8. Estimate the mixing height for the following situation: at elevations above the mixing height, the temperature-elevation behavior is given by the "standard atmosphere" shown in Fig. 5.5. At elevations below the mixing height, the temperature-elevation behavior is given by the adiabatic lapse rate for dry air. The surface temperature is 20°C = 68°F.

5.9. Repeat the calculation of the adiabatic, perfect gas lapse rate for a gas that is 1.15 mol percent water; the water does not condense. The molar heat capacity C_P of water vapor may be taken as 4.1 R. How large an error do we make by ignoring this water in computing the adiabatic lapse rate?

5.10. (*a*) Show that, when moisture condenses in a parcel of air, the density of the resultant air-water mixture is less than the density of the original parcel, so that such condensation always makes the air parcel more buoyant. Suggestions: choose 1 mol of water vapor and n mol of air. For that mixture the density (i.e., mass/volume) is

$$\rho_1 = \frac{29n + 18}{(n+1)RT_1/P}$$

After the mole of water has condensed the density is

$$\rho_2 = \frac{29n + 18}{n(RT_2/P) + V_{\text{liquid}}} \approx \frac{29n + 18}{nRT_2/P}$$

Then show that

$$T_2 = T_1 + \frac{\Delta h_{\text{condensation, water}}}{nC_{P,\text{air}}}$$

Substitute this expression in the equation for ρ_2, construct the ratio of ρ_2/ρ_1, and simplify. You will find that this ratio will always be less than 1 if

$$\frac{\Delta h_{\text{condensation, water}}}{T_1 C_{P,\text{air}}} > 1$$

Then evaluate this inequality using the following values: $\Delta h_{\text{condensation, water}} = 45.0$ kJ/mol, $C_{P,\text{air}} = 33.8$ J/mol K, $T_1 = 273$ K.

(*b*) Does the same reasoning apply, and is the result the same, when liquid water is turned to ice?

5.11. Derive the equation for the moist adiabatic lapse rate. Begin by assuming that the system of 1 mol of air and X mol of water vapor is raised adiabatically a distance dz. As it is raised,

P, T, and X will all change. Assume that the volume of the liquid formed by condensing some of the water is negligible, so that

$$V = (1 + X)\frac{RT}{P} \tag{A}$$

By an energy balance on this closed system we have

$$dU = dQ - dW = 0 - P\,dV \tag{B}$$

and

$$\begin{aligned} dU &= \frac{\partial U}{\partial T}dT + \frac{\partial U}{\partial X}dX + \frac{\partial U}{\partial P}dP \\ &= C_V\,dT + \Delta l_{\text{condensation, water}}\,dX + \frac{\partial U}{\partial P}dP \end{aligned} \tag{C}$$

Noting that $dP = 0$, we eliminate dU between equations (B) and (C). Differentiate Eq. (A), substitute dV into the resulting equation, and simplify to

$$RT(1+X)\left(\frac{dT}{T} + \frac{dP}{P} + \frac{dX}{1+X}\right) = (C_{V,\text{air}} + XC_{V,\text{water}})\,dT + \Delta u_{\text{condensation, water}}\,dX \tag{D}$$

Replace dP with its value from Eq. (5.6) and rearrange to yield

$$\frac{dT}{dz} = \frac{gM_{\text{avg}}}{(C_{V,\text{air}} + XC_{V,\text{water}}) + R(1+X)} - \frac{(\Delta u_{\text{condensation, water}} + RT)}{(C_{V,\text{air}} + XC_{V,\text{water}}) + R(1+X)}\frac{dX}{dz} \tag{E}$$

Check by setting X equal to zero and seeing if the result is the same as Eq. (5.9).

Equation (E) is the complete equation. For most situations outside the tropics, $X \ll 1$, so that we may ignore the X in the $(1 + X)$ expressions, but not in the dX/dz. Make that simplification and several thermodynamic simplifications, then use Eq. (5.9) to get

$$\frac{dT}{dz} \approx \left(\frac{dT}{dz}\right)_{\text{adiabatic, dry}} - \frac{\Delta h_{\text{condensation, water}}}{C_{P,\text{air}}}\frac{dX}{dz} \tag{5.11}$$

5.12. Based on the description of morning and afternoon wind velocities in Section 5.4.1, explain why sunsets are redder than sunrises. Explain why most of the time one may look directly at a setting sun for an instant without pain, but practically never can one look at a rising sun without pain (or permanent eye damage!).

5.13. On most of the Hawaiian Islands there is no road around the northeast side because the oceanfront cliffs there are too steep for road building. Explain why these steep cliffs occur only on the northeast sides.

5.14. Why do trade winds have that name?

5.15. If individual parcels of air moving in the Northern Hemisphere turn to their right (see Fig. 5.2), why does a region with radial inflowing air (a low-pressure system, tropical storm, tornado, or dust devil) rotate **counterclockwise**? Draw a view from above, looking down the center of the low-pressure region. Draw it first without the Coriolis force, in which case the inflow lines would be straight. Then add the curvature due to the Coriolis force.

5.16. A meteorologist discussing a record-breaking hurricane said, "It had a pressure of 850 mb in the center, so it had winds of 250 miles/h!" Explain this statement in terms of Bernoulli's equation.

5.17. Meteorologists define and use the *potential temperature*, (θ), which is the temperature that a parcel of air at some elevation would have if it were brought to the surface adiabatically. Thus,

$$\text{Potential temperature} = \theta = T_{\text{actual at } z} + (z - z_{\text{surface}})\left(-\frac{dT}{dz}\right)_{\text{adiabatic}} \tag{5.12}$$

Remake the left parts of Fig. 5.6 (parts *a–d*), replacing the actual temperature with the potential temperature. *Hint:* First determine what the adiabatic lapse rate looks like in a plot of z vs. potential temperature.

REFERENCES

1. Hasse, L., and F. Dobson: *Introductory Physics of the Atmosphere and Ocean,* D. Reidel, Dordrecht, Netherlands, p. 11, 1986.
2. "U.S. Standard Atmosphere, 1976," National Oceanic and Atmospheric Administration, U.S. Government Printing Office, Washington, DC, 1976.
3. Battan, L. J.: *Fundamentals of Meteorology,* 2d ed., Prentice-Hall, Englewood Cliffs, NJ, p. 179, 1984.
4. Holzworth, G. C.: "Mixing Heights, Wind Speeds, and Potential for Urban Air Pollution throughout the Contiguous United States," U.S. Environmental Protection Agency, Report AP-101, U.S. Government Printing Office, Washington, DC, 1972.
5. Hasse, L. and F. Dobson: op. cit., p. 22.
6. Trefil, J.: *Meditations at Sunset: A Scientist Looks at the Sky,* Scribners, New York, pp. 73–88, 1987.
7. Ruffner, J. A.: *Climates of the States,* Vols. 1 and 2, 2d ed., Gale Research Company, Detroit, MI, 1980.
8. Chicago got its reputation as "The Windy City" in the 1920s, when most U.S. Weather Bureau stations were located downtown in cities. As cities built higher and higher buildings, the Weather Bureau put its wind meters on higher and higher buildings to get away from building effects. There was a period when Chicago had the highest reported average wind speed of any major city. It also had its wind speed gauge higher than any other city at that time. I thank Herschel Slater for this interesting piece of history.
9. Oke, T. R.: *Boundary Layer Climates,* 2d ed., Methuen, London, p. 311, 1987.
10. Atkinson, B. S.: *Dynamical Meteorology,* Methuen, London, 1981.
11. Slade, D. H.: *Meteorology and Atomic Energy 1968,* TID-24190, National Technical Information Service, Washington, DC, 1968.
12. Wanta, R. C., and W. P. Lowry: "The Meteorological Setting for Dispersal of Air Pollutants," in A. C. Stern (ed.), *Air Pollution,* 3d ed., Academic Press, New York, 1976.
13. Arya, S. P.: *Air Pollution Meteorology and Dispersion*, Oxford University Press, Oxford, United Kingdom (1999).

CHAPTER 6

AIR POLLUTANT CONCENTRATION MODELS

6.1 INTRODUCTION

Air pollution law in most industrial countries is based on some kind of permitted concentration of contaminants (NAAQS in the United States). To plan and execute air pollution control programs designed to meet the requirements of these laws, one must predict the ambient air concentrations that will result from any planned set of emissions. Even if we did not use this type of air pollution law, we would probably use some other kind of law that made some use of predictions of ambient contaminant concentrations. These predictions are made by way of air pollutant concentration models.

The perfect air pollutant concentration model would allow us to predict the concentrations that would result from any specified set of pollutant emissions, for any specified meteorological conditions, at any location, for any time period, with total confidence in our prediction. The best currently available models are far from this ideal. In this chapter we consider three kinds of models, beginning with the simplest (and least reliable) and proceeding to the most complex (and most reliable). All models are simplifications of reality, leading to the belief that "All models are wrong; some models are useful." The models in this chapter are useful.

All of the models presented here (and almost all others as well) are simple material balances. A material balance is an accounting in which one applies the general balance equation to some species. In our case the species being accounted for is the air pollutant under study. The general balance equation applies to some specified set of boundaries and can be written as follows:

$$\begin{pmatrix}\text{Accumulation}\\ \text{rate}\end{pmatrix} = \begin{pmatrix}\text{all flow}\\ \text{rates in}\end{pmatrix} - \begin{pmatrix}\text{all flow}\\ \text{rates out}\end{pmatrix} + \begin{pmatrix}\text{creation}\\ \text{rate}\end{pmatrix} - \begin{pmatrix}\text{destruction}\\ \text{rate}\end{pmatrix} \tag{6.1}$$

In this form it appears quite abstract; in the models discussed in the following sections, we will see its concrete application.

All such models are applied to one air pollutant at a time. Most models can be used for several different pollutants, but they must be applied separately to each. No models presented here apply to "air pollution in general."

6.2 FIXED-BOX MODELS

Consider a rectangular city as shown in Fig. 6.1. To compute the air pollutant concentration using Eq. (6.1) in this city, we make the following major simplifying assumptions:

1. The city is a rectangle with dimensions W and L and with one side parallel to the wind direction.
2. Atmospheric turbulence produces complete and total mixing of pollutants up to the mixing height H and no mixing above this height. (See Sec. 5.3.4 for a dicussion of mixing heights.)
3. This turbulence is strong enough in the upwind direction that the pollutant concentration is uniform in the whole volume of air over the city and not higher at the downwind side than the upwind side. This assumption is quite contrary to what we observe in nature but permits a great simplification of the mathematics.
4. The wind blows in the x direction with velocity u. This velocity is constant and is independent of time, location, or elevation above the ground. This is also contrary

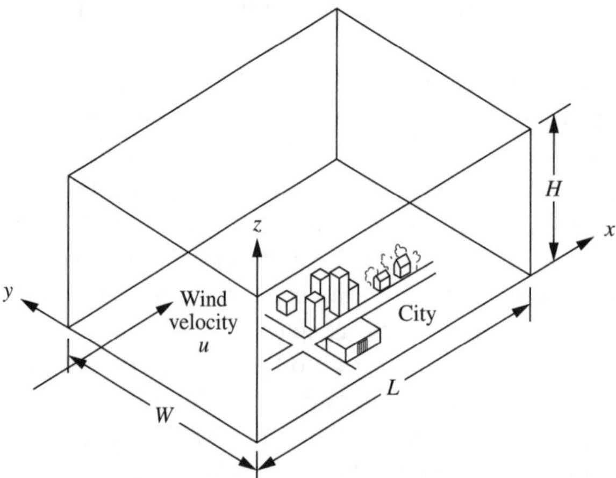

FIGURE 6.1
Rectangular city, showing meaning of symbols used in the fixed-box model.

to observation; wind speeds increase with elevation. Here we use the average u between that at the ground and that at H.

5. The concentration of pollutant in the air entering the city (at $x = 0$) is constant and is equal to b (b for "background" concentration, a term borrowed from the nuclear field, from which many of the early air pollution meteorologists came). Concentrations in this model and in most of this chapter are usually in units of g/m^3 or micrograms/m^3 (1 microgram $= \mu g = 10^{-6}$ g).

6. The air pollutant emission rate of the city is Q (typically expressed in g/s). This is normally given as an emission rate per unit area, q, in $g/s \cdot m^2$. We can convert from one to the other by

$$Q = qA \tag{6.2}$$

where A is the area of the city, which equals W times L in this case. This emission rate is constant and unchanging with time.

7. No pollutant leaves or enters through the top of the box, nor through the sides that are parallel to the wind direction.

8. The pollutant in question is sufficiently long-lived in the atmosphere that the destruction rate in Eq. (6.1) is zero (see Sec. 6.6).

With these assumptions, we can now evaluate all of the terms in Eq. (6.1). We choose as our system the volume WLH. Because all of the assumptions indicate that flows and emission rates are independent of time, we see that this is a steady-state situation in which nothing is changing with time. For any steady-state situation in any application of the general balance equation, Eq. (6.1), the accumulation rate is zero, so the term to the left of the equal sign is zero.

We may treat the emission rate Q either as a creation rate or as a flow into the box through its lower face. Either gives exactly the same result; it is more common in the air pollution literature to treat it as a flow through the lower face, so we will set the creation rate equal to zero. Thus, Eq. (6.1) has been simplified to

$$0 = \begin{pmatrix} \text{all flow} \\ \text{rates in} \end{pmatrix} - \begin{pmatrix} \text{all flow} \\ \text{rates out} \end{pmatrix} \tag{6.3}$$

There are two pollutant flow rates in. The flow rate of pollutant into the upwind side of the city is

$$\text{Flow rate in} = uWHb \tag{6.4}$$

The first three symbols constitute the volume of air that crosses the upstream boundary of the system per unit time; the student may verify that uWH has dimensions of volume/time. Multiplying it by a concentration (mass/volume), we obtain a mass flow rate (mass/time).

The second flow rate in is that of pollutant emitted by the city into the lower boundary, or face, of the system,

$$\text{Flow rate in} = Q = qWL \tag{6.5}$$

According to the preceding assumptions, the concentration in the entire city is constant and is equal to c. (Here we use c for concentration; in the older air pollution literature this is most often a χ.) The only way pollutant leaves the system is by flow out through the downwind face. The flow rate out is given by the equation

$$\text{Flow rate out} = uWHc \tag{6.6}$$

Substituting these expressions into Eq. (6.1) and solving for c, we find

$$c = b + \frac{qL}{uH} \tag{6.7}$$

which is the simple, fixed-box model. Chemical engineers will recognize that this is the same as the continuous-flow stirred-tank reactor (CSTR) model widely used in chemical engineering.

Example 6.1. A city has the following description: $W = 5$ km, $L = 15$ km, $u = 3$ m/s, $H = 1000$ m. The upwind, or background, concentration of carbon monoxide is $b = 5\ \mu\text{g/m}^3$. The emission rate per unit area is $q = 4 \times 10^{-6}\ \text{g/s} \cdot \text{m}^2$. What is the concentration c of carbon monoxide over the city?

By direct substitution into Eq. (6.7), we find

$$c = \frac{5\ \mu\text{g}}{\text{m}^3} + \left(4 \times 10^{-6} \frac{\text{g}}{\text{s} \cdot \text{m}^2}\right) \left[\frac{15{,}000\ \text{m}}{(3\ \text{m/s})(1000\ \text{m})}\right]$$

$$= 5 + 20 = 25\ \frac{\mu\text{g}}{\text{m}^3}$$ ■

In the other chapters of this book examples are shown in both English and metric units. However, air pollution models in the United States almost always use metric units, so in this chapter all examples will be shown only in metric.

Example 6.1 shows that Eq. (6.7) is simple to apply. You may already have noted that W does not enter the calculation or influence the result. This is reasonable for the model chosen; doubling the width of the city while holding q constant would not change c.

Clearly Eq. (6.7) is a great simplification of what must really occur in nature. However, all of the important variables enter, with the correct signs and powers. It correctly indicates that the upwind concentration for a long-lived pollutant is additive to the concentration produced by the city, and that the latter increases with increases in q and L and decreases with increases in u and H.

By far the worst of the foregoing assumptions is the third—that the concentrations at the upwind and downwind edges of the city are the same. Holzworth [1] developed a somewhat more complex form of Eq. (6.7) by replacing this assumption with a more realistic one.

The second-worst assumption is that the emissions are uniformly distributed over the area of the city (i.e., q is constant over the whole city). If indeed we have perfect mixing within the box, this assumption makes no difference. But if we drop the perfect-mixing assumption, this uniform emission assumption becomes the worst.

Frequently we have information on the variation of q from place to place in the city. For example, we would assume that for most pollutants q is low in the suburbs and much higher in industrial areas. Hanna [2] has presented a modification of the simple box model that incorporates the same ideas presented by Holzworth [1] and also allows one to divide the city into subareas and apply a different value of q to each.

The simple fixed-box model of Eq. (6.7) and its modifications, as well as most of the others presented in the chapter, predicts concentrations for only one specific meteorological condition. To find the annual average concentration of some pollutant, we would have to use the frequency distribution of various values of wind direction, u, and of H, compute the concentration from Eq. (6.7) for each value, and then multiply by the frequency and sum to find the annual average; that is,

$$\begin{pmatrix}\text{Annual}\\ \text{average}\\ \text{concentration}\end{pmatrix} = \sum_{\text{over all meteorologies}} \begin{pmatrix}\text{concentration}\\ \text{for that}\\ \text{meteorology}\end{pmatrix}\begin{pmatrix}\text{frequency of}\\ \text{occurrence of that}\\ \text{meteorology}\end{pmatrix} \tag{6.8}$$

Example 6.2. For the city in Example 6.1, the meteorological conditions described ($u = 3$ m/s, $H = 1000$ m) occur 40 percent of the time. For the remaining 60 percent, the wind blows at right angles to the direction shown in Fig. 6.1 at velocity 6 m/s and the same mixing height. What is the annual average concentration of carbon monoxide in this city?

First we find the concentration for the other meteorological condition using Eq. (6.7). Observe that the wind direction shift has interchanged the values of W and L (see Fig. 6.1). Thus,

$$c = 5\ \frac{\mu\text{g}}{\text{m}^3} + \left(4 \times 10^{-6}\ \frac{\text{g}}{\text{s}\cdot\text{m}^2}\right)\left[\frac{5000\ \text{m}}{(6\ \text{m/s})(1000\ \text{m})}\right] = 8.33\ \frac{\mu\text{g}}{\text{m}^3}$$

Using this value plus the one from Example 6.1 in Eq. (6.8), we find

$$\begin{pmatrix}\text{Annual}\\ \text{average}\\ \text{concentration}\end{pmatrix} = 25\ \frac{\mu\text{g}}{\text{m}^3} \times 0.4 + 8.33\ \frac{\mu\text{g}}{\text{m}^3} \times 0.6 = 15\ \frac{\mu\text{g}}{\text{m}^3}$$ ■

To make realistic application of Eq. (6.8) requires summing over several hundred meteorological conditions and their corresponding emission rates, instead of the two in Example 6.2. In cities where the major source of particulates is combustion for home heating, the emission rate per unit area is much higher in the winter than in the summer. Other pollutant emission rates vary from hour to hour and day to day. Automobile-related emissions are much higher during commuting hours than in the middle of the night. Equation (6.8) is regularly modified to take these variations in emission rate into account.

Similarly, if we wish to apply this equation to find the situation in which the highest concentration will occur, we need to know the wind speed, wind direction,

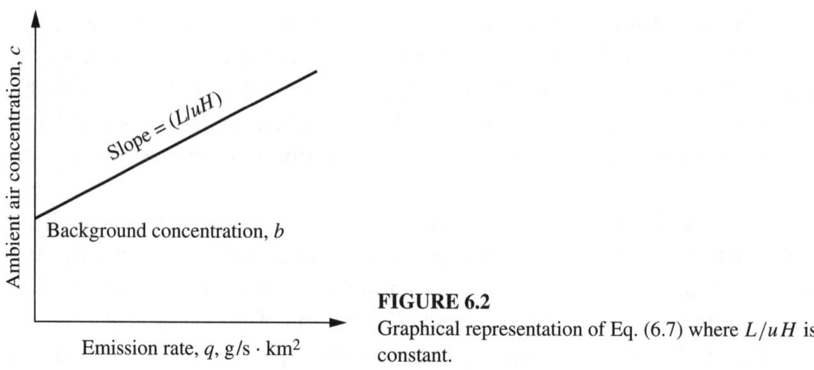

FIGURE 6.2
Graphical representation of Eq. (6.7) where L/uH is constant.

mixing height, and upwind (background) concentration that correspond to this worst case. This information is not always available. Faced with this problem, early workers in air pollution (principally Larsen [3]) proposed a simpler form of Eq. (6.7). If we hold u, L, and H constant in Eq. (6.7), we may represent it as shown in Fig. 6.2.

This plot shows that if the meteorological situation is constant, the concentration is equal to the upwind concentration b plus the concentration due to the city, which is linearly proportional to the city's emission rate. If, for example, the concentration is c_1 with emission rate q_1 and we wish to reduce the concentration to c_2, we may readily compute the allowable emission rate q_2 from Eq. (6.7) as

$$q_2 = \frac{(c_2 - b)uH}{L} \tag{6.9}$$

However, if we know q_1 and c_1, we may write an entirely analogous equation for q_1 and solve it for uH/L. If we then substitute that value of uH/L into Eq. (6.9) and rearrange, we find

$$\frac{q_2}{q_1} = \frac{(c_2 - b)}{(c_1 - b)} \tag{6.10}$$

In this formulation the meteorological parameters do not appear; they have been assumed constant between the present situation (c_1, q_1) and the future situation (c_2, q_2) and are therefore eliminated.

If the current measured pollutant concentrations exceed the applicable standard, then we must make c_2 and q_2 lower than c_1 and q_1. One can compute the fractional reduction in emission rate needed from Eq. (6.10), or

$$\begin{aligned}\begin{pmatrix}\text{Fractional reduction}\\ \text{in emission rate}\end{pmatrix} &= \frac{(q_1 - q_2)}{q_1} = 1 - \frac{q_2}{q_1} \\ &= 1 - \frac{(c_2 - b)}{(c_1 - b)} = \frac{(c_1 - c_2)}{(c_1 - b)}\end{aligned} \tag{6.11}$$

Example 6.3. The ambient air quality standard for particulates (TSP) in the United States in 1971 was 75 μg/m^3 annual average (revised when we changed from TSP to

PM_{10} and $PM_{2.5}$, see Table 2.3). In 1970 the annual average particulate concentration measured at one monitoring station in downtown Chicago was 190 $\mu g/m^3$. The background concentration was estimated to be 20 $\mu g/m^3$. By what percentage would the emission rate of particulates have to be reduced below the 1970 level in order to meet the 1971 ambient air quality standard?

Using Eq. (6.11), we find

$$\begin{pmatrix}\text{Fractional reduction}\\ \text{in emission rate}\end{pmatrix} = \frac{(190-75)}{(190-20)} = 0.67, \quad \text{or} \quad 67\%$$ ■

Equation (6.11) as well as several variants of it is known in the U.S. air pollution literature as the *proportional model* or the *rollback equation.* It has been widely used in computing the emission rate reduction needed to meet ambient air quality standards. Its virtues are that it is simple and it normally requires input data that are readily available. However, it is a great simplification of a basically complex situation, and it is unlikely to give accurate predictions except in special cases [4]. Furthermore, as we saw in Example 6.3, the equation tends to predict that high percentage reductions will be needed. More complex models generally do not make this prediction. Thus this model is more often used for cities with less severe problems than Chicago's.

Another drawback of all fixed-box models is that they make no distinction between large numbers of small sources that emit their pollutants at low elevations (autos, homes, small industry, refuse burning, etc.)—called *area sources*—and the small number of large sources that emit larger amounts per source, at higher elevations (power plants, smelters, cement plants, etc.)—called *point sources.* Both large and small sources are simply added to find the value of q. There is ample evidence that, under most circumstances, raising the release point of the pollutant will decrease the ground-level concentration due to that source, in the region near the source, although it may increase the concentration farther away. There is no easy way to deal with this drawback in fixed-box models.

What does Eq. (6.7) tell us we can do about air pollution in our own city? For an existing city we can do nothing about u, H, and L. If we are laying out a new city, we should lay it out to be long and thin, perpendicular to the wind direction (L as small as possible), or else pick a place where H and u are large. These choices will minimize air pollutant concentrations. (That generally means not to put your city in a valley; many major cities are in valleys.) For an existing city the manipulatable variables are b and q. We can reduce b by having our upwind neighbors reduce their emissions. To reduce q we must reduce our own emissions. This set of choices is illustrated by the proposal several years ago to build a large power plant in northeastern Nevada, directly upwind of Salt Lake City, Utah. The Utah politicians tried to elbow each other out of the way so they could get in front of the TV cameras to denounce this project (which would increase b). The same politicians were not willing to call for serious efforts to reduce their constituents' own emissions (which would reduce q).

6.3 DIFFUSION MODELS

In the air pollution literature this next class of models is usually called *diffusion models*. Most engineers would call them *dispersion models* because engineers reserve the word *diffusion* for molecular diffusion, which is not the principal mechanism described by these models. However, the preceding name is so common that it will be used here.

6.3.1 The Gaussian Plume Idea

Most diffusion models use the Gaussian plume idea, which also is a material balance model. In it, one considers a point source such as a factory smokestack (which is not really a point but a small area that can be satisfactorily approximated as a point) and attempts to compute the downwind concentration resulting from this point source. The schematic representation and nomenclature are shown in Fig. 6.3, where the origin of the coordinate system is placed at the base of the smokestack, with the x axis aligned in the downwind direction. The contaminated gas stream (normally called a *plume*) is shown rising from the smokestack and then leveling off to travel in the x direction and spreading in the y and z directions as it travels.

Such plumes normally rise a considerable distance above the smokestack because they are emitted at temperatures higher than atmospheric and with a vertical velocity. For Gaussian plume calculations the plume is assumed to be emitted from

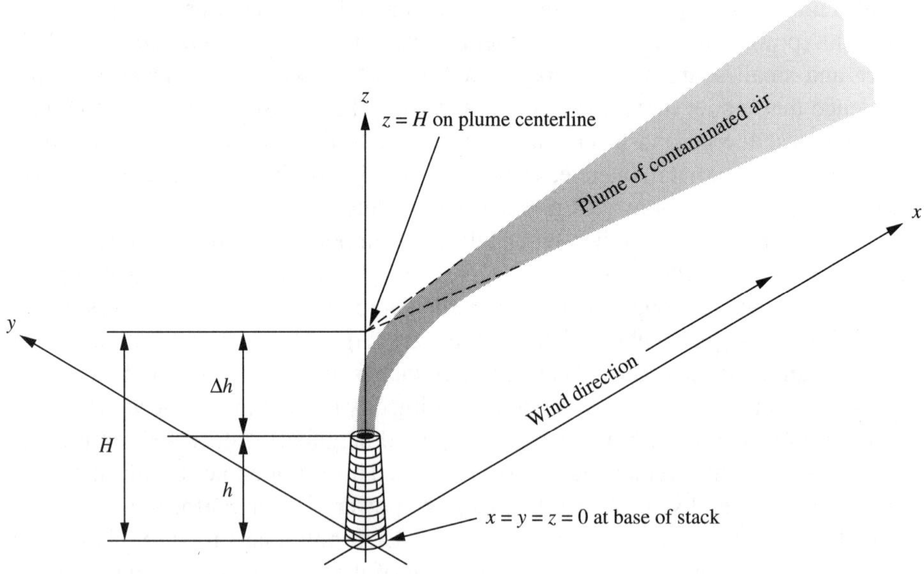

FIGURE 6.3
Coordinate system and nomenclature for the Gaussian plume idea.

a point with coordinates 0, 0, H, where H is called the *effective stack height*, which is the sum of the *physical stack height* (h in Fig. 6.3) and the *plume rise* (Δh in Fig. 6.3). Physical stack height for any existing plant can be determined with ordinary measuring instruments. Plume rise is discussed in Sec. 6.4. For the moment we will assume that we are dealing with a point source located at 0, 0, H that steadily emits a nonbuoyant pollutant at emission rate Q (normally in g/s). Let us assume the wind blows in the x direction with velocity u and that this velocity is independent of time, location, or elevation. The problem is to compute the concentration due to this source at any point (x, y, z) for $x > 0$.

If molecular diffusion alone were causing the plume to mix with the surrounding air, the plume would spread slowly and appear (if the pollutant is visible) as a thin streak moving straight down the sky (see Problem 6.33). The actual cause of the spread of plumes is the large-scale turbulent mixing that exists in the atmosphere, which may be visualized by comparing a snapshot of a plume with a time exposure of the same plume (Fig. 6.4). At any instant the plume will appear to have a twisting, snake-like shape as it moves down the sky. The twisting behavior is caused by the turbulent motion of the atmosphere that is superimposed on the plume's large-scale linear motion caused by the horizontal wind. This turbulent motion is random in nature, so that a snapshot taken a few minutes after the first would show the twists and turns in different places, but the overall form would be similar. However, time averages out these short-term variations of the plume, and thus a time exposure appears quite uniform and symmetrical. For this reason, if we placed a pollutant-concentration meter at some fixed point in the plume, we would see the concentration oscillate in an irregular fashion about some average value. The Gaussian plume approach tries to calculate only that average value without making any statement about instantaneous values. The results obtained by Gaussian plume calculations should be considered only as averages over periods of at least 10 minutes, and preferably one-half to one hour.

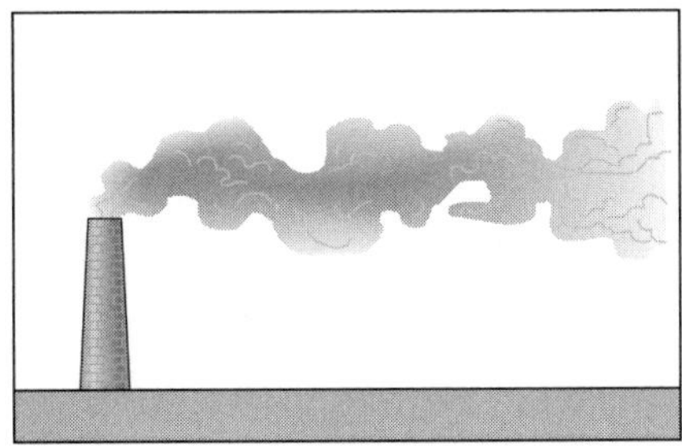

Snapshot

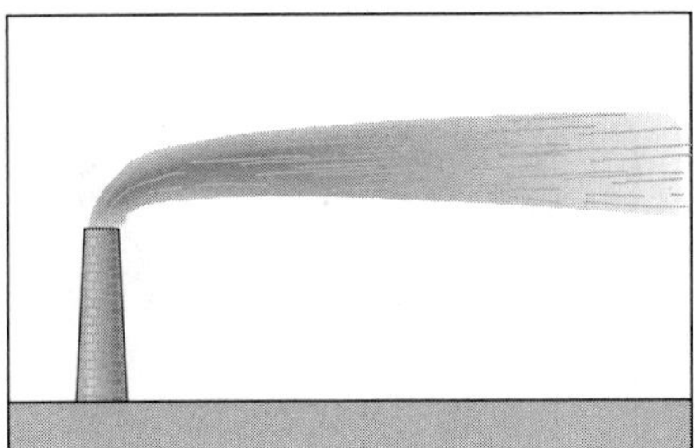

Time exposure

FIGURE 6.4
Comparison of snapshot and time exposure of a visible plume.

6.3.2 Gaussian Plume Derivation

To derive the Gaussian plume formula, we will first take the viewpoint of a person riding along with the air, the *Lagrangian viewpoint*. From this viewpoint, the ground appears to be passing below, much as the ground appears to be passing below a person in an airplane. We begin riding along upwind of the stack from which the pollutant is emitted, so we will say that the initial concentration of the pollutant is zero ($b = 0$). (If $b > 0$, we must add the value of b to the value calculated here to obtain the best estimate of the atmospheric concentration.) As we pass directly over the stack we pass into a region of high concentration. This high concentration is localized in a thin thread of contaminated air that passes directly over the stack. After we have passed the stack we will see this thread of contaminated air expand by turbulent mixing.

To find out how it expands by turbulent mixing, we will perform a material balance around some small cube of space near the center of the plume. The dimensions of this small cube are shown in Fig. 6.5. Let us consider a material that is neither created nor destroyed in the atmosphere, so that the two right-most terms of Eq. (6.1) are zero. The remaining terms are

$$\begin{pmatrix}\text{Accumulation}\\ \text{rate}\end{pmatrix} = \sum\begin{pmatrix}\text{all flow}\\ \text{rates in}\end{pmatrix} - \sum\begin{pmatrix}\text{all flow}\\ \text{rates out}\end{pmatrix} \tag{6.12}$$

The accumulation rate is the time derivative of the amount contained, which is the product of the concentration and the volume. But the volume of the cube is not changing with time, so

$$\text{Accumulation rate} = \frac{\partial}{\partial t}(cV) = V\frac{\partial c}{\partial t} = \Delta x\ \Delta y\ \Delta z\ \frac{\partial c}{\partial t} \tag{6.13}$$

There is no bulk flow (i.e., convection) into or out of the cube we are considering because the cube is moving with the local wind velocity. However, there are flows through all six faces of the cube due to turbulent mixing. We do not have a clear and complete physical or mathematical picture of the complex subject of turbulent mixing, but one may approximate it by saying that the flux of material being mixed

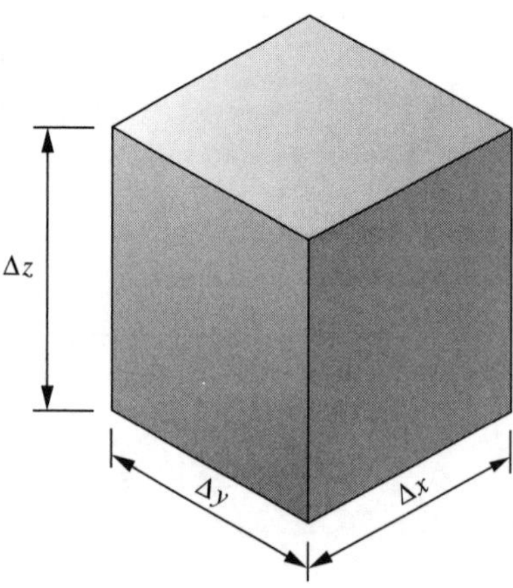

FIGURE 6.5
Dimensions of the cube used for the material balance.

across any surface is given by

$$\text{Flux} = \begin{pmatrix} \text{time rate of mass} \\ \text{flow per unit area} \end{pmatrix} = -K\frac{\partial c}{\partial n} \tag{6.14}$$

where c = concentration

n = distance in the direction considered (normally x, y, or z)

K = turbulent dispersion coefficient

Because the flux must have units of mass/time · area (e.g., g/s · m^2) and $\partial c/\partial n$ has dimensions of mass/length^4, K must have dimensions of length^2/time, e.g., m^2/s. This dimension is the same as that for molecular diffusivity or thermal diffusivity, and we will see that our equations have the same form as the equations for heat conduction or mass molecular diffusion. This does not show that the processes are the same; rather it shows that we have forced our equations into the mold of the heat and mass diffusion equations by choosing the form shown in Eq. (6.14) to represent turbulent dispersion. The minus sign in Eq. (6.14) indicates that the flow is from high concentration to low. (The approximation shown in Eq. (6.14) is called the *gradient transport* or *K-theory* or *first-order closure* approach, and the turbulent dispersion coefficient, K, is often called the *eddy diffusivity* [5].)

Our cube has two faces that look in the x direction; the one facing the reader (in Fig. 6.5) looks in the minus x direction, and the other, on the far side of the cube, looks in the plus x direction. Each of these faces has area $\Delta y\ \Delta z$. By using Eq. (6.14) twice, we can see that the net mass flow by turbulent diffusion through these two faces can be described as

$$\begin{pmatrix} \text{Net flow into the cube} \\ \text{in the } x \text{ direction} \end{pmatrix} = \left[\left(\frac{-K\,\partial c}{\partial x}\right)_{\text{at } x} - \left(\frac{-K\,\partial c}{\partial x}\right)_{\text{at } x+\Delta x}\right]\Delta y\,\Delta z \tag{6.15}$$

where the first term represents flow in through the face nearest the reader and the second represents flow out through the face away from the reader. By the same procedure we can write terms for the other four faces, giving us two terms involving $\partial c/\partial z$ and two involving $\partial c/\partial y$. These six represent the flows in or out through the six faces by turbulent mixing. From Eq. (6.12) we know that their sum is equal to the accumulation rate, Eq. (6.13). We now substitute Eqs. (6.13), (6.15), and the two analogous ones into Eq. (6.12) and divide both sides by $\Delta x\ \Delta y\ \Delta z$, finding

$$\frac{\partial c}{\partial t} = \frac{\left(\dfrac{K\,\partial c}{\partial x}\right)_{\text{at } x+\Delta x} - \left(\dfrac{K\,\partial c}{\partial x}\right)_{\text{at } x}}{\Delta x} + \frac{\left(\dfrac{K\,\partial c}{\partial y}\right)_{\text{at } y+\Delta y} - \left(\dfrac{K\,\partial c}{\partial y}\right)_{\text{at } y}}{\Delta y} + \frac{\left(\dfrac{K\,\partial c}{\partial z}\right)_{\text{at } z+\Delta z} - \left(\dfrac{K\,\partial c}{\partial z}\right)_{\text{at } z}}{\Delta z} \tag{6.16}$$

But

$$\lim_{\Delta x \to 0} \frac{\left(\dfrac{K\,\partial c}{\partial x}\right)_{\text{at } x+\Delta x} - \left(\dfrac{K\,\partial c}{\partial x}\right)_{\text{at } x}}{\Delta x} = \frac{K\,\partial^2 c}{\partial x^2} \tag{6.17}$$

so that if we take the limit of an infinitesimally small cube, Eq. (6.16) becomes

$$\frac{\partial c}{\partial t} = K\frac{\partial^2 c}{\partial x^2} + K\frac{\partial^2 c}{\partial y^2} + K\frac{\partial^2 c}{\partial z^2} \tag{6.18}$$

This is the equation for heat conduction in a solid with the variables renamed. Our immediate reaction is to factor K out of the three terms on the right, but experimental data indicate that for turbulent diffusion in the atmosphere the values of K in the three directions are not the same. So, in subsequent equations, we will write the three Ks as K_x, K_y, and K_z.

6.3.2.1 One-, two-, and three-dimensional spreading. The Gaussian plume equation is regularly applied to pollutant spreading in one, two, or three dimensions. To see how these three applications arise and the mathematics is developed, we need to consider the application of Eq. (6.18) to one-, two-, and three-dimensional spreading. An intuitively easy illustration of one-, two-, and three-dimensional spreading appears in Fig. 6.6.

In Fig. 6.6 if the medicine dropper deposits X g of dye solution instantaneously at the origin ($x = y = z = 0$) at $t = 0$, then we can solve for the dye concentration at any place and time from Eq. (6.18). This problem is entirely equivalent to the "instantaneous source" problem in the conduction of heat in solids, for which the mathematical solution is well-known [6].

The resulting concentrations calculated for one-, two-, and three-dimensional spreading are

$$c = \frac{X}{2(\pi t)^{1/2} K_x^{1/2}} \exp\left[-\left(\frac{1}{4t}\right)\left(\frac{x^2}{K_x}\right)\right] \quad \text{for one dimension} \tag{6.19}$$

$$c = \frac{X}{4(\pi t)(K_x K_y)^{1/2}} \exp\left[-\left(\frac{1}{4t}\right)\left(\frac{x^2}{K_x} + \frac{y^2}{K_y}\right)\right] \quad \text{for two dimensions} \tag{6.20}$$

$$c = \frac{X}{8(\pi t)^{3/2}(K_x K_y K_z)^{1/2}} \exp\left[-\left(\frac{1}{4t}\right)\left(\frac{x^2}{K_x} + \frac{y^2}{K_y} + \frac{z^2}{K_z}\right)\right] \quad \text{for three dimensions} \tag{6.21}$$

Comparing these three equations, we see that adding a spreading dimension multiplies the denominator of the leading fraction by $2(\pi t)^{1/2} K^{1/2}$ and adds a ($\text{dimension}^2/K$) to the exponential term on the right. We also observe that at the origin ($x = y = z = 0$) the exponential term is $\exp 0 = 1$, so that all three of

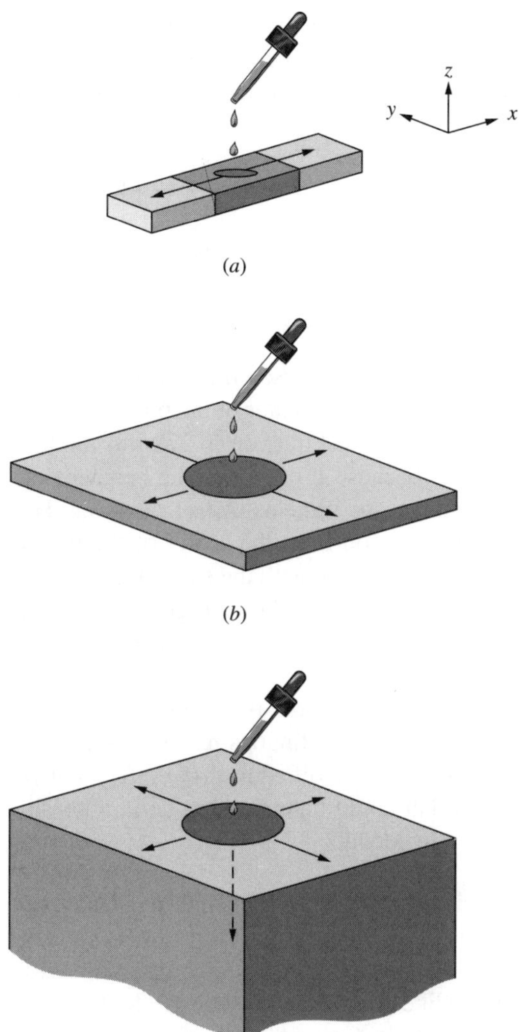

FIGURE 6.6
Illustration of one-, two-, and three-dimensional spreading. A medicine dropper puts a few drops of a dye solution onto blotting paper. (*a*) The paper is in the form of a narrow strip aligned with the x axis, and spreading is in one dimension. (*b*) The paper is a thin sheet, and spreading is in two dimensions. (*c*) The paper is a thick stack of sheets, so that spreading is in three dimensions.

these equations have a leading fraction, which is the instantaneous concentration at the origin, multiplied by an exponential term (always less than 1) that shows how much the instantaneous concentration decreases as we move away from the origin in one, two, or three dimensions. The concentration at the origin is proportional to $1/\sqrt{t}$ for one-dimensional spreading, $1/t$ for two-, and $1/t^{3/2}$ for three-dimensional spreading.

6.3.2.2 Gaussian puff, three-dimensional spreading. Consider first the application of this formulation to an instantaneous short-term release of pollutants from the chimney shown in Fig. 6.3, i.e., at $x = y = 0$ and $z = H$, as might result from a momentary breakdown in the pollution control equipment. The amount released will be $X = Q\,\Delta t$, where Q is large and Δt small. Inserting these values into Eq. (6.21), we find

$$c = \frac{Q\,\Delta t}{8(\pi t)^{3/2}(K_x K_y K_z)^{1/2}} \exp\left[-\left(\frac{1}{4t}\right)\left(\frac{x^2}{K_x} + \frac{y^2}{K_y} + \frac{(z-H)^2}{K_z}\right)\right] \tag{6.22}$$

where t = time since the release = downwind distance of the center of the pollutant cloud/wind speed

Δt = time duration of release (which is assumed small)

In Eq. (6.22), where we have taken the viewpoint of a person riding with the flow, x represents the downwind or upwind distance from the center of the pollutant cloud, which is assumed to move with local wind velocity. Equation (6.22) thus modified is often called the *Gaussian puff equation* because it describes the behavior of a "puff" of pollutants. The x in Eq. (6.22) is not the same as the x in Fig. 6.3. Here x is the distance in the up- or downwind direction from the center of the moving puff (*Lagrangian viewpoint*). In Sec. 6.3.2.3 we begin with the same definition and then change to the viewpoint of an observer standing still on the ground (*Eulerian viewpoint*), in which x will have the meaning shown in Fig. 6.3, that is, distance downwind from the base of the stack.

Equation (6.22) is only occasionally used in air pollution control calculations because we are generally much more interested in continuous releases than in puff releases. However, it is widely used (in somewhat modified form) in safety analysis where the puff of pollutants is the cloud that could be emitted in certain possible types of serious chemical plant or nuclear accidents.

6.3.2.3 Gaussian plume, two-dimensional spreading. To find the steady-state equivalent of Eq. (6.22), we make the material balance for a thin sheet of air that extends a distance 1 m in the x direction and to infinity in the y and z directions and that moves with the local wind speed (like one particular slice in a loaf of sliced bread that passes in the long direction of the loaf over the top of the stack). This sheet transfers material to the sheets immediately up- and downwind of it by turbulent dispersion, but it receives almost the same amount of material from those sheets so that the net transfer of material from the sheet in the x direction is negligible and is set equal to zero. Assuming negligible net transfer of material in the x direction makes this a two-dimensional spreading problem, for which we will use Eq. (6.20). In this case the time it takes the sheet to pass over the assumed point source is (1 m/u) so that the amount of pollutant originally injected into the slab we are considering is $X = Q/u$. (The reader may verify that Q/u has dimensions of mass/length, i.e., the amount injected per unit length of air passing over the stack.)

Making these substitutions into Eq. (6.20), we find that

$$c = \frac{Q/u}{4\pi t (K_y K_z)^{1/2}} \exp\left[-\left(\frac{1}{4t}\right)\left(\frac{y^2}{K_y} + \frac{(z-H)^2}{K_z}\right)\right] \tag{6.23}$$

where the symbols have the same meaning as before. If we had chosen our coordinates so that the pollutant source were at some arbitrary point, say (x', y', z') instead of being at $(0, 0, H)$, then the terms in the exponential part of Eqs. (6.22) and (6.23) would be $(x - x')^2$, $(y - y')^2$, etc. The choice of the origin that we made simplifies these expressions. One might choose to put the origin of the coordinate system at the top of the plume rise (which would drop the H out of Eqs. (6.22) and (6.23)), but most of us prefer that $z = 0$ be at ground level.

Although Eq. (6.23) would be perfectly satisfactory for our use, for historical reasons the form that appears in the air pollution literature is obtained by making the following three substitutions:

$$K_y = 0.5\sigma_y^2 \frac{u}{x} \tag{6.24}$$

$$K_z = 0.5\sigma_z^2 \frac{u}{x} \tag{6.25}$$

$$t = \frac{x}{u} \tag{6.26}$$

where σ_y and σ_z are called horizontal and vertical *dispersion coefficients*.* They have the dimensions of length, normally given in meters. Making these substitutions in Eq. (6.23), we find

$$\begin{aligned} c &= \frac{Q}{2\pi u \sigma_y \sigma_z} \exp\left[-\left(\frac{y^2}{2\sigma_y^2} + \frac{(z-H)^2}{2\sigma_z^2}\right)\right] \\ &= \frac{Q}{2\pi u \sigma_y \sigma_z} \exp\left(-\frac{y^2}{2\sigma_y^2}\right) \exp\left(-\frac{(z-H)^2}{2\sigma_z^2}\right) \end{aligned} \tag{6.27}$$

which is the *basic Gaussian plume equation.* This name comes from the fact that the exponential terms have the same form as the Gauss normal distribution function, which is widely used in statistics. It has many variants, a few of which we will see in subsequent paragraphs.

Equation (6.26) changes the equation from the *Lagrangian viewpoint*, in which x is at the middle of the moving cloud, to the *Eulerian viewpoint*, in which x represents some fixed distance downwind from the emission point. In most of the rest of this chapter the equations are in this Eulerian form, that is, distances are measured from the base of the stack, as in Fig. 6.3, not from the center of a moving cloud.

Equation (6.27) is the product of three terms. If we set $y = (z - H) = 0$, then the two right-most terms will be $\exp 0 = 1$, which shows that the first term is the concentration on the centerline of the plume. The two σ values increase with downwind distance, so that this centerline concentration decreases with downwind

*The Greek sigmas are used here because σ appears in the formulas in statistics that use the Gaussian distribution. Thus the sigmas here make the formulas look the same. There is no theoretical connection between the two, and some other symbol could just as well have been used, but the sigmas are used throughout the air pollution literature. The values are based on experimental data and shown in Figs. 6.7 and 6.8.

distance. The second term shows how the concentration decreases as we move in the horizontal, sidewise, $\pm y$, direction from the plume centerline. Because the second term involves y^2 it is the same for moving in the + or $-y$ direction. It is always ≤ 1.00. The third term is like the second, but it shows how the concentration decreases as we move vertically away from the elevation of the plume centerline ($z = H$). It also is symmetrical and always ≤ 1.00. The three terms are independent of each other but use the same values of the σs. This simple "product of three independent terms" formulation is the unavoidable consequence of the assumptions leading to Eq. (6.14).

Example 6.4. A factory emits 20 g/s of SO_2 at height H. The wind speed is 3 m/s. At a distance of 1 km downwind, the values of σ_y and σ_z are 30 m and 20 m, respectively. What are the SO_2 concentrations at the centerline of the plume, and at a point 60 meters to the side of and 20 meters below the centerline?

The centerline values are those for which $y = 0$ and $z = H$, so both of the terms in the exponential are zero. Since $\exp 0 = 1$, the exponential term is unity. At the centerline

$$c = \frac{20 \text{ g/s}}{2\pi(3 \text{ m/s})(30 \text{ m})(20 \text{ m})} = 0.00177\frac{\text{g}}{\text{m}^3} = 1770\ \frac{\mu\text{g}}{\text{m}^3}$$

At the point away from the centerline, we must multiply the preceding expression by

$$\exp -\left[\frac{1}{2}\left(\frac{60 \text{ m}}{30 \text{ m}}\right)^2 + \frac{1}{2}\left(\frac{-20 \text{ m}}{20 \text{ m}}\right)^2\right] = \exp -\left(2 + \frac{1}{2}\right) = 0.0818$$

so

$$c = \left(\frac{1770\ \mu\text{g}}{\text{m}^3}\right)(0.0818) = \frac{145\ \mu\text{g}}{\text{m}^3}$$ ■

The basic Gaussian plume equation predicts a plume that is symmetrical with respect to y and with respect to z. Thus if we had asked for the concentration 60 m to the other side of and 20 m above the plume centerline, we would have gotten the same answer. Different values of σ_y and σ_z mean that spreading in the vertical and horizontal directions is not equal. Most often $\sigma_y > \sigma_z$, so that at a given x a contour of constant concentration is like an ellipse, with the long axis horizontal. Close to the ground this symmetry is disturbed, as we shall discuss shortly.

To use the Gaussian plume equation one must know the appropriate values of σ_y and σ_z. From Eqs. (6.24) and (6.25) we would expect them to have the form

$$\sigma_y = \left(\frac{2\ K_y x}{u}\right)^{1/2} \quad \text{etc.} \tag{6.28}$$

However, if we reconsider our value for the Ks in Eq. (6.14), we see that we have simply assigned an arbitrary value to them, independent of atmospheric behavior. It seems reasonable to assume that they would depend on wind speed and on the degree of atmospheric turbulence, which is a function of wind speed and degree of solar heating (*insolation*) and perhaps some other factors. It is also reasonable to assume that for any given degree of insolation the value of K will be linearly proportional to the wind speed; i.e., K_y/u and K_z/u are constants. Thus from Eq. (6.28) we conclude that, for any given meteorological condition, each of the σs should be proportional to the square root of the downwind distance.

Experimental evidence does not agree well with this prediction. The available data have been correlated by Turner [7] and by others and presented in the form of plots of $\log \sigma_y$ and $\log \sigma_z$ vs. $\log x$. If the preceding calculation were correct, for each atmospheric condition such plots would be straight lines with slope $\frac{1}{2}$. The best correlations of the experimental results illustrate that on such plots the horizontal dispersion coefficient σ_y forms a family of straight lines (for various atmospheric conditions), but these have a slope of 0.894 instead of the 0.50 that we would expect from the preceding derivation (Fig. 6.7). The vertical dispersion coefficient σ_z forms a fan-shaped pattern for various atmospheric conditions (Fig. 6.8 on page 136).

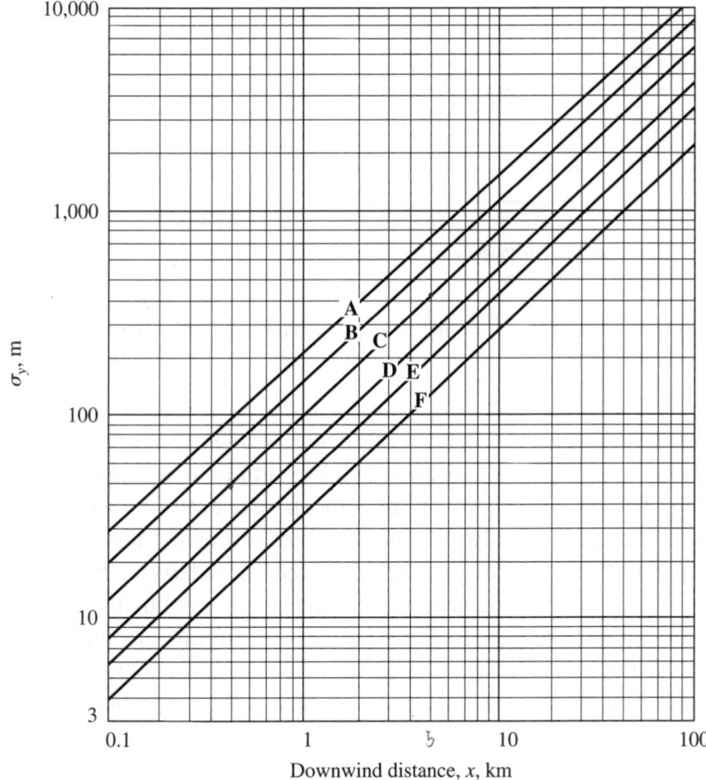

FIGURE 6.7
Horizontal dispersion coefficient σ_y as a function of downwind distance from the source for various stability categories. See Problem 6.16. (From Turner [7].)

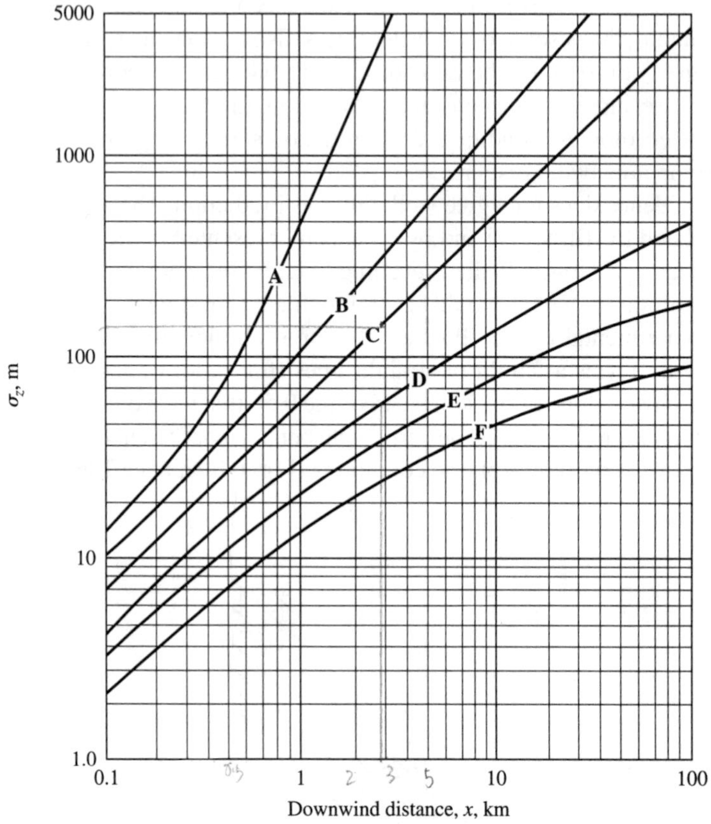

FIGURE 6.8
Vertical dispersion coefficient σ_z as a function of downwind distance from the source for various stability categories. See Problem 6.16. (From Turner [7].)

The experimental data disagree with our neat theory because the equation we assumed for atmospheric mixing, Eq. (6.14), is much too simple to account for all the complicated things that actually go on in the atmosphere, even on days with simple wind patterns, which are the only ones on which experimental tests of Eq. (6.27) are ever attempted. Thus, we can say that the preceding derivation shows us a way to obtain a logical material balance for dispersion of a pollutant from a point source in the atmosphere, subject to some strong simplifying assumptions; but that we must regard the values of σ_y and σ_z as experimental quantities that we cannot yet compute from theory. However, if we accept Figs. 6.7 and 6.8 as adequate representations of the experimental results, we can use them, along with Eq. (6.27), to make predictions of concentrations downwind from point sources. This is currently the most widely used method for routine calculations of air pollutant dispersion from point sources. The experimental data on which Figs. 6.7 and 6.8 are based are limited and not necessarily directly applicable to cities. Most of the data were taken for steady flow of winds over grasslands (the Salisbury Plain in England

TABLE 6.1
Key to stability categories

Surface wind speed (at 10 m), m/s	Day: Incoming solar radiation: Strong	Moderate	Slight	Night: Thinly overcast or $\geq \frac{4}{8}$ cloud	Clear or $\leq \frac{3}{8}$ cloud
0–2	A	A–B	B	—	—
2–3	A–B	B	C	E	F
3–5	B	B–C	C	D	E
5–6	C	C–D	D	D	D
≥ 6	C	D	D	D	D

Source: Ref. 7.
Note: The neutral class D should be assumed for overcast conditions during day or night.

and the grasslands of Nebraska). We use them for cities because we have nothing better. These plots are based on measurements for $x \leq 1$ km. The values beyond that distance are extrapolations [8]. However, comparison with experiments shows that advanced versions of this model predict observed concentrations fairly well [9].

So far we have said nothing about the lines labeled A through F on Figs. 6.7 and 6.8. These correspond to differing levels of atmospheric stability. On a clear, hot summer morning with low wind speed, the sun heats the ground, which in turn heats the air near it, causing that air to rise and thus to mix pollutants well. The atmosphere is unstable, and the values of σ_y and σ_z will be large. On a cloudless winter night, the ground cools by radiation to outer space and thus cools the air near it. The air forms an inversion layer, making the atmosphere stable and inhibiting the dispersion of pollutants, so the values of σ_y and σ_z will be small.

Atmospheric stability is one of the principal topics in meteorology (Chapter 5). In this chapter, we will use the atmospheric stability-category classification given by Turner [7], which considers only the incoming solar radiation and the wind speed (see Table 6.1). There are other systems for estimating σs; this one is simple and widely used.

Example 6.5. Estimate the values of σ_y and σ_z at a point 0.5 km downwind from a pollutant source on a bright summer day with a wind speed greater than 6 m/s. From Table 6.1 we conclude that for a bright summer day the incoming solar radiation is "strong," so we use stability category C. Then, using Figs. 6.7 and 6.8, we read (for $x = 0.5$ km) $\sigma_y = 56$ m and $\sigma_z = 32$ m. (See Problem 6.16.) ■

6.3.3 Some Modifications of the Basic Gaussian Plume Equation

6.3.3.1 The effect of the ground. At present Eq. (6.27) is our best simple prediction method for the concentration in plumes considerable distances above the ground. However, we are generally most interested in concentrations at ground level because that is where most people and property are exposed. The blind application of Eq. (6.27) at or near ground level gives misleadingly low results. It indicates that pollutants continue to disperse at any value of z, even at z less than zero. (Using it alone, we could continue Example 6.4 and compute the concentration underground; the result would bear no relation to what we would observe in nature.)

The ground damps out vertical dispersion. The upward and downward random atmospheric eddies that spread the plume in the vertical direction cannot penetrate the ground. Thus, vertical spreading terminates at ground level. To account for this in calculations it is commonly assumed that any pollutants that would have carried below $z = 0$ if the ground were not there are "reflected" upward as if the ground were a mirror. Thus, the concentration at any point is due to the plume itself plus what is reflected upward from the ground. This method is equivalent to assuming that a mirror-image plume below the ground transmits as much up through the ground surface as the aboveground plume would transmit down through the ground surface if the ground were not there.

The concentrations due to the mirror-image plume are exactly the same as those shown by Eq. (6.27), except that $(z - H)^2$ is replaced by $(z + H)^2$. This substitution shows that at the ground, or $z = 0$, both the main plume and the mirror-image plume have identical values. High in the air, for example at $z = H$, the main plume has a high concentration ($\exp 0 = 1$), whereas that for the mirror-image plume [e.g., $\exp -\frac{1}{2}(2H/\sigma_z)^2$] is a small number. The combined contribution of both plumes is obtained by writing Eq. (6.27) and the analogous equation for the mirror-image plume, adding the values for the two plumes, and factoring out the common terms to obtain

$$c = \frac{Q}{2\pi u \sigma_y \sigma_z} \exp -0.5\left(\frac{y}{\sigma_y}\right)^2 \left[\exp -0.5\left(\frac{z-H}{\sigma_z}\right)^2 + \exp -0.5\left(\frac{z+H}{\sigma_z}\right)^2\right] \tag{6.29}$$

Example 6.6. In Example 6.4 we computed the concentration at a point 20 meters below the plume centerline, ignoring the effect of the ground. Repeat the calculation for the cases where $H = 20$ m and where $H = 30$ m. For $H = 20$ m we are computing the concentration at the ground level itself. From Eq. (6.29) we see that at $z = 0$ the two terms in the brackets at the right are identical, and each is equal to the value that that term had in Example 6.4. Thus our answer is exactly twice that in the second part of Example 6.4; viz., $145 \times 2 = 290\ \mu\text{g/m}^3$. This is a general result; for $z = 0$, Eq. (6.29) always gives exactly twice the value given by Eq. (6.27).

For $H = 30$ m we have

$$c = 1770\ \frac{\mu\text{g}}{\text{m}^3}\left(\exp -0.5\left(\frac{60\text{ m}}{30\text{ m}}\right)^2\right) \times \left[\exp -0.5\left(\frac{10\text{ m} - 30\text{ m}}{20\text{ m}}\right)^2 + \exp -0.5\left(\frac{10\text{ m} + 30\text{ m}}{20\text{ m}}\right)^2\right]$$

$$= 1770\ \frac{\mu\text{g}}{\text{m}^3}\ (\exp -2)\left[\exp\left(-\frac{1}{2}\right) + \exp(-2)\right]$$

$$= 1770\ \frac{\mu\text{g}}{\text{m}^3}(0.135)(0.605 + 0.135) = 177\ \frac{\mu\text{g}}{\text{m}^3}$$

We see that at a point one-third as far off the ground as the centerline of the plume, Eq. (6.29) gives a value 22 percent greater than Eq. (6.27) (which does not take ground reflection into account). ■

Equation (6.29) is correct for ground level or any elevation above it. For large values of z, the contribution of the $(z + H)^2$ term becomes negligible and the result is practically identical with that from Eq. (6.27). Most often we are interested in ground-level concentrations. If we substitute $z = 0$ into Eq. (6.29) and simplify, we find

$$c = \frac{Q}{\pi u \sigma_y \sigma_z} \exp -0.5\left(\frac{y}{\sigma_y}\right)^2 \exp -0.5\left(\frac{H}{\sigma_z}\right)^2 \qquad \text{for } z = 0 \qquad (6.30)$$

We may consider this the "ground level modification of Eq. (6.27), taking reflection at the ground surface into account." Although Eq. (6.27) is the basic Gaussian plume equation, Eq. (6.30) is the single most widely used estimating equation because it applies directly to the problem of greatest practical interest.

As the previous examples show, hand solution of Eqs. (6.27) and (6.30) is straightforward and tedious. For this reason, numerous ways have been found to simplify their use. Here we will consider only one of these, which is probably the most useful. For conditions of $y = 0$ and $z = 0$, which correspond to the line on the ground directly under the centerline of the plume, the exponential term in y drops out of Eq. (6.30). Multiplying both sides by u/Q gives

$$\frac{cu}{Q} = \frac{1}{\pi \sigma_y \sigma_z} \exp -0.5\left(\frac{H}{\sigma_z}\right)^2 \qquad \text{for } z = 0,\ y = 0 \qquad (6.31)$$

The function on the right depends only on H and the two dispersion coefficients.

Example 6.7. Compute the value of the term on the right in Eq. (6.31) for C stability, a distance downwind of $x = 0.5$ km, and $H = 50$ m. From Example 6.5 we know that for C stability and $x = 0.5$ km, $\sigma_y = 56$ m and $\sigma_z = 32$ m. Thus,

$$\frac{cu}{Q} = \frac{1}{(\pi)(56\text{ m})(32\text{ m})} \exp -0.5\left(\frac{50\text{ m}}{32\text{ m}}\right)^2 = 5.24 \times 10^{-5}\ \text{m}^{-2}$$ ■

If we were to repeat this calculation for a wide range of distances and effective stack heights, we could make up a plot of cu/Q vs. distance with stack height as a parameter. Turner [7] has done this for the six stability categories shown in Table 6.1. Figure 6.9 on page 140 shows a plot of this type, for C stability. The reader should check to see that the value of $5.24 \times 10^{-5}/\text{m}^2$ just given is indeed the value plotted for $x = 0.5$ km and $H = 50$ m on Fig. 6.9.

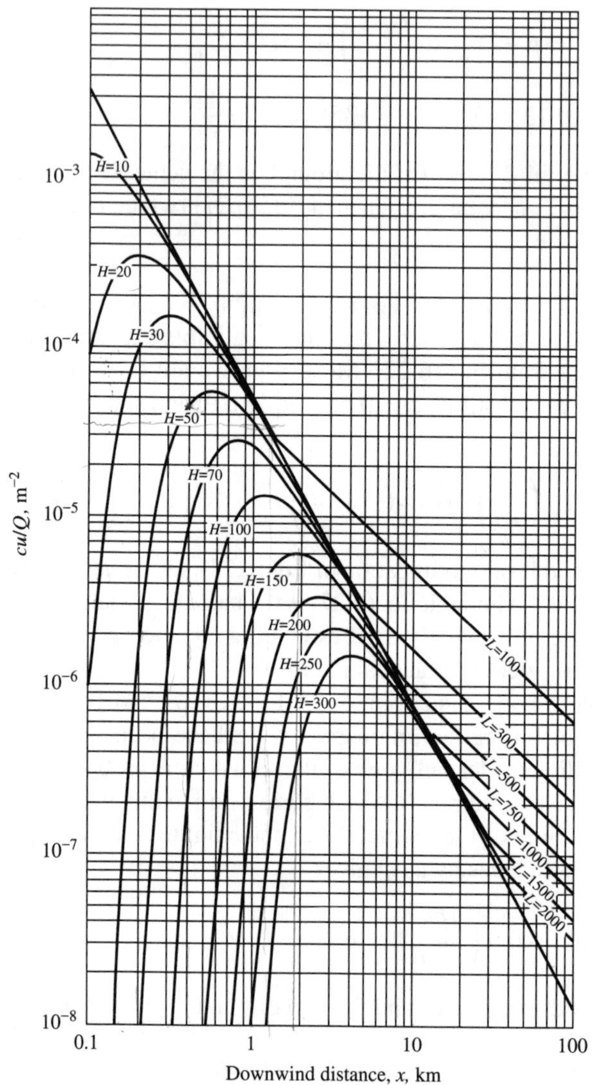

FIGURE 6.9
Ground-level cu/Q, directly under the plume centerline, as a function of downwind distance from the source and effective stack height, H, in meters, *for C stability only.* (From Turner [7].) Here L is the atmospheric mixing height, also in meters.

Example 6.8. A plant emits 100 g/s of SO_2 from a stack that has an effective stack height $H = 50$ m. The wind is blowing 3 m/s, and the stability category is C. Estimate the ground-level concentrations directly below the centerline of the plume at distances of 0.2, 0.4, 0.5, 1, 5, and 10 km downwind.

From Fig. 6.9 we may read directly that at 0.2 km, cu/Q is $1.7 \times 10^{-6}/\text{m}^2$. Thus,

$$c = \frac{cu}{Q}\frac{Q}{u} = \frac{1.7 \times 10^{-6}}{\text{m}^2}\frac{100 \text{ g/s}}{3 \text{ m/s}} = \frac{5.7 \times 10^{-5} \text{ g}}{\text{m}^3} = \frac{57 \ \mu\text{g}}{\text{m}^3}$$

We can then look up the other values of cu/Q and tabulate the results:

Distance, km	cu/Q, m^{-2}	c, $\mu g/m^3$
0.2	1.7×10^{-6}	57
0.4	4.4×10^{-5}	1467
0.5	5.3×10^{-5}	1767
1	3.6×10^{-5}	1200
5	2.7×10^{-6}	83
10	7.8×10^{-7}	24

■

If we repeat Example 6.8 for a different emission rate or a different wind speed, we can see that because of the form of the (cu/Q) factor, we can make the changes by simple multiplications. If we want to know the maximum ground level concentration and its distance downwind of the source, we can find it by inspection from Fig. 6.9 (see Problem 6.13).

6.3.3.2 Mixing height limits, one-dimensional spreading. As the plume flows downwind, it will eventually grow until it is completely mixed below the mixing height H, shown in Fig. 6.1. After that it will no longer spread vertically, but only horizontally, so a two-dimensional spreading plume has converted to a one-dimensional spreading plume. On Fig. 6.9 the mixing height is called L, and lines are drawn for long transport distances, indicating that the observed concentrations are higher than one would compute by continuing the two-dimensional spreading calculation to those distances. Observe that H and L appear with two sets of meanings in these calculations. In box models, H is the mixing height and L is the downwind length of the city. In Gaussian plume models, H is effective stack height and L is the mixing height. Alas, this usage is common.

Returning to Eq. (6.19), we see that the amount being dispersed horizontally is

$$X = \frac{Q}{uL} \tag{6.32}$$

which accounts for the fact that the X for the two-dimensional Gaussian plume is now uniformly spread over a height L. Substituting this value in Eq. (6.19) and making the substitutions of Eqs. (6.24)–(6.26), we find

$$c = \frac{Q}{\sqrt{2\pi}\,uL\sigma_y} \exp\left(-\frac{y^2}{2\sigma_y^2}\right) \tag{6.33}$$

This equation (with $y = 0$) is used to make up the sloping lines at the right edge of Fig. 6.9.

Turner [7] also gives several other representations of Eqs. (6.27), (6.29), (6.30), etc., in convenient graphical and tabular form. All serious air pollution workers have copies of "Turner's workbook," which is available at a low cost from the U.S. EPA.

Computer programs to do this type of calculation are widely available (see Problem 6.16). Note the actual title of "Turner's workbook": "Workbook of Atmospheric Dispersion Estimates." The key word is **Estimates.** In all air pollution modeling we are making best estimates, not finding scientific truth.

Turner's workbook also suggests simple methods for estimating the effects of inversions aloft, the magnitude of short-term fluctuations about the mean values computed by the basic Gaussian plume equation, inversion breakup fumigations, and other topics. All of these additional topics are explored in greater detail in more recent publications, but Turner's simple graphical and hand calculation approaches are still useful.

This treatment also explains the common observation that a plume becomes less and less opaque as it flows downwind, and finally becomes invisible. The reason is that a typical plume is spreading in two dimensions (y and z), but an observer looking across the plume, whether horizontally, vertically, or at an angle, is seeing it along one-dimensional lines of sight. The opacity (visual thickness) of the plume is given by an equation of the form

$$\begin{pmatrix}\text{Opacity or}\\ \text{visual}\\ \text{thickness}\end{pmatrix} = \begin{pmatrix}\text{some constant that takes}\\ \text{particle size and optical}\\ \text{properties into account}\end{pmatrix} \int_{\substack{\text{across the}\\ \text{plume}}} c\,dy \tag{6.34}$$

If the plume were spreading in only one dimension (e.g., if it flowed between parallel plates) then the integral on the right would be a constant, and the opacity (visual thickness) of the plume would be independent of downwind distance. But with the normal two-dimensional spreading of the plume, the concentration falls faster than the width of the plume increases, so this integral decreases in value with downwind distance, and the plume becomes less and less opaque. For a plume that has spread vertically to fill the whole space up to the mixing layer (to whichever value of L is dictated by the local, current meteorology) on the far right of Fig. 6.9, we would expect the plume to remain at a constant opacity as it continues to flow downwind.

The Gaussian plume model applies only to point sources. Various methods have been developed for applying it to area sources [7, 8].

6.4 PLUME RISE

Figure 6.3 shows the plume rising a distance Δh, called the plume rise, above the top of the stack before leveling out. Most of us have observed that the visible plumes from power plants, factories, and smokestacks tend to rise and then become horizontal, as sketched in Fig. 6.3.

Plumes rise buoyantly because they are hotter than the surrounding air and also because they exit the stack with a vertical velocity that carries them upward. They stop rising because, as they mix with the surrounding air, they lose velocity and cool by mixing. Finally, they level off when they come to the same temperature as the atmosphere.

We employ plume rise calculations to estimate the value of Δh and hence of H to use in Gaussian plume and other more complex pollutant-concentration calculations. Holland's formula for plume rise is

$$\Delta h = \frac{V_s D}{u}\left(1.5 + 2.68 \times 10^{-3} PD \frac{(T_s - T_a)}{T_s}\right) \tag{6.35}$$

where Δh = plume rise in m
V_s = stack exit velocity in m/s
D = stack diameter in m
u = wind speed in m/s
P = pressure in millibars
T_s = stack gas temperature in K
T_a = atmospheric temperature in K

Example 6.9. Estimate the plume rise for a 3-m diameter stack whose exit gas has a velocity of 20 m/s when the wind velocity is 2 m/s, the pressure is 1 atm, and the stack and surrounding temperatures are 100°C and 15°C (373 and 288 K), respectively.

$$\Delta h = \frac{20 \times 3}{2}\left(1.5 + \frac{2.68 \times 10^{-3} \times 1013 \times 3 \times (373 - 288)}{373}\right) = 101 \text{ m} \quad \blacksquare$$

Equation (6.35) is a dimensional equation, which is only correct for the dimensions shown; the other formulas in this chapter are all correct for any consistent set of dimensions. This formula is frequently corrected for atmospheric stability by multiplying the result by 1.1 or 1.2 for A and B stability or 0.8 or 0.9 for D, E, or F stability. Although this formula has some theoretical basis, it is not universally applicable. All plume rise formulas work well for some cases, but none seems to handle all cases. For plume rise calculations involved in important decisions, e.g., to permit or not permit the location of a new facility at a specific location, consult the monograph by Briggs [10].

6.5 LONG-TERM AVERAGE USES OF GAUSSIAN PLUME MODELS

The Gaussian plume formulas in this chapter allow one to estimate the concentration at a receptor point due to a single emission source for a specific meteorology. In this form they are frequently used to estimate maximum concentrations to be expected from single isolated sources. For instance, can a large single point source (e.g., a power plant or smelter) legally be placed in a given location? By how much must emissions from an existing source be reduced to meet some applicable standard?

Gaussian plume models are also applied to estimate multisource urban concentrations. The procedure is the same as in Example 6.2, using Gaussian plume calculations to determine the receptor concentration at various locations for each of the point and area sources in the city for each meteorological condition. In one typical model of this type (the Implementation Planning Program, or IPP), the summation in Eq. (6.8) is written as

$$\begin{pmatrix}\text{Annual average}\\ \text{concentration}\\ \text{at a point}\end{pmatrix} = \sum_{\substack{\text{all}\\ \text{sources}}} \sum_{\substack{\text{16 wind}\\ \text{directions}}} \sum_{\substack{\text{5 wind}\\ \text{speeds}}} \sum_{\substack{\text{6 stability}\\ \text{categories}}} (\text{frequency} \cdot c) \tag{6.36}$$

where frequency is the frequency of occurrence of a specific wind speed, wind direction, and stability category combination, and c is the concentration expected at a specific location from one specific source for that meteorology calculated by the Gaussian plume method [11].

In these models the area sources (autos, homes, small businesses) are represented by equivalent point sources, each of which represents the emissions from some small area of the region being modeled. Such programs require large amounts of input information and consume large amounts of computer time. But in principle they are no more complex than Example 6.2 of this chapter, using the concentration calculation method in Example 6.8.

6.6 POLLUTANT CREATION AND DECAY IN THE ATMOSPHERE

All of the preceding parts of this chapter have dealt with pollutants that are assumed to remain in the atmosphere forever. No pollutant really behaves that way; all pollutants have natural removal mechanisms. However, for pollutants like suspended mineral particles or carbon monoxide, it is a satisfactory approximation, because their removal rates are slow enough to ignore in most urban areas.

In contrast, sulfur dioxide, hydrocarbons, nitrogen oxides, and oxidants all undergo reactions in the atmosphere, and their reaction times may be comparable to travel times across a city. For these pollutants the simple box and Gaussian plume methods, as presented so far, predict values much higher than the observed values. They are generally modified as shown below.

Many of the early workers in Gaussian plume calculations were interested in radioactive contaminants. These contaminants convert spontaneously to other, frequently less radioactive materials. The rate of disappearance is given by the rate law for first-order decay,

$$\frac{d(\text{amount})}{dt} = -k(\text{amount}) \tag{6.37}$$

which integrates readily to

$$\frac{(\text{Amount})}{(\text{Original amount})} = \exp(-kt) \tag{6.38}$$

where k is the rate constant for decay, which has the dimension (1/time). For nuclear decays, k for any reaction is an unvarying constant, independent of temperature, other chemicals present, etc. If one evaluates the time for one-half of the material present to disappear, one finds that

$$t_{1/2} = \frac{\ln 2}{k} = \frac{0.693}{k} \tag{6.39}$$

This time for one-half to disappear is called the *half-life* and is one convenient way to discuss such decays.

If we consider the Gaussian puff formula, Eq. (6.22), and assume that the material emitted is a radioactive material with a half-life of $t_{1/2}$, then we can say that the concentration at any point shown by Eq. (6.22) should be multiplied by a decay factor,

$$\text{Decay factor} = \exp\left(-0.693\frac{t}{t_{1/2}}\right) = \exp\left(\frac{-0.693x}{ut_{1/2}}\right) \tag{6.40}$$

Example 6.10. A nuclear reactor accident releases a cloud that contains iodine-133, which has a half-life of 22 h. We have calculated the concentration-time behavior at every point using Eq. (6.22), which assumes that the material does not decay in the atmosphere. Now we wish to include the effect of decay. What is the expected decay factor at a point 10 km downwind if the wind velocity is 1 m/s?

$$\text{Decay factor} = \exp\frac{(-0.693)(10\text{ km})(1000\text{ m/km})}{(1\text{ m/s})(22\text{ h})(3600\text{ s/h})} = 0.916$$

Thus, the values previously calculated at this distance downwind should all be multiplied by 0.916. ■

The first-order decay law is a very accurate representation of nuclear decays, so for the nuclear release in Example 6.10 this correction to the much less accurate Gaussian plume calculation should be quite reliable. A much less reliable application of the same formula is frequently made for air pollutants like SO_2. For example, many workers have applied Gaussian plume formulations for SO_2, multiplying the resulting computed concentration by a decay factor from Eq. (6.40), using a half-life of 1 to 10 hours—typically 3 hours. The processes for removal of SO_2 from the atmosphere are much more complex and variable than nuclear decays, so this approach can only be considered an approximation of what happens in nature. (The SO_2 decay rate depends on temperature, light intensity, humidity, the presence of other particles, and the ozone concentration; nuclear decays depend on none of these.) While nuclear decays generally produce harmless materials, SO_2 decay produces fine sulfate particles, which are part of a different air pollution problem from the SO_2 problem.

For photochemical oxidants, which are formed in the atmosphere from hydrocarbons and nitrogen oxides, no chemical formulation as simple as the decay factor in Eq. (6.40) seems useful. Most photochemical oxidant models use predictive schemes with 10 to 30 simultaneous reactions in the atmosphere. The typical course of such reactions is shown in Fig. 1.2; the reaction times are indeed comparable to transport times in urban areas.

Figure 6.9 shows that, *according to the standard Gaussian plume assumptions*, raising the point of emissions (increasing H) lowers the ground-level concentration near the source and does not increase the ground-level concentration at any downwind point. However, much of the removal of pollutants occurs at the ground surface, where the pollutants interact with or are deposited on the ground, vegation, etc. The amount deposited is more or less proportional to the local ground-level concentration. Raising the point of emission lowers the concentrations near the emission point, reduces this removal, and leads to increased concentrations far downwind. This is certainly correct in theory, and probably correct in practice, but hard to measure.

6.7 MULTIPLE CELL MODELS

No one has yet suggested any reasonable way of incorporating the kind of complex simultaneous reaction rate expressions that describe the reaction progress shown in Fig. 1.2 into a Gaussian plume model. Currently, the most widely used approach to such problems is the multiple cell model, e.g., the Urban Airshed Model, or UAM [12, 13], for which the airspace over a city or region is divided into multiple cells, as shown in Fig. 6.10. Each cell is treated separately from the others. (This type of model is mostly used for ozone, but could be used for other secondary pollutants produced in the atmosphere.)

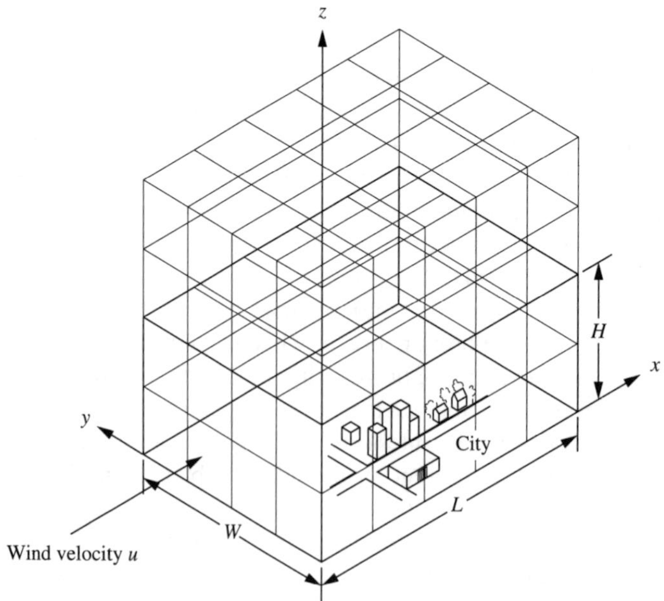

FIGURE 6.10
Division of the airspace over a city into cells for the Urban Airshed Model (UAM).

In the UAM model the division in the *x* and *y* directions has uniform grid sizes, normally 2 to 5 km each way for the whole city. In the vertical direction there are normally four or six layers, half below the mixing height and half above. The boundaries of these move up and down with the variation of the mixing height over the day, and from location to location within the city. In Fig. 6.10 there are $5 \times 4 \times 4 = 80$ cells. For a large city with 2 km grid spacing there can be thousands of cells.

All of the terms in Eq. (6.1) are retained for each cell. A model simulation for a city and some time period begins with an assumed initial distribution of pollutants and pollutant precursors in all of the cells. Then for a time step of typically 3 to 6 minutes, the program calculates the change in concentration of the pollutant of interest and its precursors in each of the cells by numerically integrating Eq. (6.1). This computation requires data or an estimating procedure for the wind velocity and direction at the center of each cell (to calculate the flows in and out across the boundaries) plus emissions estimates for each of the ground-level cells, plus a subprogram to compute the chemical transformations during the time step in any cell, plus a subprogram for deposition of the pollutant from the ground-level cells. Rather than try to solve for all of the terms in Eq. (6.1) simultaneously, UAM first computes the changes in concentration due to flows with the winds across the cell boundaries, using the concentrations from the end of the previous time step, and then computes the changes due to chemical reactions in the cell. The results of these two steps are added to estimate the concentration in each cell at the end of the time step.

To simulate a day or a few days in an urban area, this model requires a complete history of the wind pattern, solar inputs, and emissions. If these data are not available, the program has ways of estimating them. A common procedure is to choose a day on which the measured pollutant (usually ozone) concentration was the maximum for the past year or past few years. The model is run using the historical record of the wind speeds and directions, solar inputs, and estimated emissions for that day. The model's adjustable parameters are modified until the calculated concentrations match well with the observed ambient concentrations for that day. Then the model is rerun with different emission rates or distributions, corresponding to proposed or anticipated future situations, and the meteorology for that day. In this way the model performs a prediction of the *worst day* situation under the proposed future emission pattern.

Many air pollution research meteorologists believe that multiple cell models are the only models that show promise for being able to give useful guidance on the photochemical oxidant problems in places like Los Angeles. There, for example, the highest ozone concentrations regularly occur in places like Riverside, on summer afternoons, six or seven hours after the oxidant precursors were emitted 30 miles to the west. However, this type of model has not yet found much use for areas with less difficult problems, or for pollutants that do not change rapidly in the atmosphere (e.g., SO_2), because of its enormous requirements for data that are rarely available and its enormous appetite for computer time and modeler effort. This kind of model and the other models used for ozone were reviewed by Seinfeld [14].

6.8 RECEPTOR-ORIENTED AND SOURCE-ORIENTED AIR POLLUTION MODELS

The types of models in the previous parts of this chapter are called *source-oriented models.* In them one uses best estimates of the emission rates from various sources and of the meteorology to estimate the concentration of various pollutants at various downwind points. If one had perfect information about the emission rates and the meteorology as well as perfect models, these models should be totally accurate. But since our data and models are imperfect, our predictions are not nearly as accurate as we would like.

An alternative approach to air pollution modeling is called *receptor-oriented modeling.* In this approach one examines the pollutants collected at one or more monitoring sites, and from a detailed analysis of what is collected attempts to determine which sources contributed to the concentration at that receptor.

If the pollutant of interest is chemically uniform (e.g., CO, O_3, SO_2), then there is no way to distinguish between sources. But if the pollutant is particulate matter (either TSP, PM_{10}, or $PM_{2.5}$) that consists of a wide variety of chemical species, then by analyzing the chemical composition one can make some inferences about the sources. The result of such an analysis is called a *source apportionment* or *chemical mass balance* [15–18]. It normally says that of the particulates found at monitor #1, x percent are due to source #1, y percent due to source #2, z percent due to source #3, etc.

For example, clays are mostly complex compounds of aluminum and silicon. If there are no other nearby sources of aluminum and silicon compounds, and analysis of the particulate filters shows that most of the particulates are compounds of aluminum and silicon with the same elemental ratios as those in the clay component of the local dirt roads, then it is a fair assumption that road dust is the major contributor to the particulates at that location. Fortunately it is possible to determine the ratios of metallic elements in the emissions from many sources, e.g., steel plants, electric generating plants, pulp mills, etc. It is relatively easy and relatively cheap to analyze the particulate matter on a monitoring filter for metallic elements (normally by atomic absorption spectroscopy), so with this method one can make source apportionments for particles at various sampler locations at prices that are significant, but not exorbitant.

One can do the same with hydrocarbons, which are a family of compounds, rather than one uniform chemical species (Chapter 10). The distribution of species within this family depends on the emission source, e.g., oil refineries, evaporative losses in fuel transfers, tailpipe emissions of autos. Thus this approach can be used to estimate which of those (or other) sources contributed to the observed hydrocarbons at a given sampling site [19, 20]. One might also do source apportionment with isotope ratios of sulfur in SO_2, which are different for coals and sulfide ores, but that is not being done regularly. To date this method has been applied mostly to particles, occasionally to hydrocarbons.

If we had perfect knowledge of sources, meteorology, and the like, the source apportionment calculated from a receptor-oriented model should agree with the source apportionment we would calculate from a source-oriented model for the same time and place. Often they disagree significantly. When this happens, we tend to believe the receptor-oriented model more than the source-oriented model, because we have more confidence in the chemical distribution data than we have in the emission and meteorological data.

Source-oriented models can be used to estimate the effects of proposed new sources, for example, in the permitting process for new sources. Receptor-oriented models cannot be used this way. They are mostly used to test the estimates made by source-oriented models, and simultaneously to test the accuracy of the emissions estimates that are used in those models.

6.9 OTHER TOPICS

This short chapter has discussed the fundamental ideas of air pollution concentration models that will be considered by a local air pollution control official or air pollution control engineer of a company. Several other pertinent topics can be found in books on air pollution meteorology and are also discussed in the following subsections.

6.9.1 Building Wakes

A plume may get sucked into the low-pressure wake behind a building, leading to a high local concentration. Figure 6.11 on page 150 shows a dramatic example of this effect. This wake is caused by the wind flow over the building and is analogous to the low-pressure wake behind a rapidly moving truck or auto. (Stock car racing fans know that the winner of the race is the one who enters the final lap in second place. That driver uses the low-pressure wake behind the first car to "slingshot" past the first car just before the finish.) An ample literature exists on trapping of emissions in building wakes [8, 21]. The simple rule of thumb for avoiding this problem is to make the stack height at least 2.5 times the height of the tallest nearby building.

6.9.2 Aerodynamic Downwash

If one inserts an open framework like a TV tower into a plume, the tower will not disturb the air flow much, so we would expect the Gaussian plume equations to work well downstream of the tower. If one inserts a mountain into a plume, it will disturb the air flow a great deal; and we would not expect the simple Gaussian plume equations to work well for flow either toward it or directly downstream of it. Figure 6.12 on page 151 shows how a mountain can make a plume behave differently from what one would predict from the Gaussian plume equations.

The *aerodynamic downwash* shown in Fig. 6.12 is a major problem for any facility located near a mountain. At the moment shown in this figure, the ground-level concentrations from that stack were certainly many times what would be calculated with the simple (flat world) Gaussian plume equations.

After many years of effort we still do not have a good method for predicting the concentrations to be expected on the sides of such a mountain. By U.S. air pollution

FIGURE 6.11
The plume from a smoke bomb being captured in the low-pressure wake behind a building. (Courtesy of Professor J. E. Martin, University of Michigan, Ann Arbor, MI.)

law, which regulates the pollutant concentration anywhere at the ground-air interface, concentrations on the upwind side of such mountains often decide whether a plant may be located in a nearby valley or not.

6.9.3 Transport Distances

The models shown in this text and the experiments to verify them are mostly for distances less than 20 km. In the acid rain problem the transport distances are hundreds of kilometers. Work on developing the corresponding models for the acid rain problem has shown that these models do not predict accurately, nor do any others. Long-distance transport models are currently a principal research topic in air pollution modeling.

6.9.4 Initial Dispersion

Very close to the pollutant source, the standard Gaussian plume approach produces calculated concentrations significantly higher than those observed experimentally. The reason is that with Gaussian plume equations all the pollutants are emitted from a point, at $x = y = (z - H) = 0$. Real sources always are larger than a point, e.g., the

FIGURE 6.12
Wind flowing over the 455-m hill toward the ≈200-m stacks of a power plant has a strong enough downward component to carry the plume down, leading to high ground-level concentrations [22]. (Courtesy of the Air and Waste Management Association and Professor M. M. Millán.)

area perpendicular to flow of a chimney. Often in safety analyses we are concerned with maximum concentrations very close to buildings and with emissions through windows, doors, or explosion relief panels. In these studies we use the Gaussian plume method (often the three-dimensional Gaussian puff equation) but take the initial value of the σs equal to the dimension of the opening. If we add that value to the value computed from Fig. 6.7 or 6.8, we see that very near the source the values of the σs are close to those of the opening, but at long distances they become practically the same as those read from Fig. 6.7 or 6.8.

6.9.5 EPA-Recommended Models

The U.S. EPA has developed a variety of air pollution models. Most of these can be downloaded from the EPA web site, along with manuals for their use. Their guidance on which model to use for a given situation (*40CFR51*, Appendix W) is 99 pages long. Some state regulations require the use of these EPA-recommended models (or their equivalent) in permit-application modeling. The current "top of the line" EPA single-source model, ISC3, includes building wakes, a variety of plume models, ground deposition, and many other refinements on the basic models shown here. Some meteorological consultants sell their versions of these models, in which they have modified the input procedures to make them more user-friendly without changing their structure or calculated results.

6.10 SUMMARY

1. Pollutant concentration models are based on known emission rates and meteorology. These models play a crucial role in the Air Quality Management type of air pollution control strategy currently used in the United States and much of the rest of the world.
2. Fixed-box models are the simplest pollution concentration models for cities, but they have severe drawbacks. They are easily understood and used, but their numerical predictions, while qualitatively correct, are not of much quantitative use.
3. Gaussian plume models are widely used for point sources. They rest on severe simplifying assumptions but have been reasonably successful in predicting experimental results for single, elevated point sources.
4. Multiple cell models demand vast amounts of input data and computer time, but they are considered by many experts to be the only models likely to be successful for photochemical pollutants.
5. All of the models discussed here are great simplifications of the real behavior of nature. The Gaussian plume models work reasonably well in flat terrain when the instantaneous meteorology is simple. They do not work well in mountainous country, nor for long distances, nor at times when the meteorology is complex.
6. Receptor-oriented models are not predictive models like the source-oriented models described in this chapter. Rather, they are experimental source apportionment methods. They are widely used, and are often called models.
7. Building wakes, local mountains, and other sources of air flow perturbations complicate air pollution modeling.

PROBLEMS

See Common Units and Values for Problems and Examples, inside the back cover.

6.1. In the fixed-box model (see Fig. 6.1), we make many assumptions. The worst of these is #3, that there is complete mixing in the upwind direction. Here, let us remove that assumption but keep all the others. In this case the concentration will not be uniform across the city but will vary with downwind distance x. Keeping all the other assumptions in that part of the chapter unchanged, derive the equivalent of Eq. (6.7) for this case. Here c will not be a single value for the whole city but will be a function of x. We take $x = 0$ at the upwind edge of the city. Your solution should show the value of c not only at the downwind edge of the city ($x = L$) but also at any point in the city (any $L > x > 0$).

6.2. Estimate the concentration of carbon monoxide at the downwind edge of a city. The city may be considered to consist of three parallel strips, located perpendicular to the wind. For all of the strips the wind velocity u equals 3 m/s. The properties of each of the strips are described in the following table:

Name of strip	Length, km	Emission rate, q, g/s · km^2	Mixing height, H, m
Upwind suburbs	5	100	400
Downtown	2	500	500
Downwind suburbs	5	100	400

Assume that the fixed-box model applies to each of the strips. The background concentration b in the air entering the upwind suburbs is 1 mg/m^3.

6.3. For Eq. (6.29), show the simplifications that result in each of these cases:

(*a*) We are only interested in concentrations on the centerline of the plume, i.e., at $y = 0$, $z = H$. Here assume that $(z + H)$ is very large.

(*b*) We are only interested in concentrations at the same elevation as the plume centerline, i.e., at $z = H$. Here assume that $(z + H)$ is very large.

(*c*) We are only interested in concentrations directly below the plume centerline at ground level, i.e., at $z = 0$, $y = 0$.

(*d*) We are only interested in ground-level sources, i.e., $H = 0$.

6.4. Equation (6.22) is the Gaussian puff equation in the form with the Ks. Show the form that Eq. (6.22) takes if one substitutes in Eqs. (6.24), (6.25), and (6.26), plus the corresponding equation for K_x. This is the form most often actually used.

6.5. Highways are normally modeled as a *line source* (as opposed to point or area sources). The highway is aligned on the y axis, and the wind blows in the x direction. For ground-level highways, $H = 0$. For elevated highways the effect of source height must be included in the model.

(*a*) For a line source spreading is one-dimensional in the vertical direction. Show why. A simple sketch and a few words will do.

(*b*) Show the equivalent of Eq. (6.29) for a line source. Assume that the emission rate is given as Q/length, e.g., g/s · mile.

(*c*) Suggest what modification of your answer in (*b*) would be needed if the wind is not blowing at a 90° angle to the highway.

6.6. A large, poorly controlled copper smelter has a stack 150 m high and a plume rise of 75 m. It is currently emitting 1000 g/s of SO_2. Estimate the ground-level concentration of SO_2 from this source at a distance 5 km directly downwind when the wind speed is 3 m/s and the stability class is C.

6.7. The management of the smelter in Problem 6.6 has been informed that the concentration calculated in that problem at that location and for those conditions is twice the allowable. They propose to remedy this situation by installing a higher stack. How high must this stack be so that the estimated concentration will be exactly one-half that in Problem 6.6? (The plume rise is the same as in Problem 6.6.)

6.8. The Dogpatch Skunk Works emits 10 g/h of trimethylamine from a stack 10 m high with zero plume rise. The lowest concentration of trimethylamine that the average human being can detect is about 5×10^{-7} g/m^3. If the wind is blowing at 2 m/s on a totally overcast night, what is the maximum distance in the exact downwind direction at which one can smell the Dogpatch Skunk Works? (This concentration corresponds to 0.02 part per billion; it is as low a concentration as humans can smell for any substance for which the human smell threshold has been measured. Animals like bloodhounds can obviously smell much smaller concentrations than this.)

6.9. A ground-level source ($H = 0$) is emitting pollutants at an unknown rate. At 1 km directly downwind of the source the measured ground-level concentration of the pollutant is 10 $\mu g/m^3$. The stability category is A. Estimate the emission rate of this source.

6.10. For the smelter in Problem 6.6 what is the maximum calculated ground-level concentration, and at what distance downwind does it occur?

6.11. In Problem 6.6 you computed the ground-level concentration under the plume centerline at a distance 5 km directly downwind of the source for C stability.

(*a*) Now, for the same situation, calculate the concentration at the plume centerline ($x =$ 5 km, $y = 0$, $z = H$).

(*b*) Also calculate the ground-level concentration 5 km downwind of the source and 500 m to the side of the plume centerline ($x = 5$ km, $y = 500$ m, $z = 0$).

✓**6.12.** A plant is emitting 750 g/s of particulates. The stack height is 100 m and the plume rise is 50 m. The wind speed is 7 m/s and the stability category is C.

(*a*) What is the maximum estimated ground-level concentration?

(*b*) How far downwind does it occur?

6.13. If the ratio of (σ_y/σ_z) is constant, independent of x, then the maximum ground-level concentration, predicted by Eq. (6.31), will occur at the x for which $\sigma_z = H/\sqrt{2}$.

(*a*) Show this by substituting $\sigma_y = \alpha\sigma_z$ (where α is a proportionality constant) into Eq. (6.31), taking the derivative of c with respect to σ_z, setting that derivative equal to zero, and solving for σ_z.

(*b*) From Figs. 6.7 and 6.8 one can see that (σ_y/σ_z) is not really a constant. Check to see how good an approximation this is by comparing the values of σ_z at the maxima for the various values of H on Fig. 6.9 with $\sigma_z = H/\sqrt{2}$. (You will find that this is an excellent approximation for C stability because, as shown in Problem 6.16, (σ_y/σ_z) is practically a constant for C stability. For the other stability categories it is a poorer approximation because (σ_y/σ_z) is not as close to constant.)

6.14. The Kennecott Copper Corporation's Magna Smelter (near Salt Lake City, Utah, before the smelter renovation of the late 1970s) emitted approximately 300 tons/day of SO_2 from two stacks. These may be approximated for calculational purposes as one stack with $H = 300$ m. If that stack were located on a flat plain with no mountains nearby, then its behavior would be reasonably well approximated by Eq. (6.29). Based on these assumptions, calculate the maximum ground-level concentration for A stability and a wind speed of 3 m/s.

6.15. A stack with physical stack height + plume rise = 100 m is emitting 1000 g/s of SO_2. The stability category is C, the wind speed is 3 m/s, and the mixing height, L, is 500 m. Normally one can smell SO_2 at any concentration equal to or greater than about 0.5 ppm. What is the farthest downwind distance at which one could expect to be able to smell the SO_2 from this stack at ground level? Here the true situation is that with the fluctuating air currents caused by atmospheric turbulence one would smell it for some part of the time and not for others. For this problem we will ignore that fact and make the standard Gaussian plume assumptions, which indicate that the calculated concentrations are constant and nonfluctuating.

✓ **6.16.** Figures 6.7 and 6.8 are useful for hand calculations and help one visualize the behavior of the σs, but they are not useful for computer calculations. Martin [23] represents them by

$$\sigma_y = ax^{0.894} \quad \text{and} \quad \sigma_z = cx^d + f$$

where x is the downwind distance, expressed in km; the sigmas are in m; and a, c, d, and f

are constants found in the following table:

Stability category	a	$x \le 1$ km c	d	f	$x \ge 1$ km c	d	f
A	213	440.8	1.941	9.27	459.7	2.094	−9.6
B	156	106.6	1.149	3.3	108.2	1.098	2.0
C	104	61	0.911	0	61	0.911	0
D	68	33.2	0.725	−1.7	44.5	0.516	−13.0
E	50.5	22.8	0.678	−1.3	55.4	0.305	−34.0
F	34	14.35	0.740	−0.35	62.6	0.180	−48.6

Check to see how well these equations reproduce the figures by computing σ_y and σ_z for the stability category and downwind distance in Example 6.5, and comparing the results. (All of the examples in this chapter and values in the Answers to Selected Problems for this chapter were actually computed as shown in this problem. That does not change the calculated values because the values here and in Figs. 6.7 and 6.8 are the same. But it removes the uncertainty due to chart-reading inaccuracy.)

6.17. A highway carries 10 000 cars per hour (5000 per hour each way) at an average speed of 50 mi/hr. The emission factor for CO is 3.4 g CO/mi (this is only a fair approximation for cars goung 50 mi/hr, but should be used for this problem). The wind is blowing at a velocity of 1 m/s at right angles to the highway. Assume the emissions from the autos occur at ground level and all at the centerline of the highway. The stability category is C. Estimate the concentration of CO 200 m downwind of the centerline of the highway. Assume that the concentration upwind of the highway is zero.

6.18. The Huntington Canyon Power Plant of the Utah Power and Light Company releases its exhaust gases through a stack with an inside diameter of 22 ft at a velocity of 80 ft/s and a temperature of 254°F. The plant site is at 6400 ft and the stack is 600 ft high, so the point of release is about 7000 ft above sea level. At this elevation, the average pressure is 790 mb and the annual average temperature about 50°F. What is the estimated plume rise for wind velocities of 1, 3, 10, and 30 m/s?

6.19. The Huntington Canyon Power Plant (see Problem 6.18) has installed a wet scrubber that reduces the stack gas temperature from 254°F to 120°F. The stack diameter, pressure, and exhaust velocity are unchanged from Problem 6.18. Repeat Problem 6.18 for this revised stack temperature.

6.20. It is estimated that a burning dump emits 3 g/s of NO_x. What is the concentration of NO_x directly downwind from this source at distances of 0.5, 1, 3, and 5 km on an overcast night with a wind speed of 7 m/s? Assume the dump to be a ground-level point source with no plume rise.

6.21. The owner of the dump in Problem 6.20 suggests things would be much better if a 20-m stack were erected on the dump; it could then be considered a point source at an elevation of 20 m. Repeat Problem 6.20 for this condition. Is the owner right?

6.22. Problems 6.20 and 6.21 deal only with the concentration directly downwind of the source. For each of them, now sketch (using appropriate numerical values) the concentration-distance plot in the crosswind direction at a distance 1 km downwind from the source.

6.23. In Fig. 6.9, check whether the long distance lines were indeed made up from Eq. (6.33) by calculating the value of cu/Q for C stability, a mixing height of 300 m, and a downwind distance of 100 km and then comparing the calculated result with that shown on Fig. 6.9.

6.24. In Problem 6.6, what is the maximum downwind distance at which the ground level concentration, directly under the plume centerline, is greater than or equal to 30 $\mu g/m^3$,

(*a*) If the mixing height $L = 1000$ m?

(*b*) If the mixing height L is infinite?

6.25. A power plant is emitting its stack gas from a 50-m high stack. For its conditions the plume rise is given by

$$\text{Plume rise} = \frac{200\ \text{m}^2/\text{s}}{u}$$

where u is the wind speed. The stability category is C. At what wind speed will the maximum calculated ground-level concentration occur? (*Hint*: This requires a trial-and-error solution. If you use Fig. 6.9, it is fairly easy.)

6.26. A power plant emits 100 g/s of NO_x from a stack with physical stack height 100 m and plume rise 150 m. The stability category is C and the wind speed 2 m/s.

(*a*) What is the estimated maximum ground-level concentration of NO_x due to this source?

(*b*) How far downwind of the source does the maximum occur?

(*c*) If the wind speed is not necessarily fixed at 2 m/s but is taken as a variable, then there is some wind speed that causes the highest estimated ground-level concentration of NO_x. What is that wind speed?

6.27. The maximum CO concentrations normally measured in downtown Salt Lake City (early 1990s) are about 35 000 $\mu g/m^3$. These values occur during strong inversions, for which we may estimate the values of u and H as 0.5 m/s and 100 m, respectively. The background concentration for this situation is estimated to be 5000 $\mu g/m^3$. The downtown area of Salt Lake City may be approximated as a 3-km by 3-km square. Estimate the emission density ($g/s \cdot m^2$) for CO for downtown Salt Lake City.

6.28. At the 1986 Chernobyl nuclear accident, there was a very large release of radioactivity in a few minutes, followed by a slowly declining release rate over several months. The total release is estimated to have been about 5×10^7 curies. The winds carried the released materials all the way to Sweden, from which the first reports came that there had been a nuclear accident in the USSR.

For the purposes of this problem assume that there was an instantaneous release of 10^7 curies of radioactive gases (and fine particles, which are assumed to remain in the atmosphere and not settle out). Then estimate the maximum ground-level concentration of radioactive gases (curies/m^3) when the radioactive cloud from the accident got to Sweden. Make the following assumptions:

1. Ignore decay of the radioactive gases (i.e., assume their half-lives were infinite).
2. Assume the distance between Chernobyl and Sweden is 1000 km.
3. Assume that the wind speed was 3 m/s and the stability class C.
4. Assume that the mixing height was 2000 m.
5. Assume that mixing in the x direction (up and down the direction of the wind) has the same intensity as mixing in the y (crosswind) direction.

6.29. A terrorist releases 1000 g of nerve gas as a single instantaneous emission at ground level at point $x = y = z = 0$, at time $t = 0$. The wind speed is 3 m/s and the stability class is C and the mixing height, L, is 2000 m. Estimate the maximum instantaneous value of the nerve gas concentration that would be observed at a point 5 km directly downwind of the emission point ($x = 5$ km, $y = z = 0$).

6.30. Figure 6.6 shows one-, two-, and three-dimensional spreading of a dye into stationary pieces of blotting paper.

(*a*) Sketch the equivalent plots for the following situation: a wide, shallow river is flowing steadily in a straight channel with constant, rectangular cross section. We release a small amount of dye into the river, halfway from side to side and halfway from top to bottom. Show the evolution of the dye cloud as it flows downstream with the river.

(*b*) How many dimensions is the initial spreading? How many dimensions is the spreading far downstream?

(*c*) Is there a corresponding air pollution problem? Discuss!

6.31. A smelter is located near an airport. The smelter stack is 300 m high and has a plume rise of 100 m. It is emitting 5000 g/s of SO_2. Assume that the stability class is always C, and that the wind speed is always 3 m/s.

The flight path for the airport is perpendicular to the plume and 5 km downwind of the smelter. The airport safety office has determined that it is unsafe for planes to go through any portion of the plume that has an average SO_2 concentration higher than 500 $\mu g/m^3$. They have also decided that flying under the plume is unsafe, so the planes must always fly over it. What is the minimum altitude at which they can fly under these circumstances and not be exposed to SO_2 concentrations $\geq 500\ \mu g/m^3$?

6.32. In Problem 6.31 a light plane flies through the plume, 5 km downwind from the smelter, perpendicular to the plume axis, at an elevation exactly equal to that of the plume centerline. The plane's speed is 100 mi/h = 44.7 m/s. The cabin ventilation system replaces the air in the cabin with outside air at a steady rate of 10 air changes per hour. The air in the cabin is perfectly mixed at all times. What is the maximum SO_2 concentration expected in the plane's cabin? Ignore the fact that the concentrations calculated by the Gaussian plume model are averages for times longer than the time the plane spends in the plume. (This problem has no analytical solution; a numerical solution using a spreadsheet is recommended.)

6.33. Although Gaussian plume models are widely called diffusion models in the meteorological literature, the speed of dispersion of the pollutants is much faster than ever occurs by molecular diffusion. To show that this is so, observe that the smallest value of either of the sigmas on Figs. 6.7 and 6.8 is 2.2 m for σ_z at 0.1 km. Compute the value of K_z corresponding to this value, with the assumption that $u = 1$ m/s. Compare this to the typical gas-gas molecular diffusivities of about 0.1 cm^2/s.

6.34. In the text and in most air pollution modeling we treat point sources by Gaussian plume modeling and area sources by box models or their equivalent. One way to improve box models would be to continue to treat the emissions as if they were uniformly spread over the surface [i.e., emission rate $= q$ in g/($m^2 \cdot$ s) is a constant at an emission elevation of zero over the entire surface of some city], but to have the horizontal and vertical dispersion of the pollutants occur by turbulent mixing, the same as is assumed in the Gaussian plume calculations.

Show the appropriate formula for doing this, based on the following assumptions:

1. The city is the rectangular city shown in Fig. 6.1.
2. The background concentration is zero ($b = 0$) and $q = 0$ for all values of downwind distance less than zero (i.e., the upwind suburbs have no emissions, and the emissions begin at the city's upwind boundary).
3. The width of the city is practically infinite (W in Fig. 6.1 $\approx$ infinity).
4. The height of the mixing layer is so large that it plays no role in the problem (H in Fig. 6.1 $\approx$ infinity).
5. The wind is steady and blows in the direction shown in Fig. 6.1, with velocity u.

6. Vertical and horizontal dispersion of pollutants occurs as in Gaussian plume modeling, with σ_y and σ_z both increasing with increasing distance downwind and depending on stability category, as shown in Figs. 6.7 and 6.8.
7. As in the box and standard Gaussian plume models, we want to know the steady-state concentration at some point, in a situation in which nothing is changing with time.
8. You may show the formula either for c as a function of (x, y, and z) or for ground level only ($z = 0$).

REFERENCES

1. Holzworth, G. C.: "Mixing Heights, Wind Speeds, and Potential for Urban Air Pollution throughout the Contiguous United States," U.S. Environmental Protection Agency Report AP-101, U.S. Government Printing Office, Washington, DC, 1972.
2. Hanna, S. R.: "A Simple Model for Calculating Dispersion from Urban Area Sources," *J. Air Pollut. Control Assoc.,* Vol. 21, pp. 774–777, 1971.
3. Larsen, R. I.: "A Method for Determining Source Reduction Required to Meet Air Quality Standards," *J. Air Pollut. Control Assoc.,* Vol. 11, pp. 71–76, 1961.
4. de Nevers, N., and J. R. Morris: "Rollback Modeling—Basic and Modified," *J. Air Pollut. Control Assoc.,* Vol. 25, pp. 943–947, 1974. See also Air Pollution Control Association Paper, 73–139.
5. Arya, S. P.: *Air Pollution Meteorology and Dispersion*, Oxford University Press, Oxford, p. 137, 1999.
6. Carslaw, H. S., and J. C. Jaeger: *Conduction of Heat in Solids,* 2d ed., Clarendon Press, Oxford, p. 255 et seq., 1959.
7. Turner, D. B.: "Workbook of Atmospheric Dispersion Estimates," U.S. Environmental Protection Agency Report AP-26, U.S. Government Printing Office, Washington, DC, 1970. (Also available, at greater cost, as Turner, D. B.: *Workbook of Atmospheric Dispersion Estimates; an Introduction to Dispersion Modeling*, 2d ed., Lewis, Boca Raton, FL, 1994.)
8. Hanna, S. R., G. A. Briggs, and R. P. Hosker, Jr.: "Handbook on Atmospheric Diffusion," DOE/TIC-11223 (DE82002045), 1982.
9. Turner, D. B., J. S. Irwin, and A. D. Busse: "Comparison of RAM Model Estimates with 1976 St. Louis RAPS Measurements of Sulfur Dioxide," *Atmos. Environ.,* Vol. 19, pp. 247–253, 1985.
10. Briggs, G. A.: *Plume Rise,* U.S. Atomic Energy Commission, Washington, DC, 1969.
11. "Implementation Planning Program" developed for the U.S. Environmental Protection Agency by TRW Systems Group, 1970.
12. Morris, R. E., and T. C. Myers: "User's Guide for the Urban Airshed Model," U.S. Environmental Protection Agency Report EPA-450/4-90-007A and B, 1990.
13. Scheffe, R. D.: "A Review of the Development and Application of the Urban Airshed Model," *Atmos. Environ.,* Vol. 27B, pp. 23–39, 1993.
14. Seinfeld, J. M.: "Ozone Air Quality Models: A Critical Review," *J. Air Pollut. Control Assoc.,* Vol. 38, pp. 616–645, 1988.
15. Miller, M. S., S. K. Friedlander, and G. M. Hidy: "A Chemical Element Balance for the Pasadena Aerosol," *J. Colloid Interface Sci.*, Vol. 39, pp. 165–175, 1972.
16. Friedlander, S. K.: "Chemical Element Balances and Identification of Air Pollution Sources," *Environ. Sci. Technol.*, Vol. 7, pp. 235–240, 1973.
17. Gartrell, G., Jr., and S. K. Friedlander: "Relating Particulate Pollution to Sources: The 1972 California Aerosol Characterization Study," *Atmos. Environ.*, Vol. 9, pp. 279–299, 1975.
18. Gordon, G. E., W. R. Pierson, J. M. Daisey, P. J. Lioy, J. A. Cooper, J. G. Watson, and G. R. Cass: "Considerations for Design of Source Apportionment Studies," *Atmos. Environ.,* Vol. 18, pp. 1567–1582, 1984.
19. Doskey, P. V., J. A. Porter, and P. A. Scheff: "Source Fingerprints for Volatile Non-Methane Hydrocarbons," *J. Air Waste Manage. Assoc.,* Vol. 42, pp. 1437–1445, 1992.
20. National Research Council: *Rethinking the Ozone Problem in Urban and Regional Air Pollution,* National Academy Press, Washington, DC, p. 237, 1991.

21. Slade, D. H., ed.: *Meteorology and Atomic Energy 1968,* U.S. Atomic Energy Commission, Washington DC, 1968. Available as TID-24190 from NTIS.
22. Millán, M. M., E. Otamendi, L. A. Alonso, and I. Ureta: "Experimental Characterization of Atmospheric Diffusion in Complex Terrain with Land-Sea Interactions," *J. Air Pollut. Control Assoc.,* Vol. 37, pp. 807–811, 1987.
23. Martin, D. O.: "Comment on 'The Change of Concentration Standard Deviations with Distance,'" *J. Air Pollut. Control Assoc.,* Vol. 26, pp. 145–147, 1976.

CHAPTER 7

GENERAL IDEAS IN AIR POLLUTION CONTROL

In this chapter, we consider some general ideas that apply to all of the following chapters.

7.1 ALTERNATIVES

If we have an air pollution problem there are three control options available.

7.1.1 Improve Dispersion

As discussed in Chapter 2, if the true dose-response curve has a threshold value, then we can remedy the problem if we can improve the dispersion of our emissions and thereby lower the concentrations to which people are exposed to less than that threshold value. If our region regularly has pollutant concentrations above the NAAQS, we can certainly use the dispersion methods discussed shortly to reduce those concentrations. At present in the United States, this approach is *strongly disapproved* of for use by industry, but is widely used by local and regional governmental air pollution control agencies.

Fifty years ago this was the most widely used approach to pollution problems (air or water). The motto of the pollution control engineer was "Dilution is the solution to pollution." Many municipalities had regulations *requiring* air pollution sources to use tall stacks to dilute their pollutants before they came to ground. Most municipalities dumped their sewage, untreated, into the nearest river, lake, or ocean, counting on its dilution to render the sewage harmless, or at least to carry it away from them. Some dumped their solid waste (garbage) there as well.

In a sparsely populated world dilution would still be an acceptable approach. When the population density is 1 person/km^2, people may dispose of their wastes any way they like without causing any damage to their neighbors or causing long-term environmental damage. But in a densely populated world it is not a satisfactory approach. The next city downriver or downwind of us may not want to drink our sewage, breathe our air pollutants, or have our garbage wash up on their shores. When the population density is 29,000 persons/km^2 (Manhattan), there must be strict rules limiting how people dispose of their wastes just to prevent public health disasters. In the first case dilution is probably the best solution; in the second it is simply not acceptable.

For the past 20 years the thrust of U.S. environmental law regarding air pollution, water pollution, and solid waste disposal has been to prevent the emission of harmful effluents rather than to deal with them by dilution. Those who oppose the dilution solution argue that dilution merely transfers the problem somewhere else. Others argue that there are some effluents that we must emit (e.g., our human breath is high in carbon dioxide and contains some carbon monoxide). We will minimize harm to others if we minimize the amount of these effluents that other people breathe, drink, or eat. If the effluent materials have natural removal mechanisms in the environment, so that they will not accumulate, then diluting or dispersing them as a way to prevent them entering the bodies of other humans is still a prudent thing to do. It should not be a substitute for emission reduction, but may supplement it—treatment followed by dilution. Without entering further into that argument (which has largely been taken over by Congress and the courts), we can indicate three logical approaches to improving dispersion (dilution).

7.1.1.1 Tall stacks. Figure 6.9 shows that, for any one stability category, raising the point of emission (increasing the value of H) lowers the calculated ground-level concentrations for all points near the stack. For points far enough away for the plume to be well mixed up to the mixing height (the right side of Fig. 6.9), the calculated concentration becomes independent of the stack height. Thus if the assumptions behind that plot were correct (see following paragraphs), then raising the height of emissions would lower all nearby concentrations and not change the concentrations at a distance.

Figure 7.1 on page 162 shows an example of the observed effectiveness of this approach. The ground-level SO_2 concentrations at two measuring stations near a large coal-fired power plant are shown before and after the effluent was switched from five short stacks (83 to 133 m) into one tall stack (251 m). As the plot shows,

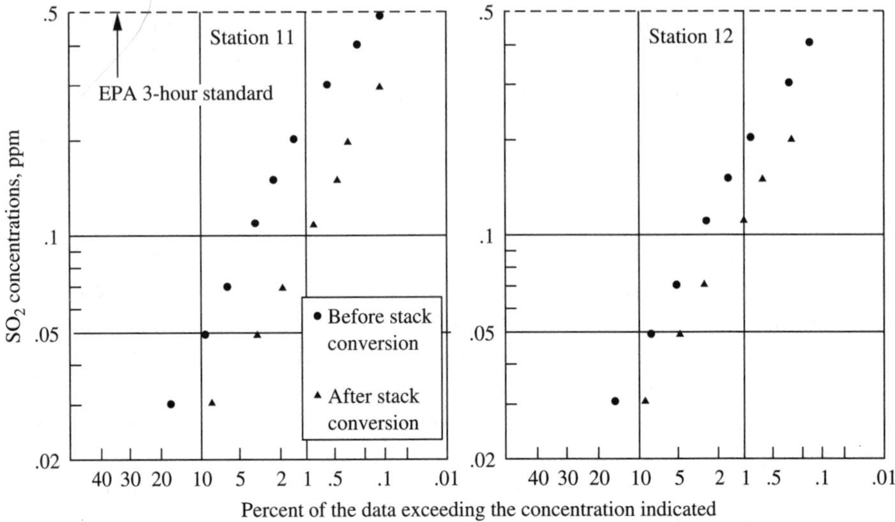

FIGURE 7.1
Comparison of observed hourly sulfur dioxide concentrations at two monitoring stations near a coal-fired power plant, before and after replacement of five short stacks (83 to 113 m) with one tall stack (251 m). Station 11 is 5.3 km southeast of the plant and station 12 is 4.4 km north-northeast of the plant. (From Ref. 1.)

at all levels of frequency of occurrence the observed ground-level concentrations were reduced. For example, at station 11, before the conversion, approximately 10 percent of the readings exceeded 0.05 ppm; after installation of the tall stack, only about 3 percent of the readings exceeded this value.

These experimental results support the calculation in Fig. 6.9 that raising the stack height lowers all ground-level concentrations near the plant. There is no comparable demonstration for long distances, and there are some grounds to believe that raising the stack may increase the concentrations at long distances. The calculations leading to Fig. 6.9 are based on the assumption that there is no natural removal of the pollutant in question. That is obviously false; pollutants like SO_2 ultimately come to the ground, mostly with rain or snow, partly by dry deposition. If the ground-level concentration is higher near the stack, then the rate of removal by the ground will be higher near the stack, and thus less pollutant will remain in the air to be transported for long distances. Using a tall stack will certainly decrease all the ground-level concentrations near the stack, but it may increase some concentrations far from the stack.

This issue has been the subject of much debate in the acid-rain controversy. Whether raising the points of emission significantly increases the concentration far downwind is an open question. There is no question that if raising the emission point is used *as a substitute for reducing the emissions,* then the concentrations far downwind will be increased compared to those that would be observed if the emissions were reduced.

7.1.1.2 Intermittent control schemes. At certain times of the year (or times of the day) emissions are more likely to come to ground in high concentrations and in populated areas than at other times. Intermittent control schemes attempt to reduce emissions then, allowing emissions to return to normal rates at other, less critical times. In most cases, the short-term emission reduction is brought about by a plant shutdown, fuel switching, or production curtailment during the period of control.

Intermittent control schemes are predictive, or observational, or combined predictive-observational. Predictive schemes are based on the knowledge that the atmospheric conditions likely to call for an emission reduction occur regularly and can be predicted with (some) accuracy. Frequently, the damaging situation is caused by a morning inversion breakup fumigation (Chapter 5) in which the pollutants that cause the NAAQS violation are emitted several hours before the violation occurs. For emission reduction to be effective in this case, it is necessary to curtail emissions several hours before the predicted violation.

The most famous, and apparently the first documented, predictive intermittent control scheme was instituted in 1941 at the lead-zinc smelter at Trail, British Columbia. This smelter is located in a narrow portion of the Columbia River valley, seven miles north of the Canada–United States border. During nighttime, emissions carried by downvalley air flow passed over agricultural lands on the United States side of the border; during morning inversion breakup, these emissions were brought to ground level, causing crop damage to orchards. This situation led to the formation of an international arbitration tribunal to award monetary damages. The tribunal instituted a study that ultimately led to the adoption of an intermittent control scheme to minimize crop damage [2]. It considered the following variables: growing and nongrowing season, wind direction, turbulence intensity, and time of day. Regulations required most stringent controls for the period from 3 A.M. to three hours after sunrise during the growing season; this is the time when turbulence is low and winds cause downvalley flow.

An entirely analogous predictive control scheme is that of the Paradise, Kentucky, steam-power complex of the Tennessee Valley Authority [3]. When meteorological conditions predict that the dispersion of the plant's plume will be severely restricted, threatening a violation of an NAAQS, the plant's power output (and thus fuel consumption and pollutant emission rate) is reduced several hours before the infraction is predicted to occur.

In an observational intermittent control scheme, emissions are promptly curtailed when an air quality sensor or network of sensors indicates that air quality is deteriorating unacceptably. This approach has limited application in many situations, particularly in circumstances where the emissions do not affect the sensor until hours after they have been emitted. A great many sensors are required if all areas in the vicinity of the emission source are to be protected. The observational scheme is most useful when it supplements a predictive scheme. It can then serve as a fail-safe backup for the predictive scheme.

In addition to the predictive part just described, the Trail, BC, intermittent control procedure also provided for continuous monitoring of SO_2 at an agricultural location in the United States and for curtailment of emissions whenever this monitor showed a continued high value [2]. Similarly, the scheme for the ASARCO smelter at El Paso, Texas, contains both predictive and observational components [4]. Apparently, the same kind of control, on a less formal basis, was practiced in the smelter industry as early as the 1920s [5].

Many western mountain communities that have large numbers of homes heated with wood stoves have instituted intermittent control systems. When the measured concentration of PM_{10} exceeds some value, a public notice is made (blowing a siren, etc.) that requires all wood-burning appliances be shut off promptly. This is a totally observational intermittent control system. Many cities have regulations curtailing certain activities during times of observed poor air quality.

The current U.S. federal regulations requiring oxygenated motor fuels in winter months are a wider-scale version of intermittent controls. High CO concentrations are observed in many U.S. cities, but generally only in the winter months. Oxygenated motor fuels reduce motor vehicle CO emissions. For that reason current U.S. federal regulations require the use of oxygenated fuels, but only during that part of the year in which high ambient CO concentrations are expected.

In all of these cases, the intermittent control operates in addition to controls that reduce emissions all the time. They are "dilution solutions" in that they take advantage of the greater dispersive capacity of the atmosphere during times when the extra, intermittent control is not applied.

7.1.1.3 Relocate the plant. It is hard to move an existing plant, but a new plant can be located where its emissions will have their greatest impact in nonpopulated areas. This reasoning is the basis for most industrial zoning and land-use planning regulations. When society first instituted those regulations we assumed that we could not have, for example, a slaughterhouse that didn't stink. So we tried to put the slaughterhouse away from residential areas, and generally downwind. (The zoning generally let poor people live near the slaughterhouse, but not the rich.) Now we believe that we can make any plant odorless, and hence a good neighbor, so we are rethinking this idea. But if, for example, a region has a severe current problem with some pollutant, we will generally not allow a new source of that pollutant to locate in that region, even if it has the best available controls, because even a well-controlled new source could add to the current problem. Instead we will try to locate the plant where any problem with that pollutant is less severe.

This siting decision may bring us into conflict with the Prevention of Significant Deterioration regulations (Chapter 3). Nonetheless, if one wished to build a new coal-fired power plant in the United States, one would not waste time trying to get the air pollution permits to build it in or near a major city. Instead, one would look for an area away from major population centers and then promise to install the most stringent currently available control methods in order to get the necessary permits.

Although improving dispersion (the dilution solution) is not allowed as a *substitute* for industrial emission reduction, it still plays a major role in U.S. air pollution control regulations, mostly in terms of intermittent control schemes that ban some activities in times of poor dispersion, and in terms of siting criteria that prevent the location of new emitting facilities in regions with severe air quality problems.

7.1.2 Reduce Emissions by Process Change, Pollution Prevention

There are many historical examples in which the most economical air pollution control solution was to modify the process to reduce the emissions. For example, when factories that applied large quantities of paint to the goods they produce (e.g., automobiles, refrigerators) were required to limit the emission of hydrocarbon solvents (paint thinners), some found that they could substitute water-based paints for some of their oil-based paints and greatly reduce their hydrocarbon emissions problem. Some copper smelters have replaced reverberatory furnaces, which produce high-volume, low-concentration SO_2 waste gases, with other smelting processes that produce lower-volume, higher-concentration SO_2 waste gases. The latter still require downstream treatment, but they are much easier and more economical to treat than the dilute waste gases from reverberatory furnaces. Open burning of municipal or industrial waste is normally smoky and sooty. Most air pollution control districts now require that such burning be carried out in closed incinerators, which have much better fuel-air mixing, fuel predrying, and heat conservation than open burning. The resulting emissions are much less than from open burning of similar wastes.

Process modification to reduce emissions is wider than one might think. One of the major reasons for installing basic oxygen process (BOP) steelmaking furnaces is that the emissions are more concentrated and thus easier to control than those from the open-hearth furnaces they replace. Most countries are forcing the owners of mercury-cell chlorine-caustic plants to switch to diaphragm-cell plants because of the toxicity of mercury. Most uses of asbestos have been banned in industrial countries, and replacements found, because of the toxicity of asbestos.

Switching fuels is also a process change to reduce emissions. The biggest improvement in air pollutant concentrations in most cities of the United States and Western Europe came about when coal was replaced by natural gas as a home and business heating fuel. Switching vehicles from gasoline to compressed natural gas, propane, or ethanol greatly reduces the vehicles' air pollutant emissions. Adding oxygenated compounds to motor fuels (typically about 2 weight percent oxygen) lowers CO emissions significantly. Requiring the use of low-sulfur fuels reduces sulfur dioxide emissions. These are all variants on the theme of process change to reduce emissions.

Getting people to carpool, to ride buses or bicycles, or to walk to work is a form of process change. If the process is "get people from home to work," then changing from the one-passenger auto to any of these alternatives is a process change that reduces emissions from the process. Replacing low-efficiency incandescent lights with higher-efficiency fluorescent lights is a process change that reduces emissions.

The process is "provide some amount of light"; fluorescent lamps require less electricity for the same amount of light, so less fuel is burned in power plants, and hence less air pollutants are emitted at the plant that produces the electricity. Any process change in any industry that reduces the consumption of fuels or other raw materials reduces air pollutant emissions, because the production, distribution, and use of raw materials and fuels all produce air pollutant emissions.

In current U.S. environmental law there is a major effort to *prevent* pollution rather than control it [6]. The goal of this effort, mostly directed at solid and hazardous waste, is never to produce such waste. The provisions of the RCRA and CERCLA acts that make it impossible for the producer of a hazardous waste ever to escape legal liability for its future misuse are strong incentives not ever to produce it.* The same pollution prevention idea applies to air pollution, although currently not as vigorously as for solid or liquid wastes.

7.1.3 Use a Downstream Pollution Control Device

A downstream pollution control device (often called a tailpipe or end-of-the-pipe control device) accepts a contaminated gas stream and treats it to remove or destroy enough of the contaminant to make the stream acceptable for discharge into the ambient air. Most of the rest of this book is about such devices. Many people think only of them when they think about air pollution control because they are widely applied and important. However, they appear third in this list of alternatives, because a prudent engineer will always first examine the previous two options to see if they are more practical and economical than a downstream control device. In the current regulatory climate, the air pollution control engineer will receive more credit for devising a process change that prevents the formation of the pollutant, than for designing an excellent device to control it once it is formed.

These three approaches need not be applied separately. In its 1977 renovation of its Magna smelter, the Kennecott Copper Corporation used a tall stack for improved dispersion, intermittent controls for dealing with particularly difficult weather situations (strong south winds, blowing over the mountains just to the south of their smelter, produced strong downdrafts that would bring their plume to ground with little dilution [see Section 6.9.2]), process changes to concentrate the off-gas, and downstream controls to collect the sulfur oxides in the off-gas. They selected this combination because they believed that no one of the options, applied singly, would have been adequate to meet the applicable NAAQS.

7.2 RESOURCE RECOVERY

If the pollutant is a valuable material or a fuel, it may be more economical to collect and use it than to discard it. Generally, reclamation is only possible if the concentra-

*RCRA (pronounced "reck-ra") stands for Resource Conservation and Recovery Act, and CERCLA stands for Comprehensive Environmental Response, Compensation, and Liabilities Act.

tion is high enough in the waste stream. This is frequently an incentive to modify the process to increase the concentration by decreasing the flow of waste gas. A clear example of this is SO_2, which can be reacted with oxygen over a vanadium catalyst to produce sulfur trioxide (SO_3). The latter, dissolved in water, forms sulfuric acid (H_2SO_4), a marketable product. (Its principal use is for the production of phosphate fertilizer; its price fluctuates with the demand for phosphate fertilizer. It has many other uses, e.g., battery acid or as a permitted food additive; see Problem 14.10.)

Those who have studied the economics of using this method to limit SO_2 emissions (see Chapter 11) have generally concluded that it is economically prudent to do so (i.e., the sulfuric acid sales will pay for the sulfuric acid plant) if the concentration of SO_2 in the waste stream is 4 percent by volume or greater. Hence smelters that extract metals from sulfide ores (copper, lead, zinc, molybdenum, nickel, and some others) can economically use this recovery process if they have a nearby market for the acid, but coal-fired electric power plants cannot because the SO_2 content of their waste gases is normally about 0.1 percent.

Other examples of resource recovery in air pollution control are the use of catalytic cracker regenerator off-gas and blast-furnace gas. Both of these waste streams generally contain enough CO to make them valuable fuels. Properly tuning engines, burners, and furnaces of all kinds both reduces air pollutant emissions and increases fuel efficiency, so such tuning is an air pollution control activity that also saves resources.

Finally, many organic solvents can be collected from waste streams and reused. This step is only economical if the concentrations are large. For this reason and for the reasons shown in the following sections, the air pollution control engineer should always examine ways to prevent the mixing of concentrated streams with dilute ones. Systems are designed to prevent the introduction of any more air than necessary into streams from which it may be possible to recover valuable products or that must be treated to minimize their effluent concentrations.

A competent pollution engineer *always* looks for opportunities to convert waste streams to profitable products or valuable raw materials. Most of the obvious possibilities have already been exploited; less obvious ones are waiting to be discovered.

7.3 THE ULTIMATE FATE OF POLLUTANTS

If possible, we prevent the formation of pollutants. If we cannot do that, we hope to capture them and put them to some good use. For most pollutants we cannot do that. If the pollutants will burn, we often destroy them by burning; this is true for most organic compounds. Most other pollutants cannot be burned. For them the most common ultimate fate is to be captured and placed in a landfill. That is the fate of most particulate pollutants, which are generally fine dusts. Most sulfur pollutants are ultimately converted to $CaSO_4 \cdot 2H_2O$, an innocuous solid, and are landfilled.

In designing any air pollution control system one should plan for the ultimate disposal of any wastes produced, because the cost of that disposal can often be a significant fraction of the total cost of air pollution control. If the collected material is classified as a *hazardous waste* [7], then its disposal cost is many, many times that of an ordinary waste.* Mixing a hazardous waste with a nonhazardous waste generally makes both hazardous. Prudent engineers try **never** to do that. Current U.S. solid waste disposal law is stringent in assigning financial responsibility to the originator of any hazardous waste. For this reason air pollution control processes that produce a solid waste, particularly one that may be classified as hazardous, are rarely chosen if there is any alternative process that produces no such solid waste.

7.4 DESIGNING AIR POLLUTION CONTROL SYSTEMS AND EQUIPMENT

Figure 7.2 shows a typical pollution control system consisting of some kind of contaminated gas capture device (a hood in this example); some kind of control device; some kind of gas mover, such as a fan or blower; some system for recycling or disposing of the collected material; and some kind of a stack. It would be most unusual for one person to design all of these pieces of equipment. Most likely for a small installation the fan and the control device would be selected from suppliers' catalogs. Standard-size equipment is much cheaper and more reliable than custom-designed equipment. (The same is true of automobiles and clothing!) For large installations (e.g., a large electric power plant) the control device would be custom-designed, but made up by assembling the proper number of standard components in a custom-designed enclosure. The designer of the whole system would be expected to specify the gas flow rate, the concentration and chemical nature of the pollutants in the gas, the required control efficiency, and the disposal method for the collected pollutant (if any). The designers at the control equipment company use much more complex and detailed design procedures than those shown in this book. Often these design procedures are trade secrets. The control equipment buyer should use the simple methods in this book to check the proposed equipment designs for gross errors, but not for precise design values.

7.4.1 Air Pollution Control Equipment Costs

The costs of air pollution control equipment are estimated by the same procedures used in all engineering cost estimation; one collects and summarizes recent purchase price information of major pieces of equipment and then uses historic data to estimate the labor costs and other costs to make a complete cost estimate [8]. Major contractors have enough historical data to make such estimates accurate to ±1 to 3% (in times of low inflation!). Students and professors, who do not have such complete data, can make them to within perhaps ±30 to 50% using published cost data. The U.S. EPA maintains an electronically available cost-estimating file, which is regularly updated

*Hazardous waste should not be confused with the Hazardous Air Pollutants classification (discussed in Chapter 15). Hazardous wastes are solids or liquids, whose disposal is the subject of very stringent and expensive federal regulations. Some materials can be both hazardous waste and Hazardous Air Pollutants, but most materials in one category are not in the other.

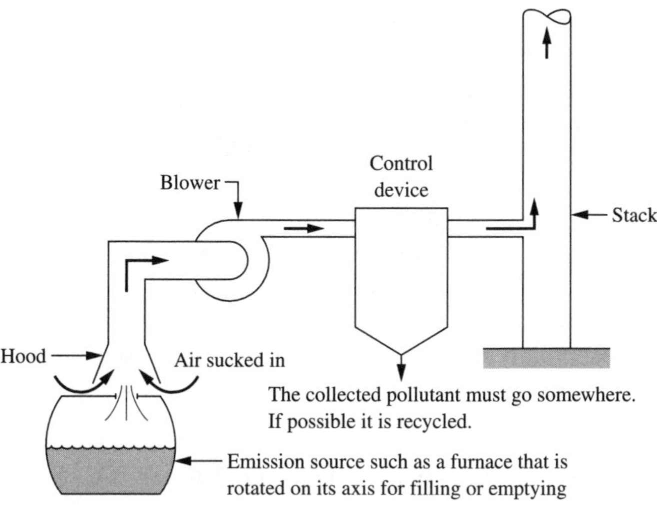

FIGURE 7.2
Typical hood/blower/control device/stack arrangement for an emission source that cannot be connected by a closed duct system to the control device. In this flowsheet the fan is between the emission source and the control device; it can also be placed between the control device and the stack. Both arrangements have advantages and disadvantages.

[9]. The cost of an individual piece of major equipment (e.g., a cyclone separator or fabric filter) is normally a function of its size, with a cost relation of the form

$$\begin{pmatrix}\text{Purchase price}\\ \text{of equipment}\\ \text{without auxiliaries}\end{pmatrix} = a + b\,(\text{size})^c \tag{7.1}$$

where a, b, and c are arbitrary values, obtained from log-log plots of historic cost data. The size parameter varies from one kind of equipment to another, e.g., filter surface area for a filter, entrance area for a cyclone separator. However, in almost every case the size dimension is proportional to the volumetric flow rate of the gas to be treated.

Example 7.1. (This example is a simplification of the example on p. 111 of [8]). We wish to purchase and install a fabric filter to collect fine particles. Using the methods shown in Chapter 9, we conclude that we will need 13 400 ft^2 of filter area. What will the complete purchased filter cost? What will the whole installation cost?

From [8] we have that the purchase price of the filter, bags included, is

$$\begin{pmatrix}\text{Purchase price}\\ \text{of equipment}\\ \text{in this size range}\end{pmatrix} = \$41\ 000 + \frac{\$8.75}{\text{ft}^2}(\text{filter area})^{1.00}$$

$$= \$41\ 000 + \$8.75 \cdot 13\ 400 = \$158\ 250$$

This estimate is based on 1988 prices. Using the *Chemical Engineering* cost index, we estimate that the 1998 price will be 1.12 times the 1988 price or \$177 000.

The total cost of the whole installation, including the foundations, site preparation, ductwork, fan, electrical connections, taxes, insurance, shipping, and erection cost [8] is $\approx$ 2.17 times the cost of the purchased filter, or \$385 000. ■

In Chapter 9 we will see that the required filter area is linearly proportional to the volumetric flow rate of the gas treated, so that the cost data in the above example could have been stated in terms of volumetric flow rate, rather than filter area. That is true of almost all pollution equipment control cost equations. In this case the size exponent $c = 1.00$. For most pollution control equipment it is in the range 0.6 to 1.00.

7.5 FLUID VELOCITIES IN AIR POLLUTION CONTROL EQUIPMENT

Most contaminated gas streams are either air or combustion gases at nearly atmospheric pressure, and at a range of temperatures from room temperature to combustion temperatures. The fluid mechanical properties of combustion gases are close enough to those of air that approximate or preliminary fluid mechanical calculations are normally made as if combustion gases were air. The same is not true for the chemical properties, discussed in Sec. 7.12 and Chapter 12.

Almost all industrial-sized flows of air or gases are turbulent. The velocity in most air conditioning and other gas-flow ducts is about 40 to 60 ft/s ($\approx$ 12 to 18 m/s), for economic reasons. As shown in any fluid mechanics book [10] there is an "economic velocity" for pumped fluid flows. This velocity minimizes the sum of pumping costs and the capital charges for the equipment. If we make the ducts or pipes bigger, the pumping cost is reduced, but the capital cost of the pipes or ducts increases. For ordinary steel construction and ordinary electric power costs, the optimal velocities are about 6 ft/s for water and 40 ft/s for air.

Example 7.2. Air at 68°F is flowing at 40 ft/s in a 2-ft diameter pipe. Estimate the Reynolds number. (See any fluid mechanics book for a discussion of the Reynolds number.)

Here

$$\mathcal{R} = \frac{DV\rho_{\text{fluid}}}{\mu} = \frac{2\text{ ft}\cdot 40\,\dfrac{\text{ft}}{\text{s}}\cdot 0.075\,\dfrac{\text{lbm}}{\text{ft}^3}}{0.018\text{ cp}\cdot 6.72\cdot 10^{-4}\,\dfrac{\text{lbm}}{\text{ft}\cdot\text{s}\cdot\text{cp}}} = 5\cdot 10^5$$ ■

This is $\approx$ 100 times the Reynolds number at the end of the transition region; thus we are quite safe in assuming that the flow of air and gases in ordinary ducts of any kind is turbulent.

Under what circumstances will the velocities be substantially different from 40 ft/s ($\approx$ 12 m/s)?

1. In some particulate control devices (see Chapter 9) we use the inertia of the particle or of a droplet for collection purposes; velocities up to 400 ft/s ($\approx$ 120 m/s) are used.
2. In other particulate control devices we want the gas to remain as long as practical in the collecting device, in order to allow time for the control process to occur. In electrostatic precipitators (Sec. 9.1.3) the normal gas velocity is 3 to 5 ft/s ($\approx$ 1 to 1.5 m/s).
3. If a gas stream is transporting a high specific gravity dust (e.g., heavy metal oxides), duct velocities up to 60 to 80 ft/s (18–23 m/s) are used, to prevent settling of the dust in the duct.
4. In countercurrent gas-liquid contacting devices, discussed in Chapters 9–11, the vertical upward gas velocity must be low enough that liquid drops can fall by gravity through the gas. This limits the upward velocity to the settling velocity of the drops (Chapter 8), normally $\leq$ 10 to 20 ft/s (3 to 6 m/s).
5. Flow-through filters (of the Surface type, see Sec. 9.2.1) and granular adsorbents (see Sec. 10.4.2) and some catalysts (see Sec. 7.13) are the exception to the above statement that the flows in air pollution control are turbulent. In these flows the actual flow passages are the spaces between the individual particles making up the filter cake or the adsorbent or catalyst bed. These are thousands of times smaller than the typical gas flow duct, so that the Reynolds number is very small. In typical air pollution surface filters ("Baghouses," Sec. 9.2.1) the superficial velocities are 1 to 3 ft/min (0.3 to 1 m/minute $\approx$ 0.016 to 0.05 ft/s) and the flow is normally laminar.

7.6 MINIMIZING VOLUMETRIC FLOW RATE AND PRESSURE DROP

All waste gas streams must be propelled through the control device and the associated ductwork and exhaust stack. Normally a fan or blower accomplishes this. The power to run this fan or blower can be one of the significant costs of the air pollution control system.

For any steady, adiabatic fan or blower that is processing an ideal gas the power input to the fan or blower is calculated by

$$\text{Power} = \frac{\dot{n}RT_1}{\eta}\left(\frac{k}{k-1}\right)\left[\left(\frac{P_2}{P_1}\right)^{(k-1)/k} - 1\right] \tag{7.2}$$

where $\dot{n}$ = molar flow rate (lbmol/s)

R = universal gas constant (see Appendix A)

T_1 = inlet absolute temperature

k = heat capacity ratio (practically a constant equal to 1.4 for air and most waste gases)

η = fan or blower efficiency

P_1 = inlet absolute pressure of the fan

P_2 = outlet absolute pressure of the fan

This relation assumes ideal gas behavior and adiabatic performance. However,

$$P_2 = P_1 + \Delta P \tag{7.3}$$

so that

$$\text{Power} = \frac{\dot{n}RT_1}{\eta}\left(\frac{k}{k-1}\right)\left[\left(1+\frac{\Delta P}{P_1}\right)^{(k-1)/k} - 1\right] \tag{7.4}$$

The quantity $(\Delta P/P_1)$ is generally $\ll 1$, and in general $(1+x)^n$ can be represented by the series expansion $1 + nx + n(n+1)x^2/2! + \cdots$. If $x \ll 1$, then all of the terms beyond the nx term can be dropped with negligible error. In Eq. (7.4), $(\Delta P/P_1)$ plays the role of x, and $[(k-1)/k]$ plays the role of n, so we may make the substitution and simplify, finding

$$\text{Power} \approx \frac{\dot{n}RT_1}{\eta}\left(\frac{k}{k-1}\right)\left[1+\left(\frac{\Delta P}{P_1}\right)\left(\frac{k-1}{k}\right) - 1\right] = \frac{\dot{n}RT_1\,\Delta P}{P_1\eta} \tag{7.5}$$

But $\dot{n}RT_1/P_1$ is equal to the inlet volumetric flow rate Q_1 so that

$$\text{Power} \approx \frac{Q_1\,\Delta P}{\eta} \tag{7.6}$$

This approximation is only valid for small $\Delta P/P_1$, which is normally the case in air pollution engineering.

Equation (7.6) shows that if one wishes to minimize the power required to drive the waste gas stream through the control system, one should minimize Q_1, minimize ΔP, and maximize η.

Example 7.3. A typical electric power plant with a net power output of 1000 MW produces a stack gas flow of approximately 2 million ft^3/min (943.9 m^3/s) at stack conditions. If we install a pollution control device that has a pressure drop of 1 psi (6895 Pa), what fraction of the power from the plant will be consumed by the 90 percent efficient fan that overcomes this pressure drop?

$$\text{Power} = \frac{(2\times 10^6\ \text{ft}^3/\text{min})(1\ \text{lbf/in.}^2)}{0.90}\cdot\frac{144\ \text{in.}^2}{\text{ft}^2}\cdot\frac{\text{hp}\cdot\text{min}}{33\,000\ \text{ft}\cdot\text{lbf}} = 9697\ \text{hp}$$

or

$$\text{Power} = \frac{(943.9\ \text{m}^3/\text{s})(6895\ \text{N/m}^2)(\text{W}\cdot\text{s/N}\cdot\text{m})}{0.90} = 7.23\times 10^6\ \text{W} = 7.2\ \text{MW} \quad \blacksquare$$

This example shows that a control device with a pressure drop of 1 psi would consume 7.2 MW or 0.7 percent of the net power output from the power plant. This fraction is significant, and considerable efforts are made to minimize the pressure drops through control devices to minimize such power consumption.

In addition to minimizing power costs, one normally minimizes other costs by reducing the volumetric flow rate. Typically (see Example 7.1) the capital cost of a pollution control device is roughly proportional to the volumetric flow rate through it and practically independent of the pressure at which it must operate. Therefore, if the gas stream is available at a high pressure, it is generally more economical to install a high-pressure gas cleaner to work when the stream's Q is small (because of the high pressure) than to reduce the pressure to atmospheric before doing the gas cleaning.

In many air pollution applications, the pollutant is emitted by an open source such as a furnace or a machine in a factory. In this case, a hood over the opening is normally connected to a vacuum system that collects the emissions. A fan then boosts this air flow to a pressure above atmospheric for cleaning and discharge (see Fig. 7.2). From Eq. (7.6) and from the preceding discussion of the relation of equipment cost to Q, it should be clear that there is a great financial saving if the hood connections in Fig. 7.2 can be made tight so that the volume of air sucked in is kept as small as possible. Ideally, this type of arrangement should be replaced by one in which there is a permanent, closed connection between the emission source and the collection system; often this is economically infeasible. Crocker has presented some interesting examples of how to minimize the volume of gas treated with resulting cost savings [11].

7.7 EFFICIENCY, PENETRATION, NINES

Consider a downstream control device that has a volumetric flow rate Q, inlet contaminant concentration c_0, and outlet contaminant concentration c_1. For such a device, the mass flow rate of contaminant into the device is Qc_0, and the contaminant flow out of it is Qc_1. If the volumetric flow rate changes as the waste stream passes through the device (which will happen if the temperature or humidity changes), then we will have to consider the inlet volumetric flow rate as Q_0 and outlet rate as Q_1. But the Qc product still represents the mass flow rate of contaminant.

Given these definitions, we can now define two new terms:

$$\text{Control efficiency or, simply, efficiency} = \eta = \frac{Q_0c_0 - Q_1c_1}{Q_0c_0} = 1 - \frac{Q_1c_1}{Q_0c_0} \tag{7.7}$$

$$\text{Penetration} = p = 1 - \text{efficiency} = \frac{Q_1c_1}{Q_0c_0} \tag{7.8}$$

Obviously, if $Q_0 = Q_1$, the Qs cancel out of these two definitions.

Why the two definitions? The efficiency is the same as the efficiencies we define in many other engineering disciplines: the ratio of what was done to the maximum that could be done. It is simple and intuitive. The penetration is the fraction not collected. We will see in the rest of this book that many calculations are easier and simpler in terms of the penetration than in terms of the efficiency. In the current air pollution literature it is becoming common to refer to the high efficiencies required for waste incinerators as "four nines," i.e., a control efficiency of 99.99 percent. New regulations are being proposed that will require "five nines," or "six nines" for very toxic materials.

If we have more than one control device in series, the mathematics of calculating their joint effect is much simpler if we use penetrations than if we use efficiencies.

Example 7.4. We wish to use four collectors in series. Each of the collectors has an efficiency of 93 percent. What is the overall efficiency of the group of four in series?

We really want to know overall efficiency,

$$\eta_{\text{overall}} = 1 - \frac{Q_4 c_4}{Q_0 c_0}$$

We can write Eq. (7.7) four times and eliminate the intermediate values as shown here:

$$\eta_{\text{overall}} = 1 - \frac{Q_4 c_4}{Q_0 c_0}$$

$$\eta_1 = 1 - \frac{Q_1 c_1}{Q_0 c_0}$$

$$\eta_2 = 1 - \frac{Q_2 c_2}{Q_1 c_1}; \qquad Q_2 c_2 = Q_1 c_1 (1 - \eta_2) = Q_0 c_0 (1 - \eta_1)(1 - \eta_2)$$

$$\eta_3 = 1 - \frac{Q_3 c_3}{Q_2 c_2}; \qquad Q_3 c_3 = Q_2 c_2 (1 - \eta_3) = Q_0 c_0 (1 - \eta_1)(1 - \eta_2)(1 - \eta_3)$$

$$\eta_4 = 1 - \frac{Q_4 c_4}{Q_3 c_3}; \qquad Q_4 c_4 = Q_3 c_3 (1 - \eta_4) = Q_0 c_0 (1 - \eta_1)(1 - \eta_2)(1 - \eta_3)(1 - \eta_4)$$

$$\eta_{\text{overall}} = 1 - \frac{Q_4 c_4}{Q_0 c_0} = 1 - (1 - \eta_1)(1 - \eta_2)(1 - \eta_3)(1 - \eta_4)$$

Here the ηs are all equal, so that we can solve easily, finding

$$\eta_{\text{overall}} = 1 - (1 - 0.93)^4 = 0.999976$$

We can solve this same problem by asking what is the penetration of the series of collectors. Here we know that the penetration of each individual collector is $(1 - 0.93) = 0.07$. Then,

$$p_{\text{overall}} = \frac{Q_4 c_4}{Q_0 c_0} = \frac{Q_4 c_4}{Q_3 c_3} \cdot \frac{Q_3 c_3}{Q_2 c_2} \cdot \frac{Q_2 c_2}{Q_1 c_1} \cdot \frac{Q_1 c_1}{Q_0 c_0}$$

$$p_{\text{overall}} = p_1 \cdot p_2 \cdot p_3 \cdot p_4$$

In this case, all of the ps are equal, so

$$p_{\text{overall}} = p^4 = (0.07)^4 = 2.40 \times 10^{-5}$$

This example shows that when we have collectors in series, the penetration is generally more practical and simpler to use than the efficiency. ■

7.8 HOMOGENEOUS AND NONHOMOGENEOUS POLLUTANTS

Some pollutants, like SO_2 and CO, are homogeneous. Every CO molecule is identical to every other CO molecule. Other pollutants such as particles with various sizes and hydrocarbons are not homogeneous. Fine particles are harder to capture, and more likely to cause health damage than coarse ones. Benzene is harder to destroy in an incinerator than hexane and is probably a more serious health threat; both are hydrocarbons. In both these cases, the regulations apply to and the control devices operate on the mixture, not on individual particle sizes or individual members of the hydrocarbon family.

Efficiency and penetration cause no confusion or difficulty when applied to homogeneous pollutants. However, when applied to heterogeneous pollutants, they are not always adequate. Typically, the efficiency of a particle-collecting device is a strong function of particle size. For most such devices (see Chapter 9), the efficiency is high for large particles and less for smaller ones.

Example 7.5. A waste stream contains particles of three sizes: large, medium, and small. These are present in equal quantities by weight in the gas stream. We pass this gas stream through a collector that is 99 percent efficient on large particles, 75 percent efficient on medium particles, and 30 percent efficient on small particles. What is the overall weight percent efficiency of this collector?

If we consider that the mass of gas that contains 0.999 kg of particles, we can compute the following:

Particle size	Incoming amount	x	Penetration	=	Outgoing amount
Large	0.333 kg		0.01		0.0033 kg
Medium	0.333 kg		0.25		0.0833 kg
Small	0.333 kg		0.70		0.2331 kg
Total	0.999 kg				0.3197 kg

And the overall efficiency $= 1 - p = 1 - (0.3197/0.999) = 0.680$. ■

Example 7.6. If in Example 7.5 we add another collector, identical to the first collector, downstream of it, what will the overall collection efficiency be?

Here again, we use a table, but the penetration for each particle size is the square of the penetration for one collector.

Particle size	Incoming amount	x	Penetration	=	Outgoing amount
Large	0.333 kg		$(0.01)^2$		0.0000 kg
Medium	0.333 kg		$(0.25)^2$		0.0208 kg
Small	0.333 kg		$(0.70)^2$		0.1633 kg
Total	0.999 kg				0.1841 kg

And the overall, two-collector efficiency $= 1 - p = 1 - (0.1841/0.999) = 0.816$. ■

Example 7.6 shows that the second collector is not as efficient as the first. For the second collector the efficiency is $1 - (0.1841/0.3197) = 42.4\%$. The reason is that it is treating a much more difficult gas stream than the first collector. The first collector took out most of the large particles, which are easy to collect. The stream leaving it contained mostly the small particles, which are hard to collect. For an individual particle size (e.g., for small particles) each collector is as effective as the other; but the ratio of particle sizes in the inlets is different. This is simply the consequence of the law of diminishing returns, applied to air pollution control.

Mathematically, the previous examples used

$$p_{\text{overall}} = \sum_{\substack{\text{over whole range}\\ \text{of particle sizes}}} p\begin{pmatrix}\text{for that}\\ \text{size range}\end{pmatrix} \Delta \begin{pmatrix}\text{total weight in}\\ \text{that size range}\end{pmatrix} \tag{7.9}$$

If we let the number of particle size intervals increase from the three in these examples to an infinite number, we can replace the summation by an integration, or

$$p_{\text{overall}} = \int_{\substack{\text{over whole range}\\ \text{of particle sizes}}} p\begin{pmatrix}\text{for that}\\ \text{size range}\end{pmatrix} d\begin{pmatrix}\text{total weight in}\\ \text{that size range}\end{pmatrix} = \int p(\text{size})\, dw \tag{7.10}$$

where p = overall penetration

$p(\text{size})$ = penetration for that particular particle size

w = weight fraction in that particle size range with values going from zero (smallest particle) to one (largest particle)

In Chapters 8 and 9 we will see examples of this kind of integration, following our discussion of particle size distributions.

For some purposes, this overall weight efficiency is really what we want to know; for many others it is not. For instance, small particles represent a greater health or visibility problem—pound for pound—than large ones. Different hydrocarbon types have different smog-forming tendencies and health effects. In rating hydrocarbon-control devices, the overall weight percentage control efficiency may not be an accurate measure of the reduction in damaging pollutants emitted to the atmosphere. For some situations, like hazardous waste incinerators, the regulations require some high destruction percentage of some one hydrocarbon component (called the *principal organic hazardous component,* or POHC), generally one that is hard to destroy. If, for example, the incinerator can destroy 99.99 percent of the benzene in the waste being incinerated, it is likely to have an even higher efficiency for the other components, like hexane, that are easier to destroy.

7.9 BASING CALCULATIONS ON INERT FLOWRATES

In most air pollution control applications the concentration of the pollutant is small enough that removing the contaminant makes a negligible change in the flow rate of the contaminated stream. However, some SO_2 streams from smelters (Chapter 11)

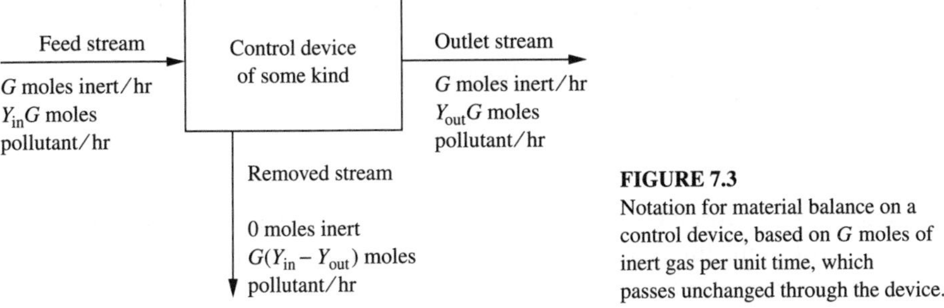

FIGURE 7.3
Notation for material balance on a control device, based on G moles of inert gas per unit time, which passes unchanged through the device.

have up to 40% SO_2, almost all of which is removed in the control device; and some air-hydrocarbon streams (Chapter 10) are also up to 40% hydrocarbon, almost all of which is removed in the control device. In calculating the behavior of such devices it is common practice to base all the calculations on the nonremoved (inert) part of the stream, because that does not change as the gas passes through the control device.

Schematically the situation is as illustrated in Fig. 7.3. The gas passing through the control device represents G mol of inert (nontransferred, noncollected) material per unit time. This amount is the same in the inlet and the outlet. The concentration of pollutant to be collected (or destroyed) is shown as a ratio to the inert, e.g.,

$$Y = \frac{\text{Mol or lb of pollutant}}{\text{Mol or lb of inert gas}}; \qquad G = \frac{\text{Mol or lb of inert gas}}{\text{Time}}$$

$$\text{Flow rate of pollutant} = YG = \left(\frac{\begin{array}{c}\text{Mol or lb}\\ \text{of pollutant}\end{array}}{\begin{array}{c}\text{Mol or lb}\\ \text{of inert gas}\end{array}} \cdot \frac{\begin{array}{c}\text{Mol or lb}\\ \text{of inert gas}\end{array}}{\text{time}} \right) \tag{7.11}$$

This approach is regularly used both in terms of mol/mol and terms of pound/pound. By common convention we use Y for the ratio of transferred material to inert material if the stream is a gas and X for the same quantity if the stream is a liquid. For humidity calculations the same function (the weight of moisture per unit weight of dry air) is given the symbol H.

We will see this same concept in the discussion of condensation of volatile organic compounds (VOC) (Chapter 10), the adsorption of VOC (Chapter 10 and Appendix E), and the absorption of sulfur compounds (Chapter 11 and Appendix F). It also appears in all humidity and evaporative cooling calculations and appears briefly in Sec. 5.3.5.

7.10 COMBUSTION

Most air pollutants are created and/or released in processes involving combustion. In Table 1.1 we saw that the categories of emission source are: transportation (mostly from combustion of motor fuels), fuel combustion, industrial processes (many of which are combustion processes), solid waste disposal (mostly solid waste incineration), and miscellaneous (of which the most important type is forest fires). Thus, at least a rudimentary understanding of combustion is needed to understand pollution sources, as well as some pollution control devices.

7.10.1 What Burns?

Appendix C discusses fuels, providing much more detail than this section. There we see that most of the things that burn are compounds of carbon and hydrogen. Most combustion reactions are of the form

$$C_xH_y + \left(x + \frac{y}{4}\right) O_2 \rightarrow x\ CO_2 + \frac{y}{2} H_2O \tag{7.12}$$

where x is zero for pure hydrogen, y is zero for pure carbon, and common fuels have both carbon and hydrogen. For example, natural gas is practically pure methane, CH_4, for which $x = 1$ and $y = 4$. Combustion means reaction with oxygen, normally from the air. In a few cases it means reaction with pure oxygen.

The other substances that will burn are sulfur, phosphorus, metals like magnesium, and iron at high temperatures. Compounds like ammonia, NH_3, contain no carbon but have enough H that they can burn. There are some other materials that burn, but at least 99% of the combustion in the world is some form of conversion of carbon and hydrogen to carbon dioxide and water.

7.10.2 Heat of Combustion

If we start with a known amount of fuel and the appropriate oxidizer (normally oxygen), react them, and then cool the products of combustion to the starting temperature, we will have removed a finite amount of energy in the form of heat, called the heat of combustion. Table 7.1 lists heat of combustion values for many common fuels, as well as many other useful data about these fuels.

For methane, the principal component of natural gas, we see from Table 7.1 the heat of combustion is 21 502 Btu/lb. There are two common definitions of heat of combustion. The *higher heating value* definition assumes that the water produced by combustion is condensed, thus giving up its latent heat of condensation. The *lower heating value* assumes that the water leaves the combustor as a gas, and hence the lower value is less than the higher heating value by the amount of that latent heat of condensation. Table 7.1 shows lower heating values. For common hydrocarbon fuels it is roughly 19 000 Btu/lb.

7.10.3 Explosive or Combustible Limits

If we start with 99 percent methane and 1 percent air and supply a spark, the mixture will not burn. There is not enough air present, and the mixture is spoken of as being "too rich." Similarly, a mixture of 1 percent methane and 99 percent air will not burn. There is not enough methane present, and the mixture is spoken of as being

TABLE 7.1
Combustion data for hydrocarbon fuels

Fuel	Molecular weight	Heat of vaporization, Btu/lb	Heat of combustion gas-gas, Btu/lb	Stoichiometric mixture, vol %	Explosive limit, % stoichiometric,[a] Lean LEL	Rich UEL	Spontaneous ignition temperature, °F[b]	Adiabatic flame temperature, °F
Acetone	58.1	224		4.97	59	233	1042	3820
Acetylene	26.0		20 734	7.72	31		581	
Benzene	78.1	169	17 446	2.71	43	336	1097−	4150
n-Butane	58.1	166	19 655	3.12	54	330	807	4060
1-Butene	56.1	168	19 475	3.37	53	353	830	4175
Carbon monoxide	28.0	91		29.5	45	251	1128	
Cyclohexane	84.2	154	18 846	2.27	48	401	518	4050
n-Decane	142.3	119	19 175	1.33	45	356	449	4115
Ethane	30.1	210	20 416	5.64	50	272	882	4040
Ethene	28.1	208	20 276	6.52	41	610	914	4275
Ethyl alcohol	46.1	368		6.52			738	
n-Heptane	100.2	136	19 314	1.87	53	450	477−	3985
n-Hexane	86.2	144	19 391	2.16	51	400	501−	4030
Hydrogen	2.0	194	51 571	29.5			1060−	
Hydrogen sulfide	34.1	237		12.24			554−	
Isopropyl alcohol	60.1	286		4.44			852	
Methane	16.0	219	21 502	9.48	46	164	1170−	
Methyl alcohol	32.0	473		12.24	48	408	878	
n-Nonane	128.3	124	19 211	1.47	47	434	453	
n-Octane	114.2	129	19 256	1.65	51	425	464−	
n-Pentane	72.1	154	19 499	2.55	54	359	544	4050
Propane	44.1	183	19 929	4.02	51	283	940−	4050
Toluene	92.1	156	17 601	2.27	43	322	1054−	4220

Source: Ref. 12.

[a]LEL, lower explosive limit; UEL, upper explosive limit.

[b]The minus signs following values indicate that slightly lower values have been reported.

"too lean." The borders between mixtures that will burn and those that will not are called the "lean limit" or *lower explosive limit* (LEL) and the "rich limit" or *upper explosive limit* (UEL). (The terms *combustible limits* and *explosive limits* are used interchangeably.)

Example 7.7. What are the stoichiometric mixture, lean limit (LEL), and rich limit (UEL) for the combustion of methane in air?

By using the values in Table 7.1, we can estimate that for methane and air the stoichiometric mixture contains 9.48 volume % methane and that the range of combustible mixtures for methane and air is

$$9.48\% \cdot 0.46 = 4.36 \text{ volume } \% \qquad \text{to} \qquad 9.48\% \cdot 1.64 = 15.55 \text{ volume } \%$$

If we wished to know the weight percents corresponding to these volume percents, we could compute, for example, at the stoichiometric mixture,

$$\text{wt } \% \text{ methane} = \frac{y_i M_i}{\Sigma y_i M_i} = \frac{0.095 \cdot 16}{0.095 \cdot 16 + 0.905 \cdot 29} = 0.055 = 5.5 \text{ wt } \%$$

The vast majority of gas compositions in combustion calculations are stated as volume % (same as mol %), but in a few cases the weight % is used, and in discussion of automotive engines most often one sees the (air/fuel) ratio (A/F), in lb/lb. For methane at the stoichiometric ratio we compute that ratio by assuming that we have one mol of fuel-air mixture, for which

$$\frac{\text{A}}{\text{F}} = \frac{n_{\text{air}} M_{\text{air}}}{n_{\text{fuel}} M_{\text{fuel}}} = \frac{(1 - 0.0948) \cdot 29}{0.0948 \cdot 16} = 17.3 \frac{\text{lb air}}{\text{lb fuel}}$$ ■

No one is able to calculate combustible limits without using experimental data. The measured combustible limits are, to some extent (e.g., ±1 percent), a function of the geometry of the device in which the test is made. They are not a thermodynamic property like density or temperature, which can be measured in a device at equilibrium; rather they are an inherently kinetic property, which can only be measured in a device in which the rapid and complex chemical reactions of combustion are taking place.

These relations are shown on Fig. 7.4. The curve is the calculated adiabatic flame temperature for various mol fractions of methane in air. For methane mol fractions less than stoichiometric, all the methane burns, and the temperature rise is dependent on the amount of methane present. For methane mol fractions more than stoichiometric, all the available oxygen is used, and the temperature rise depends on the amount of oxygen present. For concentrations below the LEL (4.36%) the mixture is too lean to burn, and for concentrations above the UEL (15.55%) the mixture is too rich to burn. The computed temperatures at the two limits are quite different, about 2000 and 3000°F. Lewis and von Elbe [13] report that for most hydrocarbons the temperature at the LEL is about 2300 to 2400°F, and that the temperatures at the UEL are generally higher and less uniform. The flammable region becomes wider if the gas is preheated; the values regularly shown are for mixtures at room temperature.

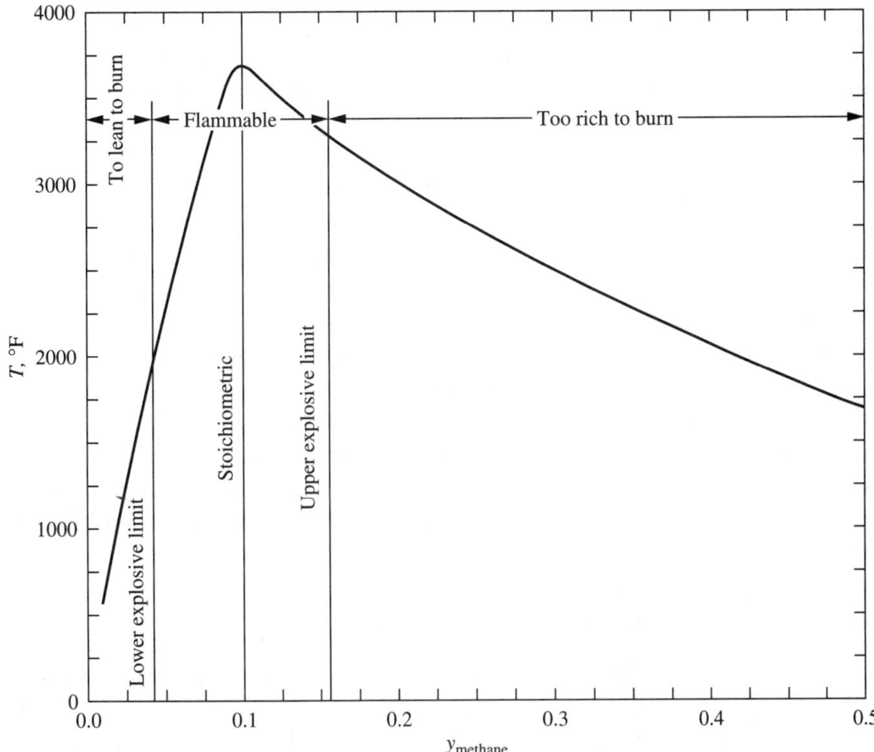

FIGURE 7.4
Calculated adiabatic flame temperatures for various mixtures of methane and air, starting at 32°F = 0°C, and explosive limits. The calculated temperatures depend on the values chosen for the heat capacities of the combustion products; these use the high temperature values in Ref. 14.

Stoichiometric concentrations and combustible limits are almost always reported in volume % = mol %, and vary widely with the molecular weight of the fuel. But on a weight basis they vary much less. On a weight basis aliphatic hydrocarbons have stoichiometric values between 5.5 and 6.4 wt%, and LELs between 2.7 and 3.4 wt%. The UELs are slightly more variable, between 14 and 22 wt% for the C_2 to C_{10} range; see Problem 7.11.

7.10.4 Equilibrium in Combustion Reactions

No chemical reactions go 100 percent to completion; there is always some unreacted material. For the industrially important reaction

$$SO_2 + \tfrac{1}{2}O_2 \rightleftarrows SO_3$$

one can show by straightforward chemical thermodynamic calculations that if the oxygen is present at a pressure of 1 atmosphere, then at a temperature of about 1500°F the conversion of SO_2 to SO_3 at equilibrium is 50 percent. The methods of making this calculation are described in detail in most books on thermodynamics, e.g., Hougen, Watson, and Ragatz [15]. (See Fig. 11.2.)

Fortunately, for most combustion reactions, the reaction equilibrium is so strongly in favor of products that one may assume that the reactions *at equilibrium* are complete. For the most careful work we must reconsider this simplification. Flagan and Seinfeld [16] show the computation of the adiabatic flame temperature and exhaust gas composition for a fuel oil, first assuming that only CO_2 and H_2O are present in the combustion products (complete reaction) and then taking into account equilibrium (and the presence of CO and H_2 in the combustion products). They find that (1) the computed adiabatic flame temperature is 95 K less in the second case than the first and (2) the calculated mole fraction ratios (CO/CO_2) and (H_2/H_2O) are 0.092 and 0.018, compared to zero for assumed complete combustion. This result was computed for a temperature of 2261 K = 3610°F; for lower temperatures this effect is much smaller and the assumption of complete conversion at equilibrium is much better.

7.10.5 Combustion Kinetics, Burning Rates

Equation (7.12) shows the general reaction for combustion. This reaction is not instantaneous. All reactions, even explosions and nuclear detonations, require some amount of time to take place. If the reaction is stopped before it is complete, then not all the fuel will be consumed (nor will all the pollutants be burned up). For almost all chemical reactions the reaction rate increases very rapidly with increasing temperature. Thus, the general way to carry out the destruction of pollutants by combustion is to bring the mixture of pollutants and air to a high enough temperature and hold them at this temperature for a long enough time so that the reaction shown in Eq. (7.12) occurs.

One of the basic theoretical problems of physical chemistry is to compute the reaction rates of chemical reactions on the basis of studies of the underlying mechanisms. Although considerable progress has been made in this direction, the results are generally complex and it is difficult to predict the rate of one reaction from data taken on another. For a few, well-studied reactions (e.g., $H_2 + \frac{1}{2}O_2 \rightarrow H_2O$) we have fairly complete descriptions of the reaction rate and the underlying mechanism [13]. For most other reactions we use empirical rate equations (for which we do not have underlying mechanistic explanations) or rules of thumb to estimate how hot the material must be and how long it must be held at that temperature to complete the reaction.

Furthermore, the balanced chemical equation for a combustion reaction is unlikely to be a correct detailed description of how the reaction actually proceeds. For example, the balanced chemical reaction equation for burning carbon monoxide is ($CO + \frac{1}{2}O_2 \rightarrow CO_2$). From that equation one would assume that a mixture of carbon monoxide and oxygen would burn easily; it does not. It is practically impossible to burn carbon monoxide in the absence of trace amounts of water. The explanation is that the hydroxyl free radical (OH) plays a crucial role in the reaction. Free radicals have unpaired electrons; the most important ones for combustion are OH, H, O, N, CH_3, and those produced by breaking up higher hydrocarbons.* None of these has a significant concentration (more than parts per billion) at room temperature, but they exist in significant quantity and are very chemically active at flame temperatures. Only a small number of OH free radicals, acting as catalysts for rapid chain reactions, control the rate of oxidation of carbon monoxide to carbon dioxide; without them the rate is effectively zero. All high-temperature gas reactions are apparently free radical reactions, as discussed further in Chapter 12.

Another illustration of this idea is provided by the burning of methane. In even the cleanest methane burner, some small amounts of chemicals with higher molecular weights than methane, e.g., ethane, are found in the gases leaving the burner. The first step in methane combustion is probably the removal of one of methane's four hydrogens. The remaining CH_3 radical can react with another methane molecule to produce higher-molecular-weight materials. The CH_3 radical does not appear in the balanced equation for burning methane, but it is certainly present in methane flames. Thus, the balanced stoichiometric equation is the correct bookkeeping of the overall reaction, but it tells us little about the details of how the reaction actually proceeds or what influences its rate.

7.10.6 Mixing in Combustion Reactions

For any combustion reaction to proceed, the fuel and the oxidizer (normally oxygen from the air) must be mixed. If the fuel and oxidizer are not properly mixed, then even if there is enough air, combustion will not be complete because some of the fuel will not get together with air in the high-temperature combustion zone. The importance of mixing is illustrated by Fig. 7.5 on page 184, which shows the theoretical composition of flue gas for burning a hydrocarbon fuel as a function of the air-fuel ratio. Two sets of curves are shown: one for perfect mixing, the other for poor mixing. With sufficient excess air, all the CO and H_2 will be used up, even with poor mixing. If the temperature were high enough (so that the reaction rate were high enough) or the retention time at high temperature long enough, then we could get to complete usage of the fuel (i.e., no unburned CO or H_2) at all mixture conditions higher than stoichiometric. The real situation is the poor mixing one shown. In all industrial furnaces we transfer heat away from the flame. That is the purpose of the furnace—to transfer heat from the flame to the substance being heated. Also the time for combustion is limited in industrial furnaces. More time requires a bigger furnace and a higher capital cost. Thus, in practice one tries to get as good mixing as possible, to minimize the amount

*Many books add a dot to the symbol for a free radical, e.g., OH·, to remind us that this is a free radical with an unpaired electron, and not an ordinary chemical. Most of the air pollution literature does not, and those dots will not be shown in this book.

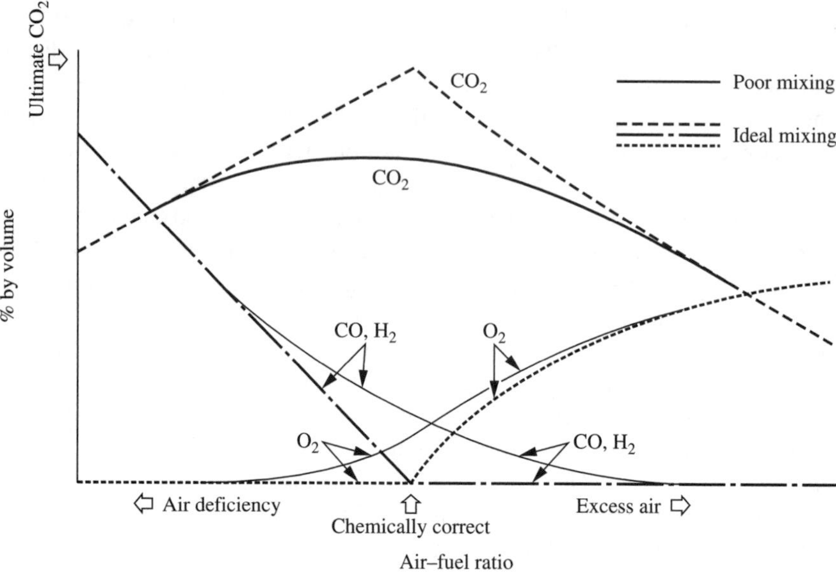

FIGURE 7.5
Effect of air-fuel ratio and quality of mixing on the composition of combustion gases. (From Ref. 17.)

of valuable fuels (CO and H_2) in the exhaust gas, and simultaneously to minimize the amount of air pollution caused by these unburned fuels. (Good mixing can increase NO_x emissions! See Chapter 12.)

7.10.7 Flame Temperature

Flames do not have uniform temperatures. Small flames like those in candles and cigarette lighters can have differences of 400°C = 720°F from one part of the visible flame to another. Larger flames are more uniform, but none is really uniform. The *adiabatic flame temperatures* shown in Table 7.1 correspond to combustion with no heat loss (as, for example, inside an insulated ceramic sleeve, which takes up the same temperature as the flame). The peak temperature of open flames depends on the following:

1. The fuel and oxidizer used
2. The size of the flame
3. The degree of fuel-air premixing
4. The amount of fuel-air preheat

Fuel-oxygen flames are much hotter than fuel-air flames, because the nitrogen in the combustion air absorbs heat, thus dividing the heat released by the combustion among more molecules, with less heat available for each molecule than in a fuel-oxygen flame. This is true for any fuel. The highest temperature flames are produced by acetylene and oxygen. They are hotter than the flames of ordinary fuels like methane with oxygen because of the extra energy stored in the acetylene triple bond, which is released on combustion, in addition to the energy released by converting carbon and hydrogen to carbon dioxide and water.

For any fuel, the combustion temperature depends on the size and shape of the flame. The flame generates heat by chemical reaction and loses it to the surroundings, partly by thermal radiation, partly by conduction, and partly by mixing with the surrounding air. The larger the surface area of the flame per unit of heat release, the faster the flame will lose energy, and thus the lower the flame peak temperature will be. For a steady-state, continuous flame (like a burning candle or Bunsen burner, but not like an explosion or the combustion in an automobile engine), we may write

$$\begin{gathered}\text{Heat generated} = \text{heat transferred to surroundings}\\ \dot{m}_{\text{fuel}}\,\Delta h_{\text{combustion}} = UA(T_{\text{flame}} - T_{\text{surroundings}})\\ T_{\text{flame}} = T_{\text{surroundings}} + \frac{\dot{m}_{\text{fuel}}\,\Delta h_{\text{combustion}}}{UA}\end{gathered} \tag{7.13}$$

Here the heat transfer coefficient U is a combined radiation, conduction, and convection coefficient, and $\Delta h_{\text{combustion}}$ is the enthalpy change of combustion.

The surface area of the flame, A, depends strongly on how much premixing of fuel and oxidizer there is. There is practically no fuel-air premixing for a device like a butane cigarette lighter, Fig. 7.6. In these lighters a low-velocity jet of butane

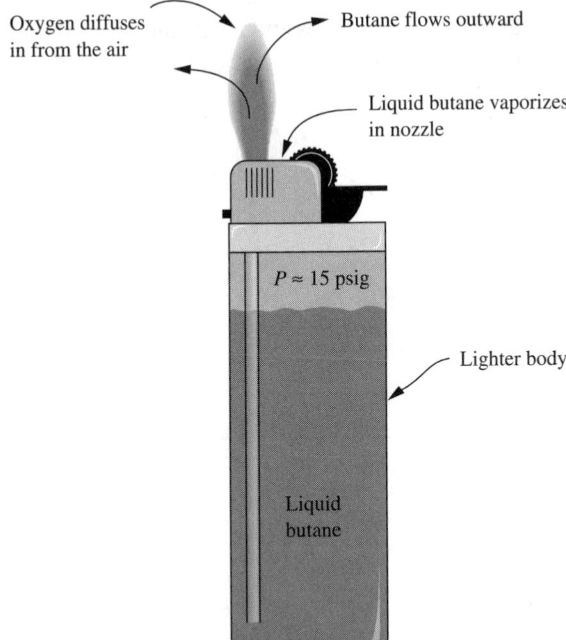

FIGURE 7.6
Butane cigarette lighter: an example of a diffusion flame.

vapor flows into the atmosphere and is lighted by a spark from a flint. The flame is large for the amount of fuel being burned, because at the surface of the flame the butane is flowing outward while the oxygen needed for combustion diffuses in, counter to the butane and combustion product flow. The flame locates where the inward flow of oxygen by diffusion is enough to support combustion. The observed peak temperatures in such flames are about 2000°F. A flame like that in Fig. 7.6, in which there is practically no fuel-air premixing, is called a *diffusion flame.* The flames of candles, wood matches, and campfires are almost pure diffusion flames; their temperatures are roughly 1900 to 2000°F. All of them are yellow, as discussed shortly.

The opposite of a diffusion flame is a *premixed flame,* in which the air and fuel are completely mixed before ignition. The flames in automobile engines are totally premixed. In them the liquid gasoline is first formed into a spray of fine droplets and is then mixed with air, either in a carburetor or a fuel injector. The mixture is then heated by compression (discussed shortly) to vaporize the liquid fuel before it is ignited. When the spark ignites the fuel, the flame front passes through the premixed air and fuel. Combustion is much more rapid than in a diffusion flame because there is no delay waiting for the air to diffuse in to mix with the fuel. As a result the maximum temperatures are higher than for a diffusion flame. The temperatures in internal combustion engines are estimated at about 4000°F.

Totally premixed burners are seldom used in industrial furnaces because the partly premixed burners discussed shortly are less expensive and work very well. The flames in most household gas appliances are partly premix and partly diffusion. The fuel and about 25 percent of the air are mixed in the venturi and mixing tube (Fig. 7.7). Then that mixture flows out of the burner holes and meets the surrounding air. The inner part of the flame is a fuel-rich premixed flame; the rest of the flame is

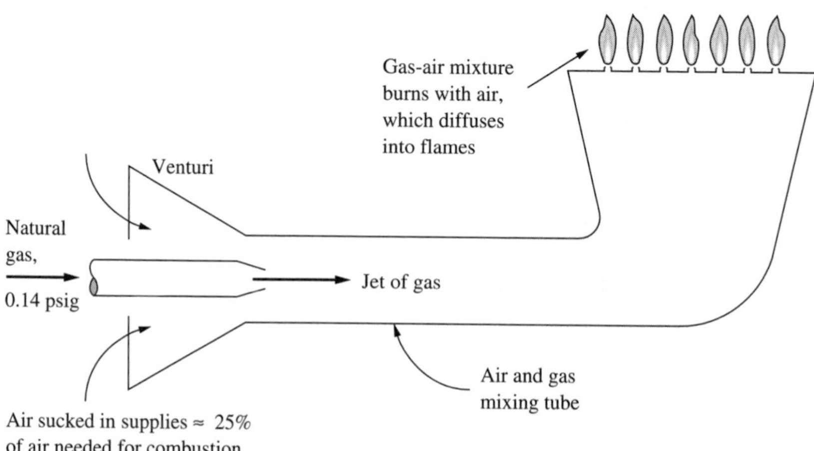

FIGURE 7.7
Gas burner of the type used in residential gas stoves, water heaters, and furnaces: an example of a partly premixed flame.

a diffusion flame, with the oxygen diffusing into the flame from the surrounding air against the net outflow of combustion products. The large industrial versions of the burner in Fig. 7.7 have highly turbulent flows that make the mixing much faster than the diffusional mixing in the individual flames of a gas stove or other residential gas appliance. The temperatures reached in burners like that in Fig. 7.7 are intermediate between those of a pure diffusion flame and a pure premixed flame, typically about 2400°F. Partly premixed flames are blue, not yellow.

The student has probably observed that flames on gas stoves do not put soot on the bottom of a pot, but those from a campfire or a candle do. The reason is that in a diffusion flame there are very small carbon particles in the center of the flame which are completely burned up at its periphery, whereas in a 25 percent premixed gas flame enough oxygen is in the inside of the flame that no such carbon particles are present. These very small carbon particles, glowing in the heat of the flames, emit the yellow light that is characteristic of candles, butane lighters, and campfires. A cold pot placed within a yellow flame (not above the flame) cools the flame enough to prevent complete combustion of these carbon particles; they deposit as soot on the pot.

In tuning a household stove, water heater, or furnace, the service person first closes the air inlet shutters gradually to find the shutter setting at which yellow tips appear on the flames, indicating that there are some carbon particles in the flames, and then opens the shutters to make the yellow tips just disappear, indicating that there is just enough premix air for clean combustion. The reader can safely try that on any gas stove. (In an oxyacetylene flame the intense white light is caused by carbon particles glowing at the very high temperature of these flames. Even though this a premixed flame, these carbon particles appear because acetylene, C_2H_2, is very carbon-rich, compared with most other gaseous fuels.)

A household propane torch of the type used to solder plumbing fittings uses the high pressure ($\approx$ 100 psig) in the propane container and a jet mixing arrangement somewhat similar to that in Fig. 7.7 to provide almost complete premixing of the propane and air. As a result of this premixing, the flame is much smaller than the butane lighter flame per unit of fuel burned and the observed temperatures in such flames are about 3000°F in the small, bright inner cone. The difference of 600°F between the butane lighter and the propane torch, for practically the same fuel, derives solely from the much higher value of $\dot{m}_{\text{fuel}}/A$ due to the difference in fuel-air mixing.

With a diffusion flame, increasing the size of the burner and fuel flow rate will not change the peak temperature much, because the flame must increase its surface area as much as needed to get the necessary oxygen to flow into it by diffusion. Thus a diffusion flame will have roughly the same heat release rate *per unit surface area,* $(\dot{m}_{\text{fuel}}/A)$, independent of the fuel flow rate. The flame size will increase or decrease with increasing or decreasing fuel flow rate, to keep this value constant. But a premixed flame has no such need, so that it will have roughly the same heat release rate *per unit volume* ($\dot{m}_{\text{fuel}}/V$ is more or less independent of its size), and $(\dot{m}_{\text{fuel}}/A)$ increases with increasing fuel flow rate.

The flames in most large industrial furnaces are partly premixed. A premixed flame is much smaller, for equal heat release, than a diffusion flame. Thus a premixed flame furnace is smaller and less expensive than a comparable output diffusion flame furnace. In the industrial equivalent of the burner in Fig. 7.7, both the air premixed with the fuel, called *primary air,* and the air that mixes in after ignition, called *secondary air,* are supplied by fans or blowers, which provide considerable turbulence to improve the mixing. Large industrial furnaces have temperatures up to 3500°F. Household furnaces, water heaters, etc. do not use fan-driven air supplies. This feature makes them somewhat larger than they would be with fan-driven air, but the extra cost of the larger furnace is offset by the low cost, simplicity, and inherent safety of the low-pressure, gas-driven, 25 percent premix burners normally used.

Both large industrial furnaces and motor vehicles preheat their air-fuel mixtures before igniting them; most other combustion users do not. That preheating causes the maximum temperatures reached in these types of combustion to be higher than those reached in other types of combustion.

In a standard gasoline engine, the air-gasoline mixture is compressed to about $\frac{1}{7}$ to $\frac{1}{10}$ of its initial volume before the spark ignites it. In an engine with a 7 : 1 compression ratio, if the compression were reversible and adiabatic, the computed temperature increase would be just over 600°F; observed temperature increases are somewhat less because the process is not completely adiabatic. The temperature increase on burning the fuel is only slightly affected by this increase in starting temperature, which means that the expected final combustion temperature is 600°F higher than would be expected burning the same fuel-air mixture in a noncompressed, totally premixed system.

Figure 7.8 compares the flows in an ordinary residential hot-air furnace to the flows in a large, modern industrial furnace. In the typical residential furnace the combustion products leave the furnace at about 750°F. They are mixed with air to keep the temperature in the chimney at a safe value, about 250°F (using a *draft hood*), and sent to the chimney. In a modern large industrial furnace the combustion gases leave the furnace at about 750°F and pass through a heat exchanger, where they preheat the incoming combustion air and reduce their own temperature to about 250°F. In so doing they heat the incoming air to about 570°F. (The heat exchangers are most often rotating wheel recuperators, known as *Ljungstrom preheaters* after their inventor.) Preheating the air increases the average and the peak flame temperatures in the industrial furnace by about 500°F compared with those in the residential furnace, assuming the same fuel and air-fuel ratio (see Problem 7.17).

7.10.8 Combustion Time

The combustion times for small flames are quite short. For larger flames they are longer.

Example 7.8. Estimate the time for complete combustion (*a*) in a gas stove, (*b*) in an auto engine, and (*c*) in a large coal-fired boiler.

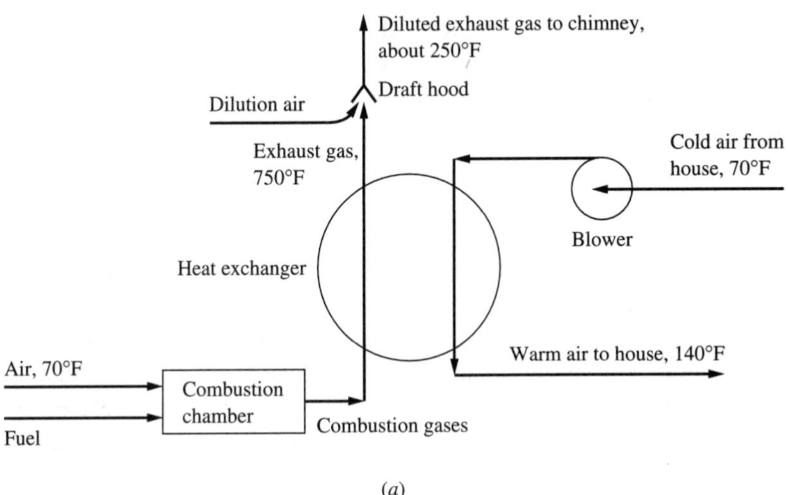

(*a*)

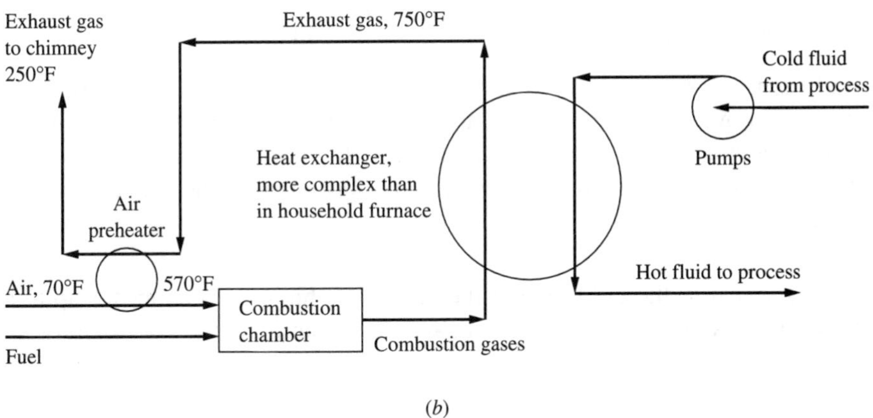

(*b*)

FIGURE 7.8
Comparison of air and combustion gas flows in (*a*) a household hot air furnace and (*b*) a large, preheated industrial furnace, showing why the combustion temperatures are about 500°F hotter in a preheated industrial furnace than in a furnace of comparable size without preheat.

(*a*) In a gas stove the flames stand still while gas and premix air pass through them and oxygen diffuses in from the air to complete the combustion. The gas velocity is comparable to the *laminar flame speed* (the speed at which a flame propagates into a fuel-air mixture that has no turbulence), which for methane is about 1 ft/s. The thickness of the flame is about $\frac{1}{16}$ of an inch = 0.005 ft, so the estimated combustion time is

$$t = \frac{L}{V} = \frac{0.005 \text{ ft}}{1 \text{ ft/s}} = 0.005 \text{ s} = 5 \text{ ms}$$

(b) In an automobile engine at 2000 RPM the time for one revolution is 0.03 s. The combustion takes place in perhaps one-twelfth of this time (see Example 13.5), so the estimated time of the combustion is 0.0025 s = 2.5 ms. Here the distance traveled by the flame front is about 2 inches, so that the flame speed must be

$$V_{\text{flame}} \approx \frac{2\text{ in.}}{0.0025\text{ s}} = 800\ \frac{\text{in.}}{\text{s}} = 66.7\ \frac{\text{ft}}{\text{s}} = 20.3\ \frac{\text{m}}{\text{s}}$$

This speed is reasonable for a premixed flame, but would not be plausible for a diffusion flame.

(c) In a typical 500-MW coal-fired power plant that uses pulverized coal [18], the gas flow through the system is about 1.2 M standard cubic ft per minute, which is equivalent to 5.3 M actual cubic ft per minute at the firebox outlet temperature of 1800°F. The main firebox is roughly 46 ft × 46 ft × 165 ft high. The flow is vertically upward with the average velocity given by

$$V_{\text{avg}} = \frac{Q}{A} = \frac{5.3 \times 10^6\ \text{ft}^3/\ \text{min}}{(46\ \text{ft})^2} = 2505\ \frac{\text{ft}}{\text{min}} = 42\ \frac{\text{ft}}{\text{s}} = 13\ \frac{\text{m}}{\text{s}}$$

and the time to traverse the 165-ft-high firebox is 165/42 = 4 s. This value somewhat overstates the combustion time, because much of the gas is admitted above the bottom of the furnace and because this value is based on the outlet temperature. At an average temperature of 3000°F one would calculate a time of 2.6 s. ■

We see that in the utility boiler the gas is at a high temperature for up to a thousand times as long as in a kitchen stove or an auto engine. The reasons are twofold:

1. The larger the burner, the longer the gas stays at high temperature. In a premixed flame the heat generated per cubic foot of flame is more or less independent of the equipment size for a given fuel; instead it depends on the burning rate of the fuel-air mix. The rate of heat removal from the flame is dependent on the surface area of the flame (or the firebox). So the time taken to remove the heat, from the peak temperature to some appropriate outlet temperature, is roughly proportional to (volume/area) or the length of the flame.
2. Although the coal particles in a modern utility boiler are ground to the consistency of face powder, they still take longer to burn than do molecules of natural gas or gasoline, so a longer *residence time* in the firebox must be provided. Coal-fired boilers are somewhat larger than natural gas or oil-fired boilers of the same capacity.

Table 7.2 summarizes the estimated times and temperatures for various kinds of burners.

TABLE 7.2
Estimated peak temperatures and combustion times in various kinds of burners

Type of flame or burner	Estimated approximate temperature, °F	Estimated approximate combustion time, s
Candle, campfire, butane lighter	1900–2000	0.005–0.01
Kitchen gas stove, hot water heater, furnace	2400	0.005
Propane torch	3000	0.001
Medium size industrial furnace without preheating	3000	1–2
Large coal-fired furnace with preheating	3500	2–4
Oxyacetylene torch	5000	0.001
Automobile engine (at 2000 RPM)	4000	0.0025

7.10.9 The Volume and Composition of Combustion Products

In air pollution control engineering we often need to know the volumetric flow rate and/or composition of gases produced by combustion. The computational scheme is summarized in Fig. 7.9 on page 192, and discussed in the next two examples.

The computation is easiest for the hydrocarbons for which we can write a simple molecular formula, e.g., CH_4, C_6H_6, or C_6H_{14}. These can all be written as C_xH_y, where, for methane, $x = 1$ and $y = 4$, etc. For all such fuels, if the fuel fed to the combustor in Fig. 7.9 is 1 mol, then, *assuming complete combustion,* the outlet gas will contain x mol of CO_2 and $y/2$ mol of H_2O. In addition, it will contain nitrogen and moisture that came in with the combustion air. The stoichiometric oxygen requirement is

$$n_{\text{stoichiometric oxygen}} = x + \frac{y}{4} \tag{7.14}$$

which we abbreviate n_{stoich}. (Here we have $y/4$, because each H atom requires $\frac{1}{4}$ mol of O_2 but produces $\frac{1}{2}$ mol of H_2O.) The dry air flow to the burner is

$$n_{\text{dry}} = n_{\text{stoich}} \left(\frac{1+E}{0.21} \right) \tag{7.15}$$

where the 0.21 accounts for the fact that air is only 21 percent oxygen, and E is the fraction of excess air* introduced to the burner. The total amount of air introduced is

*Almost all combustion devices introduce more than the stoichiometric amount of air, to make sure that there is enough oxygen for complete combustion. The amount of air beyond the stoichiometric requirement is called *excess air.* A typical value is 20%.

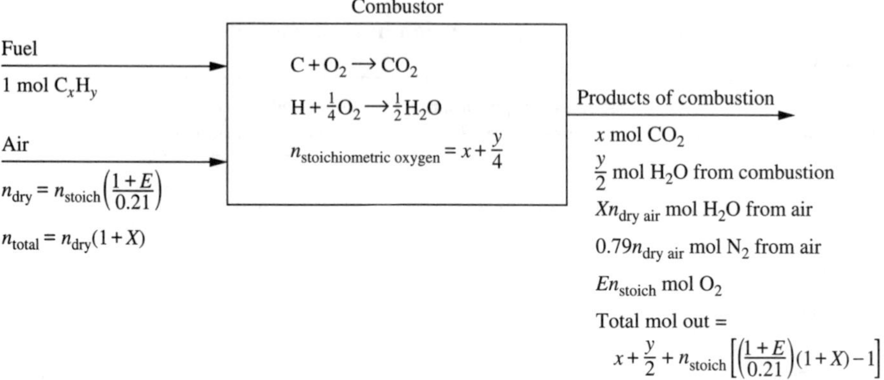

FIGURE 7.9
Illustration of the process of calculating the volume and composition of the gas from combustion of a simple hydrocarbon, C_xH_y. Complete combustion with excess air is assumed. See Example 7.9.

$$n_{total} = n_{dry}(1 + X) = n_{stoich}\left(\frac{1 + E}{0.21}\right)(1 + X) \tag{7.16}$$

where X is the humidity, expressed as (mol H_2O/mol dry air). Adding all the flows out and canceling like terms, we find that

$$n_{total\ out} = x + \frac{y}{2} + n_{stoich}\left[\left(\frac{1 + E}{0.21}\right)(1 + X) - 1\right] \tag{7.17}$$

Example 7.9. Methane is burned in air with 20 percent excess air and air moisture of 0.0116 mol/mol dry air. What are the flow rate of combustion air and the composition and flow rate of the combustion products?

Here we choose as our basis 1 mol of methane so that all calculated flows are per mol of methane. $E = 0.20$ and $X = 0.0116$, so that

$$n_{CO_2} = x = 1$$

$$n_{H_2O,combustion} = \frac{y}{2} = 2$$

$$n_{stoich} = x + \frac{y}{4} = 1 + \frac{4}{4} = 2$$

$$n_{dry} = n_{stoich}\left(\frac{1 + E}{0.21}\right) = 2\left(\frac{1.2}{0.21}\right) = 11.43$$

$$n_{total} = n_{dry}(1 + X) = 11.43 \cdot 1.0116 = 11.56$$

$$n_{N_2} = 0.79n_{dry} = 0.79 \cdot 11.43 = 9.03$$

$$n_{O_2} = En_{stoich} = 0.2 \cdot 2 = 0.4$$

$$n_{H_2O,total} = n_{H_2O,combustion} + Xn_{dry} = 2 + 0.0116 \cdot 11.43 = 2.13$$

$$n_{total\ out} = 1 + \frac{4}{2} + 2\left[\left(\frac{1 + 0.2}{0.21}\right)(1 + 0.0116) - 1\right] = 12.56$$

One may check this answer by summing the individual mols out, $1 + 2.13 + 9.03 + 0.4 = 12.56$. Thus for one mol of methane there are 11.56 mols of combustion air and 12.56 mols of product gases. The mole fraction of any component of the product gases is

$$y_i = \frac{n_i}{n_{\text{total out}}} \tag{7.18}$$

(Observe that here y_i is the mol fraction of component i, not to be confused with the y in C_xH_y.) For example, the mol fraction for CO_2 is $(1/12.56) = 0.0796$. ■

This case, in which we can write a molecular formula for the fuel, which contains only C and H, is the easiest. For solid fuels (coal, wood, biomass) the fuel composition is normally expressed **by weight** of C, H, N, S, O, and ash on a dry basis. In principle one could convert all these to a formula in terms of $C_xH_yN_z$, etc., but that is rarely done. The common procedure, similar to that preceding, is shown next.

Example 7.10. A "typical Pittsburgh seam coal" has the following *ultimate analysis* by weight: hydrogen, 5.0%; carbon, 75.8%; nitrogen, 1.5%; sulfur, 1.6%; oxygen, 7.4%; ash, 8.7% (see Appendix C). It is burned with 20% excess air with humidity 0.0116 mol/mol dry air, and combustion is complete. Determine the amount and composition of the gas produced.

Instead of choosing one mol of feed, we choose 100 g of dry coal. Then the mols of the individual components are these: $n_C = 75.8/12 = 6.32$, $n_H = 5.0/1 = 5.0$, $n_{N_2} = 1.5/28 = 0.054$, $n_S = 1.6/32 = 0.050$, and $n_{O_2} = 7.4/32 = 0.231$. Then we can see by inspection that the mols of CO_2, H_2O, and SO_2 formed by combustion are 6.32, $5/2 = 2.50$, and 0.050. The nitrogen is assumed to exit as N_2 (although some of it actually exits as NO or NO_2, see Chapter 12) so that the mols of nitrogen, from the fuel, are 0.054.

The mols of oxygen needed for combustion are those needed to oxidize the C, H, and S, less that contained in the fuel. In this case

$$n_{\text{stoich}} = n_C + \frac{n_H}{4} + n_S - n_{O_2} = 6.32 + \frac{5}{4} + 0.050 - 0.231 = 7.39 \ \frac{\text{mol}}{100 \text{ g dry coal}}$$

and

$$n_{\text{dry}} = n_{\text{stoich}} \left(\frac{1+E}{0.21} \right) = 7.39 \cdot \frac{1.2}{0.21} = 42.23 \frac{\text{mol}}{100 \text{ g dry coal}}$$

$$n_{\text{total}} = n_{\text{dry}}(1+X) = 42.23 \cdot 1.0116 = 42.72 \frac{\text{mol}}{100 \text{ g dry coal}}$$

$$n_{N_2} = 0.79 n_{\text{dry}} + n_{\text{fuel nitrogen}}$$

$$= 0.79 \cdot 42.23 + 0.054 = 33.42 \frac{\text{mol}}{100 \text{ g dry coal}}$$

$$n_{O_2} = E n_{\text{stoich}} = 0.2 \cdot 7.39 = 1.48 \frac{\text{mol}}{100 \text{ g dry coal}}$$

$$n_{H_2O,total} = n_{H_2O,combustion} + Xn_{dry} = 2.50 + 0.0116 \cdot 42.23$$
$$= 2.99 \frac{\text{mol}}{100 \text{ g dry coal}}$$

$$n_{total\ out} = n_C + \frac{n_H}{2} + n_S + n_{fuel\ N_2} + n_{stoich}\left[\left(\frac{1+E}{0.21}\right)(1+X) - 1\right]$$

$$n_{total\ out} = 6.32 + 2.50 + 0.05 + 0.054 + 7.39\left[\left(\frac{1.2}{0.21}\right) \cdot 1.0116 - 1\right]$$
$$= 44.25 \frac{\text{mol}}{100 \text{ g dry coal}}$$

One may check this by adding up the mols of individual components in the exit stream, finding the same result. We can compute the mol fraction (volume percentage) of any component in this combustion gas by dividing its number of mols by the total number of mols. Thus, for example

$$y_{SO_2} = \frac{n_{SO_2}}{n_{total\ out}} = \frac{0.05}{44.25} = 0.001130 = 1130 \text{ ppm}$$ ■

Most coals are delivered with 1 to 5 percent moisture; some subbituminous coals have up to 35 percent moisture. Coals can also pick up moisture from rain and snow in coal storage piles. Some coals are dried before burning; most are fed to the furnace wet. The foregoing example would be modified in that case to add the moisture contained in the coal, which leaves the furnace as water vapor in the exhaust gas.

One can see that this type of calculation is tedious, but it is easily written into computer programs and spreadsheets. We will use it, and the results of these two examples, several times later in this book.

7.11 CHANGING VOLUMETRIC FLOW RATES

Gas volumes change significantly with changes in temperature and/or chemical composition. These changes complicate the design of all types of air pollution control equipment.

Example 7.11. Figure 7.10 shows some kind of device with a single gas flow in and out. The gas has the properties of air, and a flow rate of 100 lbm/min. What are the volumetric flow rates in and out?

Here we first compute the molar flow rate

$$\begin{pmatrix}\text{Molar flow}\\ \text{rate}\end{pmatrix} = \frac{\text{Mass flow rate}}{\text{Molecular weight}} = \frac{100 \text{ lb/min}}{29 \text{ lb/lbmol}} = 3.45 \frac{\text{lbmol}}{\text{min}}$$

Then we note (see inside the back cover) that at standard conditions (20°C, 1 atm), 1.0 lbmol has a volume of $[1/(2.59 \times 10^{-3})] = 385.3 \text{ ft}^3$. If this flow were at standard

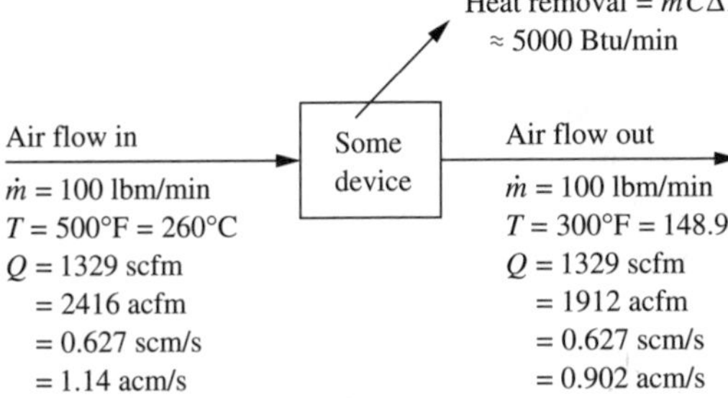

FIGURE 7.10
Example of the changes in volumetric flow rate through a device with a change in temperature.

conditions, the flow rate would be

$$\begin{pmatrix}\text{Standard}\\ \text{flow rate}\end{pmatrix} = \begin{pmatrix}\text{molar}\\ \text{flow rate}\end{pmatrix}\begin{pmatrix}\text{standard}\\ \text{molar volume}\end{pmatrix}$$

$$= 3.45\ \frac{\text{lbmol}}{\text{min}} \cdot 385.3\ \frac{\text{scf}}{\text{lbmol}} = 1329\ \frac{\text{scf}}{\text{min}}$$

where the abbreviation scf means *standard cubic feet.* This would also be written 1329 scfm, where scfm stands for *standard cubic feet per minute.* The actual molar volume is given by

$$\begin{pmatrix}\text{Molar}\\ \text{volume}\end{pmatrix} = \begin{pmatrix}\text{standard molar}\\ \text{volume}\end{pmatrix}\left(\frac{T}{T_{\text{std}}}\right)\left(\frac{P_{\text{std}}}{P}\right) \tag{7.19}$$

In most air pollution control applications the pressure is close to atmospheric, so the rightmost term is close to one and is ignored. Thus we can write

$$\begin{pmatrix}\text{Actual flow}\\ \text{rate}\end{pmatrix} = \begin{pmatrix}\text{standard flow}\\ \text{rate}\end{pmatrix}\left(\frac{T}{T_{\text{std}}}\right) = 1329\ \text{scfm} \cdot \frac{960°\text{R}}{528°\text{R}} = 2416\ \text{acfm}$$

where acfm stands for *actual cubic feet per minute.* For the flow out of the device we can see that the standard flow is unchanged, but the actual flow is the preceding, multiplied by (760°R/960°R), or 1912 acfm. ■

Here the metric equivalents are shown as *scms* and *acms* to match the English usage. One often sees Nm^3/s meaning *normal cubic meters per second.* Here *normal* is the equivalent of *standard* so if the standard and normal temperatures and pressures are the same (you must check), then $Nm^3/s = scms$. There does not seem to be an equally common metric equivalent for *acms;* sometimes one sees am^3/s. One also often sees DNm^3/s or dNm^3/s where the D or d stands for "dry." This latter form most often appears in permitted emission regulations, to put emission streams with varying moisture contents on a common basis.

In Example 7.11 the actual flows are almost twice the standard flows; in combustion gases the ratio can be three or four. Because of these changes in Q with T it is common in flow diagrams like Fig. 7.10 to show both the standard and the actual flows at various points in the system.

As we already discussed, the size of the control devices and the power to pump the gas through them are roughly proportional to the volumetric flow rate. Thus it will be economical to cool gases as much as possible before sending them to air pollution control devices, and it will be economical to locate the fan or blower at the place in the flow where the temperature is lowest. Hot gases are commonly cooled either by water sprays or by passing them through uninsulated ductwork before the gases come to control devices or fans or blowers. The acid dew point, however, limits how much can be accomplished this way (see Sec. 7.12).

Example 7.12. It is proposed to cool the inlet air stream in Fig. 7.10 from 500 to 300°F by spraying liquid water into it, not by the heat removal device shown in the figure. Water is available at 20°C = 68°F. How much water must we spray to accomplish this result? Assume that there is no heat transfer to the surroundings.

By straightforward energy balance (first law of thermodynamics) calculations we find that, for the steady flow, adiabatic mixing of hot gases and liquid water,

$$\Delta H_{\text{gas}} + \Delta H_{\text{water}} = 0 = (\dot{m} C_P \, \Delta T)_{\text{gas}} + [\dot{m}(\lambda + C_P \, \Delta T)]_{\text{water}}$$

$$\frac{\dot{m}_{\text{water}}}{\dot{m}_{\text{gas}}} = -\frac{(C_P \, \Delta T)_{\text{gas}}}{(\lambda + C_P \, \Delta T)_{\text{water}}} \approx -\frac{(C_P \, \Delta T)_{\text{gas}}}{\lambda}$$

$$= -\frac{(0.25 \text{ Btu/lb} \cdot {}^\circ\text{F})\,(-200^\circ\text{F})}{1055 \text{ Btu/lb}} = 0.047 \, \frac{\text{lb water}}{\text{lb gas}} \qquad \blacksquare$$

The amount of water calculated in Example 7.12 is small, but it produces a significant cooling effect, mostly due to the high latent heat of vaporization, λ, of water. The flow rate values on the right of Fig. 7.10 would have to be increased to take into account the water vapor that has now been added to the gas stream.

The saturation humidity of the cooled air determines how much we can cool hot gases this way. Figure 7.11 is a simplified version of a *psychrometric chart,* found in engineering handbooks and air-conditioning books. On it we see that the process in Example 7.12 can be represented as an arrow, passing from the initial state at 500°F and an assumed humidity of zero to 300°F and $H = 0.047$ lb water/lb dry air.

That figure also shows that if we continued adding water to the gas stream (dotted extension of the process path) we would reach the saturation curve at 122°F (called the *adiabatic saturation temperature*) and a humidity of 0.087 lb water/lb air. If we added still more water, the additional water would not evaporate because the air can hold no more water. We would then have a mixture of air and water drops at the adiabatic saturation temperature. In most cases we do not wish, for corrosion reasons, to add enough water to come close to the adiabatic saturation temperature, as discussed next.

7.12 ACID DEW POINT

Most gas streams that are treated for air pollution control contain moisture. If they are cooled enough to condense this moisture, the liquid thus produced may clog or

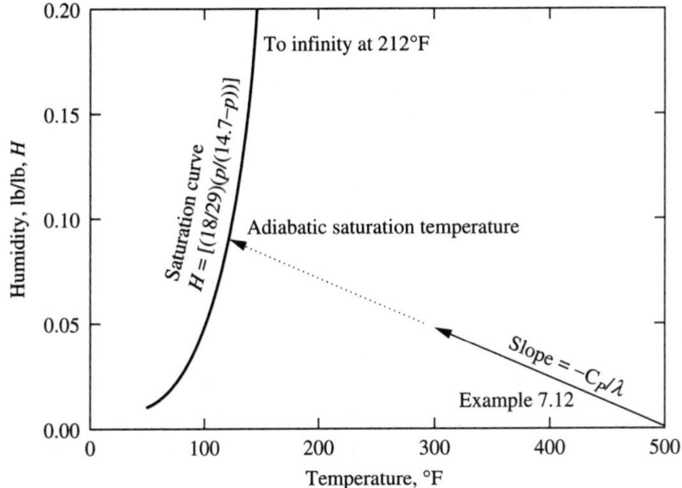

FIGURE 7.11
Path followed by the cooling in Example 7.12, shown on a psychrometric chart. Detailed charts of this type for various gas–liquid mixtures are found in many reference books, such as *Perry's Chemical Engineer's Handbook.* This chart is for air and water at one atmosphere pressure; p is the vapor pressure of water at this temperature. H is the humidity, lb/lb dry air.

plug control devices. Even worse, that water can dissolve acidic components out of the gas stream, thus forming acid liquids with impressive ability to corrode and thus destroy parts of the control equipment.

If there are no acid gases present in the gas stream, then the temperature at which the contained moisture will begin to condense, the *dew point,* is the temperature at which the ratio of the vapor pressure of water to the atmospheric pressure is equal to the mol fraction of water vapor in the gas, or

$$\left(\frac{p_{\text{water vapor}}}{P}\right)_{T=T_{\text{DEW}}} = \begin{pmatrix}\text{Mol fraction of}\\ \text{water in gas}\end{pmatrix} = y_{\text{water}} \tag{7.20}$$

Example 7.13. Methane is burned with 20 percent excess air. Estimate the dew point.

From Example 7.9 we found that the mol fraction of water in the combustion products was $y_{\text{water}} = (2.13/12.56) = 0.183$. The dew point temperature (at one atmosphere pressure) for this y_{water} is that for which the vapor pressure of water is 0.183 atm = 2.69 psia. From any table of the thermodynamic properties of steam we can find that this corresponds to $T_{\text{DEW}} \approx 137°\text{F}$. One could also compute this value using the vapor pressure equation for water in Appendix A. ■

This says that if we keep the products of combustion from natural gas above 137°F, they will not condense. Readers may have observed that if a large pot of cold water is placed over a natural gas flame, water initially condenses on its bottom and sides. As the pot heats up, the condensation disappears. At first the pot cools the products of combustion in the gas flame below 137°F. The dew point for methane is higher than for any other common fuel, because methane has the highest hydrogen/carbon ratio of any hydrocarbon fuel. Thus, if acid gases were not involved, the temperature requirement to avoid condensation would not be difficult. However, small amounts of acids, particularly sulfuric acid, can raise the dew point dramatically.

Example 7.14. A combustion gas at one atmosphere pressure contains 11.0 percent water and 1 ppm H_2SO_4. What would its dew point temperature be if there were no H_2SO_4? What is its dew point temperature with the H_2SO_4?

By a similar calculation to that in Example 7.13 we find that without the H_2SO_4, the dew point temperature of the gas would correspond to a vapor pressure of 0.11 atm = 1.62 psia, or $T_{DEW} \approx 118°F = 48°C$. Figure 7.12 is a summary of the observed dew points for this situation. From it we read an acid dew point of $\approx 230°F = 110°C$. ■

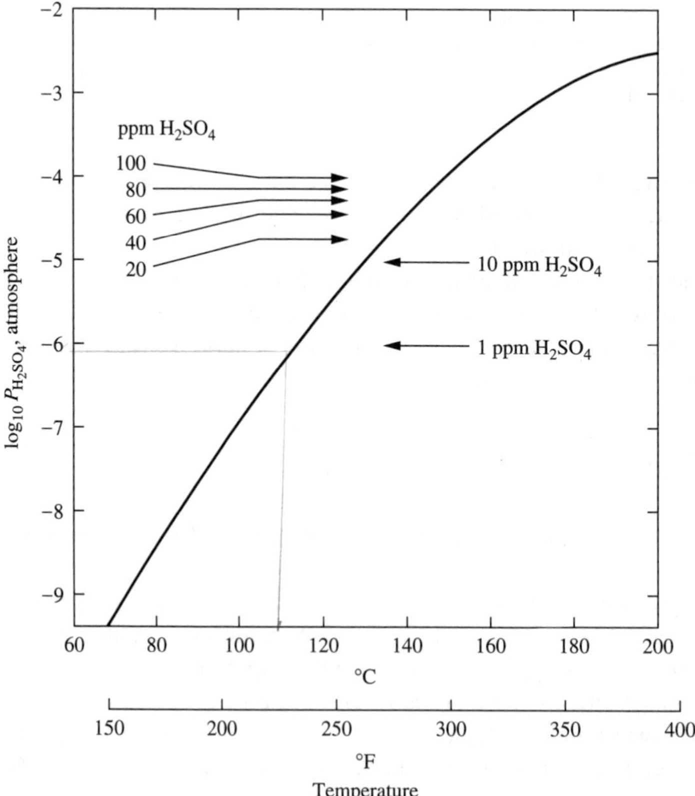

FIGURE 7.12
Acid dew point curve for flue gases containing 11.0 percent water vapor at one atmosphere pressure. (From Ref. 19.)

This result indicates that 1 ppm of H_2SO_4 in the gas raises its dew point by 112°F = 62°C. More than one major piece of equipment has been destroyed by corrosion because its designers overlooked or did not believe this startling result. Unfortunately, there is likely to be at least 1 ppm of H_2SO_4 in most combustion gases, except for those from unodorized natural gas. Most of the sulfur in coal and high-boiling oils is bound into their molecules. Propane and natural gas have sulfur compounds added as odorants for safety reasons. When fuels burn, this sulfur is converted to SO_2, which then reacts with oxygen and water vapor to form H_2SO_4. How much of the SO_2 is converted depends on the moisture content of the gas, the time spent in the combustion environment, and the catalytic effect of ash particles and/or the metal parts of the combustion device. A typical estimate is that 5 percent is converted in a large furnace. This problem is important enough (painful enough, expensive enough) that it has a substantial literature [19].

Consequently, any air pollution control device that treats a gas containing SO_2 (and therefore some H_2SO_4) must either be made of corrosion-resistant materials, which are expensive, or be protected against acid dew point corrosion. Common protective measures are to insulate the device so that it will not be cooled by the surroundings and/or to provide an auxiliary supply of inert gas or air to purge out the SO_2-containing gas whenever the device is shut down. Other acid gases (HNO_3, HCl, H_2CO_3) can also cause this problem, but not as spectacularly as H_2SO_4. The reason is that H_2SO_4 has a very strong affinity for atmospheric water vapor and will collect it and form drops at the high temperature shown in Fig. 7.12. The other acids do the same thing, but not nearly as strongly.

Acid dew point condensation is also a problem in sampling stack gases. The problem is not acid corrosion of the equipment, but change of the gas composition by condensation and removal of acid components. Stack gas sampling probes and lines are most often heated electrically to keep the sample above the acid dew point on its way through the probe and sampling line to the detector, or else enough dry dilution air is introduced near the sample inlet to reduce the acid dew point temperature below the ambient temperature. This dilution is measured and accounted for in the pollutant concentration calculations.

7.13 CATALYSTS FOR AIR POLLUTION CONTROL

A catalyst is a substance that makes a chemical reaction proceed faster without itself being consumed in the chemical reaction. Because it is not consumed itself, one catalyst molecule can influence billions of molecules in the catalyzed reaction; small amounts of catalyst produce large effects! There are many types of catalysts, of which the most important are the enzymes that regulate the chemistry of our bodies; without them life would be impossible. Most catalysts are selective; instead of speeding up all possible reactions, they speed up one reaction, without speeding up others. This means that if A + B could react to form C or D, a selective catalyst can cause the formation of almost pure C or almost pure D. Some air pollution

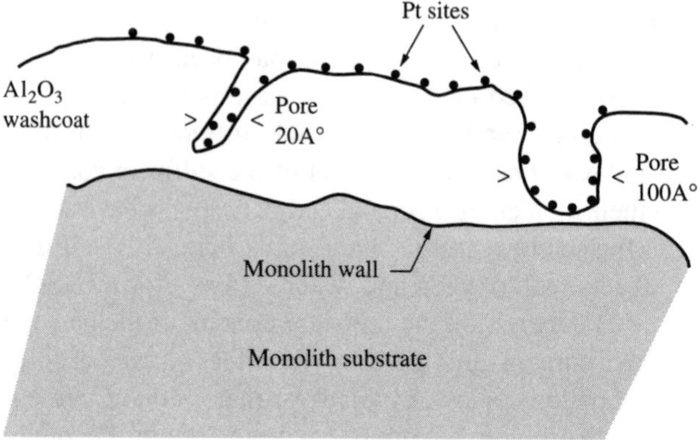

FIGURE 7.13
Conceptual view of a supported platinum catalyst. The active metal sites are on the external surface and the internal pore surface of the washcoat, which is attached to the nonactive monolith substrate. Observe the small size of the pores; they range from 5 to 25 times the diameter of a typical gas molecule. (From R.M. Heck, and R.J. Farrauto, *Catalytic Air Pollution Control: Commercial Technology*, Van Nostrand-Reinhold, New York, 1995, p. 6. [20]. Copyright ©1995, Van Nostrand-Reinhold. Reprinted by permission of John Wiley & Sons, Inc.)

control catalysts are chosen because they are selective, only promoting a desired reaction, and not promoting others. In Chapters 10–13 we will encounter examples of catalysts used for air pollution control; this section describes the general properties of the catalyst class most often used in air pollution control, supported solid catalysts.

Figure 7.13 shows, in schematic form, a highly magnified view of an active part of such a supported solid catalyst.

To prepare such a catalyst one first prepares the support, whose shape is discussed later in this section. The support is generally a nonporous ceramic; for some automotive applications it is a thin sheet of metal. The washcoat is a highly engineered ceramic which has a high internal porosity. The values of this porosity are startling.

Example 7.15. A typical catalyst support washcoat has a surface area, internal plus external, of 100 m^2/g. If this were in the form of a sheet, like a piece of paper, and had a density of 2 g/cm^3, how thick would the sheet be?

Assuming the sheet is a square with sides L long and with thickness t, we see that its area is

$$\text{Area (both sides)} = 2L^2$$

Its mass is

$$\text{Mass} = \text{volume} \cdot \text{density} = L^2 t\rho$$

Eliminating L between these equations and solving for t, we find

$$t = \left(\frac{\text{Mass}}{\text{Area}}\right)\left(\frac{2}{\rho}\right) = \left(\frac{\text{g}}{100\ \text{m}^2}\right)\left(\frac{2}{2\ \text{g/cm}^3}\right)\left(\frac{\text{m}}{100\ \text{cm}}\right)^2 = 10^{-6}\ \text{cm}$$
$$= 10^{-8}\ \text{m} = 100\text{Å}$$ ■

This value is startlingly low, about forty times the interatomic spacing in crystals! If ceramics have this much surface area, then they must have internal walls only forty atoms thick! Apparently they do. To make materials with this much surface area, one starts with a material, from which part can be removed on an atomic scale. To make porous alumina or silica (sometimes called *silica gel*), one heats a hydrous alumina or silica to drive off the water, leaving behind an alumina or silica with very small pores where the water was. If one could examine a piece of the ceramic with a microscope with resolution equal to the size of atoms, one would see that the apparently solid ceramic is really like an irregular honeycomb. If one could then examine the walls of that honeycomb, one would see that the wall itself was a smaller honeycomb, and so on down to honeycombs so small that their walls average only forty atoms thick. By the honeycomb-within-honeycomb structure one can make a strong solid out of sheets only forty atoms thick.

The catalyst is made by first preparing the inert (ceramic or metallic) substrate. This is then dipped in a dilute suspension of the raw materials for the washcoat, and removed, leaving a thin film of wet material. This is then baked to coagulate the washcoat which causes it to dehydrate on the molecular level, forming the internal, very small pores. The whole assembly is then immersed in a dilute solution of a soluble compound of the metal (e.g., chloroplatinic acid to deposit platinum). The assembly is then dried, leaving the nonvolatile metallic compound on the surface of the pores. Finally the whole assembly is heated in a hydrogen environment, which reduces the metal from its compound to metallic form. The resulting catalyst contains only a fraction of a percent of the metal, which is often expensive like platinum, palladium, rhodium, or rhenium, but all of that is on the surface of the cheaper washcoat which provides mechanical strength and the high surface area of the multiple pores. There are variants on this procedure; for example, some catalysts use the same material for support and washcoat, and petroleum cracking catalysts use highly specialized ceramics, without metal. But most air pollution catalysts are prepared in some variation of this procedure.

In the chemical and petroleum industries the catalyst support is most often prepared in the form of pellets, slightly larger than an aspirin tablet. For air pollution catalysts the most common support shape is the honeycomb, shown in Fig. 7.14 on page 202. The principal advantage of the honeycomb structure over a bed of pellets is that the honeycomb requires a lower pressure drop to force the gas through it, typically about 5–10% as much as would a comparable bed of pellets (Problem 7.28). For applications in the chemical and petroleum refining industries, the small pressure drops through catalyst beds are unimportant, and the convenience of the pellet type outweighs its pressure drop disadvantages. For air pollution applications

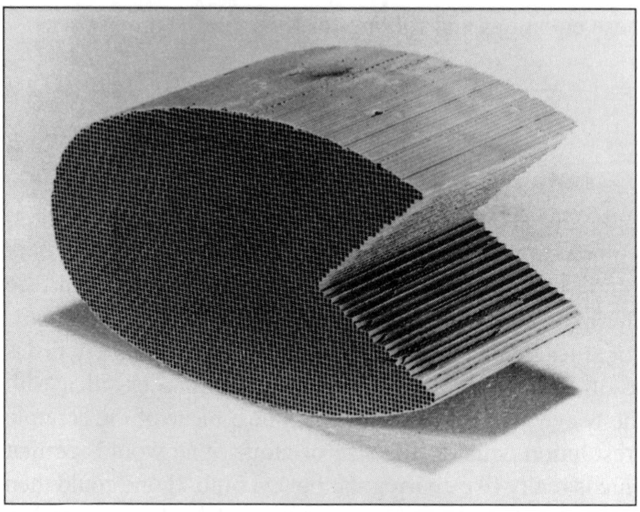

FIGURE 7.14
Honeycomb-type catalyst support. The individual flow channels pass straight through the whole structure. They are typically about a tenth of an inch square. The cutaway section at the right shows the internal structure. This support is about 10 inches wide and 6 inches long, in the shape used for automotive catalysts (see Chapter 13.). For other services the supports are rectangular or circular. (Courtesy of Dr. Ron Heck and the Engelhard Corporation.)

the reverse is true; the lower pressure drop of the honeycomb catalyst supports makes their use practically universal.

7.14 SUMMARY

1. If one is faced with an air pollution problem, the alternatives for alleviating it are improved dispersion, process change, or downstream ("tailpipe") control devices. In current U.S. law and practice the first choice is process change; the second is a downstream control device; and only if these, taken together, cannot meet the applicable standard may enhanced dispersion be used.
2. If the pollutant is valuable itself or as a fuel, and if its concentration is high enough, the best control solution may be to recover the pollutant or use it as a fuel.
3. It is almost always economical to minimize the volumetric flow rate of the stream to be treated and to minimize the pressure drop in the control equipment.
4. In most calculations, the penetration is a more convenient measure of control equipment performance than is the control efficiency.
5. Most air pollutants are the direct or indirect result of combustion or processes using combustion. This chapter provides a brief introduction to combustion.
6. The volumetric flow in control devices changes significantly with changes in temperature. For this reason process flow sheets normally show the flow rate both at standard conditions and at actual conditions (scfm, acfm).

7. All pollution control devices that treat gases with even trace amounts of SO_2 must be protected against acid dew point corrosion.
8. Catalysts are widely used in air pollution control. The catalyst support in the form of a honeycomb is most widely used because its pressure drop is less than that of alternative forms.

PROBLEMS

See Common Units and Values for Problems and Examples, inside the back cover.

7.1. (*a*) In Example 7.3, how large a percentage error do we make by using Eq. (7.6) instead of Eq. (7.4) (assume $P_1 = 14.7$ psia)?
(*b*) Sketch the paths corresponding to Eqs. (7.4) and (7.6) on a pressure-volume plot. Show what area corresponds to the error computed in part (*a*).

7.2. We pass a gas stream through a fiber filter that collects 85 percent of the particles present. If we were to use three such filters in series and if we assume that each of them has an 85 percent efficiency, what would the expected overall collection efficiency be?

7.3. Repeat Example 7.5 for three identical collectors in series, instead of the two in the example.

7.4. In Examples 7.4 and 7.5, how many identical collectors in series would we need to get 95 percent overall collection efficiency?

7.5. A gas stream contains two sizes of particles, 50 wt% large and 50 wt% small. We pass this gas stream through two collection devices in series. The collection efficiencies are shown in the following table. What overall weight fraction of the particles is collected?

	First collection device	Second collection device
Collection efficiency for small particles	0.50	0.25
Collection efficiency for large particles	0.75	0.40

7.6. A new type of particle collector consists of five identical units in series. The gas stream we are treating contains two sizes of particles, 50 wt% large and 50 wt% small. From theory we know that the collection efficiency in each of the individual units for the large particles is three times that for the small particles. The overall collection efficiency of the five units in series, for both kinds of particles combined, is 90% by weight. What is the collection efficiency of each individual unit for the large particles? For the small particles?

7.7. List as many examples as you can of industrial processes that recover a potential air pollutant from a waste gas stream and put this recovered pollutant to economic use.

7.8. Figure 7.15 on page 204 shows the frequency distribution of various values for the product of H times u for Peoria, Illinois. If the average daily emission rate for carbon monoxide for this city is 2×10^7 g/(h$\cdot$ mi^2) and the standard to be met is 10 mg/m^3, what fraction of the time must we implement intermittent control to meet this standard? Assume that the simple box model (Sec. 6.2) applies, that the size of the city is 5 mi by 5 mi, and that the background concentration is 2 mg/m^3.

7.9. In Example 7.1, what are the values of a, b, and c in Eq. (7.1)? If we wish to rewrite that equation in terms of the volumetric flow rate instead of the filter area, and if the flow velocity through the filter is 2 ft/min, what would the values of a, b, and c be?

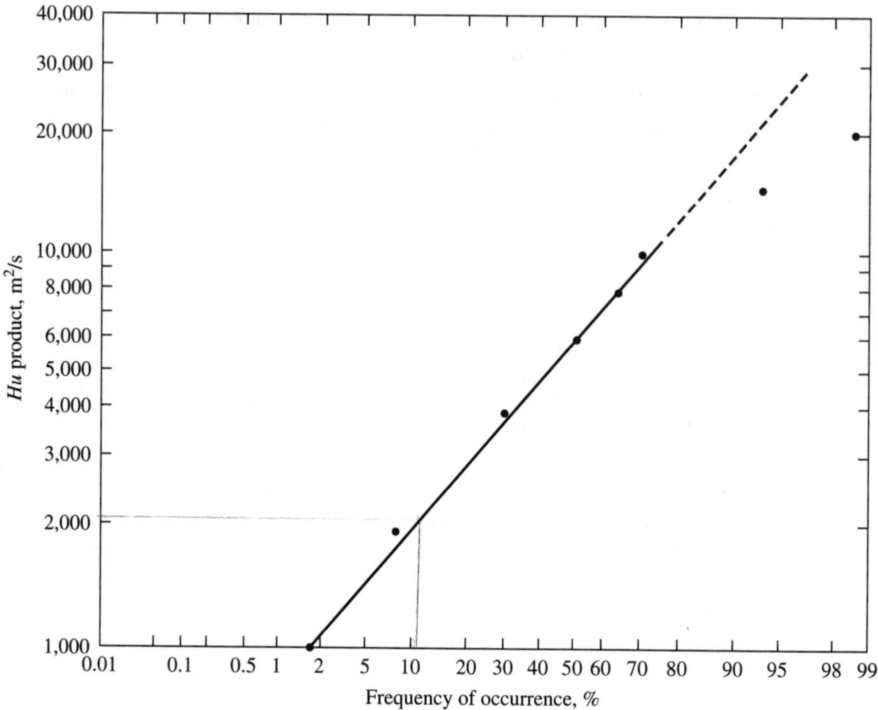

FIGURE 7.15
Frequency of occurrence of the *Hu* product for Peoria, Illinois [21].

7.10. Estimate the flammability limits (vol %) for propane in air.

7.11. Using the values in Table 7.1, calculate the LEL and UEL values by weight for the aliphatic hydrocarbons. Discuss the possible reasons why the UEL for methane is so different from the others.

7.12. Example 7.7 shows that the stoichiometric air-fuel ratio (A/F) for methane is 17.3 lb/lb.
(*a*) Compute the stoichiometric A/F for hydrogen and for carbon.
(*b*) Write an equation for the stoichiometric A/F of a fuel as a function of its molecular (hydrogen/carbon) ratio, assuming that the fuel consists only of hydrogen and carbon.

7.13. Several industrial processes produce waste gases containing carbon monoxide. Normally these waste gases contain no oxygen. They may be considered as mixtures of CO and N_2. If they contain enough CO, it is common practice to mix them with air and burn them, using the heat released to generate steam or for some other practical purpose.
(*a*) Using the values in Table 7.1, estimate the lowest concentration of CO in the waste gas for which burning is possible. Here the condition we are seeking is one where the amount of oxygen supplied is just the amount needed to burn up all the CO, with no surplus oxygen, and the concentration of CO in the final mix of waste gas and air is at the lower explosive limit. Using the LEL value from Table 7.1 is not exactly correct, because the lower explosive limit data in Table 7.1 is for CO in the presence of a mixture of N_2 and O_2 which has the ratio (79/21) while in this case the ratio will be much higher. But experimental data show that the result calculated this way is close to the observed value.

(*b*) List industrial processes that might produce such waste gases. These can be inferred by considering what kind of chemistry would produce such gases.

7.14. The explosive limit data in Table 7.1 (and all similar tables) are based on mixtures of the combustible material with air, which is a mixture of approximately 21 percent oxygen, 79 percent nitrogen. If we substitute pure oxygen for air, we normally find that the lean limit is unchanged, but the rich limit is greatly increased. Why? What are the safety consequences of this observation?

7.15. Table 7.1 lists the spontaneous ignition temperature for methane as 1170°F. This is the temperature for a stoichiometric mixture. Sketch on a copy of Fig. 7.4 the relation that must exist between spontaneous ignition temperature and y_{methane}.

7.16. Show the calculations leading to the statement in Section 7.10.7 that compression of the gas in the cylinder of an automobile engine with a compression ratio of 7 leads to a temperature increase of about 600°F.

7.17. By how much does preheating the air, as shown in Fig. 7.8*b*, increase the thermal efficiency of the preheated furnace, compared to one without preheating as shown in Fig. 7.8*a*? Assume the fuel is methane, the heating value is 21 502 Btu/lb, and there is 20 percent excess air. The heat capacity of air and combustion gases is roughly 7 Btu/(lbmol·°F). See Example 7.9.

7.18. Show the complete derivation of Eq. (7.17).

7.19. The "typical" coal in Example 7.10 is being burned under the conditions in that example, at a rate of 1000 kg/h of dry coal. The combustion products enter a treatment device at 400°F. What is the gas flow rate in scfm and acfm?

7.20. Repeat Example 7.10 for the case where the coal has the same dry analysis as shown, but has a moisture content of 5 percent as burned.

7.21. Air at 2200°F and 1 atm is flowing at a rate of 1000 lb/min out of the afterburner on a hazardous waste incinerator.

(*a*) How many scfm is this?

(*b*) How many acfm is this?

7.22. In Example 7.12, what are the values of Q in scfm and acfm in the cooled gas?

7.23. Repeat Example 7.13 for 0 percent excess air, instead of the 20 percent value shown there. If you do not have a table of the thermodynamic properties of steam, you may estimate the vapor pressure of water from the Antoine equation, Appendix A.

7.24. In Example 7.12:

(*a*) What is the dew point of the cooled gas? Assume that the hot gas is air, with zero water content.

(*b*) We now wish to cool the gas as much as possible by mixing it with a water spray, but we have a company rule not to cool below 25°F above the dew point of the cooled gas. How far can we reduce the temperature of this gas by mixing it with a water spray?

(*c*) Sketch how one would find the dew point graphically, and how the process in part (*b*) would look on a figure like Fig. 7.11.

7.25. In Example 7.12, how large a percentage error do we cause by making the approximation that, for water, $\lambda \gg C_P \, \Delta T$?

7.26. (*a*) Write the general equation for the mol fraction of water in a combustion stack gas, on the assumption that the fuel is a hydrocarbon with formula C_xH_y, that the air moisture is X mol/mol of dry air, and that the excess air is E percent.

(*b*) Using that formula, calculate the value of (y/x) that corresponds to 11.0 percent water vapor in the stack gas, for dry, stoichiometric air ($X = E = 0$).

(*c*) Compare that value with typical (y/x) values of ≈ 1 for coal, ≈ 1.8 for most fuel oils, and 4 for natural gas. What kind of fuel is probably intended in Fig. 7.12?

(*d*) Show how much the answer to part (*b*) changes if we assume $E = 0.20$ and $X = 0.0116$.

7.27. Using Fig. 7.12, estimate the acid dew point for the exhaust gases from a coal-burning power plant. The coal to be burned (this is actually the design coal analysis for the Emery, Utah, power plant) has the following analysis:

Component	% by weight
Moisture	6.4
Carbon	68.3
Hydrogen	4.9
Nitrogen	0.8
Oxygen	10.2
Sulfur	0.6
Ash	8.8
Total	100.0

Assume that $E = 0.2$, that there is 5 percent conversion of SO_2 to H_2SO_4, and that the combustion air is at 20°C and 50 percent relative humidity.

7.28. This problem follows closely an example calculation, supported by experiment in [24]. A typical automotive catalyst has a gas flow rate of 136 kg/hr at 590°C and 1 atm. The frontal area of the catalyst is 0.0182 m^2 and a total length of 0.0762 m. The exhaust gas has $\approx$ the same properties as air.

(*a*) Show that the gas density is ≈ 0.41 kg/m^3, and that the volumetric flow rate is ≈ 0.092 m^3/s. Based on these values show that the vapor hourly space velocity (VHSV), which is the volumetric flow rate divided by the external volume of the catalyst package, is $\approx$ 24,000/hr. (This is typical of automobile catalysts at low engine speeds; at full speed the flow rate is $\approx$ 4 times this amount.)

(*b*) If the catalyst is a monolith, like the one shown in Fig. 7.14, with square channels 1.5 mm on a side and an open area of 0.62 of the total area, show that the hydraulic radius, HR, is 0.375 mm, and the exposed area/volume is 1654 m^2/m^3.

(*c*) Calculate the Reynolds number, finding ≈ 131, which shows that the flow in the channels is laminar. The pressure drop is estimated [24] by

$$\Delta P = \frac{\rho V^2}{2}\left(K_c + K_e + A_g f \frac{L}{\text{HR}}\right) \tag{7.21}$$

where K_c and K_e are the contraction and expansion loss coefficients for the entrance and exit from the monolith, whose values are both ≈ 0.15, and A_g is the correction factor for the fact that this is a short tube $(L/(4\text{HR}\cdot\mathcal{R})) = 0.4$ for which one must account for the flow development. Based on a correlation shown in [24], for this problem $A_g \approx 1.1$. For laminar flow in square channels Mondt [24] suggests that the friction factor, f, can be computed from

$$f = \frac{14.23}{\mathcal{R}} \tag{7.22}$$

Graduate students should show the relationship of this equation to Poiseuille's equation for flow in a circular tube; undergraduates should simply accept it. Using it in Eq. (7.21), estimate the pressure drop needed to move this gas flow through this monolith.

(*d*) Show that if the monolith is replaced by a bed of spherical beads with diameter 3.2 mm, with the same frontal area and length and with a porosity of $\varepsilon = 0.38$, the bed of beads will have $\approx$ twice the external surface as the monolith and a hydraulic radius of 0.32 mm.

(*e*) Show that the Reynolds number, based on the interstitial velocity, is $\approx$ 185, which corresponds to the transition region between laminar flow and turbulent flow for a packed bed. For this $\mathcal{R}$ a plot in [24] suggests that $f = 0.8$. Substitute that value in the pressure drop correlation for packed beds

$$\Delta P = \frac{\rho V^2}{2}\left(f\frac{L}{\text{HR}}\right) \tag{7.23}$$

and estimate the pressure drop.

(*f*) Compare the calculated pressure drops with the statement in Sec. 7.13 that for equal flow geometries the pressure drop is much greater in a packed bed than in a honeycomb monolith. Currently almost all automotive catalysts use the honeycomb form. Earlier some used pellet beds, in which the shape was chosen to have the cross-sectional area perpendicular to flow much larger and the length in the flow direction much smaller. Repeat the calculation in (*e*) for the same total packed bed volume, but for the frontal area increased by a factor of 2 and the length decreased by a factor of 2. Here assume that f does not change. How much does that reduce the pressure drop?

REFERENCES

1. Smith, M. E., and T. T. Frankenberg: "Improvement of Ambient Sulfur Dioxide Concentrations by Conversion from Low to High Stacks," *J. Air Pollut. Control Assoc.,* Vol. 25, pp. 595–601, 1975.
2. Hewson, E. W.: "The Meteorological Control of Atmospheric Pollution by Heavy Industry," *Quarterly J. Royal Meteorolog. Soc.,* Vol. 71, pp. 266–283, 1945.
3. Leavitt, J. M., S. B. Carpenter, J. P. Blackwell, and T. L. Montgomery: "Meteorological Program for Limiting Power Plant Stack Emissions," *J. Air Pollut. Control Assoc.,* Vol. 21, pp. 400–405, 1971.
4. Nelson, K. W., M. A. Yeager, and C. K. Guptil: *Closed-Loop Control System for Sulfur Dioxide Emissions from Non-Ferrous Smelters,* American Smelting and Refining Co., Salt Lake City, UT, 1973.
5. Hill, G. R., M. D. Thomas, and J. N. Abersold: "High Stacks Overcome Concentrations of Gases," *Mining Congr. J.,* Vol. 31, Issue 4, pp. 21–34, 1945.
6. Freeman, H., T. Harten, J. Springer, P. Randall, M. A. Curran, and K. Stone: "Industrial Pollution Prevention: A Critical Review," *J. Air Waste Manage. Assoc.,* Vol. 42, pp. 618–656, 1992.
7. Wentz, C. A.: *Hazardous Waste Management,* McGraw-Hill, NY, 1989.
8. Vatavuk, W. M.: *Estimating Costs of Air Pollution Control,* Lewis Publishers, Ann Arbor, MI, 1990.
9. The cost data are currently available at ⟨www.epa.gov/ttn/catc/⟩.
10. de Nevers, N.: *Fluid Mechanics for Chemical Engineers,* 2d ed., p. 218, 1991.
11. Crocker, B. B.: "Minimizing Air Pollution Control Costs," *Chem. Eng. Prog.*, Vol. 64, pp. 79–86, 1968.
12. "Basic Considerations in the Combustion of Hydrocarbon Fuels with Air," in W. T. Olson (ed.), *NACA Report 1300,* p. 451, 1957. Table 7.1 is a much condensed version of the table in the Appendix of this report.
13. Lewis, B., and G. von Elbe: *Combustion, Flames and Explosions of Gases,* 2d ed., Academic Press, New York, 1961.
14. Hougen, O. A., K. M. Watson, and R. A. Ragatz: *Chemical Process Principles, Part I, Material and Energy Balances,* 2d ed., John Wiley, New York, p. xxiii, 1959.
15. Hougen, O. A., K. M. Watson, and R. A. Ragatz: *Chemical Process Principles, Part II, Thermodynamics,* 2d ed., John Wiley, New York, Chapter 26, 1959.
16. Flagan, R. C., and J. H. Seinfeld: *Fundamentals of Air Pollution Engineering,* Prentice Hall, Englewood Cliffs, NJ, p. 88, 1988.

17. Smith, W. S., and C. W. Gruber: "Atmospheric Emissions from Coal Combustion—An Inventory Guide," U.S. Department of Health, Education and Welfare, Public Health Service Publication No. AP-24, p. 24, 1966.
18. Bauer, T. K., and R. B. Spendle: "Selective Catalytic Reduction for Coal Fired Power Plants: Feasibility and Economics," Electric Power Research Institute, EPRI Report CS-3603, EPRI, Palo Alto, CA, 1985.
19. Holmes, D. R., ed.: *Dewpoint Corrosion,* Ellis Horwood, Ltd., Chichester, England, p. 19, 1985.
20. Heck, R. M., and R. J. Farrauto: *Catalytic Air Pollution Control: Commercial Technology,* Van Nostrand-Reinhold, NY, 1995, p. 6.
21. Holzworth, G. C.: "Mixing Heights, Wind Speeds and Potential for Urban Air Pollution throughout the Contiguous United States," U.S. Environmental Protection Agency Publication AP-101, U.S. Government Printing Office, Washington, DC, 1972.
22. Tilton, J. N.: "Fluid and Particle Dynamics," in *Perry's Chemical Engineers' Handbook,* 7th ed., D. W. Green and J. O. Maloney, eds., McGraw-Hill, NY, 1997, pp. 6–12.
23. de Nevers, *op. cit.* Chapter 12.
24. Mondt, J. R.: "Adapting the Heat and Mass Transfer Analogy to Model Performance of Automotive Catalytic Converters," *Trans. ASME, J. Eng. Gas Turb. Power*, Vol. 109, 200–206, 1987.

CHAPTER 8

THE NATURE OF PARTICULATE POLLUTANTS

If a contaminated air stream is visible, the particles it contains make it so. If the air mass over a city is hazy, the particles in the air cause the haze. Particulate pollutants are not chemically uniform (as is, for example, CO; one CO molecule is identical to another) but rather come in a wide variety of sizes, shapes, and chemical compositions. Some are much more harmful to health, property, and visibility than others. In this chapter we discuss the nature of atmospheric particles and make the distinction between primary and secondary particles. In the next chapter we discuss the emission control of primary particles.

8.1 PRIMARY AND SECONDARY PARTICLES

Most of us have an intuitive idea that particulate pollutants are like sand, gravel, or dust; i.e., there are large numbers of small individual particles, each one hard and distinct like beach sand. This is only partly right. To see why, we must consider Fig. 8.1 on page 210, which presents an overview of particles and their properties. (This figure is due to C. E. Lapple, who made many useful contributions to the practice of engineering.) The horizontal scale is particle diameter. Diameter is an obvious property of a spherical particle, but not as obvious a property of a cubical or rod-like particle. Some air pollution particles depart radically from spherical shape (e.g., rod-like asbestos). For the purposes of Fig. 8.1 and most of the remainder of this book, we will understand "diameter" for a nonspherical particle to mean "diameter of a sphere of equal volume;" that is, diameter $= (6 \cdot \text{Volume}/\pi)^{1/3}$. Particle diameters are given in microns (micron $= 10^{-6}$ m $= 10^{-3}$ mm); this size unit is used almost exclusively

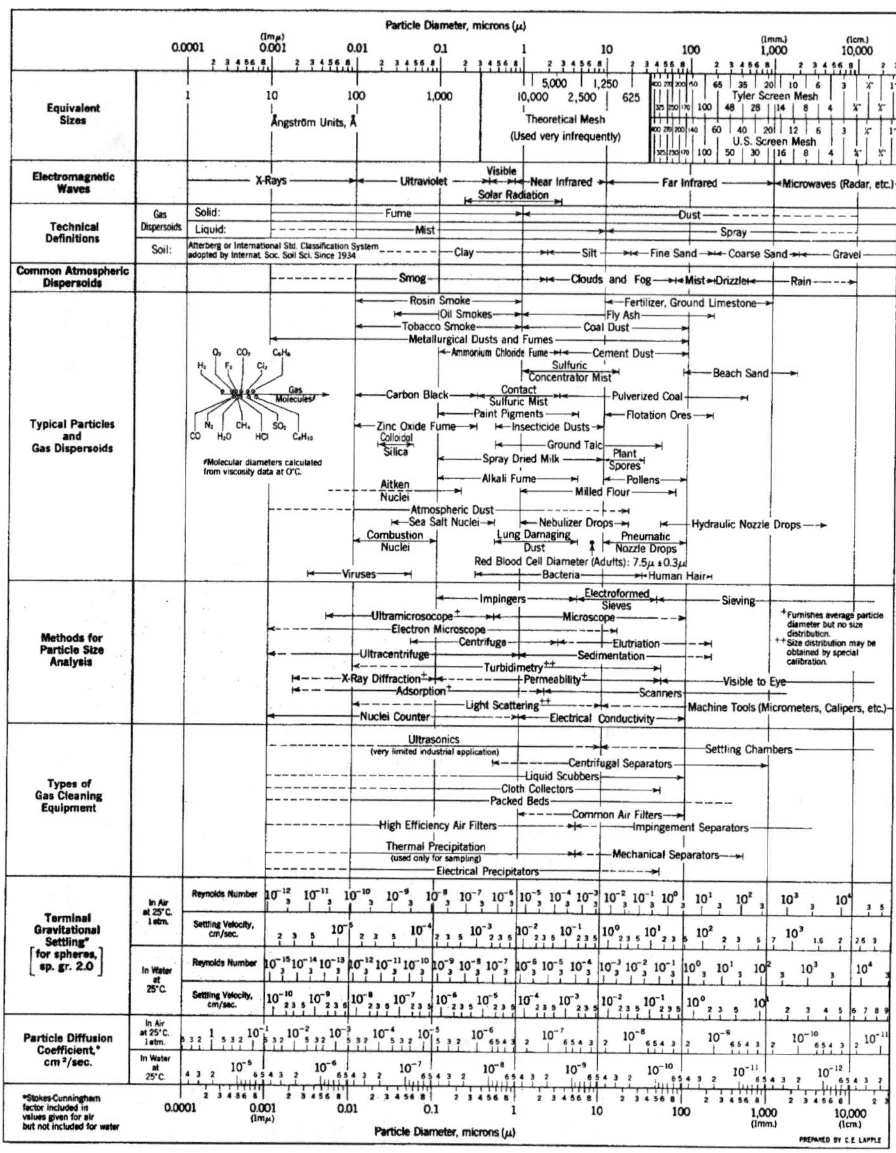

FIGURE 8.1
Sizes and characteristics of airborne particles. (From C. E. Lapple, *Stanford Res. Inst. Jour.,* Vol. 5, p. 94 (Third Quarter 1961). Reprinted by permission).

in the particle literature and is usually given the symbol μ. In SI units a micron is called a *micrometer* (μm). In the upper right of the figure we see that gravel, as we commonly understand the term, has sizes of 2000 μ or greater and sand has diameters from about 20 to 2000 μ. (A 2000-μ sand grain is very coarse.) A typical human hair has a diameter of about 50 μ. The pages of this textbook are about 100 μ thick. The particles that cause significant air pollution problems are generally in the size range 0.01 to 10 μ, much smaller than the finest sand or the diameter of a human hair. For the rest of this book all particle diameters will be stated in μ, a common practice in the United States, even when the rest of the problem or example is in English units. Particle diameters are also stated in terms of Tyler screen size in the United States (upper right, Fig. 8.1). For example, particles larger than about 100 μ will not pass through a Tyler 150 mesh screen. We will not use this terminology here, but it is also common in U.S. engineering writing.

Most of us probably recognize that sand and gravel are made by mechanically breaking up bigger rocks. (Normally this occurs in streams, but it can also occur in rock-crushing plants.) Some industrial particulate pollutants (e.g., pulverized coal, which has a size range of 3 to 400 μ) may be created mechanically but most crushing and grinding processes do not produce particles smaller than about 10 μ. The only exceptions shown on Fig. 8.1 are paint pigments and ground talc, both of which undergo extreme grinding and milling operations in order to obtain the required fine particle size.

Instead, most of the fine particles (0.1 to 10 μ) shown on the figure are obtained by combustion, evaporation, or condensation processes. One example the reader has most likely personally observed is the formation of tobacco smoke, which is shown in Fig. 8.1 as having a size range from 0.01 to 1 μ. Figure 8.2 on page 212 shows a smoking cigarette. Directly above the burning tobacco is a transparent zone 1 or 2 mm wide, above which the visible smoke plume forms. The smoke consists of droplets of condensed hydrocarbons (oils, tars) in the size range 0.01 to 1 μ. In the transparent zone, the temperature is high enough that these hydrocarbons are transparent gaseous molecules. As the combustion gases rise, they mix with colder air and reach the condensation temperature, at which the hydrocarbon gases form the very small drops that make the visible smoke.

The smoke from the exhaust pipe of an "oil-burning" car or from a smoldering campfire, and the puff of white smoke from the tires of an airliner as it lands are all high-molecular-weight hydrocarbons that were turned into vapors (or gases) by heating, and then condensed upon cooling to form fine droplets. The finest particles that have been made for research purposes are made by heating a metal or a salt to its vaporization temperature (usually by sudden, massive electric heating) and then condensing the resulting gaseous metal or salt by quickly cooling it so that many small particles form rather than a few large ones. In this way it is possible to make particles with sizes of about 0.01 μ in a fairly reproducible way [1]. It is practically impossible to produce such fine particles by any mechanical (crushing or grinding) process.

It may seem counterintuitive to think of particulate pollutants like smokes as liquids instead of the solid (sand-grain) materials you assumed. However, it is uncommon in the air pollution literature to distinguish between fine particles that

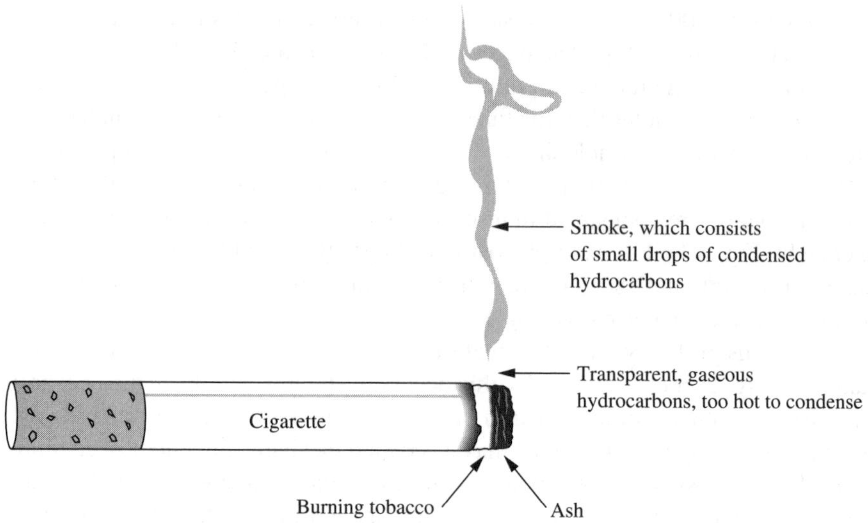

FIGURE 8.2
Burning cigarette, showing clear (gaseous) region and visible (smoke) region.

are solids and fine particles that are liquids (or tars). In the atmosphere and in common collecting devices they frequently behave alike. Furthermore, if the relative humidity of the atmosphere is high, it is common for even rock-like particles to have a film of condensed water on their surfaces that causes them to behave in a liquid way.

Most of the fine particles shown on Fig. 8.1 are obtained by condensation of gases. The others are obtained by removal of part of a larger particle by combustion or evaporation. Most fuels contain some incombustible materials, which remain behind when the fuel is burned, called *ash*. The ash left behind by the combustion of wood, coal, or charcoal contains mostly the oxides of silicon, calcium, and aluminum, with traces of other minerals. If the fuel is finely ground (or produced as a spray of fine droplets) and then burned, the remaining unburned ash particles may be quite small.

This distinction between mechanical and condensed particles is illustrated by tests in which pulverized lignite was burned (see Appendix C) in a laboratory furnace [2]. The ash particles in the exhaust gases consisted of two groups. One group had an average diameter of about 0.02 μ, the other about 10 μ. The smaller particles contained a much higher percentage of the more volatile materials in the ash (P, Mg, Na, K, Cl, Zn, Cr, As, Co, and Sb) than did the larger particles. Almost certainly the fine particles were formed by condensation, in the furnace, of materials that had been vaporized during the combustion process; the larger particles were formed from the remaining mineral matter in the fuel that was not vaporized.

Figure 8.3 shows a sample of particles collected from a coal-fired furnace operated at fuel-rich conditions, resulting in significant unburned carbon in the fly ash. The top photo shows several different types of particles. The spherical particles

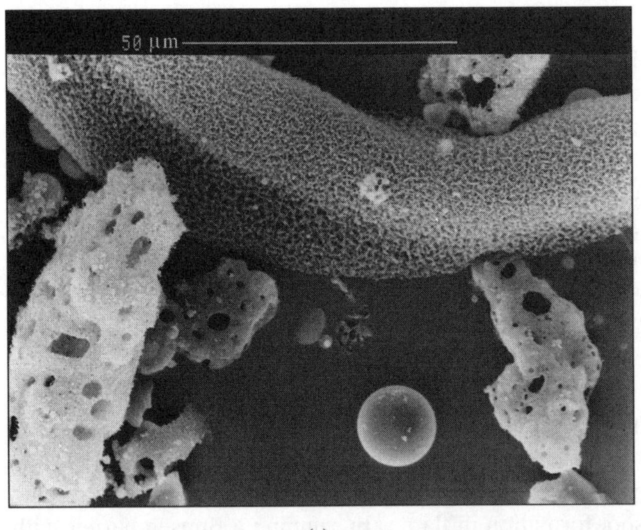

(a)

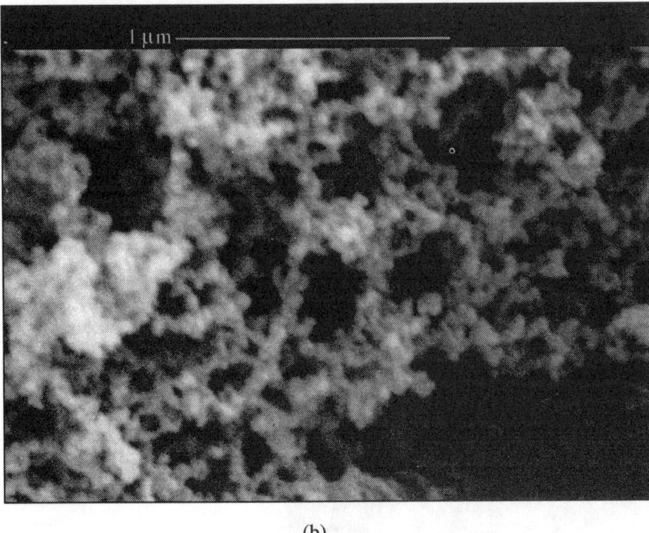

(b)

FIGURE 8.3
Scanning electron microscope (SEM) images of a variety of particles collected from a pulverized coal furnace. (*a*) Wide-angle view, (*b*) close-up of the upper part. The marks at the top of each panel indicate the distances corresponding to 50 microns and 1 micron, respectively. See the text for a description of these photos. In SEM photos one should pay attention to shape but not color; colors are not faithfully represented. (Courtesy of Dr. John Veranth, University of Utah and The Combustion Institute.)

(e.g., the $\approx 15\ \mu$ sphere near the bottom) are parts of the coal's mineral ash (mostly oxides of silicon, aluminum, calcium, and other metals) which melted in the furnace and were drawn into spherical shapes by surface tension. The photo shows one large particle of that type and many small ones. In some cases (not shown here) those particles are hollow (called *cenospheres*), suggesting that the ash particles which were initially dispersed in the coal melted and then agglomerated on the surface of a burning molten coal particle to form a spherical crust [3].

The irregular, porous particles (like the large one at the lower left) are pieces of coal *char*. Many coals burn just as wood in a campfire burns. The heat decomposes the wood or coal, driving off hydrocarbon gases, which burn in a yellow diffusion flame, leaving behind a porous residue, which is mostly carbon plus the mineral ash. In a campfire this residue forms the glowing coals, which burn slowly with no visible flame. In this furnace there was not enough time and oxygen to burn up this char, so char particles appear in the ash. The large piece at the top is *soot*, which forms long, lacy filaments up to a few millimeters long. At high magnification (Fig. 8.3*b*) one can see that the soot is an agglomerate of spherical particles, typically 0.01 to 0.03 microns in diameter. These very small particles were formed in the flames from the vaporized hydrocarbons produced by decomposing the coal. One may easily demonstrate soot formation in flames by running a Bunsen burner with the air inlet closed and letting the flame play on a cool surface; the black deposit (lampblack, soot, carbon black) is made of these very small particles, formed in the flame, mostly by polymerization of CH_3 radicals. Without the cool surface, lacy soot filaments form in the air and drift away. The lacy filaments are easily broken up into smaller ones and can be dispersed into individual very small particles. Black diesel exhaust is mostly soot.

If enough excess air had been supplied and enough time allowed, we would expect that all the char and soot would have burned, and only the spherical metal oxide ash particles would have remained. Figure 8.4 shows an example of that type

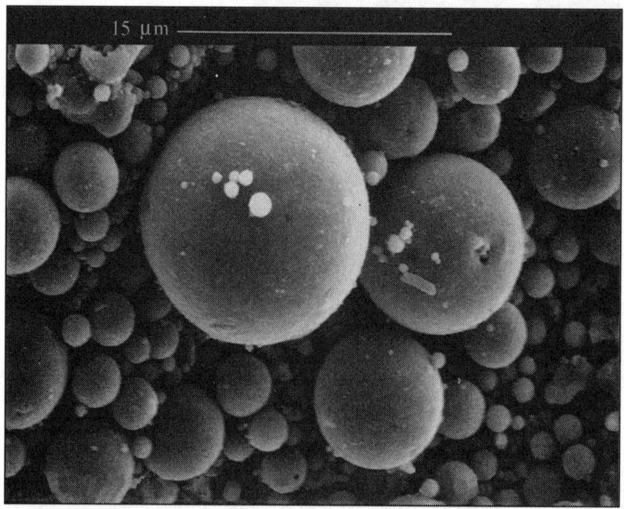

FIGURE 8.4
SEM image of a group of particles collected from a furnace that burns pulverized coal. See the text for a description of these photos. (Courtesy of Dr. Uschi Graham, University of Kentucky.)

of ash; only the spherical particles of metal oxides are present. The largest particle is roughly 20 μ in diameter, the smallest visible particles are about 0.3 μ in diameter. There are many more small particles than large. Urban atmospheres contain all the types of particles shown in Fig. 8.3, plus some others. Air pollution control devices must collect a wide variety of types of particles.

Another property of fine particles that is different from our experience with particles as large as sand grains is that when two fine particles are brought into direct physical contact, they generally will stick together by electrostatic and van der Waals bonding forces.* You have probably observed that you can pick up a seed or a grain of salt or sugar by pushing down on it with your finger, and then lifting. You can do this with any very small particle, because the surface attractive forces between your finger and any small particle are strong enough to overcome the gravity force on the small particle. You can probably not pick up a baseball this way unless your hands are very sticky!

Electrostatic and van der Waals forces are, in general, proportional to the surface area of the particle. Most of the particles we are used to are large enough that gravity or inertia will overcome electrostatic or van der Waals forces; and we all know that unless they are wet, sand grains do not stick together. But gravity and inertia forces are proportional to the particle mass, which is proportional to D^3, whereas the surface area (and hence the electrostatic and van der Waals forces) are proportional to D^2. Thus, as the particle size decreases, D^3 goes down much faster than D^2 so that the ratio of electrostatic and van der Waals to inertia and gravity forces becomes larger. As a result, if you had a handful of 1-μ particles that had been brought together into intimate contact and threw them into the air (as you probably have thrown a handful of sand at one time or another), they would not fragment into individual 1-μ particles but rather would break up into agglomerates that are the size of ordinary sand. With a microscope you could see that these agglomerates were really masses of much smaller particles held together by the forces just described.

For this reason, the basic strategy of control for particulate pollutants is to agglomerate them into larger particles that can be easily collected. This can be done by forcing the individual particles to contact each other (as in settling chambers, cyclones, electrostatic precipitators, or filters) or by contacting them with drops of water (as in wet scrubbers), all of which are discussed in Chapter 9. Agglomeration also occurs spontaneously in the atmosphere, as discussed in Sec. 8.4.

An additional peculiarity of particulate pollutants is that they can be formed in the atmosphere from gaseous pollutants. This means that if, for example, we could prevent the emission of all particulate pollutants, we would still find particles in our atmosphere. These latter particles are often called *secondary particles,* to distinguish them from those found in the atmosphere in the form in which they were emitted,

---

* Van der Waals forces are the intermolecular forces that hold ordinary liquids like gasoline and oils together. They are not as strong as the covalent forces that hold individual molecules together, nor the ionic forces that hold crystals together, nor even the hydrogen bonds that hold polar liquids like water and aqueous solutions together [4].

which are called *primary particles* [5]. These secondary particles are formed mostly from hydrocarbons, oxides of nitrogen, and oxides of sulfur.

From the theory of light scattering [6] we know that the particles that are most efficient (per unit mass or unit volume) in scattering light are those that have diameters close to the wavelength of light. From the "Electromagnetic Waves" row on Fig. 8.1, we see the wavelengths of visible light are about 0.4 to 0.8 μ. Particles in this size range are the most efficient light-scatterers. The hazy days and visible smog that occur in our cities are largely caused by secondary particles that tend to form in this size range. The Great Smoky Mountains National Park of North Carolina and Tennessee got its name from secondary particles formed from the hydrocarbons emitted from the beautiful forests on those mountains.

Near the middle of Fig. 8.1 we see "Lung-damaging dust," which has sizes from about 0.5 to 5 μ. Tests show that particles larger than about 10 μ are removed in our noses and throats; very few get into the trachea or bronchi. Particles in the size range 5 to 10 μ are mostly removed in the trachea and bronchi and do not get to the lungs [7]. Some authors use the term *inhalable particles* to refer to all particles smaller than 10 μm and *respirable particles* to refer to those smaller than 3.5 μ [8]. One also regularly sees the term *fine particles,* which has several definitions, most often particles smaller than 2.5 μ.

8.2 SETTLING VELOCITY AND DRAG FORCES

The second row from the bottom of Fig. 8.1 shows the "Terminal gravitational settling" (velocity) for spheres of specific gravity 2.0, meaning the velocity with which a particle settles through the atmosphere or through water. Ignoring for the moment where these numbers come from, we can see that the value for a coarse sand grain with a diameter of 1000 μ (= 1 mm) in air is 600 cm/s (or 6.0 m/s = 19.7 ft/s). This is much higher than the common vertical wind velocities of the atmosphere, so that it is rare for the wind to blow such particles up or to hold them up once they are in the air. (Sandstorms do occasionally occur in desert areas; they are destructive but, fortunately, infrequent and brief.) For this reason, although a factory that emitted large quantities of 1000-μ sand-sized particles into the air would be a nuisance to its neighbors, it would not contribute much to regional air pollution because almost all of the particles would settle to the ground near the plant.

The same row on Fig. 8.1 shows that the terminal settling velocity of a 1-μ diameter particle is 0.006 cm/s (or 0.00006 m/s or 0.000197 ft/s). The vertical movements of outdoor air (and even the air in most rooms) normally exceed this value, so particles this size do not quickly settle out of the atmosphere, as coarse sand would, but rather move with the gas and remain in suspension for long periods.

Thus, we distinguish between *dust,* which settles out of the atmosphere quickly because of its high gravitational settling velocity, and *suspendable particles,* which settle so slowly that they may be considered to remain in the atmosphere until they are removed by precipitation. There is no clear and simple dividing line between the two categories, but if we must make such an arbitrary distinction, it would be made somewhere near a particle diameter of 10 μ. (Particles small enough to remain suspended in the atmosphere or in other gases for long times are called *aerosols,* which indicates that they behave *as if* they were *dissolved* in the gas.)

Because the basic strategy of most particulate collection devices is to bring the particles into contact with each other so that they can coalesce and grow in size, we must have some knowledge of the drag forces that the surrounding air or gas exerts on such particles when we try to move them in order to evaluate such devices. It is easiest to examine these drag forces and settling velocities together, which we do here.

8.2.1 Stokes' Law

Figure 8.5 shows the forces acting on a spherical particle settling through a fluid under the influence of gravity. Writing Newton's law for the particle, we obtain

$$ma = \rho_{\text{part}} \left(\frac{\pi}{6}\right) D^3 g - \rho_{\text{fluid}} \left(\frac{\pi}{6}\right) D^3 g - F_d \tag{8.1}$$

The ma term represents the sum of the forces acting on the particle, equal to the downward acceleration of the particle. The three terms on the right represent, respectively, the gravity, buoyant, and drag forces acting on the particle. As we shall see later (and all know from personal experience), these drag (or air resistance) forces increase with increasing speed and are zero for zero speed. If the particle starts from rest, its initial velocity is zero, so the drag force in this equation is initially zero. The particle accelerates rapidly; as it accelerates, the drag force increases as the velocity increases, until it equals the gravity force minus the buoyant force. At this *terminal settling velocity,* the sum of the forces acting is zero, so the particle continues to move at a constant velocity. To find this velocity, we set the acceleration to zero in Eq. (8.1) and find

$$F_d = \left(\frac{\pi}{6}\right) D^3 g(\rho_{\text{part}} - \rho_{\text{fluid}}) \tag{8.2}$$

To find the velocity, we need the relation between F_d and the velocity. Stokes worked this out mathematically for a set of assumptions that are generally quite good for

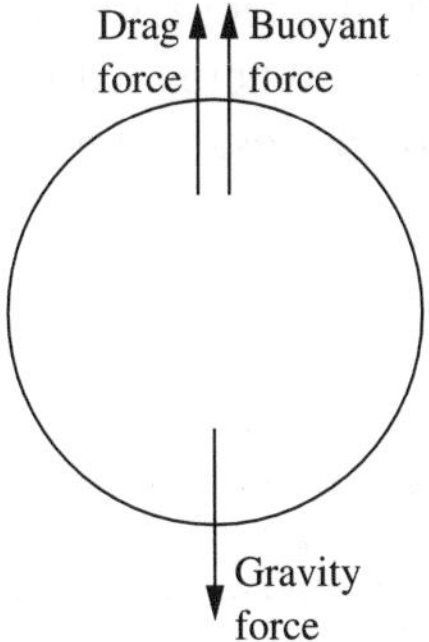

FIGURE 8.5
The forces acting on a particle in a fluid.

most of the problems in this book, finding

$$F_d = 3\pi\mu DV \tag{8.3}$$

where μ = the viscosity of the fluid (see inside the back cover). If we substitute Eq. (8.3) into Eq. (8.2) and solve for V, we find

$$V = gD^2 \frac{(\rho_{\text{part}} - \rho_{\text{fluid}})}{18\mu} \tag{8.4}$$

which is commonly referred to as Stokes' law.*

Example 8.1. Compute the terminal settling velocity in air of a sphere with diameter 1 μ. See inside the back cover for the properties of air and of particles assumed in all examples in this book. Substituting those values in Eq. (8.4), we find

$$V = \frac{(9.81\ \text{m/s}^2)(10^{-6}\ \text{m})^2(2000\ \text{kg/m}^3 - 1.20\ \text{kg/m}^3)}{(18)(1.8 \times 10^{-5}\ \text{kg/m} \cdot \text{s})}$$

$$= 6.05 \times 10^{-5}\ \frac{\text{m}}{\text{s}} = 0.00605\ \frac{\text{cm}}{\text{s}} = 1.99 \times 10^{-4}\ \frac{\text{ft}}{\text{s}} \quad ■$$

Comparing this result with the 6×10^{-3} cm/s that we read off Fig. 8.1 in the previous section, we see that they agree to within chart-reading accuracy. (The difference between the 20°C here and the 25°C on that chart is less than our uncertainty in chart reading.) That agreement should not surprise us; that part of Fig. 8.1 was made up from Eq. (8.4). We also see that the air density in the $(\rho_{\text{particle}} - \rho_{\text{fluid}})$ term contributed little to the answer. If we had set it equal to zero, our answer would have been 1.0006 times the answer shown above. We rarely know the actual particle diameters to this accuracy, so that for most air pollution applications of Eq. (8.4) we drop the ρ_{fluid} term. For gases under high pressure this omission might lead to significant error, but in most air pollution applications it will not. For gravitational settling in liquids we seldom use this simplification.

Stokes' law has been well verified for the range of conditions in which its assumptions hold good. However, for both very large and very small particles these assumptions break down. The situation is illustrated in Fig. 8.6, which is a logarithmic plot of settling velocity in air as a function of particle diameter for spheres with a specific gravity of 2. For the range of values in which Stokes' law applies, the result is a straight line with slope 2 on log paper.

*Stokes' law can be derived mathematically without the aid of experimental data. In doing so, we must assume (1) the fluid is continuous, (2) the flow is laminar, (3) Newton's law of viscosity holds, and (4) in the resulting equations the terms that involve velocities squared are negligible. The latter condition is called *creeping flow*. Even with these assumptions the derivation takes several pages [9]. The complexity is due to the three-dimensionality of the flow around a sphere.

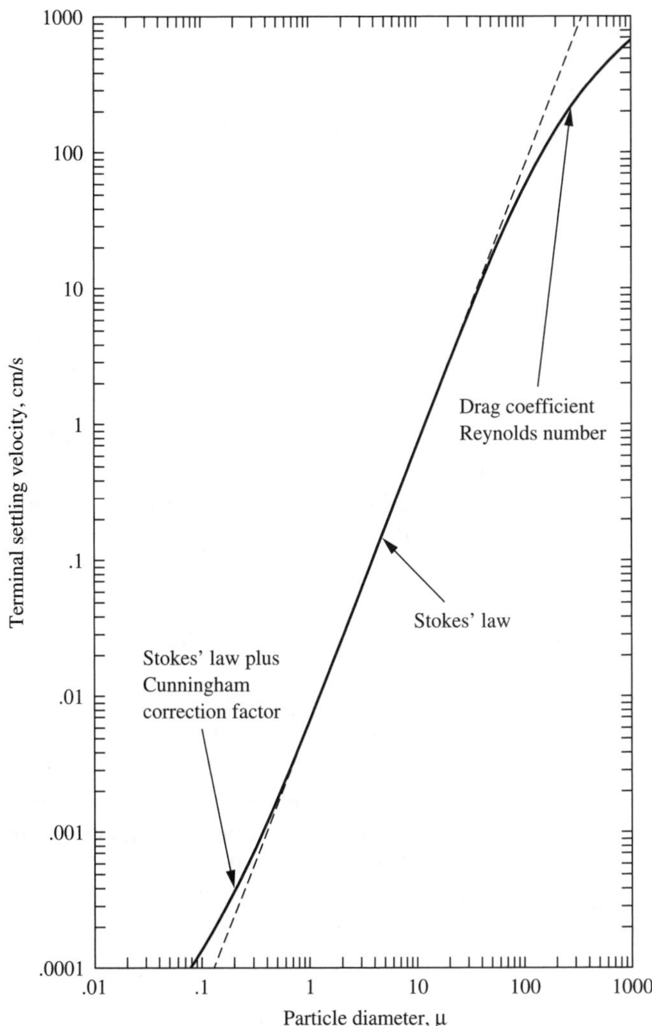

FIGURE 8.6
Terminal settling velocities for spherical particles with specific gravity = 2, in standard air.

8.2.2 Particles Too Large for Stokes' Law

As we go to larger and larger particles, we eventually find that the flow of the fluid around the sphere no longer obeys the approximation that the terms involving velocities squared are negligible. Thus the Stokes' drag equation, which is based on that assumption, becomes inaccurate. At still larger particle sizes, the flow of fluid around the sphere becomes turbulent and the principal assumptions of Stokes' law then become inapplicable.

Although various efforts have been made to derive a formula equivalent to Eq. (8.3) for larger particles, no theoretical formula represents experimental data over more than a modest range of values. However, the experimental data can be easily correlated by a nondimensional relationship. A new parameter, called the drag coefficient C_d, is defined by Eq. (8.5):

$$C_d = \frac{F_d}{(\pi/4)D^2 \rho_{\text{fluid}}(V^2/2)} \tag{8.5}$$

In addition, we introduce the Reynolds number for a particle:

$$\mathcal{R}_p = \frac{DV\rho_{\text{fluid}}}{\mu} \tag{8.6}$$

As discussed in all fluid mechanics texts, the Reynolds number is a dimensionless ratio of the inertial forces acting on a mass of fluid to the viscous forces acting on the same mass of fluid in the same flow. There are theoretical grounds for believing that, for smooth spheres in uniform, subsonic flow in constant-density Newtonian fluids, the drag coefficient (C_d) should depend on the Reynolds number alone (i.e., on a plot of C_d versus $\mathcal{R}_p$, all the data for all sizes of spheres and all constant-density Newtonian fluids should fall on a single curve). Experimentally, this has been shown to be true within the range of experimental accuracy. Such plots appear in most fluid mechanics books.

The reader may verify that the Stokes' drag term [F_d in Eq. (8.3)] can be substituted in Eq. (8.5) and the result rewritten as $C_d = (24/\mathcal{R}_p)$. This is an alternative way of writing the Stokes' drag term. Experimentally, it has been found that Stokes' law represents the observed behavior of particles satisfactorily for Reynolds numbers less than about 0.3. For larger values of the Reynolds number, the experimental value of C_d is larger than the $(24/\mathcal{R}_p)$ predicted by Stokes' law. For the $0.3 \le \mathcal{R}_p \le 1000$, the experimental drag coefficient data can be represented with satisfactory accuracy by the following empirical, data-fitting equation [10] (see Problem 8.9):

$$C_d = \frac{24}{\mathcal{R}_p}(1 + 0.14\mathcal{R}_p^{0.7}) \tag{8.7}$$

Example 8.2. A spherical particle with diameter 200 μ is falling in air. If Stokes' law were correct for this particle, how fast would it be falling, and what would its Reynolds number be?

We can take our result from Example 8.1, and observe that, if Stokes' law applies, then the velocity is proportional to the diameter squared, so

$$V = \left(0.00605\ \frac{\text{cm}}{\text{s}}\right)\left(\frac{200\ \mu}{1\ \mu}\right)^2 = 2.42\ \frac{\text{m}}{\text{s}} = 7.94\ \frac{\text{ft}}{\text{s}}$$

$$\mathcal{R}_p = \frac{DV\rho_{\text{fluid}}}{\mu} = \frac{(200 \times 10^{-6}\ \text{m})(2.42\ \text{m/s})(1.20\ \text{kg/m}^3)}{(1.8 \times 10^{-5}\ \text{kg/m}\cdot\text{s})} = 32.3 \qquad \blacksquare$$

This result is clearly greater than the Reynolds number of 0.3, which is the normal upper limit for reliable use of Stokes' law.

Example 8.3. Estimate the true settling velocity of the 200-μ diameter particle in Example 8.2 using the experimental drag coefficient correlation in Eq. (8.7).

In this case we must use a trial-and-error solution because we do not know the real Reynolds number. In principle we could combine Eqs. (8.2), (8.5), (8.6), and (8.7) and solve for V, but the resulting transcendental equation cannot be solved analytically (see Problem 8.12).

We begin by assuming that the particle Reynolds number is 20. Then from Eq. (8.7) we see that

$$C_d = \frac{24}{20}(1 + 0.14 \times 20^{0.7}) = 2.57$$

Solving Eq. (8.5) for the velocity and substituting, we find

$$V_t = \left(\frac{F_d}{C_d(\pi/8)D^2 \rho_{\text{fluid}}} \right)^{1/2}$$

Here, at the terminal settling velocity, $F_d = mg = (\pi/6)D^3 \rho_{\text{part}} g$. Thus

$$V_t = \left[\frac{(4/3)D\rho_{\text{part}} g}{C_d \rho_{\text{fluid}}} \right]^{1/2}$$

Inserting values, we find

$$V_t = \left[\frac{(4/3)(200 \times 10^{-6}\ \text{m})(2000\ \text{kg/m}^3)(9.81\ \text{m/s}^2)}{2.57(1.20\ \text{kg/m}^3)} \right]^{1/2} = 1.30\ \frac{\text{m}}{\text{s}} = 4.3\ \frac{\text{ft}}{\text{s}}$$

Now we must check our assumption that $\mathcal{R}_p = 20$. Using this value of the velocity, we find

$$\mathcal{R}_p = \frac{DV\rho_{\text{fluid}}}{\mu} = \frac{(200 \times 10^{-6}\ \text{m})(1.30\ \text{m/s})(1.20\ \text{kg/m}^3)}{(1.8 \times 10^{-5}\ \text{kg/m} \cdot \text{s})} = 17.3$$

From Eq. (8.7) we compute that this corresponds to $C_d = 2.82$. Since the calculated V_t is proportional to $(1/C_d)^{1/2}$, we can compute the new estimate of the terminal velocity as

$$V_t = 1.30\ \frac{\text{m}}{\text{s}} \left(\frac{2.57}{2.82} \right)^{1/2} = 1.24\ \frac{\text{m}}{\text{s}}$$

Repeating the $\mathcal{R}_p$ calculation, we find $\mathcal{R}_p = 16.5$. For this $\mathcal{R}_p$ Eq. (8.7) shows a C_d of about 2.90, leading to a new velocity estimate of 1.22 m/s = 4.0 ft/s and a new $\mathcal{R}_p$ of 16.2. At this point, we may consider the problem solved, because further cycles of trial-and-error will make changes in the answer smaller than our uncertainty in the drag coefficient correlation. ■

Example 8.3 shows the following:

1. For a 200-μ particle of specific gravity 2 settling in air at 20°C, the true velocity is only 50 percent (= 1.22/2.42) of the velocity calculated by Stokes' law.
2. This type of calculation by trial and error is tedious.

For the latter reason it is common practice to make up plots of velocity versus particle diameter and read directly from them, e.g., Fig. 8.6. One may also read the velocity for this case from the bottom of Fig. 8.1, finding (as closely as one can read it) the same value shown here. Figure 8.7 is similar to Fig. 8.6, but it covers a range of particle densities and also shows particle settling velocities in water.

The conclusion from these two examples is that for particles much larger than about 50 μ, one would make a serious error by using Stokes' law instead of the relations based on experimental drag coefficients. The situation is shown on the right side of Fig. 8.6, where the experimental settling velocities, labeled Drag coefficient Reynolds number, are significantly less than those shown by the extrapolated straight line, which represents Stokes' law.

8.2.3 Particles Too Small for Stokes' Law

When the particle becomes very small, another of the assumptions leading to Stokes' law becomes inaccurate. Stokes' law assumes that the fluid in which the particle is moving is a continuous medium. We know that real gases, liquids, and solids are not truly continuous but are made up of atoms and molecules. As long as the particle we are considering is much larger than the spaces between the individual gas molecules or atoms, the fluid interacts with the particle as if it were a continuous medium. When a particle becomes as small as or smaller than the average distance between molecules, then its interaction with molecules changes. When a particle has a large number of molecular collisions per unit time, most of the molecules bounce off the particle in a *specular* (mirror-like) way, with angle of reflection equaling angle of incidence. This is the way billiard balls collide. If the number of collisions is small, then some significant fraction of the colliding gas molecules are adsorbed onto the surface of the particle and remain long enough to "forget" what direction they came from. In this case their direction of leaving is *diffuse,* meaning random, subject to some statistical rules.

The effect of the change from specular to diffuse reflection is to lower the drag force, which causes the particle to move faster. The most widely used correction factor for this change has the form

$$F_d = \frac{F_{d\text{-Stokes}}}{1 + A\lambda/D} \tag{8.8}$$

where A = an experimentally determined constant

λ = mean free path (the average travel distance of a gas molecule between successive collisions)

$F_{d\text{-Stokes}}$ = the drag force computed according to Stokes' law

The $(1 + A\lambda/D)$ term used here is commonly called the *Cunningham correction factor*. It is only applicable for values of λ/D of order of magnitude one. For much

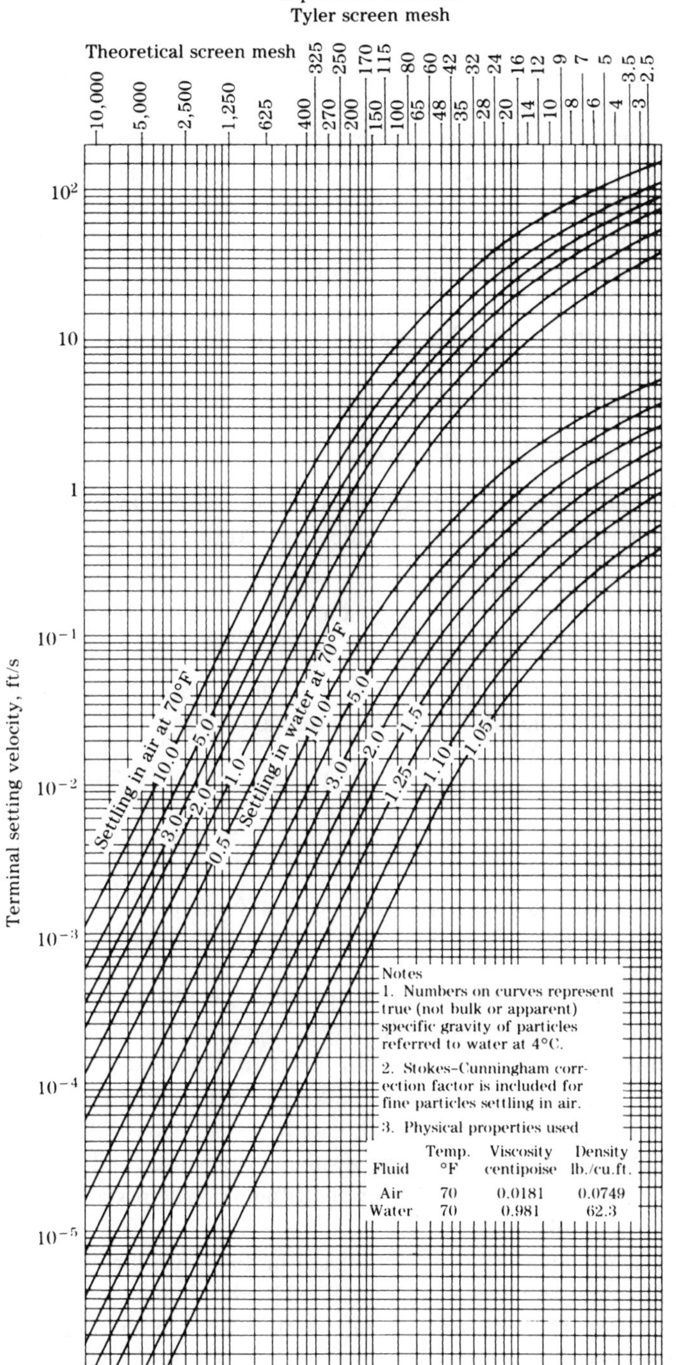

FIGURE 8.7

Terminal settling velocities of spherical particles of different densities settling in air and water at 70°F under the influence of gravity. (From C. E. Lapple, et al., *Fluid and Particle Mechanics,* University of Delaware, Newark, 1951, p. 292.) (Observe that the scale is 1, 1.5, 2, 2.5, 3, 3.5, 4, 5....)

larger values of λ/D, more complex formulae are used [11]. Although formulae for the precise calculation of λ are reasonably well-known (based on the measured viscosity of gases), the A in Eq. (8.8) is not a universal constant for all particles but varies from one kind of particle to another. Most workers use the value found by Millikan for oil droplets settling in air, $A = 1.728$; this is not derived theoretically, nor is it necessarily applicable to other kinds of particles or other gases but it is widely used because we do not have better information.

Example 8.4. A spherical particle with diameter 0.1 μ is settling in still air. What is its terminal settling velocity?

By combining Eqs. (8.2) and (8.8), we find that the velocity is given by

$$V = V_{\text{Stokes}}(1 + A\lambda/D) \tag{8.9}$$

where V_{Stokes} is the terminal settling velocity calculated from Stokes' law. We find that V_{Stokes} is our answer from Example 8.1, divided by 100 (due to the smaller diameter, squared), or

$$V_{\text{Stokes}} = 6.05 \times 10^{-7}\ \frac{\text{m}}{\text{s}} = 6.05 \times 10^{-5}\ \frac{\text{cm}}{\text{s}} = 1.98 \times 10^{-6}\ \frac{\text{ft}}{\text{s}}$$

The mean free path λ depends on T, P, and M. For air at one atmosphere and room temperature $\lambda \approx 0.07$ μ (see Problem 8.14), so the correction term is

$$1 + A\frac{\lambda}{D} = 1 + 1.728\left(\frac{0.07\ \mu}{0.1\ \mu}\right) = 2.21$$

and

$$V = 6.05 \times 10^{-7}\ \frac{\text{m}}{\text{s}}(2.21) = 1.34 \times 10^{-6}\ \frac{\text{m}}{\text{s}} = 4.39 \times 10^{-6}\ \frac{\text{ft}}{\text{s}} \quad \blacksquare$$

This example shows that for a 0.1-μ diameter particle, the estimated settling velocity is 2.21 times faster than one would predict from Stokes' law. Because of the uncertainties of determining the right value of A for this particular kind of particle as well as the shortage of experimental data for this kind of experiment (which is tedious and difficult to make), the calculated result is only an estimate. But it does show that we expect such small particles to have a smaller drag force and thus to settle more rapidly and to be moved more rapidly by centrifugal or electrostatic force than one would predict from Stokes' law.

For oil droplets Fuchs suggested the limits listed in Table 8.1 for errors in the applicability of Stokes' law, and Stokes' law modified by the simple form of the Cunningham correction factor. The table shows for Example 8.4 that the simple form of the Cunningham correction factor we used is probably in error by about 10 percent, and that for the most precise work we should use the more complex versions, which include terms involving (λ/D) to powers higher than one. Since most of the particle collectors discussed in Chapter 9 work on particles with diameters in the range of 1.6 to 70 μ, most are designed (at least in the early design stages) on the basis of Stokes' law. For the final design calculation, the Cunningham correction factor is used as

TABLE 8.1
Calculated applicability range of Stokes' law, alone, and with the simple form of the Cunningham correction factor

Formula	Permissible error	
	1%	10%
Stokes' law [Eq. (8.4)]	$16 < D < 30$ μ	$1.6 < D < 70$ μ
Stokes' law with the simple form of the Cunningham correction [Eq. (8.9)]	$0.36 < D < 30$ μ	$0.1 < D < 70$ μ

Source: Ref. 11.

well. Referring to Fig. 8.6 we see on the left that the curve marked Stokes' law plus Cunningham correction factor lies above the straight line representing Stokes' law. As Example 8.4 shows, for 0.1-μ particles the calculated settling velocity is 2.2 times the Stokes' law velocity.

8.2.4 Stokes Stopping Distance

Example 8.5. A 1-μ diameter spherical particle with specific gravity 2.0 is ejected from a gun into standard air at a velocity of 10 m/s = 32.8 ft/s. How far does it travel before it is stopped by viscous friction? Here we ignore the effect of gravity.

Here we apply Newton's law $F = ma$. The drag force is the only force acting on the particle after it leaves the gun. It operates in the direction opposite the direction of motion and is given by the Stokes drag resistance, Eq. (8.3), modified by the Cunningham correction factor, Eq. (8.8). Inserting these, we have

$$F = -\frac{3\pi \mu D V}{C} = ma = \frac{\pi}{6} D^3 \rho_{\text{part}} \frac{dV}{dt}$$

$$\frac{dV}{dt} = -\frac{18\mu V}{D^2 \rho_{\text{part}}\, C} \tag{8.10}$$

Here C is the Cunningham correction factor, normally taken as $(1 + A\lambda/D)$. Substituting $dt = dx/V$, separating variables, canceling the two V terms, and integrating, we find

$$\int_{V=V_o}^{V=0} dV = -\frac{18\mu}{D^2 \rho_{\text{part}}\, C} \int_{x=0}^{x=x} dx \tag{8.11}$$

and

$$x_{\text{Stokes stopping}} = \frac{V_0 D^2 \rho_{\text{part}}\, C}{18\mu} \tag{8.12}$$

This expression defines the *Stokes stopping distance*. Inserting values [including the value of C from Eq. (8.9)] into Eq. (8.12), we have

$$x_{\text{Stokes stopping}} = \frac{(10 \text{ m/s})(10^{-6} \text{ m})^2 (2000 \text{ kg/m}^3) 1.12}{18(1.8 \times 10^{-5} \text{ kg/m} \cdot \text{s})} = 6.9 \times 10^{-5} \text{ m} = 69 \text{ μ} \quad \blacksquare$$

This value is surprisingly small; the particle stops in 0.07 mm = 0.0027 inch. This makes clear that for particles of this size, and most particles of air pollution interest, the air is a very viscous fluid indeed. Intuitively this is comparably viscous to a baseball thrown into a pot of cold maple syrup.

In Chapter 9 we will see that the Stokes stopping distance is a natural distance scale for the behavior of particles. Several of the control-efficiency relations we develop there include one term that is the ratio of the Stokes stopping distance to some dimension of the piece of control equipment.

8.2.5 Aerodynamic Particle Diameter

Equation (8.12) also shows that any two particles that have the same value of $D^2 \rho_{\text{part}}\, C$ will have the same Stokes stopping distance for any initial velocity (in air with the same viscosity). We will see in Chapter 9 that any two particles with the same value of this set of properties will behave identically in several kinds of control devices. They have the same aerodynamic behavior. For that reason, we define a new property, the *aerodynamic particle diameter*:

$$\text{Aerodynamic particle diameter} = D_a = D(\rho_{\text{part}}\, C)^{1/2} \tag{8.13}$$

Often one sees this definition with the C omitted. This is a peculiar diameter, because it has the dimensions [(length) (mass/length^3)$^{1/2}$], e.g., [(m)(kg/m^3)$^{1/2}$]. It is strange to speak of a diameter with this kind of dimension, but that is the common usage. Thus the particle in Example 8.5 would have an aerodynamic particle diameter, D_a, of

$$D_a = 0.1\ \mu \left(2\ \frac{\text{g}}{\text{cm}^3} \cdot 2.21\right)^{0.5} = 0.21\ \mu \left(\frac{\text{g}}{\text{cm}^3}\right)^{0.5} = 0.21\ \mu_a$$

where the symbol μ_a stands for "microns, aerodynamic." In SI units this should be stated as 0.21 μm (1000 kg/m^3)$^{0.5}$, but that usage is seldom seen.

8.2.6 Diffusion of Particles

Small particles move by Brownian motion, which we describe according to the equations for diffusion. If a particle is large, then in any short period of time (e.g., 1 s) it will experience many collisions with the surrounding gas molecules that are hitting it from all sides, and the resulting net force will be quite small relative to the mass of the particle. Thus we do not see houses, desks, or marbles being moved about by Brownian motion. If the particle is small enough that it can only expect a few collisions per second and its inertia is small because of its small size, then the force of an individual collision is enough to make it move. Subsequent collisions, whose directions are random, will move it in other directions, so that its time-series path will be a series of short jumps in one direction and then another. One must use a microscope to observe such behavior, because the particle size at which it becomes important is too small for our eyes to distinguish.

In a uniform solution or suspension, Brownian motion does not cause any net change in the concentration with time in any part of the solution. As many particles move one way as move another. But if the concentration is not uniform, then Brownian motion tends to equalize the concentration. In so doing, it makes the particles move by diffusion, just as molecules in nonuniform solutions do. From diffusion theory, we know that for three-dimensional, nonsteady-state diffusion

$$\frac{\partial c}{\partial t} = \mathcal{D}\left(\frac{\partial^2 c}{\partial x^2} + \frac{\partial^2 c}{\partial y^2} + \frac{\partial^2 c}{\partial z^2}\right) \tag{8.14}$$

where $\mathcal{D}$ = diffusivity (normal units m^2/s)

c = concentration

For steady-state, one-dimensional diffusion Eq. (8.14) reduces to the well-known *Fick's law* of diffusion,

$$\text{Flux} = \frac{\text{Diffusive flow rate}}{\text{Unit area}} = -\mathcal{D}\frac{dc}{dx} \tag{8.15}$$

For spherical particles suspended in a perfect gas, $\mathcal{D}$ may be estimated from the kinetic theory of gases as

$$\mathcal{D} = \frac{kTC}{3\pi\mu D} \tag{8.16}$$

where k = Boltzmann constant

C = Cunningham correction factor from Eq. (8.8)

Example 8.6. Estimate the diffusivity of a 1-μ diameter particle in air at 20°C and 1 atm.

For a 1-μ diameter particle the Cunningham correction factor can be shown from Eq. (8.8) to be about 1.16, so

$$\mathcal{D} = \frac{(1.38 \times 10^{-23}\ \text{kg} \cdot \text{m}^2/\text{s}^2 \cdot \text{K})(293.15\ \text{K})(1.16)}{(3\pi)(1.8 \times 10^{-5}\ \text{kg/m} \cdot \text{s})(10^{-6}\ \text{m})} = 2.8 \times 10^{-11}\ \frac{\text{m}^2}{\text{s}} \quad \blacksquare$$

Most gases diffuse in air with diffusivities of about 10^{-5} m^2/s, and diffusivities of solutes in liquids are typically about 10^{-9} m^2/s. Thus, particles on the order of a few microns do not diffuse rapidly. We may now turn back to Fig. 8.1 and observe that along the bottom of the page the values of $\mathcal{D}$ are shown for particles in air and water; the student may verify that the result in Example 8.6 is the same value shown on that figure.

We will see that for some collection devices diffusion plays a measurable role, and also that the coalescence behavior of fine particles in the atmosphere is governed by diffusion, with the diffusivity values shown in Example 8.6 and on Fig. 8.1.

8.3 PARTICLE SIZE DISTRIBUTION FUNCTIONS

So far we have considered a single particle, or a group of particles, all with the same size. But in particulate air pollution problems we are concerned with groups of particles having a variety of sizes (see Fig. 8.4). To discuss such groups and to make useful calculations about their behavior in collection devices, we need some way of describing the particle size distributions. This section discusses *distribution functions* and their application to groups of particles. Students who are familiar with distribution functions can skip this section.

8.3.1 A Very Simple Example: The Population of the United States

Table 8.2 shows the age distribution of the population of the United States, taken from the 1990 census. On the basis of the first line, 18.35 million people had not reached their fifth birthday by the date of the census. Dividing this number by the total at the bottom of the second column, we see that this was 7.38 percent of the total population. The next line shows a similar set of numbers; we also see in the rightmost column that the cumulative total population in the age range zero to nine was 14.66 percent. Every number in the rightmost column is the sum of the number above it (the cumulative percent up to the previous age group) and the number to its left (the incremental percent in this age group).

Figures 8.8 and 8.9 on pages 229 and 230, show the same information as Table 8.2, plotted in two different ways, integral and differential. Figure 8.8 shows a smooth curve drawn through the values in the rightmost column of Table 8.2 plotted vs. the age corresponding to the end of each interval. The resulting plot is a cumulative (integral) distribution of ages in the population of the United States.

TABLE 8.2
Population of the United States, 1990

Age range, years, Δn	Population, millions, ΔN	% in this age range, $\Delta\Phi$	Cumulative % of the total population up to age n, Φ
0–4	18.35	7.38	7.38
5–9	18.10	7.28	14.66
10–14	17.11	6.88	21.54
15–19	17.75	7.14	28.68
20–24	19.02	7.65	36.32
25–29	21.31	8.57	44.89
30–34	21.86	8.79	53.68
35–39	19.96	8.03	61.71
40–44	17.62	7.08	68.79
45–49	13.87	5.58	74.37
50–54	11.35	4.56	78.94
55–59	10.53	4.23	83.17
60–64	10.62	4.27	87.44
65–69	10.11	4.07	91.51
70–74	7.99	3.21	94.72
75–79	6.12	2.46	97.18
80–84	3.93	1.58	98.76
85+	3.08	1.24	100.00
Total	$248.7 = N$		

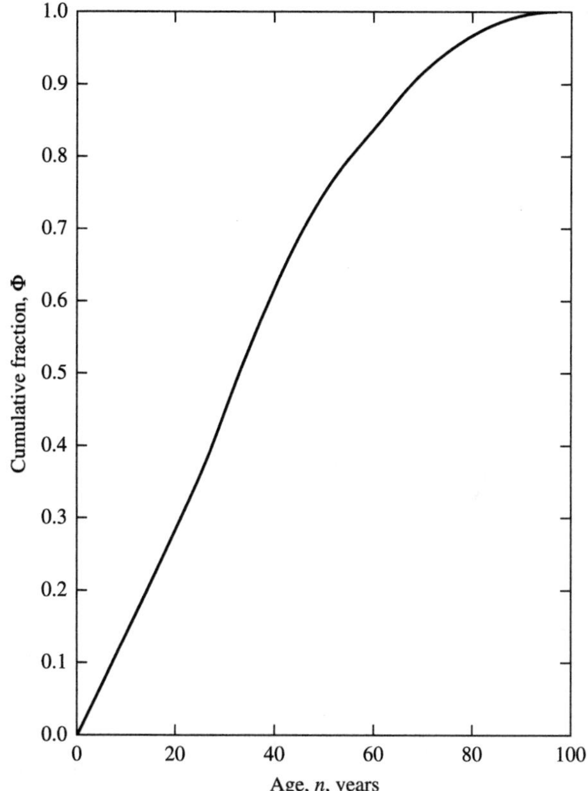

FIGURE 8.8
Age distribution for the United States in 1990 in integral form, based on Table 8.2.

If the total number of people in any age range, column 2 in Table 8.2, is ΔN and the total number of people in the whole population is N, and n represents some specific age, then the fraction of the population with ages n or less is Φ, defined by

$$\Phi = \frac{\sum_0^n \Delta N}{\sum_0^\infty \Delta N} \tag{8.17}$$

Φ has values from 0 (actually $1/248.7$ million, practically zero, corresponding to the most recently born baby) to 1.00 ([actually 1.00 minus $1/248.7$ million], practically 1.00, corresponding to the oldest person in the population). Figure 8.8 and Table 8.2 show Φ going from 0 to 1 (or 0 to 100 percent). In the language of statistics, the Φ curve is called a *normalized* curve, which means that all values have been divided by a suitable total so that the value of the variable ranges from 0 to 1. *Normalization* is common in statistics and practically universal in the study and use of distribution functions.

Figure 8.9 shows a plot of $d\Phi/dn$ vs. n. It is much more informative than Fig. 8.8. On it we can see that the birthrate reached a peak about 1960 (1990 minus 30) and a low about 1932 (1990 minus 58). These dates correspond to the baby boom following World War II and the birthrate decline corresponding to the Great

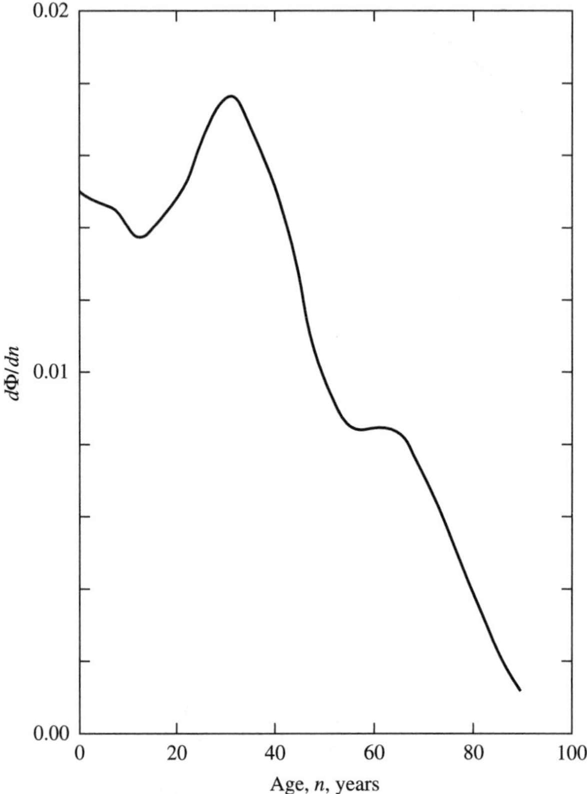

FIGURE 8.9
Age distribution for the United States in 1990 in differential form, based on Table 8.2.

Depression. This information could also be found from a careful examination of Fig. 8.8, but the information is much clearer on Fig. 8.9. (Plotting the integral in Fig. 8.8 smooths out most of the interesting details.) How can we relate these two figures? Table 8.2 shows the $\Delta\Phi$ values that correspond to each five-year period. If we divided these values by the time interval, five years, we would get the values of $\Delta\Phi/\Delta n$. So, for example, for the first five-year period, we would have

$$\frac{\Delta\Phi}{\Delta n} = \frac{\left(\dfrac{18.35 \text{ million}}{248.7 \text{ million}}\right)}{5 \text{ yr}} = 0.0148/\text{yr} = 1.48\%/\text{yr}$$

In making Fig. 8.9 this value (in decimal form) was plotted at 2.5 years, the corresponding values for each subsequent interval were plotted at the midpoint of those intervals, and a smooth curve was drawn through the points. For the last interval, 85+, Δn was arbitrarily selected as 15 years.

Since Fig. 8.8 is a plot of Φ vs. n, the slope of the curve at any value of n must be $d\Phi/dn$ at that n, and hence the smooth curve drawn through the values on Fig. 8.9 must represent the derivative of the curve in Fig. 8.8. From this it follows that the area under the curve on Fig. 8.9 (plotted as $d\Phi/dn$ vs. n) from n equals zero to infinity must have the value 1.0. Patient students might try counting squares to see if this is correct; it is.

Table 8.2 gives the most detailed information, but it is not very intuitive. The two figures are much more helpful for visualizing the situation. However, if we want to represent the data in the most compact form and to make mathematical manipulations with it, we would like to have some mathematical distribution function to represent the data. This would be of the form Φ = function of n. If two populations of different sizes have a similar distribution of ages (e.g., the United States and Canada), then they will have similar values of Φ as a function of n, even though the total populations' Ns are very different. If we know Φ as a function of n, and we know the total population N, then we can easily deduce the number of persons in any age segment of the population for either country.

Example 8.7. The simplest population distribution function we can think of is of the form $\Phi = (\text{constant}) \cdot n$, over the age range from 0 to n_{max}. We would expect this distribution in a population in which the birthrate was constant and everyone lived to the age n_{max}, but no longer. Such populations occasionally occur in science fiction, e.g., *Brave New World,* by Aldous Huxley.

Let us assume $n_{\text{max}} = 50$ years. Then, because we know that the Φ corresponding to n_{max} must be 1.0, we can determine the value of the constant, i.e.,

$$\text{Constant} = \frac{\Phi_{\text{max}}}{n_{\text{max}}} = \frac{1.00}{50\ \text{yr}} = \frac{0.02}{\text{yr}}$$

The figure analogous to Fig. 8.8 would be a straight line passing through the origin, with slope 0.02/yr, reaching 1.00 at 50 years. The figure corresponding to Fig. 8.9 would be a horizontal line with value $d\Phi/dn = 0.02/\text{yr}$ from 0 to 50 years, then dropping to 0 for all ages > 50 years. If we wanted to know the number of people in the age range 18 to 23 years we would need to know the total population N, which we here assume to be 500,000. Then

$$\begin{aligned} \Delta N &= N\,\Delta\Phi = N(\Phi_{\text{final}} - \Phi_{\text{initial}}) \\ &= 500\,000\left(\frac{0.02}{\text{yr}}\right)(23 - 18)\ \text{yr} = 50\,000\ \text{people} \end{aligned}$$

Here we would have to be clear that "people in the age range 18 to 23" means those who have had their eighteenth birthday, and have not yet had their twenty-third birthday. This distinction can be important in more complex cases. ■

This example is given in much more detail than such a simple distribution function requires; but the manipulations are the same for the more mathematically complex distribution functions that follow. Returning to the U.S. population example, we see there is no simple mathematical equation for Φ as a function of n. By brute-force curve fitting we can compute

$$\begin{aligned} \frac{d\Phi}{dn} &= 1.544 \times 10^{-2} - 5.7907 \times 10^{-4} n + 5.9344 \times 10^{-5} n^2 - 1.9224 \times 10^{-6} n^3 \\ &\quad + 2.3849 \times 10^{-8} n^4 - 1.2072 \times 10^{-10} n^5 \end{aligned} \tag{8.18}$$

which has no theoretical significance, but which represents the data fairly well (correlation coefficient = $R^2 = 0.966$).

Table 8.2, Figs. 8.8 and 8.9, and Eq. (8.18) are all different ways of representing the same set of experimental data. Given any one of them, and the value of N, one could reproduce all the others. The equation has the least intuitive content but is most satisfactory if we wish to use the data in a computer. The table also has little intuitive content, but is the most precise representation (although census data are only estimates of the true population). The two figures give the most intuitive picture of the data. In the following parts of this section, we look at the relation between distribution equations and their corresponding plots and tables.

The complexity of human behavior is great and so variable over time that one can seldom find a simple mathematical description of human behavior. (Equation 8.18 is brute-force and ugly!) However, for phenomena that do not involve individual human decisions, often we can find a satisfactory mathematical description of Φ as a function of some suitable variable. For example, if we measure the diameter of 1000 grains of beach sand and make a plot of Φ vs. diameter, we will probably find that the resulting curve can be satisfactorily represented by some relatively simple mathematical relation. Many distribution functions have been found to represent natural phenomena, e.g., the Gaussian or normal distribution, the log-normal, the gamma, the Weibull, the Poisson, etc. All of these are of the form Φ = some function of some parameter like age, or diameter, or wind speed, etc.

8.3.2 The Gaussian, or Normal, Distribution

The most famous and most widely used distribution function is the *Gaussian,* or *normal,* or *error distribution function.* It represents a great variety of observed distribution data well and is described by

$$\frac{d\Phi}{dx} = \frac{1}{\sigma\sqrt{2\pi}} \exp - \left[\frac{(x - x_{\text{mean}})^2}{2\sigma^2}\right] \tag{8.19}$$

Here Φ has the same meaning as before (i.e., the fraction of the cumulative total in the size range of interest), x is some suitable dimension or measure (e.g., age, diameter, etc.), x_{mean} is the average value of x (a suitably chosen average; there are several choices), and σ^2 is a quantity called the variance, which can be considered as a constant for the purposes of Eq. (8.19).

Equation (8.19) shows that if we plot $d\Phi/dx$ vs. x as in Fig. 8.9, the plot must be symmetrical about x_{mean} because $(x - x_{\text{mean}})$ enters squared. The maximum value of $d\Phi/dx$ must occur at $x = x_{\text{mean}}$ because for that value the exponential term is 1.0, and for any other value of x it is smaller. When x is a large positive or negative number the exponential term will approach zero asymptotically, so the curve must approach zero in both directions moving from the center, producing a symmetrical, bell-shaped curve. A small value of σ makes the argument of the exponential term larger, so that the values are concentrated near x_{mean}; a large value of σ spreads the values out over a wide range of xs. Therefore, for the same x_{mean}, a small σ will give a tall, narrow bell, whereas a large σ will give a low, broad bell. For any value of σ the area under the bell-shaped curve from $x = -\infty$ to $x = +\infty$ is equal to 1.0. [The $(1/\sigma\sqrt{2\pi})$ term ahead of the exponential makes this integration come out right; it is called the *normalizing factor*].

So far, no one has found a way to integrate Eq. (8.19) analytically to get the explicit equation that we would like for Φ as a function of x, x_{mean}, and σ. (Many great mathematicians have tried; fame and fortune await the clever student who can do it!) But although there is no available analytical solution, the integration has been performed numerically, and tables of its values are widely available. Rather than treat x, x_{mean}, and σ as separate variables, all of these tables combine them by defining a new variable z as

$$z = \text{number of standard deviations from the mean} = \frac{(x - x_{\text{mean}})}{\sigma} \tag{8.20}$$

(z is sometimes called the number of *probits* from the mean.)

Substituting this definition into Eq. (8.19) and simplifying, we find

$$\frac{d\Phi}{dz} = \frac{1}{\sqrt{2\pi}} \exp\left(-\frac{z^2}{2}\right) \tag{8.21}$$

In this equation we can separate variables and integrate numerically to obtain a table of Φ as a function of z. Table 8.3 on page 234 presents the results of such a numerical integration. Much more detailed tables are available in mathematical handbooks. Table 8.3 does not contain any negative values of z because, as one can see from Eq. (8.21), the value of $d\Phi/dz$ is the same for a positive or negative value of z. It also shows that for $z = 0$ the value of Φ is 0.5; the distribution is symmetrical about $z = 0$, $\Phi = 0.5$.

Example 8.8. An investigator reports that the height of adult males in the United States is well represented by the normal, or Gaussian, distribution, with x = height, $x_{\text{mean}} = 5.75$ ft, and $\sigma = 0.8$ ft.

If this is correct, what fraction of this population is taller than 6 ft? shorter than 4 ft? Are there any men taller than 10 ft? Are there any shorter than 1 ft?

For 6 ft we have

$$z = \frac{6\text{ ft} - 5.75\text{ ft}}{0.8\text{ ft}} = 0.31$$

From Table 8.3 we see that this value of z corresponds to a Φ of approximately 0.62, which indicates that $(1 - 0.62) = 0.38 = 38$ percent of this population is predicted to be taller than 6 ft. For 4 ft we find

$$z = \frac{4\text{ ft} - 5.75\text{ ft}}{0.8\text{ ft}} = -2.19$$

Here we use the symmetry property of the normal distribution, shown at the bottom of Table 8.3, to calculate that

$$\Phi(-2.19) = 1 - \Phi(2.19) = 1 - 0.986 = 0.014 = 1.4\%$$

TABLE 8.3
Values of the cumulative frequency integral Φ as a function of z

z	Φ	z	Φ
0.0	0.5000	2.1	0.9821
0.1	0.5398	2.2	0.9861
0.2	0.5793	2.3	0.9893
0.3	0.6179	2.4	0.9918
0.4	0.6554	2.5	0.9938
0.5	0.6915	2.6	0.9953
0.6	0.7258	2.7	0.9965
0.7	0.7580	2.8	0.9974
0.8	0.7881	2.9	0.9981
0.9	0.8159	3.0	0.9986
1.0	0.8413	3.1	0.9990
1.1	0.8643	3.2	0.9993
1.2	0.8849	3.3	0.9995
1.3	0.9032	3.4	0.9997
1.4	0.9192	3.5	0.9998
1.5	0.9332	3.6	0.9998
1.6	0.9452	3.7	0.9999
1.7	0.9554	3.8	0.9999
1.8	0.9641	3.9	See Prob. 8.29
1.9	0.9713	4.0	See Prob. 8.29
2.0	0.9772		

Note: For negative values of z use $\Phi(-z) = 1 - \Phi(z)$. For example, $\Phi(-0.2) = 1 - \Phi(0.2) = 1 - 0.5793 = 0.4207$.

We would expect 1.4 percent of this population to have a height less than 4 ft. For 10 ft we compute $z = (10 - 5.75)/0.8 = 5.31$. Using the approximation in Problem 8.29, we find that $(1 - \Phi)$ is 6×10^{-8}. So, if this distribution truly represents the population, and if there are approximately 10^8 adult males in the United States, then we would expect to find about six men with a height above 10 ft. For the 1-ft-tall man, $z = -5.94$ and (see Problem 8.29) $\Phi = 0.14 \times 10^{-8}$, so we would expect to find about 0.14 adult male less than a foot tall in the population (or to find one 14 percent of the time). ■

This example shows how one uses the normal distribution function and Table 8.3. We also see that although the normal distribution is easy to use, it cannot be an absolutely correct description of this particular population, because we can be quite certain from observation that there are no adult males taller than 10 ft or shorter than 1 ft. One can carry this calculation out to even taller and shorter values, even to negative heights, and find a very small but nonzero probability that we will find a man with negative height. This should help the student realize that these mathematical distribution functions are *useful approximations* of experimental reality but not exact descriptions of nature. Generally, mathematical distribution functions like the normal, or others (log-normal, Weibull, gamma), do a satisfactory job of representing experimental data in the middle of the data range (where most of the data are) but become unreliable at representing the experimental data at the extreme values (tails) of the distributions.

Students are sometimes confused by the fact that x_{mean} and σ, which have exact and unambiguous definitions, appear in these distributions, which are approximations. For any sample with n members,

$$x_{\text{mean}} = \frac{1}{n}\sum x_i \quad \text{and} \quad s = \frac{1}{(n-1)}\left[\sum (x_i - x_{\text{mean}})^2\right]^{1/2} \tag{8.22}$$

These expressions are independent of whether the sample is best represented by the normal distribution function or some other distribution function, or is not well represented by any simple distribution function. In the limit, as n becomes large, the s in the preceding definition (the *sample standard deviation*) becomes σ, or (the *variance*)$^{0.5}$. Often people speak of statistics with the hidden assumption that the measurements we are discussing are taken from a population that is well represented by the normal distribution. That is generally a good guess, but not always right, as shown later.

Now we are ready to talk about particle size distributions. We can presumably obtain a sample of the particles in a gas stream by catching them on a filter or by some other technique; and we can count the particles of various sizes using a microscope and make up a table just like Table 8.2, with diameter replacing age range. However, we generally find that data obtained from this kind of experiment are *not* well represented by the normal distribution function of Eq. (8.19).

8.3.3 The Log-Normal Distribution

If we let x in Eq. (8.19) represent, not the particle diameter but its natural logarithm, we will obtain the following *log-normal distribution,* which is almost as widely used as the normal distribution:

$$\frac{d\Phi}{d\ln D} = \frac{1}{\sigma\sqrt{2\pi}}\exp -\left[\frac{(\ln D - \ln D_{\text{mean}})^2}{2\sigma^2}\right] \tag{8.23}$$

or, alternatively,

$$\frac{d\Phi}{d\ln D} = \frac{1}{\sigma\sqrt{2\pi}}\exp -\left[\frac{[\ln(D/D_{\text{mean}})]^2}{2\sigma^2}\right] \tag{8.24}$$

Many authors write Eqs. (8.23) and (8.24) with the σ replaced by $\ln\sigma$, i.e.,

$$\frac{d\Phi}{d\ln D} = \frac{1}{\ln\sigma\sqrt{2\pi}}\exp -\left[\frac{[\ln(D/D_{\text{mean}})]^2}{2(\ln\sigma)^2}\right] \tag{8.25}$$

Since σ is a constant for any particular distribution, this change makes no difference, except that the σ one finds using Eq. (8.25) is the exponential of the σ one finds using Eqs. (8.23) and (8.24). Typical values of σ in Eqs. (8.23) and (8.24) for particle distributions are 0.5 to 2, which correspond in Eq. (8.25) to σs of 1.64 to 7.39. The latter are often called *logarithmic standard deviations* or *geometric standard deviations* and are sometimes written σ_g. The smallest possible value (σ_g) is 1.0, corresponding to a σ of zero.

The value of z that we defined in Eq. (8.20) for the normal distribution is converted to the log-normal distribution of particle diameters by replacing every x by ln D, or

$$z = \left[\frac{(x - x_{\text{mean}})}{\sigma}\right]_{\text{normal distribution}}$$
$$= \left[\frac{(\ln D - \ln D_{\text{mean}})}{\sigma} = \frac{\ln(D/D_{\text{mean}})}{\sigma}\right]_{\text{log-normal distribution}} \tag{8.26}$$

The student may verify that substituting Eq. (8.26) into Eq. (8.23) converts the latter into Eq. (8.21). Thus we can use Table 8.3 for the log-normal distribution, just as we did for the normal distribution, with the proper value of z from Eq. (8.26).

Returning now to the problem of the particle distribution function, we see that if Eq. (8.23) or (8.24) is a satisfactory representation of the distribution and if we could plot ln D vs. z (which is a function of D), we should obtain a straight-line plot. Fortunately, graph papers are available that make this easy. On them one simply plots D vs. Φ, and if the data are log-normally distributed, the result is a straight line. Figure 8.10 shows such a representation on log-normal paper (most often called *log-probability* paper) of particle sizes normally encountered in the exhaust gas from pulverized-coal furnaces. This paper is plotted so that z proceeds linearly across the bottom of the paper; the values of Φ corresponding to any z (looked up on Table 8.3 or its equivalent) are shown instead of z itself. As a result, the scale is compressed in the middle and expanded greatly near the right and left edges. This kind of representation is practically universal in the air pollution literature. No other way of presenting particle size data seems to be nearly as successful or as widely used.*

Because the representation in Fig. 8.10 is common in air pollution work, let us familiarize ourselves with its properties (which are explored in much more detail in Ref. 12). First, we observe that the axes are reversed compared to Fig. 8.9; diameter is plotted vertically and Φ horizontally. Log-probability paper is always laid out that way. The line, and any straight line on log-probability paper, is a representation of Eq. (8.23). That equation contains only two constants, D_{mean} and σ. Thus, if we specify the line, we have specified these two values, and conversely if we specify these two values, we have specified one and only one line on log-probability paper. To find the value of D_{mean} from the line on Fig. 8.10, we need only read the diameter that corresponds to the 50 percent "less than stated" size. On Fig. 8.10 this is

*Many other natural phenomena are also well represented by the log-normal distribution. Most weather data—e.g., distribution of hourly wind speeds over a year—are better represented by the log-normal distribution than by any other distribution function.

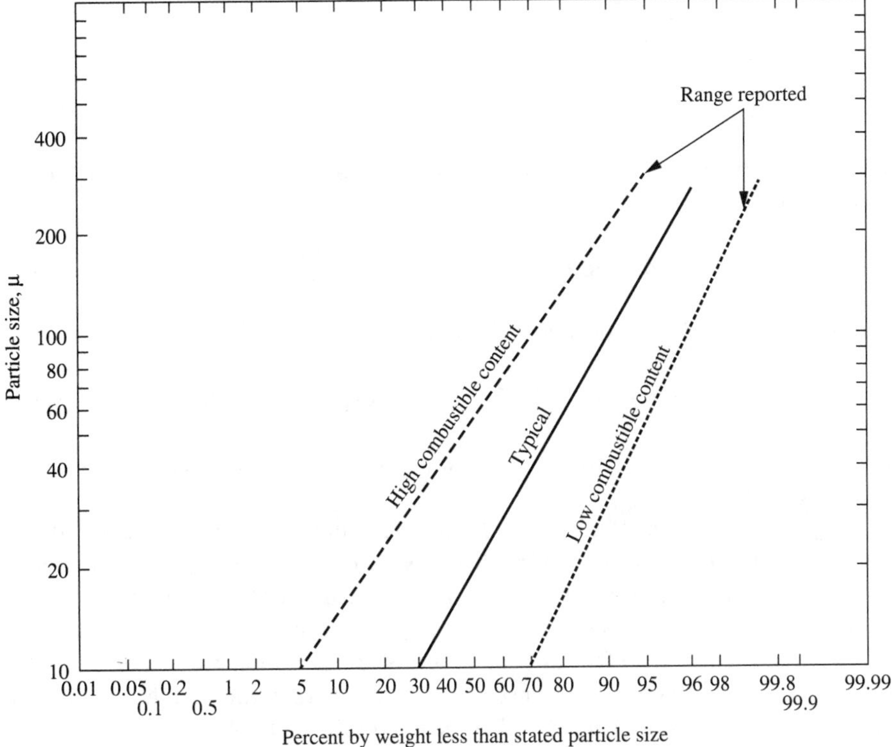

FIGURE 8.10
Example from the air pollution literature of the representation of particle size data in log-normal form. (From Ref. 13.) This is the distribution of the particles collected at the outlet of a pulverized-coal furnace. The particles in Fig. 8.4 would be approximately described by the "Typical" line on this figure.

approximately $D_{\text{mean}} = 20\ \mu$. Observe that this is not the arithmetic mean we are all used to. For a group of N particles with diameter D_i

$$\text{Arithmetic mean diameter} = \frac{1}{N}\sum D_i \tag{8.27}$$

and

$$\text{Log mean diameter} = \exp\left(\frac{1}{N}\sum \ln D\right) = (D_1 \cdot D_2 \cdot \cdots \cdot D_N)^{1/N} \tag{8.28}$$

The latter mean, called the *geometric* or *logarithmic mean,* is the value we obtained by reading the 50 percent point on Fig. 8.10. The reader may verify that these two means are not the same by considering a particle sample with only two particles, one with a diameter of 1 μ and the other with a diameter of 9 μ. Using Eqs. (8.27) and (8.28), we see that the arithmetic mean diameter is 5 μ and the log mean is 3 μ. For most particle size groupings encountered in nature this difference is not important, but when σ becomes large, it becomes more important.

To find σ from Fig. 8.10, we observe in Table 8.3 that $z = 1$ corresponds to $\Phi = 0.8413$, so that, in Eq. (8.20), we have

$$z = 1 = \frac{x_{0.84} - x_{\text{mean}}}{\sigma} \tag{8.29}$$

but in the distribution we are considering, the xs are the natural logs of the diameters, so we can solve for σ, writing

$$\sigma = \ln D_{0.84} - \ln D_{\text{mean}} = \ln \frac{D_{0.84}}{D_{\text{mean}}} \tag{8.30}$$

Reading the value of $D_{0.84}$ from Fig. 8.10 as about 70 μ, we find

$$\sigma = \ln \frac{70\ \mu}{20\ \mu} = 1.25; \qquad \sigma_g = \exp 1.25 = 3.49 \tag{8.31}$$

Thus, the complete characterization of the straight line drawn on Fig. 8.10 is $D_{\text{mean}} =$ 20 μ, $\sigma = 1.25$.

Because of the symmetry about D_{mean}, we could just as well have found the value of σ from D_{mean} and $D_{0.16}$ if we had wished (except that it is off-scale in this figure). Because of the utility of the values at 16 percent and 84 percent for estimating σ, some log-probability papers have heavy lines drawn in at those percentages.

All the discussion so far has been in terms of natural logarithms, or ln. Since $\log_{10} x = (\ln x)/2.303$, we can convert all the formulae in this chapter that are in terms of ln to $\log_{10}$ by inserting 2.303 at the appropriate places.

8.3.4 Distributions by Weight and by Number

If we determine the distribution by catching the particles on a greased microscope slide and measuring the diameter of a suitable number of particles, our results will be presented as the percent by number at various size ranges. That is not the most common way of representing the data.

Example 8.9. A group of particles consists of three members, one with a diameter of 1 μ, one with a diameter of 4 μ, and one with a diameter of 10 μ. All three are spheres, and all have the same density. What percent by number of the particles have diameters less than 5 μ? What percent by length, by surface area, and by mass have diameters less than 5 μ?

Here, by number, we have $2/3 = 66.6$ percent of the particles have diameters less than 5 μ. By length we see that if we were to line the particles in a row, the length of those less than 5 μ would be $(1 + 4)$ μ, whereas the total length would be $(1 + 4 + 10)$ μ; so the percent by length less than 5 μ is $(5/15) = 33.3\%$. The surface area of each particle is πD^2, so the surface area of the particles less than 5 μ is $\pi(1^2 + 4^2)\ \mu^2$. Taking the ratio of this sum to the total, and noting that the πs cancel, we find the percentage of the surface area in particles less than 5 μ is $(1 + 16)/(1 + 16 + 100) = 14.5\%$. Proceeding the same way for mass we observe that the mass of any particle is $(\rho\pi/6)D^3$, and that the $(\rho\pi/6)$ terms will cancel, so that the fraction of the mass in particles less than 5 μ in diameter is $(1+64)/(1+64+1000) = 6.1\%$. ■

This example shows that if one asks what percent of the particles is smaller than some value, without specifying which percent one means, one can get widely varying answers, all correct. In Fig. 8.10 the axis label makes clear that the percent shown there is percent by weight. That is the most commonly used percent in such distributions. Percent by number is also common. Percent by area is widely used in discussing sprays (e.g., spray dryers and paint sprayers) and sometimes in air pollution work. The percent by length has no common application.

A general—and very useful—property of log-normal distributions is that if Φ of D^a is log normal, then Φ of D^b is also log normal, the values of σ are the same for both distributions, and the mean of the new distribution is

$$D_{\text{new mean}} = D_{\text{old mean}} \exp[(b-a)\sigma^2)] \tag{8.32}$$

Example 8.10. Compute the D_{mean} by number that corresponds to the distribution given in Fig. 8.10, for which we know that in the distribution by weight we have $D\text{mean} = 20\ \mu$, and $\sigma = 1.25$.

In the distribution by weight $a = 3$ (because the weight of a particle is proportional to D^3) and in the distribution by number $b = 0$ (because the number of a particle is independent of its diameter, $D^0 = 1$). Substituting into Eq. (8.32), we find

$$D_{\text{mean, number}} = 20\ \mu \exp[(0-3)1.25^2] = 0.18\ \mu$$ ■

This difference in mean diameters by weight and by number appears startling but is correct. The big particles have almost all the weight, so the mean by weight is close to the diameter of the largest-size particles that are present in significant numbers. But there are many more small particles than large, so the number mean is much smaller than the weight (or mass) mean.

Because both distributions have the same value of σ, the lines representing them on log-normal paper are parallel. Thus, once we have computed the logarithmic mean by number in the preceding example, we could in principle draw a line parallel to the line on Fig. 8.10, passing through the $D_{\text{mean by number}}$, and have the complete distribution by number. (In Fig. 8.10 that line would run off the plot at the bottom). The saving in time and effort afforded by using this set of properties of the log-normal distribution is very great and is one of the principal reasons why almost all workers in the air pollution field have selected this distribution to represent particle size data.

8.4 BEHAVIOR OF PARTICLES IN THE ATMOSPHERE

Much of what we discussed in this chapter is illustrated by Fig. 8.11 on page 240, which describes the behavior of particles in the atmosphere. This is a plot of $d\Phi_{\text{by area}}/dD$, similar to Fig. 8.9. If $d\Phi_{\text{by mass}}/dD$ were plotted, the peak to the

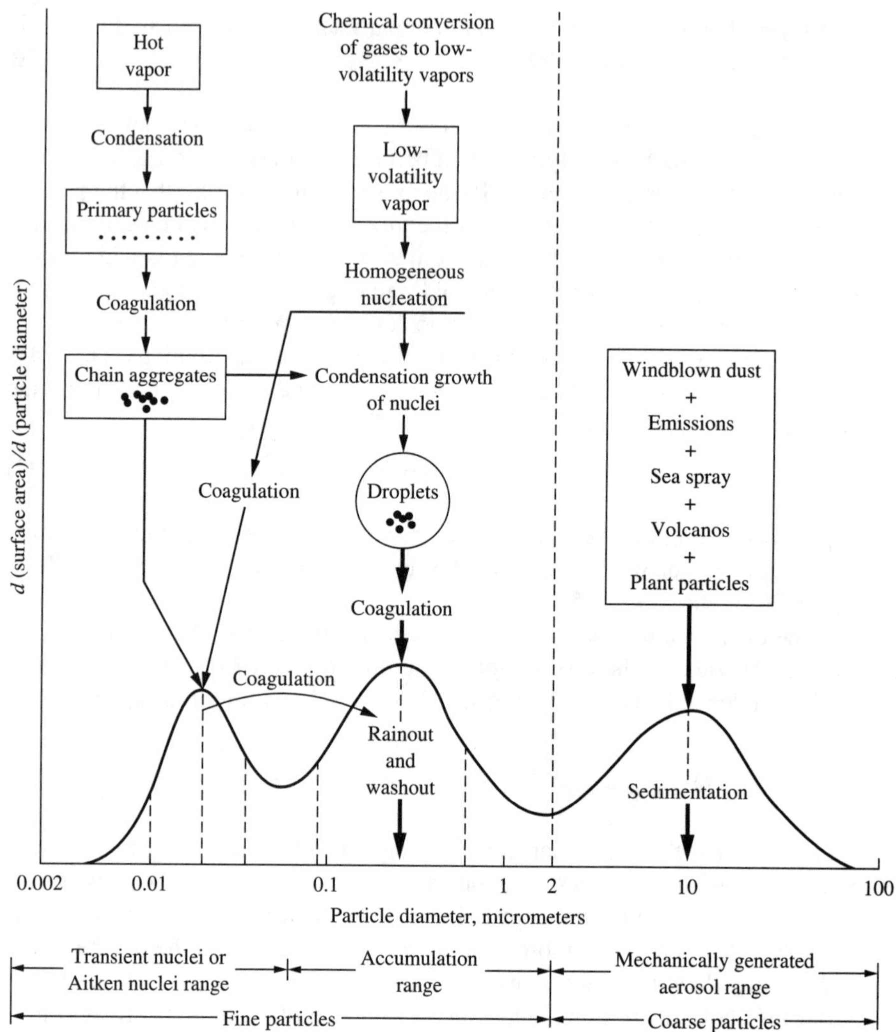

FIGURE 8.11
An estimate of the distribution of particles, *by surface area,* in an industrial atmosphere, after Whitby [14]. (Courtesy of EPRI.)

right would be much larger than the others, or if $d\Phi_{\text{by number}}/dD$ were presented, the peak at the left would be much larger than the others. It shows that the finest particles, with diameters 0.005 to 0.1 μ, enter the atmosphere mostly by condensation of hot vapors from combustion sources. Over time (usually several hours) these smallest particles grow, mostly by agglomeration onto each other. Some of this agglomeration occurs in the gas phase, caused by Brownian motion (diffusion) bringing them into contact; some occurs inside cloud or fog droplets.

Midsized particles (0.1 to 1 μ) are formed partly by the agglomeration of finer particles and partly by chemical conversion of gases and vapors to particles in the atmosphere. These particles are large enough to be removed by *rainout* (capture by drops in clouds) or *washout* (capture by falling raindrops). Although they do grow by agglomeration to form larger particles, this process is slow compared to rainout and washout. The larger particles (2 to 100 μ) are, as shown, mechanically generated; some are derived from industrial particle sources, whose control is discussed in the next chapter. These larger particles are mostly removed by gravity settling, with or without the action of clouds and rain.

The first two peaks in Fig. 8.11 represent almost exclusively secondary particles, formed in the atmosphere from gaseous precursors; the third peak represents for the most part primary particles, emitted to the atmosphere in particulate form. There is some deposition of smaller particles onto these primary particles, but it is not the major method of removal of these smaller particles.

The gaseous precursors of secondary particles are primarily SO_2, NO_x, NH_3, and hydrocarbons. The control of emissions of hydrocarbons, sulfur oxides, and nitrogen oxides are discussed in Chapters 10–12. Ammonia (NH_3) is widely distributed in the atmosphere, coming mostly from biological sources, rather than from human sources.

Figure 8.12 summarizes this chapter. On it we see a truck hauling sand down the road. Sand blows off the truck and falls to the ground, causing a local nuisance. The truck stirs up road dust and generates tire wear particles that are local air pollutants but that do not remain long in the atmosphere. The truck's exhaust contains fine

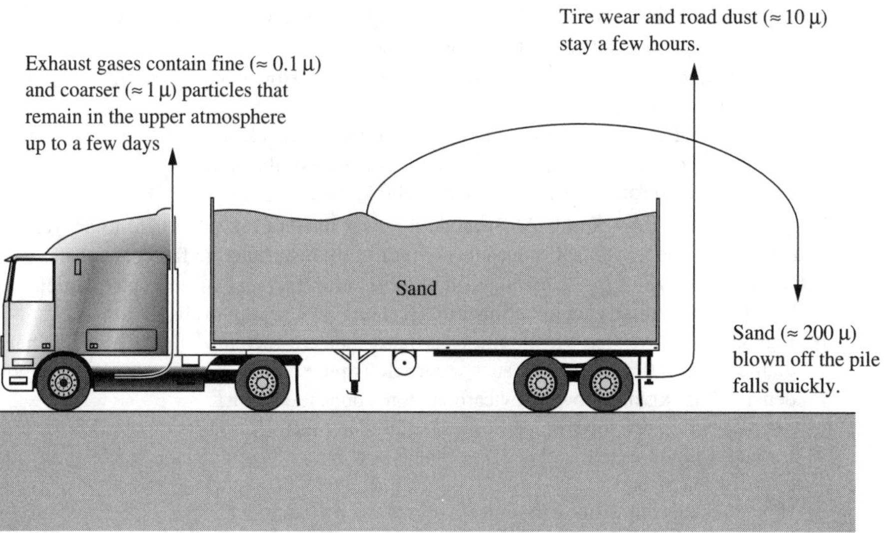

FIGURE 8.12
A truck, loaded with sand, puts three different sizes of particles into the atmosphere.

particles, generated by combustion, that remain in the atmosphere for several days and contribute to the regional fine particulate problem.

8.5 SUMMARY

1. The particles of air pollution interest are mostly in the size range 0.01 to 10 μ.
2. Particles smaller than about 2 μ are rarely produced by mechanical means; they are primarily produced by condensation or chemical reaction of gases or vapors.
3. These small particles behave quite differently from the particles with which we are familiar, like sand and gravel. Their high surface area per unit mass makes them adhere to one another if they are brought into contact.
4. Most particles of air pollution interest are in the size range where the Stokes' equation for the drag force on the particle can be used with satisfactory accuracy.
5. Because particles of air pollution interest are rarely present in the air or in a gas stream as a uniform particle size set, we normally have to deal with the distribution of particle sizes.
6. The fine particles in the atmosphere are largely secondary particles, formed in the atmosphere from gaseous precursors. Most of the coarser particles in the atmosphere are primary particles, which enter the atmosphere as particles.

PROBLEMS

See Common Units and Values for Problems and Examples, inside the back cover.

8.1. Determine the thickness of the pages in this textbook, in microns, by measuring the thickness of the text (excluding covers) and dividing that by the number of pages. Take into account the fact that page numbers go on both sides of the page.

8.2. One lbm of water is dispersed in droplets of diameter D. How many square feet of surface does the water have when $D = 1$ cm, 1 μ, 0.01 μ?

8.3. (*a*) If a solid material has a density of 1000 kg/m^3 and a particle of this material has a mass of 1 microgram and is a cube, how long is each side of the cube?
(*b*) Repeat part (*a*) for a particle of density 2000 kg/m^3.

8.4. A typical coal is 10 wt % ash. Most modern power plants grind their coal to an average particle size of about 100 μ. If the ash were uniformly distributed in the coal, what would be the expected size of the remaining ash particles after the coal was burned? Particles as small as 1 μ are regularly found in this ash. Explain how they are probably formed.

8.5. (*a*) Figure 8.1 shows that the smallest particles that are recognizable as particles have diameters of 0.01 μ. Suppose such a particle is pure carbon, atomic weight 12 g/mol, density 2000 kg/m^3. How many carbon atoms does it contain?
(*b*) Why does Fig. 8.1 show no particles smaller than this?

8.6. (*a*) Based on Fig. 8.1 (extrapolated!), estimate how far an SO_2 molecule would settle due to gravity in a year.
(*b*) How does this compare with typical vertical wind velocities?
(*c*) Is there any industrial process that separates gases by gravity?

8.7. Dustfall rates (sediment accumulations from the air) can be up to 100 tons/square mile · month.

(*a*) Is that number big or small?
(*b*) How many pounds per square foot per day is that?
(*c*) If the dust, in a settled condition, has a bulk density of 30 lbm/ft^3, how thick a layer will accumulate in a month?

8.8. In the U.S. air pollution literature and regulations, particle concentrations in gas streams are often expressed in grains/ft^3 (1 lbm = 7000 grains = 7000 gr; 1 gr = 0.065 g). These concentrations are normally abbreviated as gr/acf (grains per actual cubic ft) and often referred to as *grain loadings*.
(*a*) For a typical concentration of 100 gr/ft^3 in a dirty gas stream, what is the weight percentage of solids?
(*b*) What is the metric equivalent (g/m^3) of 100 gr/ft^3?
(*c*) If the particles are 10 μ spherical particles, how many are there in a ft^3?
(*d*) What is the most likely historical origin of the grain as a unit of mass?
(*e*) What other common materials normally have their masses expressed in grains?

8.9. Particles with a diameter of 1 mm (which corresponds roughly to coarse beach sand) are emitted from a tall stack. The wind is blowing at a velocity of 10 mi/h. The distance from the centerline of the stack to the plant's property line is five stack heights. What is the likelihood that most of the sand will fall on the plant's property?

8.10. To determine the diameter of a small spherical particle, we let it settle by gravity in air in the field of view of a microscope. The settling velocity was 0.001 ft/s. Estimate the diameter of the particle.

8.11. A particle is a hollow sphere of a metal oxide. The density of the metal oxide is 2000 kg/m^3. The hollow portion in the center of the sphere is full of air that has the same density as the surrounding air through which the sphere is falling at its terminal velocity. The outside diameter of the sphere is 10 μ and the thickness of its walls is 0.1 μ (i.e., the bubble in the center has a diameter of 9.8 μ). How fast is it falling?

8.12. (*a*) What value of C_d does Eq. (8.7) give for $\mathcal{R}_p = 0.3$?
(*b*) What is the percentage difference between this value and the Stokes' law value at this Reynolds number?

8.13. Example 8.3 shows the trial-and-error solution to a particle settling problem. Most students now have hand calculators with a "solve" routine that will do that trial-and-error calculation easily. Rework this problem on that kind of hand calculator:
(*a*) Combine Eqs. (8.2) and (8.5)–(8.7) and rearrange to

$$V + V^{1.7} \cdot 0.14 \left(\frac{D\rho_{\text{fluid}}}{\mu} \right)^{0.7} - \frac{D^2 g \rho_{\text{part}}}{18\mu} = 0$$

(*b*) Evaluate the constants, finding

$$V + V^{1.7} \cdot 0.8582 \text{ (s/m)}^{0.7} - 2.422 \text{ m/s} = 0$$

(*c*) Solve, using a solve routine, finding $V = 1.219$ m/s.

8.14. From the kinetic theory of gases we know that

$$\lambda = \frac{1}{\sqrt{2}\pi n \sigma^2} \tag{8.33}$$

where λ is the mean free path (the average distance a molecule travels between collisions), n is the number concentration of molecules (molecules/volume), and σ is the collision diameter of an individual molecule. This latter is determined by experimental viscosity measurements and has values in the range of 2 to $4 \cdot 10^{-10}$ m for common gases (see Fig. 8.1). For air the value is approximately $3.48 \cdot 10^{-10}$ m. The number concentration of molecules is Avogadro's number ($6.02 \cdot 10^{23}$ molecules/mol) times the molar density, which for ordinary gases at modest pressures is given by the idela gas law, $\rho = RT/P$.

Using these values, estimate the mean free path of air at 1 atm and 20°C.

8.15. A Crookes radiometer is an evacuated glass tube, with a vertical shaft, to which are attached small plates at a radius of a few centimeters. One side of each plate is polished like a mirror, and the other is painted flat black. The mirrors all point in one direction around the shaft, the black sides in the other direction. When the radiometer is placed in a bright light, the shaft rotates; the brighter the light, the faster it rotates.

(*a*) Which direction does it rotate, i.e., do the mirrored surfaces go forward or backward? Why?

(*b*) Would it behave the same way in a perfect vacuum? Why?

(*c*) How does this relate to the Cunningham correction factor?

8.16. A spherical particle with diameter 1 μ and specific gravity 4.0 is settling in still air.

(*a*) What is the terminal settling velocity of this particle, according to Stokes' law?

(*b*) What is the terminal settling velocity of this particle, according to Stokes' law, taking the Cunningham correction factor into account?

8.17. In Example 8.5 we saw that a particle with a 1-μ diameter, specific gravity of 2, and an initial velocity of 10 m/s would be stopped by air in a travel distance of 69 diameters.

(*a*) If we inject a baseball ($D = 2.9$ inches, $m = 0.32$ lbm) into a tank of some viscous fluid (molasses or honey or lube oil) at the same velocity and it is stopped in 69 diameters, what is the viscosity of the fluid? Assume Eq. (8.12) applies.

(*b*) Is the Reynolds number small enough for the Stokes' stopping distance to be applicable? If not, estimate what the observed stopping distance would be.

8.18. In Example 8.5.

(*a*) How long does it take the particle to come to zero velocity?

(*b*) How long does it take the particle to come to 1 percent of its initial velocity?

(*c*) What is the initial value of $\mathcal{R}_p$?

(*d*) If this is too large for the Stokes' drag force to be applicable, will the observed stopping distance be larger or smaller than that calculated in Example 8.5? By what percentage?

8.19. Figure 8.8 is a Φ vs. age plot for the United States in 1990. Assume that a childhood influenza had killed all the people in the United States born during the five-year period 1970 through 1975 and none of the subsequent immigrants to the United States had been born during that period.

(*a*) Sketch what Fig. 8.8 would look like in this circumstance.

(*b*) Sketch what Fig. 8.9 would look like in this circumstance.

8.20. Sketch the equivalents of Figs. 8.8, 8.9, and 8.10 for a particle group that is log-normally distributed, with each of the following sets of parameters (rough sketches with no numerical values will be satisfactory):

(*a*) $D_{\text{mean}} = 0.250$ in., $\sigma = 0.0001$

(*b*) $D_{\text{mean}} = 5.0$ in., $\sigma = 10$

(*c*) $D_{\text{mean}} = 10$ μ, $\sigma = 2$

(*d*) What physical systems might these distributions correspond to?

8.21. For the group of particles in Example 8.9:

(*a*) Sketch a plot of Φ *by mass* vs. particle diameter for this distribution, indicating all the important numerical values on your sketch. (The sketch can be quite rough and need not be to scale).

(*b*) Repeat (*a*) using Φ *by number*.

8.22. A group of particles is described by the log-normal distribution with $D_{\text{mean by weight}} = 5$ μ, and $\sigma = 0.8$.

(*a*) What fraction by weight of the particles have diameters less than 1 μ?
(*b*) What fraction by number of the particles have diameters less than 1 μ?

8.23. In Example 8.10 we computed the mass-mean diameter from the count-mean diameter by using a value of $b = 0$ in Eq. (8.32). What would be the physical significance of the distributions we would obtain if we had repeated the calculation in Example 8.10 using values of $b = 1$ and $b = 2$?

8.24. As described in Chapter 5, average wind velocities in the United States are about 10 mi/h. The highest values are about 100 mi/h, and the lowest about 1 mi/h. Assume for this problem only that 100 mi/h and 1 mi/h winds occur with equal frequency.
(*a*) Would this distribution of wind speeds be well represented by the normal distribution?
(*b*) Would it be well represented by the log-normal distribution?
(*c*) Sketch the equivalent of Fig. 8.9 for wind speeds, both in the normal and the log-normal form.

8.25. The emission factors table gives the following data for the particle size distribution in the waste gas from a mass-burn municipal waste incinerator [15]:

Particle diameter, μ	Φ, cumulative weight % to this diameter
0.625	14
1.0	18
2.5	24
5.0	32
10.0	37
15.0	47

No values are given for particles larger than 15 μ because they are of little air pollution interest.

Can these data be satisfactory represented by the normal distribution? by the log-normal distribution?

8.26. For the "Typical" line on Fig. 8.10, estimate the diameter that corresponds to 10 percent by weight and to 1 percent by weight.

8.27. (*a*) We now pass the "typical" particle group shown on Fig. 8.10 through a particle collector that is 100 percent efficient for particles larger than or equal to 10 μ in diameter, and zero percent efficient for particles with diameters less than 10 μ. For the particles that pass through this collector, sketch the equivalents of Figs. 8.8 and 8.9.
(*b*) Calculate the mass mean diameter and the number mean diameter of the particles that pass through.

8.28. If a population of particles is log normal with $D_{\text{mean by weight}} = 10$ μ and $\sigma = 1$, what is the diameter that has 99.9 percent of the weight smaller than it? What is the diameter that has 0.01 percent of the weight smaller than it?

8.29. Table 8.3 is easy to use with hand calculations, but not with a computer. Furthermore, it is not easily used for values of z greater than 3.8. For these purposes it is common to use the following algebraic approximation [16]:

$$z \approx t - \frac{a_0 + a_1 t}{1 + b_1 t + b_2 t^2}$$

where $t = \sqrt{\ln[1/(1-\Phi)^2]}$

$a_0 = 2.30753$

$a_1 = 0.27061$

$b_1 = 0.99229$

$b_2 = 0.04481$

This approximation can only be used for $\Phi > 0.5$ and has a maximum error in z of ± 0.003.

(*a*) Test the accuracy of this approximation by computing the value of z corresponding to a Φ of 0.9772 and comparing that estimate with the value in Table 8.3.

(*b*) Using this approximation, estimate the Φ corresponding to $z = 6.0$.

8.30. A gas stream contains a group of particles whose size distribution is given by the "rectangular distribution," which is

$$d\Phi/dD = C_1 \text{ for particle diameters from 0 to } D_{\max}$$

$$d\Phi/dD = 0 \text{ for particle diameters greater than } D_{\max}$$

Here Φ is the cumulative fraction by mass of particles with diameter less than D, and $D_{\max}$ is the diameter of the largest particle. C_1 is equal to $(1/D_{\max})$.

We pass this gas stream through a particle collector in which the collection efficiency is proportional to the particle diameter squared and is equal to 1.0 for a particle diameter of $D_{\max}$. What is the overall collection efficiency of this collector?

8.31. A gas stream has particles whose distribution is represented by the "triangular distribution function," which is

$$\frac{d(\text{weight fraction})}{d(\text{particle diameter})} = b(\text{particle diameter})$$

for sizes 0 to 10 μ and

$$\frac{d(\text{weight fraction})}{d(\text{particle diameter})} = b(e - \text{particle diameter})$$

for sizes 10 to 20 μ. Here, $b = 0.01/\mu^2$ and $e = 20$ μ.

A particle collection device has collection efficiency represented by the equation

$$\text{Efficiency} = a(\text{particle diameter})^2$$

over the range of 0 to 20 μ. Here a has the value $0.0025/\mu^2$.

We now pass this gas stream through this collector. What fraction by weight of the particles is collected?

8.32. A gas stream contains a group of particles whose particle size distribution by weight is given by the "quadratic distribution function," which is

$$\text{Weight fraction with diameter less than } D = k_1 D^2 \qquad \text{for} \qquad 0 < D < \sqrt{1/k_1}$$

We now pass this gas stream through a collector whose efficiency as a function of particle size is given by these equations:

$$\text{Fraction collected} = k_2 D \qquad \text{for} \qquad 0 < D < 1/k_2$$

$$\text{Fraction collected} = 1.0 \qquad \text{for} \qquad 1/k_2 < D$$

What weight fraction of the particles in the gas stream is caught by this collector? Here, $k_1 = 0.01/\mu^2$ and $k_2 = 0.1/\mu$.

8.33. A contaminated air stream contains particles that follow the log-normal distribution by mass with $D_m = 10\ \mu$, $\sigma = 1.5$. We now pass this gas through a separator that removes all particles $D \geq 5\ \mu$. All particles $D < 5\ \mu$ pass through.
(*a*) What fraction by mass of the particles is removed?
(*b*) Sketch the equivalents of Figs. 8.8, 8.9, and 8.10 for the remaining particles (i.e., those still in the gas stream).
(*c*) What is the mass mean diameter of the particles that are captured? What is the mass mean diameter of the particles that pass through uncollected?

8.34. A contaminated air stream contains particles that follow the log-normal distribution by mass, with $D_m = 5\ \mu$ and $\sigma = 1.5$. We pass this contaminated air stream through a particle collector that removes all the particles larger than 4 μ, and which is 50% efficient for particles in the size range 2 to 4 μ. What is the overall weight percent collection efficiency of this collector for these particles?

8.35. The particles in an air stream are described by the log-normal distribution, with $D_{\text{mean by mass}} = 10\ \mu$ and $\sigma = 1.5$. We now pass this dirty air stream through a collector that is 100% efficient for particles with $D \geq 40\ \mu$, 50% efficient for particles 10 to 40 μ in diameter, and 0% efficient for particles smaller than 10 μ.
(*a*) What fraction by mass is collected by this collector?
(*b*) What is the mass median diameter of the particles that pass through uncollected?

8.36. Figure 8.11 suggests that in a typical atmosphere, about one-third of the surface area of the particles is contained in particles with diameter centered about 0.02 μ, about one-third in particles with diameter centered about 0.3 μ, and about one-third in particles with diameter centered about 10 μ. If the true situation were that the distribution by area was exactly one-third in each of these diameter ranges and if, instead of the broad distributions shown in Fig. 8.11, all of the particles were exactly either 0.02, 0.3, or 10 μ in diameter, then
(*a*) What would the fraction by weight be for each of the three particle sizes?
(*b*) What would the fraction by number be for each of the three particle sizes?

REFERENCES

1. Fuchs, N. A., and A. G. Sutugin: *Highly Dispersed Aerosols,* Ann Arbor Science, Ann Arbor, MI, 1970.
2. Quann, R. J., M. Neville, M. Janghorbani, C. A. Mims, and A. F. Sarofim: "Mineral Matter and Trace Element Vaporization in a Laboratory Pulverized Coal Combustion System," *Environ. Sci. Technol.*, Vol. 16, pp. 776–781, 1982.
3. Sarofin, A. F., J. B. Howard, and A. S. Padia: "The Physical Transformation of the Mineral Matter in Pulverized Coal under Simulated Combustion Conditions," *Combust. Sci. Tech.*, Vol. 16, pp. 187–204, 1977.
4. Moore, W. J.: *Physical Chemistry,* 3d ed., Prentice-Hall, Englewood Cliffs, NJ, p. 675, 1962.
5. Fennelly, P. F.: "Primary and Secondary Particulates as Pollutants—A Literature Review," *J. Air Pollut. Control Assoc.*, Vol. 25, pp. 697–704, 1975.
6. Butcher, S. S., and R. J. Charlson: *An Introduction to Air Chemistry*, Academic Press, New York, p. 184, 1972.
7. Hatch, T. F., and P. Gross: *Pulmonary Deposition and Retention of Inhaled Aerosols,* Academic Press, New York, 1964.
8. Lippmann, M.: "Size-Selective Health Hazard Sampling," in S. V. Hering (ed.), *Air Sampling Instruments for Evaluation of Atmospheric Contaminants,* 7th ed., American Conference of Governmental Industrial Hygienists, Cincinnati, OH, p. 163, 1989.
9. Lamb, H.: *Hydrodynamics,* Dover, New York, pp. 597–598, 1932.
10. Sakiadis, B. C.: "Fluid and Particle Mechanics," in D. W. Green and J. O. Maloney (eds.), *Perry's Chemical Engineers' Handbook*, 6th ed., McGraw-Hill, New York, pp. 5–63, 1984.

11. Fuchs, N. A.: *The Mechanics of Aerosols*, Pergamon-Macmillan, New York, Chapter 11, 1964.
12. Aitchison, J., and J. A. C. Brown: *The Log-Normal Distribution*, Cambridge University Press, Cambridge, 1957.
13. Smith, W. S., and C. W. Gruber: "Atmospheric Emissions from Coal Combustion—An Inventory Guide," U.S. Department of Health, Education, and Welfare Publication No. AP-24, 1966.
14. Whitby, K. T.: "Modeling of Atmospheric Aerosol Particle Size Distributions," University of Minnesota–Minneapolis Mechanical Engineering Department, Particle Technology Laboratory Report No. 253, 1975.
15. "Compilation of Air Pollutant Emission Factors," AP-42, 4th ed., p. 2.1–9 (9/91 supplement), 1985. (See Ref. 7 of Chapter 4.)
16. Abramowitz, M., and I. A. Stegun (eds.): *Pocketbook of Mathematical Functions,* Verlag Harri Deutsch-Thun, Frankfurt/Main, p. 409, 1984.

CHAPTER 9

CONTROL OF PRIMARY PARTICULATES

As discussed in Chapter 8, most of the fine particles in the atmosphere are secondary particles. Nonetheless, the control of primary particles is a major part of air pollution control engineering. Many of the primary particles, e.g., asbestos and heavy metals, are more toxic than most secondary particles. Although primary particles are generally larger than secondary particles, many primary particles are small enough to be respirable and are thus of health concern. The average engineer is more likely to encounter a primary particle control problem than any other type of air pollution problem. If possible the collected particles are recycled to somewhere in the process that generates them. Most often (e.g., ash and soot from coal combustion), the collected particles go to a landfill.

9.1 WALL COLLECTION DEVICES

The first three types of control devices we consider—gravity settlers, cyclone separators, and electrostatic precipitators—all function by driving the particles to a solid wall, where they adhere to each other to form agglomerates that can be removed from the collection device and disposed of. Although these devices look different from one another, they all use the same general idea and are described by the same general design equations.

9.1.1 Gravity Settlers

A gravity settler is simply a long chamber through which the contaminated gas passes slowly, allowing time for the particles to settle by gravity to the bottom. It is an old, unsophisticated device that must be cleaned manually at regular intervals. But it is simple to construct, requires little maintenance, and has some use in industries treating very dirty gases, e.g., some smelters and metallurgical processes. Furthermore, the mathematical analysis for gravity settlers is very easy; it will reappear in modified form for cyclones and electrostatic precipitators.

Figure 9.1 shows a gravity settler. Its cross-sectional area (WH) is much larger than that of the duct approaching it or leading the gas away from it, so that the gas velocity inside is much lower than in either of those two ducts. Baffles of some kind are used to spread the incoming flow evenly across the settling chamber; without baffles most of the flow will go through the middle and poor particle collection will result.

To calculate the behavior of such a device, chemical engineers generally rely on one of two models. Either we assume that the fluid going through is totally unmixed *(block flow* or *plug flow model)* or we assume total mixing, either in the entire device or in the entire cross section perpendicular to the flow *(backmixed* or *mixed model)*. Each of these sets of assumptions leads to simple calculations. The observed behavior of nature most often falls between these two simple cases, so that with these two models we can set limits on what nature probably does. Both models are widely used in air pollution control device calculations. We will calculate the behavior of a gravity settler both ways.

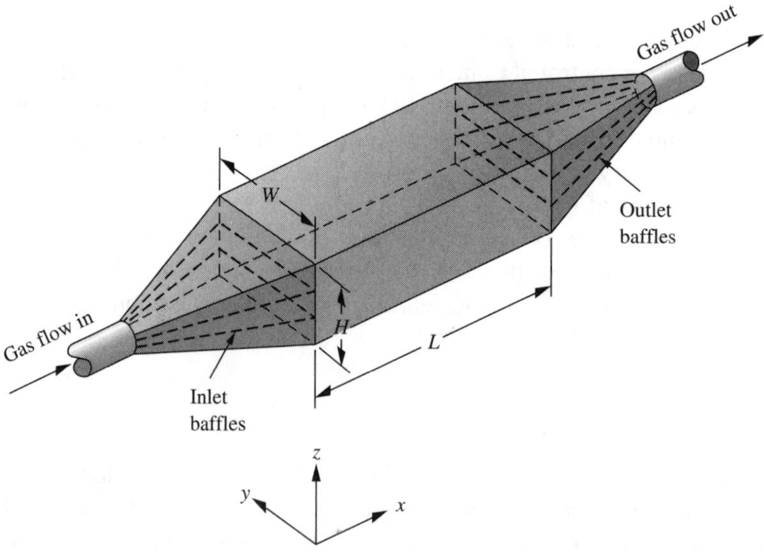

FIGURE 9.1
Schematic of a typical gravity settler.

For either block or mixed flow, the average horizontal gas velocity in the chamber is

$$V_{\text{avg}} = \frac{Q}{WH} \tag{9.1}$$

For the block flow model, we will assume

1. The horizontal velocity of the gas in the chamber is equal to V_{avg} everywhere in the chamber (but see Problem 9.1).
2. The horizontal component of the velocity of the particles in the gas is always equal to V_{avg}.
3. The vertical component of the velocity of the particles is equal to their terminal settling velocity due to gravity, V_t.
4. If a particle settles to the floor, it stays there and is not re-entrained.

With these assumptions we can compute the behavior of a gravity settling chamber according to the block flow model.

Consider a particle that enters the chamber some distance h above the floor of the chamber. The length of time the gas parcel it entered with will take to traverse the chamber in the flow direction is

$$t = \frac{L}{V_{\text{avg}}} \tag{9.2}$$

During that time the particle will settle by gravity a distance,

$$\text{Vertical settling distance} = tV_t = V_t \frac{L}{V_{\text{avg}}} \tag{9.3}$$

If this distance is greater than or equal to h (its original distance above the floor), then it will reach the floor of the chamber and be captured. If all the particles are of the same size (and hence have the same value of V_t), then there is some distance above the floor (at the inlet) below which all of the particles will be captured, and above which none of them will be captured. If we now further assume that all of the particles are the same size, that they are distributed uniformly across the inlet of the chamber, and that they do not interact with one another, then we can say that the fraction of particles that will be captured, which is the fractional collection efficiency, is

$$\text{Fraction captured} = \eta = \frac{LV_t}{HV_{\text{avg}}} \quad \text{for block flow} \tag{9.4}$$

To compute the efficiency-particle diameter relationship, we replace the terminal settling velocity in Eq. (9.4) with the gravity-settling relations described in Sec. 8.2.2. For most air pollution applications, Stokes' law [Eq. (8.4) with the air density ignored] is appropriate; substituting it in Eq. (9.4), we find

$$\eta = \frac{LgD^2\rho_{\text{part}}}{HV_{\text{avg}}18\mu} \quad \text{for block flow} \tag{9.5}$$

Now to consider the mixed flow model, we assume that the gas flow is totally mixed in the z direction but not in the x direction. (Most real gas flows are turbulent, leading to internal mixing in process equipment.) This makes sense, because mixing in the x direction moves particles both up- and downstream, with little effect on collection efficiency, whereas mixing in the z direction leads to a decrease in collection efficiency. We then consider a section of the settler with length dx. In this section the fraction of the particles that reach the floor will equal the vertical distance an average particle falls due to gravity in passing through the section, divided by the height of the section, which we may write as

$$\text{Fraction collected} = \frac{V_t\,dt}{H} \tag{9.6}$$

The change in concentration passing this section is

$$dc = -c \cdot (\text{fraction collected}) = -\frac{cV_t\,dt}{H} \tag{9.7}$$

The time the average particle takes to pass through this section is

$$dt = \frac{dx}{V_{\text{avg}}} \tag{9.8}$$

Combining these equations and rearranging, we have

$$\frac{dc}{c} = -\frac{V_t}{HV_{\text{avg}}}\,dx \tag{9.9}$$

which we may integrate from the inlet ($x = 0$) to the outlet ($x = L$), finding

$$\ln\frac{c_{\text{out}}}{c_{\text{in}}} = -\frac{V_t L}{HV_{\text{avg}}} \qquad \text{mixed flow} \tag{9.10}$$

or

$$\eta = 1 - \left(\frac{c_{\text{out}}}{c_{\text{in}}}\right) = 1 - \exp-\left(\frac{V_t L}{HV_{\text{avg}}}\right) \tag{9.11}$$

Finally we can substitute for V_t from Stokes' law, finding

$$\eta = 1 - \exp-\left(\frac{LgD^2\rho_{\text{part}}}{HV_{\text{avg}}18\mu}\right) \qquad \text{mixed flow} \tag{9.12}$$

Comparing this result with that for the block or plug flow assumption, Eq. (9.5), we see that Eq. (9.12) can be rewritten as

$$\eta_{\text{mixed}} = 1 - \exp(-\eta_{\text{block flow}}) \tag{9.13}$$

Example 9.1. Compute the efficiency-diameter relation for a gravity settler that has $H = 2$ m, $L = 10$ m, and $V_{\text{avg}} = 1$ m/s for both the block and mixed flow models, assuming Stokes' law.

Here we can get the result using only one computation and then using ratios. First we compute the block flow efficiency for a 1-μ particle, viz.,

$$\eta = \frac{LgD^2\rho_{\text{part}}}{18\mu H V_{\text{avg}}} = \frac{(10\text{ m})(9.81\text{ m/s}^2)(10^{-6}\text{ m})^2(2000\text{ kg/m}^3)}{(18)(1.8\times10^{-5}\text{ kg/m}\cdot\text{s})(2\text{ m})(1\text{ m/s})} = 3.03\times10^{-4}$$

For 1-μ particles the block flow assumption leads to an efficiency of 3.03×10^{-4}. The mixed assumption leads to practically the same result, viz.,

$$\eta_{\text{mixed}} = 1 - \exp(-3.03\times10^{-4}) = 3.029\times10^{-4}$$

To find the efficiencies for other particle diameters, we observe that the block efficiency is proportional to the particle diameter squared, so we make up a table of block flow efficiencies by simple ratios to the value for 1 μ, and then compute the corresponding mixed flow efficiencies as just shown.

Particle diameter, μ	η_{block}	η_{mixed}
1	0.000303	0.000303
10	0.0303	0.0298
30	0.273	0.239
50	0.76	0.53
57.45	1.00	0.63
80	—	0.86
100	—	0.95
120	—	0.99

These values are shown in Fig. 9.2 on page 254. ■

For small particles, for which the calculated collection efficiencies are small, the mixed and block flow models give practically the same answer. For larger particles the calculated collection efficiencies become larger, and the two models give different answers. The block flow model shows the efficiency reaching 100 percent for a particle diameter of 57.45 μ, whereas the mixed flow model shows the efficiency asymptotically approaching 100 percent for particles larger than about 100 μ. If one substitutes a diameter of 100 μ in the block flow equation one finds an efficiency of 303 percent, which is meaningless.

One may gain some insight into these two models by asking what the dust pile on the floor would look like if we ran a gravity settler with a single-size dust for a long period of time and then shut it down. In the block flow model, we would expect a pile of absolutely uniform height ending abruptly at that length for which $L = HV_{\text{avg}}/V_t$. For the mixed model we would expect a pile that is deepest at the inlet end and whose depth falls exponentially, approaching zero depth asymptotically as L becomes large.

This type of device would be useful for collecting particles with diameters of perhaps 100 μ (fine sand) but not for particles of air pollution interest, whose diameters go down to fractions of a micron. We could increase the efficiency by

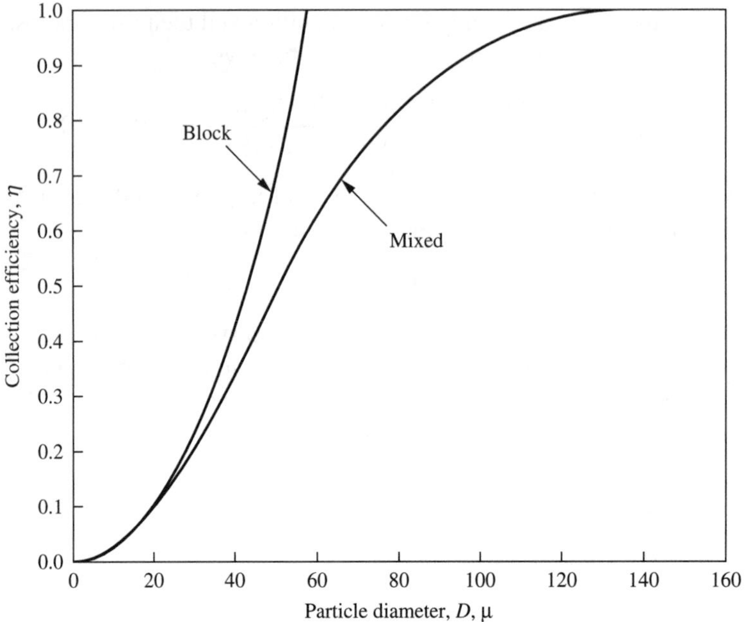

FIGURE 9.2
Comparison of the efficiencies for a gravity settler, calculated by the block and mixed models (see Example 9.1).

making L larger (which makes the device very long and expensive), by making H smaller (which is sometimes done by subdividing the chamber with horizontal plates, which makes the cleanup much more difficult), by lowering V_{avg} (which requires a larger cross-sectional area and hence larger and more costly device), or by increasing g. The latter is the only practical alternative; it requires substituting some other force for the force of gravity in driving the particles from the gas stream to the collecting surface.

Small gravity settlers used for particle sampling are sometimes called *horizontal elutriators.* In them air flow is very slow, and particles are collected by gravity on greased plates for subsequent microscopic examination [1] (see Problem 9.3).

9.1.2 Centrifugal Separators

We have spent considerable time on gravity settlers because it is easy to see what all their mathematics mean. But they have little practical industrial use because they are ineffective for small particles. If we are to use them or devices like them, we must find a substitute that is more powerful than the gravity force they use to drive the particles to the collection surface. Physics and mechanics books usually show that *centrifugal force* is a pseudoforce that is really the result of the body's inertia carrying it straight while some other force makes it move in a curved path. It is convenient to use this pseudoforce for calculational purposes.

If a body moves in a circular path with radius r and velocity V_c along the path, then it has angular velocity $\omega = V_c/r$, and

$$\text{Centrifugal force} = \frac{mV_c^2}{r} = m\omega^2 r \tag{9.14}$$

Example 9.2. A particle is traveling in a gas stream with velocity 60 ft/s (18 m/s) and radius 1 ft. What is the ratio of centrifugal force to the gravity force acting on it?

$$\frac{\text{Centrifugal force}}{\text{Gravity force}} = \frac{mV_c^2/r}{mg} = \frac{(60\text{ ft/s})^2/(1\text{ ft})}{32.2\text{ ft/s}^2} = 111.8 \quad \blacksquare$$

At even modest velocities and common radii, the centrifugal forces acting on particles can be two orders of magnitude larger than the gravity forces. For this reason centrifugal particle separators are much more useful than gravity settlers.

For further work we will use a centrifugal equivalent of Stokes' law, given in Eq. (8.4). We obtained Stokes' law by equating the (gravitational minus buoyant) force to the Stokes' form of the drag force. Normally we drop the buoyant term for particles in gases because it is small. To obtain the centrifugal equivalent, we need only substitute the centrifugal force for the gravitational force (or the centrifugal acceleration for the gravitational acceleration, since the masses are equal). In Eqs. (8.2) and (8.4) we replace g by V_c^2/r or by $\omega^2 r$. Doing this poses a problem, because now there are two velocities in the equation that are not the same. To save confusion we will call the terminal settling velocity in the radial direction V_t and the velocity along the circular path V_c. The relation of these two is sketched in Fig. 9.3.

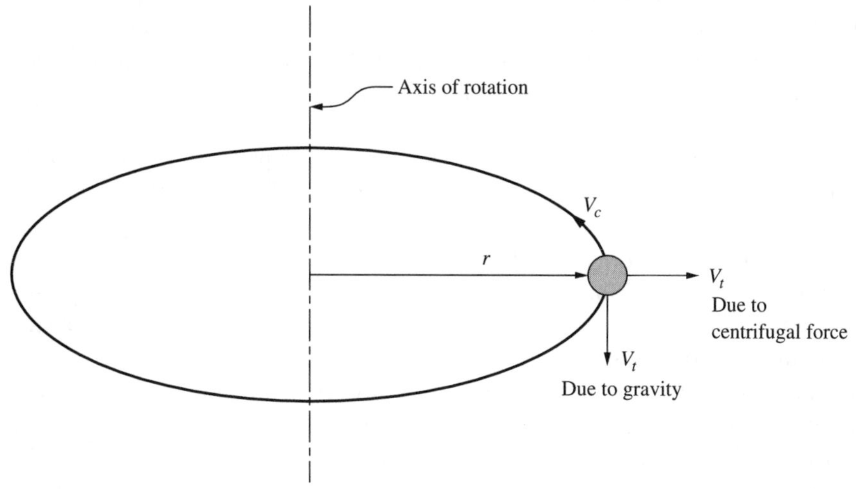

FIGURE 9.3
Relation of defined terms for rotational motion.

Figure 9.3 also shows another V_t due to gravity (assuming the axis of the circle is vertical). If, as shown previously, the centrifugal force is normally more than 100 times the gravitational force, then this gravitational settling velocity will be less than a hundredth of the centrifugal one and can be left out of consideration. V_t, the terminal settling velocity we calculate, is a velocity in the *radial direction* at right angles to the main circular motion of the particle.

Now substituting this centrifugal acceleration for the gravitational one in Eq. (8.4) and dropping the ρ_{fluid} term, we find

$$V_t = \frac{V_c^2 D^2 \rho_{\text{part}}}{18\mu r} \tag{9.15}$$

Example 9.3. Repeat the computation of the terminal settling velocity shown in Example 8.1 for a particle in a circular gas flow with velocity $V_c = 60$ ft/s (18.29 m/s) and radius 1 ft (0.3048 m). The density of the fluid can be ignored. By direct substitution in Eq. (9.15), we find

$$V_t = \frac{(18.29\ \text{m/s})^2 (10^{-6}\ \text{m})^2 (2000\ \text{kg/m}^3)}{(18)(1.8 \times 10^{-5}\ \text{kg/m} \cdot \text{s})(0.3048\ \text{m})} = 0.0068\ \frac{\text{m}}{\text{s}}$$

$$= 0.68\ \frac{\text{cm}}{\text{s}} = 0.022\ \frac{\text{ft}}{\text{s}} \qquad \blacksquare$$

This answer is 112 times as large as the value found in Example 8.1, indicating again that much greater settling velocities can be obtained this way. One may compute the particle Reynolds number here, finding that it is about 0.00046. Hence the assumption of a Stokes' law type of drag seems reasonable. In centrifugal devices the settling velocities are higher than those due to gravity, so that if we were to make up a centrifugal equivalent to Fig. 8.6 we would find that the drag-coefficient Reynolds number curve would begin at smaller particle diameters than for gravity settling. The Cunningham correction factor is unaffected by how fast the particles move, and thus that part of the curve would be unaffected by the switch from gravity settling to centrifugal settling.

At this point let us reconsider the Stokes' law assumption. If we consider the overall gas flow, with velocities on the order of 60 ft/s, the Reynolds numbers are on the order of a half million. The flow is highly turbulent. How can we apply Stokes' law, which requires that the particle Reynolds numbers be less than about 0.3 and that the fluid flow around the particle be laminar? If we take the view of a person riding on the particle, we can see that the patch of fluid surrounding us is in turbulent, rapid circular motion, with one turbulent eddy moving us toward the center, then another moving us away from the center, etc. However, in the immediate locality of the particle there is a small net movement of the particle relative to the surrounding gas caused by centrifugal force. This net movement is so slow that the gas molecules can easily move out of the particle's way in a laminar fashion. It is this net particle movement, superimposed on the overall turbulent gas flow, that causes the average radially outward movement of the particle and is the movement discussed in this section.

After all this theoretical discussion, how does one construct a practical centrifugal particle collector? There are many types, but the most successful is sketched in Fig. 9.4. It is universally called a *cyclone separator,* or simply a *cyclone.* It is probably the most widely used particle collection device in the world. In any industrial district of any city, a sharp-eyed student can find at least a dozen of these outside various industrial plants.

As the sketch shows, a cyclone consists of a vertical cylindrical body, with a dust outlet at the conical bottom. The gas enters through a rectangular inlet, normally

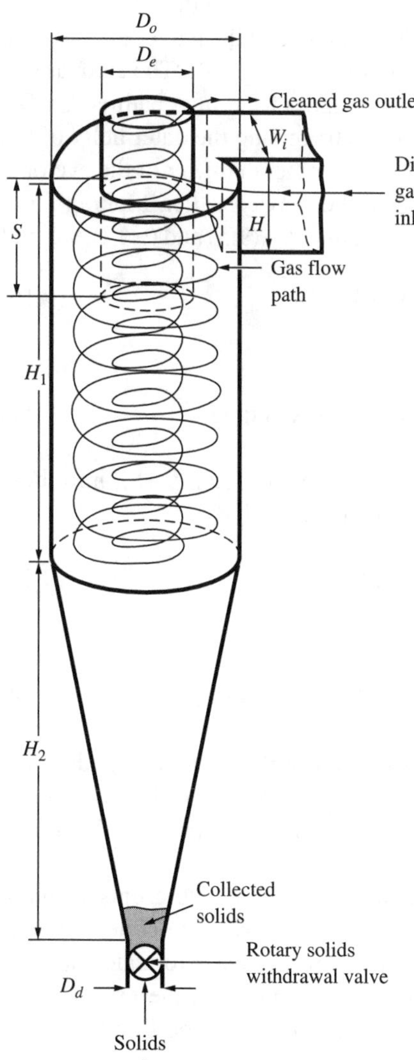

FIGURE 9.4
Schematic of a cyclone separator. Dimensions are typically based on the overall diameter D_o. Taken as ratios to that dimension, $W_i = 0.25$, $H = 0.5$, $H_1 = 2$, $H_2 = 2$, $D_e = 0.5$, $S = 0.625$, $D_d = 0.25$. For example, if $D_o = 1$ ft, then $W_i = 0.25$ ft, etc. Ashbee and Davis [2] show a table with six sets of values for these dimension ratios. The principal differences are that high-efficiency cyclones have smaller values of W_i whereas high-throughput cyclones have larger values of W_i and of D_e. The dimension ratios here are for the "conventional" design.

twice as high as it is wide, arranged tangentially to the circular body of the cyclone, so that the entering gas flows around the circumference of the cylindrical body, not radially inward. The gas spirals around the outer part of the cylindrical body with a downward component, then turns and spirals upward, leaving through the outlet at the top of the device. During the outer spiral of the gas the particles are driven to the wall by centrifugal force, where they collect, attach to each other, and form larger agglomerates that slide down the wall by gravity and collect in the dust hopper in the bottom.

Clearly the cyclone separator sketched in Fig. 9.4 is merely a gravity settler that has been made in the form of two concentric helices. Only the outer helix contributes to collection; particles that get into the inner helix, which flows upward to the gas outlet, escape uncollected. Thus the outer helix is equivalent to the gravity settler. The inlet stream has a height W_i in the radial direction, so that the maximum distance any particle must move to reach the wall is W_i (defined on Fig. 9.4). The comparable distance in a gravity settler is H (Fig. 9.1). The length of the flow path is $N\pi D_o$, where N is the number of turns that the gas makes traversing the outer helix of the cyclone, before it enters the inner helix, and D_o is the outer diameter of the cyclone. This length of the flow path corresponds to L in the gravity settler. Making these substitutions directly into the gravity settler equations, Eqs. (9.5) and (9.12), we find

$$\eta = \frac{N\pi D_o V_t}{W_i V_c} \qquad \text{block flow} \tag{9.16}$$

and

$$\eta = 1 - \exp - \left(\frac{N\pi D_o V_t}{W_i V_c} \right) \qquad \text{mixed flow} \tag{9.17}$$

If we then substitute the centrifugal Stokes' law expression, Eq. (9.15), into these two equations, and make the appropriate cancellations, we find

$$\eta = \frac{\pi N V_c D^2 \rho_{\text{part}}}{9 W_i \mu} \qquad \text{block flow} \tag{9.18}$$

and

$$\eta = 1 - \exp - \left(\frac{\pi N V_c D^2 \rho_{\text{part}}}{9 W_i \mu} \right) \qquad \text{mixed flow} \tag{9.19}$$

Here D is the particle diameter. The outside diameter of the cyclone, D_o, does not appear directly but only indirectly through W_i, which is proportional to it. Observe also that the right side of Eq. (9.18) is the Stokes' stopping distance (Section 8.2.4) divided by $W_i/2\pi N$.

Equations (9.18) and (9.19) contain a parameter N, which represents the number of turns the gas makes around the cyclone before it leaves the collecting area near the wall. There seems to be no satisfactory theoretical basis for calculating N from fluid mechanical principles. A value of $N = 5$ represents the experimental data best. Unless one has specific information to the contrary, one should assume that $N = 5$ throughout this book.

Example 9.4. Compute the efficiency-diameter relation for a cyclone separator that has $W_i = 0.5$ ft, $V_c = 60$ ft/s, and $N = 5$, for both the block and mixed flow assumptions, assuming Stokes' law.

Here, as in Example 9.1, we can get the result with one numerical computation, using ratios. First we compute the block flow efficiency for a 1-μ particle, viz.,

$$\eta = \frac{\pi N V_c D^2 \rho_{\text{part}}}{9 W_i \mu}$$

$$= \frac{(\pi)(5)(60 \text{ ft/s})(10^{-6} \text{ m})^2 (3.28 \text{ ft/m})^2 (124.8 \text{ lbm/ft}^3)}{(9)(0.5 \text{ ft})(1.8 \times 10^{-2} \text{ cp})[6.72 \times 10^{-4} \text{ lbm/(ft} \cdot \text{s} \cdot \text{cp})]} = 0.0232$$

Then, as we did in Example 9.1, we can use this number, plus the fact that the particle diameter enters the equation to the second power, to make up the following table:

Particle diameter, μ	η_{block}	η_{mixed}
0.1	0.000232	0.000232
1	0.0232	0.0230
2	0.0930	0.0888
3	0.209	0.189
4	0.372	0.311
5	0.582	0.441
6.559	1.00	0.632
10	—	0.902
15	—	0.995

■

Comparing this result to that for gravity settling chambers in Example 9.1, we see the form of the result is the same, but the maximum particle size for which the device is effective is much smaller. If we plotted these data as in Fig. 9.2, we would find an identical plot, but with the diameter scale multiplied by a factor of $(6.559/57.45) = 0.114$. This occurs because the models and their resulting equations are truly the same except for the substitution of centrifugal force for gravity, and the change in dimensions.

Next we introduce a new term, the *cut diameter,* which is widely used in describing particle collection devices. This definition gives us a measure of the size of particles caught and the size passed for a particle collector. A kitchen colander—a sheet metal dish with uniform, circular holes—has a cut diameter; all the particles that can pass through the holes in any direction will do so (if we shake long enough), whereas those larger than the holes will not. If we considered only spherical peas in a colander with uniform circular holes, then the cut diameter would be the diameter of the holes. For peas larger than the cut diameter the collection efficiency would be 100 percent, and for those smaller it would be 0 percent. For all practical particle collection devices the separation is not that sharp; there is no single diameter at which the efficiency goes suddenly from 0 percent to 100 percent. The universal convention in the air pollution literature (and the particle technology literature in general) is to define *cut diameter* as the diameter of a particle for which the efficiency curve has the value of 0.50, i.e., 50 percent.

We can substitute this definition into Eq. (9.18) and solve for the cut diameter that goes with Stokes' law, block flow model, finding:

$$D_{\text{cut}} = \left(\frac{9 W_i \mu}{2\pi N V_c \rho_{\text{part}}} \right)^{1/2} \quad \text{block flow} \tag{9.20}$$

Although one might logically expect that Eq. (9.19), with its more realistic mixed flow model, would better represent experimental data, Eq. (9.18) appeared in the literature earlier [3] and has been more widely used. It is widely known as the *Rosin-Rammler equation* and is reasonably accurate in estimating the performance of cyclones.

Example 9.5. Estimate the cut diameter for a cyclone with inlet width 0.5 ft, $V_c = 60$ ft/s and $N = 5$.

$$D_{\text{cut}} = \left(\frac{(9)(0.5\ \text{ft})(1.8 \times 10^{-5}\ \text{kg/m} \cdot \text{s})}{2\pi(5)(60\ \text{ft/s})(2000\ \text{kg/m}^3)} \right)^{1/2} = 4.63 \times 10^{-6}\ \text{m} \approx 5\ \mu \quad \blacksquare$$

This example shows that for a typical cyclone size and the most common cyclone velocity and gas viscosity, the cut diameter is about 5 μ. Comparing this calculation with that in Example 9.4 shows that the cut diameter we would calculate by the mixed model is somewhat larger, but not dramatically so. It is an industrial rule of thumb that if a gas stream contains few particles smaller than 5 μ then a cyclone is probably the only collector one should consider. It works well on most particles that size and larger (e.g., sawdust from wood shops and wheat grains from pneumatic conveyers), and is a low-cost, easy-maintenance device. It is not satisfactory for sticky particles, like tar droplets.

Suppose we wish to apply a cyclone separator for even smaller particles. What are our options? From Eq. (9.20) we can see that the alternatives are to make W_i smaller or V_c larger. (Generally we cannot alter the gas viscosity or the particle density.) Making V_c larger is generally too expensive because, as we shall see later in this section, the pressure drop across a cyclone is generally proportional to the velocity squared. To make W_i smaller, we must make the whole cyclone smaller if we are to keep the same ratios of dimensions. But the inlet gas volumetric flow is proportional to W_i squared, so that a small cyclone treats a small gas flow. Very small cyclones have been used to collect small particles from very small gas flows for research and gas-sampling purposes, but the industrial problem is to treat large gas flows. Several practical schemes have been worked out to place a large number (up to several thousand) small cyclones in parallel, so that they can treat a large gas flow, capturing smaller particles. The most common of these arrangements, called a **multiclone**, is sketched in Fig. 9.5.

The many small cyclones in the multiclone are mass-produced and inserted into sheet metal supporters. In the device shown, the circular gas motion in each

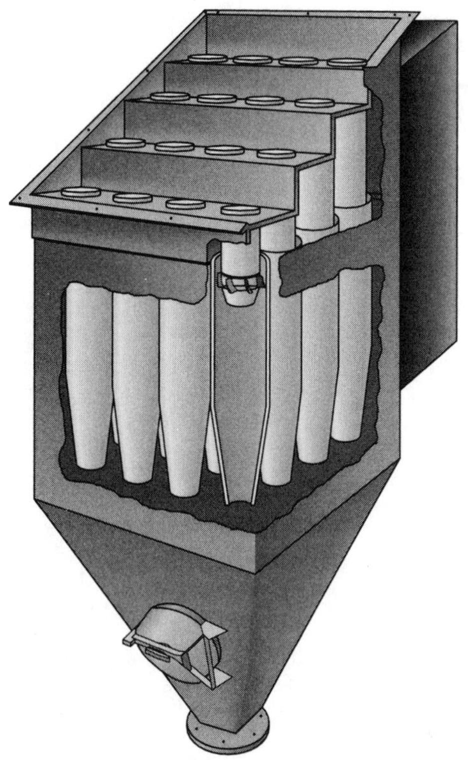

FIGURE 9.5
A multiclone, which places a large number of small cyclones in parallel. The dirty gas flows through an entrance duct, the edge of which is shown in the sketch at the rear, into the chamber shown in the cutaway, then flows downward into the individual tubes, getting its spiral motion from the turning vanes shown. The cleaned gas flows up the central tubes and out through the top of the device (through an outlet flue, not shown, which bolts to the slanting top of the device). The collected particles fall to the conical bottom. (Courtesy of Joy Environmental.)

cyclone is caused by a set of sheet metal turning vanes that replace the solid top of an ordinary cyclone. The gas outlet tubes are connected to a common gas outlet header. If the individual cyclone were one-half foot in diameter, the W_i in Eq. (9.20) would be about 0.125 ft. Repeating Example 9.5 for a W_i of 0.125 ft, we find a predicted cut diameter of 2.3 μ, which is about the actual cut diameter of these devices.

Although Eq. (9.20) is a fair predictor of cut diameters, Eq. (9.18), upon which it is based, is a poor predictor of the relation of collection efficiency to diameter. Equation (9.19), which takes mixing into account, is a better predictor, but neither is really good. Figure 9.6 on page 262 compares the predictions that Eqs. (9.18) and (9.19) make with a curve representing a summary of experimental data [4] that can be represented with satisfactory accuracy by the following totally empirical data-fitting equation:

$$\eta = \frac{(D/D_{\text{cut}})^2}{1 + (D/D_{\text{cut}})^2} \tag{9.21}$$

(See also Problem 9.6.)

Example 9.6. A gas stream contains particles with a particle size distribution by mass that is given by the log-normal distribution, with $D_m = 20$ μ and $\sigma = 1.25$

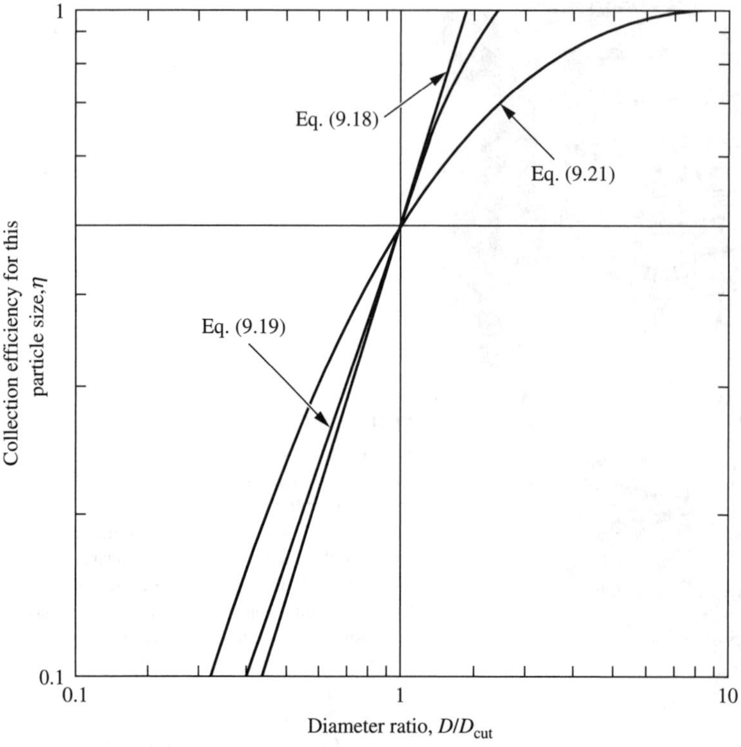

FIGURE 9.6
Collection efficiency vs. particle diameter curves for cyclones. Here, all three curves must pass through 0.5 at $D = D_{cut}$ because of the definition of D_{cut}. Equation (9.21) is very close to the experimental results for typical cyclones.

(see Fig. 8.10). We pass this through a cyclone separator whose cut diameter is 5 μ, and whose efficiency-diameter relation is given by Eq. (9.21) (and shown in Fig. 9.6). What is the percentage by weight of the particles caught? What is the mass mean diameter of the particles that pass through?

We cannot solve this problem analytically but must instead divide the particle distribution into size fractions and compute the penetration for each one, as illustrated in Section 7.8. The result is shown in Table 9.1. In the first column we have divided the distribution into 10 fractions, those from 0 to 0.1 of the mass of the particles, those from 0.1 to 0.2, etc. The second column shows the z corresponding to the Φ at the end of this interval, such as 0.1, 0.2, etc. These values are found from a table like Table 8.3, but arranged for even values of Φ instead of even values of z. The third column shows the value of (D/D_{mean}) at the end of the size interval, found by solving Eq. (8.19) for a log-normal distribution. The first value is

$$\frac{D}{D_{mean}} = \exp(z\sigma) = \exp(-1.282 \times 1.25) = 0.2014$$

This calculation shows that 0.1 = 10 percent of the particles have diameters less

TABLE 9.1
Performance computation for a cyclone separator

Φ	z	$\left(\dfrac{D}{D_{\text{mean}}}\right)_{\text{end}}$	$\left(\dfrac{D}{D_{\text{mean}}}\right)_{\text{mid}}$	η	$p\,\Delta\Phi$	$\sum p\,\Delta\Phi$
0.1	−1.282	0.2014	0.1007	0.1396	0.0860	0.0860
0.2	−0.842	0.3491	0.2752	0.5479	0.0452	0.1312
0.3	−0.524	0.5194	0.4343	0.7511	0.0249	0.1561
0.4	−0.253	0.7289	0.6242	0.8617	0.0138	0.1700
0.5	0	1.0000	0.8644	0.9228	0.0077	0.1777
0.6	0.253	1.3720	1.1860	0.9575	0.0043	0.1819
0.7	0.524	1.9251	1.6486	0.9775	0.0022	0.1842
0.8	0.842	2.8648	2.3950	0.9892	0.0011	0.1853
0.9	1.282	4.9654	3.9151	0.9959	0.0004	0.1857
1		1.0000	4.9654	0.9975	0.0003	0.1859

than (0.2014 × 20 μ) = 4.02 μ. The average diameter of the smallest 10 percent of the particles is approximately half of this, or 2 μ. The fourth column shows this average diameter ratio, listed as $(D/D_{\text{mean}})_{\text{mid}}$. For the first entry this is the average of the end value and zero. For the next eight values it is the average of the value at the end of the range and at the end of the previous range. The final value is taken as the end value of the preceding range, which introduces only a small error.

The fifth column of Table 9.1 shows the collection efficiency η for the midrange diameter, computed by Eq. (9.21):

$$\eta = \frac{(D/D_{\text{cut}})^2}{1 + (D/D_{\text{cut}})^2}$$

$$= \frac{[(D/D_{\text{mean}})(D_{\text{mean}}/D_{\text{cut}})]^2}{1 + [(D/D_{\text{mean}})(D_{\text{mean}}/D_{\text{cut}})]^2} = \frac{(0.1007 \cdot 20\ \mu/5\ \mu)^2}{1 + (0.1007 \cdot 20\ \mu/5\ \mu)^2} = 0.1396$$

In the sixth column is $p\,\Delta\Phi$, the amount of mass in this size interval that passes through uncollected, e.g.,

$$p\,\Delta\Phi = (1 - 0.1396)(0.1 - 0) = 0.0860$$

We see that 86 percent by mass of the particles in this size range (8.6 percent of the total particle mass) pass through the cyclone uncollected. The final column is the sum of the values in column 6, showing the cumulative fraction uncollected. The lower right-hand value shows that 0.186 = 18.6 percent of the particles are not collected, so that the overall collection efficiency is 0.814 ≈ 81 percent.

The mass mean diameter of the particles that pass through the cyclone is the diameter that corresponds to half of the value at the bottom of column 7, or 0.0930. This is slightly more than the value at the end of the 0 to 0.1 weight fraction interval, so from the third column in Table 9.1 we know that it corresponds to a diameter of about 0.2014 ≈ 0.2 of the mean diameter or about 4 μ. At the end of this long example, the reader is encouraged to compare it with Example 7.5. This is simply that example repeated, using a real particle size distribution and a real collector efficiency relation. For all the devices discussed later in this chapter, final design calculations are made by the equivalent of this table.

One may repeat this example using 20 size intervals instead of the 10 here, and find that the final penetration value is 0.1836 instead of 0.1859. We rarely have size distribution or control efficiency data good enough to justify that extra computation. ■

The low collection efficiency, 81 percent, of Example 9.6 shows that a typical cyclone cannot meet modern control standards (usually > 95 percent required control efficiency) for any particle group that has a substantial fraction smaller than 5 μ in diameter.

Although Eq. (9.20) and Example 9.5 show that the practical cut diameter is limited, the physical reason is hidden in the mathematics. To get a high value of V_t, we need a high value of V_c; but a high value of V_c means that the gas stream is in the cyclone for only a very short time and has little time to be acted on by the high centrifugal force.

Example 9.7. In Example 9.5 how long does the gas spend in the high centrifugal force field near the wall where a particle has a good chance of being captured?

Here, following the assumptions leading to Eqs. (9.16) and (9.17),

$$t = \frac{L}{V} = \frac{N\pi D_o}{V_c} = \frac{5\pi \cdot 2\text{ ft}}{60\text{ ft/s}} = 0.525\text{ s}$$ ■

The distance the particle can move toward the wall is equal to the product of this time and V_t, but V_t is proportional to V_c squared, so that to get better collection efficiencies we must go to lower and lower times in the cyclone.

Previously we stated that the typical velocity at a cyclone inlet is 60 ft/s (18.29 m/s) and that this velocity is selected for pressure drop reasons. If one measures the pressure in the pipe leading the gas to the cyclone and the pressure in the pipe leaving the cyclone, one will find that the inlet pressure is higher. For a given cyclone one will generally find that the pressure drop, for various conditions, can be represented by an equation of the form

$$\text{Pressure drop} = P_{\text{in}} - P_{\text{out}} = K\left(\frac{\rho_g V_i^2}{2}\right) \tag{9.22}$$

where ρ_g is the gas density and V_i is the velocity at the inlet to the cyclone. (V_i is not the same as the velocity in the duct approaching the cyclone; typically it is about 1.5 times as high.) Designers who work regularly with air-conditioning or other piping systems have observed that most pressure drop data for their kinds of systems can be represented in the form of Eq. (9.22), with each particular kind of device having its own K. (All sudden expansions have a K of 1.0, all sudden contractions have a K of 0.5, etc. Tables of Ks for various types of pipes and fittings are widely published [5].) Most cyclone separators have Ks of about 8. It is also common in air-conditioning design to refer to the quantity ($\rho_g V^2/2$) as a *velocity head*, so one could say that most cyclones have pressure drops of about 8 velocity heads.

Example 9.8. A cyclone has an inlet velocity of 60 ft/s and a reported pressure loss of 8 velocity heads ($K = 8$). What is the pressure loss in pressure units?

Applying Eq. (9.22), we find

$$\begin{aligned}\text{Pressure drop} &= 8\left(0.075\,\frac{\text{lbm}}{\text{ft}^3}\right)\left(60\,\frac{\text{ft}}{\text{s}}\right)^2\left(\frac{1}{2}\right)\left(\frac{\text{lbf}\cdot\text{s}^2}{32.2\ \text{lbm}\cdot\text{ft}}\right)\left(\frac{\text{ft}^2}{144\ \text{in.}^2}\right)\\ &= 0.23\,\frac{\text{lbf}}{\text{in.}^2}\\ &= 8\left(1.20\,\frac{\text{kg}}{\text{m}^3}\right)\left(18.29\,\frac{\text{m}}{\text{s}}\right)^2\left(\frac{1}{2}\right)\left(\frac{\text{N}\cdot\text{s}^2}{\text{kg}\cdot\text{m}}\right)\\ &= 1606\,\frac{\text{N}}{\text{m}^2} = 1.61\ \text{kPa} = 0.23\ \text{psi} = 6.4\ \text{in. H}_2\text{O}\end{aligned}$$

■

Typically this pressure drop must be overcome by a fan or blower somewhere in the system. If the system is already under pressure, this poses no problems for the designer. However, if it is a new system that must consist of cyclone, blower, and associated ductwork, then the designer has two options, both of which have disadvantages. The first of these is shown in Fig. 7.2. There the blower is located before the cyclone, which is the pollution control device in this case. The disadvantage of this arrangement is that the blower is exposed to the dirty gas. The particles will get into its bearings and collect on its blades, throwing it out of balance. The alternative arrangement is to put the blower downstream of the pollution control device (cyclone), in which case the blower works on cleaned gas and has fewer maintenance problems. The disadvantage of this arrangement is that the cyclone now operates under a weak vacuum, and if the seal at the solids removal valve is not very good, air will be sucked in and re-entrain the collected particles, degrading the overall collection efficiency. Both systems can be made to work with adequate attention to engineering detail.

There are many other variants on the centrifugal collector idea, but none approaches the cyclone in breadth of application. These devices are simple and almost maintenance-free. Because any medium-sized welding shop can make one, the big suppliers of pollution control equipment, who have test data on the effects of small changes in the internal geometry, have been unwilling to make these data public. However, there do not appear to be designs that are substantially better than the simple one shown in Fig. 9.4. The alternative dimension ratios shown in Ref. 2 allow a smaller cut diameter at the price of a higher pressure drop, or a higher throughput at the price of a larger cut diameter, but not an improvement of one performance measure without a cost in terms of some other performance parameter. There is no reason why one cannot obtain better collection efficiency by placing one cyclone downstream of another; the standard design of catalytic cracker regenerators has two cyclones in series to remove catalyst particles from the waste gas (see Problem 9.8).

The same basic device as the cyclone separator is used in other industrial settings where the goal is not air pollution control, but some other kind of separation. When it is used to separate solids from liquids it is generally called a *hydroclone*. A cyclone called an *air-swept classifier* is attached to many industrial grinders. It passes those particles ground fine enough, and collects those that are too coarse, returning them to the grinder.

9.1.3 Electrostatic Precipitators (ESP)

If gravity settlers and centrifugal separators are devices that drive particles against a solid wall, and if neither can function effectively (at an industrial scale) for particles below about 5 μ in diameter, then for wall collection devices to work on smaller particles, they must exert forces that are more powerful than gravity or centrifugal force. The *electrostatic precipitator* (ESP) is like a gravity settler or centrifugal separator, but electrostatic force drives the particles to the wall. It is effective on much smaller particles than the previous two devices.

In all three kinds of devices, the viscous (Stokes' law) resistance of the particle to being driven to the wall is proportional to the particle diameter [see Eq. (8.3)]. For gravity and centrifugal separators, the force that can be exerted is proportional to the mass of the particle, which, for constant density, is proportional to the diameter cubed. Thus the ratio of driving force to resisting force is proportional to (diameter cubed/diameter) or to diameter squared. As the diameter decreases, this ratio falls rapidly. In ESPs the resisting force is still the Stokes viscous drag force, but the force moving the particle toward the wall is electrostatic. This force is practically proportional to the particle diameter squared, and thus the ratio of driving force to resisting force is proportional to (diameter squared/diameter) or to the diameter. Thus it is harder for an ESP to collect small particles than large ones, but the difficulty is proportional to $(1/D)$ rather than to $(1/D^2)$, as in gravitational or centrifugal devices.

The basic idea of all ESPs is to give the particles an electrostatic charge and then put them in an electrostatic field that drives them to a collecting wall. This is an inherently two-step process. In one type of ESP, called a two-stage precipitator, charging and collecting are carried out in separate parts of the ESP. This type, widely used in building air conditioners, is sometimes called an electronic air filter. However, for most industrial applications the two separate steps are carried out simultaneously in the same part of the ESP. The charging function is done much more quickly than the collecting function, and the size of the ESP is largely determined by the collecting function.

Figure 9.7 shows in simplified form a wire-and-plate ESP with two plates. The gas passes between the plates, which are electrically grounded (i.e., voltage = 0). Between the plates are rows of wires, held at a voltage of typically −40 000 volts. The power is obtained by transforming ordinary alternating current to a high voltage and then rectifying it through some kind of solid-state rectifier. This combination of charged wires and grounded plates produces both the free electrons to charge the

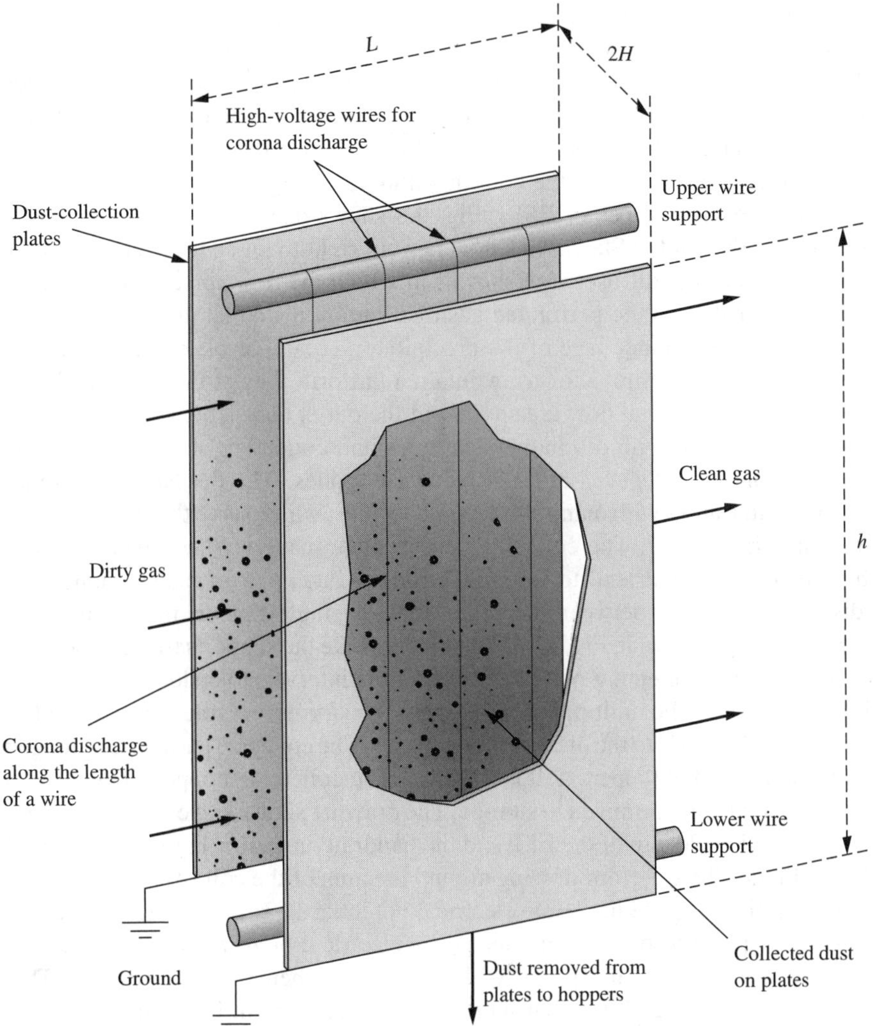

FIGURE 9.7
Diagrammatic sketch of a simplified ESP with two plates, four wires, and one flow channel. Industrial-size ESPs have many such channels in parallel; see Fig. 9.8.

particles and the field to drive them against the plates. On the plates the particles lose their charge and adhere to each other and the plate, forming a "cake." The cleaned gas then passes out the far side of the precipitator as shown in Fig. 9.7.

Solid cakes are removed by rapping the plates at regular time intervals with a mechanical or electromagnetic rapper that strikes a vertical or horizontal blow on the edge of the plate. Through science, art, and experience designers have learned to make rappers that cause most of the collected cake to fall into hoppers below the plates (not shown on Fig. 9.7). Some of the cake is always re-entrained, thereby lowering the efficiency of the system. If the collected particles are liquid, e.g., sulfuric acid mist, they run down the plate and drip off. For liquid droplets the plate is often replaced by a circular pipe with the wire down its center. Some ESPs (mostly the circular pipe variety) have a film of water flowing down the collecting surface, to carry the collected particles to the bottom without rapping.

There are many types of ESPs; Fig. 9.8 shows one of the most common in current use in the United States. Gas flow is from right to left. The gas enters at the right through an inlet diffuser (not shown) in which the flow spreads out from the much narrower duct to the perforated gas distribution plate that distributes the gas evenly across the entrance face of the precipitator. A similar plate and converging nozzle on the left side (not shown) maintain a uniform flow at the outlet and then reduce the cross-sectional flow area to that of the outlet duct. The whole interior of the structure is filled with discharge electrodes and collecting plates; the cutaway shows only one set of plates and discharge electrodes. The discharge electrodes consist of rigid frames with many short, pointed stubs, which serve the same function as the wires in Fig. 9.7. The collecting surfaces are made of sheet metal sections with vertical joints that tend to trap the particles. Each pair of plates, along with the discharge electrode between them, acts like the single channel in the simplified version of an ESP shown in Fig. 9.7. The rappers strike the supports for the discharge electrodes and the collecting plates at regular time intervals to dislodge the cake of collected particles. The multiple power supply transformer-rectifier sets supply DC current at $\approx -40\,000$ V to the discharge electrodes. The collected particles, dislodged from the plates by the rappers, fall into the particle collecting hoppers, from which they are automatically removed to storage. The drawing shows some of the structural steel frame and enclosure of the ESP and the handrail on its top, but not the internal seals that hinder the gas from flowing around the area of the collecting plates.

Each point in space has some electrical potential V. If the electrical potential changes from place to place, then there is an electrical field, $E = \partial V/\partial x$, in that space. If we connect two such points with a conductor, then a current will flow. This V is the voltage we are all familiar with, and E is its gradient in any direction; the units of E are V/m.

In a typical wire-and-plate precipitator, as sketched in Fig. 9.7, the distance from the wire to the plate is about 4 to 6 in., or 0.1 to 0.15 m. With a voltage difference of 40 kV and 4-in. spacing, one would assume a field strength of $40\text{ kV}/0.1\text{ m} = 400\text{ kV/m}$. This is indeed the field strength near the plate. However, all of the electrical flow that reaches the plate comes from the wires, and the surface area of the wires is much lower than that of the plate; thus, by conservation of charge, the driving potential near the wires must be much larger. Typically it is 5 to 10 MV/m. (The first person to utilize this fact was presumably Benjamin Franklin, who invented the sharp, pointed lightning rod.)

When a stray electron from any of a variety of sources encounters this strong a field, it is accelerated rapidly and attains a high velocity. If it then collides with a gas molecule, it has enough energy to knock one or more electrons loose, thus

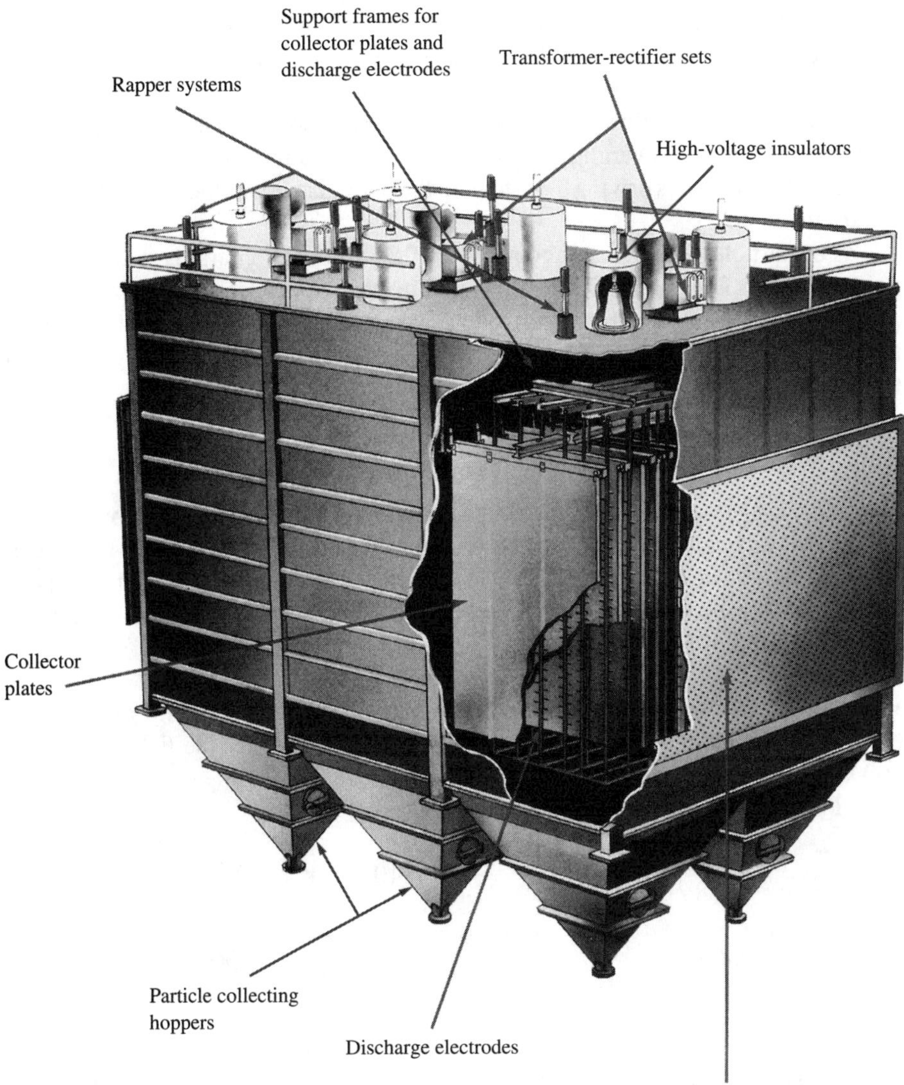

FIGURE 9.8
Cutaway view of a large, modern ESP showing the various parts. In this design the wire discharge electrodes have been replaced by rigid frames with many short, pointed stubs. (Courtesy of The Babcock and Wilcox Company, Barberton, Ohio.)

ionizing the gas molecule. These electrons are likewise accelerated by the field and knock more electrons loose, until there are enough free electrons to form a steady **corona discharge**. In a dark room this discharge appears as a dim glow that forms a circular sheath about the wire. The positive ions formed in the corona migrate to the wire and are discharged. The electrons migrate away from the wire, toward the plate. Once they get far enough away from the wire for the field strength to be too low to accelerate them fast enough to ionize gas molecules, the visible corona ceases and they simply flow as free electrons.

As the electrons flow toward the plate, they encounter particles and can be captured by them, thus charging the particles. Then the same electric field that created the electrons and that is driving them toward the plate also drives the charged particles toward the plate.

For particles larger than about 0.15 μ, the dominant charging mechanism is *field charging*. This is practically equivalent to the capture of any electron by any particle lying in its path. However, as the particles become more highly charged, they bend the paths of the electrons away from them. Thus the charge grows with time, reaching a steady state value of

$$q = 3\pi\left(\frac{\varepsilon}{\varepsilon + 2}\right)\varepsilon_0 D^2 E_0 \tag{9.23}$$

Here q is the charge on the particle, and ε is the dielectric constant of the particle—a dimensionless number that is 1.0 for a vacuum, 1.0006 for air, and 4 to 8 for typical solid particles. The permittivity of free space ε_0 is a dimensional constant whose value in the SI system of units is $8.85 \times 10^{-12}\ \text{C}/(\text{V}\cdot\text{m})$. D is the particle diameter, and E_0 is the local field strength.

Example 9.9. A 1-μ diameter particle of a material with a dielectric constant of 6 has reached its equilibrium charge in an ESP at a place where the field strength is 300 kV/m. How many electronic charges has it?

From Eq. (9.23) we can write

$$q = 3\pi\left(\frac{6}{8}\right)\left(8.85 \times 10^{-12}\ \frac{\text{C}}{\text{V}\cdot\text{m}}\right)(10^{-6}\ \text{m})^2\left(300\ \frac{\text{kV}}{\text{m}}\right)$$

$$= 1.88 \times 10^{-17}\ \text{C} \times \left(\frac{1.602 \times 10^{19}\ \text{electrons}}{\text{C}}\right) = 300\ \text{electrons} \qquad \blacksquare$$

The value computed in Example 9.9 is typical for this large a particle. The charge is proportional to diameter squared, so that a $\frac{1}{3}$-μ diameter particle would be expected to have about 33 electronic charges.

This charge is the steady-state value, reached after the particles have been in the precipitator a long time. Theoretical calculations show that for most particles in most precipitators this "long time" is much less than the average time a particle spends in the precipitator, so we can use this steady-state value as if the particle had it from the moment of its entry without serious error. If the particle is smaller than about 0.15 μ, then we would make a serious error computing its charge by Eq. (9.23); we must consider the additional charge that is acquired by *diffusion charging*. This latter results from particle-electron collisions caused not by the net motion of the electrons due to the electric field, but by the random motion superimposed on that motion by electron-gas molecule collisions, which make the electron behave like a gas molecule with a Boltzmann velocity distribution. Readers interested in field charging, the time necessary to charge a particle, the mathematics leading to Eq. (9.23), or a thorough treatment of all aspects of ESPs should consult Ref. 6.

The electrostatic force on a particle is

$$F = qE_p \tag{9.24}$$

Here E_p is the local electric field strength causing the force. Why do we use E_p in this equation and E_0 in the previous one? A particle may acquire its charge in a region of high E (near the wire) and then move into a region of lower E (near the plate). If we substitute for q from Eq. (9.23), we find

$$F = 3\pi\left(\frac{\varepsilon}{\varepsilon+2}\right)\varepsilon_0 D^2 E_0 E_p \tag{9.25}$$

The two subscripts on the Es remind us that one represents the field strength at the time of charging, the other the instantaneous (local) field strength. For all practical purposes we use an average E; and in the rest of this chapter we will use $E_0 = E_p = E$ and write subsequent equations with an E. If the particle's resistance to being driven to the wall by electrostatic forces is given by the Stokes drag force, Eq. (8.3), we can set the resistance force equal to the electrostatic force in Eq. (9.25) and solve for the resulting velocity, finding

$$V_t = \frac{D\varepsilon_0 E^2\left(\dfrac{\varepsilon}{\varepsilon+2}\right)}{\mu} = w \tag{9.26}$$

This velocity is called the *drift velocity* in the ESP literature, and is given the symbol w. We will use that symbol here although it is clearly the same as the V_t we found for gravity or centrifugal terminal settling velocities.

Example 9.10. Calculate the drift velocity for the particle in Example 9.9.

$$w = \frac{(10^{-6}\ \text{m})(8.85\times10^{-12}\ \text{C/V}\cdot\text{m})(3\times10^{5}\ \text{V/m})^2(6/8)\times(\text{N}\cdot\text{m/C}\cdot\text{V})}{(1.8\times10^{-5}\ \text{kg/m}\cdot\text{s})(\text{N}\cdot\text{s}^2/\text{kg}\cdot\text{m})}$$

$$= 0.033\ \frac{\text{m}}{\text{s}} = 0.109\ \frac{\text{ft}}{\text{s}}$$ ■

Since the calculated drift velocity is proportional to the particle diameter, one would compute larger values for the larger particles present in the gas stream.

Equation (9.26) shows that the drift velocity is proportional to the square of E, which is approximately equal to the wire voltage divided by the wire-to-plate distance. If we could raise the voltage or lower the wire-to-plate distance, we should be able to achieve unlimited drift velocities. The limitation here is sparking. The conditions between the wire and the plate are the same ones that exist between a thundercloud and the ground during a thunderstorm. Occasionally an ionized conduction path will be formed between the wire and the plate; this ionized path is then a good conductor and forms a continuous standing spark, which is in every way equivalent to a lightning stroke. The power supply to the wire must sense this sudden increase in current and stop the flow into it to prevent a burnout of the transformer. Normally the current is shut off for a fraction of a second, the lightning stroke ends, and then the field is reestablished. As one raises the values of E, the frequency of sparks increases. These sparks are energetic events that disrupt the cake on the plate (just as lightning strokes cause damage where they touch the earth), thus reducing the collection efficiency, so a large number of sparks are bad. Experimentally it has also been found that setting the voltage low enough to have zero sparks results in too low an E for optimum efficiency. Most ESP control systems are set for about 50 to 100 sparks per minute, which seems to be the optimum balance between the desire to increase E and the desire not to have too many sparks. Furthermore, it is common practice to subdivide the power supply of a large precipitator into many subsupplies so that each part of the precipitator can operate at the optimum voltage for its local conditions, and so that during the fraction of a second in which the system is shut down to neutralize a spark, only a small part of the whole ESP is shut down. (The multiple transformer-rectifiers are shown on the roof of the ESP in Fig. 9.8.)

When we compare the drift velocity here with the terminal settling velocity computed for the same particle in a cyclone separator in Example 9.3, we see that this is only about five times as fast. Why then is an ESP so much more effective than a cyclone for fine particle collection? As mentioned before, the drift velocity is proportional to D for an ESP and to D^2 for a cyclone. But to obtain a high drift velocity in a cyclone, one must use a high gas velocity. Thus, as shown in Example 9.7, the length of time the particle is exposed to centrifugal force in a cyclone is very short. On the other hand, the gas velocity does not enter Eq. (9.26), and the velocity with which the particle approaches the wall is independent of gas velocity. We can make the precipitator large enough that the particle spends a long time in it and has a high probability of capture. Typical modern ESPs have gas velocities of 3 to 5 ft/s (1 to 2 m/s), and the gas spends from 3 to 10 seconds in them. This is in marked contrast to the high gas velocities (and low residence times) necessary to make centrifugal separators work.

Since a precipitator is really a gravity settler in which we have replaced the gravitational force with an electrostatic force as the mechanism for driving the particles to the wall, it seems reasonable to assume that we can predict the behavior of ESPs by using Eqs. (9.5) and (9.11) and substituting the drift velocity w for V_t. Figure 9.7 is, in effect, two gravity settlers back to back, with one being the space between the wires and the far plate, and the other being the space between the wires and the near plate. The particles are driven from the wires toward both of the plates, in opposite horizontal directions. The maximum distance perpendicular to the flow that a particle must travel is the distance from the wire to the plate, which is the equivalent of H in Fig. 9.1.

If we now consider the section between the row of wires and one plate on Fig. 9.7, we see its collecting area is

$$A = Lh \tag{9.27}$$

and the volumetric flow through the section is

$$Q = HhV_{avg} \tag{9.28}$$

Making these substitutions in Eqs. (9.4) and (9.11), we find

$$\eta = \frac{wA}{Q} \quad \text{block flow} \tag{9.29}$$

and

$$\eta = 1 - \exp\left(-\frac{wA}{Q}\right) \quad \text{mixed flow} \tag{9.30}$$

In the literature, Eq. (9.29) is occasionally called the "theoretical laminar flow equation," which would hold if we had block flow of gas with no mixing [6, 7]. It has no practical use. Equation (9.30) is the *Deutsch-Anderson equation*, the most widely used simple equation for design, analysis, and comparison of ESPs. It is the same equation we have used for gravity settlers and cyclones, with the terms renamed.

Example 9.11. Compute the efficiency-diameter relation for an ESP that has particles with a dielectric constant of 6 and $(A/Q) = 0.2$ min/ft (≈ 0.060 s/m). We will use only the mixed flow equation.

Using the results of Example 9.10 we know that a 1-μ diameter particle will have a drift velocity of 0.109 ft/s, and that the drift velocity will be linearly proportional to the particle diameter. Thus for a 1-μ particle we may compute

$$\eta = 1 - \exp\left(-\frac{wA}{Q}\right) = 1 - \exp\left[-\left(0.109\ \frac{\text{ft}}{\text{s}}\right)\left(0.2\ \frac{\text{min}}{\text{ft}}\right)\left(\frac{60\ \text{s}}{\text{min}}\right)\right] = 0.73$$

As in Examples 9.1 and 9.4, we make up a table using this one computed value by taking advantage of the fact that the computed drift velocity is proportional to the particle diameter to the first power.

Particle diameter, μ	η
0.1	0.12
0.5	0.48
1	0.73
3	0.98
5	0.998

■

This example shows that this fairly typical precipitator has a cut diameter of about 0.5 μ, one-tenth of the cut diameter of a typical cyclone. If we plotted these values on an efficiency-diameter plot like Fig. 9.2, we would find a somewhat different shape, because the drift velocity in an ESP depends on D, whereas the terminal settling velocity in a gravity settler or cyclone depends on D^2.

One might hope to calculate the value of w from the theory previously presented and thus design precipitators by Eq. (9.30) with confidence. If every particle that got to the wall stayed there, then the performance calculated that way would be observed in working precipitators. Unfortunately, the rapping that loosens the particles from the wall also re-entrains some of them in the gas, and various particles have various re-entrainment properties. However, Eq. (9.30) suggests that if we pass a particle-laden gas stream through various precipitators, all of the data for this stream will form a straight line on a plot of log p vs. A/Q. Figure 9.9 is such a plot, in which the third variable is percent sulfur in the coal burned (explained later). Since the re-entrainment process is likely the same kind of random statistical process as the turbulence process, and since the amount re-entrained is likely to be proportional to the local cake thickness, which in turn should be a function of the local particle

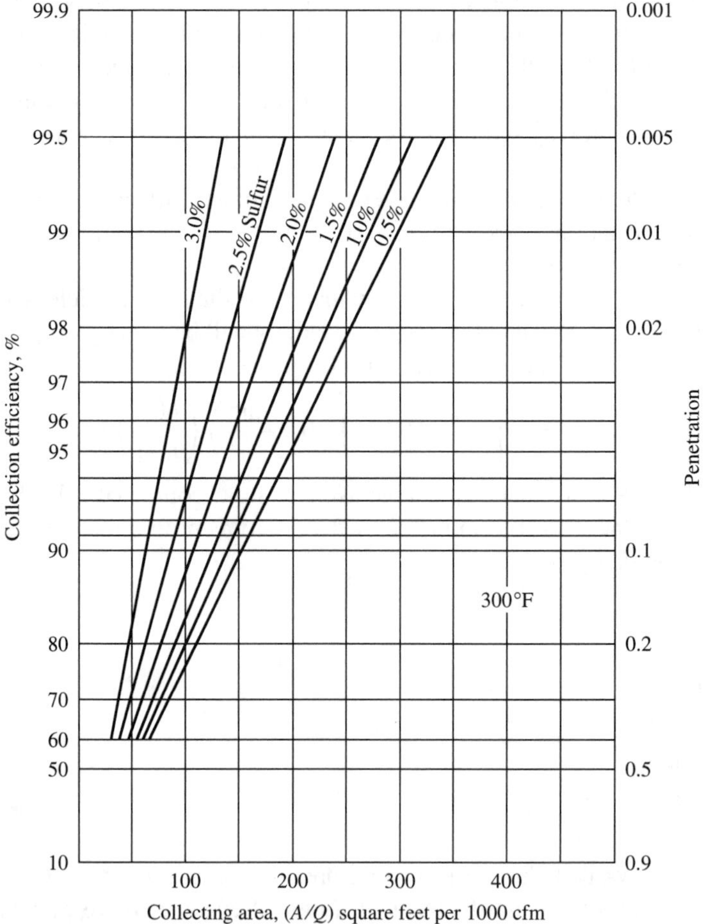

FIGURE 9.9
Summary of size-efficiency data for coal-fired power plant precipitators. Each line represents coal of a specified sulfur content. The precipitator size is specified in square feet of collecting area per 1000 cfm of gas flow; this is the common usage in the ESP industry. (From Ref. 8.) (Reprinted with permission of American Power Conference.)

concentration, we are not surprised that these experimental data for any fixed sulfur content fall on a straight line.

Example 9.12. From Fig. 9.9, estimate the value of w for coal containing 1 percent sulfur. From that figure at 99.5 percent efficiency we read that for 1 percent sulfur coal,

$$\frac{A}{Q} = \frac{310 \text{ ft}^2}{1000 \text{ ft}^3/\text{min}} = 0.31 \, \frac{\text{min}}{\text{ft}}$$

From Eq. (9.30), we calculate

$$w = -\frac{\ln p}{A/Q} = -\frac{\ln 0.005}{0.31 \text{ min/ft}} = 17.09 \, \frac{\text{ft}}{\text{min}} = 0.28 \, \frac{\text{ft}}{\text{s}} = 0.086 \, \frac{\text{m}}{\text{s}} \qquad \blacksquare$$

The different lines for different coal sulfur contents on Fig. 9.9 are caused by sulfur's indirect effect on fly ash resistivity (discussed later). ESPs work well with medium-resistivity solids, but poorly with low-resistivity or high-resistivity solids. We can see why by referring to Fig. 9.10, which shows three situations. In each situation the voltage at the wire is −40 kV and the voltage at the plate is zero; these

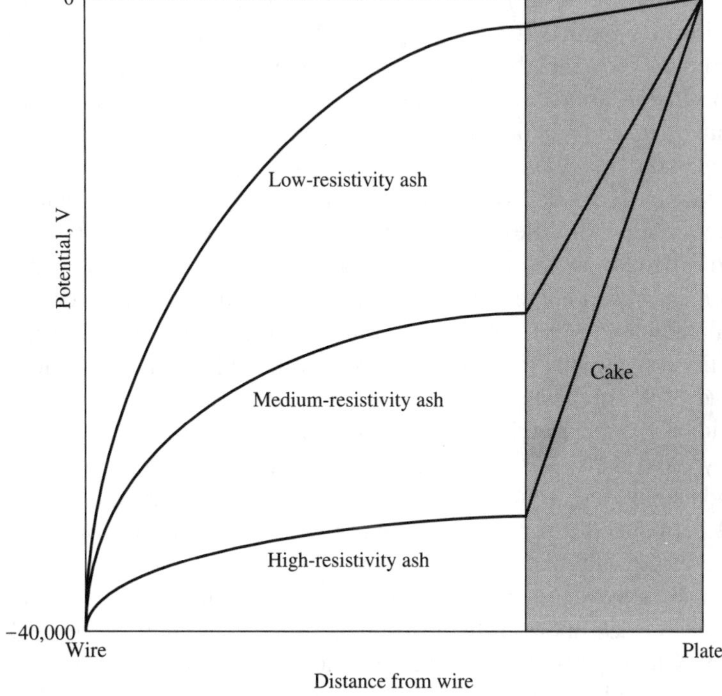

FIGURE 9.10
Voltage-distance relation between plate and wire for a low-, a medium-, and a high-resistivity ash. $E = \partial V/\partial x$ is the slope of the voltage–distance curve.

are common conditions for precipitators. For the case of a low-resistivity solid, e.g., carbon black, the material forms a cake that is a good conductor of electricity. The voltage gradient in the cake is small. On reaching the plate the particles are discharged and hence there is very little electrostatic force holding the collected particles to the plate. The collected particles do not adhere and are easily re-entrained; the overall collection is poor. (In one instance an ESP was used to agglomerate fine carbon particles although it could not collect them; they were subsequently collected in a cyclone, which could not collect the unagglomerated particles [7].) Because of this low cake adhesion for low-resistivity particles, ESPs are generally not used to collect particles with a resistivity less than 10^7 ohm · cm.

Figure 9.10 also shows a cake of medium-resistivity particles on the collecting plate. The voltage gradient across the cake is adequate to provide electrostatic force to hold the cake in place, but not enough to cause trouble. Figure 9.10 also shows particles of very high resistivity, e.g., elemental sulfur, on the plate. Here most of the voltage gradient occurs through the cake, causing at least two problems. First, the voltage gradient near the wire has now fallen so much that it cannot produce a good corona discharge. Thus, the particles are not properly charged. Second, the voltage gradient inside the cake is so high that in the gas spaces between the particles stray electrons will be accelerated to high velocities and will knock electrons off of gas molecules and form a *back corona* inside the cake. This back corona is a violently energetic conversion of electrostatic energy to thermal energy that causes minor gas explosions, which blow the cake off the plate and make it impossible to collect the particles. It is considered impractical to collect particles with resistivities greater than 2×10^{10} ohm · cm. The practical resistivity range is greater than 10^7 and less than 2×10^{10} ohm · cm.

If the resistivity of the particles is too low, little can be done. If the resistivity is too high, there are some possibilities. The resistivity of many coal ashes is too high at 300°F for good collection, but satisfactorily low at 600°F. (The resistivity change is due to improvement in conduction of some minerals in the ash with temperature increase.) Thus a precipitator operating after the air preheater at 300°F (the normal power plant location for the precipitator) might not work well on this ash, but an ESP located ahead of the preheater at 600°F, called a *hot-side precipitator,* might work well. Hot-side ESPs are used in some coal-fired power plants.

An ash may have high resistivity because its surface is a poor conductor. If one could condense on its surface a hygroscopic, conducting material, the ash resistivity would be reduced. Such condensation is reflected by the various lines on Fig. 9.9. Some of the sulfur in coal is converted in the furnace to SO_3, which collects on the ash, absorbs water, and makes the ash more conductive. Hence low-sulfur coal produces an ash more difficult to collect than does high-sulfur coal, as seen in Fig. 9.9. One logical cure is to add SO_3 to the gas stream approaching the precipitator to "condition" the ash. This works well sometimes. Coal ash is basic, so an acid conditioner seems best. Portland cement is acidic, and a basic conditioner like ammonia seems to work best for it. There are many proprietary conditioners on the market, working in the area between art and science. (SO_3 added as a conditioner increases the sulfur oxide emissions, but normally by a negligible amount.)

Another approach to the ash resistivity problem is to separate the charging and collecting functions. If the particles are charged in a separate charger, one can use a higher voltage and not worry much about the resulting sparks, because they do not pass through the cake and disrupt it. This idea has been tested on a pilot scale [9], with results positive enough that it is expected to be tried at full scale in the near future.

To calculate an appropriate value of w to use in Eq. (9.30), one would want to know the particle size distribution, the dielectric properties of the material, its resistivity, and whether the particles formed a coherent cake. Generally not all of that information is available in advance for a new material, so trials are made. Table 9.2 shows some representative values of w for industrial precipitators [6]. One should not think that these are truly the average particle velocities in the direction of collection. The rapping process re-entrains some fraction of the collected particles into the gas, so they must be collected again. Some of the gas bypasses the collecting zone in each section of the precipitator, in spite of the baffles that try to force it all through the collecting zone. The combination of these effects plus other effects discussed next causes the overall collection efficiency to be less than what we would calculate from Eq. (9.30) if we substituted the true drift velocity (if we could measure or calculate it). The values in Table 9.2 are those which, when substituted into Eq. (9.30), reproduce the observed ESP efficiencies in those industries.

TABLE 9.2
Typical values of the drift velocity encountered in industrial practice

Application	Drift velocity w, ft/s
Pulverized coal (fly ash)	0.33–0.44
Paper mills	0.25
Open-hearth furnace	0.19
Secondary blast furnace (80% foundry iron)	0.41
Gypsum	0.52–0.64
Hot phosphorus	0.09
Acid mist (H_2SO_4)	0.19–0.25
Acid mist (TiO_2)	0.19–0.25
Flash roaster	0.25
Multiple-hearth roaster	0.26
Portland cement manufacturing (wet process)	0.33–0.37
Portland cement manufacturing (dry process)	0.19–0.23
Catalyst particles	0.25
Gray iron cupola (iron-coke ratio = 10)	0.10–0.12

Source: Ref. 6.

Example 9.13. Our ESP has a measured efficiency of 90 percent. We wish to upgrade it to 99 percent. By how much must we increase the collecting area?

Using Eq. (9.30), we calculate

$$p_{\text{existing}} = 1 - \eta_{\text{existing}} = 0.1 = \exp\left(\frac{-wA_{\text{existing}}}{Q}\right)$$

$$p_{\text{new}} = 1 - \eta_{\text{new}} = 0.01 = \exp\left(\frac{-wA_{\text{new}}}{Q}\right)$$

$$\frac{\ln 0.1}{\ln 0.01} = 0.5 = \frac{(-wA_{\text{existing}}/Q)}{(-wA_{\text{new}}/Q)} = \frac{A_{\text{existing}}}{A_{\text{new}}}$$

$$\frac{A_{\text{new}}}{A_{\text{existing}}} = 2$$

■

This example shows that if the Deutsch-Anderson equation were obeyed exactly, then going from 90 percent to 99 percent efficiency requires that we double the collecting area; 90 percent to 99.9 percent, that we triple it, etc. Unfortunately life is harder than that. In Eq. (9.26) we found that the drift velocity is proportional to the particle diameter (down to the very small particles, where diffusion charging becomes important). Thus the big particles, which contain most of the mass, are removed first and, as the percentage efficiency by weight increases, the remaining particles become smaller and smaller and harder and harder to collect. To take this phenomenon into account, some designers use a modified Deutsch-Anderson equation with the form

$$p = 1 - \eta = \exp -(wA/Q)^k \tag{9.31}$$

where k is an arbitrary exponent, typically about 0.5 [10].

Example 9.14. Rework Example 9.13 using Eq. (9.31) instead of (9.30), taking $k = 0.5$.

Here for the existing unit $(wA/Q) = (-\ln p)^{1/k} = (-\ln 0.1)^2 = 5.30$. For the upgraded precipitator we need $(wA/Q) = (-\ln 0.01)^2 = 21.20$, so the new value of (A/Q)—assuming constant w—is $(21.20/5.30) = 4.0$ times the old value of (A/Q). We must quadruple the size of the precipitator instead of doubling it. ■

Equation (9.31) has no theoretical basis; it is a simple way to deal with the fact that as penetrations are reduced to lower and lower values the remaining particles become smaller and smaller, and collecting them becomes harder and harder. Theoretically we should make w a function of p (w gets smaller as p gets smaller) but using Eq. (9.31) is simpler. (See Problems 9.36 and 9.37.)

Uniform distribution of gas through a precipitator is very important. Nonuniform distribution of the gas flow in an ESP lowers its collection efficiency. We can see why by applying Eq. (9.30) to ESPs with uniform and nonuniform flows.

Example 9.15. A precipitator consists of two identical sections in parallel, each handling one-half of the gas. It is currently operating at 95 percent efficiency. We now hold the total gas flow constant, but maldistribute the flow so that two-thirds of the gas goes through one of the sections, and one-third through the other. What is the predicted overall collection efficiency?

For the existing situation, we calculate

$$p = 0.05 = \exp\left(\frac{-wA}{Q}\right); \qquad \frac{wA}{Q} = -\ln 0.05 = 2.995$$

For the new situation, we have

$$p_1 = \exp(-2.995)\left(\frac{1/2}{2/3}\right) = 0.1057$$

$$p_2 = \exp(-2.995)\left(\frac{1/2}{1/3}\right) = 0.0111$$

$$Q_1 p_1 = \frac{2}{3}Q(0.1057) = 0.070\,Q$$

$$Q_2 p_2 = \frac{1}{3}Q(0.0111) = 0.004\,Q$$

Adding these two equations, we find

$$(Q_1 + Q_2)p = 0.074\,Q$$

$$p = 0.074; \qquad \eta = 1 - p = 92.6\%$$

■

This example shows mathematically, for a very simple case, how maldistribution degrades precipitator performance. In a maldistributed flow, most of the gas passes through the high-velocity part, where it spends less than the average amount of time in the precipitator; and hence the collection efficiency is lower. Thus the gas passing through the high-velocity part contributes more to the penetration than it would in the uniform flow case. Considerable efforts are made to distribute the flow evenly through the precipitator. In Fig. 9.8 the gas enters through a set of perforated plates that even out the flow. In a new installation where there are long straight ducts to and from the ESP, this is normally satisfactory. In a retrofit, an ESP must often be fit into a plant near other big pieces of equipment and connected to the other equipment by short pieces of ducting that have frequent sharp bends. These bends introduce nonuniformities into the flow, which propagate through the ESP and result in poor performance. Inlet and outlet screens and baffles can even out the flow, but they cause a pressure drop, which is expensive for large gas flows. This trade-off between the desire to keep the pressure drop low (normally a few inches of water) and the need to have a uniform flow is important enough that for many large retrofit ESP installations in plants with limited space, a fluid mechanical model at 1/4 to 1/16 scale is built and lab tested with models of the associated ductwork to ensure that there will be adequate uniformity of flow without excessive pressure drop.

In a settling chamber or cyclone, all of the gas to be treated passes through the collecting zone. The same is not true for an ESP. In Fig. 9.8 we see that the wires and plates cannot reach completely to the top and bottom of the collecting volume; some space must be allowed for one or the other, because of the high voltage on one. Thus, some of the gas must pass through a region with poor collection. This is called *sneakage* in the ESP literature. Serious efforts are made to minimize it, using seals and baffles. The overall turbulence in the precipitator mixes the gas between the poor treatment regions and the major region, which has good treatment. Experts who are troubleshooting a poorly-operating precipitator always consider excessive sneakage due to worn or damaged seals as a likely cause.

The typical linear velocity of the gas inside an ESP is 3 to 5 ft/s, much lower than that in a cyclone. The typical pressure drop is 0.1 to 0.5 in. H_2O, again much less than in a cyclone. The pressure drop in the ducts leading to and from the precipitator is generally more than in the ESP itself.

The ESP industry is now well established. Standard package units are available for small flows (down to the size of home air conditioners), and large power plants have precipitators costing up to $30 million. The design shown in Fig. 9.8 is widely used, but other designs are widely used also. The collection requirements have been pushed from the 90–95% range typical in 1965 to the 99.5%-plus range now commonly specified. Faced with this challenge, and with the problem of upgrading existing precipitators to meet more stringent control requirements, ESP manufacturers have continued to use designs like Fig. 9.8 for collection of the first 90 or 95% of particles from large gas streams, but then often substitute other designs for the final collection stage. Vatavuk [11] presents a table similar to Table 9.2, showing values of w for various industries, both for ordinary dry ESPs and for wet ESPs in which the collected particles are continually removed by a film of fluid (normally water) flowing down the collecting surface instead of by intermittent rapping. In his table the values of w for wet ESPs are two to three times those for dry precipitators, mostly reflecting that there is no particle re-entrainment by rapping. Wet ESPs are more complex, and the collected particles are not in the convenient form of a dry powder. But for the final 5% cleanup these problems seem a modest price to pay for the greatly improved collection efficiency. Another approach is to make the final 5% collection in a filter, as described next. Sometimes the ESP-filter combination is more economical than an equivalent-performance ESP or filter [12].

9.2 DIVIDING COLLECTION DEVICES

Gravity settlers, cyclones, and ESPs collect particles by driving them against a solid wall. Filters and scrubbers do not drive the particles to a wall, but rather divide the flow into smaller parts where they can collect the particles. In this section we shall first consider the two types of filters used in air pollution control, *surface filters* and *depth filters*. Then we shall discuss scrubbers.

The public often refers to any kind of pollution control device as a filter, giving the word *filter* the meaning "cleaning device." Technically, a filter is one of the devices described in this section. Other devices (e.g., the "biofilters" described in Chapter 10) are not truly filters. Engineers must live with the difference between the technical meaning and that used by nonprofessionals.

9.2.1 Surface Filters

Most of us have personal experience with surface filters, as exemplified by those in a coffee percolator or a kitchen sieve. The principle of operation is simple enough; the filter is a membrane (sheet steel, cloth, wire mesh, or filter paper) with holes smaller than the dimensions of the particles to be retained.

Although this kind of filter is sometimes used for air pollution control purposes, it is not common because constructing a filter with holes as small as many of the particles we wish to collect is very difficult. One only needs to ponder the mechanical problem of drilling holes of 0.1-μ diameter or of weaving a fabric with threads separated by 0.1 μ to see that such filters are not easy to produce. It can be done on a laboratory scale by irradiating plastic sheets with neutrons and then dissolving away the neutron-damaged area. The resulting filters have analytical uses but are not used for industrial air pollution control (although they are used industrially to filter some beers and other products, removing trace amounts of bacteria). Figure 9.11 shows asbestos crystals captured on such a filter. Although these filters are very useful in determining the chemical identity and size distribution of air pollution particles, they are much too expensive and fragile for use as high-volume industrial air cleaners.

Although industrial air filters rarely have holes smaller than the smallest particles captured, they often act as if they did. The reason is that, as fine particles are caught on the sides of the holes of a filter, they tend to bridge over the holes and make them smaller. Thus as the amount of collected particles increases, the cake of collected material becomes the filter, and the filter medium (usually a cloth) that originally served as a filter to collect the cake now serves only to support the cake,

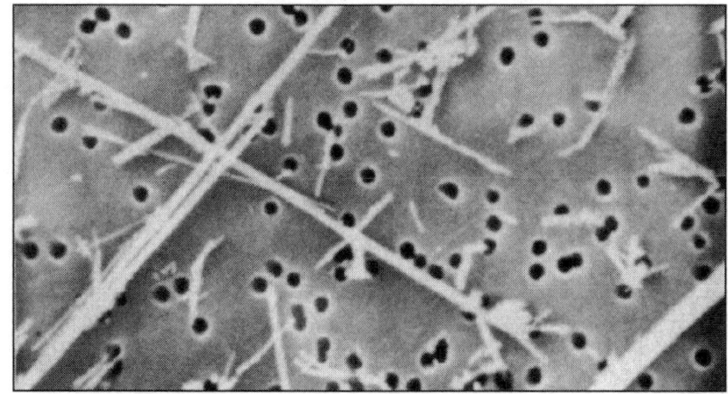

FIGURE 9.11
Scanning electron micrograph of chrysotile asbestos crystals collected on a Nucleopore® polycarbonate analytical filter with holes approximately 0.4 μ in diameter. (Courtesy of the Costar Corporation.)

and no longer as a filter. This cake of collected particles will have average pore sizes smaller than the diameter of the particles in the oncoming gas stream, and thus will act as a sieve for them. The particles collect on the front surface of the growing cake. For that reason this is called a *surface filter.*

One may visualize this situation with a screen having holes 0.75 in. (1.91 cm) in diameter. We could collect a layer of Ping-Pong balls easily on this screen. Once we had such a layer, we could then collect cherries, which, by themselves, could pass through the holes in the screen but cannot pass through the spaces between the Ping-Pong balls. Once we have a layer of cherries, we could put on a layer of peas, then of rice, then of sand. In that way we could collect sand on a screen with holes 0.75 inch in diameter. In typical industrial filters the particles are of a wide variety of sizes, so they do not go onto the screen in layers, but all at once. The effect is the same; very small particles are collected by the previously collected cake on a support whose holes are much larger than the smallest particles collected.

The theory of cake accumulation and pressure drop for this type of device is well-known from industrial filtration. The flow through a simple filter is shown schematically in Fig. 9.12. A fluid containing suspended solids (in this case a dirty gas stream) flows through a *filter medium,* which is most often a cloth, but sometimes a paper, porous metal, or bed of sand. The solid particles in the stream deposit on the face of the filter medium forming the *filter cake.* The cleaned gas, free from solids, flows through both cake and filter medium. If we follow the gas stream from point 1 to point 3 we see that the flow is horizontal and has a small change in velocity because the pressure drops, causing the gas to expand, and because the gas is leaving behind its contained particles. For most filters of air pollution interest, the combined effect of these changes is negligible. Therefore, the only fluid mechanical effect of interest is the decrease in pressure due to the frictional resistance to flow through the filter cake and the filter medium. In most industrial filters, both for gases and liquids, the flow velocity in the individual pores is so low that the flow is laminar. Therefore, we may use the well-known relations for laminar flow of a fluid in a porous medium

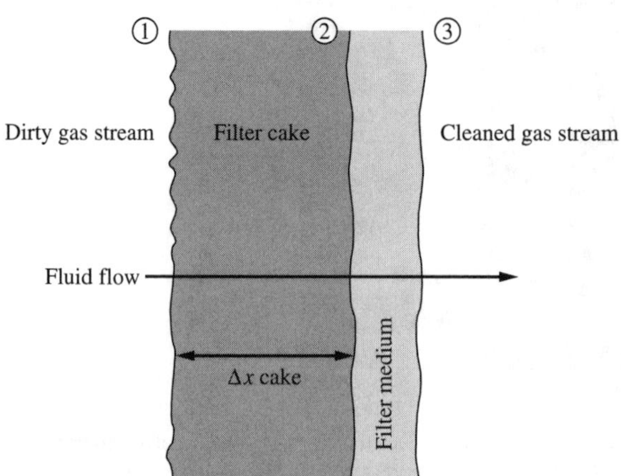

FIGURE 9.12
Flow through a surface filter.

[13], which indicate

$$V_s = \frac{Q}{A} = \left(\frac{-\Delta P}{\mu}\right)\left(\frac{k}{\Delta x}\right) \tag{9.32}$$

Here, k is the *permeability,* a property of the bed (or of the filter medium). For very simple beds, like stacked spheres, k can be calculated with fair accuracy from fluid mechanical principles [13]. For all beds of industrial interest k is determined experimentally, although the values calculated for spheres of comparable size may be used for rough estimates. For a steady fluid flow through a filter cake supported by a filter medium, there are two resistances to flow in series, but the flow rate is the same through each of them. Writing Eq. (9.32) for this flow rate (see Fig. 9.12), we find

$$V_s = \left(\frac{P_1 - P_2}{\mu}\right)\left(\frac{k}{\Delta x}\right)_{\text{cake}} = \left(\frac{P_2 - P_3}{\mu}\right)\left(\frac{k}{\Delta x}\right)_{\text{f.m.}} \tag{9.33}$$

where the subscript "f.m." indicates "filter medium." Solving for P_2, we get

$$P_2 = P_1 - \mu V_s\left(\frac{\Delta x}{k}\right)_{\text{cake}} = P_3 + \mu V_s\left(\frac{\Delta x}{k}\right)_{\text{f.m.}} \tag{9.34}$$

and then solving for V_s, we get

$$V_s = \frac{(P_1 - P_3)}{\mu[(\Delta x/k)_{\text{cake}} + (\Delta x/k)_{\text{f.m.}}]} = \frac{Q}{A_{\text{filter}}} \tag{9.35}$$

This equation describes the instantaneous flow rate through a filter; it is analogous to Ohm's law for two resistors in series. The $\Delta x/k$ terms are called the *cake resistance* and the *cloth resistance*.

The resistance of the filter medium is usually assumed to be a constant that is independent of time, so $(\Delta x/k)_{\text{f.m.}}$ is replaced with a constant α. If the filter cake is uniform, then its resistance is proportional to its thickness. However, this thickness is related to the volume of gas that has passed through the cake by the following material balance:

$$\begin{aligned}\Delta x_{\text{cake}} &= \left(\frac{\text{Mass of cake}}{\text{Area}}\right)\left(\frac{1}{\rho_{\text{cake}}}\right)\\ &= \left(\frac{1}{\rho_{\text{cake}}}\right)\left(\frac{\text{volume of gas}}{\text{area}}\right)\left(\frac{\text{mass of solids removed}}{\text{volume of gas}}\right)\end{aligned} \tag{9.36}$$

Customarily we define

$$W\eta = \left(\frac{\text{Mass of solids removed}}{\text{Volume of gas}}\right)\left(\frac{1}{\rho_{\text{cake}}}\right) = \frac{\text{volume of cake}}{\text{volume of gas processed}} \tag{9.37}$$

Here W is the volume of cake per volume of gas processed, which corresponds to a collection efficiency, η, of 1.00. For most surface filters $\eta \approx 1.00$, so the η is normally dropped when we use Eq. (9.37). Thus

$$\Delta x_{\text{cake}} = \left(\frac{V}{A}\right)W \quad \text{and} \quad \frac{d(\Delta x_{\text{cake}})}{dt} = V_s W \tag{9.38}$$

Here V is the volume of gas cleaned ($V = \int Q\,dt$). Substituting Eq. (9.38) for the cake thickness in Eq. (9.35), we find

$$V_s = \frac{Q}{A} = \frac{1}{A}\left(\frac{dV}{dt}\right) = \frac{(P_1 - P_3)}{\mu[(VW/kA) + \alpha]} \tag{9.39}$$

For most industrial gas filtrations the filter is supplied by a centrifugal blower at practically constant pressure, so $(P_1 - P_3)$ is a constant, and Eq. (9.39) may be rearranged and integrated to

$$\left(\frac{V}{A}\right)^2\left(\frac{\mu W}{2k}\right) + \left(\frac{V}{A}\right)\mu\alpha = (P_1 - P_3)t \qquad \text{[constant pressure]} \tag{9.40}$$

For many filtrations the resistance α of the filter medium is negligible compared with the cake resistance, so the second term of Eq. (9.40) may be dropped; in such cases the volume of gas processed is proportional to the square root of the time of filtration [14]. See Problems 9.42 and 9.43.

For some industrial gas filtrations a positive displacement blower, which is practically a constant-flow-rate device, feeds the filter at a pressure that steadily increases during the filtration. From Eq. (9.40) we see that for constant k and negligible α the pressure increases linearly with time, because the cake thickness increases linearly with time.

The theory presented here is equally applicable to the filtration of solids from gases or from liquids. In typical gas cleaning applications, k is practically a constant and is independent of pressure. In many filtrations from liquids, particularly filtration of soft or flocculant materials like water-treatment chemicals, k decreases as pressure increases, so the previous integrations that considered k as a constant must be redone with k taken as a function of P.

The two most widely used designs of industrial surface filters are shown in Figs. 9.13 and 9.14 on pages 285 and 286. Because the enclosing sheet metal structure in both figures is normally the size and roughly the shape of a house, this type of gas filter is generally called a *baghouse*. The design in Fig. 9.13, most often called a *shake-deflate* filter, consists of a large number of cylindrical cloth bags that are closed at the top like a giant stocking, toe upward. These are hung from a support. Their lower ends slip over and are clamped onto cylindrical sleeves that project upward from a plate at the bottom. The dirty gas flows into the space below this plate and up inside the bags. The gas flows outward through the bags, leaving its solids behind. The clean gas then flows into the space outside the bags and is ducted to the exhaust stack or to some further processing.

For the baghouse in Fig. 9.13 there must be some way of removing the cake of particles that accumulates on the filters. Normally this is not done during gas-cleaning operations. Instead the baghouse is taken out of the gas stream for cleaning. When the gas flow has been switched off, the bags are shaken by the support to loosen the collected cake. A weak flow of gas in the reverse direction may also be added to help dislodge the cake, thus deflating the bags. The cake falls into the hopper at the bottom of the baghouse and is collected or disposed of in some way. Often metal

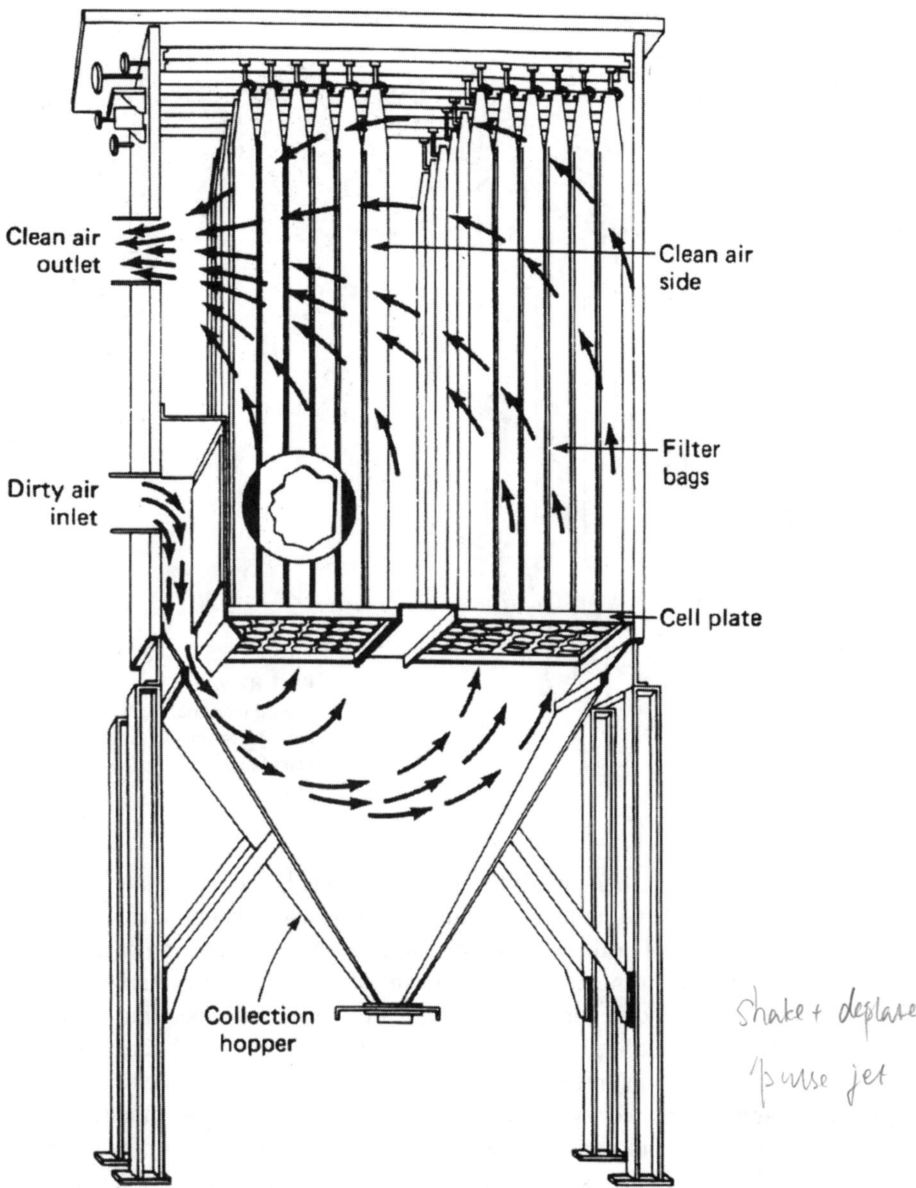

FIGURE 9.13
Typical industrial baghouse of the shake-deflate design. (Courtesy of Wheelabrator Air Pollution Control, Inc.)

rings are sewn into filter bags at regular intervals so that the bag will only partly collapse when the flow is reversed, and a path will remain open for the dust to fall to the hopper.

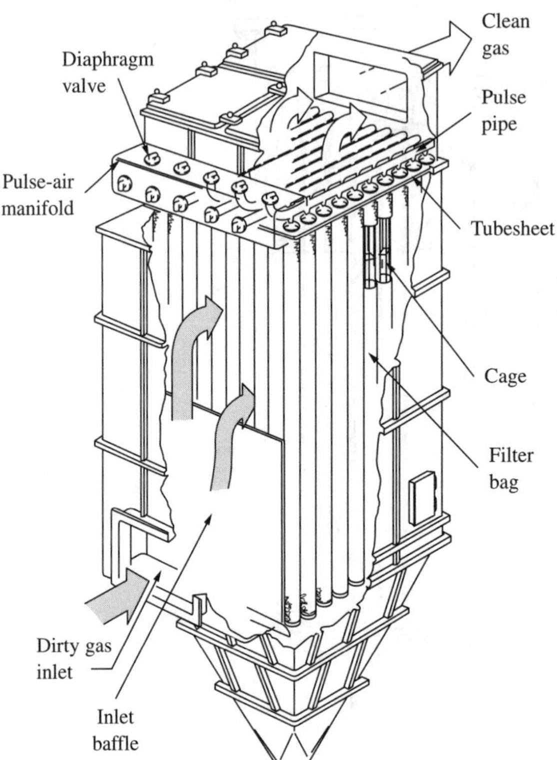

FIGURE 9.14
Typical industrial baghouse of the pulse-jet design. (Courtesy of ABB Fläkt Industriella Processer AB, Sweden.)

Because it cannot filter gas while it is being cleaned, a shake-deflate baghouse cannot serve as the sole pollution control device for a source that produces a continuous flow of dirty gas. For this reason, one either uses a large enough baghouse so that it can be cleaned during periodic shutdowns of the source of contaminated gas or installs several baghouses in parallel. Typically, for a major continuous source like a power plant, about five baghouses will be used in parallel, with four operating as gas cleaners during the time that the other one is being shaken and cleaned. Each baghouse might operate for two hours and then be cleaned for 10 minutes; at all times one baghouse would be out of service for cleaning or waiting to be put back into service. Thus the baghouse must be sized so that four of them operating together provide adequate capacity for the expected gas flow rate.

The other widely used baghouse design, called a *pulse-jet* filter, is shown in Fig. 9.14. In it the flow during filtration is inward through the bags, which are similar to the bags in Fig. 9.13 except their ends open at the top. The bags are supported by internal wire cages to prevent their collapse. The bags are cleaned by intermittent jets of compressed air that flow into the inside of the bag to blow the cake off. Often these baghouses are cleaned while they are in service; the internal pulse causes much of the collected solids to fall to the hopper, but some are drawn back to the filter cloth. Just after the cleaning the control efficiency will be less than just before the next cleaning, but the average efficiency meets the legal control requirements.

Example 9.16. The shake-deflate baghouse on the Nucla Power Station has six compartments, each with 112 bags that are 8 in. in diameter and 22 ft long, for an active area of 46 ft^2 per bag [15]. The gas being cleaned has a flow rate of 86,240 ft^3/min. (This very small power plant had one of the first baghouses on a coal-fired power plant and was the subject of extensive testing.) The pressure drop through a freshly cleaned baghouse is estimated to be 0.5 in. H_2O. The bags are operated until the pressure drop is 3 in. H_2O, at which time they are taken out of service and cleaned. The cleaning frequency is once per hour. The incoming gas has a particle loading of 13 grains/ft^3. The collection efficiency is 99 percent, and the filter cake is estimated to be 50 percent solids, with the balance being voids. Estimate how thick the cake is when the bags are taken out of service for cleaning. What is the permeability, k, of the cake?

First we compute the average velocity coming to the filter surface, V_s.

$$V_s = \frac{Q}{A} = \frac{86\,240 \text{ ft}^3/\text{min}}{(5)(112)(46 \text{ ft}^2)} = 3.35 \frac{\text{ft}}{\text{min}} = 1.02 \frac{\text{m}}{\text{min}}$$

The 5 is used here because one of the six compartments is always out of service for cleaning. In the baghouse literature, V_s is commonly referred to as the *air-to-cloth ratio* or *face velocity*. The dimension of (ft/min) is commonly dropped, so this filter would be referred to as having an air-to-cloth-ratio of 3.35 in countries using English units. In fluid mechanics this would be called a *superficial velocity* to indicate that it is total volumetric flow divided by total cross-sectional area of the filter. It is the same just before the filter cake as inside the filter cake. The velocity inside the pores of the filter is called the *interstitial velocity,* to distinguish it from this superficial velocity. It is larger, because Q is the same for both, but the A is less inside the cake. In most cases the superficial velocity is well-known, because the projected area of the cake is known; but the interstitial velocity is not known, because the interstitial projected area is not known. If the filter remains in service for 1 hour before cleaning and V_s is constant, then 1 square foot of bag will collect the following mass of particles:

$$\frac{m}{A} = cV_s\eta t = \left(13 \frac{\text{gr}}{\text{ft}^3}\right)\left(3.35 \frac{\text{ft}}{\text{min}}\right)(0.99)\left(\frac{60 \text{ min}}{\text{h}}\right)\left(\frac{\text{lbm}}{7000 \text{ gr}}\right)$$

$$= 0.369 \frac{\text{lbm}}{\text{ft}^2} = 1.80 \frac{\text{kg}}{\text{m}^2}$$

The thickness of the cake collected in 1 hour is

$$\text{Thickness} = \frac{m/A}{\rho}$$

$$= \frac{0.369 \text{ lbm/ft}^2}{(2 \text{ g/cm}^3)(0.5)(62.4 \text{ lbm} \cdot \text{cm}^3/\text{ft}^3 \cdot \text{g})}$$

$$= 5.9 \times 10^{-3} \text{ ft} = 0.071 \text{ in.} = 1.8 \text{ mm}$$

Taking $\alpha = 0$, we can solve Eq. (9.35) for k, and find

$$k = \frac{V_s\,\Delta x\,\mu}{(-\Delta P)}$$

$$= \frac{(3.35\text{ ft/min})(0.071\text{ ft}/12\text{ in.})(0.018\text{ cp})(2.09\times10^{-5}\text{ lbf}\cdot\text{s/ft}^2\cdot\text{cp})(\text{min}/60\text{ s})}{(3\text{ in. H}_2\text{O})(5.202\text{ lbf/ft}^2\cdot\text{in. H}_2\text{O})}$$

$$= 7.96\times10^{-12}\text{ ft}^2 = 7.40\times10^{-13}\text{ m}^2$$

Those familiar with the flow of fluids in porous media can compare this with values found in groundwater and underground oil flows by converting this to the conventional unit of permeability,

$$k = (7.96\times10^{-12}\text{ ft}^2)\left(\frac{\text{darcy}}{1.06\times10^{-11}\text{ ft}^2}\right) = 0.75\text{ darcies}$$

The calculated permeability of this material is roughly the same as that of a highly permeable sandstone. ■

The flow velocities through such filters are very low, typically a few feet per minute. In contrast, in devices like cyclones the flow is about 60 feet per second. A wind velocity equal to the typical flow through such a filter is so low that someone standing in it could not tell in which direction it was blowing and would report that there was no wind at all.

This calculation shows that the collected cake is about 0.07 in. thick, the average increase during one cycle. If the cleaning were perfect, this would be the cake thickness. However, it is hard or impossible to clean the bags completely, and in power plant operation it is common for the average cake thickness on the bags to be up to 10 times this amount. During each cleaning cycle some part of the cake falls completely away, leaving bare patches on the bag; and most of the cake does not come off at all. If one could examine a bag after a cleaning, one would probably see nine-tenths of the surface covered with a cake perhaps 0.7 in. thick, and one-tenth of the bag with a bare surface. The operators would like to clean the bags more thoroughly, but more vigorous cleaning procedures (harder shaking, faster reverse gas flow) tend to wear out the bags faster and lead to more frequent maintenance shutdowns. Most operators have used mild cleaning cycles, leading to long bag lives and low maintenance costs but higher pressure drops than would be needed if all the cake came off the bag at each cleaning cycle [16]. One of the advantages of the pulse-jet design is that it cleans the bags more thoroughly, allowing a higher V_s, at the cost of a somewhat shortened bag life.

Figure 9.15 is a set of typical results from tests of collection efficiency for this kind of filter. The individual lines represent different values of the superficial velocity (face velocity, air-to-cloth ratio). Consider first the curve for 0.39 m/min. We see that at zero fabric loading (new or freshly cleaned cloth) the outlet concentration is high and practically equal to the inlet concentration of about 0.8 g/m^3. As the cake builds up, the outlet concentration declines, finally stabilizing at a value about 0.001 times

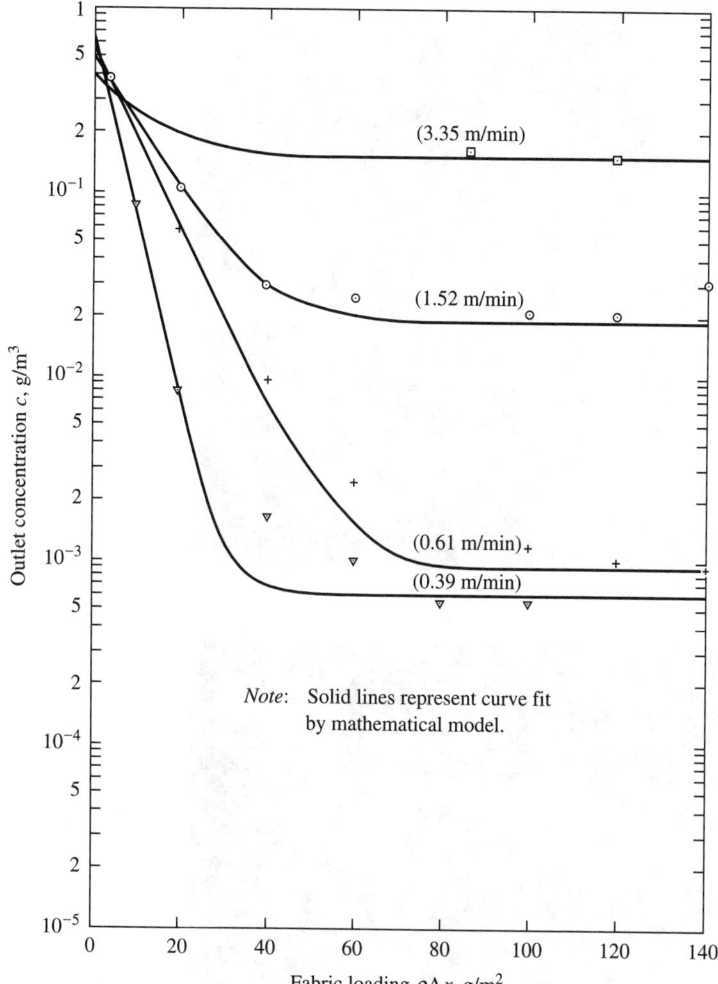

FIGURE 9.15
Effect of fabric loading (mass of collected particles per unit area) and face velocity on filter outlet concentration. For all the tests, the inlet concentration was about 0.8 g/m^3. (From Ref. 17.)

the inlet concentration (i.e., $\eta \approx 99.9$ percent). Once the cake has been properly established, the filtration efficiency remains constant. But why do any particles at all get through? The simple picture presented previously would suggest complete particle capture. Furthermore, by comparing the four curves in this figure, we see that if the superficial velocity increases, the efficiency falls; for a superficial velocity of 3.35 m/min the outlet concentration is about 20 percent of the inlet concentration. The particles that pass through such a filter do not pass through the cake but through *pinholes,* which are regions where the cake did not establish properly. Figure 9.16 on page 290 shows several such pinholes. They are apparently about 100 μ in diameter,

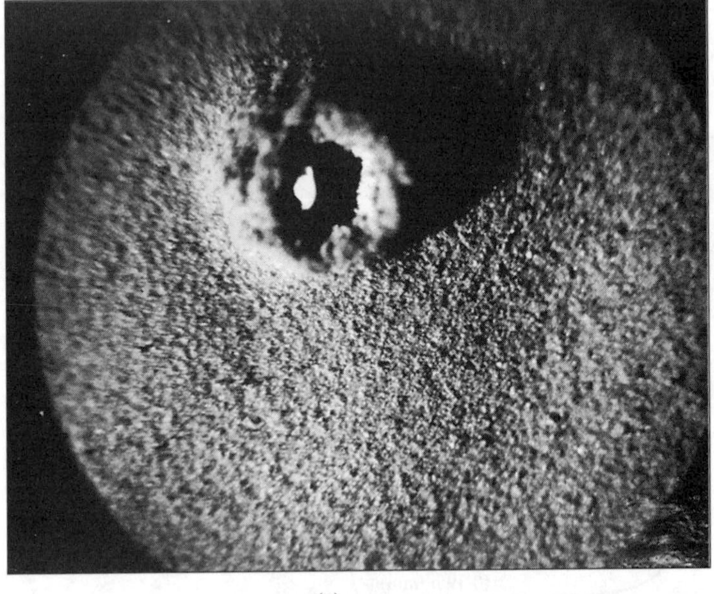

(*a*)

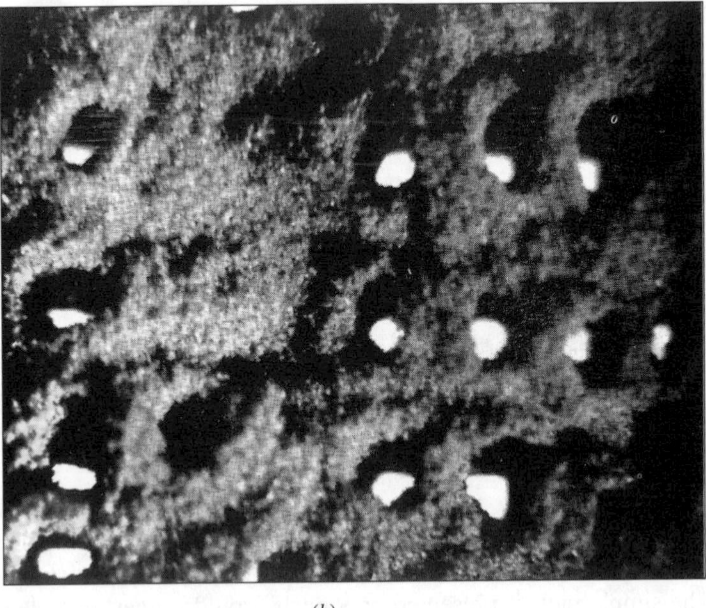

(*b*)

FIGURE 9.16
Photos of pinhole leaks in surface filters, with the filter surface lighted from below: (*a*) pinhole leak showing characteristic mound (20× magnification) and (*b*) massive pinhole leaks with monofilament screen, without loose fibers. (Most air pollution filter cloths are made from fibers with many loose ends, which establish the filter cake. These lead to fewer pinholes and much better collection efficiency than does the monofilament cloth, shown below, which has no such loose ends.) (From Ref. 17.)

much too large for a single particle to block because there are rarely 100-μ particles in the streams being treated. When the superficial velocity is high, more pinholes form, and thus a higher fraction of the flow passes through the pinholes than when it is low. The observed particle size distribution in the gas passing through this type of filter is practically identical to the particle size distribution in the gas entering the filter. That observation is only possible if the particles were in a part of the gas stream that passed through the filter practically unprocessed.

Example 9.17. Estimate the velocity through a pinhole in a filter with a pressure drop of 3 in. of water. Assuming that this is the pressure drop corresponding to the curve for 0.39 m/min on Fig. 9.15, that the steady-state penetration at that velocity is 0.001, and that the pinholes have a diameter of 100 μ, estimate how many pinholes per unit area there are in the cake.

One may show that, even though the pinhole is small, the flow through it is best described by Bernoulli's equation, from which we find the average velocity as

$$V = C\left(\frac{2\,\Delta P}{\rho}\right)^{1/2}$$

$$= 0.61\left[\frac{2(3 \text{ in. H}_2\text{O})}{(1.20 \text{ kg/m}^3)}\left(\frac{249 \text{ Pa}}{\text{in. H}_2\text{O}}\right)\left(\frac{\text{kg}}{\text{Pa}\cdot\text{m}\cdot\text{s}^2}\right)\right]^{1/2} = 21.5\,\frac{\text{m}}{\text{s}} = 70.6\,\frac{\text{ft}}{\text{s}}$$

Here the (area · velocity) of the pinholes must be 0.001 times the (area · velocity) of the rest of the cake. Hence

$$\frac{A_{\text{pinholes}}}{A_{\text{cake}}} = (0.001)\frac{V_s}{V_{\text{pinholes}}} = \frac{0.001(0.39 \text{ m/min})}{(21.5 \text{ m/s})(60 \text{ s/min})} = 3.0 \times 10^{-7}\,\frac{\text{m}^2}{\text{m}^2}$$

Each pinhole has an area of $A = (100 \times 10^{-6}\text{m})^2(\pi/4) = 7.85 \times 10^{-9}\ \text{m}^2$, so there must be

$$\frac{(3.0 \times 10^{-7})}{(7.85 \times 10^{-9}\ \text{m}^2)} = 38\,\frac{\text{pinholes}}{\text{m}^2}$$

■

The calculated velocity through a pinhole is $(21.6 \times 60)/0.39 = 3300$ times the velocity through the cake. The pinhole area need not be large to carry much of the flow with this high a velocity ratio. For the assumed conditions each pinhole is surrounded by an area of 0.026 m^2 ($= \text{m}^2/38$), which could be a square 0.16 m (6.4 inches) on a side. Unless the filter was illuminated from below (as in Fig. 9.16), one would probably not see this small number of very small holes.

If surface or cake-forming filters are operated at low superficial velocities, they can have very high efficiencies, and they generally collect fine particles as efficiently as coarse ones. For these two reasons they have found increasing application, particularly in electric power plants, as particle emission regulations have become steadily more stringent, making it necessary to collect particles in the size range from 0.1 to 0.5 μ, which are difficult for ESPs to collect.

9.2.2 Depth Filters

Another class of filters, widely used for air pollution control, does not form a coherent cake on the surface, but instead collects particles throughout the entire filter body. These are called *depth filters* to contrast them to the *surface filters* discussed in Sec. 9.2.1. The examples with which the student is probably familiar are the filters on filter-tipped cigarettes and the lint filters on many home furnaces. In both of these a mass of randomly oriented fibers (not woven to form a single surface) collects particles as the gas passes through it.

In Fig. 9.17, we see a particle-laden gas flowing toward a target, which we may think of as a cylindrical fiber in a filter. In Fig. 9.17 we are looking along the length of the fiber. The gas flow must bend to flow around the fiber, just as the wind bends to flow around a building or a river bends to flow around a rock in its middle. However, the contained particles, which are much denser (typically 2000 times) than the gas, are carried by their inertia, which makes them tend to continue going straight. Thus some of them hit the target rather than following the gas around it.

To determine whether a particle bumps into the target (and presumably adheres to it by electrostatic or van der Waals forces) or flows around it, we can compute its path, using the known flow fields for flow around various obstacles, computing the relative velocity between particle and gas using the appropriate equivalents of Stokes' law. That task was apparently first undertaken by Langmuir and Blodgett [18], who were working on the problem of ice formation on the leading edge of airplane wings. In Fig. 9.17, we may think of the target as an airplane wing moving to the left through still air and the particles as water drops in a cloud. If they contact the wing they may adhere and freeze, causing problems. Langmuir and Blodgett's mathematical solution is too long to include here, but Fig. 9.18 conveniently summarizes it for the small particles of interest in air pollution work. To see how they obtained it, consider a single particle in a turning part of the gas stream as shown in Fig. 9.19.

The particle, if it moves directly to the right, will run into the target. The force moving it upward on the figure (and hence around the target) is given by the Stokes drag force, Eq. (8.3). However, the appropriate velocity to use in Stokes' law is not

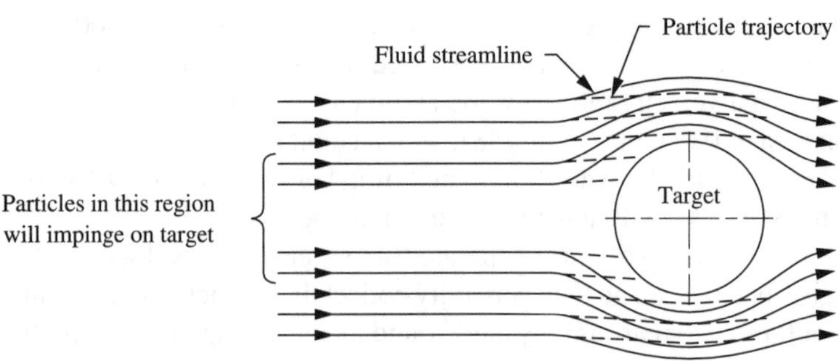

FIGURE 9.17
Flow of gas and particles around a cylinder.

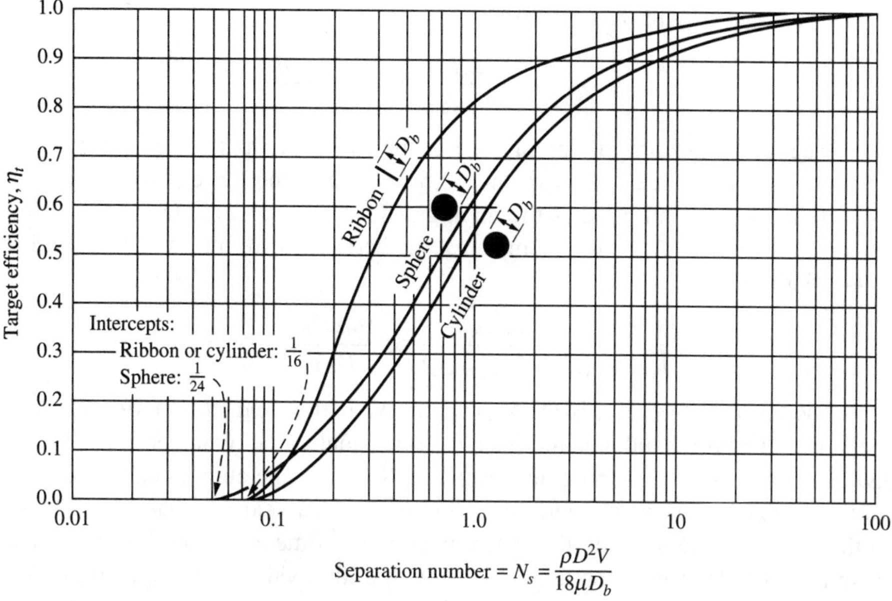

FIGURE 9.18
Target efficiency as a function of separation number, for cylinders, ribbons, and spheres. (From Ref. 18.)

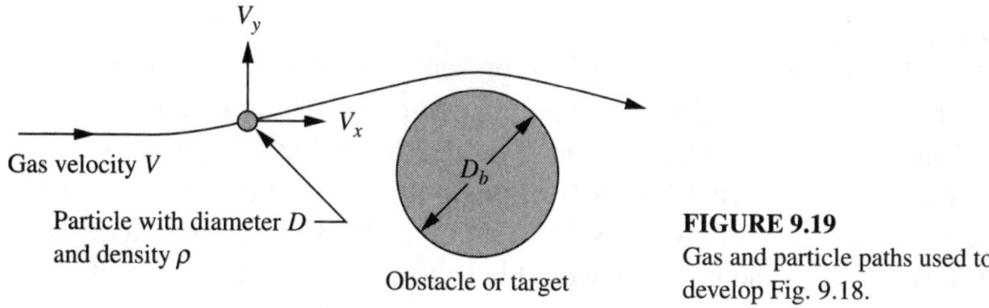

FIGURE 9.19
Gas and particle paths used to develop Fig. 9.18.

the overall stream velocity but rather the difference in y-directed velocity between the particle and the gas stream. Generally, the gas stream will have a larger velocity in this direction than the particle, so we may write

$$F_{y\text{-drag}} = 3\pi\mu D(V_{y\text{-gas}} - V_{y\text{-particle}}) \tag{9.41}$$

The resisting (inertial) force of the particle is

$$F_{y\text{-inertial}} = ma = \frac{\pi}{6}\rho D^3 \frac{dV_{y\text{-particle}}}{dt} \tag{9.42}$$

If there are no electrostatic, magnetic, or other forces acting, then these are equal and opposite, so we may solve for the particle's acceleration in the y direction,

$$\frac{dV_{y\text{-particle}}}{dt} = \frac{18\mu(V_{y\text{-gas}} - V_{y\text{-particle}})}{\rho D^2} \tag{9.43}$$

We can separate variables and integrate to find

$$\int \frac{dV_{y\text{-particle}}}{(V_{y\text{-gas}} - V_{y\text{-particle}})} = \frac{18\mu\,\Delta t}{\rho D^2} \tag{9.44}$$

We may further simplify by saying that the Δt on the right is the time available for the y-directed forces to move the particle around the target, which must be proportional to the time it takes the main gas flow to go past the target. This time $\Delta t = D_b/V$, where D_b is the diameter of the barrier, so we may substitute in Eq. (9.44), finding

$$\int_{y_1}^{y_2} \frac{dV_{y\text{-particle}}}{(V_{y\text{-gas}} - V_{y\text{-particle}})} \propto \frac{18\mu D_b}{\rho D^2 V} = \frac{1}{N_s} \tag{9.45}$$

The term on the right is $(1/N_s)$, where N_s is the *separation number,* which appears on the horizontal axis in Fig. 9.18. It is equal to the diameter of the barrier divided by the Stokes stopping distance (Sec. 8.2.4). Some authors call N_s the *impaction parameter* or *inertia parameter*. In Eq. (9.45) we can see that if the integral on the left is a large number, then there is plenty of time and force for the flow to move the particle around the target. Thus with a high value of the integral (a low value of N_s) there is a great likelihood that the particle will be swept around the obstacle and the target efficiency (η_t in Fig. 9.18, see next paragraph) will be low. Conversely, if the integral is small (a large value of N_s) then the time and force are inadequate to move the particle around the target, and most of the particles will contact the target.

The *target efficiency* η_t in Fig. 9.18 represents the number of particles that actually contact the target, divided by the number that would have contacted it if all particles had moved perfectly straight and none had been drawn around the target by the gas stream. To construct Fig. 9.18, Langmuir and Blodgett integrated Eq. (9.45) using the known fluid flow paths for various kinds of targets and taking into account the various paths the particles would have to follow. For example, at an N_s of 0.8 for a cylindrical target, the target efficiency in Fig. 9.18 is roughly 0.5. This value means that those particles whose trajectories, if continued perfectly straight, would have hit the outer 50 percent of the target will be carried around it and not contact it. We have presented the integration limits in Eq. (9.45) as y_1 and y_2, with no explanation. In Langmuir and Blodgett's work, these values were calculated and used to make up Fig. 9.18.

Example 9.18. A single, cylindrical fiber 10 μ in diameter is placed perpendicular to a gas stream that is moving at 1 m/s. The gas stream contains particles that are 1 μ in diameter and the particle concentration is 1 mg/m^3. What is the rate of collection of particles on the fiber?

If all the particles that start moving directly toward the fiber hit it (i.e., a target efficiency of 100 percent), then the collection rate would be equal to the volumetric flow rate approaching the fiber times the concentration of particles, i.e.,

$$\text{Maximum possible rate} = VD_b c = \left(1\frac{\text{m}}{\text{s}}\right)(10^{-5}\ \text{m})(10^{-3}\ \text{g/m}^3)$$

$$= 10^{-8}\ \frac{\text{g}}{\text{m}\cdot\text{s}} = 6.7 \times 10^{-12}\ \frac{\text{lbm}}{\text{ft}\cdot\text{s}}$$

If we catch them all, we will collect 10^{-8} g/s for every meter of fiber length. The actual amount caught will be this number times the target efficiency. The separation number is

$$N_s = \frac{(2000\ \text{kg/m}^3)(10^{-6}\ \text{m})^2(1\ \text{m/s})}{(18)(1.8 \times 10^{-5}\ \text{kg/m}\cdot\text{s})(10^{-5}\ \text{m})} = 0.617$$

From Fig. 9.18, we see that for cylinders this value of N_s corresponds to a target efficiency of about 0.42, so we would expect to collect about 0.42×10^{-8} g/m · s. ■

Example 9.19. A filter consists of a row of parallel fibers across a flow, as described in Example 9.18, with the center-to-center spacing of the fibers equal to five fiber diameters. What collection efficiency will the filter have for the particles? Assume that the fibers are far enough apart that each one behaves as if it were in an infinite fluid, uninfluenced by the other fibers.

Here, we can use the preceding results to see that 42 percent of the particles that were traveling directly toward the fibers are collected. If the fibers are spaced five fiber diameters apart, then the open area is 80 percent $[(5-1)/5]$, and the blocked area is 20 percent $(1/5)$. The target efficiency, as just described, applies only to those particles that were flowing toward the blocked area, so the overall collection efficiency is

$$\text{Collection efficiency} = (\text{target efficiency})\ (\text{percentage blocked})$$
$$= 0.42 \times 0.2 = 0.084 = 8.4\%$$

■

Example 9.20. A filter consists of 100 rows of parallel fibers as described in Example 9.19, arranged in series. They are spaced far enough apart that the flow field becomes completely uniform between one row and the next (i.e., the rows do not interact). What is the collection efficiency of the entire filter?

Here, we calculate

$$\eta_{\text{overall}} = 1 - p_{\text{overall}} = 1 - (p_{\text{individual}})^n = 1 - (1 - 0.084)^{100} = 0.9998$$ ■

These three examples show, in idealized form, what goes on within depth filters. Most such filters do not have an orderly array of parallel fibers; the filter medium consists of a tangled jumble of fibers in a random orientation, making up a thick mat. The mat resembles the felt material used to make hats, line pool tables, etc., or steel wool or fiberglass building insulation. (The student should examine a piece of any of these materials to see how different it is from woven cloth.) The idealization that individual fibers do not interact is clearly an approximation. But these thick fiber mats do operate almost entirely by impaction, sometimes called *impingement,* as calculated here. The individual particles have many chances to contact an individual fiber on their path through the mat, and their likelihood of being caught on any one is shown in Fig. 9.18.

Such filters are often used where the particles to be caught are fine drops of liquids that are only moderately viscous. Such drops will coalesce on the fibers and then run off as larger drops, leaving the fibers ready to catch more fine drops. If the particles were solid, then this type of filter would require regular cleaning; for the liquid application it does not. The most widespread air pollution control use of depth filters is in the collection of very fine liquid drops, *sulfuric acid mist,* produced in sulfuric acid plants. Similar devices are used in many gas-liquid contacting devices to catch fine droplets; one brand uses the trade name Demister. This kind of device is also used for removing solid particles from gas streams that contain few of them, e.g., for cleaning the air of industrial clean rooms or hospital surgical suites and in personal protection dust masks. The filters are thrown away when they have collected enough particles that their pressure drop begins to increase. The depth filters used in those applications are normally called *high-efficiency, particle-arresting* (HEPA) or *absolute filters*. The air filters on household furnaces operate this way as well; typically the fibers are coated with a sticky substance to improve the retention of the collected dust and lint.

Depth filters collect particles mostly by impaction. Some older types of particle collectors also used impaction, to catch particles on solid walls, but they are seldom used now. Some size-specific particle analyzers (*impactors* or *cascade impactors*) use impaction on collecting surfaces to collect specific sizes of particles. In liquid scrubbers (discussed later), one of the principal collection mechanisms is the collision between the particle and a moving drop of liquid (usually water). We will have further use of Fig. 9.18 when we discuss scrubbers.

As discussed in Sec. 8.2.6, small particles move in gases by diffusion. In depth filters that diffusion leads to particle collection in addition to that computed above by impaction. We can use previously developed solutions for mass transfer in gases to compute the efficiency with which particles will diffuse to a collecting surface. In Fig. 9.17 consider the case of a very small particle, for which the separation number is so small that it has practically zero chance of impacting the target. If, however, it is in the stream of gas that passes close to the target and Brownian motion at right angles to the main flow moves it against the target, it will probably adhere. In this case, we would say that it was collected by diffusion (see Sec. 8.2.6) rather than by impaction.

Using this idea, Freidlander developed a theoretical equation, with constants determined by experiment, for the case of diffusional collection of particles from a gas stream flowing past a cylinder under circumstances where impaction was negligible [19]. Most of the published data could be represented by

$$\eta_t = \frac{6\mathcal{D}^{2/3}}{\nu^{1/6} D_b^{1/2} V^{1/2}} + \frac{3D^2 V^{1/2}}{\nu^{1/2} D_b^{3/2}} \tag{9.46}$$

where all the terms are as defined previously, and ν is the kinematic viscosity. The first term on the right is for diffusional collection, whereas the second is for collection by noninertial contact. The calculations of Langmuir and Blodgett are based on point masses; the final term in Eq. (9.46) takes into account the fact that the particles have finite diameters and hence will contact the target if their center passes within $D/2$ of it. This behavior is called *interception.*

Example 9.21. Repeat Example 9.18 for particles having a diameter of 0.1 μ. Take into account impaction, diffusion, and interception.

In this case N_s is $(0.1)^2 = 0.01$ times the previous value, or 0.062, for which, from Fig. 9.18 we can read η_t = practically zero. Hence, a particle of this size will not be collected by impaction.

From Fig. 8.1 we can read that the diffusivity is about 6×10^{-6} cm²/s (6×10^{-10} m²/s). So

$$\eta_t = \frac{6(6 \times 10^{-10}\ \text{m}^2/\text{s})^{2/3}}{(1.49 \times 10^{-5}\ \text{m}^2/\text{s})^{1/6}(10^{-5}\ \text{m})^{1/2}(1\ \text{m/s})^{1/2}} + \frac{3(10^{-7}\ \text{m})^2(1\ \text{m/s})^{1/2}}{(1.49 \times 10^{-5}\ \text{m}^2/\text{s})^{1/2}(10^{-5}\ \text{m})^{3/2}}$$

$$= 0.0086 + 0.00025 = 0.0088 \approx 0.9\%$$

■

The diffusion term is $(0.0086/0.00025) = 34.4$ times the interception term. As the particles become smaller the diffusion term becomes relatively more important, whereas the interception term increases in importance as the particles increase in size. The interception and impaction mechanisms respond in the same general way to changes in velocity and particle diameter. The mechanisms are compared in Table 9.3.

There is some particle size at which there is a minimum collection efficiency (Problem 9.57). Typically, this size is in the range 0.1 to 1 μ, which is the size most likely to be deposited in the human lung. We would like to have a particle collection device that was most efficient for this size particle; no such device is known.

It has also been observed that if the particles are charged before they enter the filter, they will be collected with a higher efficiency than if they are not. This has led to the ESP–baghouse combination, in which an old ESP that does not meet new

TABLE 9.3
Comparison of collection mechanisms

	Impaction and interception	Diffusion
Increasing particle size causes efficiency to	Increase	Decrease
Increasing gas velocity causes efficiency to	Increase	Decrease
Increasing target diameter causes efficiency to	Decrease	Decrease

emission standards has a baghouse attached to its downstream side. The particles passing from the ESP to the baghouse are mostly the smallest of the particles that entered the ESP, and many of them are charged. The measured performance of this combination is often better than one would predict for an ESP plus a baghouse treating uncharged particles.

9.2.3 Filter Media

Whether a filter behaves as a surface or a depth filter depends on the type of filter medium used. For shake-deflate baghouses (Fig. 9.13) the filter bags are made of tightly woven fibers, much like those in a pair of jeans. (The reader is invited to look at the sun through a single layer of such fabric, seeing that it has some pinholes, allowing light to come through, and to blow into such a fabric, observing that one can breathe in and out through one.) Pulse-jet baghouses (Fig. 9.14) use high-strength felted fabrics, so that they act partly as depth filters and partly as surface filters. This allows them to operate at superficial velocities (air-to-cloth ratios) two to four times those of shake-deflate baghouses; in recent years this higher capacity per unit size has allowed them to take market share away from the previously dominant shake-deflate type baghouses.

Filter fabrics are made of cotton, wool, glass fibers, and a variety of synthetic fibers. The choice depends on price and suitability for the expected service. Cotton and wool cannot be used above 180 and 200°F, respectively, without rapid deterioration, whereas glass can be used to 500°F (and short-term excursions to 550°F). The synthetics have intermediate service temperatures. In addition the fibers must be resistant to acids or alkalis if these are present in the gas stream or the particles as well as to flexing wear caused by the repeated cleaning. Typical bag service life is 3 to 5 years. Generally fibers that have many small microfibers sticking out their sides form better cakes than those that do not. The student should examine under a microscope a thread of cotton, which has such microfibers, and one of monofilament fishing line, which does not.

9.2.4 Scrubbers for Particulate Control

Just as filters work by separating the flow of particle-laden gas into many small streams, so also *scrubbers* effectively divide the flow of particle-laden gas by sending many small drops through it.

In air pollution control engineering, the term *scrubber* originally meant a device for collecting fine particles on liquid drops. Then when liquid drops were used to collect sulfur dioxide (see Chapter 11), the devices that did that were also called *scrubbers*. Recently, alas, some other types of devices have been marketed as *dry scrubbers*. In this chapter, we will use the original meaning of the term: a scrubber is a device that collects particles by contacting the dirty gas stream with liquid drops.

Most fine particles will adhere to a liquid drop if they contact it. So if we can make the drop and the particle touch each other, the particle will be caught on the drop. Particles 50 μ and larger are easily collected in cyclones. If our problem is to collect a set of 0.5-μ particles, cyclones will not work at all. However, if we were to introduce a large number of 50-μ diameter drops of a liquid (normally water) into the gas stream to collect the fine particles, then we could pass the stream through a cheap, simple cyclone and collect the drops and the fine particles stuck on them. This idea is the basis of almost all scrubbers for particulate control.

A complete scrubber has several parts, as sketched in Fig. 9.20. Most often, the gas–liquid separator is a simple cyclone of the type discussed in Sec. 9.1.2; water drops of the size encountered in most scrubbers pose few difficulties for such cyclones. The liquid–solid separator can be of many kinds although gravity settlers seem to be the most common. If possible, the engineer should try to save money by finding a place where the contaminated water stream can be recycled inside the plant without first removing the solids. There are many examples where that has been done successfully. Obviously, if there is no good way to deal with the contaminated water stream, then the scrubber has merely changed an air pollution problem into a water pollution problem.

For the rest of this chapter, we will assume that the gas–liquid and the liquid–solid separations are relatively easy; we will only concern ourselves with the gas–liquid contactor, in which the particles are caught on the drops. Most of that capture takes place by impaction or impingement, as described in Figs. 9.18 and 9.19, to which we will refer often.

9.2.4.1 Collection of particles in a rainstorm. We will begin with a collection device that all students have witnessed—a rainstorm. From that we will work toward the more complex geometries of industrial interest.

Figure 9.21 on page 300 shows the geometry for which we will make a material balance on the particles and on the drops. We consider a space with dimensions Δx, Δy, Δz. The concentration of particles in the gas in this space is c (lbm/ft^3 or kg/m^3).

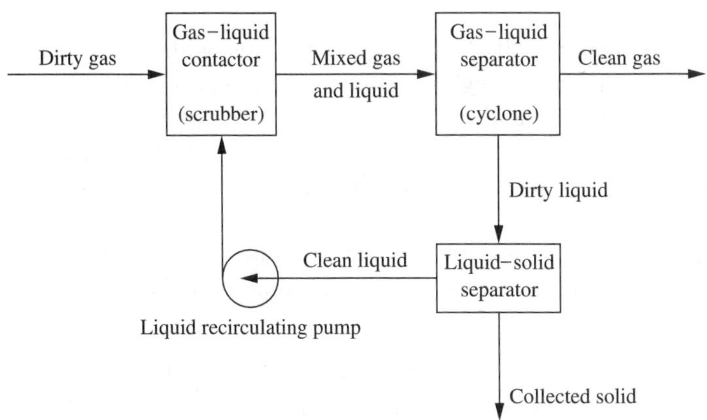

FIGURE 9.20
Component parts of a scrubber installation.

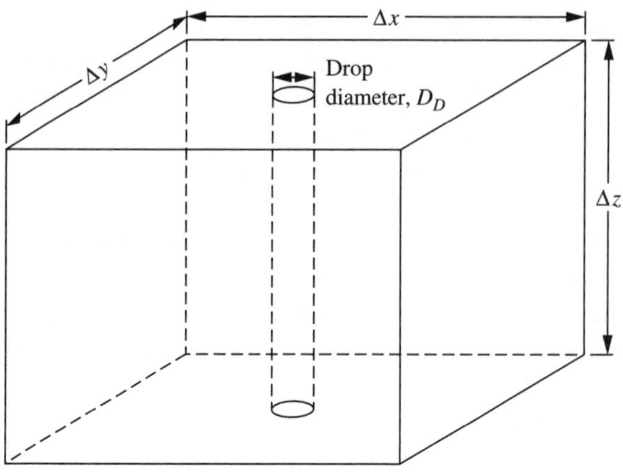

FIGURE 9.21
Region considered in the material balance for a rainstorm.

Now we let one spherical drop of water of diameter D_D pass through this space. How much of the particulate matter in the space will be transferred to the drop? We can see that the volume of space swept out by the drop is the cylindrical hole shown in Fig. 9.21, whose volume is

$$V_{\text{swept by one drop}} = \frac{\pi}{4} D_D^2 \, \Delta z \tag{9.47}$$

The total mass of particles that was originally in that swept volume is that volume times the concentration c. The fraction of these that will be collected by the drop is the target efficiency η_t, which we can determine from Fig. 9.18 or its equivalent. So the mass of particles transferred from the gas to the drop is

$$\begin{aligned} \begin{pmatrix} \text{Mass transferred} \\ \text{to one drop} \end{pmatrix} &= \begin{pmatrix} \text{swept} \\ \text{volume} \end{pmatrix} (\text{concentration}) \begin{pmatrix} \text{target} \\ \text{efficiency} \end{pmatrix} \\ &= \frac{\pi}{4} D_D^2 \, \Delta z c \eta_t \end{aligned} \tag{9.48}$$

Next we consider a region of space (still $\Delta x \, \Delta y \, \Delta z$) that is large with respect to the size of any one individual raindrop through which a large number of raindrops are falling at a steady rate N_D, expressed as drops/time. Each of the drops stirs the region of gas around it so that there is no distinction between volume "swept by a drop" and volume "not swept by a drop," as there would be for the foregoing single-drop example.

For the region $\Delta x \, \Delta y \, \Delta z$ we wish to know how the concentration of particles in the air changes during the rainstorm. From a material balance on the particles in the space, we can say that

$$\begin{aligned} \frac{dc}{dt} &= -\frac{(\text{mass transferred to each drop})(\text{number of drops/time})}{(\text{volume of the region})} \\ &= -\frac{(\pi/4)(D_D^2 \, \Delta z c \eta_t N_D)}{\Delta x \, \Delta y \, \Delta z} = -\frac{\pi}{4} D_D^2 c \eta_t \left(\frac{N_D}{\Delta x \, \Delta y}\right) \end{aligned} \tag{9.49}$$

We multiply top and bottom of the equation by the volume of a single spherical drop and simplify to obtain

$$\frac{dc}{dt} = -\frac{\pi}{4} D_D^2 c \eta_t \left(\frac{N_D}{\Delta x \, \Delta y} \right) \frac{(\pi/6) D_D^3}{(\pi/6) D_D^3} = -1.5 \frac{c \eta_t}{D_D} \left(\frac{N_D (\pi/6) D_D^3}{\Delta x \, \Delta y} \right) \tag{9.50}$$

The final term in parentheses in Eq. (9.50) represents the volume of rain that fell per unit time (the number of drops per unit time times their individual volume) divided by the horizontal area through which they fell. (Weather reports often tell of rain falling at a rate of one inch per hour, a rapid rate indeed. One may also think of this as the rate at which the level in a container like a glass will rise if the rain falls in at its open top and none exits.) For the rest of this chapter the total liquid volumetric flow rate going to a scrubber (or to this region of space) will have the symbol Q_L (m^3/s or equivalent) so the rightmost term is Q_L/A, where A is the horizontal projection of the region of interest. Substituting this into Eq. (9.50), we can rearrange and integrate to find

$$\begin{aligned} \frac{dc}{dt} &= -\frac{1.5}{D_D} c \eta_t \frac{Q_L}{A} \\ \frac{dc}{c} &= -\frac{1.5}{D_D} \eta_t \frac{Q_L}{A} \, dt \end{aligned} \tag{9.51}$$

$$\ln p = \ln \frac{c}{c_0} = -\frac{1.5}{D_D} \eta_t \frac{Q_L}{A} \, \Delta t \tag{9.52}$$

Example 9.22. A rainstorm is depositing 0.1 in./h, all in the form of spherical drops 1 mm in diameter. The air through which the drops are falling contains 3-μ diameter particles at an initial concentration 100 $\mu g/m^3$. What will the concentration be after one hour?

Solving Eq. (9.52) for c, we find

$$c = c_0 \exp - \left(\frac{1.5 \eta_t Q_L \, \Delta t}{D_D A} \right)$$

We know all of the quantities on the right except η_t. From Fig. 8.7 we can read the terminal settling velocity of a 1-mm diameter drop of water in still air is about 14 ft/s = 4.2 m/s, so we can compute N_S from Eq. (9.45) as

$$N_s = \frac{\rho D_p^2 V}{18 \mu D_b} = \frac{(2000 \text{ kg/m}^3)(3 \times 10^{-6} \text{ m})^2 (4.2 \text{ m/s})}{(18)(1.8 \times 10^{-5} \text{ kg/m} \cdot \text{s})(10^{-3} \text{ m})} = 0.23$$

Here D_b (the barrier diameter in the definition of N_S) = D_D (the drop diameter). From Fig. 9.18 we can read $\eta_t \approx 0.23$, so

$$c = 100 \frac{\mu\text{g}}{\text{m}^3} \exp \left[-\frac{(1.5 \cdot 0.23)(0.1 \text{ in./h})(1 \text{ h})}{10^{-3} \text{ m}} \cdot \frac{\text{m}}{39.37 \text{ in.}} \right] = 43 \frac{\mu\text{g}}{\text{m}^3} \quad \blacksquare$$

This example shows that the result depends on the total amount of rain that fell, $Q_L \, \Delta t/A$, which is 0.1 inch in this case, not on the time or rainfall rate separately.

The fractional removal is independent of the initial concentration; c/c_0 does not depend on the value of c_0. Although rainfall collects large particles well, it does poorly for small particles. If the example had asked for the collection efficiency for particles of 1-μ diameter, we would have calculated an N_s one-ninth as large, and from Fig. 9.18 we would have computed an η_t of zero. Recall that Fig. 9.18 only describes the impaction mechanism, which would be zero in this case; the diffusional mechanism would have led to some collection, but the efficiency would have been very small.

This calculation suggests that, contrary to popular opinion, a rainstorm does not clean the air well. The rainstorm will remove large particles but have little effect on those smaller than 1-μ, which are of the greatest health concern and which are the most efficient light scatterers. It is a common observation in the northern and western United States that the air is much clearer after a rainstorm than before. The reason is not that the raindrops cleared the air, but that rainstorms in this region are normally followed by a flow of polar air, or air from over the Pacific Ocean; the incoming air is generally cleaner than the air it replaces.

9.2.4.2 Collection of particles in crossflow, counterflow, and co-flow scrubbers. To get good removal of small particles, we must find some way to increase the value of N_s for the drop-particle interaction to get a higher value of η_t. We will consider several scrubber geometries to see what the possibilities are.

Crossflow scrubbers. Consider the crossflow scrubber sketched in Fig. 9.22, which shows the overall dimensions and some of the notation. This is a large box with multiple spray nozzles that disperse the incoming liquid, Q_L, uniformly over the horizontal surface and a floor drain that collects the liquid at the bottom. The gas is assumed to move through the scrubber in uniform, blocklike flow at a total volumetric flow rate of Q_G.

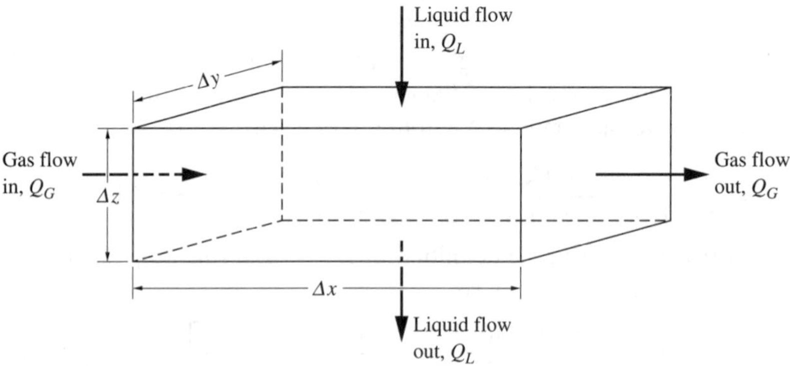

FIGURE 9.22
Schematic of a crossflow scrubber.

A parcel of air moving through this scrubber behaves just like a parcel of air standing still in a rainstorm. If we can compute the time it takes such a parcel of air to travel through the scrubber, we can use it in Eq. (9.52) to compute the collection efficiency. The linear velocity of the gas is $(Q_G/\Delta y\ \Delta z)$, and hence the time it takes a parcel of gas to pass through is the length of the scrubber divided by the linear velocity, or

$$\text{Travel time} = \Delta t = \frac{\Delta x\ \Delta y\ \Delta z}{Q_G} \tag{9.53}$$

One may also think of this as the volume of the scrubber divided by the volumetric flow rate of gas, and hence as the quantity (1/the number of scrubber volumes of gas admitted per unit time). Substituting this value for travel time into Eq. (9.52), we find

$$\ln p = \ln \frac{c}{c_0} = -\frac{1.5\eta_t Q_L\ \Delta x\ \Delta y\ \Delta z}{D_D A Q_G} = -1.5 \cdot \frac{\eta_t}{D_D} \cdot \frac{Q_L}{Q_G} \cdot \Delta z \tag{9.54}$$

This equation says that the smaller the drop and the taller the scrubber, the more efficient it will be in removing particles. However, we must consider the path taken by an individual drop in the scrubber. A very large drop will fall almost straight down, because its vertical velocity due to gravity is much larger than the horizontal velocity of the gas. But a small drop has a much lower vertical velocity, so it will be carried along in the flow direction by the gas. If we try to get a good collection efficiency (a low value of $p = c/c_0$) by increasing Δz or decreasing D_D, we see that the drops will pass out with the gas and not be collected in the scrubber. For this reason, this type of scrubber is not widely used. There are some applications; for example, one wishes to capture a valuable dust, of fairly large particle size, in an aqueous solution. In such cases it is common practice to locate the spray heads only in the most upstream part of the roof of the scrubber. The distance between the most downstream spray head and the outlet of the scrubber is calculated to allow most of the drops to reach the bottom of the scrubber before they reach the outlet.

Counterflow scrubbers. The next geometry to consider is the counterflow scrubber, sketched in Fig. 9.23 on page 304. Liquid enters the top of the scrubber through a series of spray nozzles that distribute it uniformly and falls by gravity. The gas enters the bottom of the scrubber and flows upward in uniform, blocklike flow.

We might be tempted to proceed as we did for the crossflow scrubber and simply compute the gas transit time and substitute it into Eq. (9.52). Alas, there is a complication. In the rainstorm problem and in the crossflow scrubber, the distance that a drop travels relative to fixed coordinates is the same distance it travels relative to the gas (Δz in both cases). Here that is no longer the case, because if the drop is at its terminal settling velocity V_t relative to the gas *that surrounds it,* but that gas is moving upward with velocity $V_G = Q_G/\Delta x\ \Delta y$, then the velocity of the drop relative to the fixed coordinates of the scrubber is $V_{D\text{-Fixed}} = V_t - V_G$.

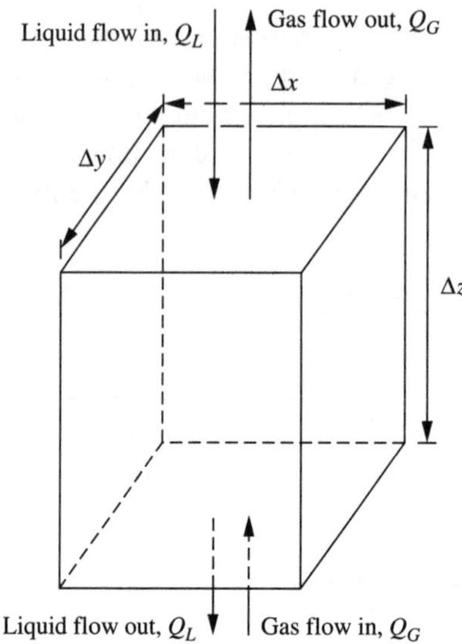

FIGURE 9.23
Schematic of a counterflow scrubber.

We remake our previous material balance as in the following:

$$\begin{pmatrix}\text{Mass of particles transferred}\\ \text{to drops per unit time}\\ \text{per unit volume}\end{pmatrix} = -\begin{array}{l}\text{mass of particles transferred out of}\\ \text{the gas per unit time per unit volume}\end{array}$$
$$= (\text{volume swept/time})(\text{particle concentration}) \times (\text{target efficiency})$$
$$= -(\text{gas volumetric flow rate}) \times (\text{change in particle concentration}) \tag{9.55}$$

To compute the quantity (volume swept by drops/time) we must compute the instantaneous number of drops per unit volume. The liquid flow into the system is Q_L (m^3/s), and this consists of N_D drops/time, each of volume $(\pi/6)D_D^3$. The average time each such drop spends in the scrubber is the vertical distance divided by the vertical velocity *relative to fixed coordinates*, or

$$\text{Average time} = \frac{\Delta z}{(V_t - V_G)} \tag{9.56}$$

so at any time the number of drops in the system is

$$\text{Drops present at any time} = \frac{N_D\,\Delta z}{(V_t - V_G)} \tag{9.57}$$

The volume of gas that these drops sweep out per unit time is their number times their cross-sectional area times the velocity at which they move *relative to the gas,* which is V_t. So we can compute

$$\frac{\text{Volume swept}}{\text{Time}} = \left(\frac{N_D\,\Delta z}{V_t - V_G}\right)\left(\frac{\pi D_D^2}{4}\right)V_t$$

$$= \left(\frac{Q_L}{\pi D_D^3/6}\right)\left(\frac{\Delta z\,\pi}{4}\right)\left(\frac{D_D^2 V_t}{V_t - V_G}\right) \tag{9.58}$$

$$= Q_L\left(\frac{1.5}{D_D}\right)(\Delta z)\left(\frac{V_t}{V_t - V_G}\right)$$

We substitute Eq. (9.58) into Eq. (9.55), finding

$$Q_L\left(\frac{1.5}{D_D}\right)(\Delta z)\left(\frac{V_t}{V_t - V_G}\right)c\eta_t = -Q_G\,\Delta c \tag{9.59}$$

If we now let the scrubber height be infinitesimally small, so that Δz and Δc become dz and dc, we can separate the variables and integrate, finding

$$\frac{dc}{c} = -1.5 \cdot \frac{\eta_t}{D_D} \cdot \frac{Q_L}{Q_G} \cdot \frac{V_t}{(V_t - V_G)}\,dz \tag{9.60}$$

$$\ln p = \ln\frac{c}{c_0} = -1.5 \cdot \frac{\eta_t}{D_D} \cdot \frac{Q_L}{Q_G} \cdot \frac{V_t}{(V_t - V_G)}\,\Delta z \tag{9.61}$$

Comparing Eq. (9.61), for counterflow scrubbers, to Eq. (9.54), for crossflow scrubbers, we see that the only difference is the addition of a $[V_t/(V_t - V_G)]$ term, which accounts for the fact that each drop moves farther relative to the gas than it moves relative to the fixed geometry of the scrubber.

Equation (9.61) also allows us to see the limitation of this kind of scrubber. We can get 100 percent efficiency ($c/c_0 = 0$) if we let $V_t = V_G$, because that makes the value of the right side negative infinity. Physically, that means that if the upward velocity of the gas equals the terminal settling velocity of the liquid, then the individual drop will stand still in the scrubber and will collect from an infinitely long column of gas as the gas passes. However, if we continue to put liquid into the scrubber (Q_L not equal to zero) and no liquid leaves, we will fill the scrubber with liquid. It will become *flooded* and will cease to operate as a scrubber. Since we want to use the smallest practical size drops in order to get high values of N_s and thus of η_t, flooding sets a very strong practical limitation on this kind of scrubber. There are some important applications where they are used (Chapter 11), but they do not play a major role in *particulate* air pollution control.

Co-flow scrubbers. Clearly we need a geometrical arrangement in which we can get very small drops to move at high velocities relative to the gas being scrubbed, to get a high N_s and high η_t, without blowing the drops out the side or top of the scrubber. The solution to this problem is the co-flow scrubber, shown schematically in Fig. 9.24 on page 306. In it, both gas and liquid enter at the left and exit at the right. However, the liquid enters at right angles to the gas flow; it comes in with

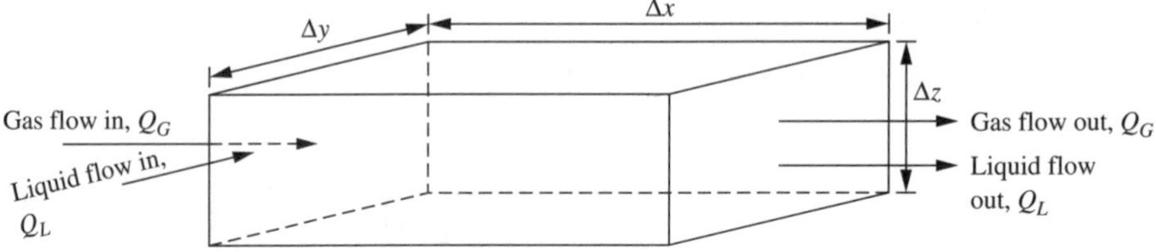

FIGURE 9.24
Schematic of a co-flow scrubber.

much less velocity in the x direction than does the gas. Since both liquid and gas flow in the same direction, there is no problem with the gas blowing the drops in the wrong direction.

Very high gas velocities can be used in this type of scrubber, as much as 400 ft/s (122 m/s). The liquid enters with zero or negligible velocity in the x direction so that at the inlet the *relative* velocity may be as high as 400 ft/s. This may be 100 times the maximum tolerable relative velocity in a crossflow or counterflow scrubber.

To make up a material balance for a co-flow scrubber we begin with Eq. (9.55). Here our problem is harder, because the velocity of the liquid drops changes as they move in the x direction. At the inlet they have zero or practically zero velocity, but by the time they reach the outlet they normally will have come to the same velocity as the gas. If we write the material balance for a differential length dx of the scrubber, we can say that the average time it takes a drop to pass that section is

$$\text{Average time} = \frac{dx}{V_{D\text{-Fixed}}} = \frac{dx}{V_G - V_{\text{Rel}}} \tag{9.62}$$

where $V_{D\text{-Fixed}}$ = drop velocity referred to fixed coordinates
V_G = local gas velocity
V_{Rel} = velocity of the drops relative to the gas = $V_G - V_{D\text{-Fixed}}$

Then we can write out the analog of Eq. (9.58), using V_{Rel} in place of V_t (both are the drop velocity relative to the gas for their appropriate scrubber) and Eq. (9.62) in place of Eq. (9.57). We find

$$\begin{aligned} d\left(\frac{\text{volume swept}}{\text{time}}\right) &= \left(\frac{N_D\,\Delta x}{V_G - V_{\text{Rel}}}\right)\left(\frac{\pi D^2}{4}\right) V_{\text{Rel}} \\ &= Q_L\left(\frac{1.5}{D_D}\right)\left(\frac{V_{\text{Rel}}}{V_G - V_{\text{Rel}}}\right) dx \end{aligned} \tag{9.63}$$

and

$$\frac{dc}{c} = -\frac{1.5}{D_D}\cdot \eta_t \frac{Q_L}{Q_G}\cdot \frac{V_{\text{Rel}}}{V_G - V_{\text{Rel}}}\,dx \tag{9.64}$$

For the previous examples we could integrate the equations corresponding to Eq. (9.64) because all the terms on the right were independent of distance or time. For a co-flow scrubber they are not. Figure 9.25 shows how some of these variables change for a simple, constant cross-sectional area scrubber.

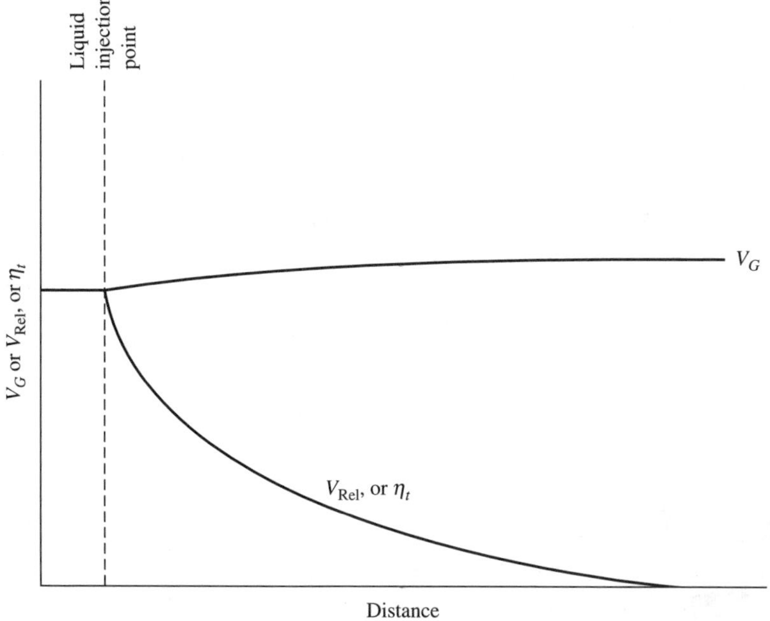

FIGURE 9.25
Sketch of behavior of V_G, V_{Rel}, and η_t vs. distance in a constant cross-sectional area co-flow scrubber.

At the liquid injection point there is a small increase in gas velocity. This appears odd but is real. The gas must transfer momentum to the liquid to speed it up. That causes a fall in pressure and hence a fall in gas density. But the gas mass flow rate is constant along the scrubber, so the velocity must increase to offset the decreasing density.

Initially the relative velocity is almost equal to the gas velocity because $V_{Rel} = V_G - V_{D\text{-Fixed}}$, but V_{Rel} declines rapidly as the drops speed up. If the scrubber is long enough, V_{Rel} will approach zero as the liquid drops reach the velocity of the gas. Because N_s is linearly proportional to V_{Rel}, and η_t is a monotonically increasing function of N_s, it seems clear that η_t will change more or less proportionately to V_{Rel}, as sketched on Fig. 9.25.

If these changes with distance were all that was involved, we could presumably use the drag force relationships previously introduced to calculate V_{Rel} as a function of distance, calculate η_t at every point, and then integrate Eq. (9.64) to find the predicted performance of any such scrubber. Two other factors make the problem even more complex. First, when a jet of fluid is injected into a fast-moving stream, it breaks up into droplets of widely varying sizes. Then these, being acted on by the fast-moving gas stream, break up into even smaller drops. The breakup does not happen all at once, but over time. So the assumption of constant drop size (or constant drop size distribution) that was tolerable for the rainstorm and the crossflow and counterflow scrubbers cannot be used here. No attempt to integrate Eq. (9.64) that does not take into account the changing drop size distribution has much chance of success. Second, although the simple geometry shown in Fig. 9.24 is occasionally used (most often in a small-diameter pipe treating a small flow of dirty gas), for large flows the venturi design shown in Fig. 9.26 is much more economical.

The co-flow venturi scrubber shown in Fig. 9.26 has a rectangular cross section, chosen for ease of fabrication. Liquid is injected into the throat of the venturi; the discharge passes immediately to a large cyclonic separator. The venturi design is widely used because it saves fan power. We know from fluid mechanics that for such a design the velocity at any point is equal to the volumetric flow rate Q_G, which is practically constant through the device, divided by the cross-sectional area. For the venturi shown in Fig. 9.26 the throat cross-sectional area is about one-fifth that at the inlet or outlet, so the velocity there must be about five times the velocity at the inlet or outlet. To achieve high velocities in a gas flow we must have a drop in pressure;

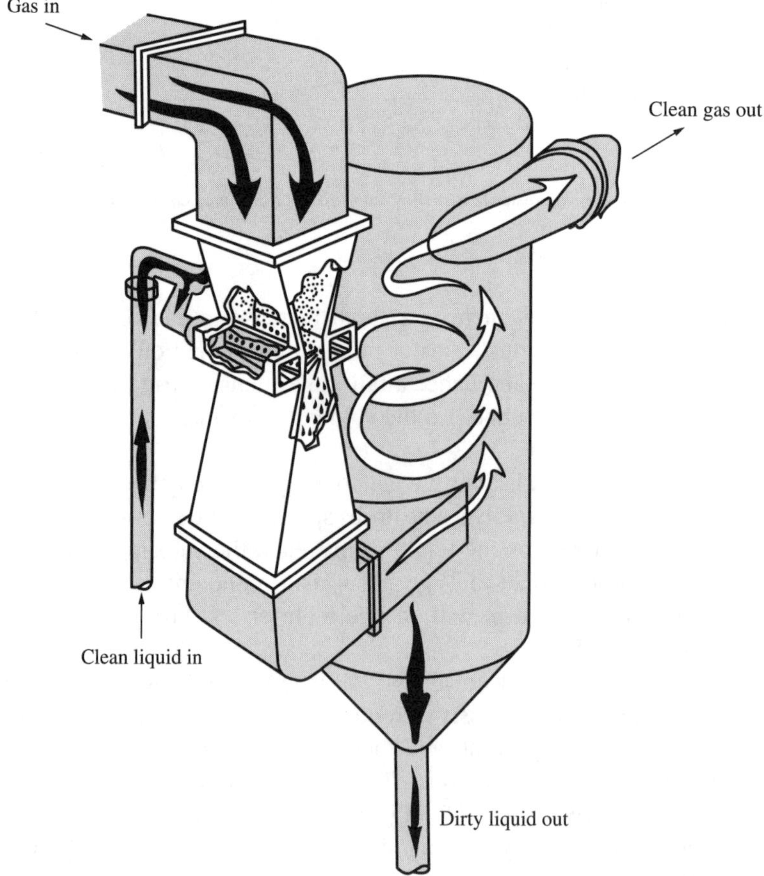

FIGURE 9.26
Schematic of a typical downflow venturi scrubber, with liquid injection at the throat and the discharge passing to a large cyclone separator. (Courtesy of Wheelabrator Air Pollution Control Inc.)

for steady, horizontal frictionless flow we know that

$$P_2 - P_1 = \left(\frac{\rho}{2}\right)\left(V_1^2 - V_2^2\right) \tag{9.65}$$

The flow in the converging section of a venturi obeys this frictionless equation fairly well. In the diverging section downstream of the throat, we would find that Eq. (9.65) still applied if we had frictionless flow, and we would get back all the pressure we lost in speeding up the flow. Real venturis never work that well, but in practice we can get a gas flow of a specified high velocity with much less overall pressure loss in a venturi arrangement like that shown in Fig. 9.26 than we can in any other simple device. A venturi co-flow scrubber seems the most economical way to get a high velocity for rapid liquid breakup into drops and high collection efficiency with minimum fan power.

Example 9.23. In a venturi scrubber the throat velocity is 400 ft/s = 122 m/s. The particles to be collected have diameters of 1 μ, and the droplet diameter is 100 μ. We are feeding 10^{-3} m^3 of liquid per m^3 of gas to the scrubber ($Q_L/Q_G = 10^{-3}$). At a point where V_{Rel} is 0.9 V_G, what is the rate of decrease in particle concentration in the gas phase?

We begin by evaluating N_s; the appropriate velocity to use in N_s is V_{Rel}.

$$N_s = \frac{\rho D_p^2 V}{18\mu D_b} = \frac{(2000\ \text{kg/m}^3)(10^{-6}\ \text{m})^2(0.9 \times 122\ \text{m/s})}{(18)(1.8 \times 10^{-5}\ \text{kg/m}\cdot\text{s})(10^{-4}\ \text{m})} = 6.78$$

From Fig. 9.18 we see that $\eta_t = 0.92$. Then from Eq. (9.64), we may write

$$\frac{dc/c}{dx} = -\left(\frac{1.5}{10^{-4}\ \text{m}}\right)(0.92)\left(10^{-3}\right)\left(\frac{0.9V_G}{V_G - 0.9V_G}\right) = -\frac{124}{\text{m}} = -\frac{0.124}{\text{mm}}$$

In this part of the scrubber the concentration of particles in the gas is decreasing by 12.4 percent for every millimeter of gas travel in the flow direction. ■

Example 9.24. How rapidly is V_{Rel} changing for the drop in Example 9.23? The particle Reynolds number for the droplet is

$$R_p = \frac{D\rho V}{\mu} = \frac{(10^{-4}\ \text{m})(1.20\ \text{kg/m}^3)(0.9 \cdot 122\ \text{m/s})}{1.8 \times 10^{-5}\ \text{kg/m}\cdot\text{s}} = 732$$

This Reynolds number is very high because of the very high assumed speed. In any fluid mechanics text we can look up the corresponding drag coefficient for a spherical droplet, finding that it is about 0.7. Then we can compute the acceleration as

$$a = \frac{dV}{dt} = \frac{F}{m} = \frac{(\pi/4)D_D^2 C_d \rho_{\text{air}}(V^2/2)}{(\pi/6)D_D^3 \rho_D} = 1.5 C_d \rho_{\text{air}} \frac{V^2}{2D_D\rho_D}$$

$$= \frac{(1.5)(0.7)(1.20\ \text{kg/m}^3)(106.7\ \text{m/s})^2}{(2)(10^{-4}\ \text{m})(1000\ \text{kg/m}^3)} = 7.2 \times 10^4 \frac{\text{m}}{\text{s}^2} = 2.4 \times 10^5 \frac{\text{ft}}{\text{s}^2}$$

= 3700 times the acceleration of gravity! ■

These two examples show that things happen very fast near the throat of the venturi, where the scrubbing is very efficient, and the droplet accelerations are very rapid. It is also the region where the cross-sectional area perpendicular to the flow is changing fastest (percentage change per unit length is largest).

These two examples also show that, if we knew the relation of drop size to the other parameters operating, we could in principle solve Eq. (9.64) numerically and thus predict venturi scrubber performance for any set of conditions. There is no very simple published solution to that problem. Calvert [20, 21] made several simplifications and thus was able to perform the integration numerically; the results of that integration, for a typical venturi scrubber, are summarized in Fig. 9.27. The figure's use will be illustrated later.

9.2.4.3 Pressure drop in scrubbers. The crossflow and counterflow scrubbers shown previously have very small pressure drops (and very poor efficiencies). Venturi scrubbers have higher pressure drops and higher efficiencies. The power cost of the fan that drives the contaminated gas through a venturi scrubber often is much more important than the purchase cost of the scrubber.

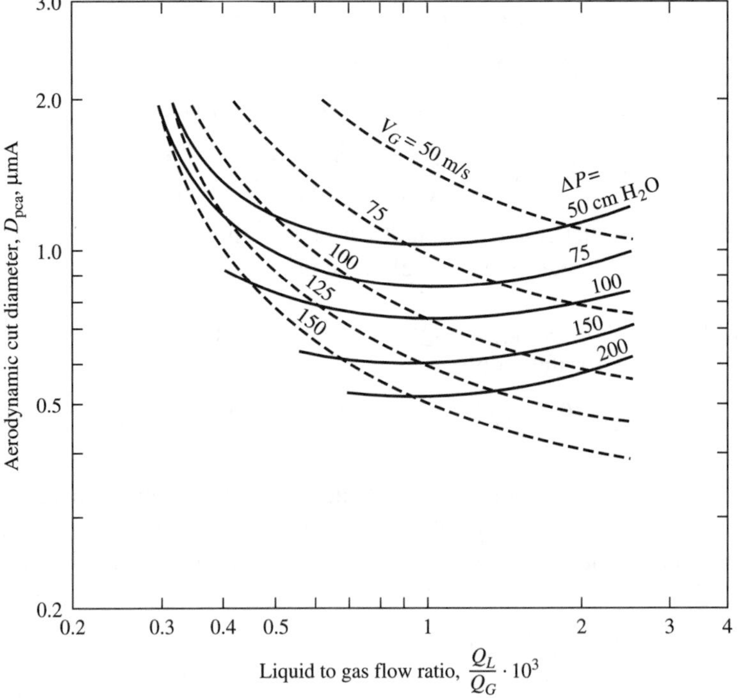

FIGURE 9.27
Aerodynamic cut diameter and pressure drop predictions for a typical venturi scrubber. V_G is the velocity at the throat. (From Ref. 21.)

Example 9.25. A typical venturi scrubber has a throat area of 0.5 m^2, a throat velocity of 100 m/s, and a pressure drop of 100 cm of water = 9806 N/m^2. If we have a 100 percent efficient motor and blower, what is the power required to force the gas through this venturi?

From Eq. (7.6) we calculate

$$\text{Po} = Q_G\,\Delta P = (0.5\text{ m}^2)\left(100\,\frac{\text{m}}{\text{s}}\right)\left(9806\,\frac{\text{N}}{\text{m}^2}\right)\left(\frac{\text{kW}\cdot\text{s}}{10^3\text{ N}\cdot\text{m}}\right) = 245\text{ kW} = 328\text{ hp}$$

If the fan and scrubber operate 8760 h/yr and the electricity costs 5 cents/kWh, the annual power cost will be

$$\text{Power cost} = (245\text{ kW})\left(8760\,\frac{\text{h}}{\text{yr}}\right)\left(\frac{\$0.05}{\text{kWh}}\right) = \frac{\$107{,}300}{\text{yr}}$$ ■

We can see why the pressure drop is so high by performing a simple momentum balance on the system shown in Fig. 9.24. For steady flow and negligible wall friction, we have (letting 1 stand for inlet conditions and 2 stand for outlet conditions)

$$(P_1 - P_2)(\Delta x\,\Delta y) = Q_G\rho_G(V_{G_2} - V_{G_1}) + Q_L\rho_L(V_{L_2} - V_{L_1}) \tag{9.66}$$

In most cases the gas velocity changes very little from inlet to outlet, although it may have a very high value at the throat, so the first term on the right is negligible. Normally the liquid inlet velocity V_{L_1} is negligible, and the liquid outlet velocity V_{L_2} is equal to the gas velocity. So Eq. (9.66) normally is simplified to

$$P_1 - P_2 = \frac{Q_L\rho_L V_G}{\Delta x\,\Delta y} = \frac{Q_L Q_G \rho_L}{A^2} = V_G^2\rho_L\frac{Q_L}{Q_G} \tag{9.67}$$

Example 9.26. For a scrubber using water as the scrubbing liquid, estimate the pressure drop for

$$V_G = Q_G/\Delta x\,\Delta y = 100\text{ m/s},\ \text{and}\ Q_L/Q_G = 0.001.$$

Substituting in the rightmost form of Eq. (9.67), we find that

$$P_1 - P_2 = \left(100\,\frac{\text{m}}{\text{s}}\right)^2\left(1000\,\frac{\text{kg}}{\text{m}^3}\right)(0.001)\left(\frac{\text{N}\cdot\text{s}^2}{\text{kg}\cdot\text{m}}\right)$$

$$= 10^4\,\frac{\text{N}}{\text{m}^2} = 10^4\text{ Pa} \approx 0.1\text{ atm} = 102\text{ cm H}_2\text{O}$$ ■

Unfortunately, the very properties that cause wet scrubbers to have high pressure drops are the same ones that make them efficient particle collectors: the rapid acceleration of liquid by the fast-moving gas produces both effects. Figure 9.28 on page 312 shows test results taken by Lapple and Kamack in which they used one kind of particle, a talc, in a variety of scrubber designs both in the laboratory and semi-works scales [22]. They found that, although the larger equipment worked better than the small, still there seemed to be a unique relation between penetration and

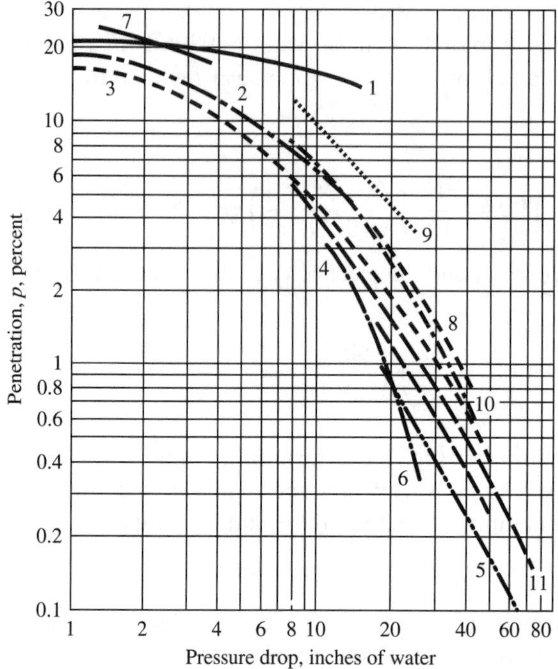

FIGURE 9.28
Lapple and Kamack's test results for penetration vs. pressure drop for a single kind of particle in a variety of equipment types and sizes. The numbers on the curves refer to 11 types and sizes of scrubbers, identified in the original article. (From Ref. 22.)

pressure drop, with all the data for a wide variety of scrubbers falling in a narrow band on a plot of penetration vs. pressure drop. Fame and fortune await the engineer who can devise a scrubber whose performance lies significantly to the lower left of the experimental data shown in Fig. 9.28.

Returning to Fig. 9.27, we can illustrate its use by one more example.

Example 9.27. We wish to treat a gas stream to remove most of the particles. We conclude that if we have a cut diameter of 0.5 μ we will have made a satisfactory particle removal. If we use $Q_L/Q_G = 0.001$, what gas velocity at the throat will we need, and what will the expected pressure drop be?

Figure 9.27 shows cut diameters as *aerodynamic diameters,* discussed in Sec. 8.2.5. Using the results in Example 8.4 for a 0.5-μ particle, we can compute a Cunningham correction factor of 1.24, so the aerodynamic cut diameter, D_{pca}, is

$$D_{\text{pca}} = (0.5\ \mu)\left(2\,\frac{\text{gm}}{\text{cm}^3}\cdot 1.24\right)^{1/2} = 0.79\ \mu\left(\frac{\text{gm}}{\text{cm}^3}\right)^{1/2}$$

From Fig. 9.27 it is clear that this would require a throat velocity of about 90 m/s and a pressure drop of about 80 cm of water. ■

The venturi scrubber, discussed above, is the most widely used particulate scrubber. In the early days of air pollution control a wide variety of scrubber designs were tested, manufactured, and used. Most had low pressure drops and low collection efficiencies for fine particles. As required control efficiencies have increased, these low-efficiency scrubbers have fallen out of favor. Some are still in use, but few new ones are being installed. The venturi scrubber seems to have found numerous niches in particulate control. It is relatively cheap to fabricate and install, does a much better job on fine particles than anything else except a baghouse, but it has a high pressure drop cost. If the particles to be collected are tarry, they will coalesce on and plug up a filter or an ESP. The logical control choices in this instance are a wet ESP, a throwaway filter, or a venturi scrubber. For a large application the venturi scrubber is almost always used. If a liquid stream enters the process and the particles in the gas can be reused in the process, then using the inlet liquid stream to capture the particles and thus return them to the process is often very economical.

9.3 CHOOSING A COLLECTOR

In choosing a primary particle collection device one must consider the size of the particles to be collected, the required collection efficiency, the size of the gas flow, the allowed time between cleanings, and details of the nature of the particles. The following rules of thumb may be helpful:

1. Small or occasional flows can be treated by throwaway devices, e.g., cigarette and motor oil filters, in which the collected particles remain in the device. Large and steady flows require collection devices that operate continuously or semicontinously, and from which the collected particles can be removed continuously or semicontinuously. A throwaway device may be used as a final cleanup device, e.g., a high-efficiency filter may remove the last few particles from the air flowing to a microchip production clean room.
2. Sticky particles (e.g., tars) must be collected either on throwaway devices or into a liquid, as in a scrubber or cyclone, filter, or wet ESP whose collecting surfaces are continually coated with a film of flowing liquid. There must be some way to process the contaminated liquid thus produced.
3. Particles that adhere well to each other but not to solid surfaces are easy to collect. Those that do the reverse often need special surfaces, e.g., Teflon-coated fibers in filters that release collected particles well during cleaning.
4. Electrical properties of the particles are of paramount importance in ESPs, and they are often significant in other control devices where friction-induced electrostatic charges on the particles can aid or hinder collection.
5. For nonsticky particles larger than about 5 μ, a cyclone separator is probably the only device to consider.
6. For particles much smaller than 5 μ one normally considers ESPs, filters, and scrubbers. Each of these can collect particles as small as a fraction of a micron.
7. For large flows the pumping cost makes scrubbers very expensive; other devices are chosen if possible.
8. Corrosion resistance and acid dew point (Sec. 7.12) must always be considered.

9.4 SUMMARY

1. Gravity settling chambers, cyclones, and ESPs work by driving the particles to a solid wall where they form agglomerates that can be collected. These three devices have similar design equations.
2. Filters and scrubbers divide the flow. They have different design equations from wall collection devices and from each other.
3. Both surface and depth filters are used for particle collection. Surface filters are used to collect most of the particles in a heavily laden gas stream. Depth filters are mostly used for the final cleanup of air or gas that must be very clean or for fine liquid drops, which coalesce on them and then drop off.
4. To collect small particles, a scrubber must have a very large relative velocity between the gas being cleaned and the liquid drops. For this reason co-flow scrubbers are most often used. The venturi scrubber is the most widely used type of co-flow scrubber.

PROBLEMS

See Common Units and Values for Problems and Examples, inside the back cover.

9.1. Equations (9.5) and (9.12) for gravity settling chambers, as well as the corresponding equations for cyclones and ESPs, are based on the assumption of blocklike flow (i.e., V_{gas} is independent of distance from the collecting surface). How much difference would it make if we calculated efficiency for a laminar velocity profile? To find out, derive the formula for the settling chamber length needed for 100 percent collection efficiency [equivalent to Eq. (9.5) with $\eta = 1.0$] for the following assumptions: H = height; L = length; V_t = terminal settling velocity, which is assumed constant; and V, instead of being a constant, is given by

$$V = V_{avg}\left(\frac{6}{H^2}\right)(zH - z^2)$$

where z is the vertical distance above the lower plate. (Ignore any entrance effects, i.e., assume V is independent of downstream distance. Assume that the horizontal particle velocity is always the same as the local gas velocity, which is equivalent to assuming the particles have negligible inertia.)

9.2. Equations (9.5) and (9.12) for gravity settlers are based on Stokes' law. Now assume that we wish to design a gravity settling chamber for particles that are so large that Stokes' law no longer applies. Instead, each particle may be assumed to have a drag resistance given by Eq. (8.5). Derive the equivalents of Eqs. (9.5) and (9.12) for that situation.

9.3. We wish to design a particle sampling device that will have a greased, horizontal microscope slide 75 mm long, over which a contaminated gas stream will flow in a narrow slit 2 mm high. What velocity should we choose if we want 90% of the 1-μ diameter particles to fall to (and stick to) the surface of the microscope slide? (This device, called a *horizontal elutriator,* has been widely used for particle sampling in England, and the name is regularly used in industrial hygiene licensing multiple-choice exams.)

9.4. In Stokes' law we commonly drop the ρ_f because it is negligible compared with ρ_{part}. This assumption is equivalent to ignoring buoyant forces in comparison to gravity forces. That is a good assumption for almost all air pollution problems, but not for particles settling in liquids. In our treatment of cyclone separators, we have dropped the buoyant term as well. Is this an approximation, as it was for gravity settlers, or is it strictly correct because the buoyant force is at right angles to the centrifugal force as shown in Fig. 9.3?

9.5. (*a*) Repeat Example 9.2 for the following combinations of velocity and radius: $V = 60$ ft/s, $r = 0.1$ ft; $V = 120$ ft/s, $r = 1$ ft.
(*b*) At the equator, what is the ratio of centrifugal force due to the earth's rotation to gravitational force? The radius of the earth is about 4000 miles, and its rotational velocity at the equator is about 1000 mi/h.
(*c*) How much would the angular velocity of the earth have to increase so that, at the equator, the combination of gravity and centrifugal force would sum to zero (and we would be "weightless")?

9.6. Check to see if Fig. 9.6 was made up correctly by calculating the values of D/D_{cut} for collection efficiencies of 10 percent and 90 percent by Eqs. (9.18), (9.19), and (9.21) and comparing them with the values shown on that figure.

9.7. Our cyclone separator is operating with $D_{\text{cut}} = 5\ \mu$. It is now necessary to increase the flow rate to the cyclone (and hence the inlet velocity) by 25 percent (i.e., the new velocity will be 1.25 times the old velocity). Nothing else will change. The cyclone is believed to obey Eq. (9.18). Estimate the new cut diameter.

9.8. Equation (9.18) is based on a set of assumptions that are discussed in the text. We have one cyclone that obeys those assumptions; and its cut diameter is D_1. We now connect three identical cyclones, each with an individual cut diameter of D_1, in series, with complete remixing of the gas stream between them. We will call the cut diameter for this combination of three cyclones D_3.
(*a*) What is the value of (D_3/D_1)?
(*b*) Is this the same answer we would get for (D_3/D_1) if, instead of having three cyclones in series with remixing, we had one cyclone with $N = 15$ instead of the customary $N = 5$? [Here it may help to sketch η vs. (D_3/D_1), all on the same plot, for one cyclone that obeys Eq. (9.18), for three cyclones with remixing, and for one cyclone with $N = 15$.]

9.9. We wish to use a cyclone to remove 50 percent of particles 1 μ in diameter from an air stream with an inlet velocity of 50 ft/s. Using Eq. (9.18), estimate the maximum allowable value of W_i.

9.10. A cyclone separator has an inlet width of 0.25 ft and $V_c = 60$ ft/s.
(*a*) Estimate the cut diameter for this cyclone separator.
(*b*) Estimate the collection efficiency of this cyclone separator for particles with diameter 1 μ.

9.11. A cyclone separator is operating in conditions where $D_{\text{cut}} = 10\ \mu$. We are offered another cyclone that is of the same design, but all the dimensions are one-half as big as the present one. If we feed the same air stream (same total volumetric flow rate, same particle loading, same particle size distribution) to this new cyclone, what will the new value of D_{cut} be? Assume that Eq. (9.18) is applicable.

9.12. Equation (9.18) was worked out for particles small enough for the Stokes' law form of the drag force to be applicable. For large spherical particles moving at the high speeds encountered in cyclones, the Stokes' law assumption is not correct. Rather, the drag force will be given by Eq. (8.5). Derive the equivalent of Eq. (9.18), based on this assumed drag-force relationship.

9.13. A gas stream contains particles with diameter 1 μ. It passes around one circular turn of a collection device with radius of the streamline = 10 cm and tangential velocity = 20 m/s. How far is each of the particles displaced toward the outside of the circle during the single circular turn?

9.14. Manufacturer X offers a standard-size cyclone with $D_{cut} = 10\ \mu$. We wish to install enough of these in series to collect 99.5 percent of the particles 10 μ or larger in diameter. How many in series must we install? [Assume that Eq. (9.18), and all the assumptions that went into it, apply. Assume complete remixing between each individual cyclone in series.]

9.15. The human nose and nasal passages remove particles from the air destined for our lungs. Consider it as a cyclone separator, with $N = 0.25$; its average dimensions perpendicular to flow are 1 cm × 1 cm. Assume that a typical breath is 1 liter, drawn in over a period of 1 s. Estimate the cut diameter of the human nose for particles. (The observed behavior of the nasal system is that few particles larger than about 5 μ reach the lungs, so the calculation here is only approximately correct.)

9.16. We are passing a gas stream through a cyclone with $D_{cut} = 10\ \mu$. We now must treat twice as much gas in the same cyclone, so the average gas velocity will be increased by a factor of 2. We are told by the cyclone manufacturer that we should have no special troubles with re-entrainment or disturbance of the flow patterns if we double the flow rate. What will the cut diameter be at the new flow rate?

9.17. A particle stream has 33 weight percent particles with diameter 1 μ, 33 percent with diameter 5 μ, and 34 percent with diameter 10 μ. We now pass this stream through a cyclone whose efficiency is described by Eq. (9.21), with $D_{cut} = 5\ \mu$. What fraction by weight of the particles will the cyclone collect?

9.18. In any cyclone separator there is a slow-moving boundary layer adjacent to the outer wall of the cyclone body. Using the nomenclature and ideas of boundary layer theory, estimate the thickness of this layer for a cyclone with $V_i = 60$ ft/s, $D_0 = 1$ ft, $N = 5$. See any fluid mechanics textbook for a discussion of boundary layer theory. List your assumptions.

9.19. Figure 9.29 shows the tangential and radial gas velocity distributions measured in a typical cyclone separator. It appears that in the main body of the cyclone, the radial velocity is practically constant, while the tangential velocity increases from the wall to the diameter of the gas outlet, and then falls rapidly. Why? (*Hint:* See any fluid mechanics textbook on the subject of potential flow.)

9.20. One assumption used in making Eq. (9.18) is that when the gas speeds up at the cyclone inlet, the particles do not lag behind (i.e., $V_{c\text{-particles}} = V_{c\text{-gas}}$). How good an assumption is that? At the inlet of the cyclone, the gas is rapidly accelerated by the decreasing pressure it encounters. Does the particle velocity lag behind? How much?

(*a*) Write the general differential equation for $(V_{gas} - V_{particle})$, assuming that the only force acting between the two is the drag force, which is given by Stokes' drag force. Then assume that at $t = 0$, $V_{gas} = V_{particle} = V_0$ and that the gas velocity is instantly increased to V_1. Solve the equation for this case.

(*b*) Then evaluate the constants in the equation for $V_0 = 30$ ft/s, $V_1 = 60$ ft/s, $D_{part} = 1\ \mu$. Estimate how much time elapses before $(V_{gas} - V_{particle})$ is 1 percent of its value at $t = 0$.

9.21. We wish to design a cyclone for sampling purposes that will have a cut diameter of 1 μ. The inlet velocity will be 20 m/s.

(*a*) Estimate the outside diameter of this cyclone.

(*b*) Estimate the volumetric flow rate through this cyclone.

9.22. We wish to design a cyclone for sampling purposes that will have a cut diameter of 1 μ and will process 10^{-5} m^3/s of gas. The outlet will be at atmospheric pressure; what inlet pressure will be needed?

9.23. We wish to design a multiclone (a group of cyclone separators that operate in parallel; see Fig. 9.5). The specifications for the project include the total volumetric flow rate Q of gas to be treated and the cut diameter D_{cut} required. We must select the size of individual cyclones

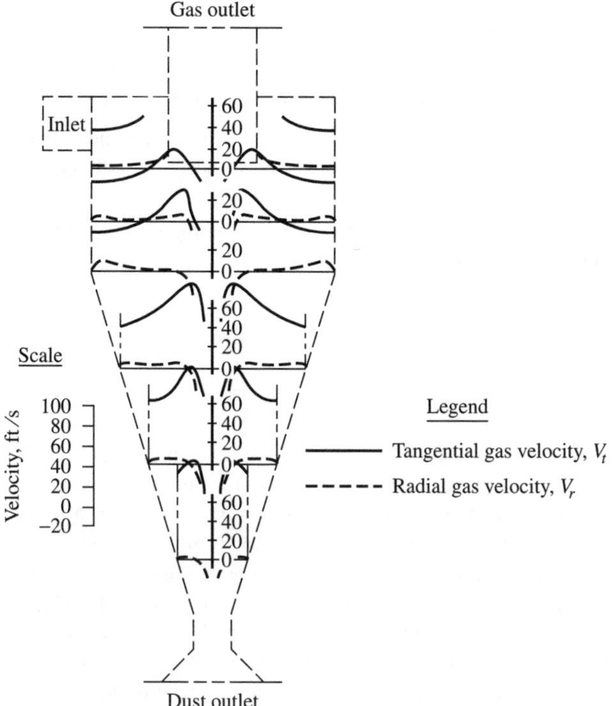

FIGURE 9.29
Variation of tangential and radial gas velocity at different points in a cyclone. (From Ref. 23. Courtesy of the Institution of Mechanical Engineers.)

to use, i.e., select the right value of D_o in Fig. 9.4. Here it is easier to solve for the most economic inlet velocity V_{econ}, from which we can compute W_i in Eq. (9.18) and then D_o from the ratio on Fig. 9.4. Derive the formula for V_{econ} based on the following information.

The individual cyclones can be assumed to obey Eq. (9.18) and to have the dimension ratios shown in Fig. 9.4. The value of N in Eq. (9.18) = 5 for any size cyclone. The pressure drop through the cyclone is equal to 8 inlet velocity heads.

The power cost to drive the air through the cyclones is given by

$$\text{Annual power cost} = C_1 Q\,\Delta P$$

where Q = volumetric flow rate

ΔP = pressure drop

The annual cost of owning the cyclone installation (sum of interest, depreciation, taxes, and insurance) is

$$\text{Annual cost of owning} = C_2(\text{installed cost})$$

(One may think of "annual cost of owning" as the equivalent rental cost one would pay if one did not own the equipment, but rented it. The person renting it to us would decide how much to charge per year as some fixed fraction of his or her cost of buying it; i.e., C_2 times the installed cost.)

The installed cost of a bank of cyclones in parallel is

Installed cost

$= C_3$(number of cyclones in parallel)(outer diameter D_o of individual cyclone)^2

(This expression makes the cost proportional to the surface area of the sheet metal from which the individual cyclones are fabricated.) C_1 has dimensions of \$/kWh, C_2 of 1/yr and C_3 of \$/ft^2.

9.24. Leith and Licht present a much more complex approach to computing the efficiency of cyclone separators [24]. Their final relationship is

$$\eta = 1 - \exp[-2(C\Psi)^{1/(2n+2)}] \tag{9.68}$$

where n and C are functions of cyclone dimensions and

$$\Psi = \frac{\rho_p D_p^2 V_c}{18\mu D_o}(n+1) \tag{9.69}$$

(Here some of their symbols have been changed to match those used in this book.) Test this relation by repeating Example 9.4 using it, and compare your results with those shown in that example. For the cyclone dimensions in Example 9.4, $n = 0.57$ and $C = 50$. The predictions of Eqs. (9.68) and (9.69) for a variety of cyclone sizes are shown in Ref. 2.

9.25. Example 9.6 is direct but tedious. It can be easily programmed on a computer. For hand calculations one can find practically the same result using Fig. 9.30, [21] in which Eq. (7.10) has been integrated numerically, with the assumptions that the particle size distribution is log-normal and that $p = \exp(-AD_a^2)$. This plot was devised for scrubbers but it is applicable to any device whose penetration-size curve can be represented in the above form.

(*a*) Show that Eq. (9.21) can be rewritten and then expanded by binomial expansion as

$$p = \frac{1}{1 + (D/D_{\text{cut}})^2} = 1 - (D/D_{\text{cut}})^2 + (D/D_{\text{cut}})^4 + \cdots \tag{9.70}$$

(*b*) Then show that $p = \exp(-AD_a^2)$ can be represented by the series

$$p = 1 - (AD_a^2) + \frac{(AD_a^2)}{2!} + \cdots \tag{9.71}$$

in which the two first terms have the same form as Eq. (9.70), so that Eq. (9.21) is *approximately* the same as the assumed form for Fig. 9.30.

(*c*) Show that in Example 9.6 $(D_{\text{cut}}/D_{\text{mean}}) = 0.25$, that $\sigma_g = \exp\sigma = 3.49$, and that, based on those values, one reads an overall penetration from Fig. 9.30 of ≈ 0.14. Given the approximate nature of Eq. (9.21) this is a satisfactory agreement with the result in Example 9.6. It does not show the distribution of sizes in the outlet stream, which that examples does.

(*d*) Show why we could use D instead of D_{aero} in this case without error.

(*e*) Repeat the spreadsheet calculation leading to Table 9.1, using $p = \exp(-AD_a^2)$ instead of Eq. (9.21), and showing that Fig. 9.30 does indeed reproduce the spreadsheet solution.

9.26. Equation (9.30) does not contain plate-to-wire distance as an explicit variable. Why? Does it contain it implicitly?

9.27. An ESP is treating a particle-laden air stream, collecting 95 percent of the particles. We now double the air flow rate, keeping the particle loading constant. What is the new percent recovery? Assume Eq. (9.30) is applicable.

9.28. Supercolossal Enterprises's ESP is recovering 95 percent of the particles. The new pollution laws require them to recover 99 percent of the particles. Their present precipitator consists of 10 standard, factory-built units, each with area A.

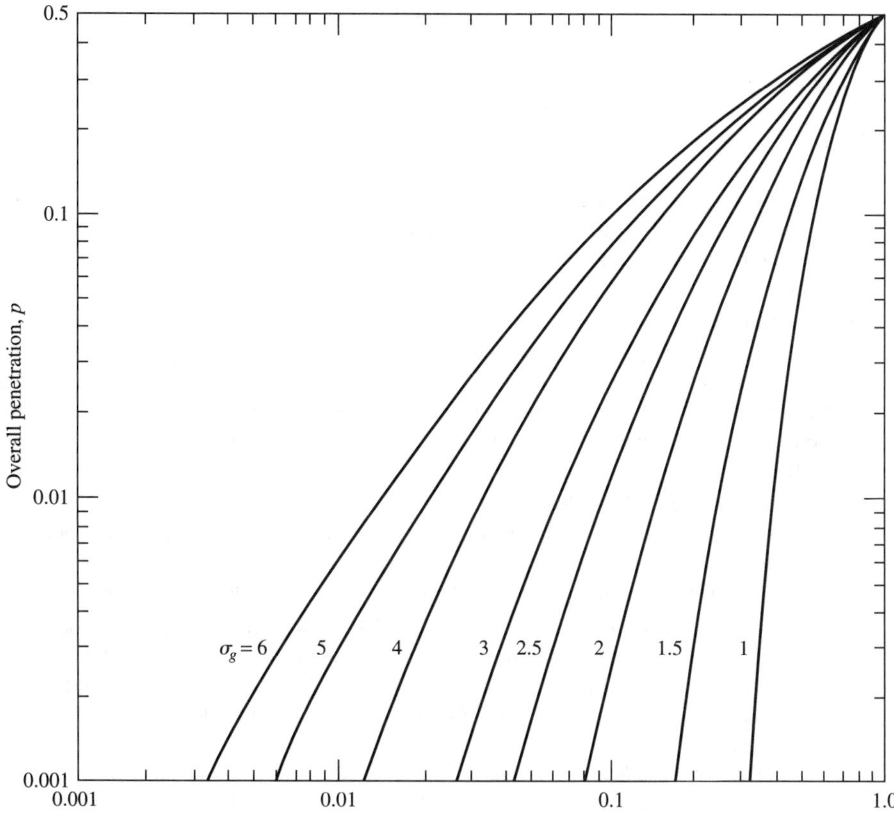

FIGURE 9.30
Integrated form of Eq. (7.10), for the assumption that the particle size distribution is log-normal and that the penetration-size relationship is given by $p = \exp(-AD_a^2)$. Here $\sigma_s = \exp\sigma$ [21]. (Reproduced by permission of Van Nostrand Reinhold Publishing Co.)

(*a*) How many more, just like these 10, must they add to meet the new specification? Assume Eq. (9.30) is correct.

(*b*) Should they add the new units in parallel or in series?

9.29. Our ESP is collecting 95 percent of the particles in our waste gas. A salesperson now offers us an additive to add to the gas that will change the resistivity of the collected cake of particles, thus doubling the value of w, the effective drift velocity. If we use this additive and these statements are true, what will the collection efficiency be?

9.30. We are constructing a new multiple hearth roaster, for which we plan to install an ESP to control the particles in the exhaust gas. The gas flow will be 10,000 acfm. We believe that the values of the drift velocity shown in Table 9.2 for multiple hearth roasters will be applicable to the particles coming from this roaster. Regulations require us to capture 99.5 percent of the particles in the exhaust gas. Estimate the required ESP collecting area A.

9.31. Our ESP is collecting 95 percent of the particles entering it. We must collect 99 percent of the particles. Our engineers suggest that we could do this by increasing the voltage applied to the precipitator. With new solid-state controls we can presumably keep the sparking rate within manageable limits. If we assume that all the particles are the same size and that Eqs. (9.26) and (9.30) apply, by what percent must we increase the applied voltage to increase the collection efficiency from 95 percent to 99 percent?

9.32. Our ESP is collecting 95% of the particles in our exhaust stream. We are under pressure to improve its performance. It has been suggested that if we were to cool the inlet gas from the present 600°F to 250°F, the performance would improve. What is the estimated collection efficiency at 250°F, if we keep the mass flow rate constant, do not change any of the dimensions of the ESP, and if changing the temperature does not change the drift velocity, w?

9.33. The Deutsch-Anderson equation does not explicitly show the time that the gas spends in the ESP. Show that A/Q can be expressed as some function of the time the gas spends in the ESP and the plate spacing. (Although there is some variation in plate spacing, most large ESPs have about the same plate spacing, 8 to 12 inches between plates. Thus a big ESP can be thought of as having a large value of A/Q or as having the gas spend long time in the ESP.)

9.34. The Geneva (Utah) Steel Works of US Steel Company in 1976 upgraded their ESPs by changing the electric power supply to raise the operating voltage. Assume that the original efficiency was 95 percent and the original voltage was 40 kV. If they increased the voltage to 50 kV and held everything else constant, what is the estimated new collection efficiency?

9.35. A gas stream contains particles of three sizes—10 μ, 7 μ, and 3 μ. The particle density is the same for all three sizes, and the weight concentration in the gas stream of each of the three sizes of particles is the same (0.33333). We pass this gas stream through an ESP that obeys Eq. (9.30). Assume that Eq. (9.26) is correct and the drift velocity is proportional to the diameter.

The overall collection efficiency of the precipitator (weight fraction of all the entering particles captured) is 95 percent. What are the individual percent collection efficiencies for each of the three sizes of particle?

9.36. In the Deutsch-Anderson equation for ESPs, Eq. (9.30), it is assumed that all the particles entering the precipitator have the same diameter, and hence the same value of the drift velocity w. We now wish to develop a modification of this equation to take into account the fact that for all real ESPs there is a distribution of particle sizes in the inlet gas stream.

Derive the equivalent of Eq. (9.30) with the following changes:

1. The inlet particles are described by the triangular distribution,

$$\frac{d\Phi}{dD} = C_1 D \qquad \text{for diameters from 0 to } D_{\text{mean}}, \text{ and}$$

$$\frac{d\Phi}{dD} = C_1(2D_{\text{mean}} - D) \qquad \text{for diameters between } D_{\text{mean}} \text{ and } 2D_{\text{mean}}$$

There are no particles with diameters larger than $2D_{\text{mean}}$. Here Φ is the cumulative fraction by *mass* of particles with diameter less than D, and D_{mean} is 50 percent of the diameter of the largest particle. C_1 is equal to $(1/D^2_{\text{mean}})$.

2. The drift velocity for various diameter particles is given by

$$w = C_2 D$$

where D = particle diameter

C_2 = a constant

Here we can safely assume that at the collection efficiencies of interest practically all of the particles larger than D_{mean} are collected, so they can be ignored.

9.37. Equation (9.31) and Problem 9.36 show two ways to modify the Deutsch-Anderson equation, Eq. (9.30), to take into account the change in average particle size as most of the particles are removed. Another approach is to keep the same form of equation, but to let w decrease as p decreases. Turner et al. suggest the following values for the collection of fly ash from burning bituminous coal [25]:

Design efficiency, %	w, ft/s
95	0.41
99	0.33
99.5	0.31
99.9	0.27

This table can be represented with satisfactory accuracy ($R^2 = 0.96$) by

$$w = 0.506 + 0.0356 \ln p$$

(*a*) Rework Example 9.14 using this equation and Eq. (9.30). Compare the results with those in Examples 9.13 and 9.14.

(*b*) The A/Q values computed this way can be reasonably well represented by an equation of the form of Eq. (9.31), but with a different value of k from that shown in Example 9.14. What value of k makes the values computed this way practically the same as the value computed from Eq. (9.31)?

9.38. (*a*) An ESP has three identical chambers operating in parallel. When the flow is distributed equally (one-third to each), the particle removal efficiency is 95 percent. Now, as a result of maldistribution, the flows become 50 percent, 30 percent, and 20 percent to the three chambers. The total flow is unchanged. What is the overall particle collection efficiency under this flow condition?

(*b*) Two cyclone separators are operating in parallel. They have identical dimensions, and are processing equal flows of dirty gas. The particles in the gas all have the same diameter and their concentration in the gas is uniform. The overall collection efficiency of the combination of two cyclones, which is the same as the individual efficiency of each of the two cyclones, is 70 percent. Now, because of a valve blockage, the flow rates to the two cyclones are perturbed so that $\frac{2}{3}$ of the flow goes to one of them and $\frac{1}{3}$ of the flow goes to the other. The total flow is unchanged. (This change will raise the pressure required to drive the gas, but that is of no concern for this problem.) The two streams leaving the cyclones are joined and mixed. What is the overall collection efficiency for the two cyclones in parallel under these conditions? Assume that Eq. (9.18) applies.

(*c*) Do the answers from parts (*a*) and (*b*) show the same trend? Should they?

9.39. Example 9.15 treats the problem of nonuniform flow in a two-chamber precipitator. Assume we have such a two-chamber precipitator, each chamber with area A. Also assume that the total flow through the two combined precipitators is a constant, equal to Q. (Here $Q = Q_1 + Q_2$, where Q_1 is the flow in the first chamber and Q_2 is the flow in the second.) The efficiency for equal flow ($Q_1 = Q_2$) is η_0. Write the equation for η/η_0 as a function of $Q_1/(Q_1 + Q_2)$. Sketch this function on a plot of η/η_0 vs. $Q_1/(Q_1 + Q_2)$. On the sketch show two curves, one for a low value and one for a high value of wA.

9.40. Typically 10 to 30 ppm of SO_3 are added to the gas going to an ESP to condition the gas, thereby lowering the resistivity of the collected solid and improving the collection efficiency

[25]. If the gas stream is produced by burning a coal with properties as shown in Example 7.10, what fraction of the SO_2 in the gas would have to be converted to SO_3 to produce this concentration of SO_3?

9.41. Monroe *et al.* [26] present test data on a wet ESP treating the exhaust gas from a low-efficiency dry ESP at a coal-fired electric power plant. They report average wet ESP penetrations of 5%, with $A/Q = 0.035$ min/ft.

(*a*) Estimate the drift velocity for this wet ESP.

(*b*) The above value is roughly three times the typical value for dry ESPs in this service (see Table 9.2). This appears to be a common ratio for wet and dry ESPs [12]. Explain this experimental finding.

(*c*) Explain why anyone uses dry precipitators when this result indicates they could use much smaller wet precipitators to get the same collection efficiency.

9.42. Equation (9.39) shows the instantaneous pressure drop through a filter.

(*a*) Sketch ΔP vs. t for V_s = constant and $\Delta x_{\text{cake}} = 0$ at $t = 0$.

(*b*) Sketch Δx vs. t for ΔP = constant and $\Delta x_{\text{cake}} = 0$ at $t = 0$.

(*c*) For (*b*), sketch V_s vs. t.

9.43. Equations (9.32) to (9.40) are presented in the traditional nomenclature of filtration, shown in standard filtration books. It is common in the air pollution filtration literature [13] to rewrite Eq. (9.35) as

$$\frac{\Delta P}{V_s} = \mu\left[\left(\frac{\Delta x}{k}\right)_{\text{cake}} + \left(\frac{\Delta x}{k}\right)_{\text{f.m.}}\right] = \left(\begin{array}{l}\text{"filter drag," normally}\\ \text{expressed as in. H}_2\text{O/(ft/min)}\end{array}\right) \tag{9.72}$$

and then make the following substitutions,

$$\mu\left(\frac{\Delta x}{k}\right)_{\text{f.m.}} = K_r = K_f + W_r K_2 \quad \text{and} \quad \mu\left(\frac{\Delta x}{k}\right)_{\text{cake}} = W K_2 \tag{9.73}$$

where K_r, K_f, and K_2 are the "dirty fabric resistance," the "clean fabric resistance," and the "specific cake resistance coefficient." Here W and W_r are the instantaneous cake loading (mass per unit area) and the residual cake loading on the imperfectly cleaned filter fabric. These Ws are not the same as the W in Eq. (9.38), as they are mass of solids per unit of filter surface. Making these substitution, we find

$$\frac{\Delta P}{V_s} = W K_2 + K_f + W_r K_2 \tag{9.74}$$

which seems to be the most common formulation. One also sees the entire right side of Eq. (9.73) replaced by an S [27], which is called the "filter drag," a parameter correlated with cake and cloth properties. In Eq. (9.74) the gas viscosity has been absorbed into the Ks, so that these are a function of temperature. The reported values normally state the temperature at which they apply, so one may estimate behavior at other temperatures, by taking the changes in viscosity into account.

In [12] an example is given for a high-velocity reverse jet air filter, operating downstream of an existing low-efficiency precipitator, with $V_s = 11.3$ ft/min, $c_{\text{inlet}} = 0.13$ gr/acf, cleaning 3.3 times/h, $T = 340°$F. The drag is reported to be

$$\frac{\Delta P}{V_s} = 140.59W + 0.1288$$

with the drag expressed in (in. H_2O/(ft/min)) and W in lbm/ft^2. The V_s shown is the average over the cycle. The actual operation is at a constant pressure drop, with the flow rate through any section of the filter decreasing from its highest value just after cleaning to its lowest value just before the next cleaning. The cleaning cycles of various sections are not simultaneous, so that the flow is continually redistributing itself. For this problem assume that V_s is a constant and:

(*a*) Estimate the value of $K_f + W_r K_2$.
(*b*) Estimate the value of K_2.
(*c*) Estimate the value of the increase in W during one cycle between cleanings.
(*d*) Estimate the pressure drop just before the bags are cleaned.
(*e*) Estimate the pressure drop just after the bags are cleaned. Then, using the value from part (*b*) and the assumption that $K_f \approx 0$, estimate the value of W_r.
(*f*) Estimate the permeability of the cake, with the assumptions in Example 9.16.
(*g*) The reported mean particle diameter is 4 μ. Estimate the permeability of a bed of spheres with this diameter and a porosity of 0.5 by the standard methods of fluid mechanics [13], and, based on that, estimate K_2. Compare your result to that shown in part (*f*). Do not expect more than order-of-magnitude agreement.

9.44. In Problem 9.43 we simplified the calculations by assuming that V_s was constant and independent of time. That is true for the average over all the bags. For any one bag V_s will be a maximum just after cleaning and a minimum just before the next cleaning. Using the values from Problem 9.43 and the assumption that the pressure drop is constant at 7 in. H_2O,
(*a*) Estimate V_s for a freshly cleaned bag.
(*b*) Estimate W for a bag about to be cleaned. [Here you must integrate Eq. (9.74) to find the equivalent of Eq. (9.40).]
(*c*) Estimate V_s for a bag about to be cleaned.

9.45. (*a*) If $V_s = 5$ ft/min, how many square feet of filter surface would be needed for a 750-MW coal-fired power plant that produces 1.5 million acfm of stack gas?
(*b*) If this filter area is in the form of cylindrical bags 40 ft long and 1 ft in diameter, how many of them will be needed?

9.46. Most real blowers are neither constant-pressure nor constant-flow-rate but somewhere in between. Assume a blower has the relation

$$\Delta P = \Delta P_0 - bQ$$

where
ΔP = pressure rise across the blower
ΔP_0 = pressure rise at zero flow rate (i.e., with the downstream valve closed)
Q = volumetric flow rate through the blower
b = an arbitrary constant

We now connect such a blower to a process whose exhaust gas contains c lbm/ft^3 of particles and pass the stream through a filter with surface area A. The filter behaves according to Eq. (9.39) with $\alpha = 0$. The collection efficiency is 100 percent. At time zero, the mass of particles on the filter cloth is zero. What is the algebraic relation between mass per unit area on the filter and time?

9.47. In Fig. 9.15, at a fabric loading of 140 g/m^2,
(*a*) What is the collection efficiency for each of the four velocities shown?
(*b*) What is the estimated cake thickness, assuming a cake porosity of 0.3?
(*c*) For the lowest curve, approximately how long must the filter have operated to reach this loading?

9.48. We plan to treat a particle-laden gas stream with volumetric flow rate Q in a baghouse. The pressure drop through the baghouse must be overcome by the electrically driven blower attached to the baghouse. The annual power cost is $C_1 Q \Delta P$ where C_1 is a constant proportional to the price of electricity. The pressure drop through the baghouse is $\Delta P = C_2 Q/A$, where C_2 is a constant and A is the filter surface area. The annual cost of owning the baghouse (which includes interest, depreciation, taxes, insurance, and maintenance) is $C_3 A$ where C_3 is a constant.

(*a*) Find the formula for the economic optimum filter surface area.

(*b*) Compare your answer to the answers you got to Problem 9.23. Are they the same? If not, why not?

(*c*) Here, if $C_1 = \$300/(\text{kw} \cdot \text{yr})$, $C_2 = 0.025$ psi/(ft/min), and $C_3 = \$2/(\text{ft}^2 \cdot \text{yr})$, what is the optimum velocity?

9.49. In Problem 9.48, we computed the economic optimum filter surface in terms of various constants. Redo part (*a*) of that problem with the following change: the annual cost of owning the baghouse, which was previously $C_3 A$ (where C_3 was a constant), is now $C_3 A^{0.6}$, i.e., the cost is proportional to the 0.6 power of the filter area.

9.50. (*a*) If the design pressure on a baghouse is of the order of 10 inches of water, what is the force per unit area on the wall of the baghouse container, in lb/in.^{2}? For a 10 ft × 10 ft wall, what is the total force?

(*b*) Is this force more difficult to resist if it is internal or external?

(*c*) What, if any, kind of safety device would be used to protect against overpressure in the case of an upset?

9.51. Using Fig. 9.18, estimate the percent target efficiency for a 10-μ particle in air moving at 100 ft/s against a 1-cm diameter cylinder.

9.52. A single fiber 10 μ in diameter is placed perpendicular to a gas stream that is moving 1 m/s. The gas stream contains particles with a diameter of 10 μ. The particle concentration is 1 mg/m^3. What is the rate of collection of particles on the fiber in $\text{g}/(\text{m} \cdot \text{s})$?

9.53. A typical household furnace air filter consists of a mass of randomly oriented collecting fibers. An idealized model of such a filter would have multiple screens in series, each perpendicular to the flow and looking to the oncoming gas like a venetian blind, i.e., a series of parallel obstructions with openings between them. Assume that such a filter can be represented as 10 such sets of obstructions in series, that each such obstruction has an area that is 80 percent open and 20 percent blocked by the fibers, that for the individual fibers the collection mode is entirely by impaction, and that the target efficiency as shown on Fig. 9.18 is 25 percent. What is the overall collection efficiency of this entire filter?

9.54. A collector consists of a sieve made of cylindrical wire fibers with diameter 100 μ. These are arranged in a parallel array, like a venetian blind, with spaces of 100 μ between the fibers. We are passing through them an air stream containing particles with diameters of 10 μ and of 0.01 μ. The relative velocity at which the air approaches the sieve is 1 m/s. Estimate the fraction of each of the two types of particles that is collected by impaction and by diffusional deposition.

9.55. We are passing a gas stream with velocity 1 m/s over a single array of cylindrical collectors with diameter 100 μ. This array has 50 percent open area, 50 percent area blocked by the collectors.

(*a*) What is the smallest particle that will be collected with 40 percent efficiency?

(*b*) What is the smallest particle for which the collection efficiency by impaction is greater than zero?

9.56. The Lone Ranger is riding through a sandstorm. Does his mask (a bandanna that covers his nose) significantly decrease the amount of sand he breathes in? Assume the bandanna is made of fibers, each 300 μ in diameter. They are woven in a simple weave that blocks 75 percent of the projected area; the remaining 25 percent is open, with holes up to 75 μ in diameter. For the purpose of this problem, the bandanna may be considered equal to a single row of parallel fibers, each 300 μ in diameter, with 75 percent of the flow area blocked and 25 percent open. The gas velocity approaching the cloth as he breathes in is 1 m/s.

(*a*) Estimate both the collection efficiency for 10-μ particles and the cut diameter for the bandanna used as a filter.

(*b*) The Lone Ranger now decides that his bandanna will work better as a dust mask if he folds it so that there are two layers of cloth across his nose, instead of one layer. All other parts of the problem are unchanged. What is the overall collection efficiency for particles 10 μ in diameter for these two layers of cloth?

(*c*) What is the cut diameter for two layers of cloth?

(*d*) The Lone Ranger now decides that he wants to use as many layers of bandanna as are needed to have the cut diameter for the combined multiple layers of bandana be 2.5 μ. How many layers of bandanna does he need?

9.57. Freidlander shows that if one computes diffusivities using Eq. (8.16) with the Cunningham correction factor set equal to 1, then one can solve Eq. (9.46) for the diameter at which there is a minimum in the collection efficiency [19]. Defining $\gamma = kT/3\pi\mu$, one finds

$$D_{\text{minimum efficiency}} = \frac{(0.859\gamma^{1/4}\nu^{1/8}D_b^{3/8})}{V^{3/8}} \tag{9.75}$$

Show the calculations leading to this value and then compute the diameter with the minimum collection efficiency for Example 9.21.

9.58. Iinoya and Orr [28] reported (p. 425) that sand beds 7 ft deep were used as fine-particle collectors by the U.S. Atomic Energy Commission (AEC). Calvert [21] reports (p. 277) that for beds of that type the penetration can be estimated by

$$p = \exp\left(-\frac{7zV_sD_{\text{pa}}^2}{9D_c^2\mu_g\varepsilon}\right) \tag{9.76}$$

where z = length of the bed in the flow direction

V_s = superficial gas velocity

D_{pa} = aerodynamic particle diameter

ε = porosity or void fraction in the bed

D_c = diameter of the collecting particles (in this case the practically spherical sand grains)

μ_g = gas viscosity

(*a*) Estimate the collection efficiency of the AEC sand beds for the following assumptions: particle diameter = 0.5 μ; diameter of sand grains = 1 mm; gas velocity V_s = 0.2 ft/s; porosity ε = 0.3.

(*b*) Estimate how thick a sand bed would be needed for 99.9 percent collection efficiency.

9.59. Derive Eq. (9.76). *Hint:* This is a remixed set of cyclones in series, even if it doesn't look like it.

9.60. A 2-mm diameter raindrop falls at its terminal velocity through 1000 ft of air that contains 80 μg/m^3 of 3-μ spherical particles.

(*a*) How many particles does the drop collect?

(*b*) What is the percentage increase in mass of the drop due to these particles?

9.61. Rain is falling steadily at a rate of 1 in./h. The raindrops are all 2 mm in diameter and are falling at their terminal velocity. What is the concentration of raindrops in the air in drops per cubic foot?

9.62. A rainstorm is depositing 0.1 in./h of rain over a large area. The drops have an average diameter of 2 mm, for which the target efficiency for the particles in the air is estimated to be 0.1. How long must the rainstorm continue at this rate to collect 90 percent of the particles in the air?

9.63. A rainstorm deposits 1 in. of water uniformly over a large area. Of the rain that falls, one-half of the volume is in the form of drops with a diameter of 2 mm, for which the target efficiency for particles is 0.1, and one-half of the volume is in the form of drops with diameter 1 mm, for which the target efficiency for particles is 0.15. Estimate what fraction of the particles originally present will be collected.

9.64. A rainstorm deposits 1.0 in. of water uniformly over a large area. The drops come in two sizes, 2 mm and 0.5 mm. They are distributed so that 0.9 of the mass of water is in 2-mm drops and 0.1 of the mass is in 0.5-mm drops. Their estimated terminal velocities and target efficiencies for atmospheric particles are shown here:

Drop diameter, mm	2	0.5
Estimated terminal velocity, ft/s	21	7.5
Estimated target efficiency for atmospheric particles	0.1	0.3

(*a*) What mass fraction of the atmospheric particles is collected by this rainstorm?

(*b*) Of the particles that are collected by this rainstorm, what fraction is collected by the 0.5-mm drops? What fraction is collected by the 2-mm drops? Assume that the rate of deposition of the two sizes of drops is constant over the time period.

9.65. Our crossflow scrubber is collecting 90 percent of the particles passing into it. We now must double the gas flow rate. If we hold everything else constant, what is the anticipated collection efficiency at the new flow rate?

9.66. Our crossflow scrubber is collecting 90 percent of the 3-μ particles entering. The water drops are all of the same diameter, 400 μ. We now install new spray nozzles that make all the drops 200 μ in diameter. Q_L is not changed. What is the new collection efficiency?

9.67. We are collecting 90 percent of the particles from a gas stream in a crossflow scrubber. We wish to increase the collection efficiency to 95 percent by increasing the water flow to the scrubber, holding everything else constant. By what percentage must we increase the water flow to the scrubber?

9.68. A co-flow wet scrubber has $Q_G = 10^4$ cfm, $Q_L = 10$ cfm, $A = 0.6$ ft^2, and an inlet liquid velocity of zero. The gas has the same properties as air, and the liquid the same properties as water.

(*a*) Estimate the pressure drop of the scrubber.

(*b*) Estimate the aerodynamic cut diameter.

(*c*) Estimate the physical cut diameter, if the particle density is 2 g/cm^3 and $C = 1.3$.

9.69. In Example 9.27, if the particles are log-normally distributed with $D_{\text{mean}} = 2\ \mu$ and $\sigma = 1.5$, what is the expected overall collection efficiency, based on Fig. 9.30 (Problem 9.25)?

REFERENCES

1. Hering, S. V.: "Inertial and Gravitational Collectors," in S. V. Hering (ed.), *Air Sampling Instruments for Evaluation of Atmospheric Contaminants,* 7th ed., American Council of Governmental Industrial Hygienists, Cincinnati, OH, p. 361, 1989.
2. Ashbee, E., and W. T. Davis: "Cyclones and Inertial Separators," in A. J. Buonicore and W. T. Davis (eds.), *Air Pollution Engineering Manual,* Van Nostrand Reinhold, New York, pp. 73–78, 1992.
3. Rosin, P., E. Rammler, and W. Intelmann: "Grundlage und Grenzen der Zyklonentstaubung ("Basis and limits of cyclone dust removal")," *VDI,* Vol. 76, pp. 433–437, 1932.
4. Semrau, K. T.: "Air Pollution" section of "Solids Drying and Gas-Solids Systems," in R. H. Perry, D. W. Green, and J. O. Maloney (eds.), *Perry's Chemical Engineers' Handbook,* 6th ed., McGraw-Hill, New York, pp. 20–86, 1984.
5. Anonymous: "Flows of Fluids through Valves, Fittings, and Pipes," Tech. Paper No. 410, Crane Company, 475 N. Gary Ave., Carol Stream, Ill. 1957 (regularly reprinted).
6. Oglesby, S., Jr., and G. B. Nichols: *Electrostatic Precipitation,* Marcel Dekker, New York, 1978.
7. Danielson, J. A. (ed.): *Air Pollution Engineering Manual,* U.S. Environmental Protection Agency Report PB 225-132/OAS, U.S. Government Printing Office, Washington, DC, p. 149, 1967.
8. Ramsdell, T. G., Jr.: "Design Criteria for Precipitators for Modern Central Station Power Plants," *Proc. Amer. Power Conf.,* Vol. 30, pp. 444–449, 1968.
9. Rinard, G: "Proof of Concept Testing of ESP Retrofit Technology for Low and High Resistivity Flyash," in *Proceedings of the 7th Particulate Control Symposium,* R. F. Altman (ed.), Volume I, pp. 19-1 to 19-14, Electric Power Research Institute, Pub. GS 6208, 1988.
10. Lloyd, D. A.: *Electrostatic Precipitator Handbook,* Adam Hilger, Bristol, England, p. 34, 1988
11. Vatavuk, W. M.: *Estimating Costs of Air Pollution Control,* Lewis, Chelsea, MI, p. 117, 1990.
12. Miller, R. L., W. A. Harrison, D. B. Prater, and R. Chang: "Alabama Power Company E. C. Gaston 272 MW Electric Stream Plant-Unit No. 3 Enhanced COPAC Installation," *EPRI-DOE-EPA Combined Utility Air Pollutant Control Symposium, TR-108683-V3,* Electric Power Research Institute, Palo Alto, CA, August 1997.
13. de Nevers, N.: *Fluid Mechanics for Chemical Engineers,* 2d ed., McGraw-Hill, New York, Chapter 12, 1991.
14. de Nevers, N.: " 'Product in the Way' Processes," *Chem. Eng. Educ.,* Vol. 12, pp. 146–151, 1992.
15. Ensor, D. S., R. Hooper, R. W. Scheck, and R. C. Carr: "Performance and Engineering Evaluation of the Nucla Baghouse," in *Symposium on Particulate Control in Energy Processes,* U.S. Environmental Protection Agency Report EPA-600/7-76-010, U.S. Government Printing Office, Washington, DC, 1976.
16. Carr, R. C., and W. B. Smith: "Fabric Filter Technology for Utility Coal-Fired Power Plants, Parts I–VI." These six articles appeared in the 1984 *Journal of the Air Pollution Control Association* and were reprinted by the Electrical Power Research Institute as *EPRI* CS-3724-SR.
17. Dennis, R., and N. F. Surprenant: "Particulate Control Highlights: Research on Fabric Filtration Technology," U.S. Environmental Protection Agency Report EPA-600/8-87-005d, U.S. Government Printing Office, Washington, DC, 1978.
18. Langmuir, I., and K. B. Blodgett: "A Mathematical Investigation of Water Droplet Trajectories," U.S. Army Air Forces Technical Report No. 5418, 1946. (Available from the U.S. Department of Commerce as PB-27565.)
19. Friedlander, S. K.: "Theory of Aerosol Filtration," *Ind. Eng. Chem.,* Vol. 50, pp. 1161–1164, 1958.
20. Calvert, S., J. Goldshmid, D. Leigh, and D. Mehta: *Scrubber Handbook,* NTIS No. PB-213-016, NTIS, Springfield, VA, 1972.
21. Calvert, S.: "Scrubbing," in *Air Pollution,* 3d ed., A. C. Stern (ed.), Vol. 4, Academic Press, New York, 1977.
22. Lapple, C. E., and H. J. Kamack: "Performance of Wet Dust Scrubbers," *Chem. Eng. Prog.,* Vol. 51, pp. 110–121, 1955.
23. ter Linden, A. J.: "Investigations into Cyclone Dust Collectors," *Inst. Mech. Engrs. Proc.,* Vol. 160, pp. 233–251, 1949.
24. Leith, D., and W. Licht: "The Collection Efficiency of Cyclone Type Particle Collectors—A New Theoretical Approach," *AIChE Symp. Series,* Vol. 126, pp. 196–206, 1972.

25. Turner, J. H., P. A. Lawless, T. Yamamoto, D. W. Coy, G. P. Greiner, J. P. McKenna, and W. M. Vatavuk: "Electrostatic Precipitators," in *Air Pollution Engineering Manual,* A. J. Buonicore and W. T. Davis (eds.), Van Nostrand Reinhold, New York, pp. 89–114, 1992.
26. Monroe, L. S., K. M. Cushing, W. A. Harrison, and R. Altman: "Testing of a Combined Dry and Wet Electrostatic Precipitator for Control of Fine Particle Emissions from a Coal-Fired Boiler," *EPRI-DOE-EPA Combined Utility Air Pollutant Control Symposium, TR-108683-V3,* Electric Power Research Institute, Palo Alto, CA, August 1997.
27. Donovan, R. P.: *Fabric Filtration for Combustion Sources,* Marcel Dekker, New York, p. 153, 1985.
28. Iinoya, K., and C. J. Orr: "Source Control by Filtration," in *Air Pollution,* 2d ed., A. C. Stern (ed.), Vol. 3, Academic Press, New York, pp. 409–435, 1968.

CHAPTER 10

CONTROL OF VOLATILE ORGANIC COMPOUNDS (VOCs)

Volatile organic compounds (VOCs) are liquids or solids that contain organic carbon (carbon bonded to carbon, hydrogen, nitrogen, or sulfur, but not carbonate carbon as in $CaCO_3$ nor carbide carbon as in CaC_2 or CO or CO_2), which vaporize at significant rates. VOCs are probably the second-most widespread and diverse class of emissions after particulates.

VOCs are a large family of compounds. Some (e.g., benzene) are toxic and carcinogenic, and are regulated individually as hazardous pollutants (see Chapter 15). Most VOCs are believed not to be toxic (or not very toxic) to humans. Our principal concern with VOCs is that they participate in the "smog" reaction [Eq. (1.2), see also Appendix D] and also in the formation of secondary particles in the atmosphere. These latter are mostly in the fine particle size range. Some VOCs are powerful infrared absorbers and thus contribute to the problem of global warning (see Sec. 14.1).

Table 10.1 on page 330 shows the estimated U.S. emissions of VOC for 1997. We see that more than 80% came from solvent usage (e.g., paint thinners and other similar solvents), the transportation and storage of VOC, and motor vehicles (including autos, airplanes, boats, and railroad engines). This reflects the fact that our principal uses of VOCs are for motor fuels and for solvents. The table also shows a wide variety of other sources, such as incomplete combustion in fireplaces and

TABLE 10.1
National emissions estimates for VOC for 1997 (see Table 1.1)

Source category	Emissions, thousands of tons/yr	Percent of total
Residential wood combustion	527	2.74
Chemical and allied processing	461	2.40
Petroleum and related industries	538	2.80
Other industrial processes	458	2.38
Solvent utilization	6483	33.74
VOC storage and transport	1377	7.17
Waste disposal and recycling	449	2.34
Motor vehicles	7660	39.86
Forest fires and wood waste combustion	767	3.99
All other sources	496	2.58
Total	19 216	100.00

Source: Ref. 1.

forest fires, and many others as small as fingernail polish remover and paint spray cans. In addition to the uses as solvents and fuels, many VOCs are intermediates in the production of plastics and other chemicals, e.g., vinyl chloride (the principal raw material for PVC plastics), which is also regulated as a hazardous air pollutant (see Chapter 15). Solvents and motor fuels are mostly derived from petroleum, so most of the emissions listed here are ultimately based on petroleum. Some small fraction is based on wood (e.g., turpentine and wood smoke) and coal (some small amount due to coal combustion) but most VOC emissions are of refined petroleum products, used as fuels or solvents. The total emissions shown in Table 10.1 are roughly 2% of the total petroleum usage in the United States.

10.1 VAPOR PRESSURE, EQUILIBRIUM VAPOR CONTENT, EVAPORATION

To understand which chemicals are volatile (evaporate at significant rates) we must consider the idea of *vapor pressure.* Figure 10.1 shows vapor pressures as a function of temperature for a variety of compounds. Consider first the line for water. At 212°F, its *normal boiling point* (NBP), water has a pressure of 14.696 psia (= 760 torr = 1 atmosphere = 101.3 kPa ≈ 14.7 psia). The normal (atmospheric) boiling point is the temperature at which the vapor pressure equals the atmospheric pressure and the liquid converts to a vapor by the vigorous bubble formation we call boiling. At room temperature (68°F = 20°C) the vapor pressure of water is 0.339 psia = 17.5 torr = 0.023 atm. At this temperature water does not boil. But it does evaporate if the surrounding air is not saturated. (Wet clothes and swimsuits dry slowly at this temperature, faster in a dry climate than in a wet one.) One cannot easily read Fig. 10.1 to three significant figures; the preceding values are from more extensive tables of the vapor pressure of water. But if one could read Fig. 10.1 to

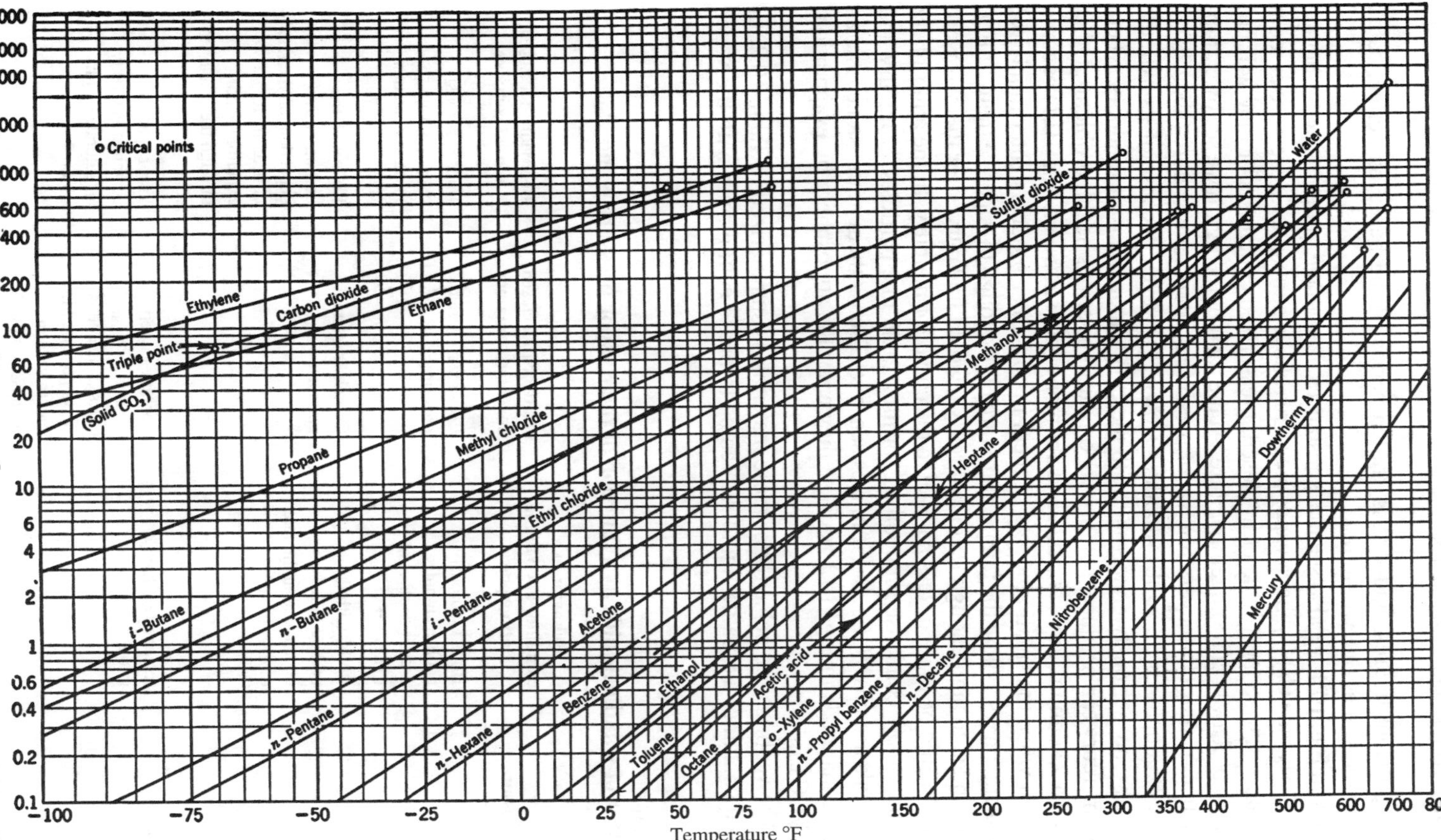

FIGURE 10.1
Vapor pressures of 27 compounds as a function of temperature. From G. G. Brown et al., *Unit Operations,* Wiley, New York, p. 583, 1951. (Reproduced by permission of the publisher.)

three significant figures (and if it is drafted perfectly and based on perfect data), those are the values one would read.

From Fig. 10.1 we also see that substances like ethane, propane, and n-butane (C_2H_6, C_3H_8, and C_4H_{10}) have vapor pressures above atmospheric pressure at room temperature. These must be kept in closed, pressurized containers or they will immediately boil away at room temperature. At the right of the figure we see by extrapolation that mercury has a very low vapor pressure at room temperature (about 10^{-6} atm; see Example 10.3), so it evaporates slowly. However, because of its extreme toxicity, we are seriously concerned with mercury vapor exposures indoors, where spilled mercury can produce mercury vapor, causing serious health problems. Other metals like cadmium, zinc, arsenic, and antimony (not shown on Fig. 10.1) become a problem when the small amounts of them in consumer products enter municipal waste incinerators, where the temperatures are high enough to vaporize them. If Fig. 10.1 extended to high enough temperatures, we would find curves for them as well, and they would be more or less parallel to the curve for mercury. In the atmosphere the metal vapors condense on cooling, forming toxic fine particles. (Mercury and other metals are not VOCs, but it makes sense to discuss them at the same time as the other materials in Fig. 10.1 because they enter the atmosphere by evaporation, as do VOCs.)

The vaporization behavior of these materials is summarized in Table 10.2. In a closed container a volatile liquid will come to phase equilibrium with the vapor

TABLE 10.2
Behavior of volatile liquids as a function of their vapor pressure p and P_{atm}

Vapor pressure p	Behavior in a container open to the atmosphere	Behavior in a closed, unvented container	Behavior in a closed, vented container
$p > P_{atm}$	Boils vigorously, cools as it boils, until it cools enough to have $p = P_{atm}$	Container internal pressure $= p$	Boils vigorously expelling vapor through container vent
$p = P_{atm}$	Boils, with the boiling rate dependent on the rate of heat input	Container internal pressure $= P_{atm}$	Boils with the boiling rate dependent on the rate of heat input, expelling vapor through the container vent
$p < P_{atm}$	Evaporates slowly into air	Container internal pressure $< P_{atm}$ unless some other gas is present. May lead to vacuum collapse of vessel	Vapor space in container is mostly air saturated with vapor

above it. If it is a pure liquid, then, as shown in the table, the pressure in the container will be the vapor pressure of the liquid, and the vapor will have the same chemical composition as the liquid. If the container also contains a gas like air, then at equilibrium that air will be saturated with the vapor evaporated from the liquid. For the low pressures of interest for air pollution control, we make only small errors if we assume that the vapor mix behaves as a perfect gas, and that we can estimate the content of volatile liquid in the vapor mix by *Raoult's law* (which is not a law like Newton's laws, but rather a useful approximation):

$$y_i = x_i \frac{p_i}{P} \tag{10.1}$$

where y_i = mol fraction [= (volume %/100) for perfect gases] of component i in the vapor

x_i = mol fraction of component i in the liquid

p_i = vapor pressure of pure component i

P = total pressure

Example 10.1. Estimate the water content of air that is in equilibrium with pure water at 68°F = 20°C.

Using the value read from Fig. 10.1, we find

$$y_i = x_i \frac{p_i}{P} = 1.00 \frac{0.023 \text{ atm}}{1 \text{ atm}} = 0.023$$

Here we have taken $x_i = 1.00$, ignoring the small amount of air dissolved in the water. ■

Example 10.2. Repeat Example 10.1 for a liquid mixture of 50 mol percent benzene and 50 mol percent toluene in equilibrium with air in a closed container. From Fig. 10.1 we read the vapor pressures of benzene and toluene at 68°F as about 1.5 and 0.4 psia, respectively. Using more extensive tables, we find that the values are 1.45 and 0.42 psi. Then applying Eq. (10.1) twice, we have

$$y_{\text{benzene}} = x_{\text{benzene}} \frac{p_{\text{benzene}}}{P} = 0.5 \frac{1.45 \text{ psia}}{14.7 \text{ psia}} = 0.049$$

$$y_{\text{toluene}} = x_{\text{toluene}} \frac{p_{\text{toluene}}}{P} = 0.5 \frac{0.42 \text{ psia}}{14.7 \text{ psia}} = 0.014$$

$$y_{\text{air}} = 1 - 0.049 - 0.014 = 0.937$$

Here we have assumed that benzene and toluene form an *ideal solution*. That is a good assumption for this mixture, and most of the VOC mixtures of air pollution interest. For more complex mixtures, consult any book on chemical engineering thermodynamics, which will show how one solves this type of problem for *nonideal solutions*. ■

These examples show how we estimate the equilibrium concentrations of volatile liquids, either pure liquids or mixtures, in a closed container that contains air. If the same liquids were exposed to the atmosphere, and the atmosphere contained water or benzene or toluene mol fractions less than those computed here, then the liquid would evaporate into the air. Conversely, if the air contained mol fractions higher than those calculated here, then there would be condensation of liquid out of the air. That is rare with organic vapors, but occurs regularly with water, when the air temperature near the ground is lowered enough that dew forms on the ground, leaves, etc. If we cooled the container in Example 10.2, the two values of p_i would decline, and some of the benzene and toluene in the vapor would condense into the liquid.

We use VOCs as fuels (propane, gasoline, jet fuel, diesel fuel) because liquid hydrocarbon fuels have a better combination of ease of production, transportation, storage, and use in small quantities than any competing fuels. Large stationary sources like power plants mostly use coal and uranium as fuels. These fuels have a much lower cost per unit of energy than common liquid fuels, but they require more expensive and complex pollution control and safety equipment than seem practical on motor vehicles, lawn mowers, and other garden tools. Natural gas is an excellent fuel but so far it has proven difficult to store enough of it in a moving vehicle and to refuel such a vehicle quickly for it to compete with the VOC fuels in private vehicles (it is often used for centrally fueled fleet vehicles). Solar, wind, geothermal, and other alternative energy sources do not currently compete with VOCs as fuels for vehicles.

VOC solvents are widely used because they evaporate into the air, leaving either no residue (dry-cleaning solvent, nail polish remover) or a thin layer of previously dissolved solid (paint, fingernail polish, floor wax, inks). The rate of evaporation (pounds evaporated per square foot of exposed surface per hour) is roughly proportional to the vapor pressure. If we want quick evaporation (spray paints, nail polish), we use a solvent with a high vapor pressure at room temperature; if we want slower evaporation (brushed-on paints, cleaning solvents) we choose a solvent with a lower vapor pressure at room temperature.

Figure 10.1 has a high visual content, but for computer programs we need an algebraic representation of the same information. Figure 10.1 is approximately a plot of log p vs. $1/T$ (K or °R), with the values of $1/T$ not shown, but plotted from right to left, and the corresponding values of T shown, running from left to right and printed on the horizontal axis. (High values of $1/T$ correspond to low values of T, etc.) One can show from thermodynamics that at low pressures the vapor pressure of all pure compounds is represented with good (but not perfect) accuracy by the Clausius-Clapyron equation,

$$\log p = A - \frac{B}{T} \tag{10.2}$$

where A and B are constants determined by experiment for each individual chemical compound. If this were exactly true for all pressures, then the vapor pressure data for any compound would form a straight line on a plot of $\log p$ vs. $1/T$. This is approximately true, even to pressures of thousands of psi. However, the experimental data can often be represented with better accuracy by the Antoine equation,

$$\log p = A - \frac{B}{(T - C)} \tag{10.3}$$

P559

in which A, B, and C are totally empirical constants, determined from the experimental data. Even more complex vapor pressure equations are used to make up tables of thermodynamic properties. Values of Antoine equation constants for a variety of substances have been published; the constants for 23 substances are shown in Appendix A. [Equations (10.2) and (10.3) are written for temperatures in K. However many tables, including those in Appendix A, use T in °C, and show the denominator in Eq. (10.3) as $(T + C)$, showing C as a positive number normally less than 273.15. See Problem 10.2(*b*).] The horizontal axis in Fig. 10.1 is slightly perturbed from being exactly $1/T$ to accomplish the same improvement of the representation of the experimental data as inserting C in Eq. (10.3) does (see Problem 10.3).

Example 10.3. Estimate the vapor pressure of mercury at 20°C by extrapolating the line on Fig. 10.1. Assume that Eq. (10.2) applies.

From that figure, the vapor pressure is 0.1 psia at about 330°F and 20 psia at about 700°F. We could work the example with those values, but we will find a more interesting and useful answer if we look up more precise values available in the widely available tables of the vapor pressure of mercury [2], finding a pressure of 0.1 psia at 329°F and of 18.93 psia at 700°F. We write Eq. (10.2) twice, working in °R, finding

$$\log 0.1 = A - \frac{B}{(329 + 460)} \quad \text{and} \quad \log 18.93 = A - \frac{B}{(700 + 460)}$$

These equations can be solved to find $A = 6.11993$, $B = 5617.6$°R.

Then we may estimate that for 68°F = 20°C

$$\log p = 6.11993 - \frac{5617.6}{(68 + 460)} = -4.5195$$

$$p = 3.02 \times 10^{-5} \text{ psia} = 2.06 \times 10^{-6} \text{ atm} = 0.00156 \text{ torr}$$

The commonly reported value, based on a more complex vapor pressure equation than Eq. (10.2), is 1.6×10^{-6} atm [2], so the estimate here is 29 percent too large. ■

This example shows that the Clausius-Clapyron equation is a good, but not perfect, representation of vapor pressures. Here we have extrapolated downward in vapor pressure by a factor of 3000 with an error of only 29 percent.

10.2 VOCs

We may now state, as an approximate rule, that VOCs are those organic liquids or solids whose room temperature vapor pressures are greater than about 0.01 psia (= 0.0007 atm) and whose atmospheric boiling points are up to about 500°F (= 260°C), which means most organic compounds with less than about 12 carbon atoms. Materials with higher boiling points evaporate quite slowly into the atmosphere unless they are heated, and hence they are less likely to become part of our VOC problems. (If vaporized, they condense in the atmosphere, forming part of our fine particle problem. A lighted cigarette produces a gaseous mixture of high-boiling organic compounds; when this mixture is cooled on leaving the cigarette it forms a smoke of fine particulate droplets. They are part of our particulate problem, but not our VOC problem.) Figure 10.1 contains data for only a few of the millions of organic chemicals in that vapor pressure range. The Clean Air Act Amendments of 1990 list 189 compounds that are considered to be health hazards and that are to be regulated to prevent or minimize emissions [3]; most are VOCs (see Sec. 15.3).

The legal definition used for regulatory purposes [4] does not set a lower vapor pressure limitation and excludes a large variety of compounds that have negligible photochemical reactivity, including methane, ethane, and most halogenated compounds. Compounds with boiling points above about 500°F will have negligible emission rates under normal circumstances, so the absence of a lower vapor pressure limitation causes little problem.

The terms VOC and *hydrocarbon* (HC) are not identical, but often are practically identical. Strictly speaking, a hydrocarbon contains only hydrogen and carbon atoms. But gasoline is normally called a "hydrocarbon fuel" because it contains mostly hydrogen and carbon atoms, but also some oxygen, nitrogen, and sulfur atoms. Acetone, CH_3–CO–CH_3, the principal ingredient of nail polish remover, is a VOC but is not strictly speaking a hydrocarbon because it contains an oxygen atom. In common usage it would often be grouped with the hydrocarbons.

Hydrocarbons are only slightly soluble in water, so we can normally separate liquid HCs from liquid water by simple phase separation and decantation. However, the water left behind often contains enough dissolved hydrocarbon that it cannot be discharged to the sewer or natural body of water without additional treatment. Polar VOCs, which almost all contain an oxygen or nitrogen atom in addition to carbons and hydrogens (alcohols, ethers, aldehydes and ketones, carboxylic acids, esters, amines, nitriles) are much more soluble in water. This difference in solubilities makes the polar VOCs easier to remove from a gas stream by scrubbing with water, but harder to remove from water once they dissolve in it. Table 10.3 shows some typical values of these solubilities.

We see that the polar organics are generally about 100 times more soluble than the hydrocarbons (compounds of H and C only) of the same molecular weight, and that within each chemical family the solubility decreases with increasing molecular weight.

TABLE 10.3
Solubilities of various categories of VOC in water at 25°C

Chemical class	Individual compound	*M*, g/mol	Solubility in water, wt %
HC, linear	*n*-Pentane	72	0.0038
	n-Hexane	86	0.00095
HC, cyclic	Cyclohexane	84	0.0055
HC, aromatic	Benzene	78	0.18
	Toluene	92	0.052
	Ethyl benzene	106	0.020
Alcohols	Methyl, ethyl, *n*-propyl, isopropyl,	32, 46, 60, 60,	Totally miscible
	ethylene glycol	62	Totally miscible
	n-butanol	74	7.3
	Cyclohexanol	100	4.3
Ketones	Acetone	58	Totally miscible
	Methyl ethyl ketone	72	26
	Methyl isobutyl ketone	100	1.7
Ethers	Diethyl ether	74	6.9
	Di-isopropyl ether	102	1.2
Esters	Methyl acetate	74	24.5
	Ethyl acetate	88	7.7
	n-Butyl acetate	116	0.7

Source: Ref. 5.

10.3 CONTROL BY PREVENTION

If possible, we prevent the formation of a VOC-containing air or gas stream, which we must treat by some kind of tailpipe control device. The ways of doing this for VOCs are substitution, process modification, and leakage control.

10.3.1 Substitution

Oil-based paints, coatings, and inks harden by the evaporation of VOC solvents such as paint thinner into the atmosphere. Water-based paints are concentrated oil-based paints, emulsified in water. After the water evaporates, the small amount of organic solvent in the remaining paint must also evaporate for the paint to harden. Switching from oil- to water-based paints, coatings, and inks greatly reduces but does not totally eliminate the emissions of VOCs from painting, coating, or printing. For many applications, e.g., house paint, the water-based paints seem just as good as oil-based paints. But water-based paints have not yet been developed that can produce auto body finishes as bright, smooth, and durable as the high-performance oil-based paints and coatings now used.

There are numerous other examples where a less volatile or nonvolatile solvent can be substituted for the more volatile one. This replacement normally reduces but does not eliminate the emission of VOCs. In addition, a less toxic solvent can often be substituted for a more toxic one, although the more toxic solvents often have special solvent properties that are hard to replace.

Replacing gasoline as a motor fuel with compressed natural gas or propane is also a form of substitution that reduces the emissions of VOCs, because those fuels can be handled, metered, and burned with fewer VOC emissions than can gasoline. The petroleum industry is working hard to improve the burning properties, handling, and use of gasoline, to make it as low-emission a fuel as compressed natural gas and propane, so that gasoline can keep its dominant position in the auto fuel market.

10.3.2 Process Modification

Process modification to prevent or reduce the formation of the VOC stream may be more economical than applying the control options discussed below. Often substitution (Sec. 10.3.1) and process modification are indistinguishable. (Changing fuels or solvents without changing their use is clearly substitution. Changing from standard solvent-based painting to fluidized-bed powder coating could be considered process modification or substitution.)

Replacing gasoline-powered vehicles with electric-powered vehicles is a form of process modification that reduces the emissions of VOCs, as well as emission of carbon monoxide and nitrogen oxides, in the place the vehicle is. On the other hand, it causes other emissions where the electricity is generated. If we consider the process as "get workers from their homes to their place of employment," then improved public transport, mandatory ride pools, etc., are modifications of the process that reduce emissions of VOCs (and of CO and NO_x).

Many coating, finishing, and decoration processes that at one time depended on evaporating solvents have been replaced by others that do not, e.g., fluidized-bed powder coating and ultraviolet lithography.

Finding alternatives to VOC solvents and fuels can be difficult, but it is often the most cost-effective way to reduce VOC emissions.

10.3.3 Leakage Control

10.3.3.1 Filling, breathing, and emptying losses. Tanks containing liquid VOCs can emit VOC vapors because of filling and emptying activities as well as changes in temperature and atmospheric pressure. These emissions are called *filling* or *displacement losses, emptying losses,* and *breathing losses,* or, collectively, *working losses*. Figure 10.2 shows a simple tank of some kind being filled with liquid from a pipeline. As the liquid enters the tank and the liquid level rises, the vapor space above the tank must decrease in volume.* Normally that vapor space (called *headspace*) is connected by a vent to the atmosphere so the vapor, which is mostly air, will

**Vapor* means a substance in gaseous form (below its critical temperature) or more commonly a mixture of such a gas with air.

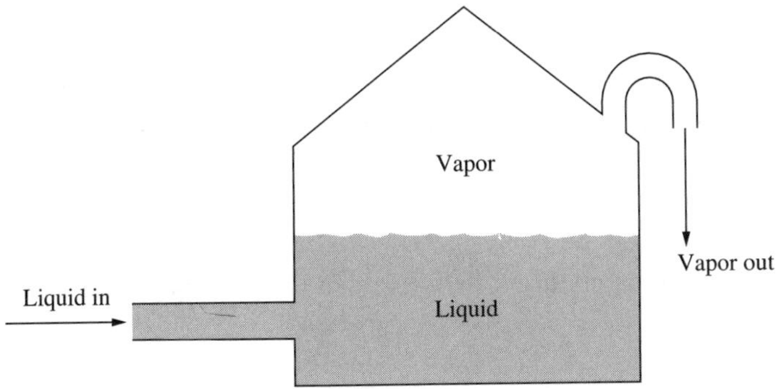

FIGURE 10.2
Displacement losses occur when a vented tank is filled with liquid, thus displacing vapor from the tank's headspace. The tank walls support the roofs of small cone-roof tanks; large ones have internal supports. All have some kind of vent on the roof.

be expelled. When liquid is withdrawn from the tank, air will flow in through the vent to fill the space made available by the fall in liquid level. If the tank were not vented, changing the liquid level would cause an overpressure during filling or a vacuum during emptying. (When large tank vents are plugged, e.g., by an ice storm or blizzard, emptying the tank sometimes causes a vacuum that causes the tank to collapse [6]!)

For all three kinds of working losses, we can write

$$\text{VOC emission} = \begin{pmatrix}\text{volume of air-VOC mix}\\ \text{expelled from the tank}\end{pmatrix}\begin{pmatrix}\text{concentration of}\\ \text{VOC in that mix}\end{pmatrix} \tag{10.4}$$

$$m_i = \Delta V c_i$$

where m_i = mass emission of component i

c_i = concentration (lb/ft^3, kg/m^3, or equivalent) in the displaced gas

Concentration c_i can be expressed as

$$c_i = \frac{y_i M_i}{V_{\text{molar, gas}}} \tag{10.5}$$

Replacing the vapor mol fraction by Raoult's law, Eq. (10.1), replacing the gas molar volume by the perfect gas law, and substituting in Eq. (10.4), we find

$$\frac{m_i}{\Delta V} = \frac{x_i p_i M_i}{P} \cdot \frac{P}{RT} = \frac{x_i p_i M_i}{RT} \tag{10.6}$$

Example 10.4. The tank in Fig. 10.2 contains pure liquid benzene at 68°F which is in equilibrium with the air-benzene vapor in its headspace. If we now pump in liquid benzene, how many pounds of benzene will be emitted in the vent gas per cubic foot of benzene liquid pumped in? What fraction is this of the liquid benzene pumped into the tank?

Using the benzene vapor pressure from Example 10.2, we find

$$\frac{m_i}{\Delta V} = \frac{x_i p_i M_i}{RT} = \frac{1.00 \cdot 1.45 \text{ psia} \cdot 78 \text{ lb/lbmol}}{(10.73 \text{ psi} \cdot \text{ft}^3/\text{lbmol} \cdot {}^\circ\text{R})(528^\circ\text{R})} = 0.020 \frac{\text{lb benzene}}{\text{ft}^3\text{vapor}}$$

$$= 0.32 \frac{\text{kg benzene}}{\text{m}^3\text{vapor}}$$

The density of liquid benzene is about 54 lb/ft^3 so that the fraction of the filling emitted in the vapor is

$$\frac{\begin{pmatrix}\text{Mass of benzene}\\ \text{emitted in vapor}\end{pmatrix}}{\begin{pmatrix}\text{Mass of liquid}\\ \text{benzene filled}\end{pmatrix}} = \frac{0.020 \text{ lb benzene/ft}^3}{54 \text{ lb benzene/ft}^3} = 3.7 \times 10^{-4} = 0.037\%$$

See Problem 10.8. ■

Example 10.5. The tank in Example 10.4 is now heated by the sun to 100°F; both vapor and liquid are heated to this temperature. How many pounds of benzene are expelled per cubic foot of tank? Assume that initially the tank was 50 percent by volume full of liquid, 50 percent by volume full of vapor.

Here there are two contributors to the emission—the vapor expelled because of simple thermal expansion of the vapor and liquid in the tank, and the vapor expelled because of the vaporization of benzene as the liquid temperature is raised. We simplify by assuming these processes take place in sequence—heating then equilibration. To calculate the first we compute the volume of vapor expelled due to simple thermal expansion;

$$\begin{pmatrix}\text{Volume of}\\ \text{vapor expelled}\end{pmatrix} = \begin{pmatrix}\text{increase in}\\ \text{vapor volume}\end{pmatrix} + \begin{pmatrix}\text{increase in}\\ \text{liquid volume}\end{pmatrix} - \begin{pmatrix}\text{increase in}\\ \text{tank volume}\end{pmatrix} \tag{10.7}$$

The fractional change in volume caused by heating is normally expressed as

$$\frac{dV}{V} = \alpha dT \tag{10.8}$$

where α is the coefficient of thermal expansion at constant pressure. Substituting α three times into Eq. (10.7), we have

$$\left(\frac{dV}{V}\right)_{\text{expelled}} = \left(\frac{V_{\text{vapor}}}{V_{\text{tank}}}\alpha_{\text{vapor}} + \frac{V_{\text{liquid}}}{V_{\text{tank}}}\alpha_{\text{liquid}} - \alpha_{\text{tank}}\right) dT \tag{10.9}$$

For the tank, α_{tank} is the coefficient of the volume expansion, which is three times the coefficient of linear expansion for the material of which the tank is made; for a steel tank α is $3 \cdot 6.5 \times 10^{-6}/{}^\circ\text{F} = 1.95 \times 10^{-5}/{}^\circ\text{F}$. For liquids like benzene α is typically about 6×10^{-4}/°F. For perfect gases at room temperature α is

$$\alpha = \left(\frac{dV/V}{dT}\right)_P = \frac{1}{T} = \frac{1}{528°\text{R}} \approx 1.9 \times 10^{-3}/°\text{F} \qquad \text{perfect gases} \qquad (10.10)$$

Substituting these values into Eq. (10.9), we find that

$$\left(\frac{dV}{V}\right)_{\text{expelled}} = (0.5 \cdot 1.9 + 0.5 \cdot 0.6 - 0.0195)(10^{-3}/°\text{F})dT$$

$$= (1.23 \times 10^{-3}/°\text{F})dT$$

$$\left(\frac{\Delta V}{V}\right)_{\text{expelled}} \approx (1.23 \times 10^{-3}/°\text{F})\ \Delta T = (1.23 \times 10^{-3}/°\text{F})(100 - 68)°\text{F}$$

$$= 0.039 = 3.9\%$$

Next we look up the vapor pressure of benzene at 100°F (= 37.78°C) on Fig. 10.1 or in suitable tables, finding 3.22 psia. For every cubic foot of benzene evaporated in this step, 1 cubic foot of benzene-air mixture is displaced from the container vent. The volume of benzene vaporized is the volume of the vapor in the tank times the change in mol fraction (= volume fraction). Thus we find

$$\frac{\Delta V_{\text{expelled}}}{V_{\text{tank}}} = \left(\frac{V_{\text{vapor}}}{V_{\text{tank}}}\right)(y_{\text{benzene final}} - y_{\text{benzene initial}})$$

$$= 0.5\left[\frac{(3.22 - 1.45)\ \text{psia}}{14.7\ \text{psia}}\right] = 0.076 = 0.076\ \frac{\text{ft}^3}{\text{ft}^3 \text{of tank}} \qquad (10.11)$$

And the total fraction of the tank volume expelled = 0.039 + 0.076 = 0.115. If we assume that there is plug flow displacement of this vapor, then it would be expected to have the benzene mol fraction corresponding to 68°F, and

$$\frac{\text{lb benzene}_{\text{expelled}}}{V_{\text{tank}}} = \left(0.115\frac{\text{ft}^3}{\text{ft}^3 \text{of tank}}\right)\left(0.020\frac{\text{lb benzene}}{\text{ft}^3}\right)$$

$$= 0.0023\ \frac{\text{lb benzene}}{\text{ft}^3 \text{of tank}}$$

■

This is called the *breathing loss* because the tank must "breathe" in and out whenever its temperature changes, normally out every day and in every night. The assumption of plug flow displacement of the vapor is plausible for a stationary tank, but not for the fuel tank of a moving vehicle, where sloshing of the liquid will keep the vapor close to equilibrium at all times. In that case we would have to integrate the emission over the temperature change, assuming equilibrium vapor content at all temperatures; the resulting calculated emissions would be substantially higher than those shown here. The breathing loss due to changes in atmospheric pressure is normally much smaller than that due to changes in temperature (see Problems 10.11 and 10.12).

The third kind of loss, sometimes called *emptying loss,* arises from the slow vaporization of the contents of the tank after partial emptying.

Example 10.6. The tank in Example 10.4 contains liquid benzene at 68°F. We now rapidly pump out some liquid. Air at 68°F enters the tank to replace the liquid withdrawn. During the pumpout process none of the benzene evaporates into the fresh air. After we have finished pumping out the liquid benzene, some of the remaining liquid benzene slowly evaporates into the fresh air, eventually saturating it with benzene. How much benzene escapes this way?

Combining Eqs. (10.4) and (10.11), we write

$$m_i = (c_i V_{\text{air added to tank}})(y_{\text{benzene final}} - y_{\text{benzene initial}}) \tag{10.12}$$

Here $y_{\text{benzene initial}} = 0$. There is no obvious choice for c_i. If evaporation takes place in a pistonlike displacement from below, then the gas forced out the vent would be the fresh air just brought in and would contain no benzene. If the tank is emptied a little at a time, with the incoming air mixing well with the air already in the tank, then the gas forced out the vent will have a benzene concentration close to the saturated value. Here we assume an average of these two, or $c_{i \text{ in emitted air}} = 0.5\ c_{i \text{ saturated}}$. Making that assumption and using Eqs. (10.5) and (10.6), we find

$$\frac{m_i}{V_{\text{air admitted}}} = 0.5\frac{(y_{\text{benzene final}})(M)}{(V_{\text{molar gas}})}(y_{\text{benzene final}}) = 0.5\left(\frac{x_i^2 p^2 M}{PRT}\right) \tag{10.13}$$

or, in this example,

$$\frac{m_i}{V_{\text{air admitted}}} = 0.5\left[\frac{1^2(1.45\text{ psia})^2(78\text{ lb/lbmol})}{(14.7\text{ psia})(10.73\text{ psi}\cdot\text{ft}^3/\text{lbmol}\cdot{}^\circ\text{R})(528^\circ\text{R})}\right]$$

$$= 0.00098\frac{\text{lb benzene}}{\text{ft}^3\text{of air admitted}}$$

This value is about 5 percent as large as the direct displacement loss shown in Example 10.4. ■

Breathing, filling, and emptying losses are minimized by attaching to the vent of the tank in Fig. 10.2 (and any similar tank) a *pressure-vacuum valve,* also called a *vapor conservation valve*. These valves remain shut when the pressure difference across them is small, typically 0.5 psi positive pressure or 0.062 psi (1 oz/in.^{2}) negative. They open for the significant flows of vapor in and out that are caused by filling and emptying and by major changes in temperature or pressure.

10.3.3.2 Displacement and breathing losses for gasoline. The greatest interest in these types of losses concerns gasoline, because we use so much of it—about 350 million gallons per day in the United States. Gasoline is a complex mixture, typically containing perhaps 50 different hydrocarbons in concentrations of 0.01 percent or more, plus traces of many others. The smallest molecules have 3 carbon atoms; the largest, 11 or 12. A "typical" gasoline has an average formula of about C_8H_{17} and thus an average molecular weight of about 113. Its composition varies with season of the year and from refinery to refinery.

In the previous examples the vapor pressure and molecular weight played dominant roles. For a pure component like benzene, the molecular weight is a constant and the vapor pressure is a simple function of temperature; that choice made these examples simple. For any **mixture** of VOCs, like gasoline, both the vapor pressure and molecular weight of the vapor change as the liquid vaporizes. This behavior is sketched in Fig. 10.3.

To estimate the displacement and breathing losses for a mixture like gasoline, we first observe that only a small fraction of the gasoline is normally evaporated into the headspace of its containers, so the appropriate vapor pressure and molecular weight are ≈ those corresponding to zero percent vaporized, or roughly 6 psia and 60 g/mol at 20°C = 68°F for the gasoline in Fig. 10.3.

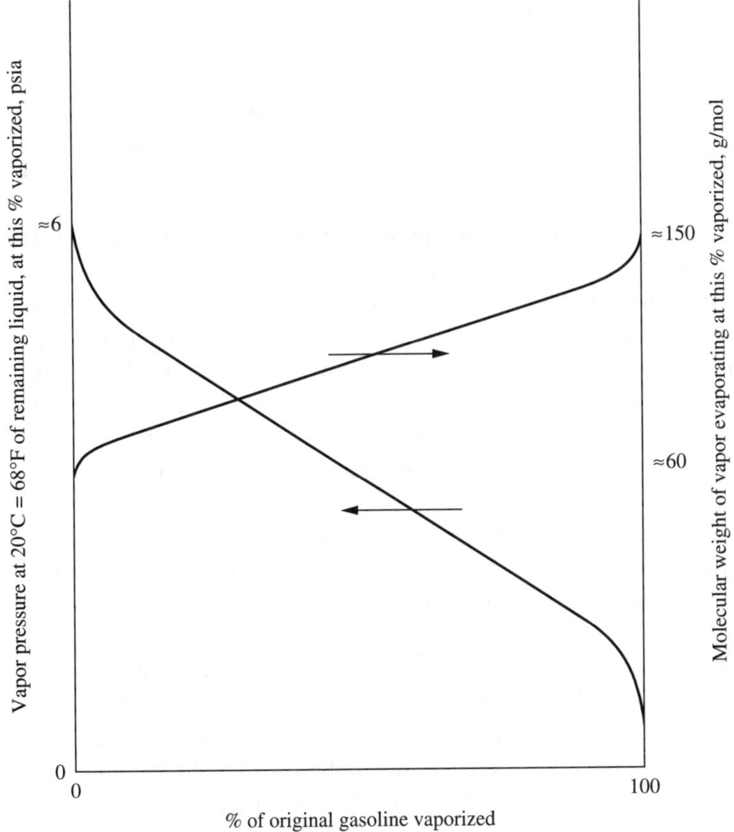

FIGURE 10.3
Change of vapor pressure of remaining liquid and of molecular weight of the vapor removed with change in % vaporized, for a typical gasoline (10 RVP, described later) at room temperature.

Example 10.7. Estimate the total volume of gasoline vapor emitted as displacement losses in the United States when gasoline is transferred from service station storage tanks to the gasoline tanks of the customers' vehicles.

From Fig. 10.3 we estimate the vapor pressure as 6 psia and the molecular weight as 60 g/mol. The density of liquid gasoline is roughly 47 lb/ft^3, so we can rework Example 10.4, finding that the concentration of gasoline in the displaced vapor is

$$\frac{m_i}{\Delta V} = \frac{x_i p_i M_i}{RT} = \frac{1.00 \cdot 6\text{ psia} \cdot 60\text{ lb/lbmol}}{(10.73\text{ psi} \cdot \text{ft}^3\text{/lbmol} \cdot {}^\circ\text{R})(528^\circ\text{R})} = 0.063\,\frac{\text{lb gasoline}}{\text{ft}^3\text{ vapor}}$$

$$= 1.02\frac{\text{kg gasoline}}{\text{m}^3\text{ vapor}}$$

and the fraction of the gasoline filled that is emitted is

$$\frac{(\text{Mass of gasoline emitted in vapor})}{(\text{Mass of liquid gasoline filled})} = \frac{0.063\text{ lb gasoline/ft}^3}{47\text{ lb gasoline/ft}^3} = 1.34 \times 10^{-3}$$

$$= 0.134\%$$

Multiplying this percentage by the approximately 3.5×10^8 gal/day of gasoline used in the United States, we find that the displacement emissions from this one transfer amount to about 470 000 gal/day. ■

This example makes clear the magnitude of the displacement emissions problem for gasoline. We know that 470 000 gal/day $\approx$ 0.48 million ton/yr $\approx$ 2.5% of the total VOC emissions in the United States (see Table 10.1). Example 10.7 only considered the final transfer of the gasoline, from the service station to the customer's vehicle. In the early days of the gasoline industry there were corresponding losses for every transfer—from one tank to another within a refinery, from the refinery to the tank truck, and from the tank truck to the tank of the service station. Much of our equipment for controlling displacement and breathing losses was developed by the petroleum industry.

For large-scale storage, the petroleum industry never puts large amounts of a liquid with as high a vapor pressure as gasoline in a simple, cone roof tank like that shown in Fig. 10.2. Instead they store large amounts of such liquids in *floating roof tanks,* as shown in Fig. 10.4.

Current EPA regulations require that a floating roof be used on any tank that stores more than 151 m^3 (40 000 gal) of fluid with a vapor pressure more than 5.2 kPa (= 0.75 psia) or stores 75 to 151 m^3 of fluid with a vapor pressure of 27.7 kPa (= 3.9 psia) at the maximum monthly temperature at the site [7]. The tank sketched in Fig. 10.4 is described as an external floating roof tank because the pontoon is open to rain and snow. Some tanks have internal floating roofs, with an external, sloping roof to divert rain and snow from the pontoon.

The transfer of gasoline from tank trucks to underground storage tanks at service stations in the United States now uses the scheme shown in Fig. 10.5. (Service

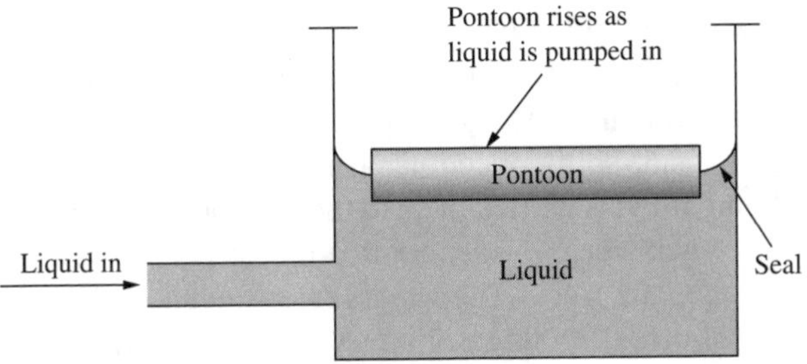

FIGURE 10.4
Floating roof tank, used to store large amounts of high vapor pressure fluids. The sealed pontoon floats on the fluid, so that there is no headspace. The pontoon moves up and down as fluid is inserted or withdrawn. This completely eliminates the displacement and breathing losses discussed above. However, the seals (normally spring-backed rubber sheets like windshield wiper blades) are not perfect, so that there are *seal losses* at them. This sketch does not show the provisions for removal of rainfall or snowmelt, or various other details.

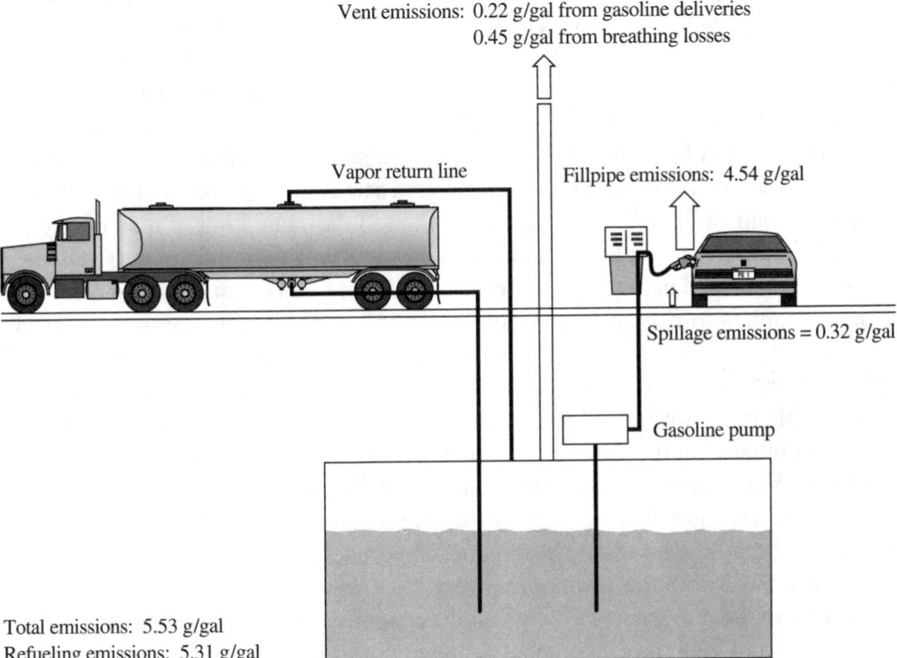

FIGURE 10.5
Loading a service station underground storage tank that has a vapor return system. This set of equipment is commonly called *Stage 1 control.* The emissions shown are a mixture of calculated and experimental values, based on work of the California Air Resources Board. (Courtesy of Sierra Research Inc.)

station tanks are placed underground both to save valuable ground space and to reduce the fire hazard of a leak or spill from such a tank. These tanks corrode and leak, polluting groundwater. This has caused conflict between the environmental engineers, who want the tanks aboveground, where leakage can be seen and corrected, and the local fire marshals, who want the tanks underground so they cannot cause a fire. In crowded urban areas the tanks are all underground; in more remote areas the new ones are aboveground, inside spill-collection basins.) In this system, the vapor displaced from the tank being filled is carried back to the tank being emptied. Withdrawing liquid from the tank truck creates a vacuum in it that sucks the vapors out of the underground tank into the tank truck. The vapors from the underground storage tank thus go back to the refinery or bulk gasoline terminal, where they are treated in one of several ways. The vent lines from all of the underground tanks are normally located at the side or rear of the service station building. The vent line shown in Fig. 10.5 remains open during tank filling to prevent any excessive pressure or vacuum in the system. This system recovers about 95 percent of the vapor from the tank being filled, the other 5 percent (0.22 g/gal in Fig. 10.5) exits from the underground tank's vent. The vent remains open between tank fillings so that, as gasoline is withdrawn and placed in the customers' vehicles, air can flow in to replace it.

The storage tank vent also handles breathing losses, which are the sum of the losses computed in Examples 10.5 and 10.6. Figure 10.5 shows the displacement loss and spillage loss as well, averaged over many tests, for fueling the auto, leading to a total loss for this system of 5.53 g/gal, or 1.4 times the value in Example 10.7.

The system in Fig. 10.5 was first applied in southern California in the 1950s and is now used in most urban areas of the United States. In regions with severe VOC problems (urban California; Washington, DC; etc.) the same kind of technology is used for the transfer of gasoline from the underground tank to the customer's vehicle; the system is sketched in Fig. 10.6. In some versions of this system the vapor from the customer's tank is forced back into the supply tank by the positive pressure caused by pumping liquid into the customer's tank, as shown in Fig. 10.6. Other versions use mechanical blowers to accomplish the same result. The blower systems bring a larger volume of air into the storage tank than the volume of liquid withdrawn, so some vapor passes out the vent. This is normally passed through an incinerator that destroys the VOC before it reaches the ambient air. The numbers in Fig. 10.6 suggest that this system reduces displacement losses by 95 percent, and spillage losses by 62 percent. The breathing losses are also reduced because much less fresh air enters the storage tank. Even with these vapor recovery systems the amount of gasoline vapor that escapes to the atmosphere is large and is a significant part of our nation's VOC problem. Starting with the 1999 model year and phasing in over several years, autos and small trucks sold in the United States will have "on board" devices to capture the fillpipe emissions shown in Fig. 10.6. For these vehicles Stage 2 is unnecessary; when they have replaced the older vehicles, the Stage 2 equipment will be removed (see Sec. 13.3).

The vapor pressure of gasoline is specified by the *Reid vapor pressure* (RVP), which is found by a standard test [8]. The value is close to the true vapor pressure at 100°F. Refiners adjust the RVP of their product by adjusting the ratio of low-boiling components (butanes and pentanes) to higher-boiling components (other hydrocarbons up to about C_{12}). In winter they raise the RVP to improve the cold-

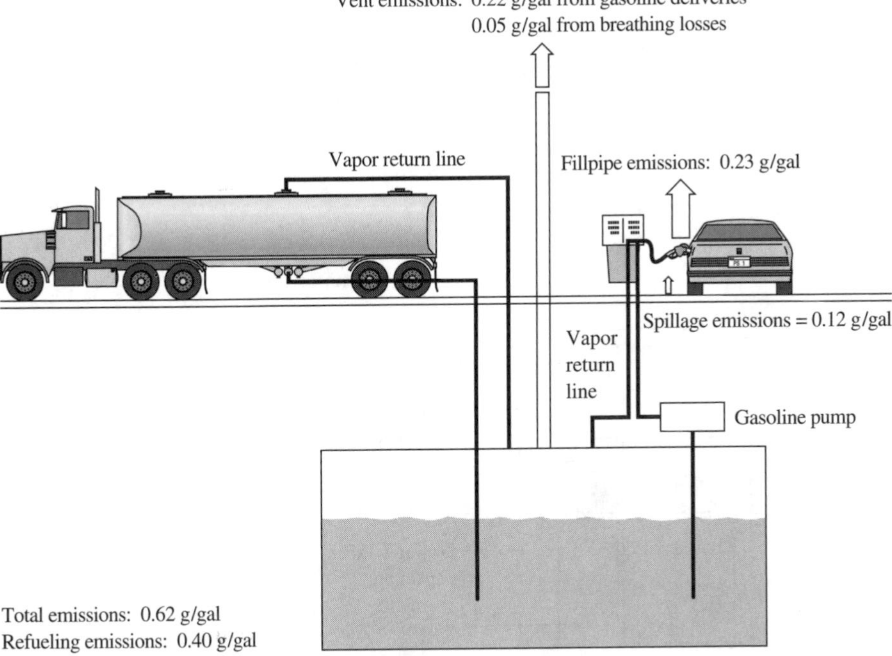

FIGURE 10.6
Loading gasoline from a service station underground storage tank to a user's vehicle using a vapor return system. This set of equipment is commonly called *Stage 2 control*. The emissions shown are a mixture of calculated and experimental values, based on work of the California Air Resources Board. (Courtesy of Sierra Research Inc.)

starting properties of the gasoline. In summer they lower the RVP because cold starting is not a problem, but vapor lock can be. Typical winter RVP values in the United States are 9 to 15 psi, with the lowest values in Florida and Hawaii and the highest values in the colder states. Typical summer values are 8 to 10 psi.

The previous examples showed that displacement and breathing losses increase with increasing vapor pressure. For this reason current EPA regulations and the Clean Air Act limit the allowable RVP of gasoline. The limitation is only applicable in the summer months, in which VOC emissions contribute to photochemical ozone formation. The rules limit RVP to 9 psia for those areas that meet the ozone standard ("ozone attainment areas") and to 7.8 psia for those that do not ("ozone nonattainment areas") [9].

10.3.3.3 Seal leaks. Many small emissions of VOCs occur as leaks at seals. In recent years these have come under regulatory control because, as the larger sources are controlled, these become a more significant part of the remaining problem.

Figure 10.7 on page 348 shows three kinds of seals. Figure 10.7*a* shows a static seal, as exists between the bottle cap and the top of a bottle of carbonated

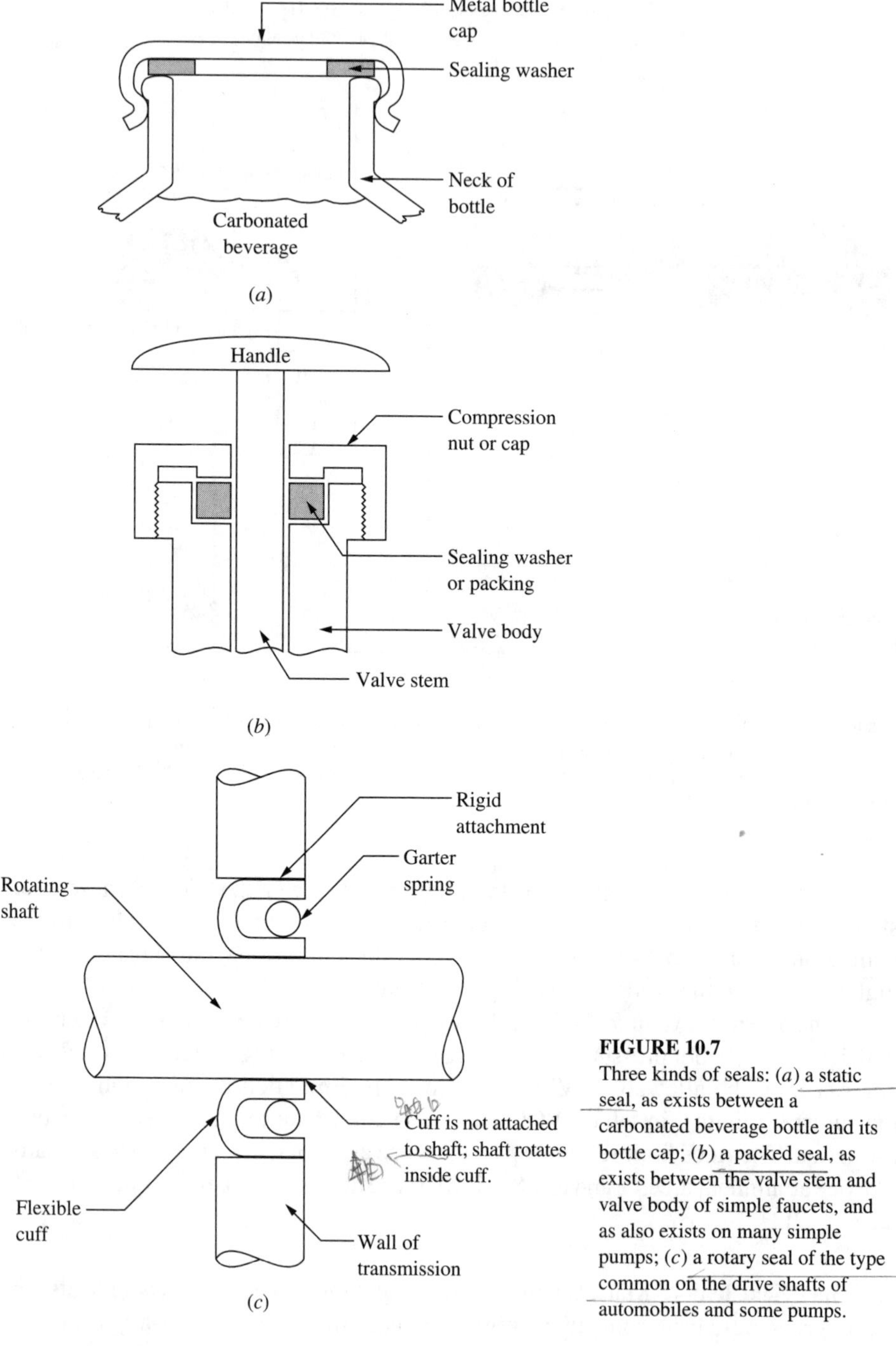

FIGURE 10.7
Three kinds of seals: (*a*) a static seal, as exists between a carbonated beverage bottle and its bottle cap; (*b*) a packed seal, as exists between the valve stem and valve body of simple faucets, and as also exists on many simple pumps; (*c*) a rotary seal of the type common on the drive shafts of automobiles and some pumps.

beverage. A thin washer of elastomeric material is compressed between the metal cap and the glass bottle top. This compressed material forms a seal that prevents the escape of CO_2 (carbonation), often for many years. Leaks through this kind of seal are generally unimportant. Sealing is more difficult when one of the two surfaces involved in the seal moves relative to the other.

Figure 10.7*b* shows a simple compression seal between a housing and a shaft. The example shown is a water faucet, in which a nut screws down over the body of the faucet to compress an elastomeric seal that is trapped between the body of the faucet and the stem of the valve. The compressed seal must be tight enough to prevent leakage of the high-pressure water inside the valve out along the edge of the stem, but not so tight that the valve cannot be easily rotated by hand. Students are probably aware from personal experience that this type of seal often leaks. If the leak is a small amount of water into the bathroom sink, that causes little problem; tightening the nut normally reduces the leak to a rate low enough that it becomes invisible (but does not become zero).

Example 10.8. A valve has a seal of the type shown in Fig. 10.7*b*. Inside the valve is gasoline at a pressure of 100 psig. The space between the seal and the valve stem is assumed to have an average thickness of 0.0001 in. The length of the seal, in the direction of leakage, is 1 in. The diameter of the valve stem is 0.25 in. Estimate the gasoline leakage rate.

From any fluid mechanics book one may find that the flow rate for the conditions described in this problem is given to a satisfactory approximation as laminar flow in a slit, for which

$$Q = \frac{P_1 - P_2}{\Delta x} \cdot \frac{1}{12\mu} \cdot yh^3 \tag{10.14}$$

Inserting values, we have

$$Q = \left(\frac{100 \text{ lbf/in.}^2}{1 \text{ in.}}\right)\left(\frac{1}{12 \cdot 0.6 \text{ cp}}\right)(\pi)(0.25 \text{ in.})(10^{-4} \text{ in.})^3$$
$$\times \left(\frac{\text{cp} \cdot \text{ft}^2}{2.09 \times 10^{-5} \text{ lbf} \cdot \text{s}}\right)\left(\frac{144 \text{ in.}^2}{\text{ft}^2}\right)$$
$$= 7.5 \times 10^{-5} \frac{\text{in.}^3}{\text{s}} = 0.27 \frac{\text{in.}^3}{\text{h}}$$

$$\dot{m} = Q\rho = 0.27 \frac{\text{in.}^3}{\text{h}} \cdot 0.026 \frac{\text{lbm}}{\text{in.}^3} = 0.007 \frac{\text{lbm}}{\text{h}} = 0.0032 \frac{\text{kg}}{\text{h}}$$

Published values indicate that the average refinery valve processing this kind of liquid leaks about 0.024 lbm/h, 3.5 times the value calculated here [10]. See Problem 10.14. ■

Figure 10.7*c* shows in greatly simplified form the seal that surrounds the drive shaft of an automobile at the point where the shaft exits from the transmission. The inside of the transmission is filled with oil. The elastomeric seal is like a shirt cuff turned back on itself with the outside held solidly to the wall of the transmission and the inside held loosely against the rotating shaft by a garter spring. If we set that spring loosely, then there will be a great deal of leakage. If we set it tightly, then the friction and wear between the cuff and the shaft that rotates inside it will be excessive. Setting the tension on that spring requires a compromise between the desire for low leakage and the desire for low friction and wear. That compromise normally leads to a low, but not a zero, leakage rate; a small amount of oil is always dripping out, and accumulating on the floor of our garages. Valves and pumps also have shafts that must rotate, and hence they have the same kind of leakage problem.

All of the pumps and valves in facilities that process VOCs have this same kind of leakage problem. The seals regularly used are more complex versions of types *b* and *c* in Fig. 10.7. There is considerable regulatory pressure for the seals to be made more and more leak-tight. Mostly this goal will be accomplished by replacing simple, low-quality seals on pumps and valves with more complex and expensive, higher-quality seals.

A truly innovative example of VOC leakage control occurred when the ARCO oil company sank and anchored a large, steel, inverted funnel over a natural gas seep at the bottom of the Santa Barbara Channel off the coast of southern California. The gas captured by the funnel is piped to shore and processed. The value of this gas is less than the cost of the equipment that captures it, but the company thereby removed VOCs from the atmosphere and gained needed VOC pollution-control credit (see Sec. 3.4), at a lower cost per pound than it could have in any of its other southern California facilities.

10.4 CONTROL BY CONCENTRATION AND RECOVERY

Most VOCs are valuable fuels or solvents; if we can recover them in pure or nearly pure form we can reuse or sell them for a profit. For large VOC-containing gas streams this is often economical, but not often for small streams. We can concentrate and recover VOC by condensation, adsorption, and absorption.

10.4.1 Condensation

One can remove most of the VOCs from an air or gas stream by cooling the stream to a low enough temperature that most of the VOCs are condensed as a liquid and then separated from the gas stream by gravity.

Example 10.9. We wish to treat an airstream containing 0.005 mol fraction (0.5%, 5000 ppm) toluene, moving at a flow rate of 1000 scfm at 100°F and 1 atm, so as to remove 99% of the toluene by cooling, condensation, and phase separation, as sketched in Fig. 10.8. To what temperature must we cool the airstream?

Ninety-nine percent recovery will reduce the mol fraction in the gas stream to 0.005% = 0.00005 mol fraction = 50 ppm. Assuming the recovered liquid is practically pure toluene ($x_{\text{toluene}} \approx 1$), we know that we must find from Eq. (10.1) the temperature at which

$$p_{\text{toluene}} = y_{\text{toluene}} \cdot P = 0.00005 \cdot 14.7 \text{ psi} = 7.35 \times 10^{-4} \text{ psia} = 5 \text{ Pa}$$

Using the Antoine equation constants for toluene in Appendix A, we find that this corresponds to a temperature of −74°F = −60°C. ■

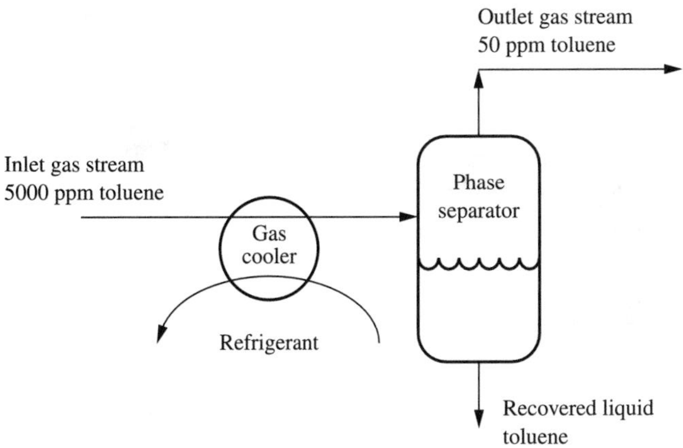

FIGURE 10.8
Simplified flow diagram of the removal of a VOC (toluene) from a gas stream by condensation and phase separation.

The difficulties with the simple condenser-phase separator in this example are these:

1. The temperatures are low enough that ordinary one-stage refrigerators cannot be used.
2. Often the temperatures required for this high a removal efficiency are below the freezing temperature of the material being removed so that the material freezes on the cooling coils, requiring frequent defrosting. In this case the freezing point of toluene is −95°C; we would not expect to encounter freezing.
3. If the gas being treated contains significant amounts of water vapor, it will condense and freeze on the cooling coils, thus requiring frequent defrosting. The water may contaminate the recovered liquid.
4. The cleaned gas leaving the system is very cold; the refrigeration work to cool it is wasted.

In spite of these difficulties such devices are used for medium-sized and/or intermittent flows of gas streams containing VOCs. Most often the cooling and condensation occur in stages, with most of the water taken out in the first stage, which operates just above 32°F (= 0°C). Figures 10.5 and 10.6 show how gasoline-laden air is transferred from the tanks of the auto to the service station's underground tanks and then back to the tanks of the gasoline delivery truck. When a truck returns to its bulk-loading terminal, that air must be displaced.

Example 10.10. The current EPA requirement for gasoline bulk-loading terminals is that the displacement loss from the returning tank trucks must not exceed 35 mg of VOCs per liter of gasoline filled [11] (see Problem 10.18). One common solution to this problem is sketched in Fig. 10.9 on page 352. This is clearly two of the systems in Fig. 10.8, in series, with the first one taking out much of the gasoline and most of

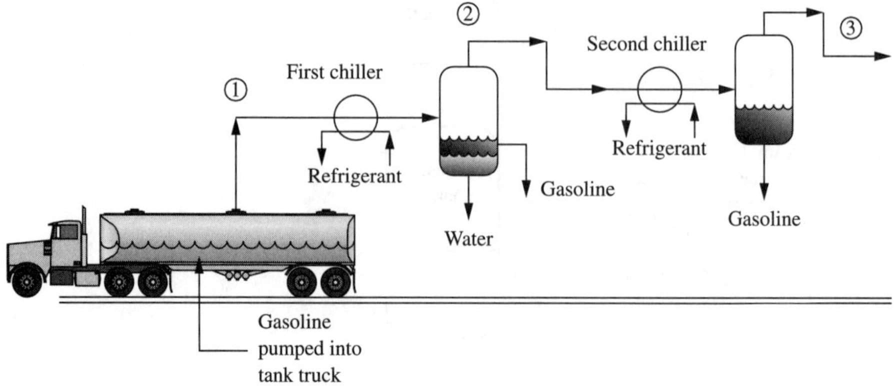

FIGURE 10.9
Simplified flow diagram of a two-stage condenser-separator for recovering VOC from the displacement vapors of a gasoline tank truck, at a bulk-loading terminal. The circled numbers refer to stream compositions, shown in the following table.

the water, and the second, at a lower temperature, taking out most of the remaining gasoline. Assuming that the vapor leaving the truck is at 20°C = 68°F, in equilibrium with the remaining gasoline in the truck, and that that vapor is 1 mol % water vapor, (*a*) how cold must the second chiller cool the displaced vapor before discharging it to the atmosphere? Assume that the discharged vapor is in equilibrium with liquid gasoline at that temperature, that the gasoline vapor in the air has a molecular weight of 60, and that its vapor pressure is given approximately by

$$\ln P \text{ (psia)} = 11.724 - (5236.5°\text{R})/T \tag{10.15}$$

which is a fair approximation for 10 RVP gasoline [12]. (*b*) What fraction of the gasoline will be removed in the first (water removal) stage, which cools the gas to about 32°F? (*c*) What is the ratio of ice formed to gasoline condensed in the second stage?

The standard chemical engineering approach to such problems is to choose as a basis 1 mol of vapor leaving the tank in stream 1 and record results as calculated in the following table. The table shows the completed solution, with the values found one at a time in the following text.

Stream #	**1**	**2**	**3**
Mols air	0.576	0.576	0.576
Mols gasoline	0.414	0.145	0.0140
Mols water	0.01	0.004	≈ 0
Total mols	1.00	0.726	0.590
Mol fraction gasoline	0.414	0.200	0.024
T, °F	68	32	−50.0
Gasoline vapor pressure, psia	6.09	2.95	0.350
Water or ice vapor pressure, psia	—	0.089	0.0008 ≈ 0
Mol fraction water	0.01	0.0060	≈ 0

We can immediately write that the total mols in stream 1 = 1.00 and that the mols of water are 1% of that, or 0.01. From Eq. (10.15) we estimate the vapor pressure of the gasoline as 6.09 psia, from which we can compute the gasoline mol fraction as (6.09 psia/14.7 psia) = 0.414. Thus for the assumed basis of 1 mol, the mols of gasoline are 0.414, and those of air (1.00 – 0.414 – 0.010) = 0.576. Thus we have a complete description of stream 1.

Next we assume (see Sec. 7.9) that the air passes unchanged through the system. This is equivalent to assuming that no air dissolves in the gasoline or water we remove, a very good (but not perfect) assumption. Thus we can fill in the row for mols of air. To find the permitted amount of gasoline in stream 3 we observe that the 1 L of gasoline displaces 1 L of vapor (stream 1), and write

$$\begin{pmatrix}\text{Permitted}\\ \text{emission}\end{pmatrix} = \frac{0.035\text{g}}{\text{L gasoline}} \cdot \frac{24.056\text{ L}}{\text{mol}} \cdot \frac{\text{mol}}{60\text{ g}} = 0.0140\frac{\text{mol gasoline}}{\text{mol stream 1}}$$

which allows us to write in the number of permitted mols of gasoline in stream 3. We will see later that the mols of water in stream 3 are negligible, so we can compute

$$\begin{pmatrix}\text{Mol fraction}\\ \text{gasoline in}\\ \text{stream 3}\end{pmatrix} = \frac{\text{Mols gasoline}}{\text{Mols (gasoline + air)}} = \frac{0.0140}{(0.0140 + 0.576)} = 0.024$$

We are now able to solve part (*a*). The vapor pressure of gasoline in stream 3 is equal to the total pressure times the gasoline mol fraction, (14.7 · 0.024) = 0.350 psia; solving Eq. (10.15) for the temperature corresponding to this pressure, we find $T_3 = 410°\text{R} = -50.0°\text{F}$.

To answer part (*b*), we estimate the vapor pressure of gasoline at 32°F from Eq. (10.15), finding 2.95 psia, and look up the vapor pressure of water at 32°F = 0.089 psia. Then we find

$$\begin{pmatrix}\text{Mol fraction}\\ \text{gasoline in}\\ \text{stream 2}\end{pmatrix} = \frac{\text{Vapor pressure gasoline}}{\text{Total pressure}} = \frac{2.95\text{ psia}}{14.7\text{ psia}} = 0.200$$

and, similarly for water, we find a mol fraction of 0.006. Then by difference we find the mol fraction of air (1.00 – 0.200 – 0.006) = 0.794, from which the total number of mols in stream 2 is (0.576/0.794) = 0.726. Thus the mols of gasoline in stream 2 are (0.726 · 0.200) = 0.145, and correspondingly for water, 0.004. Thus the amount of gasoline removed in the first stage is (0.414 – 0.145) = 0.269 mols, or (0.269/0.414) = 65% of the gasoline in stream 1.

To answer part (*c*) we must estimate the vapor pressure of ice at −50.0°F. Extrapolating the values from the steam table [13], we find $p_{\text{ice}} \approx 0.001$ psia ≈ 0 (see Problem 10.19); practically all of the water in stream 2 will be frozen in the second chiller. Then the mols of ice formed are 0.004 – 0 = 0.004, whereas the mols of gasoline condensed are (0.145 – 0.014) = 0.131, and the molar ratio of ice to gasoline is 0.004/0.131 = 0.03 = 3%. ■

This is the typical example of basing calculations on the inert material flow rate (Sec. 7.9); the total flow decreases from 1.00 to 0.590 mol from inlet to outlet. All of the calculations refer to the air flow, which does not change from inlet to outlet. In the most careful work we need to account for the change in the molecular weight of the gasoline, as part of it is condensed. We see that we can get most of the gasoline and water out in a chiller that operates just above the water freezing temperature, but that the second chiller will accumulate ice and need to be defrosted. This type of plant does not normally operate 24 hours a day, so the defrosting is done at night.

These condensers encounter a special fire hazard. Most VOC-contaminated air streams have VOC concentrations less than the lower explosive limit. Removing VOCs from them takes them even farther from a combustible condition. The vapors from gasoline tanks generally contain enough gasoline that they are above the upper explosive limit. Removing VOCs from them, by condensation or adsorption, causes them to pass through the combustible range before they pass below the lower explosive limit within the control device. Inside the device they are often combustible; special care is required to exclude all possible ignition sources and to provide flame arrestors.

For more detailed technical information on this type of VOC control, see [14]. See also Problems 10.21 and 10.22.

10.4.2 Adsorption

A**d**sorption means the attachment of molecules to the surface of a solid. In contrast, a**b**sorption means the dissolution of molecules within a collecting medium, which may be liquid or solid. Generally, absorbed materials are dissolved *into* the absorbent, like sugar dissolved in water, whereas adsorbed materials are attached *onto* the surface of a material, like dust on a wall. Absorption mostly occurs into liquids, adsorption mostly onto solids. This section deals only with a**d**sorption onto the surface of a solid adsorbent.

Adsorption is mostly used in air pollution control to concentrate a pollutant that is present in dilute form in an air or gas stream. The material collected is most often a VOC like gasoline or various paint thinners and solvents. The solid is most often some kind of activated carbon. The student is possibly familiar with cigarettes that have activated carbon filters to collect some of the harmful materials in the smoke. They are used once and thrown away. The student is probably less familiar with the activated carbon canisters used in industrial face masks. These are worn by workers exposed to solvents, as in paint spraying or solvent cleaning. The worker's lungs suck the air in through thin beds of activated carbon, contained in replaceable cartridges on the face mask. When the activated carbon is loaded (i.e., the solvent begins to come through into the worker's breathing space) the cartridge of activated carbon is discarded and a fresh one installed.

For large-scale air pollution applications, like collecting the solvent vapors coming off a large paint-drying oven or a large printing press, the normal procedure is to use several adsorption beds. As shown in Fig. 10.10, the contaminated air stream

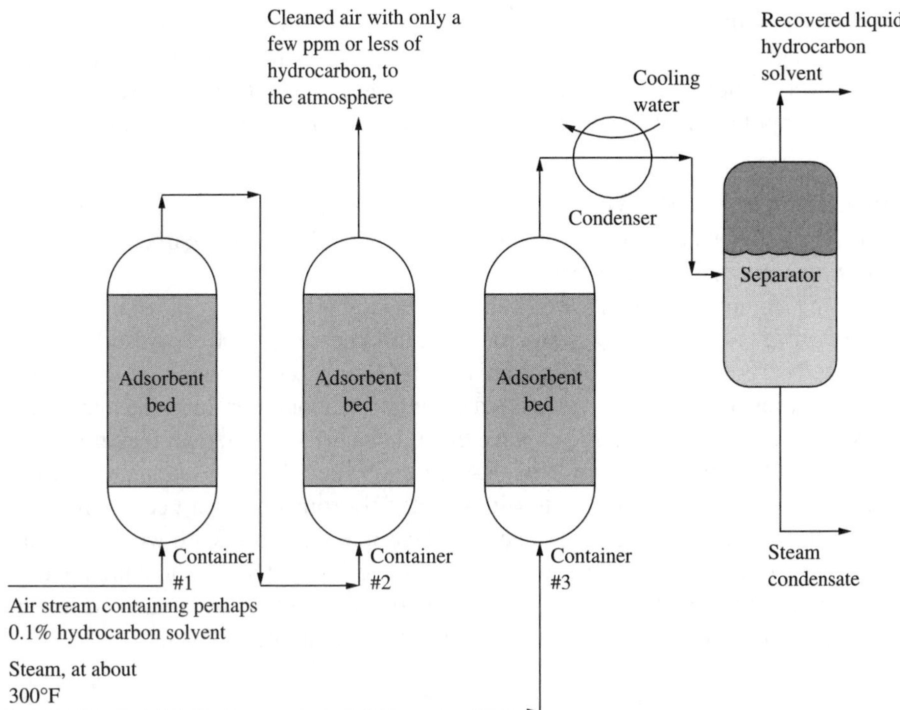

FIGURE 10.10
The typical arrangement for adsorption of a VOC from a gas stream, using three adsorbent beds; automatic switching valves; and steam desorption, condensation, and gravity separation.

passes through two vessels in series. Inside each of the vessels is a bed of adsorbent that removes the VOCs. From the second vessel the cleaned air, normally containing at most a few parts per million of VOCs, passes to the atmosphere. Meanwhile, a third vessel is being regenerated. Steam passes through it, removing the adsorbed VOCs from the adsorbent. The mixture of steam and VOCs coming from the top of the vessel passes to a water-cooled condenser that condenses both the VOCs and the steam. Both pass in liquid form to a separator, where the VOCs, which are normally much less dense than water and have little solubility in water, float on top and are decanted and sent to solvent recovery.

After a suitable time period a set of automatically programmed valves changes the position of the containers in the flow sheet. (The containers do not move; their place in the piping arrangement changes.) Container 1, which is most heavily loaded, goes to the regeneration position. Container 2, which is lightly loaded with VOCs, goes to the position where container 1 was; and container 3, which is now regenerated and very clean, goes to the position previously held by container 2, making the final cleanup on the air stream. Figure 10.10 shows the steam condensate leaving the phase separator, without specifying where it goes. As discussed in Sec. 10.2, this condensate will be saturated with dissolved VOC. The VOC concentration may be high enough to prevent its being sent back to the steam boiler, or for it to be discharged to a sewer. If there is no good way to deal with this stream, then the absorber solves a large air pollution problem but creates a small water pollution problem!

10.4.2.1 Adsorbents. The most widely used adsorbent for VOCs is activated carbon. This somewhat fancier version of the charcoal used for barbecuing has an amazing amount of surface area.

In Example 7.15 we showed that catalyst supports typically have surface areas of 100 m^2/g, corresponding to internal wall thicknesses of 100 Å. Adsorbents like activated carbon often have surface areas of 1000 m^2/g, corresponding to an internal wall thickness of 10 Å. This value is startlingly low, about four times the interatomic spacing in crystals! If adsorbents have this much surface area, then they must have internal walls only four atoms thick! Apparently they do. To make materials with this much surface area, one starts with a material, from which part can be removed on an atomic scale. In the case of activated carbon, one starts with wood (or peach pits, or coconut shells, or some other woody material) and heats it to a high enough temperature that the wood decomposes (pyrolyzes), producing a gas and leaving behind a solid carbon residue, in the form of these thin internal walls.

10.4.2.2 Adsorbent capacity. To design an adsorber of the type shown in Fig. 10.10 we must consider both the adsorbent capacity and the breakthrough performance (discussed below) of the adsorbent. Figure 10.11 shows the capacity of adsorbents in the form suggested by Polanyi, as presented in Fair [15]. For an actual design one would need the corresponding capacity curve for the material to be adsorbed and the adsorbent to be used. With Fig. 10.11 we can make estimates without such data.

Example 10.11. Using Fig. 10.11, estimate the adsorbent capacity curves for toluene on a typical activated carbon at 1.0 atm and 100°F and at 300°F.

From the legend for Fig. 10.11 we see that curves D through I represent various activated carbons. Since we do not know which curve best represents the behavior of toluene on a typical activated carbon, we select curve F, which lies near the middle of this family of curves. Then we compute the point on $w^* - P$ coordinates for 1 atm, 100°F, and an arbitrarily selected value of 1% toluene in the gas. From that point, we calculate the values for other percent toluene values by ratios.

Here $T = 560°\text{R}$ and $M = 92$ g/mol. We estimate ρ'_L, the toluene density at the normal boiling point of 110.6°C, as 0.782 g/cm^3 from the 20°C density (0.8669 g/cm^3) and the typical coefficient of thermal expansion for organic liquids (0.67×10^{-3}/°F).

At atmospheric pressure, the fugacity f can be replaced by the partial pressure (mol fraction $\cdot$ total pressure) $= 0.01$ atm and the saturation fugacity f_s can be replaced by the vapor pressure. Using the vapor pressure constants in Appendix A

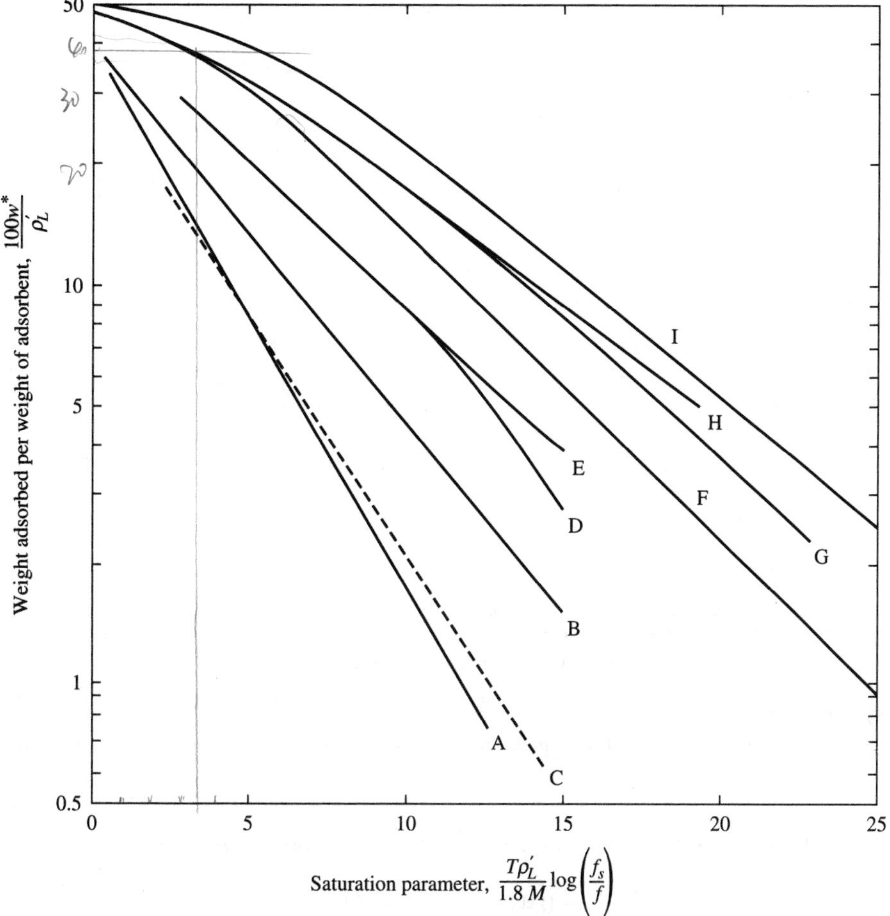

FIGURE 10.11
Adsorption curves (isotherms) for hydrocarbons and some other gases on activated carbon and silica gel. (From Fair [15].)

Curves A, B, and C are for silica gel as the adsorbent; curves D through I are for various activated carbons. Curves D and E coincide at the left, but not at the right; they represent differing results for olefins and paraffins on the same adsorbent. The same is true for curves G and H, on another adsorbent. The gases adsorbed are mostly paraffins and olefins, C_1 to C_6.

Here w^* is the weight of adsorbate per weight of adsorbent, ρ'_L is the liquid density of the adsorbate at the normal boiling point, T is the absolute temperature, M is the molecular weight of the adsorbate, f is the fugacity of the adsorbate in the gas stream, and f_s is the fugacity of the adsorbate at vapor-liquid equilibrium. The fugacity is a "corrected partial pressure"; at the low pressures of air pollution interest $(f_s/f) \approx (p/yP)$.

This plot is dimensional; T, M, w^*, and ρ'_L must be expressed in °R, g/mol, lb/lb (or equivalent), and g/cm^3, respectively.

we estimate a vapor pressure of 1.03 psia = 53 torr = 0.070 atm. Thus we write

$$\frac{T\rho'_L}{1.8\,M}\log\left(\frac{f_s}{f}\right) = \frac{560 \cdot 0.782}{1.8 \cdot 92}\log\left(\frac{0.070}{0.010}\right) = 2.23$$

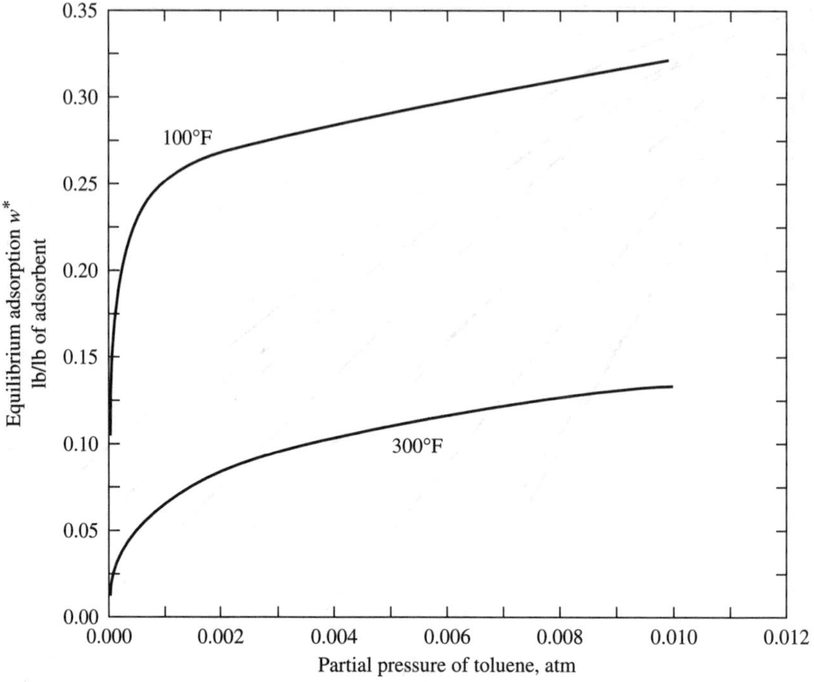

FIGURE 10.12
Calculated equilibrium curves (adsorption isotherms) for toluene on a typical activated carbon. See Example 10.11.

From curve F of Fig. 10.11 we read an ordinate of about 41, so that

$$w^* = \frac{41}{100}\rho_L' = 0.41 \cdot 0.782 = 0.31 \frac{\text{lb toluene}}{\text{lb solid adsorbent}}$$

We then repeat the calculation for other values of the mol fraction of toluene in the gas, and for $T = 300°\text{F}$, and plot the results as shown in Fig. 10.12. ■

From Fig. 10.12 we see that both curves pass asymptotically into the y-axis, indicating that we can get practically complete removal of toluene from air with this kind of adsorbent. We also see that the curve for 300°F is much lower than the curve for 100°F, so that we should be able to regenerate our adsorbent at this temperature.

Example 10.12. We wish to treat the airstream in Example 10.9 to remove practically all the toluene. If the bed must operate 8 h between regenerations, how many pounds of activated carbon must it have (a) if it is only used once and then thrown away, and (b) if it is regenerated to an outlet stream toluene content of 0.5 percent?

Here the incoming air flow is

$$\text{Flow} = \dot{n} = 1000 \text{ scfm} \cdot 2.595 \times 10^{-3} \frac{\text{lbmol}}{\text{scf}} = 2.595 \frac{\text{lbmol}}{\text{min}}$$

and the contained toluene is

$$\dot{m}_{\text{toluene}} = \dot{n}_{\text{air}} M_{\text{toluene}} y_{\text{toluene}} = 2.595 \frac{\text{lbmol}}{\text{min}} \cdot 92 \frac{\text{lb}}{\text{lbmol}} \cdot 0.005$$

$$= 1.19 \frac{\text{lb}}{\text{min}} = 0.0090 \frac{\text{kg}}{\text{s}}$$

If all the toluene is to be recovered, we must recover

$$m = \dot{m}_{\text{toluene}} \Delta t = 1.19 \frac{\text{lb}}{\text{min}} \cdot 8 \text{ h} \cdot \frac{60 \text{ min}}{\text{h}} = 572 \text{ lb} = 260 \text{ kg}$$

From Fig. 10.12 we read that for a toluene partial pressure of 0.005 atm at 100°F, $w^* = 0.29$ lb/lb, and at 300°F it equals 0.11 lb/lb. Thus, for part (*a*) we can say that the amount of adsorbent needed is

$$\text{Adsorbent needed} = \frac{m}{w^*} = \frac{572 \text{ lb}}{0.29 \text{ lb/lb}} = 1970 \text{ lb} = 895 \text{ kg}$$

For part (*b*) the adsorbent is to be reused and regenerated. The net amount adsorbed per cycle will be 0.29 minus 0.11 or 0.18 lb/lb, and the same calculation leads to an adsorbent requirement of 3180 lb = 1445 kg. ■

As discussed in the next section, this is an optimistic estimate of the amount of adsorbent needed.

10.4.2.3 Breakthrough performance. The calculation in Example 10.12 assumes that the adsorbent fills up with adsorbed material uniformly, like filling an auditorium one row at a time. Unfortunately, real adsorbers never work that well, and some of the material to be adsorbed "breaks through" before the bed has reached its maximum capacity. The reason for this early breakthrough is that there is a finite resistance to mass transfer between the gas and the solid, so some finite amount of time is needed for each particle to be loaded with adsorbate. If there were infinitely rapid mass transfer, there would be no early breakthrough.

Figure 10.13 on page 360 compares the ideal breakthrough curve with a typical real breakthrough curve. If the material being adsorbed filled the adsorbent like filling an auditorium one row at a time, then none of the material being adsorbed would appear in the outlet gas until the bed was completely filled. Then the concentration in the outlet gas would jump instantaneously from zero to the concentration of the inlet gas at time t_{ideal}.

In the real situation some of the adsorbable material appears at the outlet before the bed is filled. At some time t_b less than the ideal time the concentration in the outlet stream reaches the *breakthrough value* (typically 1 percent of the inlet value). The adsorbent bed is then switched to regeneration or, if it is a throwaway adsorber like those in workers' face masks, the adsorber cartridge is discarded and a new one installed. If that did not occur, then the outlet concentration would rise in an S-shaped curve to the inlet concentration, as shown in Fig. 10.13.

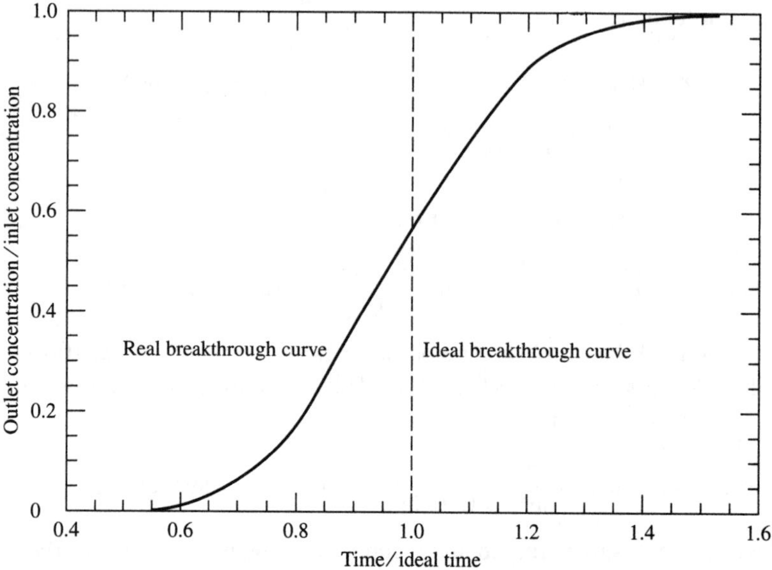

FIGURE 10.13
Ideal and real breakthrough curves for adsorption in a fixed bed.

The complete calculation of the breakthrough curve requires numerical solution of coupled partial differential equations [16] (see Appendix E). However, a reasonable estimate can be made using Fig. 10.14.

Example 10.13. Using the data from Example 10.12, estimate the breakthrough curve that would be observed if we started with clean adsorbent. Assume that the adsorbent is an activated carbon with a bulk density of 30 lb/ft^3 and a particle diameter of 0.0128 ft. The volume of the bed is [1970 lb/(30 lb/ft^3)] = 66 ft^3. If we assume a cubically shaped bed (not practical to fabricate, but easy to calculate), the sides of the bed will be 4.03 ($\approx$ 4) ft.

Then as shown in Appendix E, we may estimate that a = 14.4/ft and b = 7.4/h. Thus on Fig. 10.14, $\mathbf{N} = ax$ = 4 ft × 14.4/ft = 57.6 ≈ 60, so that the estimated breakthrough behavior of the bed will follow the curve for $\mathbf{N}$ = 60. The outlet concentration will become 1 percent of the inlet concentration at a bt value of about 36, which corresponds to a time of

$$t = \frac{bt}{b} = \frac{36}{7.4/\text{h}} = 4.9 \text{ h}$$

This value is 61% of that for perfect filling of the bed (i.e., no mass transfer resistance). ■

To summarize, adsorbers remove VOCs from air or other gas streams in which the VOCs are present in low concentration. VOCs are recovered chiefly by regener-

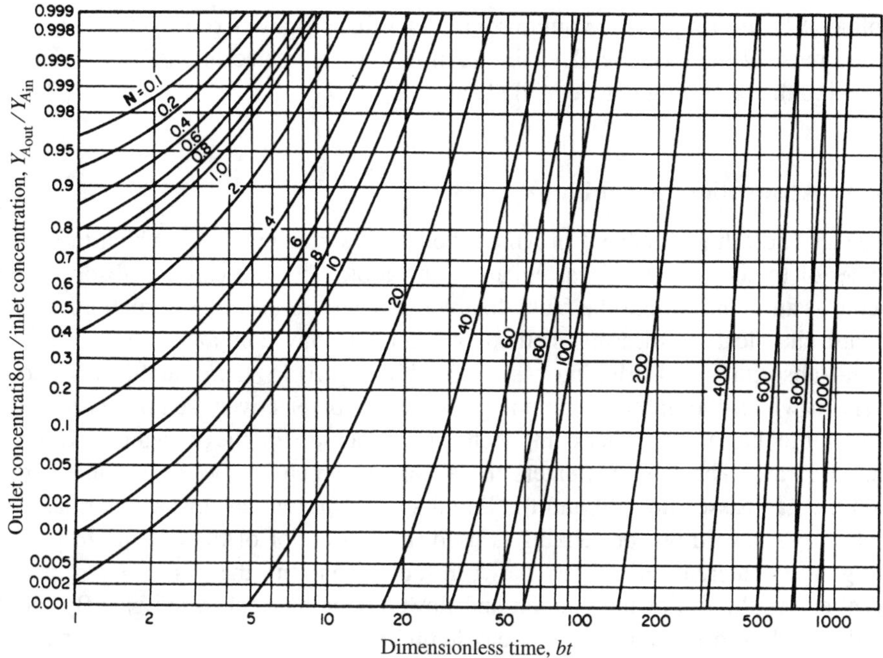

FIGURE 10.14
Dimensionless breakthrough plot for a fixed bed adsorber. (Vermeulen, T., M. D. LeVan, N. K. Hiester, and G. Klein: "Adsorption and Ion Exchange," in *Perry's Chemical Engineer's Handbook,* 6th ed., D. W. Green and J. O. Maloney, eds., McGraw-Hill, New York, Chapter 16, 1984 [17]. Reproduced by permission of McGraw-Hill.) The notation has been changed from that source to match the notation in the classic paper by Hougen and Marshall [18], which anyone interested in adsorption should read. In the latter paper the **N** parameter shown on the curves is called the dimensionless bed thickness ax. (See Appendix E.)

ating the adsorbents with steam and condensing the VOCs as liquids. The cleaned air or gas streams leaving the adsorbers contain only a few parts per million of VOC. The same principles apply to the throwaway adsorbers used in industrial face masks and to the adsorbers used to capture some of the fuel emissions in most modern autos (Chapter 13).

The curve on Fig. 10.13 was made up from Example 10.13, reading the values of bt for various values of (Outlet concentration/inlet concentration) for the $\mathbf{N} = 60$ curve. The plots look different because Fig. 10.13 is on arithmetic coordinates, whereas Fig. 10.14 is on log-normal coordinates.

Example 10.10 illustrates the use of two-stage condensers to capture the gasoline vapor breathing loss caused by filling gasoline delivery trucks at bulk terminals. That technology competes on more or less even terms with adsorption systems, which use two large carbon adsorbent beds, regenerate the loaded bed by reducing the pressure in it to about 0.1 atm, and allow a small flow of air to carry away the gasoline. A liquid-ring vacuum pump compresses that air-gas mixture to 1 atm, condensing most of the gasoline. The air stream is further treated in an absorber and returned to the bed being regenerated [19].

10.4.3 Absorption (Scrubbing)

If we can find a liquid solvent in which the VOC is soluble and in which the remainder of the contaminated gas stream is insoluble, then we can use absorption to remove and concentrate the VOC for recovery and re-use, or destruction. The standard chemical engineering method of removing any component from a gas stream—absorption and stripping—is sketched in Fig. 10.15. If we can find a liquid solvent in which the gaseous component we wish to selectively remove is much more soluble than are the other components in the gas stream, the procedure is quite straightforward. The feed gas enters the absorber, which is a vertical column in which the gas passes upward and the liquid solvent passes downward. Normally, bubble caps, sieve trays, or packing is used in the interior of the column to promote good countercurrent contact between the solvent and the gas. The stripped solvent enters the top of the column and flows countercurrent to the gas. By the time the gas has reached the top of the column, most of the component we wish to remove has been dissolved into the solvent; the cleaned gas passes on to the atmosphere or to its further uses. The loaded solvent, which now contains most of the component we are removing from the gas, passes to the stripper, which normally is operated at a higher temperature and/or a lower pressure than the absorber. At

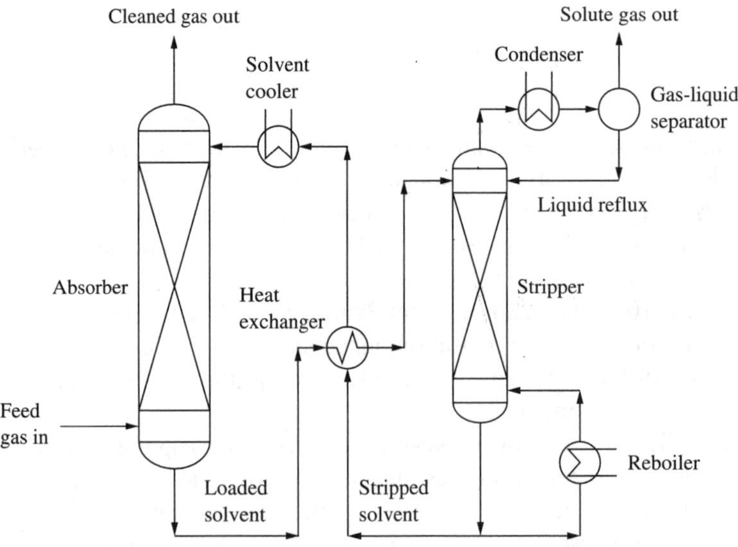

FIGURE 10.15
The flow diagram for the most common method for removing one component from a gas stream.

this higher temperature and/or lower pressure, the solubility of the gas in the selective solvent is greatly reduced so the gas comes out of solution. In Fig. 10.15 the separated component is shown leaving as a gas for use, sale, or destruction. In some cases it is condensed and leaves as a liquid. The stripped or lean solvent is sent back to the absorber column. Very large absorption-stripping systems often use tray columns, but the small ones used in most air pollution control applications use internal packings. The rest of this section assumes that we are discussing packed absorber columns.

Functionally, this is the same as the adsorption process sketched in Fig. 10.10. The chosen component is selectively removed from the gas stream onto an adsorbent or into an absorbent in one vessel and is subsequently removed at much higher concentration (often practically pure) in another vessel at a higher temperature and/or lower pressure. The absorption-stripping scheme in Fig. 10.15 is mechanically simpler because it is easy to move liquids with pumps and pipes. It is much harder to move solids the same way. The adsorption equivalents of Fig. 10.15 have been tried, but the mechanical difficulties have been severe enough that most adsorption is done with the solids remaining in place as shown in Fig. 10.10, using a semisteady-state operation.

The absorption solvent must have the following properties:

1. It must afford reasonable solubility for the material to be removed, and, if this material is to be recovered at reasonable purity, it must not dissolve and thus carry along any of the other components of the gas stream.
2. In the absorber, the gas being treated will come to equilibrium with the stripped solvent. The vapor pressure of the solvent, at absorber temperature, must be low enough that if the cleaned gas is to be discharged to the atmosphere, the emission of solvent is small enough to be permissible. Some solvent is lost this way; the cost of replacing it must be acceptable. If the solvent is water this is not a problem (unless we need the gas to be dry for its next use), but for other solvents this can be a problem.
3. At the higher temperature (or lower pressure) of the stripping column, the absorbed material must come out of solution easily, and the vapor pressure of the solvent must be low enough that it does not contaminate the recovered VOC. If the solvent vapor pressure in the stripper is too large, one may replace the stripper by a standard distillation column (combination stripper and rectifier) to recover the transferred material at adequate purity.
4. The solvent must be stable at the conditions in the absorber and stripper, and be usable for a considerable time before replacement.
5. The solvent molecular weight should be as low as possible, to maximize its ability to absorb. This requirement conflicts with the low solvent vapor pressure requirement, so that a compromise must be made.

10.4.3.1 Design of gas absorbers and strippers. The treatment here is a simplified version of the detailed discussion of the design of these devices in Refs. 20–22.

In the device shown in Figure 10.15, the basic design variables are the choice of selective reagent to be used, the system pressure, the flow rates of the gas and liquid, the gas velocity in the column, and the amount of liquid-gas contact needed to produce the separation. In air pollution control applications we will generally know the flow rate of gas to be treated, the content of the material to be removed from the gas, and the required degree of removal.

The pressure will normally be 1 atmosphere for most gases that are to be discharged to the atmosphere. For many industrial applications of the system in Fig. 10.15, the gas to be treated will be available at a pressure higher than 1 atmosphere. It will generally be economical to treat the gas at that higher pressure because the size and cost of the treating equipment are proportional to the volumetric flow rate of gas handled, which is $\approx$ proportional to $1/P$.

In Fig. 10.15 both columns operate in counterflow: the liquid flows down the column by gravity; the gas flows up the column, driven by the decrease in pressure from bottom to top. This design not only utilizes gravity efficiently in moving the liquid but also provides for very efficient contacting. Figure 10.16 shows why. In this figure, the curve at the right shows the mol fraction in the gas of the component to be absorbed, y_i, decreasing from its high value where it enters the bottom of the column to its low value where it leaves the top of the column. The path is shown curved because the removal rate is not linearly proportional to the column height. The curve at the left shows the concentration of absorbable component that would be in equilibrium with the liquid absorbent, y_i^*, which increases from top to bottom

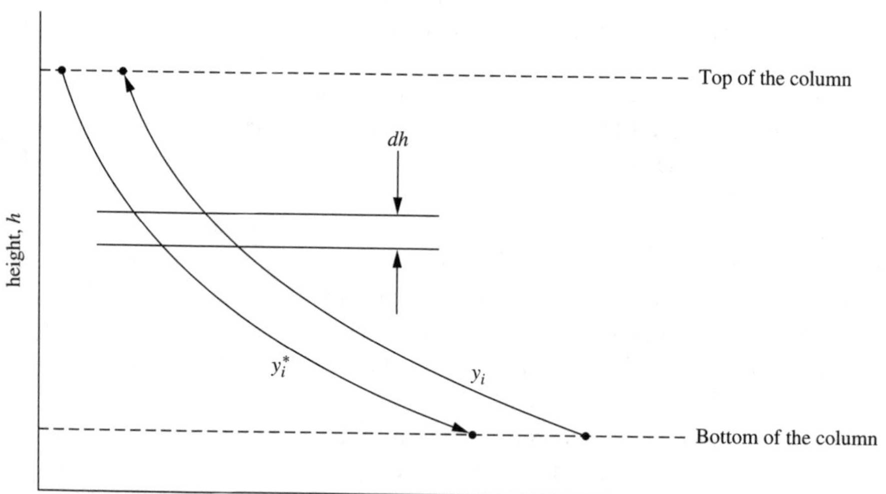

FIGURE 10.16
Simplified schematic diagram of the changes in gas composition, and of the equilibrium gas composition, which depend on liquid composition, for the absorbing (leftmost) column in Fig. 10.15.

of the column as the descending absorbent removes the absorbable component from the gas.

The relation between y_i^* and the mol fraction of absorbable component in the liquid absorbent, x_i, is complex and differs from one chemical system to another. The simplest description of such absorption equilibria, for slightly soluble gases, is *Henry's law* (which is not a law like Newton's laws but rather a crude, but simple and useful approximation of reality). Henry's law is normally written as

$$x_i = \frac{Py^*}{H_i} \tag{10.16}$$

Here P is the absolute pressure, and H_i is the *Henry's law constant* for component i, which is normally a strong function of temperature, but a weak function of pressure or concentration. (Values of Henry's law constants are widely published [23].) If we assume that this simplest relation is applicable, then we can compute the value of y_i^* from the liquid mol fraction x_i at any point in the column.

In Fig. 10.16 the transfer of the absorbable component will be from gas to liquid as long as $y_i > y_i^*$, i.e., the actual concentration in the gas at that location in the column is higher than the concentration that would be in equilibrium with the absorbing liquid. (In the stripper in Fig. 10.15 the relation is reversed, $y_i < y_i^*$, and the absorbable component flows from liquid to gas.) With the counterflow arrangement shown, the flow of absorbable component is from gas to liquid in the whole absorption column. By increasing the height of the column, we can make the concentration of the absorbable component in the outlet gas stream approach the concentration that would be in equilibrium with the inlet (stripped) solvent. In this way, if we strip the solvent thoroughly and provide enough gas-liquid contact, we can reduce the content of absorbable material in the gas to very low values.

We next perform a material balance on the transferred component for the small section dh of the column shown in Fig. 10.16, finding

$$\begin{aligned} &\begin{pmatrix}\text{Mols of } i\\ \text{transferred}\\ \text{from the gas}\end{pmatrix} = \begin{pmatrix}\text{mols of } i\\ \text{transferred}\\ \text{to the liquid}\end{pmatrix} = \begin{pmatrix}\text{mass transfer}\\ \text{capacity per}\\ \text{unit volume}\end{pmatrix} (d \text{ volume}) \\ &-G\,dY_i \qquad\qquad = L\,dX_i \qquad\qquad = (KaP)(y_i - y_i^*)A\;dh \end{aligned} \tag{10.17}$$

Here G and L are the molar flows of gas and liquid (mol/s), **excluding the flow of the transferred components**. Y_i and X_i are the gas and liquid contents of the transferred component, respectively, expressed in (mol/mol of nontransferred components). $A\,dh$ is the product of the column cross-sectional area and incremental height, equal to the column volume corresponding to dh. Ka is the product of the mass transfer coefficient and the interfacial area for mass transfer (ft^2 of transfer area per ft^3 of volume), discussed later, and P is the system pressure.

Here we base the calculations on the flows of the nontransferred components because those flows do not change from place to place within the column, while other properties and concentrations all do (see Sec. 7.9).

Example 10.14. We wish to treat the stream in Example 10.9, recovering the toluene by absorption in a suitable solvent. (See also Examples 10.11–10.13.) Select a suitable solvent, and estimate the required solvent flow rate.

The solubility of toluene in water (see Table 10.3) is low enough that we are unlikely to use water as a solvent (see Problem 10.31). Our logical choice is an HC with a higher boiling point than toluene. In Example 10.9 the permitted toluene concentration in the exhaust gas is 50 ppm. If we assume that we can emit an equal concentration of the solvent, then its vapor pressure at column temperature (100°F in that example) must be no more than $50 \cdot 10^{-6}$ atm, because the gas leaving the system will be in equilibrium with the practically pure solvent at the top of the column. From Fig. 10.1 we see that the highest-boiling HC shown, n-decane, has a vapor pressure at 100°F of about 0.06 psi $=$ 0.004 atm, which is $\approx$ 80 times too high. Using the Antoine equation constants in [24], we find that the vapor pressure falls rapidly as the molecular weight increases, and that n-tetradecane ($C_{14}H_{30}$; $M = 198$ g/mol) has an acceptable calculated vapor pressure of $47 \cdot 10^{-6}$ atm at 100°F. We would not use a pure HC as absorbent, because pure HCs are expensive. But this shows that a hydrocarbon mixture with a vapor pressure comparable to n-tetradecane could be used (if we can accept 50 ppm of it in the waste gas stream). This is comparable to the vapor pressure of diesel fuel; we would probably use a very clean diesel fuel as our absorbent. For the rest of this example we will use n-tetradecane, because its properties are well known, and doing so simplifies the calculations. We can also calculate from the Antoine equation that the atmospheric boiling point of n-tetradecane is $\approx$ 490°F, which is the temperature we would expect at the bottom of the stripper.

Next we integrate the left two terms of Eq. (10.17) from the top to the bottom of the column and rearrange, finding

$$\frac{L}{G} = \frac{-\Delta Y_i}{\Delta X_i} = \frac{Y_{i\ \text{bottom}} - Y_{i\ \text{top}}}{X_{i\ \text{bottom}} - X_{i\ \text{top}}} \tag{10.18}$$

In making this integration, observe that in Eq. (10.17) the dX and dY terms apply to flows in the opposite direction, so that in the integration one more minus sign appears. Then we observe from the definitions of Y and X that

$$Y = \frac{y}{1-y} \quad \text{and} \quad X = \frac{x}{1-x} \tag{10.19}$$

so that for small values of y and x

$$Y \approx y \quad \text{and} \quad X \approx x \tag{10.20}$$

To estimate the Henry's law constant we observe that Eq. (10.16) is equivalent to Raoult's law [Eq. (10.1)], with the vapor pressure of toluene taking the place of H. At 100°F we know from Example 10.11 that $p_{\text{toluene}} \approx 0.070$ atm, and we use this as an estimate of H.

$$x_{\text{toluene}} = \frac{P y^*_{\text{toluene}}}{p_{\text{toluene}}} = \frac{1 \text{ atm } y^*_{\text{toluene}}}{0.070 \text{ atm}} = 14.3 y^*_{\text{toluene}} \tag{10.21}$$

On Fig. 10.16 we can draw in the values for the inlet gas, $y_i = 0.5\% = 5000$ ppm, and for the outlet gas, 50 ppm. If we assume that the stripper is 100 percent effective, then the stripped solvent will have zero toluene (and the leftmost curve will go to $y_i^* = 0$ at the top of the column). The maximum conceivable liquid outlet concentration produces a y_i^* equal to the inlet value of 5000 ppm. On Fig. 10.16 that would correspond to the two curves meeting at the bottom of the column, which means that the concentration difference driving the absorption would be zero at the bottom of the column so that in Eq. (10.17), $dh = 1/0 = \infty$. To prevent this, we arbitrarily specify that the outlet liquid shall have $y_i^* = 0.8y_i$. Thus we can calculate that

$$x_{i\text{ bottom}} = 0.8 \cdot 14.3 \cdot 0.005 = 0.057$$

(Here we do not write this as 57 000 ppm because in liquids ppm always means ppm by mass, and this is a mol fraction!) Then we may write

$$\frac{L}{G} = \frac{Y_{i\text{ bottom}} - Y_{i\text{ top}}}{X_{i\text{ bottom}} - X_{i\text{ top}}} \approx \frac{y_{i\text{ bottom}} - y_{i\text{ top}}}{x_{i\text{ bottom}} - x_{i\text{ top}}} = \frac{5000\text{ ppm} - 50\text{ ppm}}{0.057 - 0} = 0.087$$

The molar flow rate of gas is

$$G = 1000\,\frac{\text{scf}}{\text{min}} \cdot \frac{\text{lbmol}}{385.3\text{ scf}} = 2.6\,\frac{\text{lbmol}}{\text{min}} = 1.25\,\frac{\text{lb}}{\text{s}} = 1800\,\frac{\text{mol}}{\text{min}}$$

so the required liquid flow rate is

$$L = 0.087 \cdot 2.6\,\frac{\text{lbmol}}{\text{min}} = 0.23\,\frac{\text{lbmol}}{\text{min}} = 44\,\frac{\text{lb}}{\text{min}} = 20\,\frac{\text{kg}}{\text{min}}$$

We can also use the Henry's law expression to estimate how thoroughly we must strip the solvent before reusing it. At the top of the column we also arbitrarily specify that

$$y^*_{\text{toluene}} = 0.8y_{\text{toluene}} = 0.8 \cdot 50\text{ ppm} = 40\text{ ppm}$$

$$x_{\text{toluene}} = \frac{1\text{ atm} \cdot 40 \cdot 10^{-6}}{0.070\text{ atm}} = 5.71 \cdot 10^{-4} = 0.000571$$

This is a difficult but not impossible stripping requirement. If we substituted this value of $x_{i\text{ top}}$ into the above calculation in the place of the assumed value of zero, it would increase the computed value of L/G by 1%, which we ignore. ■

This part of the design is easy, requiring little empirical data; based on a simple material balance and reasonable property estimates we can learn the required solvent flow rate and the purity required in the stripped solvent. The next part, the selection of the column diameter and height, is much more dependent on empirical data. The subject is treated in detail in [20–22], and only sketched here.

Example 10.15. Estimate the required column diameter in Example 10.14.

The column diameter is determined almost entirely by the gas flow rate. Although the liquid flow rate in (mol/mol) may be close to that of the gas, the liquid flow rate in (volume/volume) is seldom more than 10 percent of the gas flow rate because of the large difference in densities, and it is practically ignored in sizing the column. This difference is almost always true for such devices intended for air pollution control, which mostly operate at atmospheric pressure. For high-pressure applications it is less often true. In all such devices (as in the counterflow particulate scrubber described in Sec. 9.2.4.2), the gas velocity is chosen low enough that it is less than the terminal settling velocity of the smallest liquid droplets that are likely to be present. In addition, if the gas velocity is too high the gas flow will trap liquid in the column, leading to flooding. Typically, such columns will operate at a gas velocity that is approximately 75 percent of the flooding velocity. For most packed absorption columns, the flooding velocity is predicted from a semitheoretical, graphical correlation [20–22] that can be satisfactorily approximated by

$$\log \alpha = -1.6798 - 1.0662 \log \beta - 0.27098(\log \beta)^2$$

where

$$\alpha = \frac{G'^2 F \Psi \mu^{0.2}}{\rho_L \rho_G g_c} \quad \text{and} \quad \beta = \frac{L'}{G'}\left(\frac{\rho_G}{\rho_L}\right)^{1/2} \tag{10.22}$$

Here β is dimensionless, with G' = the gas *mass velocity* = mass flow rate of gas per unit area = GM/A, and L' = the liquid *mass velocity* = LM/A. Since α is not dimensionless, the quantities in it must be expressed in the following units (or suitable conversions must be made): G' is in lb/(ft^2 · s), F is a dimensionless *packing factor* whose values are presented in tables for various packings, Ψ is the specific gravity of the liquid, μ is the liquid viscosity in centipoise, the liquid and gas densities are in lb/ft^3, and $g_c = 32.2$.

We know that $L/G = 0.087$, from which it follows that

$$\frac{L'}{G'} = \frac{LM_L/A}{GM_G/A} = \frac{L}{G} \cdot \frac{M_L}{M_G} = 0.087\frac{198}{92} = 0.187$$

The column will operate at 1 atm and 100°F, at which temperature the density of the air stream will be about 0.071 lbm/ft^3 and that of n-tetradecane about 47 lbm/ft^3, so that

$$\beta = \frac{L'}{G'}\left(\frac{\rho_G}{\rho_L}\right)^{1/2} = 0.187\left(\frac{0.071}{47}\right)^{1/2} = 0.0073$$

and $\log \beta = \log 0.0073 = -2.14$ for the flooded condition. Substituting this value in Eq. (10.22), we find $\alpha = 0.23$.

We estimate the viscosity of n-tetradecane at 100°F [24] is 1.6 cp, and its specific gravity is 0.75. From [21] we find that $F \approx 50$ for typical packed columns, and using the units with the dimensions specified for Eq. (10.22), we compute

$$G'_{\text{at flooding}} = \sqrt{\frac{\alpha \rho_L \rho_G g_c}{F \Psi \mu^{0.2}}} = \sqrt{\frac{0.23 \cdot 47 \cdot 0.071 \cdot 32.2}{50 \cdot 0.75 \cdot 1.6^{0.2}}} = 0.77 \ \frac{\text{lb}}{\text{ft}^2 \cdot \text{s}}$$

and at 75 percent of flooding $G' = 0.58$ lb/(ft^2·s). From Example 10.14 we know that the gas flow rate is $\dot{m}_{\text{gas}} = 1.25$ lb/s. Thus

$$A = \frac{\dot{m}_{\text{gas}}}{G'} = \frac{1.25 \text{ lb/s}}{0.58 \text{ lb/ft}^2 \cdot \text{s}} = 2.18 \text{ ft}^2 = 0.203 \text{ m}^2$$

and

$$D = \sqrt{\frac{4}{\pi} A} = \sqrt{\frac{4}{\pi} 2.18 \text{ ft}^2} = 1.66 \text{ ft} = 0.51 \text{ m}$$

$$V_{\text{gas}} = \frac{\dot{m}_{\text{gas}}}{\rho_G A} = \frac{1.25 \text{ lb/s}}{(0.071 \text{ lb/ft}^3) \cdot (2.18 \text{ ft}^2)} = 8.2 \frac{\text{ft}}{\text{s}} = 2.5 \frac{\text{m}}{\text{s}}$$ ■

This calculation was based on fluid mechanics, and the result is independent of the column height. The column height required to perform the separation depends on mass transfer.

Example 10.16. Estimate the required column height for the gas absorption in Examples 10.14 and 10.15.

We return to Eq. (10.17) and rearrange the rightmost two terms to find that

$$dh = -\frac{G}{(KaP)(y_i - y_i^*)A} dY_i \tag{10.23}$$

Ka, which was not described when we introduced it with Eq. (10.17), is the product of the mass transfer coefficient and the interfacial area between liquid and gas per unit volume of the absorber. The two parts—K and a—are normally presented and correlated as a product because it is much easier to measure their product than to measure them individually. This product depends on the chemical and physical properties of the liquid and the gas and on the geometry of the column internals, whose function is to provide as large a value of Ka as possible with a minimum pressure drop and to maintain true countercurrent flow between the two phases. Kohl and Nielsen [21] present a table of values of Ka observed in industrial practice for a variety of absorption systems. These vary from 0.007 to 20 lbmol/h · ft^3 · atm.

For this example we will assume that Ka is a constant = 4.0 lbmol/h · ft^3 · atm. We have specified that $y_i^* = 0.8y_i$ at both the top and the bottom of the column. To find the values between those points we must calculate x_i by the material balance, Eq. (10.17), and then compute y_i^* from that value by Eq. (10.21). We may rearrange Eq. (10.17) so that

$$Y_{i \text{ top}} = Y_{i \text{ bottom}} - \frac{L}{G}(X_{i \text{ bottom}} - X_{i \text{ top}}) \tag{10.24}$$

Then we observe that this equation was found by integrating from bottom to top of the column. It must be correct for the integration from the bottom to any intermediate point in the column as well, so that we can drop the "top" subscripts and use the equation for any point in the column. Then we apply Eq. (10.20) twice, replacing X

and Y by x and y, solve Eq. (10.21) for y_i^*, and substitute into the right side of Eq. (10.23), finding

$$N = h\frac{KaPA}{G} = -\int_{\text{bottom}}^{\text{top}} \frac{dy_i}{(y_i - y_i^*)}$$

$$= -\int_{\text{bottom}}^{\text{top}} \left(\frac{dy_i}{[y_i(1 - HG/PL) - (H/P)x_B + (HG/PL)y_B]} \right) \tag{10.25}$$

$$= \frac{-1}{(1 - HG/PL)} \ln \left[\frac{y_T(1 - HG/PL) - (H/P)x_B + (HG/PL)y_B}{y_B(1 - HG/PL) - (H/P)x_B + (HG/PL)y_B} \right]$$

where N is the *number of transfer units,* a measure of the difficulty of the separation, quite analogous to the same term used in heat transfer; x_B and y_B are the liquid and vapor concentrations at the bottom of the column; and y_T is the vapor concentration at the top of the column. (The required column height is linearly proportional to N.) For this example

$$\frac{H}{P} = \frac{0.070 \text{ atm}}{1 \text{ atm}} = 0.070; \qquad \frac{HG}{PL} = \frac{0.070}{0.087} = 0.805;$$

$$\left(1 - \frac{HG}{PL}\right) = (1 - 0.805) = 0.195$$

$$N = \frac{-1}{0.195} \ln \left[\frac{50 \cdot 10^{-6} \cdot 0.195 - 0.070 \cdot 0.057 + 0.805 \cdot 5000 \cdot 10^{-6}}{5000 \cdot 10^{-6} \cdot 0.195 - 0.070 \cdot 0.057 + 0.805 \cdot 5000 \cdot 10^{-6}} \right] = 16.6$$

and

$$h = \frac{NG}{KaPA} = \frac{16.6 \cdot 0.23 \text{ lbmol/min}}{(4 \text{ lbmol/h} \cdot \text{ft}^3 \cdot \text{atm}) \cdot 2.18 \text{ ft}^2} \times \frac{60 \text{ min}}{\text{h}} = 26 \text{ ft}$$ ■

At the end of this long example, we observe

1. The Henry's law assumption greatly simplifies the calculation. If we must take more complex vapor-liquid equilibria into account, then the integration in Eq. (10.25) will generally have to be numerical. The Henry's law assumption is normally good for dilute gas streams, but not for concentrated ones.
2. Texts on absorption and on mass transfer [20–22] show similar examples without the many simplifications that were used here.
3. Absorption (also called scrubbing) is used not only for VOC recovery but also for capture of various other gases, discussed in the following chapter. Absorption-stripping, as shown in Fig. 10.15, is a general-purpose chemical engineering operation, widely used in industry.
4. We have not discussed the design of the stripper. That is covered in mass-transfer texts.

10.5 CONTROL BY OXIDATION

The final fate of VOCs is mostly to be oxidized to CO_2 and H_2O, as a fuel either in our engines or furnaces, in an incinerator, in a biological treatment device, or in the atmosphere (forming ozone and fine particles). VOC-containing gas streams that are too concentrated to be discharged to the atmosphere but not large enough to be concentrated and recovered are oxidized before discharge, either at high temperatures in an *incinerator* or at low temperatures by biological oxidation.

10.5.1 Combustion (Incineration)

Some air pollutants consist mostly of materials that, when burned, produce other materials that are harmless or much less harmful than the original ones. (Please review Sec. 7.10, on combustion fundamentals.) Materials in this category are largely compounds of carbon, hydrogen, oxygen, nitrogen, and sulfur. Most VOCs fall into this category, as do some others. Thus the treatment here mainly considers VOCs, but it also applies to other materials. Converting these materials from the harmful or objectionable form to a harmless or less objectionable form by combustion may be the most economical and practical way of solving a VOC emission problem.

Some examples of interest are

$$CO + \tfrac{1}{2}\,O_2 \rightarrow CO_2 \tag{10.26}$$

$$C_6H_6 + 7\tfrac{1}{2}\,O_2 \rightarrow 6\,CO_2 + 3\,H_2O \tag{10.27}$$

$$H_2S + \tfrac{3}{2}\,O_2 \rightarrow H_2O + SO_2 \tag{10.28}$$

In Eqs. (10.26) and (10.27), carbon monoxide, which has well-known harmful human health effects (see Sec. 15.1), and benzene, which is a reactive hydrocarbon, smog precursor, and carcinogen, are converted to harmless materials. Equation (10.28) may appear not to belong here because SO_2 is a significant air pollutant for which we have a national control program (see Chapter 11). However, H_2S (hydrogen sulfide) is very toxic at high concentrations and has a strong smell (rotten eggs) that most of us can detect at much lower concentrations than we can detect SO_2 (the smell of burning sulfur or of a wooden match being lighted). Typical estimates of the minimum concentration that average humans can smell are 0.0005 ppm for H_2S and 0.5 ppm for SO_2. Thus, if the problem is caused by the odor of H_2S, it can frequently be alleviated by burning the H_2S to form the less odorous SO_2. This makes sense only for low H_2S concentrations; higher concentrations are treated in an entirely different way (see Chapter 11).

(As an interesting sidelight on the last example, it is common practice to add a strong-smelling sulfur compound—called an odorant—to odorless natural gas or propane to help detect leaks. This is an absolutely necessary safety feature; without the smell for a warning, a stove burner accidentally left on but not ignited could lead to a disastrous explosion! Normally this odorant is a mercaptan, which is a near chemical relative of H_2S. In normal combustion this odorant is converted to water, carbon dioxide, and sulfur dioxide. These odorants are effective at such low concentrations that the resulting SO_2 concentration in the combustion gases is below our ability to detect, and we find the combustion products odorless. The concentration is also low enough that the SO_2 produced probably has no health effects.)

The nitrogen present in compounds being incinerated normally enters the atmosphere partly as N_2, NO, or NO_2. The latter two are pollutants for which we have control programs (Chapter 12). Thus, we would not use incineration to limit the emissions of these materials. However, many organic chemicals contain small amounts of organic nitrogen that will pass through the combustion process and emerge as NO or NO_2, e.g.,

$$2\,(CH_3)_3N + 11\tfrac{1}{2}\,O_2 \rightarrow 6\,CO_2 + 9\,H_2O + 2\,NO \tag{10.29}$$

In this case trimethylamine, one of the smelliest compounds known (decaying fish), is oxidized to harmless CO_2 and H_2O plus some NO that will contribute, after further atmospheric oxidation, to any regional NO_2 and ozone problems.

In most incinerators, the chlorine content of the material burned will leave the incinerator as hydrochloric acid, HCl. Municipal waste incinerators generally receive enough polyvinyl chloride plastic that HCl in the exhaust gas can cause corrosion in the incinerator and damage to the neighborhood. Most modern municipal waste incinerators have some kind of acid capture technology to prevent the emission of this HCl.

As discussed previously, at incinerator temperatures some metals become vapors, e.g., mercury, cadmium, zinc. Municipal waste contains some of these materials (mercury in dry cells, cadmium and zinc in metal plating); their emissions can be a problem.

The combustible pollutant can be a gas, a mist droplet, or a solid particulate. The gases most often treated by burning are CO, hydrocarbons of all kinds, and strong odor producers, which are normally VOCs containing sulfur and/or nitrogen. Almost all VOCs can be destroyed by incineration. The particulates treated by combustion are largely hydrocarbon smokes; examples are the smoke from meat smokehouses, the fumes formed in asphalt processing and paint baking, miscellaneous tars, etc. The principles of dealing with these are the same, although the details may differ.

In all combustion and incineration, incomplete combustion is a permanent problem. Many of the intermediate products produced between the original components and the final carbon dioxide and water are themselves harmful, e.g., aldehydes, dioxins, furans. Incomplete combustion of a waste stream can produce an exhaust gas that is more harmful than the input gas. All air pollution incinerators (and municipal and hazardous waste incinerators) are designed to ensure that combustion is as complete as practical, and that the emissions of products of incomplete combustion are as small as possible.

10.5.1.1 Combustion kinetics of the burning of gases.

Most combustion takes place in the gas phase. Liquids and solids mostly vaporize before they burn. For chemical reactions of any kind in any phase (gas, liquid, or solid) the reaction rates are typically expressed by equations of the form

$$\begin{pmatrix}\text{Decrease in concentration}\\ \text{of } A \text{ per unit time}\end{pmatrix} = \frac{-dc_A}{dt} = r = kc_A^n \tag{10.30}$$

where r = reaction rate

k = a kinetic rate constant whose value is strongly dependent on the temperature but is independent of the concentration of the reactants

c_A = concentration of A

n = reaction order

For combustion reactions we usually apply this equation to reactions in the gas phase.

For most chemical reactions the relation between the kinetic rate constant k and the temperature T is given to a satisfactory approximation by the Arrhenius equation

$$k = A \exp\left(-\frac{E}{RT}\right) \tag{10.31}$$

where A and E = experimental constants. (E is normally called the *activation energy,* and is related to the bond energies in the molecules. A is called the *frequency factor* and is related to the frequency of collisions of the reacting molecules.)

R = universal gas constant

T = absolute temperature

In the absence of data to the contrary, one should assume that the kinetic rate constant for an unknown reaction is represented by this equation. Table 10.4 on page 374 lists values for A and E to be used in Eq. (10.31), and the computed values of k for three temperatures, based on the strong simplifying assumption that $n = 1$ (first-order reaction) for the combustion of a variety of compounds [25].

An additional strong simplifying assumption in Table 10.4 is that the concentration of VOC to be burned is much less than the concentration of oxygen in the contaminated air stream. The true kinetic expression is presumably

$$\begin{pmatrix}\text{Decrease in concentration}\\ \text{of VOC per unit time}\end{pmatrix} = \frac{-dc_{\text{VOC}}}{dt} = r = kc_{\text{VOC}}c_{\text{O}_2} \tag{10.32}$$

However, in most cases the oxygen concentration starts out close to 21 volume percent and does not change much during the reaction because the VOC concentration is generally small. So the k in Eq. (10.30) is equivalent to $k \cdot c_{\text{O}_2}$ in Eq. (10.32).

Example 10.17. Show the calculation leading to the value of k in Table 10.4 for benzene at 1000°F.

$$\begin{aligned} k &= A \exp\left(-\frac{E}{RT}\right) \\ &= \frac{7.43 \times 10^{21}}{\text{s}} \exp\left(-\frac{95\,900 \text{ cal/mol}}{[1.987 \text{ (cal/mol)/K}](1000^\circ + 460^\circ\text{R})(\text{K}/1.8^\circ\text{R})}\right) \\ &= 0.00011/\text{s} \end{aligned}$$

■

TABLE 10.4
Thermal oxidation parameters, based on first-order kinetics

Compound	A, 1/s	E, kcal/mol	k, 1/s; at 1000°F	1200°F	1400°F
Acrolein	3.30E + 10	35.9	6.99258	102.37	841.47
Acrylonitrile	2.13E + 12	52.1	0.01946	0.96	20.34
Allyl alcohol	1.75E + 06	21.4	2.99528	14.83	52.07
Allyl chloride	3.89E + 07	29.1	0.56034	4.93	27.21
Benzene	7.43E + 21	95.9	0.00011	0.14	38.59
1-Butene	3.74E + 14	58.2	0.07760	6.02	183.05
Chlorobenzene	1.34E + 17	76.6	0.00031	0.09	8.41
Cyclohexane	5.13E + 12	47.6	0.76467	26.84	438.42
1,2-Dichloroethane	4.82E + 11	45.6	0.24851	7.51	109.11
Ethane	5.65E + 14	63.6	0.00411	0.48	19.93
Ethanol	5.37E + 11	48.1	0.05869	2.14	35.97
Ethyl acrylate	2.19E + 12	46.0	0.88094	27.44	407.99
Ethylene	1.37E + 12	50.8	0.02804	1.25	24.64
Ethyl formate	4.39E + 11	44.7	0.39562	11.18	154.04
Ethyl mercaptan	5.20E + 05	14.7	56.86353	170.64	404.29
Hexane	6.02E + 08	34.2	0.36628	4.72	35.13
Methane	1.68E + 11	52.1	0.00153	0.08	1.60
Methyl chloride	7.43E + 08	40.9	0.00708	0.15	1.66
Methyl ethyl ketone	1.45E + 14	58.4	0.02658	2.09	64.38
Natural gas	1.65E + 12	49.3	0.08565	3.41	61.61
Propane	5.25E + 19	85.2	0.00058	0.34	49.99
Propylene	4.63E + 08	34.2	0.28171	3.63	27.02
Toluene	2.28E + 13	56.5	0.01358	0.93	25.54
Triethylamine	8.10E + 11	43.2	1.85139	46.78	590.11
Vinyl acetate	2.54E + 09	35.9	0.53822	7.88	64.77
Vinyl chloride	3.57E + 14	63.3	0.00313	0.36	14.58

Source: Ref. 25.

Example 10.18. Estimate the time required to destroy 99.9 percent of the benzene in a waste gas stream at 1000°, 1200°, and 1400°F.

For $n = 1$, a first-order reaction, the calculation is simple. We can integrate Eq. (10.30) from $t = 0$ to $t = t$, finding

$$\frac{c}{c_0} = \exp[-k(t - t_0)] \tag{10.33}$$

At 1000°F, we calculate

$$t = \frac{1}{k}\ln\frac{c_0}{c} = \frac{1}{0.00011/\mathrm{s}}\ln\frac{1}{0.001} = 62\ 800\ \mathrm{s} = 17.4\ \mathrm{h}$$

Repeating the calculation at 1200° and 1400°F, we find 49 s and 0.2 s. ■

This example shows that incinerating benzene is impractical at 1000°F, but quite practical at 1400°F. From Table 10.4 we see that benzene is one of the more difficult materials to burn; it has one of the lowest values of k. We also see that it has the highest value of E, thus showing the highest rate of increase of k with an increase in T. (Conversely, ethyl mercaptan has the lowest E and hence the slowest increase of k with an increase in T.) Example 10.18 also shows that for first-order kinetics, the time needed for a given percent destruction of contaminants is independent of the starting concentration. For all other reaction orders (for $n \neq 1$) the time required for a given percent destruction depends on the starting concentration (Problem 10.42). This example also illustrates how one measures k. If c_0, c_i, and t are measured, one can compute k for that experiment. The values in Table 10.4 were found that way.

This treatment vastly simplifies the real problem, because the VOCs being oxidized in incinerators are normally mixtures. For example, if a mixture of 0.5 percent benzene and 0.5 percent hexane were treated in an incinerator at 1000°F, the percent destruction of hexane might be as predicted with Eq. (10.33) using the values in Table 10.4, whereas that of benzene would be much larger than the value calculated the same way. The reason is that the free radicals generated in the burning of hexane, which is more easily attacked, will encounter benzene molecules and attack them. Thus benzene, which is one of the more difficult materials to incinerate by itself because its structure makes it hard for it to form a free radical, is more easily incinerated in the presence of other materials that form free radicals more readily.

Barnes et al. present a much more complete and complex account of the observed kinetics of incineration of VOCs [26]. They also suggest the following as typical values of the operating conditions of industrial gas incinerators:

Gas velocity: 25–50 ft/s	
Residence time: 0.2–1 s	
Temperatures:	
Odor control	900–1350°F
Oxidize hydrocarbons	900–1200°F
Oxidize CO	1200–1450°F

The typical way of carrying out the combustion of VOCs is shown in Fig. 10.17*a* on page 376. The contaminated gas stream is mixed with fuel. If the contaminated gas stream does not contain enough oxygen to burn the fuel, additional air is also mixed in. The burning of the fuel takes place in a combustion chamber, where the high velocities of the inlet streams provide good turbulent mixing. From the combustion chamber the gases pass to an insulated retention chamber, where they remain long enough at high temperature for the reactions to complete the destruction of the VOCs begun in the combustion chamber. Then the hot gases pass to the stack.

The biggest drawback with the arrangement in Fig. 10.17*a* is the high cost of the fuel. If the contaminated gas stream contained enough VOC to burn (i.e., the VOC concentration was above its LEL) then no additional fuel would be needed, and the waste stream would be a valuable fuel source instead of an air pollution control problem. For air pollution control we almost always have to add fuel. One way to lower the fuel cost is to put a heat exchanger into the system, as shown in Fig. 10.17*b*.

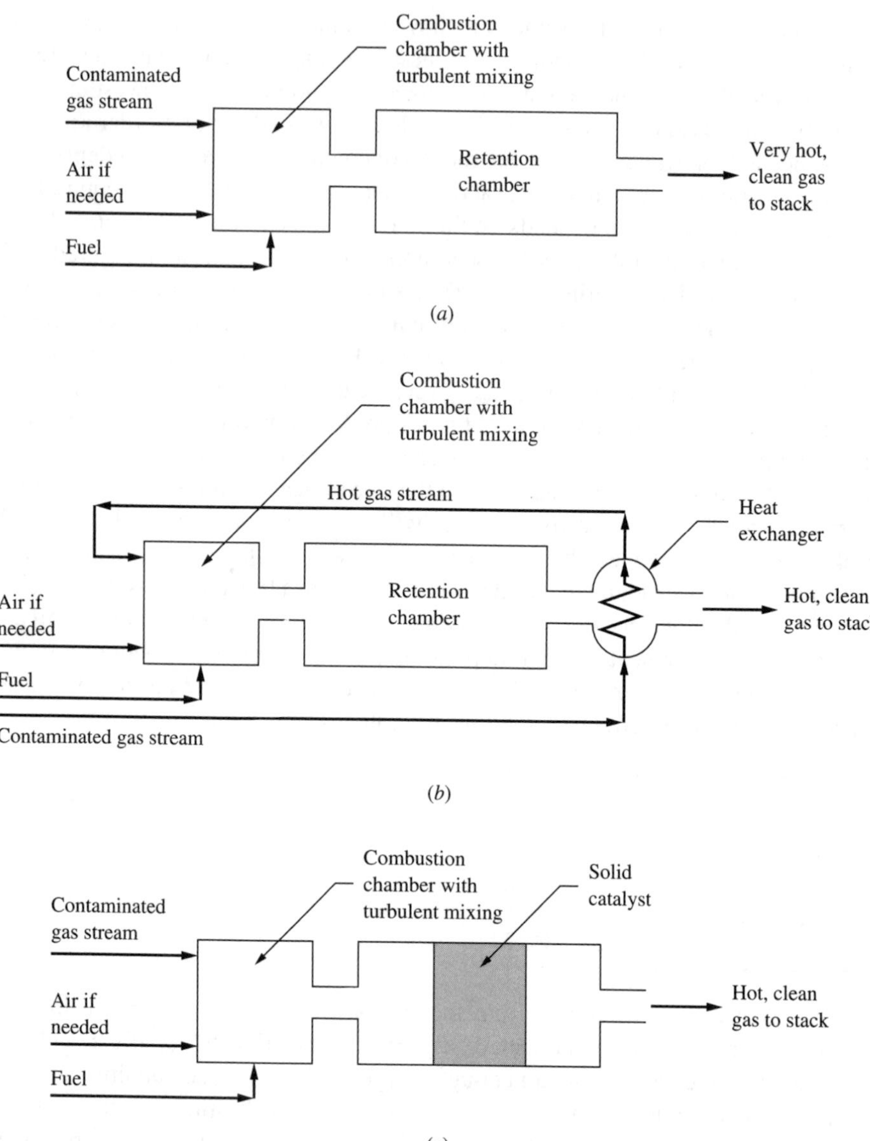

FIGURE 10.17
Arrangements for destroying a VOC in a gas stream by incineration: (*a*) simple thermal incinerator, (*b*) thermal incinerator with regenerative heat recovery, and (*c*) catalytic incinerator.

There the hot, clean gas leaving the retention chamber transfers heat to the incoming contaminated gas stream, thus reducing the outlet temperature and reducing the amount of fuel needed to bring the mixture of fuel + air + contaminated gas stream up to the temperature at which the VOC-destruction reactions proceed. Unfortunately, the hot-gas-to-cold-gas heat exchangers shown in Fig. 10.17*b* are expensive and often have severe corrosion problems. They are used in some situations, but not all.

The second modification of the basic idea is to put an oxidation catalyst in the retention chamber. (See Sec. 7.13 on catalysts.) Such catalysts can cause VOC destruction reactions to occur at much lower temperatures than they would without a catalyst. In some circumstances they greatly reduce the lower explosive limit, so that a contaminated gas stream that would not burn without additional fuel in a simple combustion chamber will burn without additional fuel over a catalyst. Barnes et al. state that the operating temperature of an afterburner can typically be reduced from about 1000° to 1200°F to about 600°F if a catalyst is used. The catalyst may be expensive, but the fuel savings are significant.

10.5.1.2 Combustion kinetics of the burning of solids. Most fuels will not burn in the solid or liquid state; they must be vaporized before they will mix with air and burn. Propane and butane come from their containers as gases and need no vaporization to burn. Gasoline is a mixture of hydrocarbons that has a much higher vapor pressure than water; these vaporize easily when gasoline is mixed with air in a carburetor or fuel injector. Diesel fuel has a lower vapor pressure than gasoline and must be heated by the hot compressed air in the diesel engine cylinder before it will vaporize, mix, and ignite. Higher-boiling fuel oils are normally preheated before use to raise their vapor pressure enough for them to vaporize, mix, and burn. The only common fuel that will burn as a solid is charcoal; most of us know that it is fairly difficult to ignite and burns slowly. The following treatment of the burning of solids is thus limited to solids like charcoal. However, the mathematics are the same for other solids, with the combustion rate at the surface being replaced by the pyrolysis and evaporation rate at the surface, which, in turn is driven by the heat flow to the surface, normally supplied by the combustion of the vaporized fuel. (Reactions of powdered metals occur in the solid state; e.g., the thermite reaction,

$$2\ \mathrm{Al} + \mathrm{Fe_2O_3} \rightarrow \mathrm{Al_2O_3} + 2\ \mathrm{Fe}\ (\text{liquid})$$

produces temperatures comparable to combustion reactions, but it is not combustion in the ordinary sense of that term.)

If a solid has a flat surface and is burning, then we would assume that the burning rate could be expressed in terms like (mass burned)/[(amount of exposed surface) × (time)]. This assumption would agree with our experience with wood-burning campfires and fireplaces; to get a high heat release rate (high mass burned/time), we increase the exposed wood surface, usually by putting on more wood or many small pieces of wood with a high surface area. Figure 10.18 on page 378 shows the measured burning rates for pure carbon; the rate in $\mathrm{g/cm^2 \cdot s}$ is plotted versus temperature.

Example 10.19. How large a spherical carbon particle can we completely oxidize in an airstream that is held at 1000 K (1340°F) for 3 s?

From Fig. 10.18 we read that the burning rate at this temperature is approximately $r = 0.018 \times 10^{-3}\ \mathrm{g/cm^2 \cdot s}$. If we write a mass balance for a spherical particle

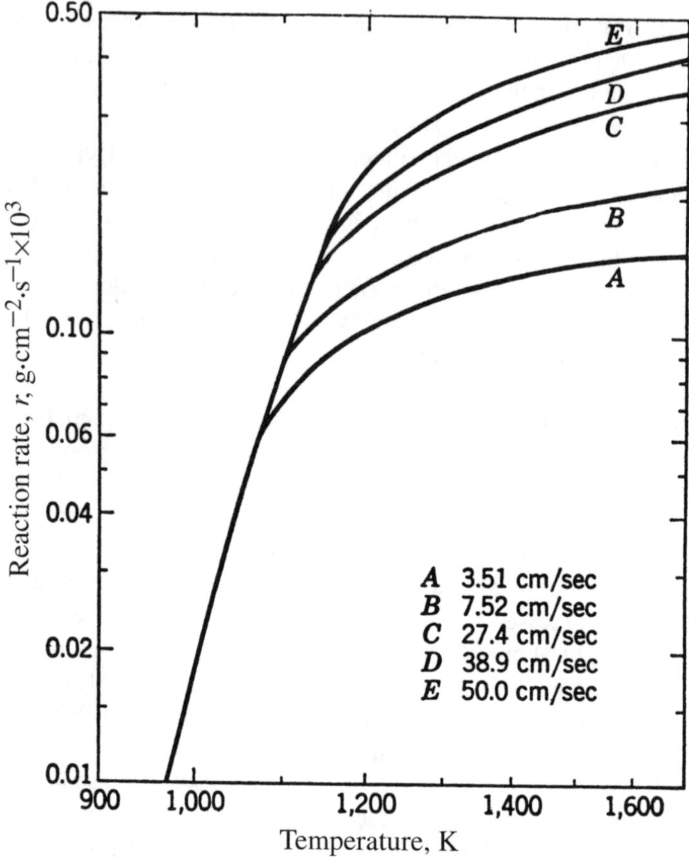

FIGURE 10.18
Effect of air velocity and temperature on combustion rate of carbon. (From Ref. 27.) The individual curves represent different velocities of air blowing across the surface of the carbon. Below about 1100 K, air velocity has no effect; above that temperature it controls the reaction rate. The ultimate data source is listed in Ref. 27 as Tu, Davis, and Hottel, *Ind. Eng. Chem.*, Vol. 26, 749, 1934.

whose surface is burning away, we find

$$-\frac{dm}{dt} = -\rho\frac{\pi}{2}D^2\frac{dD}{dt} = rA = r\pi D^2$$

$$-\frac{dD}{dt} = 2\frac{r}{\rho}$$

$$-\int dD = 2\frac{r}{\rho}\int dt$$

$$-(D - D_0) = 2\frac{r}{\rho}(t - t_0) \tag{10.34}$$

Taking $D = D_0$ at time t_0 and $D = 0$ (i.e., all burned up) at time t, we find that the largest particle that we can completely burn in time t has the following diameter:

$$D_0 = 2\frac{rt}{\rho}$$

Using the value of r given earlier and a density of 2 g/cm^3 for carbon, we find

$$D_0 = 2\frac{(0.018 \times 10^{-3}\ \text{g/cm}^2 \cdot \text{s})(3\text{s})}{2\ \text{g/cm}^3} = 5.4 \times 10^{-5}\ \text{cm} = 0.54\ \mu$$ ■

This example illustrates several things about combustion reactions. First, for combustion of a solid, the mathematical product of the reaction rate and the time of exposure determines how large a particle can be burned. Second, on Fig. 10.18, the coordinates are logarithmic on the ordinate and some other value on the abscissa. One may verify that the abscissa is (1/absolute temperature) with the origin taken at the right instead of the left (i.e., the values of $1/T$ increase from right to left). A straight line on such a plot would obey the Arrhenius equation, Eq. (10.31). On Fig. 10.18 the data collected at the lowest temperatures form a straight line on these coordinates. However, at higher temperatures they do not, as explained below.

Pure carbon burns more slowly than most other materials; we should be able to burn up much larger particles of other materials at the same temperature. Most other combustible materials (e.g., wood, tars, coal, etc.) will decompose (pyrolyze) on heating, giving off combustible gases. Students have certainly observed the flames that stand away from burning wood; they are burning the gaseous decomposition products of the wood, which gives them off as it is heated by the flames. For wood, the sequence is pyrolyze-vaporize-mix with air and oxidize to give the final products, CO_2 and H_2O. Because of its very low vapor pressure, pure carbon does not give off such gaseous decomposition products. The combustion of decomposable (technically, pyrolyzable) solid materials, such as wood, coal, and plastics, is more complex than that of carbon; Example 10.19 is about as simple a solid combustion situation as one can imagine.

10.5.1.3 Mixing in combustion reactions. See Sec. 7.10.6 for an introduction to this topic, mostly applied to gaseous combustion. Figure 10.18 makes clear that it is important for combustion of solids as well. In that figure at low temperatures the combustion rate is independent of the rate of air movement across the surface of the burning carbon; but at higher temperatures the combustion rate depends on the air flow rate. At low temperatures molecular diffusion moves the air in to the solid carbon surface and the carbon dioxide out faster than the chemical reaction can transform them, so the chemical reaction determines the overall reaction rate. (It is always the *slowest* step that determines overall rate; any experienced freeway driver knows that.) At higher temperatures the chemical reaction is so fast that it uses up the oxygen as fast as diffusion can bring it in, and the overall rate is determined not by the chemical reaction at the surface, but by how fast the oxygen can get to the burning surface (or how fast the carbon dioxide produced can get away from the surface, to make room for more oxygen to get in).

The same idea applies to combustion for VOC control. To get complete destruction of the VOC, there must be excellent mixing of the VOC to be oxidized, the oxygen to perform the oxidation, and the hot gases from burning the fuel to raise the temperature high enough for the oxidation to occur.

10.5.1.4 Application to boilers, furnaces, flares, etc. One of the first major undertakings in the history of air pollution control was the control of emissions from coal-burning boilers, furnaces, etc. Unburned coal or products of incomplete combustion formed a substantial part of these emissions. These were one of the easiest pollutants to control; all that is required for good control is sufficient excess air (see Fig. 7.5) and adequate mixing between the burning coal, its decomposition products, and the air.

As Fig. 7.5 shows, to get complete combustion with imperfect mixing, one must supply *excess air* in addition to that needed for stoichiometric combustion (see Example 7.9). The amount of excess air to be used is determined by economics. At zero excess air, some valuable fuel escapes unburned to pollute the atmosphere. Large amounts of excess air lower the combustion temperature by diluting the combustion products, and carry away more heat in the exhaust gas. This lowers the furnace's efficiency (fraction of heating value of the fuel transferred to whatever is being heated). Large industrial furnaces operate with 5 to 30 percent excess air. Autos (Chapter 13) have variable excess air, depending on engine load. The optimum amount of excess air for VOC destruction is generally higher than the optimum for fuel efficiency; air pollution control officers try to induce furnace operators to use the optimum amount for VOC destruction.

The mixing problem is especially difficult in *flares*. These are safety devices used in oil refineries and many other processing plants. All vessels containing fluids under pressure have high-pressure relief valves that open if the internal pressure of the vessel exceeds its safe operating value. All household water heaters have such a valve to prevent tank rupture in some unlikely but not impossible circumstances (see Problem 10.54). With a hot water heater, if the valve opens, hot water drops onto the floor. In the case of a large petroleum-processing vessel (distillation column, cracker, isomerizer, etc.) the material released is an inflammable VOC, which cannot safely be dropped on the floor. The outlets of a refinery's relief valves are piped to a flare (or "flare stack"), which is an elevated pipe with pilot lights to ignite any released VOCs. Many have steam jets running constantly to mix air into the gas being released. These flares handle significant amounts of VOCs only during process upsets and emergencies at the facilities they serve. When there is a small release, the steam jets can often mix the gas and air well enough that there is practically complete combustion. For a large release the mixing is inadequate, and the large, bright orange, smoky flame from the flare indicates a significant release of unburned or partly burned VOC.

In the coal combustion process one difficulty, even in well-designed modern furnaces, is that some particles of coal and some hydrocarbons pass out of the flame zone before they can be combusted. These are called *soot* (see Fig. 8.3). In modern steam boilers this soot will collect in parts of the furnace where it is too cold for soot to burn, typically on the tubes in which the water is boiled or the steam superheated. If soot is allowed to collect there, it will impede heat transfer and make the boiler less efficient. The cure for this problem is a *soot blower,* which is typically a fixed or moving steam jet that blows high-pressure steam onto the surface of the tubes to remove this soot. Normally, soot blowing is required only a few minutes per day. Soot dislodged in this way exits the furnace as short-period emissions of black smoke. Most public relations officers ask plant engineers to do all soot blowing at night.

Combustion is discussed further in Chapters 12 and 13.

10.5.2 Biological Oxidation (Biofiltration)

As discussed above, the ultimate fate of VOCs is to be oxidized to CO_2 and H_2O, either in our engines or furnaces, or incinerators, or in the environment. Many microorganisms will carry out these reactions fairly quickly at room temperature. They form the basis of most sewage treatment plants (oxidizing more complex organic materials than the simple VOCs of air pollution interest). Microorganisms can also oxidize the VOCs contained in gas or air streams. The typical *biofilter* (not truly a filter but commonly called one; better called a *highly porous biochemical* reactor) consists of the equivalent of a swimming pool, with a set of gas distributor pipes at the bottom, covered with several feet of soil or compost or loam in which the microorganisms live. The contaminated gas enters through the distributor pipes and flows slowly up through soil, allowing time for the VOC to dissolve in the water contained in the soil, and then to be oxidized by the microorganisms that live there.

Typically these devices have soil depths of 3 to 4 ft, void volumes of 50%, upward gas velocities of 0.005 to 0.5 ft/s, and gas residence times of 15 to 60 s. They work much better with polar VOCs, which are fairly soluble in water (see Sec. 10.2) than with HCs whose solubility is much less. The microorganisms must be kept moist, protected from conditions that could injure them, and in some cases given nutrients. Because of the long time the gases must spend in them, these devices are much larger and take up much more ground surface than any of the other devices discussed in this chapter. In spite of these drawbacks, there are some applications for which they are economical, and for which they are used industrially [28].

10.6 THE MOBILE SOURCE PROBLEM

Table 10.1 shows that motor vehicles are the largest source of VOC emissions in the United States, with 40% of the total. This includes all kinds of motor vehicles, autos, busses, aircraft, and boats. Although autos have a higher control efficiency than most of the others, because of the large number of autos they are still the largest source of VOC. Chapter 13 discusses the auto problem in greater detail. In general, the approaches taken to date have been control of leaks, adsorption followed by recycling for some sources, and improved combustion, both in the engine and in a catalyst in the exhaust system, to minimize the emission of VOCs. The principles are the same as those discussed here, but the application is complex and difficult.

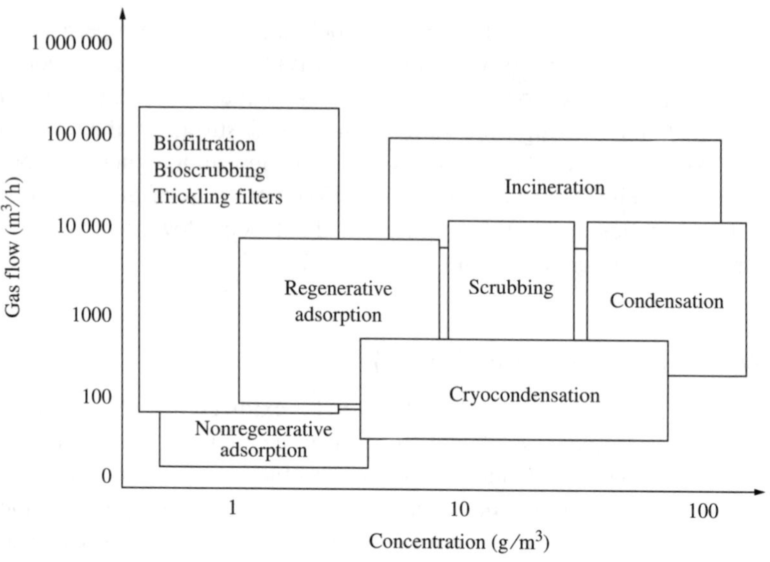

FIGURE 10.19
A guide to choosing VOC control technologies, based on flow rate and concentration only [29]. (With permission of KPMG Management Consultants, Ottawa, Ontario.) Cryocondensation is discussed in Problem 10.21. Nonregenerative adsorbers are placed in a landfill or incinerated, instead of being regenerated as shown in Fig. 10.10.

10.7 CHOOSING A CONTROL TECHNOLOGY

In choosing a VOC control technology, one must consider the permitted emission regulation, the flow rate of the gas stream, the concentration of the VOC in it, and the special properties of the contained VOC. If the VOC is water-soluble the options are different from when it is not. If the VOC is biologically fairly inactive (e.g., most straight-chain HCs) it will be treated very differently from a very biologically active VOC like dioxin. Figure 10.19 shows one author's guide to choosing technology, based only on flow rate and concentration.

10.8 SUMMARY

1. VOCs are emitted from a wide variety of sources and have a wide variety of individual components, each with its own properties. We use VOCs mostly as petroleum-based fuels and solvents. The majority of our VOC emissions come from fuel and solvent usage, transportation, and storage.
2. The control alternatives are prevention, concentration and recovery, or oxidation.
3. Some of these control options can also be used for non-VOC emissions, e.g., incineration for odor control of H_2S, adsorption for SO_2 or mercury vapor, and leakage control for any process source.

PROBLEMS

See Common Units and Values for Problems and Examples, inside the back cover.

10.1. Figure 10.1 shows only one line for every substance except CO_2, for which two lines are shown.

(*a*) Explain why.

(*b*) How low in pressure would the plot have to go to show a similar situation for water?

10.2. (*a*) Repeat Example 10.3 using the Antoine equation (Appendix A). Observe the wording that precedes Eq. (10.3), which says, "The experimental data can *often* be represented with better accuracy by the Antoine equation."

(*b*) What value of C makes the Antoine equation the same as the Clausius-Clapyron equation? How does that compare to the values of C in Appendix A?

(*c*) The Clausius-Clapyron equation is a straight line on a plot of $\ln p$ vs. $1/T$. What does the Antoine equation look like on those coordinates?

10.3. Figure 10.1 is a *Cox chart*. On it the curve for a "reference substance," water in this case, is drawn as a perfectly straight line with a logarithmic abscissa. Then the temperature scale on the ordinate is made up to match the tabulated values of the vapor pressure of the reference substance. Next the vapor pressure curves of other substances are drawn in on this plot, whose abscissa is the log of the vapor pressure and whose ordinate is a special, nonlinear scale based on the vapor pressure line of the reference substance.

The temperature scale is approximately $[1/(T - 43.15)]$, plotted from right to left. This is equivalent to using the Antoine equation with $C = 230°C$. Show that this is so by observing that the water curve on Fig. 10.1 is perfectly straight, and then that a plot of the p values for steam vs. $[1/(T - 43.15)]$ is also perfectly straight.

10.4. Some safety officials are concerned with the problem of mercury (M = 200.6 g/mol), trapped in cracks in laboratory floors, causing hazardous exposure of workers to mercury vapor. Is that plausible?

(*a*) If the mercury content in the room is exactly at the TLV of 25 μg/m^3 (TLV = Threshold Limit Value = permissible concentration for exposure of workers), how much mercury would a typical worker breathe in during an eight-hour shift? The average adult inhales about 15 kg/day of air.

(*b*) What mol fraction of mercury in air corresponds to the stated TLV?

(*c*) What mol fraction of mercury in air corresponds to equilibrium with liquid mercury? At 20°C the vapor pressure of liquid mercury is 0.0012 torr.

(*d*) How large an exposed area of liquid mercury would be needed to hold the mercury content of the room at the TLV? Here assume that the room is $5 \times 5 \times 3$ meters, that the ventilation rate is 2 air changes per hour, and that the average mass flux from liquid mercury in a floor crack to the room air is 6×10^{-10} mol/cm$^2 \cdot$ s.

10.5. Evaporation rates of hydrocarbon liquids from horizontal surfaces are **approximately** given by

$$\text{Evaporation rate} = \left(0.5 \frac{\text{mol}}{\text{m}^2 \cdot \text{s}}\right) \frac{p}{P}$$

where p is the vapor pressure of the liquid and P is the atmospheric pressure.

(*a*) We have spilled a layer of lubricating oil on a metal drip pan in our garage. It forms a layer 1 mm thick. Its density is 1 g/cm^3 and its molecular weight is 400 g/mol. Its vapor pressure is approximately 10^{-7} torr at 20°C. Estimate how long it will take all of this motor oil to evaporate at 20°C.

(*b*) Repeat the calculation for a 1-mm layer of gasoline. Use the vapor pressure, density, and molecular weight from Example 10.7.

(*c*) Repeat part (*b*) with the following changes:

The vapor pressure of gasoline is

$$p = 6 \text{ psia} - (3 \text{ psia}) \cdot (\text{weight fraction evaporated})$$

The molecular weight of gasoline is

$$M = 60 \text{ g/mol} + (60 \text{ g/mol}) \cdot (\text{weight fraction evaporated})$$

The density of liquid gasoline can be considered constant at 0.75 g/cm^3. (Please read Problem 10.6.)

10.6. The evaporation rate shown in Problem 10.5 is the average value for many circumstances.

(*a*) Welker and Cavin [30] show two more general equations for estimating evaporation rates in terms of j_m, the mass transfer factor,

$$j_m = \frac{k'_g}{G_m}(Sc)^{2/3}$$

where Sc is the Schmidt number, which is typically 1.0 to 3.0 for most VOCs diffusing in air, G_m is the air molar mass velocity (molar density · velocity), and k'_g is the mass transfer coefficient in the same units as G_m, with the driving force expressed in mol fraction [31]. They show two equations for j_m,

$$j_m = 0.036\, \mathcal{R}^{0.2}$$

which is based on extensive published measurements of water evaporation [31], and

$$j_m = 4.4\, \mathcal{R}^{0.57}$$

which is based on their own measurements with evaporating liquid propane. Here $\mathcal{R}$ is the Reynolds number based on the air physical properties, wind velocity, and the downwind length of the evaporating liquid surface. The two equations give the same value of $j_m = 2.7 \times 10^{-3}$ for $\mathcal{R} = 4.87 \times 10^5$, which corresponds to a 10 ft/s air velocity over a pool 8 ft long. For that circumstance calculate the formula equivalent to the formula in Problem 10.5. Explain any difference in the answers.

(*b*) Mackay and Matsugu [32] represent the evaporation rate as

$$N = k_m(P - P_\infty)/RT_P$$

where
N = evaporation rate in (mol/m^2 · h)
k_m = the mass transfer coefficient in m/h
P = the vapor pressure of the liquid in atm
P_∞ = the partial pressure of the evaporating material in the bulk atmosphere (atm), normally assumed to be zero
T_P = the temperature of the liquid pool, in K
R = the universal gas constant in (atm · m^3)/(mol · K)

They correlated their experimental data for the evaporation of water and hydrocarbons and additional evaporation data from the literature by

$$k_m = 0.0292 U^{0.78} X^{-0.11} Sc^{0.67}$$

where U = the wind speed in m/h

X = the evaporating liquid pool diameter in m

Sc = the Schmidt number for the evaporating vapor in air, normally in the range from 1 to 3 for hydrocarbons evaporating in air

Here the symbols and dimensions from the original paper have been used; these are not the same as those used in this book. Using the above equations, estimate the value of the constant in the evaporation rate equation in Problem 10.5 for $X = 2$ m, $U = 3\text{m/s} = 10\ 800\ \text{m/h}$, and $Sc = 2$. Compare your result with that in Problem 10.5.

10.7. The U.S. Environmental Protection Agency standard for coating beverage cans [33] allows 0.29, 0.46, and 0.89 kilograms of VOC emission per liter of solids deposited for the two kinds of outside coating and one kind of inside coating in a standard two-piece beverage can. The coatings (paints, varnishes) consist of solid polymers, plus pigments, dissolved in VOC solvents. The solids fraction, by weight, of the paints and varnishes varies from about 20 percent to about 40 percent.

(*a*) If we assume that the 0.29 kg/L restriction applies to a paint with 40 percent solids, what fraction of the solvent (thinner) VOC must be captured or destroyed to meet this standard? (Assume that both solids and solvent have a density of 1 g/cm^3.)

(*b*) VOC regulations in Los Angeles restrict the solvent emissions from painting to 0.25 kg/L of solids. If you are a manufacturer of paints for outdoor use in Los Angeles, what are your possible ways of complying with this regulation?

10.8. The U.S. Environmental Protection Agency recommends the following formula for estimating filling (displacement) losses in a vented tank [34]:

$$D_L = 12.46\left(\frac{S \cdot P \cdot M}{T}\right)$$

where D_L = vapor content of the displaced vapor, in lb/1000 gal of liquid transferred

S = saturation factor, which ranges from 0.2 to 1.45 depending on the filling method and history of the tank; its values are given in Ref. 34

P = vapor pressure in psia

M = molecular weight in g/mol

T = temperature in °R

(*a*) Rework Example 10.4 using this formula, with $S = 1$. Does the value agree with the value in that example? Should it?

(*b*) The same source suggests the following values of S: submerged loading of ships, 0.2; submerged loading of trucks and railroad cars, 0.5; splash loading of clean cargo tanks, 1.45. Sketch what must be meant by submerged and splash loading, and suggest why those values of S are as recommended there.

10.9. A vented tank contains liquid mercury at 20°C. The vapor space above the liquid is saturated with mercury vapor. We now pump 1 m^3 of liquid mercury into this tank from below. Estimate the mass of mercury emitted.

10.10. Compare tank filling losses in Example 10.7 to the permitted hydrocarbon tailpipe emission from modern autos. Assume that the auto gets 20 miles/gallon of gasoline, and that the emissions are equal to the current tailpipe standard, 0.41 g of hydrocarbon per mile. Present your results as the ratio (filling emissions/tailpipe emissions).

10.11. Estimate the fraction of the contained gasoline that is emitted from the gasoline tank of an auto when the contents of the tank are heated from 70° to 100°F. Assume that the tank is half-full of liquid, the coefficient of volume expansion of liquid gasoline is 0.0006/°F, the gasoline vapor in the air has a molecular weight of 60, and its vapor pressure is given approximately by Eq. (10.15). (This type of emission is collected by the *charcoal canister* on modern autos.)

10.12. Estimate the fraction of the contained gasoline that is emitted from the gasoline tank of an auto when the surrounding atmospheric pressure decreases by 1 percent. Assume the tank is half-full of liquid gasoline at 68°F, $p_i = 6$ psia. The coefficient of volume contraction on compression of the gasoline ≈ 0.000007/psi, the vapor behaves as a perfect gas, and the volume of the tank itself does not change. (This type of emission is collected by the charcoal canister on modern autos.)

10.13. Based on the values in Fig. 10.5,

(*a*) What fraction of the gasoline dispensed is spilled?

(*b*) For a typical filling of 12 gallons, how much liquid is this? Does that value agree with your personal observation of how much is spilled each time you fill your tank?

(*c*) The EPA mobile source emission tables [35] present a figure for the combination of displacement and spillage losses of 0.45 g/mi. That corresponds to an average fuel economy of 12.7 mi/gal. Does that lead to the same ratio as in part (*a*)? Discuss this result.

10.14. Example 10.8 shows that the calculated emission rate is less than the observed average value for refinery valves. Assume that the difference is caused by the estimate of the average thickness of the leakage path, which was assumed to be 0.0001 inch. What value of the assumed leakage path thickness makes the calculated value equal the reported value?

10.15. Show the calculation leading to the required condensation temperature of −60°C in Example 10.9.

10.16. In Example 10.9, estimate the required power to supply the refrigeration. See any elementary thermodynamics textbook for methods of making this estimate.

10.17. Rework Example 10.9 for a permitted emission of 1 ppm.

10.18. Although the current U.S. standard for Example 10.10 is 35 mg/L of gasoline, the European standard is 10 mg/L, and it is expected that the United States will soon adopt that standard. Rework Example 10.10 on the basis of meeting that standard.

10.19. The steam tables [13] show the vapor pressure of solid water (ice) for temperatures from 32 to −40°F. For Example 10.10 we needed the value at −50.0°F.

(*a*) Use the following two values to estimate the constants A and B in Eq. (10.2) for ice; at $T = 32$°F, $p = 0.0886$ psia, and at $T = -40$°F, $p = 0.0019$ psia.

(*b*) Check your constants by using them to estimate the vapor pressure at 0°F, for which the table value is 0.0185 psia.

(*c*) Then (if part *b* works out well) use Eq. (10.2) to estimate the vapor pressure at −50.0°F.

10.20. Example 10.10 assumes that the tank truck returns to the bulk plant with the vapor in the tank at equilibrium with the remaining (small amount of) gasoline at 20°C. This is plausible for spring and fall, but not for summer or winter. Estimate the required temperature of stream 3 in that example for:

(*a*) Summer, with the tank truck at equilibrium at 30°C.

(*b*) Winter, with the tank truck at equilibrium at 0°C.

(*c*) This problem is somewhat unrealistic, because refiners change the properties of gasoline from summer to winter. But ignore that and assume that the gasoline obeys Eq. (10.15) all year, and that the gasoline vapor molecular weight is 60 g/mol all year. Comment on which way the vapor pressure changes discussed in Sec. 10.3.3.2 would change the answers to parts (*a*) and (*b*).

10.21. The required temperatures for condensation are low enough that ordinary mechanical refrigeration may be too expensive to use. An alternative is "cryocondensation" [14] in which the gas stream is cooled to some intermediate temperature by mechanical refrigerators, and then cooled to the final separation temperature with liquid nitrogen. Sketch the equivalent of Fig. 10.8 for this process. Include a gas-gas heat exchanger to recover the sensible refrigeration in the cold cleaned gas.

10.22. The hydrocarbon-air mixture in the gasoline truck in Example 10.10 is above its upper explosive limit. The cleaned mixture leaving the system (stream 3 in Fig. 10.9) is below its lower explosive limit. Somewhere between these points there must exist a mixed fuel-gas stream in the combustible range. One alternative to the system shown in Fig. 10.9 [14] takes this mixture directly to an internal combustion engine, burning it to produce a clean exhaust gas, and uses the power produced by the engine to drive the refrigerators in the system.

(*a*) Sketch the flow diagram for such a system.

(*b*) Estimate the temperature to which the hydrocarbon-air mixture should be cooled to produce an ideal fuel-air mixture with an air-fuel mixture of 15 : 1 by weight.

(*c*) Discuss the advantages and disadvantages of this arrangement.

10.23. Show the calculations leading to the values on Fig. 10.12 for 1 atm and the following conditions:

(*a*) 100°F and 0.001 mol fraction of toluene.

(*b*) 300°F and 0.005 mol fraction of toluene (at 300°F, $p_{\text{toluene}} = 2.66$ atm).

10.24. Figure 10.12 shows adsorption capacity curves (normally called *adsorption isotherms*) based on Fig. 10.11 in the form suggested by Polanyi. There are other forms widely used. Sketch a plot with the same axes as Fig. 10.12 showing the shapes of the following popular adsorption isotherm equations:

(*a*) Freundlich, $w^* = \alpha p^{(1/n)}$

(*b*) Langmuir, $w^* = Kp/(1 + Kp)$

where α, n, and K are data-fitting constants.

10.25. Estimate the pressure drop through the adsorbent bed in Example 10.13. Use the Kozeny-Carman equation for estimating pressure drop for laminar flow in porous media,

$$\frac{\Delta P}{\Delta x} = 150 \frac{V_s \mu (1 - \varepsilon)^2}{D_p^2 \varepsilon^3}$$

where V_s = superficial velocity, equal to the total volumetric flow rate divided by the cross-sectional area of the bed

ε = bed porosity, which may be assumed here to be 0.3

D_p = the particle diameter, shown in Example 10.13.

10.26. Our adsorption bed is at 100°F, and contains 0.29 lb of toluene per lb of adsorbent, (see Example 10.12). We now shut this bed off from the stream it is treating and regenerate it by pulling a vacuum on it. We reduce the pressure in the bed to 0.1 atm. Then we allow a small steady flow of purge air to pass through the bed, while our vacuum pumps continue to hold the pressure at 0.1 atm.

(*a*) If that air comes to equilibrium with the toluene on the carbon in the bed, what will the mol fraction of toluene in this purge air be?

(*b*) If we continue until the concentration in this purge air falls to 0.01 mol fraction (1 mol %), what will the toluene concentration in the bed (lb toluene/lb absorbent) be?

10.27. A painter's face mask has two small charcoal canisters that protect the worker wearing the mask against the solvent fumes in a paint-spraying operation. The solvent has the same properties as toluene, and the adsorbent has the same properties as that used in Example 10.11. However, the temperature of the air and of the charcoal is 68°F (= 20°C) instead of the 100°F in that example. The vapor pressure of toluene at 68°F is 0.029 atm. The charcoal canisters originally contain no toluene, and are used once and thrown away when the toluene vapor "breaks through."

If the combined mass of charcoal in the two canisters is 100 g, the air breathed has a toluene mol fraction of 1000 ppm, and the painter breathes in at the rate of 15 kg of air per 24 hours, how long will this mask protect the painter? Assume an ideal breakthrough curve, as shown on Fig. 10.13.

10.28. The Henry's law constant for O_2 in water at 20°C is 40 100 atm. Estimate the equilibrium oxygen content of water at 20°C. What are the biological implications of this value being so small?

10.29. (*a*) Show the derivation of Eq. (10.19) from the definition of X and Y.

(*b*) How large an error do we make by substituting Eq. (10.20) for (10.19) if $x = y = 1\%$?

10.30. Compute Q_L/Q_G, the ratio of volumetric flow rates, in Example 10.14. Does the value bear out the statement that in practical systems the volumetric flow rate of liquid is rarely greater than 10 percent that of the gas?

10.31. Repeat Example 10.14 using water as the absorbing liquid.

(*a*) Show that the solubility of toluene in water at 25°C, listed in Table 10.3, corresponds to a Henry's law constant of $\approx$ 9800 atm. Assume that this value at 25°C = 77°F is applicable also at 100°F (only a fair assumption!).

(*b*) Then compute $x_{\text{toluene, bottom}}$ and from it L/G.

(*c*) Comment on the feasibility of using water in this example.

10.32. It has now been decided that the vapor pressure of n-tetradecane at 100°F is too high, resulting in an unacceptably large concentration in the vapor leaving the absorber in Example 10.14. At this temperature, n-pentadecane ($C_{15}H_{32}$, M = 212 g/mol) has a calculated vapor pressure of $15 \cdot 10^{-6}$ atm, about a third of that of n-tetradecane. If we substitute n-pentadecane for n-tetradecane in Examples 10.14, 15 and 16, what will all the changes be?

10.33. (*a*) Show the mathematics of finding Eq. (10.25).

(*b*) Colburn and Pigford [31] showed that this expression can be put in somewhat simpler form and reduced to a simple plot that appears in all mass-transfer books [20–22]. Find that form in your mass-transfer book, and show the algebra between it and Eq. (10.25).

(*c*) Repeat the calculation of N in Example 10.16, using the plots in mass-transfer books.

10.34. A paint-baking oven is to remove 1000 lbm/h of toluene from decorated items. For safety reasons the concentration of toluene must be kept below 10 percent of the lower explosive limit.

(*a*) Estimate the required air flow rate.

(*b*) Would it be practical to lower the air flow rate enough to go above the upper explosive limit?

10.35. It is a rule of thumb that for organic chemical reactions near room temperature, raising the temperature by 10°C will double the reaction rate. If we take room temperature as 20°C and assume that the rate of some reaction doubles when we raise the temperature to 30°C and that Eq. (10.31) applies,

(*a*) What is the value of E, the activation energy, for this reaction?

(*b*) How much would the reaction rate increase if we raised the temperature from 1000 to 1010°C?

10.36. In tests of incineration of herbicides [36], the incinerator temperature averaged 1500°C, and the time spent by the material being burned at this temperature was 1.0 s. The destruction of dioxin (2,3,7,8-tetrachlorodibenzo-*p*-dioxin), averaged over three testing periods, was 99.93 percent.

Assuming that this material burns according to the first-order equation, estimate how long the material would need to be held at 1500°C to get 99.999 percent destruction. How long would it take to get 99.9999% (six nines) destruction?

10.37. A widely used "rule of thumb" for VOC incinerators is that they should hold the heated gases at the peak temperature for 1 s. If the incineration reaction is first-order, and we require 99.99% destruction of the VOC, what is the required first-order rate constant, k?

10.38. We wish to design an incinerator to destroy acrolein in a 10 000 scfm waste gas stream. The acrolein in the waste gas stream must be 99.99% destroyed. The kinetics of acrolein destruction are well represented by the values shown in Table 10.4. What volume must the combustor and retention chamber have to get 99.99% destruction of acrolein at a temperature of 1200°F?

10.39. Estimate the temperature required for an afterburner that will destroy 99.5 percent of the toluene contained in an airstream with a residence time of 0.5 s using the values in Table 10.4. Compare your results with the example calculation in Cooper and Alley [37], in which they show that three different methods of making this estimate lead to answers of 1326°, 1263°, and 1331°F.

10.40. Many authors report that in VOC incinerators the steps leading to the formation of CO and H_2O are relatively fast and that the oxidation of CO to CO_2 is slower, so the size of the incinerator is determined by the size required to complete the CO oxidation. Test this idea with the following calculations:

(*a*) Estimate the time required to combust 99 percent of the toluene in a waste stream containing 5000 ppm of toluene at 1400°F; use the values in Table 10.4.

(*b*) Assume that the step

$$C_7H_8 + 5.5O_2 \rightarrow 7CO + 4H_2O$$

is instantaneous, and then estimate the time required for the step $CO + 0.5O_2 \rightarrow CO_2$. Assume the kinetics are given by

$$\frac{d[\text{CO}]}{dt} = -A\exp(-E/RT)[\text{CO}][\text{O}_2]^{0.5}[\text{H}_2\text{O}]^{0.5}$$

where [X] indicates the concentration of X in mol/cm^3, $A = 1.3 \times 10^{14}$ cm^3/mol · s, and $E = 30$ kcal/mol [26, page 29] and that we need 99 percent destruction of the CO.

10.41. We wish to treat an airstream from a paint dryer in a combustor to destroy 90 percent by weight of the contained hydrocarbons. The hydrocarbon vapor consists of components A, B, and C. Their weight percentages in the feed are 25, 25, and 50. Each is believed to combust to carbon dioxide and water according to first-order kinetics. The combustor will operate at 1500°F. At this temperature the individual first-order rate constants (1/s) are 1.0, 2.0, and 3.5.

How long must this gas be held at 1500°F to get 90 percent by weight conversion to carbon dioxide and water? Assume that the three components do not interact on burning (not a good assumption in the real world, but a satisfactory one for this problem) and that the heatup to 1500°F and cooldown from it are instantaneous.

10.42. Barnes et al. [26] report that in the gas phase the data for combustion of benzene at 1200°F are best represented by $n = 2$ (a second-order reaction) with $k = 1.8 \times 10^6$ cm^3/(mol · s). If this relation is correct, and if we start with a gas that is 0.5 mol percent (5000 ppm) benzene in air, how long will it take to burn to 0.0005 percent (5 ppm) benzene?

10.43. Barnes et al. [26] presented the following table of the temperatures required to oxidize various compounds to CO_2 and H_2O:

	Ignition temperature, °F	
Compound	**Thermal**	**Catalytic**
Benzene	1076	575
Toluene	1026	575
Xylene	925	575
Ethanol	738	575
Methyl isobutyl ketone	858	660
Methyl ethyl ketone	960	660
Methane	1170	932
Carbon monoxide	1128	500
Hydrogen	1065	250
Propane	898	500

They also suggest that the required temperatures for the catalytic combustion of various compounds decrease in the following order: methane, ethane, propane, cyclopropane, ethylene, propylene, propadiene, propene, acetylene.

Based on this information, write a set of general rules indicating which compounds are hard to oxidize thermally and which are easy. Do the values shown in this table make chemical sense?

10.44. We are designing an afterburner to destroy solid particles of a new type of plastic with density 1 g/cm^3. The particles will be spheres with diameter 10 μ. The afterburner will operate at a temperature of 1200°F. Our laboratory has tested the burning rate of this plastic and finds that at 1200°F, the burning rate in air is 0.001 g/(cm^2 · s). Based on this burning rate, estimate how long the particles must be held at 1200°F to burn up completely.

10.45. The waste stream from a textile-weaving plant contains fine fibers that are practically cylindrical (with length $\gg$ diameter), and with diameters of 10 μ. We wish to destroy these in a combustion chamber in which the temperature will be 2000°F. At this temperature the combustion rate at the surface of the fibers is estimated to be 0.2×10^{-2} g/(cm^2 · s). The fibers have a density of 1.5 g/cm^3 and zero ash. How long must they be held at this temperature to burn up completely?

10.46. This problem is the same as Problem 10.45, except that now instead of leaving the fibers in the furnace long enough to burn them up completely, we take them out after 0.2 second and cool them quickly so that the reaction stops after 0.2 second. What fraction by mass of the fibers will be burned in this furnace?

10.47. In this problem the conditions are the same as in Problem 10.45, except that now we want to burn up only 90 percent by weight of the fibers. How long must they remain at 2000°F to accomplish this?

10.48. A graphite plant produces a waste stream contaminated with fine flakes of graphite. These flakes are essentially rectangular, with $x = y \gg z$. For a typical flake, $z = 10$ μ.

We wish to destroy these in a combustion chamber in which the temperature will be 2000°F. At this temperature the combustion rate at the surface of the flakes is estimated to be 10^{-4} g/(cm^2 · s). The flakes have a density of 2.0 g/cm^3 and an ash content of zero. How long must they be held at this temperature to burn them up completely?

10.49. Repeat Problem 10.48 with particles that are cubes with an edge length of 10 μ.

10.50. We are passing a gas stream containing spherical carbon particles through an incinerator where the temperature is 1000 K. At this temperature, the burning rate at the surface of the particles is 0.015×10^{-3} g/cm^2 · s. The particle density is 2.0 g/cm^3. The particles stay in the incinerator for 2 seconds. The inlet particle set is characterized by the rectangular particle size distribution:

$$\frac{d\Phi}{dD} = C_1 \quad \text{for } 0 < D \leq 1.0\ \mu$$

Here Φ is the cumulative fraction *by number* in the gas stream. There are no particles with diameters larger than 1 μ. C_1 is a constant = 1.00/μ.

(*a*) What fraction by *weight* of the particles will be burned up in this incinerator?

(*b*) How long must the stream remain in the incinerator to get 95 percent by weight burnup of the particles?

10.51. Pulverized-coal furnaces grind their coal so that 80+ percent is smaller than 200 microns in diameter. These particles are completely burned in the approximately 2 s that they spend in the combustion zone of the furnace. Estimate their surface burning rate. Compare it to the values in Example 10.19. A "typical coal" has a density of 1.2 g/cm^3.

10.52. We plan to treat a gas stream containing hydrocarbons in an afterburner to destroy them. We can use either a simple thermal afterburner or a catalytic afterburner. Based on tests it is clear that we can destroy the hydrocarbons thermally at 1200°F. With a catalyst we can accomplish the same result at 1000°F.

We will select the catalytic unit if the fuel savings obtained by heating to only 1000°F instead of 1200°F are more than the difference between the annual cost of owning the thermal unit and the catalytic unit.

The cost of the thermal unit is \$$A$/(standard cubic foot per minute), e.g., for treating 1000 scfm we would have to pay \$1000A to purchase a thermal unit. The cost of the catalytic unit is \$$C$/(scfm).

How much more can we pay for the catalytic unit, i.e., how large is $(C - A)$, given the following conditions:

1. The annual cost of owning equipment (= depreciation + interest + taxes + insurance) is (20%/year)(purchase cost of the equipment).
2. Natural gas is used as a fuel and costs \$3.00/1000 ft^3.
3. The heat release for natural gas in the range from 1000 to 1200°F is 720 Btu/ft^3 of natural gas.
4. The heat capacity (or specific heat) of the gas we are going to treat is 0.25 Btu/lb ·°F.

10.53. An alternative to the burners shown in Fig. 10.17 is the regenerative burner shown in Fig. 10.20 on page 392. In this arrangement the flow direction is regularly reversed by rotating the valve 90 degrees. The heat of combustion is stored in one of the beds and then used to preheat the gas coming into the combustor on the next cycle. What are the advantages and disadvantages of this arrangement compared with that shown in Fig. 10.17*b*?

10.54. All household water heaters have a high-pressure relief valve. The student should examine such a heater and find that valve. What are the conditions that would cause that valve to have

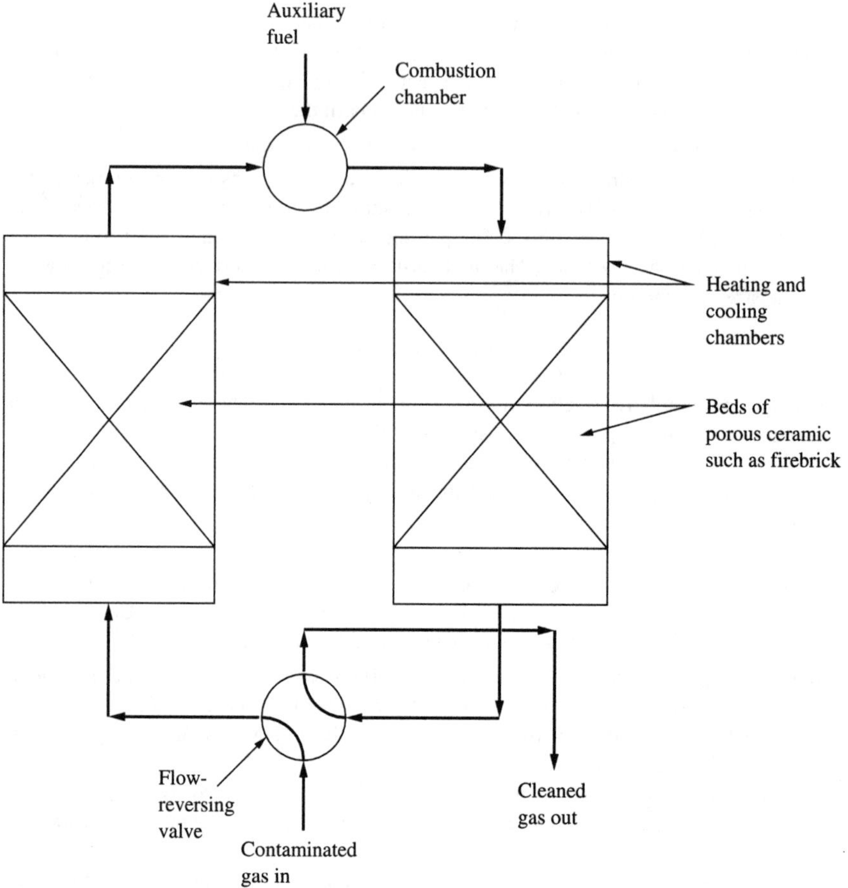

FIGURE 10.20
Flow diagram of a regenerative incinerator. These are described in some detail in Ref. 38.

to open? *Hint:* Sketch the flow diagram of a hot water heater. (This is not an air pollution problem, but it is the simplest example of why we have high-pressure relief valves. Relief valves on VOC vessels can contribute to air pollution problems.)

10.55. Numerous examples in this chapter deal with a flow of 1000 scfm of air containing 5000 ppm of toluene.

(*a*) If we wish to treat this stream by biological oxidation in a bed 3 ft deep with a residence time (based on a bed porosity of 50%) of 30 s, estimate the horizontal surface area of the required bed.

(*b*) How does this compare to the sizes of equipment we estimated for condensation, adsorption, and absorption?

10.56. Numerous examples in this chapter deal with a flow of 1000 scfm of air containing 5000 ppm of toluene. Locate this stream on Fig. 10.19, and based on that figure estimate which control technologies are likely to be most economical.

REFERENCES

1. *National Air Quality and Emissions Trends Report, 1997,* EPA-454/R-98-016.
2. Lide, D. R., ed. in chief: *CRC Handbook of Chemistry and Physics*, 71st ed., CRC Press, Boca Raton, FL, 1990, pp. 6–85.
3. Clean Air Act of 1990, *Public Law 101-549,* Section 301.
4. "Definitions, Volatile Organic Compounds (VOC)," *40CFR51.100(s).* See also the test method in *40CFR60,* Appendix A, Method 18.
5. Smallwood, I.: *Solvent Recovery Handbook,* McGraw-Hill, New York, 1993.
6. de Nevers, N.: "Vacuum Collapse of Vented Tanks," *Process Safety Progress,* Vol. 15, No. 2, pp. 74–79, 1996.
7. "Standards of Performance for Volatile Organic Liquid Storage Vessels (Including Petroleum Liquid Storage Vessels) for Which Construction, Reconstruction, or Modification Commenced after July 23, 1984," *40CFR60.110b.*
8. *Standard Test Method for Vapor Pressure of Petroleum Products (Reid Method),* ASTM, Publ. D 323-72, 1973.
9. Clean Air Act of 1990, *Public Law 101-549,* Section 216.
10. "Emission Factors for Equipment Leaks of VOC and HAP," U.S. Environmental Protection Agency, Report No. EPA-450/3-86-002, U.S. Government Printing Office, Washington, DC, 1986.
11. "Standards of Performance for Bulk Gasoline Terminals," *40CFR60.500.*
12. Shedd, S. A.: "Gasoline Marketing," in A. J. Buonicore and W. T. Davis (eds.), *Air Pollution Engineering Manual,* Van Nostrand Reinhold, New York, p. 341, 1991.
13. Keenan, J. H., et al.: *Steam Tables, Thermodynamic Properties of Water, Including the Vapor, Liquid and Solid Phases,* John Wiley, New York, 1969.
14. Anon.: *Air Pollution Control Manual for Hydrocarbon Vapor Recovery Equipment,* Edwards Engineering Corp., Pompton Plains, NJ, 1997.
15. Fair, J. R.: "Sorption Processes for Gas Separation," *Chem. Eng.,* Vol. 76, Issue 15, pp. 90–110, July 14, 1969.
16. Noll, K. E., V. Gounaris, and W.-S. Hou: *Adsorption Technology for Air and Water Pollution Control,* Lewis, Chelsea, MI, Chapter 9, 1992.
17. Vermeulen, T., M. D. LeVan, N. K. Hiester, and G. Klein: "Adsorption and Ion Exchange," in D. W. Green and J. O. Maloney (eds.), *Perry's Chemical Engineers' Handbook,* 6th ed., McGraw-Hill, New York, Chapter 16, 1984.
18. Hougen, O. A., and W. R. Marshall: "Adsorption from a Fluid Stream Flowing through a Stationary Granular Bed," *Chem. Eng. Prog.,* Vol. 43, pp. 197–208, 1947.
19. Anon.: *Vapor Recovery Systems,* John Zinc Co., Tulsa, OK, 1997.
20. Fair, J. R., M. A. Steinmeyer, W. R. Penney, and B. B. Crocker: "Gas Absorption and Gas-Liquid System Design," in D. W. Green and J. O. Maloney (eds.), *Perry's Chemical Engineers' Handbook,* 7th ed., McGraw-Hill, New York, Chapter 14, 1997.
21. Kohl, A. L., and R. B. Nielsen: *Gas Purification,* 5th ed., Gulf Publishing, Houston, TX, 1997.
22. Treybal, R.: *Mass Transfer Operations,* 3d ed., McGraw-Hill, New York, 1980.
23. Liley, P. E., G. H. Thompson, D. G. Friend, T. E. Daubert, and E. Buck: "Physical and Chemical Data," in D. W. Green and J. O. Maloney (eds.), *Perry's Chemical Engineers' Handbook,* 7th ed., McGraw-Hill, New York, Chapter 2, 1997.
24. Reid, R. C., J. M. Prausnitz, and T. K. Sherwood: *The Properties of Liquids and Gases,* 3d ed., McGraw-Hill, New York, Appendix A, 1977.
25. Lee, K. C., et al.: "Revised Model for the Prediction of the Time-Temperature Requirements for the Thermal Destruction of Dilute Organic Vapors and Its Usage in Predicting Compound Destructibility," Paper No 82-5.3, 75th Annual Meeting of the Air Pollution Control Association, New Orleans, June 20–25, 1982.
26. Barnes, R. H., M. J. Saxton, R. E. Barrett, and A. Levy: "Chemical Aspects of Afterburner Systems," U.S. Environmental Protection Agency, Report No. EPA-600/7-79-096, U.S. Government Printing Office, Washington, DC, 1979.
27. Walas, S. M.: *Reaction Kinetics for Chemical Engineers,* McGraw-Hill, New York, Figure 6-1, 1959.

28. Devinny, J. S., M. A. Deshusses, and T. A. Webster: *Biofiltration for Air Pollution Control,* Lewis, Boca Raton, FL, 1999.
29. Kostelz, A. M., A. Finkelstein, and G. Sears: "What Are the 'Real Opportunities' in Biological Gas Cleaning for North America," *Proceedings of the 89th Annual Meeting and Exhibition of the Air and Waste Management Association,* Pittsburgh, PA, 1996.
30. Welker, J. R., and W. D. Cavin: "Vaporization, Dispersion and Radiant Fluxes from LPG Spills," U.S. Department of Energy Publication No. DOE/EV/06020–1, U.S. Government Printing Office, Washington, DC, 1982.
31. Colburn, A. P., and R. L. Pigford: "General Theory of Diffusional Operations," in J. H. Perry (ed.), *Chemical Engineers' Handbook,* 3d ed., McGraw-Hill, New York, p. 545, 1950.
32. Mackay, D., and R. S. Matsugu: "Evaporation Rates of Liquid Hydrocarbon Spills on Land and Water," *Can. J. Chem. Eng.,* Vol. 51, pp. 434–439, 1973.
33. "Standards of Performance for the Beverage Can Surface Coating Industry," *40CFR60.490.*
34. "Compilation of Air Pollution Emission Factors, Volume 1: Stationary Point and Area Sources," 4th ed., AP-42, U.S. EPA, Office of Air Quality Planning and Standards, pp. 4.4-1–17, 1985.
35. "Supplement A to Compilation of Air Pollutant Emission Factors, Volume II: Mobile Sources," AP-42 Supplement A, p. H-8, 1991.
36. "At Sea Incineration of Herbicide Orange Onboard the M/T Vulcanus," U.S. Environmental Protection Agency, Report No. EPA-600/2-78-086, U.S. Government Printing Office, Washington, DC, 1978.
37. Cooper, C. D., and F. C. Alley: *Air Pollution Control: A Design Approach,* PWS Engineering, Boston, MA, p. 316, 1986.
38. Corey, R. C.: "Heat Regeneration," in D. W. Green and J. O. Maloney (eds.), *Perry's Chemical Engineers' Handbook,* 6th ed., McGraw-Hill, New York, pp. 9–89, 1984.

CHAPTER 11

CONTROL OF SULFUR OXIDES

The control of particulates and VOCs is mostly accomplished by physical processes (cyclones, ESPs, filters, leakage control, vapor capture, condensation) that do not involve changing the chemical nature of the pollutant. Some particles and VOCs are chemically changed into harmless materials by combustion. This chapter and the next concern pollutants—sulfur oxides and nitrogen oxides—that cannot be economically collected by physical means nor rendered harmless by combustion. Their control is largely chemical rather than physical. For this reason, these two chapters are more chemically oriented than the rest of the book.

Sulfur and nitrogen oxides are ubiquitous pollutants, which have many sources (see Table 1.1). SO_2, SO_3, and NO_2 are strong respiratory irritants that can cause health damage at high concentrations. We have NAAQS for SO_2 and NO_2 (see Table 2.3). The states are required to prepare SIPs for the control of NO_2 and SO_2. These gases also form secondary particles in the atmosphere, contributing to our PM_{10} and $PM_{2.5}$ problems and impairing visibility. They are the principal causes of acid rain. The Clean Air Act of 1990, Section 401—Acid Deposition Control, requires substantial reductions in our national emissions of both sulfur and nitrogen oxides over the next few decades.

11.1 THE ELEMENTARY OXIDATION-REDUCTION CHEMISTRY OF SULFUR AND NITROGEN

This chapter concerns sulfur oxides; the next, nitrogen oxides. Their sources and control methods are significantly different, but their chemistry is quite similar, as this short section shows.

Both sulfur and nitrogen in the elemental state are relatively inert and harmless to humans. Both are needed for life; all animals require some N and S in their bodies. However, the oxides of sulfur and nitrogen are widely recognized air pollutants. The reduced products also are, in some cases, air pollutants.

Table 11.1 shows, in parallel form, the oxidation and reduction products of nitrogen and sulfur. Reduction means the addition of hydrogen or the removal of oxygen. If we reduce nitrogen, we produce ammonia (which logically should be called hydrogen nitride; but because it had a common name before modern chemical naming systems were devised, it goes by its common name, ammonia). Similarly, if we reduce sulfur, we produce hydrogen sulfide. Both hydrogen sulfide and ammonia are very strong-smelling substances, gaseous at room temperature (−60°C and −33°C boiling points, respectively), and toxic in high concentrations. (High concentrations due to accidental releases often cause fatalities. These occur in the production and use of ammonia as a fertilizer and refrigerant and in the production and processing of "sour" gas and oil, which contain hydrogen sulfide.) Neither ammonia nor hydrogen sulfide has been shown to be toxic in the low concentrations that normally exist in the atmosphere.

TABLE 11.1
Elementary oxidation and reduction of sulfur and nitrogen

Reduction ←	**Elemental form**	**→ Oxidation, first step**	**→ Oxidation, second step**	**Reaction, with water**	**Reaction, with NH_4^+ or other cations**
Normally requires high pressure, high temperature, hydrogen gas and a catalyst. Occurs in many biological processes at low pressures and temperatures.		Most often accomplished by reaction with oxygen from the atmosphere, quickly at high temperatures in the case of burning, or slowly at low temperatures, e.g., rusting.	Slowly in the atmosphere, or quickly in a catalytic reactor.	Rate depends on atmospheric moisture content.	Rate depends on concentration of atmospheric cations
NH_3 ← ammonia	← N_2 → nitrogen	→ NO nitric oxide	→ NO_2 nitrogen dioxide	→ HNO_3 nitric acid	nitrate particles
H_2S ← hydrogen sulfide	← S → sulfur	→ SO_2 sulfur dioxide	→ SO_3 sulfur trioxide	→ H_2SO_4 sulfuric acid	sulfate particles

When nitrogen is oxidized, nitric oxide (NO) and then nitrogen dioxide (NO_2) form; likewise, sulfur forms sulfur dioxide (SO_2) and then sulfur trioxide (SO_3). These are all gases at room temperature or slightly above room temperature (boiling points 21°C, 34°C, −10°C, and 45°C, respectively). The oxides have higher boiling points than the hydrides. Both nitrogen and sulfur can also form other oxides, but these are the ones of principal air pollution interest.

In the atmosphere NO_2 and SO_3 react with water to form nitric and sulfuric acids, which then react with ammonia or any other available cation to form particles of ammonium nitrate or sulfate or some other nitrate or sulfate. These particles, generally in the 0.1 to 1-μ size range, are very efficient light-scatterers; they persist in the atmosphere until coagulation and precipitation remove them. They are significant contributors to urban PM_{10} and $PM_{2.5}$ problems. They are the principal causes of acid deposition (Chapter 14) and of visibility impairment in our national parks. NO and NO_2 also play a significant role in the formation of O_3 (see Chapter 12 and Appendix D).

The estimated concentrations of these materials in unpolluted parts of the world's atmosphere are SO_2, 0.2 ppb; NH_3, 10 ppb; NO_2, 1 ppb [1].

11.2 AN OVERVIEW OF THE SULFUR PROBLEM

Figure 11.1 on page 398 shows in part how sulfur moves in the environment as a result of human activities. It does not include the large amounts of sulfur emitted by volcanic eruptions nor the movement of sulfur into growing plants and then back out of decaying plants. Sulfur is the sixteenth-most abundant element in the earth's crust, with an abundance of about 260 ppm [2]. The vast majority of this sulfur exists in the form of sulfates, mostly as gypsum, $CaSO_4 \cdot 2H_2O$, the principal ingredient of plaster and wallboards, or anhydrite, $CaSO_4$. Gypsum is a chemically inert, nontoxic, slightly water-soluble mineral, found widely throughout the world.

All organic fuels used by humans (oil, coal, natural gas, peat, wood, other organic matter) contain some sulfur. Fuels like wood have very little (0.1 percent or less), whereas most coals have 0.5 percent to 3 percent (see Appendix C). Oils generally have more sulfur than wood but less than coal. If we burn the fuels, the contained sulfur will mostly form sulfur dioxide,

$$\underset{\text{(in fuel)}}{S} + O_2 \rightarrow SO_2 \tag{11.1}$$

If we put this into the atmosphere, it will eventually fall with precipitation, mostly in the ocean (because most of the world's rain falls on the ocean), and over time become part of the land mass as a result of geologic processes. Again over geologic time, it will enter into fossil fuels and sulfide minerals, which humans extract and use. These uses generally lead to the formation of SO_2. If we wish to prevent this SO_2 from getting into the atmosphere, we can use any of the methods described in this chapter, all of which have the effect of capturing the sulfur dioxide in the form of $CaSO_4 \cdot 2H_2O$ that will then be returned to the earth, normally in a landfill. Most

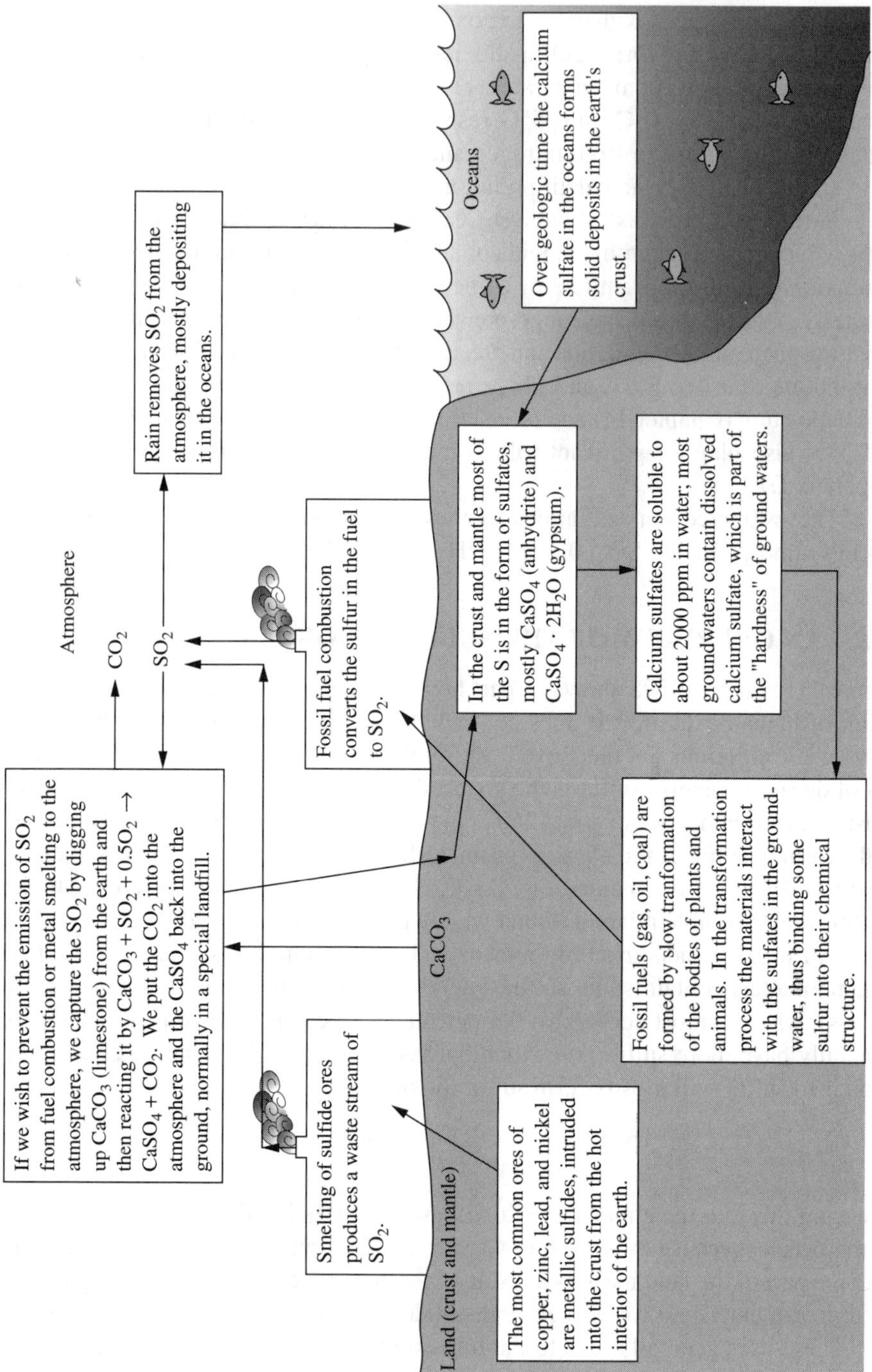

FIGURE 11.1
Principal flow of sulfur in the environment, as influenced by humans. This figure omits the large amounts of sulfur that some volcanic eruptions release, and the flow into and out of the atmosphere to growing and decaying plants, respectively.

often the overall reaction will be

$$\underset{\text{(limestone)}}{CaCO_3} + SO_2 + 0.5O_2 \rightarrow CaSO_4 + CO_2 \tag{11.2}$$

In this reaction one kind of widely available rock (limestone) is mined and used to produce another rock (anhydrite or, with $2H_2O$, gypsum), which we put back into the ground, and to release carbon dioxide to the atmosphere. We are concerned about adding to the CO_2 in the atmosphere (Chapter 14), but not nearly as much as we are about adding an equivalent amount of SO_2. Although Eq. (11.2) appears simple, the details of carrying it out on a large scale are complex, as discussed in this chapter.

In natural gas most of the sulfur is in the form of H_2S, which is easily separated from the other constituents of the gas. In oil (liquid petroleum) and also in oil shales and tar sands, the sulfur is chemically combined with the hydrocarbon compounds; normally it cannot be removed without breaking chemical bonds. In oils the sulfur is concentrated in the higher-boiling fraction of the oil, so the same crude oil can yield a low-sulfur gasoline (average 0.03% S) and a high-sulfur heavy fuel oil (e.g., 0.5 percent to 1 percent S). In coal much of the sulfur is also in the form of chemically bound sulfur, but some coals have a large fraction of their sulfur in the form of small (typically 100 μ) crystals of iron pyrite ("fools gold," FeS_2). When the fuel is burned, almost all of the sulfur in the fuel, whether chemically bound or pyritic, is converted to sulfur dioxide (SO_2) and carried along with stack gas. Some small fraction is captured in the ash, and some is converted to SO_3. Mixtures of SO_2 and SO_3 are sometimes called SO_x to remind us that some of the sulfur is in the form of SO_3. Usually the SO_3 is negligible, and we speak of these streams as if the only sulfur oxide they contained was SO_2.

The other important source of SO_2 attributable to humans is the processing of sulfur-bearing ores. The principal copper ore of the world is chalcopyrite, $CuFeS_2$. The basic scheme for obtaining copper from it is the overall high-temperature smelting reaction,

$$CuFeS_2 + \tfrac{5}{2}O_2 \rightarrow Cu + FeO + 2SO_2 \tag{11.3}$$

in which the iron is converted to a molten oxide that will float on the molten copper (with a silica flux) and thus be separated from it. The sulfur is converted to gaseous SO_2. The principal ores of lead, zinc, and nickel are also sulfides, whose processing is similar to Eq. (11.3).

Because the SO_2 liberated in the preceding process has been widely recognized as an air pollutant for many years, considerable effort has been devoted to finding other ways to process these ores that do not produce SO_2. There has been some success in developing processes that treat these ores by aqueous chemistry without producing any SO_2 at all. Currently such processes are economical for partly oxidized copper oxide ores containing smaller amounts of sulfur. However, for ores like chalcopyrite, the processes have not proven economical and most of these ores are currently smelted with air or oxygen.

Table 11.2 [3] on page 400 shows in more detail than Table 1.1 the emission sources for SO_2 in the United States in 1997. We see that coal combustion contributes

TABLE 11.2
U.S. emissions estimates for sulfur dioxide for 1997 (see Table 1.1)

Source category	Emissions, thousands of tons/yr	Percent of total	
Coal combustion, utilities	12 532	61.51	
industrial	1769	8.69	
commercial	206	1.01	
Subtotal			71.21
Oil combustion, utilities	436	2.14	
industrial	847	4.16	
commercial	414	2.03	
Subtotal			8.33
Industrial processes			
Chemical and allied	301	1.48	
Metals processing	552	2.71	
Petroleum and related	385	1.89	
Subtotal			6.08
Vehicle emissions, on-road	320	1.57	
off-road	1064	5.22	
Subtotal			6.79
All other sources	1545	7.59	7.59
Total	20 371	100.00%	

Source: Ref. 3.

71% of the total, and that coal plus oil combustion (including vehicle emissions) account for 86% of the total. The emissions from metals processing, which were major sources before 1970, have now been largely controlled and contribute only 2.7% of the total.

The sulfur-containing gas streams most often dealt with in industry belong to three categories—reduced sulfur, concentrated SO_2 streams, and dilute SO_2 streams—each with its own control method, as discussed in this chapter.

11.3 THE REMOVAL OF REDUCED SULFUR COMPOUNDS FROM PETROLEUM AND NATURAL GAS STREAMS

As discussed in Sec. 11.1, we can convert sulfur in organic compounds to various forms by oxidation or reduction. Here we discuss the technology for removing sulfur from gas streams when the sulfur is present in reduced form. These gas streams occur in many natural gas deposits and in many by-product gases produced in oil refining and in the fuel gases produced by coal gasification.

Example 11.1. A flow of 10^8 scf per day (32.8 std m^3/s) of natural gas ($\approx$ 0.2 percent of average U.S. consumption), which contains 1 percent (10 000 ppm) of H_2S, is treated in the apparatus sketched in Fig. 10.15 to reduce the H_2S concentration to 4 ppm (the maximum allowed in commercial natural gas in the United States). The gas is scrubbed at a pressure of 100 atm and 20°C. Assuming that we will use water as the scrubbing agent, estimate the required water flow rate.

From Ref. 4 we find the Henry's law constant for H_2S at 20°C = 483 atm, so that from Eq. (10.16) we see that

$$x_i = \frac{Py_i^*}{H_i} = \frac{100 \text{ atm}}{483 \text{ atm}} y_i^* = 0.207 y_i^*$$

As in Example 10.14 (please reread that example!), we arbitrarily specify that the outlet liquid shall have $y_i^* = 0.8 y_i$. Thus we can calculate that

$$x_{i \text{ bottom}} = 0.8 \cdot 0.207 \cdot 0.01 = 1.66 \times 10^{-3}$$

(Here we do not write this as 1660 ppm because in liquids ppm always means ppm by mass, and this is a mol fraction!) Then we may write

$$\frac{L}{G} = \frac{Y_{i \text{ bottom}} - Y_{i \text{ top}}}{X_{i \text{ bottom}} - X_{i \text{ top}}} \approx \frac{y_{i \text{ bottom}} - y_{i \text{ top}}}{x_{i \text{ bottom}} - x_{i \text{ top}}} = \frac{10\,000 \text{ ppm} - 4 \text{ ppm}}{1.66 \times 10^{-3} - 0} = 6.04$$

The molar flow rate of gas is

$$G = 10^8 \frac{\text{scf}}{\text{day}} \cdot \frac{\text{lbmol}}{385.3 \text{ scf}} \cdot \frac{\text{day}}{24 \cdot 3600 \text{ s}} = 3.0 \frac{\text{lbmol}}{\text{s}} = 1360 \frac{\text{mol}}{\text{s}}$$

so the required liquid flow rate is

$$L = 6.04 \cdot 3.0 \frac{\text{lbmol}}{\text{s}} = 18.1 \frac{\text{lbmol}}{\text{s}} = 326 \frac{\text{lb}}{\text{s}} = 148 \frac{\text{kg}}{\text{s}}$$ ■

This is a large liquid flow rate. To make the system shown in Fig. 10.15 practical, one must find a solvent that can absorb much more H_2S than can the water in this example. Fortunately, for many of the gases of air pollution and industrial interest, we can do that. H_2S, SO_2, SO_3, NO_2, HCl, and CO_2 are *acid gases,* which form acids by dissolving in water. For H_2S the process is

$$H_2S \text{ (gas)} \rightleftarrows H_2S \text{ (dissolved in water)} \rightleftarrows H^+ + HS^- \tag{11.4}$$

If we can add something to the scrubbing solution that will consume either the H^+ or the HS^-, then more H_2S can dissolve in the water, and much less water is needed. For acid gases, the obvious choice is some alkali, a source of OH^- that can remove the H^+ by

$$H^+ + OH^- \rightleftarrows H_2O \tag{11.5}$$

Removing the H^+ on the right side of Eq. (11.4) drives the equilibrium to the right, greatly increasing the amount of H_2S absorbed. If the solution is to be regenerated in the rightmost column in Fig. 10.15, then the alkali should be a weak alkali that can easily give back the acid gas on heating or pressure reduction. If only the leftmost column of Fig. 10.15 is used and the resulting solution is discarded (see Sec. 11.5), then a strong alkali, which could not be easily regenerated, can be used. The most common choices of alkali for H_2S removal are ethanolamines (monoethanolamine, diethanolamine, and triethanolamine) and also the sodium or potassium salts of weak acids like carbonic or phosphoric.

Example 11.2. Repeat Example 11.1, using as the absorbent a 2 N (12.2 wt %, 3.94 mol %) solution of monoethanolamine (MEA), $HO(CH_2)_2NH_2$, as the scrubbing solution, at 77°F = 25°C.

For this solution strength and temperature, Kohl and Nielsen [5] give a plot of $p_i = Py_i^*$ as a function of mols H_2S per mol of MEA that can be approximated by

$$\begin{pmatrix}\text{Vapor pressure}\\ \text{of } H_2S\end{pmatrix} = p_{H_2S} \approx Py_i^* \approx (0.00281 \text{ psi}) \exp(195x_{H_2S}) \qquad (11.6)$$

Following Example 11.1, we choose the outlet liquid concentration to have $y_i^* = 0.8y_i$ and solve Eq. (11.6) for the corresponding x, finding

$$x_{H_2S} = \frac{\ln(0.8 \cdot 1470 \text{ psi} \cdot 0.01/0.00281 \text{ psi})}{195} = 0.0427$$

If we also require that at the top of the column $y_i^* = 0.8y_i$, we can compute the maximum permitted concentration of H_2S in the regenerated solution as

$$x_{H_2S} = \frac{\ln(0.8 \cdot 1470 \text{ psi} \cdot 4 \times 10^{-6}/0.00281 \text{ psi})}{195} = 0.00264$$

and

$$\frac{L}{G} = \frac{Y_{i\text{ bottom}} - Y_{i\text{ top}}}{X_{i\text{ bottom}} - X_{i\text{ top}}} \approx \frac{y_{i\text{ bottom}} - y_{i\text{ top}}}{x_{i\text{ bottom}} - x_{i\text{ top}}} = \frac{10\,000 \text{ ppm} - 4 \text{ ppm}}{0.0427 - 0.00264} = 0.250$$

The required liquid flow rate is that in Example 11.1 multiplied by $(0.250/6.04) = 0.041$, or 13.5 lb/s = 6.1 kg/s. ■

This example shows that by changing from water to an MEA solution we can reduce the required liquid flow rate by a factor of about 24. We could repeat Examples 10.15 and 10.16 to find the required column diameter and height (see Problems 11.5 to 11.7). The procedure is the same as in those examples, but the pressure is high enough that we can no longer consider the gas to be ideal. In addition, the equilibrium relation is much more complex than the simple Henry's law relation, Eq. (10.16), in those examples, so that the integration of Eq. (10.25) can no longer be done in closed form, but must be done numerically. Nonetheless the calculation procedures are straightforward and available in the literature [6, 7]. There are hundreds of plants all over the world performing the separation in Example 11.2, not only for natural gas but also for H_2S-containing gases generated in petroleum refining. Every major petroleum refinery has at least one of them.

11.3.1 The Uses and Limitations of Absorbers and Strippers for Air Pollution Control

As shown in Chapter 10, absorber-stripper combinations are widely used to remove HCs from exhaust gas streams. This example shows that the removal of H_2S from natural gas and similar streams is simple and straightforward. The system in Fig. 10.15 also works extremely well for removing ammonia from a gas stream, because NH_3 is very soluble in water or in weak acids, forming a weak alkali by the following reaction:

$$NH_3 + H_2O \rightarrow NH_4^+ + OH^- \tag{11.7}$$

It is possible to make practically complete removal of NH_3 from gas streams with water or weak acids. The solubility of ammonia is so high that generally the simplest possible forms of this arrangement are satisfactory.

To remove SO_2 from gas streams by this method is also relatively easy *if* there are no other acid gases present. For example, SO_2 could be easily removed from N_2 by the scheme shown in Fig. 10.15 using any weak alkali (for example, ammonium hydroxide), and the solution would be easily regenerated to produce pure SO_2. The problem of removing sulfur dioxide from combustion gases is much more complex and difficult, as discussed in Sec. 11.5.

NO and NO_2 are not readily removed from gas streams by the process shown in Fig. 10.15. Although NO_2 is an acid gas that produces nitric acid by reaction with water,

$$3NO_2 + H_2O \rightarrow 2HNO_3 + NO \tag{11.8}$$

the reaction rate is slow. NO is not an acid gas, so that although we can remove NO_2 from a gas stream with an alkaline solvent, we cannot remove NO with the same solvent. For this reason, weak alkali solvents are not successful for the joint removal of NO and NO_2 or for the rapid removal of NO_2 alone. No other solvent is known that serves well for this task. (My generation has not found a suitable solvent to do this; fame and fortune await the person who finds a suitable solvent to remove NO, NO_2, and SO_2 economically from combustion gases by the scheme shown in Fig. 10.15!) The scheme in Fig. 10.15 is widely used in the chemical and petroleum industries to make separations not directly related to pollution control, e.g., the separation of CO_2 from H_2. The absorption column can also be used without regenerating the absorbent solution if the amount of material to be collected is small and there is some acceptable way of disposing of the loaded absorbent.

11.3.2 Sulfur Removal from Hydrocarbons

Once H_2S has been separated from the other components of the gas, it is normally reacted with oxygen from the air in controlled amounts to oxidize it only as far as elemental sulfur,

$$H_2S + \tfrac{1}{2}O_2 \rightarrow S + H_2O \tag{11.9}$$

and not as far as SO_2,

$$H_2S + \tfrac{3}{2}O_2 \rightarrow SO_2 + H_2O \tag{11.10}$$

The elemental sulfur is either sold for use in the production of sulfuric acid or landfilled if there is no nearby market for it. Although the chemical reaction in Eq. (11.9) for production of sulfur (the Claus process) is simple enough, there are a variety of ways of carrying it out, and the details can be complex; see Kohl and Nielsen [5, Chapter 8]. Hundreds of such plants operate successfully throughout the world; every major petroleum refinery has at least one.

Because elemental sulfur is inert and harmless and because reduced sulfur in the form of hydrogen sulfide or related compounds can be easily oxidized to sulfur or sulfur oxides, the entire strategy of the petroleum and natural gas industries in dealing with reduced sulfur in petroleum, natural gas, and other process gases is to keep the sulfur in the form of elemental sulfur or reduced sulfur (for example, H_2S). Oxygen from the air is virtually free, so we can always move in the oxidation direction at low cost. In contrast, hydrogen is an expensive raw material, so that moving in the reduction direction is expensive.

Sulfur in hydrocarbon fuels (natural gas, propane, gasoline, jet fuel, diesel fuel, furnace oil) is normally converted to SO_2 during combustion and then emitted to the atmosphere. Large oil-burning facilities can have equipment to capture that SO_2, but autos, trucks, and airplanes do not. The only way to limit the SO_2 emissions from these sources is to limit the amount of sulfur in the fuel. For this reason the Clean Air Act of 1990 (Section 217) limits the amount of sulfur in diesel fuel to 0.05 percent by weight. Crude oils vary in their sulfur contents: low-sulfur crudes are called "sweet"; high-sulfur crudes, "sour." If the fraction of the crude oil going to gasoline or diesel fuel has too high a sulfur content (which many do under current regulations), most of that sulfur is removed by catalytic hydrodesulfurization,

$$\begin{pmatrix}\text{Hydrocarbon}\\ \text{containing S}\end{pmatrix} + H_2 \xrightarrow{\text{Ni or Co catalyst promoted with Mo or W}} \text{hydrocarbon} + H_2S \tag{11.11}$$

The mixture leaving the reactor is cooled, condensing most of the hydrocarbons. The remaining gas stream, a mixture of H_2 and H_2S, is one of the streams treated in a refinery for H_2S removal by the process shown in Fig. 10.15. Some petroleum streams in refineries are treated over these catalysts to remove both sulfur and nitrogen because those elements interfere with the catalysts used for subsequent processing. The resulting gas streams contain both H_2S and NH_3.

Whether the treatment of gases with high concentrations of H_2S and NH_3 should be considered as air pollution control is an open question. For natural gas fields with H_2S, treatment is a market requirement, because the typical purchase specification for natural gas in the United States is $H_2S \leq 4$ ppm. However, at one time in oil refineries H_2S-containing gases were customarily burned for internal heat sources in the refineries if the H_2S content was modest. Current U.S. EPA air pollution regulations (NSPS, see Chapter 3) forbid the burning of such refinery waste gases if they contain more than 230 mg/dscm (dry standard cubic meter) of H_2S, so the removal of H_2S down to that concentration in oil refinery gases is done by the method shown in Fig. 10.15 to meet air pollution control regulations.

11.4 REMOVAL OF SO_2 FROM RICH WASTE GASES

The SO_2 concentrations in off-gases from the smelting of metal sulfide ores depend on which process is used and vary with time within the batch smelting cycle. However, they generally range from 2 percent to 40 percent SO_2. Such gases can be economically treated in plants that produce sulfuric acid by the following reactions:

$$SO_2 + 0.5O_2 \xrightarrow{\text{vanadium catalyst}} SO_3 \tag{11.12}$$

and

$$SO_3 + H_2O \rightarrow H_2SO_4 \tag{11.13}$$

Example 11.3. One of the largest copper smelters in the United States (Kennecott, at Salt Lake City) produces 320 000 tons of copper per year. The copper ore smelted is principally chalcopyrite. If all the sulfur were emitted to the atmosphere as SO_2, how much would be emitted? If all the sulfur in the ore were converted to sulfuric acid, how much sulfuric acid per year would the smelter produce?

From Eq. (11.3) we know that we would expect to produce 2 mols of SO_2 per mol of copper. The molecular weights are 64 for SO_2 and 63 for copper, so we would expect to produce 320 000 (ton/yr) $[(2 \cdot 64)/63] = 650\ 000$ ton/yr of SO_2. If this were all converted to H_2SO_4 (molecular weight 98), it would be 650 000 ton/yr $\times (98/64) = 996\ 000$ ton/yr.

From Table 11.2 we can see that if Kennecott did not capture any of its SO_2, this emission would have been 3.2 percent of the total U.S. SO_2 emissions. If it captured all the SO_2 and converted it to H_2SO_4, the resulting 996 000 ton/yr would be roughly 2 percent of the total U.S. production of H_2SO_4. Currently this smelter captures 99.9+ percent of that sulfur dioxide and converts it to sulfuric acid, emitting the rest to the atmosphere. ■

Equation (11.12) is an equilibrium reaction, which does not go to completion. Furthermore, this reaction is exothermic, so that the percent conversion at equilibrium is higher at low temperatures than at high temperatures. For this reason the reaction is customarily carried out in three or four separate catalyst beds with intercoolers between them. The course of the reaction is shown in Fig. 11.2 on page 406.

The preheated feed gas enters the first catalyst bed at about 420°C. The exothermic reaction heats it. All of the catalyst beds except the last are small enough (have low-enough residence time) that the gases do not come close to their equilibrium conversion. The gases leave each of the first three beds and are cooled before entering the next bed. As shown on the figure, the equilibrium conversion decreases rapidly with increasing temperature. The exit gas from the final bed is at about 425°C and close to equilibrium at this temperature. In this way about 98 percent of the incoming

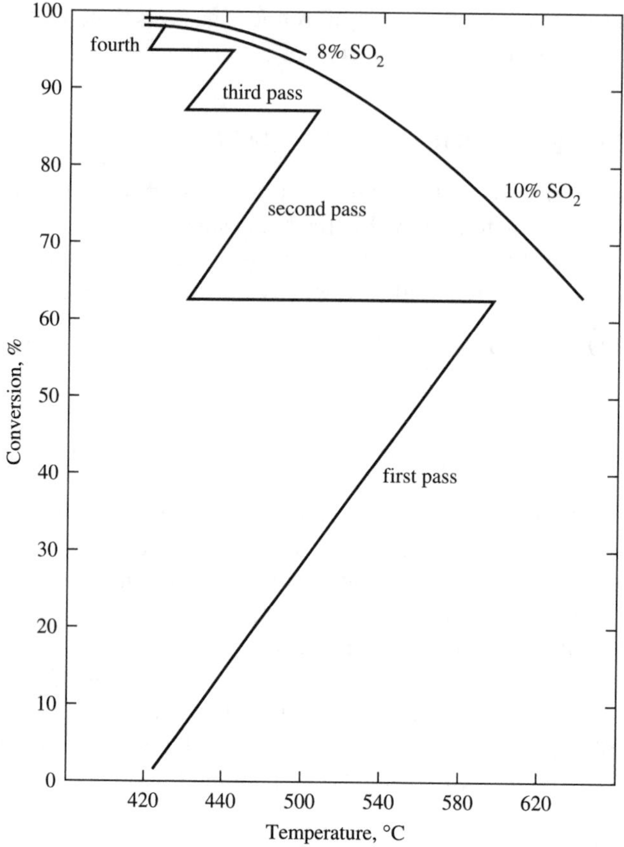

FIGURE 11.2
Temperature-conversion history for a simple four-bed SO_2 converter in a sulfuric acid plant. The sloping lines represent the conversion-temperature behavior inside the catalyst beds. The horizontal lines represent the behavior in the intercoolers. The curves at the upper right show the equilibrium conversion as a function of temperature and initial SO_2 content. (J. R. Donovan and J. M. Salamone, "Sulfuric Acid and Sulfur Trioxide," in *Kirk-Othmer Encyclopedia of Chemical Technology,* 3d ed., Volume 22, Copyright © 1983 by John Wiley & Sons, Reprinted by permission of John Wiley & Sons. [8]).

SO_2 can be converted to H_2SO_4. In a simple acid plant, the remaining 2 percent of the SO_2 is vented to the atmosphere.

In response to legal pressure to reduce this SO_2 emission, plants have been developed that are, in effect, two plants in series; i.e., equilibrium in Eq. (11.12) is obtained, the SO_3 is removed by Eq. (11.13), and then the "tail gas" is again passed over a catalyst so that it will again come to equilibrium according to Eq. (11.12). The resulting gas is again contacted with water so that the SO_3 is absorbed according to Eq. (11.13). Such plants are called *double contact* or *double absorption* plants. They can typically convert over 99.7 percent of the incoming SO_2 to H_2SO_4 [9]. Because the feed material (smelter off-gas or SO_2-bearing gas made by burning sulfur) is quite cheap, the extra cost of the second reaction and absorption step is not repaid by increased acid production and is truly a pollution control expenditure. Figure 11.3 shows simplified flow diagrams for single- and double-absorption sulfuric acid plants.

Comparing this flow diagram to Fig. 10.15, we see an SO_3 absorber but no stripper. There is no need to regenerate the absorbing liquid, because the solution of SO_3 in water, H_2SO_4, is the saleable product. The Henry's law constant for SO_3 in water at 20°C is roughly 10^{-25} atmosphere, so this absorption is very rapid and easy. In fact, it is too easy; a fine mist of sulfuric acid drops is formed, which must be removed before the waste gas passes to the atmosphere. Although there are interesting technical challenges in the design and operation of H_2SO_4 plants [10], hundreds of these plants operate satisfactorily throughout the world.

Sulfuric acid is the cheapest industrial acid, and it is used whenever an inexpensive acid is needed; it has many uses [11]. The largest use, consuming the bulk of the world's sulfuric acid, is the production of phosphate fertilizer. Naturally occurring phosphate rock contains fluorapatite, $Ca_{10}F_2(PO_4)O_6$. This mineral is quite insoluble in water and, hence, of no use as a fertilizer. Reacting it with sulfuric acid converts the insoluble fluorapatite to water-soluble phosphoric acid, which is a useful fertilizer ingredient (and also produces an HF waste gas that must be collected and a $CaSO_4$ waste that is landfilled). Generally the most economical procedure in

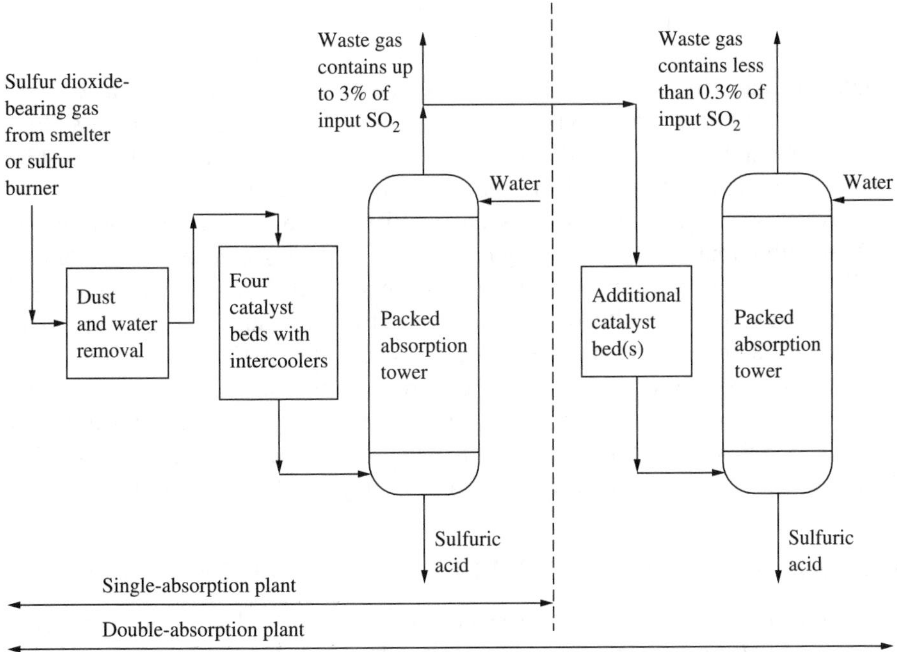

FIGURE 11.3
Block diagrams of single-absorption and double-absorption sulfuric acid plants.

the fertilizer industry is to ship solid sulfur to the phosphate rock deposit and make sulfuric acid there, rather than ship sulfuric acid to the phosphate rock deposit, or phosphate rock to the sulfuric acid source. The reason is that one pound of sulfur (atomic weight 32) makes 3.06 pounds of sulfuric acid (molecular weight 98); and sulfur is a nontoxic, noncorrosive solid that is cheaper to ship than toxic, corrosive liquid sulfuric acid. Thus large-scale production of sulfuric acid from smelter off-gases is only economical when there is a nearby deposit of phosphate rock. (Perhaps in the future we will ship metal sulfide ore concentrates to the place where the phosphate rock is and smelt them there.) These economic facts place some limitations on the application of sulfuric acid conversion as an SO_2 control technology.

An additional limitation is that production of H_2SO_4 is uneconomic when the concentration of SO_2 in the waste gas is too low. Most analysts believe that with SO_2 concentrations more than 4 percent, the acid plant can show a profit if there is a nearby market for the acid, but below this concentration it cannot. The reason is that both the capital and operating costs of the plant depend on the volumetric flow rate of gas processed [12]. To process a given volumetric flow rate Q of gas requires the same size pipes, blowers, absorbers, etc. regardless of the percent SO_2 contained in the gas; and the power required to move the gas through the plant is also proportional to Q and independent of the SO_2 content. But the acid production rate is proportional to the product of the volumetric flow rate and the percent SO_2 in the feed, $Q \cdot y_{SO_2}$, so that

$$\begin{pmatrix}\text{Cost per unit}\\ \text{of acid produced}\end{pmatrix} \propto \frac{Q}{Q \cdot y_{SO_2}} \approx \frac{1}{y_{SO_2}}$$

In addition, plants treating gases with more than about 4 percent SO_2 are *autothermal,* meaning that the heat of reaction provides all of the heat needed for the process. With less than about 4 percent SO_2 the plants are not autothermal; external heat must be supplied, adding fuel cost and increasing the plant complexity.

Traditional smelting processes are multistage, producing some waste gas streams with enough SO_2 to treat by conversion to H_2SO_4 and some too dilute to treat economically this way. Faced with pressure not to discharge these dilute gases to the atmosphere, the smelter operators have had two alternatives: (1) treat the dilute gases by the methods for lean gases described in Sec. 11.5, or (2) modify their smelting processes to eliminate or minimize the production of lean gases so that all the waste gases are rich enough to be economically treated in sulfuric acid plants. Most have decided that the second option is the more economical. Some have modified their processes to use O_2 in place of air, thus raising the SO_2 concentration to 40%.

11.5 REMOVAL OF SO_2 FROM LEAN WASTE GASES

The major source of SO_2, except near uncontrolled copper, lead, zinc, and nickel smelters, which no longer exist in the United States, but do in some developing countries, is the stacks of large coal- or oil-burning facilities. Most of the largest ones are coal-burning electric power plants (see Table 11.2). For them, the typical

SO_2 content of the exhaust gas is about 0.1 percent SO_2, or 1000 ppm (see Example 7.10), which is much too low for profitable recovery as H_2SO_4.

The most widely used procedure for controlling SO_2 emissions from these sources is scrubbing with water containing finely ground limestone; the overall effect is shown by Eq. (11.2). The whole process is called *flue gas desulfurization*, and the acronym FGD is widely used. The installation and operation of FGD devices are major costs for the electric utility industry, which is continually trying to improve FGD performance and to develop less expensive alternative processes [13]. To understand these devices, we begin with a simpler problem.

Example 11.4. A power plant produces 10^6 scfm (471.9 sm^3/s) of exhaust gas with 0.1 percent SO_2. We are required to remove 90 percent of it before the gas is discharged into the atmosphere. We propose to do so by dissolving the gas in water taken from the nearby river, using an absorber column like the left column in Fig. 10.15. How large a flow of water will be needed if we make the same requirement as in Example 11.1, that at the bottom of the absorber $y_i^* = 0.8y_i$?

This situation is clearly similar to Example 11.1. However, here the Henry's law constant for SO_2 is much smaller, leading to a greater solubility. Here the use of Henry's law is speculative, because the apparent Henry's law constant depends strongly on the acidity or alkalinity of the water (see Problems 11.26 and 11.27). But for this example a Henry's law constant of 9 atm gives close to the right result. The pressure is 1 atm instead of 100 atm; and the gas flow rate is much larger. The outlet liquid will have the following characteristics:

$$x_i = \frac{P \cdot 0.8y_i}{H_i} = \frac{1 \text{ atm} \cdot 0.8 \cdot 0.001}{9 \text{ atm}} = 0.000089 = 0.0089 \text{ mol\%}$$

$$\frac{L}{G} = \frac{Y_{i \text{ bottom}} - Y_{i \text{ top}}}{X_{i \text{ bottom}} - X_{i \text{ top}}} \approx \frac{y_{i \text{ bottom}} - y_{i \text{ top}}}{x_{i \text{ bottom}} - x_{i \text{ top}}} = \frac{1000 \text{ ppm} - 100 \text{ ppm}}{0.000089 - 0} = 10.5$$

$$G = 10^6 \text{ scfm} \cdot \frac{\text{lbmol}}{385.3 \text{ scf}} \cdot \frac{\text{min}}{60 \text{ s}} = 43.3 \frac{\text{lbmol}}{\text{s}} = 1.96 \times 10^4 \frac{\text{mol}}{\text{s}}$$

$$L = 10.1 \cdot 43.3 \frac{\text{lbmol}}{\text{s}} = 438 \frac{\text{lbmol}}{\text{s}} = 7900 \frac{\text{lb}}{\text{s}}$$

$$= 3585 \frac{\text{kg}}{\text{s}} = 126 \frac{\text{ft}^3}{\text{s}}$$

■

There are several drawbacks to this procedure for dealing with the SO_2 from an electric power plant. First, it requires a large amount of water. The computed water flow is approximately 1 percent of the flow of the Hudson River at New York City. Power plants located on the Hudson, the Mississippi, the Ohio, or the Columbia rivers could obtain such amounts of water, but most of the power plants in the world could not. Second, the waste water stream, which is 80 percent saturated with SO_2, would emit this SO_2 back into the atmosphere at ground level (river level), causing an SO_2 problem that might be more troublesome than the emission of the same amount of SO_2 from the power plant's stack. Third, in aqueous solution SO_2 undergoes Reaction (11.12), (without the catalyst) which would remove most of the dissolved O_2 in the river, making it impossible for fish to live in it. For this reason alone, simple dissolution of large quantities of SO_2 in most rivers is prohibited.

However, the first large power plant to treat its stack gas for SO_2 removal did remove SO_2 with river water. The Battersea Plant of the London Power Company is located on the banks of the Thames River, which is large enough to supply the water it needed [14]. Furthermore, the water of the Thames is naturally alkaline because its course passes through many limestone formations, so that it will absorb substantially more SO_2 than would pure water. To prevent the dissolved SO_2 from consuming O_2 in the river, the effluent from the gas washers was held in oxidizing tanks, where air was bubbled through it until the dissolved SO_2 was mostly oxidized to sulfate (SO_4^{2-}), before being discharged to the Thames. In this form the sulfur has a low vapor pressure and does not reenter the air nor kill the fish by consuming the river's dissolved oxygen. Although this pioneering plant had its problems, it was a technical success—removing over 90 percent of the SO_2—and operated from 1933 to 1940. (The SO_2 removal system was shut down in 1940 because the exhaust plume from this plant was wet due to the scrubber and, hence, very visible. It made a good navigation marker for German aircraft during the Battle of Britain.)

As we saw in Example 11.2, the amount of scrubbing water required can be substantially reduced if we add a reagent to the water that increases the solubility of the gas being removed.

Example 11.5. The power plant in Example 11.4 wishes to remove 90 percent of the SO_2 by scrubbing the exhaust gas with a dilute solution of sodium hydroxide, NaOH. How much sodium hydroxide will they need? What problems will they encounter?

The overall reaction (including the oxidation of sulfite to sulfate) will be

$$2NaOH + SO_2 + \tfrac{1}{2}O_2 \rightarrow Na_2SO_4 + H_2O \tag{11.14}$$

From Example 11.4, we know that we must remove (0.9)(0.001)(43.3 lb mol/s) = 0.039 (lbmol/s) of SO_2. Therefore, we will need, as a minimum, 2(0.039) = 0.078 lbmol/s of NaOH (35 mol/s). One lbmol of NaOH weighs 40 lb, so the annual NaOH requirement will be $0.078 \cdot 40 \cdot (3.15 \times 10^7\ \text{s/yr}) = 98.4 \times 10^6\ \text{lb/yr} =$ 49 200 t/yr = 44 700 tonne/yr. The prices of industrial chemicals fluctuate, but the price of sodium hydroxide is about \$500/ton (dry basis), so that the sodium hydroxide for this plant would cost about \$22 million per year. ■

Comparing this problem to the H_2S removal problem in Examples 11.1 and 11.2, we see that:

1. The volumetric flow rate of the gas is about 1700 times that in the H_2S removal problem (14 times because of the higher molar flow rate, and 120 times because of the lower gas density At 100 atm, methane has ≈ 1.2 times the density of a perfect gas).

2. The power cost to drive the gas through the scrubber (see Chapter 7) is thus 1700 times as large, for an equal ΔP. Thus minimizing pressure drop is much more important in this problem than in that.
3. Here there is no regenerator. If we regenerated the solution to produce a stream of practically pure SO_2 we would have no economical way of converting it to a harmless solid, as the Claus process does with H_2S.

In the previous examples we said little about the internal features of the absorbing column. For the high-pressure treatment of H_2S, either plate or packed towers are used, with little problem. For the SO_2 problem, three plausible arrangements are sketched in Fig. 11.4. The first of these is a simple bubbler, in which the gas is forced under pressure through perforated pipes submerged in the scrubbing liquid. As the bubbles rise through the liquid, they approach chemical equilibrium with it. If the liquid is deep enough and the bubbles are small enough, this kind of device will bring the gas close to chemical equilibrium with the liquid. However, it has a high pressure drop. The gas pressure must at least equal the hydrostatic head of the liquid. If, for example, the liquid is a foot deep, then the hydrostatic head will be 12 inches of liquid, which is large enough to be quite expensive (see Example 7.3). Plate-type distillation and absorption columns are, in effect, a series of such bubblers, stacked one above the other, with the gas flowing up from one to the next and the liquid flowing down from one to the next through pipes called *downcomers*. At high pressures, where pressure drops are unimportant, they are the most widely used device.

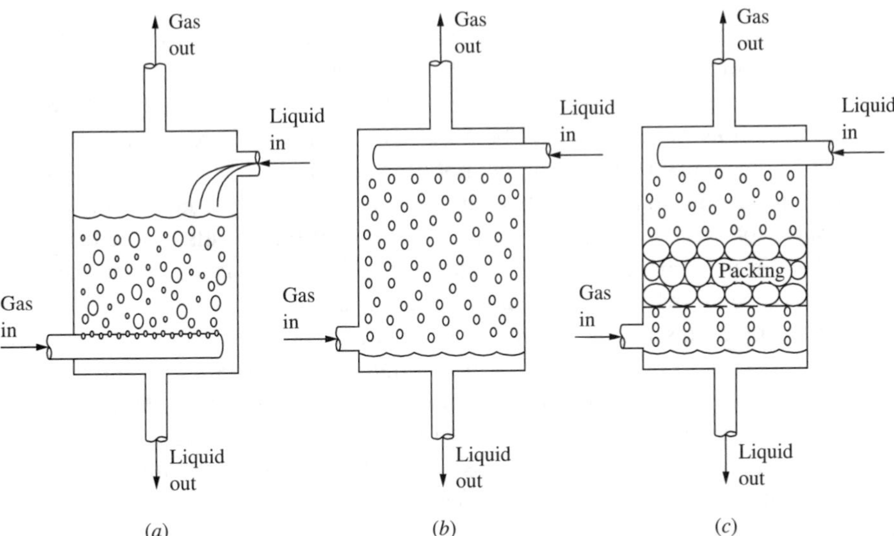

FIGURE 11.4
Three plausible arrangements for scrubbing a gas with a liquid: (*a*) bubbler, (*b*) spray chamber, and (*c*) packed column.

The second arrangement is a spray chamber. In it the gas flows up through an open chamber while the scrubbing liquid falls from spray nozzles, much like the heads in bathroom showers, through the gas. In this arrangement the gas pressure drop is small, but it is difficult to approach equilibrium because the gas does not contact the liquid as well as it does in the bubbler. Nonetheless, it is widely used because of its simplicity, low pressure drop, and resistance to scale deposition and plugging.

The third arrangement is a packed column, which is similar to the spray chamber except that the open space is filled with some kind of solid material that allows the liquid to coat its surface and run down over it in a thin film. The gas passes between pieces of solid material and comes in good contact with the liquid films. In the most primitive of these, the solid materials were gravel or crushed rocks. More advanced ones use special shapes of ceramic, plastic, or metal that are fabricated to provide the optimum distribution of liquid surface for contact with the gas. This third kind of contactor can be designed to have a better mass transfer per unit of gas pressure drop than either of the other two kinds. All three of these arrangements, plus combinations of them, plus some other arrangements are in current use for removal of SO_2 from power plant stack gases.

The gas velocities in such devices range from about 1 ft/s in a packed tower to 10 ft/s for a spray chamber. If we assume we are going to treat the gas in Example 11.5 in a spray chamber at a gas velocity of 10 ft/s, the cross-sectional area perpendicular to the gas flow will be

$$A = \frac{Q}{V} = \frac{10^6 \text{ ft}^3/\text{min}}{10 \text{ ft/s}} \cdot \frac{\text{min}}{60 \text{ s}} = 1667 \text{ ft}^2 = 155 \text{ m}^2$$

Such devices are almost always cylindrical, because that shape is easier and cheaper to fabricate than, for example, a rectangular vessel of equal cross-sectional area. For this example the diameter would be $(4 \cdot 1667 \text{ ft}^2/\pi)^{0.5} = 46 \text{ ft} = 14 \text{ m}$. A typical length in the flow direction would be 50 ft. That is a very large diameter for any piece of chemical plant equipment, but not for a power plant. Often the flow will be divided into several smaller scrubbers in parallel. This choice avoids having to ship or fabricate too large a vessel and ensures that one of the vessels can be taken out of service for maintenance while the rest are in operation. Thus the power plant can continue to operate while one part of the scrubber is out of service.

What problems might power plant operators encounter? First, there is the question of what to do with the sodium sulfate produced. Sodium sulfate (also called "salt cake") is used in detergent manufacture and in paper making, as well as in some miscellaneous uses. However, for those uses it must be quite pure. The sodium sulfate produced in this process would be contaminated with fly ash from the coal. (In most such operations the scrubber is downstream of an electrostatic precipitator, but even so some particles pass through the precipitator and are caught in the scrubber.) Thus if we wished to sell the sodium sulfate, we would have to get it out of solution (by evaporation and crystallization) and then purify it. If we did, we would find that the total amount produced in a few power plants would glut the current market, so that although a few power plants might sell their sodium sulfate, most could not. Because of its water solubility, it is not generally acceptable in landfills unless they are well protected from water infiltration.

But the real difficulty is with carbon dioxide. Here we assumed that we could treat the exhaust gas with dilute alkaline solutions and remove the SO_2, which is an acid gas. However, the exhaust gas from combustion sources contains another acid gas, CO_2. Normally its concentration is about 12 percent, or 120 times that of the SO_2. We are not generally concerned with the fate of CO_2, but if it gets into solution it will use up sodium hydroxide by the reaction

$$2\,NaOH + CO_2 \rightarrow Na_2CO_3 + H_2O \tag{11.15}$$

Any sodium hydroxide used up this way is not available to participate in Reaction (11.14). The real problem is how to absorb one acid gas while not absorbing another acid gas that is present in much higher concentration!

Fortunately, this is possible because SO_2 forms a much stronger acid than does CO_2. The reactions that occur in the liquid phase are these:

$$CO_2(\text{gas}) \rightleftarrows CO_2(\text{dissolved}); +H_2O \rightleftarrows H_2CO_3 \rightleftarrows H^+ + HCO_3^- \tag{11.16}$$

$$SO_2(\text{gas}) \rightleftarrows SO_2(\text{dissolved}); +H_2O \rightleftarrows H_2SO_3 \rightleftarrows H^+ + HSO_3^- \tag{11.17}$$

These show that each of the gases goes from the gas state to the dissolved state, then reacts with water to form the acid, which then dissociates to form hydrogen ion and the bisulfite or bicarbonate ion. If we find the right concentration of H^+ in solution, it may be possible to drive the equilibrium in Eq. (11.16) to the left while driving the equilibrium in Eq. (11.17) to the right. That is indeed possible if the concentration of hydrogen ions is between 10^{-4} and 10^{-6} mols per liter (pH$=$ 4 to 6) [15]. But this calculation shows that we cannot use an alkaline scrubbing solution at all; alkaline solutions have pH values of 7 or more. To remove SO_2 without absorbing CO_2, we must use a scrubbing solution that is a *weak acid*. Furthermore, we must be careful to control the pH of our solution so that it is acid enough to exclude CO_2 but not acid enough to exclude SO_2. As the solution absorbs SO_2 it becomes more acid, and thus less able to absorb SO_2. Controlling pH during the SO_2 absorption process is of crucial importance to the operation of these devices (see Problem 10.29).

If the problem were to use NaOH to remove SO_2 from a gas stream that contained no other acid gases, this would be a simple problem for which ordinary chemical engineering techniques would be satisfactory. The real problem is different from this one for the following reasons:

1. There is another acid gas, CO_2, present that will use up our alkali unless we keep the solution acid enough to exclude it.
2. The amount of alkali needed is high, and the cost of sodium hydroxide is enough that we would prefer to use a cheaper alkali if possible.
3. We have to do something with the waste product, either sell it or permanently dispose of it.

4. Because the volume of gas to be handled is very large, we must be very careful to keep the gas pressure drop in the scrubber low. The pressure drops that are normally used in the chemical and petroleum industry in gas absorbers are much too large to be acceptable here.

11.5.1 Forced-Oxidation Limestone Wet Scrubbers

The most widely used process to deal with these problems is forced-oxidation limestone wet scrubbing. There are a variety of flowsheets and of mechanical arrangements for this process; Figs. 11.5 and 11.6 (page 415) and Table 11.3 on page 416, show one of the most commonly used varieties. In it we see that the flue gas, from which the solid fly ash particles have been removed, passes to a scrubber module where it passes countercurrent to a scrubbing slurry containing water and limestone particles (as well as particles of other calcium salts). In principle this is the same as the H_2S scrubber in Examples 11.1 and 11.2. Figure 11.6 shows the scrubber module as a vertical spray tower column with a single gas-liquid contacting tray, and with the bottom serving as a liquid storage and oxidation tank. At the top are two levels of *entrainment separators*. (These are often called *Demisters,* which is a brand name for one type.) The separators in the figure are *chevron type*. These devices cause the fine droplets carried with the gas to collect on their surfaces, coalesce, and fall back into the scrubber as drops large enough to fall counter to the upward-flowing gas.

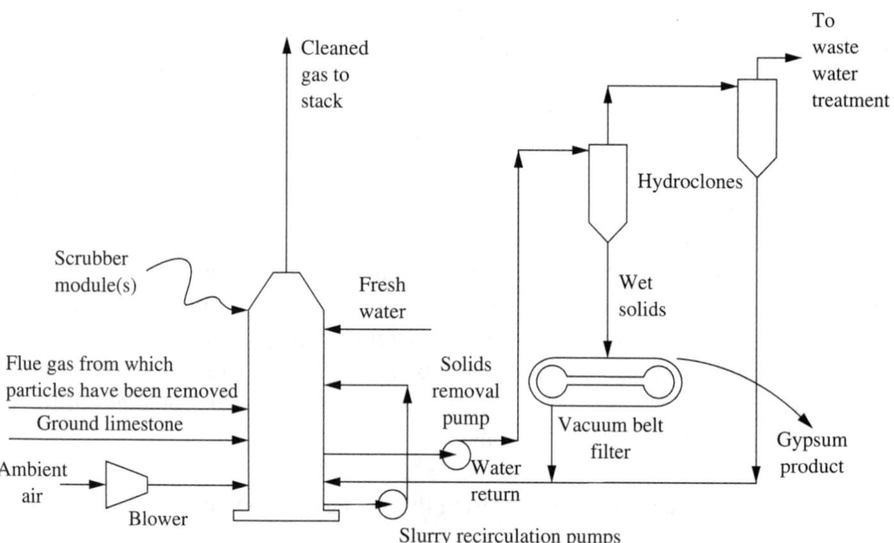

FIGURE 11.5
Flow diagram for forced-oxidation limestone wet scrubbing. The scrubber module is shown in detail in Fig. 11.6. The process is described in the text.

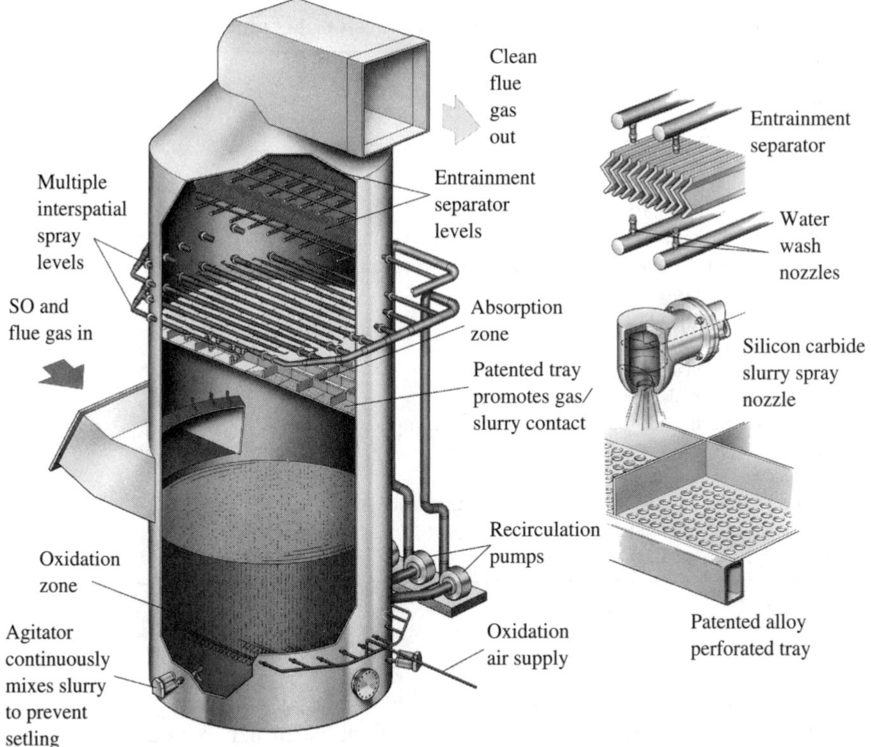

FIGURE 11.6
Internal details of the scrubber module, which has one sieve tray without downcomers, several levels of spray nozzles, two levels of entrainment separators, and a large sump that serves as a forced oxidation reactor. For large installations there will often be several such modules operating in parallel. (Courtesy of The Babcock and Wilcox Company, Barberton, Ohio.)

Some other designs use a packing with a very high open area in the tower or specialized bubbler designs. In the tower the SO_2 dissolves in the slurry and reacts with limestone (and the other dissolved and suspended calcium salts) producing CO_2, which enters the gas stream, and solid $CaSO_3$. The latter is almost entirely oxidized to $CaSO_4$, partly by the excess oxygen in the flue gases in the tower, mostly in the bottom of the scrubber module. In earlier designs the oxidation took place in a separate vessel, but most current devices use the bottom of the scrubber as a liquid oxidation reactor.

The slurry of water and solid particles ($CaCO_3$, $CaSO_4 \cdot 2H_2O$, and $CaSO_3 \cdot 0.5H_2O$) is pumped from the sump at the bottom of the module to the sprays, where it forms drops that fall through the rising flue gas and do the actual SO_2 removal. Finely ground limestone is added to the sump. A small stream of slurry is sent by the solid removal pump to a hydroclone (much like a multiclone, Fig. 9.5) from which

TABLE 11.3
Design values for a typical forced-oxidation limestone scrubber

Inlet temperature, °F	400
Outlet temperature, °F	125
Inlet SO_2 concentration, ppm	1000
Collection efficiency for SO_2	0.9
Pressure drop, inlet to outlet, inches of water	10
Vertical upward gas velocity, ft/s	10
Liquid/gas ratio in scrubber. This is almost always stated as gallons of liquid per 1000 actual cubic feet of gas, **measured at the inlet temperature**	100
Height of the scrubbing zone, ft	30
Typical drop size of slurry, mm	3
Solids content of slurry, wt %	15
Limestone feed rate, mols/mol SO_2 in the feed gas	1.1
Oxidation air feedrate, mols/mol SO_2 in the feed gas	1.5
Liquid residence time in oxidation tank, between recycle trips through scrubber, minutes	6
Solids residence time in oxidation tank, minutes	18
Water content of stream passing from first hydroclone to belt filter, wt %	30
Water content of solid gypsum leaving belt filter, wt %	10
Fraction of $CaSO_3$ oxidized to $CaSO_4$, mol/mol	0.95

the underflow passes to a belt filter, from which a semidry gypsum product leaves the system. The overflow from the first hydroclone is further treated to produce a liquid waste stream with a very low solids content, which goes to water treatment and disposal. The fresh water enters the system almost exclusively as wash water for the entrainment separators (see Fig. 11.6). The scrubber operates at or near the adiabatic saturation temperature (Sec. 7.11) of the entering flue gas, which is about 125°F. In some installations the cleaned flue gas is reheated to about 175°F to restore plume buoyancy and prevent acid corrosion of the ducts and stack downstream of the reheater. Other installations discharge the gas at scrubber temperature and use corrosion-resistant materials to deal with the small amount of acid liquid that is not removed by the entrainment separators.

Example 11.6. The spray nozzles in Figure 11.6 are normally designed to produce droplets with an average diameter of about 3 mm.

(*a*) How fast do those fall, relative to the gas and relative to the container?

(*b*) For the slurry flow rate shown in Table 11.3, what is the equivalent rainfall rate in in./s?

(*c*) What fraction of the droplets evaporates to bring the gas from its inlet temperature to its adiabatic saturation temperature?

(*d*) What fraction of the $CaCO_3$ will react as the slurry makes one pass through the scrubber?

Solution

(*a*) From Fig. 8.7 we read that a 3 mm sphere with a specific gravity (s.g.) of 1.0 would be expected to fall about 20 ft/s. (The s.g. will be slightly more than 1 because of the dissolved and dispersed solids, and the shape will vary from a sphere because the drop is fluid. The estimate of 20 ft/s is still plausible.) This velocity is relative to the gas; relative to the walls of the scrubber, the velocity will be $(20 - 10) = 10$ ft/s.

(*b*) The equivalent rainfall rate is

$$\frac{Q_L}{A} = \frac{Q_L}{Q_G} \cdot \frac{Q_G}{A} = \frac{Q_L}{Q_G} V_G = \frac{100 \text{ gal}}{1000 \text{ acf}} \cdot \frac{860°\text{R}}{585°\text{R}} \cdot 10 \, \frac{\text{ft}}{\text{s}} \cdot \frac{\text{ft}^3}{7.48 \text{ gal}}$$
$$= 0.197 \, \frac{\text{ft}}{\text{s}} = 2.36 \, \frac{\text{in.}}{\text{s}} = 60 \, \frac{\text{mm}}{\text{s}}$$

(The temperature ratio appears because the liquid rates are stated based on inlet conditions, whereas the gas velocity is based on scrubber temperature.) This is far beyond any outdoor rainfall rate. Cloudbursts sometimes deliver a few inches of rain an hour; this is 2 inches per second.

(*c*) From Example 7.12 we know that the evaporated water ratio is given by

$$\left(\frac{\dot{m}_{\text{water}}}{\dot{m}_{\text{combustion gases}}}\right)_{\text{evaporated}} \approx -\frac{C_p \, \Delta T}{\lambda}$$
$$= -\frac{(0.25 \text{ Btu/lb} \cdot °\text{F})(125 - 400)°\text{F}}{1055 \text{ Btu/lbm}}$$
$$= 0.065 \, \frac{\text{lb water}}{\text{lb air}}$$

The liquid-gas ratio passing up and down the scrubber is

$$\left(\frac{\dot{m}_{\text{water}}}{\dot{m}_{\text{gas}}}\right)_{\text{total}} \approx \frac{100 \text{ gal}}{1000 \text{ acf}} \cdot \frac{8.3 \text{ lb water}}{\text{gal}} \cdot \frac{\text{ft}^3}{0.075 \text{ lbm gas}} \cdot \frac{860°\text{R}}{528°\text{R}}$$
$$= 18.0 \, \frac{\text{lb water}}{\text{lb gas}}$$

so that about 0.4 percent of the water is evaporated inside the scrubber vessel.

(*d*) We see that 1000 acf at scrubber inlet temperature corresponds to 614 scf, which corresponds to 1.71 lbmol, containing 0.00171 lbmol of SO_2. The 90% of that which is captured corresponds to 0.00154 lbmol, which equals the number of mols of $CaCO_3$ reacted. The corresponding slurry is 100 gal $\cdot$ 8.3 lb/gal $\approx$ 830 lb, of which 15% or 124.5 lb is solids, of which 11.6% or 14.4 lb (see Problem 11.16) is $CaCO_3$. Thus the fraction of the contained $CaCO_3$ that could be converted in one pass through the scrubber is (0.00154 lbmol$\cdot$100 lb/lbmol)/14.4 lb) $\approx 1\%$. ■

This example shows that the liquid circulation rates in these scrubbers are *very, very large*. As a consequence, even though they remove most of the SO_2 from the gas, the scrubbing slurry passes through them practically unchanged. Most of the chemical reactions take place in the effluent hold tank. (The slurry spends about 3 seconds per pass in the scrubber and about 8 minutes between passes through the scrubber in the hold tank.) Observations suggest that very little of the preceding 1% possible reaction occurs while the drops are falling; there is not enough time. This results in the pH of the drops declining as they fall, reducing their absorptive capacity (see Problems 11.28 and 11.29).

11.5.1.1 The development problems with limestone scrubbers. In the 1970s and early 1980s the electric utility industry suffered through the very painful development period of limestone scrubbers. By now the major problems have largely been solved, and these devices are reasonably reliable and useful if designed and operated properly. The major development problems were these:

1. Corrosion: The exhaust gases from coal combustion contain small amounts of many chemicals, e.g., chlorides (see Problem 11.16). In an acid environment these proved much more corrosive to metals, including stainless steels, than the designers of the first systems had anticipated.
2. Solids deposition, scaling and plugging: Calcium sulfate and its near chemical relatives are slightly soluble in water and can precipitate on solid surfaces to form hard, durable scales that are very difficult to remove. These are the "boiler scales" that collect in teapots and hot water heaters. The scales formed in valves, pumps, control instruments, and generally anywhere that their effect could cause the most trouble.
3. Entrainment separator plugging: The spray nozzles shown in Fig. 11.6 do not produce totally uniform drops; some of the drops are small enough to be carried along with the gas and must be removed from the gas in the entrainment separator. If they are not removed, they will plug and corrode the ductwork downstream of the scrubber. The early entrainment separators were plugged by the solids contained in those small drops.
4. Poor reagent utilization: The product sulfates and sulfites can precipitate on the surface of the limestone particles, thus blocking their access to the scrubbing solution. This caused a high percentage of the limestone to pass unreacted into the solid waste product, raising reagent and waste disposal costs.
5. Poor solid-liquid separation: $CaSO_3 \cdot 0.5H_2O$ tends to form crystals that are small, flat plates. These are very good at trapping and retaining water. If the solid product has too many of these it will have the consistency of toothpaste and not be acceptable for landfills. $CaSO_4 \cdot 2H_2O$ forms larger, rounder crystals that are much easier to settle and filter. Flocculating agents added to the thickener improve this separation.

The solution to these problems has been found by careful attention to engineering and chemical detail. The rate of liquid rejection to waste water (Fig. 11.5) is chosen to control the chloride content of the circulating liquid. It is kept low enough to protect the very expensive materials it contacts. (The most widely used metal for lining the surfaces of the modules is alloy C-276, 55% Ni, 17% Mo, 16% Cr, 6% Fe, 4% W. It costs roughly 15 times as much as ordinary steels.) The solids deposition was caused by local supersaturation with gypsum. Enough gypsum is kept in the circulating slurry to prevent that supersaturation, vastly reducing the scale deposition. The original entrainment separators were of the woven wire variety, which plugged easily. The chevron type shown in Fig. 11.6 is much easier to keep clean. All the fresh water entering the system comes in as entrainment separator wash water, which is applied as strong jets for a few minutes of each hour. The liquid holding tanks were made larger, thus allowing more time for the reagent to dissolve. This additional time plus more vigorous application of oxidation air resulted in converting $\approx 95\%$ of the captured sulfur to gypsum, which forms large, easily filtered crystals. Some plants produce a gypsum waste stream clean enough and dry enough that wallboard manufacturers will purchase it, thus converting the plant's waste disposal cost to a by-product sale.

In principle these systems are designed by the same methods as in Examples 11.1 and 11.2. In practice the five problems just listed have dominated the design and all efforts have been devoted to overcoming these problems. Most of these problems are now solved or reduced to manageable proportions by careful control of the process chemistry, good mechanical design, and careful operation. The design and operating practices are continuously being improved [13]. But these scrubbers are still expensive and troublesome, and they generate large amounts of solid waste, which are a disposal problem. Detailed descriptions of the design, chemistry, and operating experience of these scrubbers have been published [16, 17] .

11.5.2 Other Approaches

During the period of development of the limestone scrubber, when its growing pains seemed unendurable (many believed that it would *never* work satisfactorily), many other approaches to the problem were suggested and tested. As the technical difficulties with the limestone scrubber were worked out, it became the clear economical choice for scrubbing stack gas from the combustion of medium- or high-sulfur coal. The other processes are not being used now for new installations, and some of those installed 20 years ago are being converted to forced oxidation limestone scrubbers to save operating costs. Table 11.4 on page 420 compares these processes.

11.5.2.1 Other wet systems. $Ca(OH)_2$ (hydrated lime, quicklime) is an alternative to limestone in wet throwaway processes. (*Throwaway processes* are ones in which the reagent is used once and then thrown away.) Its use is similar to that of limestone, shown in Fig. 11.6. Normally, CaO (lime, burned lime) is added to the oxidation tank and hydrates there to $Ca(OH)_2$. It is more chemically reactive than limestone, mostly because it has a much higher surface area. (CaO is prepared by heating limestone and driving off the CO_2. The result is a porous structure, as discussed in Sec. 7.13.

TABLE 11.4
Some possibilities for removing SO_2 from dilute gas streams

	Reagents				
Processes	**$CaCO_3$, limestone**	**$Ca(OH)_2$, hydrated lime (quicklime)**	**Na_2CO_3, sodium carbonate**	**$NaHCO_3$, sodium bicarbonate**	**Regenerable adsorbents or absorbents**
Wet throwaway	Limestone scrubbing	Lime scrubbing			
Double alkali			Double alkali	Double alkali	
Dry throwaway	Boiler limestone injection	Boiler lime injection	Boiler or flue injection	Boiler or flue injection	
Wet-dry		Spray dryers	Spray dryers	Spray dryers	
Regenerative					Many kinds, producing SO_2 or S or H_2SO_4. Some control both SO_2 and NO_x.

Typical surface areas are 15 m^2/g.) But to use CaO requires an extra process step to prepare it for insertion in the process shown in Fig. 11.6. In the early days of scrubber development this extra reactivity seemed necessary, but as the problems with wet limestone scrubbers have mostly been solved, the additional reactivity of lime has seemed less likely to repay its extra cost. The other reagents shown are all more expensive than limestone and would not be used in a wet, throwaway process where cheap limestone can be made to work.

Table 11.4 also lists double alkali systems as wet throwaway systems. These were devised in the early days of limestone scrubber development in response to the problems of solid deposition, scaling, and plugging that resulted from using calcium compounds in the scrubber. In double alkali processes the scrubbing step is done with a sodium carbonate or sodium bicarbonate solution in the presence of a very low concentration of calcium. The solubility of sodium salts is much higher than that of calcium salts, so that in the scrubber all the salts are in solution and the liquid is practically free of solids.

The liquid is taken out of the scrubber, the alkali is regenerated with lime or limestone in a reaction tank. The main reaction in the scrubber is

$$Na_2CO_3 + SO_2 \rightarrow Na_2SO_3 + CO_2 \tag{11.18}$$

The overall reaction in the reaction tank is

$$Na_2SO_3 + CaCO_3 + 0.5O_2 + 2H_2O \rightarrow CaSO_4 \cdot 2H_2O + Na_2CO_3 \tag{11.19}$$

which regenerates the sodium carbonate (or bicarbonate) in solution and precipitates the calcium as $CaSO_4 \cdot 2H_2O$. All the liquid from the reaction tank is sent to a thickener, where the calcium is removed, either as gypsum or as unreacted calcium carbonate (or as calcium hydroxide). Some dissolved sodium carbonate is lost in the solution in the moist solid waste stream, so additional sodium carbonate or bicarbonate is added in the thickener overflow tank, and the clear liquid from it is used as the scrubbing liquid.

The ease of operation and reliability of the double alkali systems allowed them to compete with the early limestone scrubbers. As the limestone scrubbers improved, the extra complexity (more chemicals to handle, more vessels, pumps, lines, valves) and higher reagent cost of the double alkali systems made them uncompetitive.

11.5.2.2 Dry systems. The solids handling and wet sludge handling and disposal difficulties that are integral to wet throwaway processes induced engineers to develop dry throwaway processes that would have fewer corrosion and scaling difficulties and would produce a waste product much easier to handle and dispose of. All of these systems inject dry alkaline particles into the gas stream, where they react with the gas to remove SO_2. The SO_2-containing particles are then captured in the particle collection device that the plant must have to collect fly ash (most often a baghouse, sometimes an ESP). If successful, this approach eliminates the problems with disposal of wet scrubber sludge and all the difficulties involved with the wet limestone process. It increases the volume of dry solids to be disposed of, but that is considered a less difficult problem. The flow diagrams for such systems are sketched in Fig. 11.7.

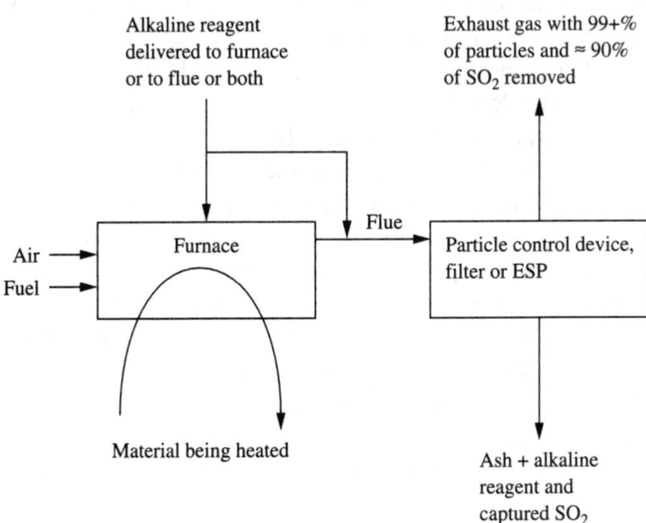

FIGURE 11.7
Flow diagram for SO_2 control systems using dry solids addition.

Table 11.4 shows four entries under "Dry throwaway." The first two call for the injection of powdered limestone or lime into the boiler. In the high-temperature part of the furnace the limestone would convert to lime, so that either way the active reagent would be CaO. The desired reaction is

$$CaO + SO_2 \rightarrow CaSO_3 \tag{11.20}$$

$CaSO_3$ would then oxidize to $CaSO_4$. In principle this should work, but most tests have shown that to get high SO_2 collection efficiencies one must put a large excess of lime or limestone into the system, thus increasing reagent costs, increasing the load on the particle collector, and increasing the volume of solid wastes to be disposed of. However, if one uses more reactive (and much more expensive) $NaHCO_3$ or Na_2CO_3, the collection efficiency is much better, mostly because of the much higher chemical reactivity of these sodium salts.

The design of such devices is, in principle, done the same way as in Examples 11.1 and 11.2. However, here we have co-flow, which is much less efficient than counterflow. Mass transfer between gases and solids is much less well understood than that between gases and liquids, so that the design of these devices is much more heavily dependent on test and empiricism than is the design of systems like that in Fig. 10.15.

11.5.2.3 Wet-dry systems. Wet-dry systems (Table 11.4) combine some features of the preceding two kinds of systems. The most widely used wet-dry systems (called dry scrubbers) are spray dryers; the flow diagram of this arrangement is shown in Fig. 11.8.

Spray dryers are widely used in the process industries. Masters [18] presents a five-page list of products that are commercially spray dried, e.g., powdered milk, instant coffee, laundry detergents, etc. In all such spray dryers a liquid (almost always water) containing dissolved or suspended solids is dispersed as droplets into a hot gas stream. The dispersion can be done by a high-pressure gas-atomizing nozzle or a rapidly rotating (about 10 000 rpm) atomizing wheel. The hot gas is well above the boiling temperature of water, so that the water in the droplets evaporates rapidly. The particles formed from the evaporating drops are dry before they reach the wall or bottom bin of the dryer, so they form a free-flowing powder that is easily removed.

In industry a spray dryer is most often used when the product is heat sensitive. The drying is done very quickly, and the powder can be cooled quickly after it leaves the dryer. In addition, by controlling the solids concentration in the feed and the size of the droplets, one may control the size of the particles produced, often producing a particle size distribution not easily obtained any other way. With soluble solids one can often produce particles that are hollow spheres. The student should study some powdered coffee (not freeze-dried coffee) or laundry detergent as an example of products made this way.

In treating SO_2-containing flue gases, as shown in Fig. 11.8, the hot gas enters the spray dryer chamber, usually from the side and/or top and flows out most often at the bottom or side or through an outlet tube that dips down into the dryer vessel. The

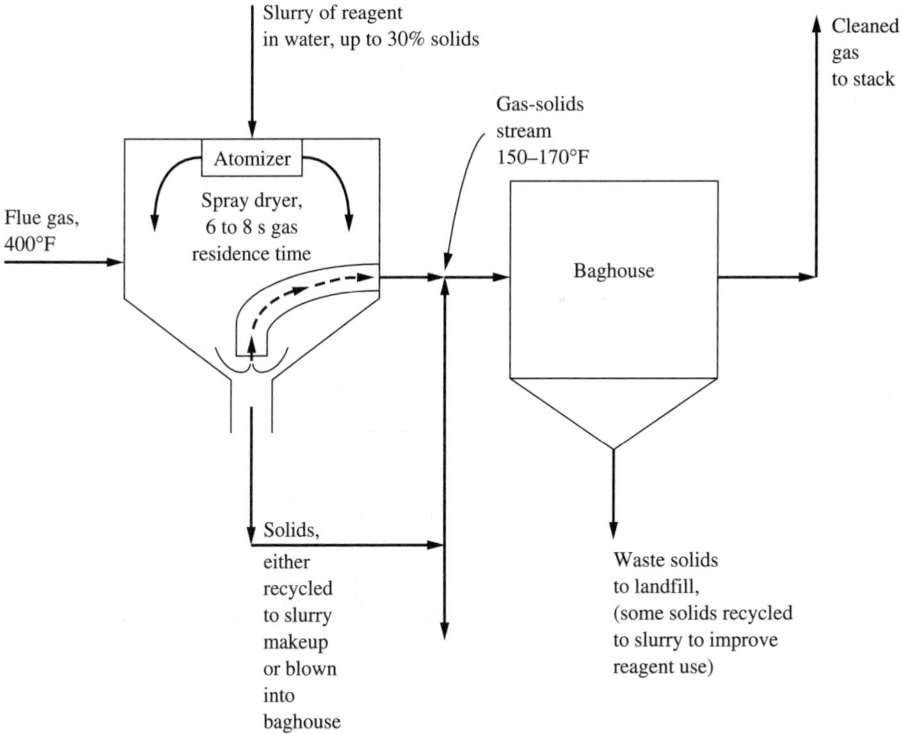

FIGURE 11.8
Flow diagram for spray dryer process for removal of SO_2 from stack gases.

reagent slurry is dispersed as 10- to 50-μ drops, containing about 30 weight percent solids. The resulting dry particles are small enough that most are carried along with the gas stream; this feature is different from most spray dryers for consumer products, in which the particles are large enough (or the air velocity small enough) that most of the particles settle to the bottom of the dryer. Some test results from a large pilot plant spray dryer designed to control SO_2 are shown in Fig. 11.9 on page 424.

This is called a wet-dry system because part of it is like a wet lime scrubber and part is like dry sorbent injection. The freshly formed drops behave very much like the drops in a wet lime scrubber. The SO_2 dissolves in the water and reacts there with the dissolved $Ca(OH)_2$. As the water evaporates from the drops, the individual fine particles in them coalesce to form a single porous particle from each drop. This particle then behaves like the dry sorbent particles injected as shown in Fig. 11.7.

Example 11.7. On Fig. 11.9, at a reagent ratio of 1.1:

(*a*) What is the percent efficiency of the spray dryer alone for SO_2 capture?

(*b*) What is the percent efficiency of the filter alone for SO_2 capture, based on the inlet concentration to the filter?

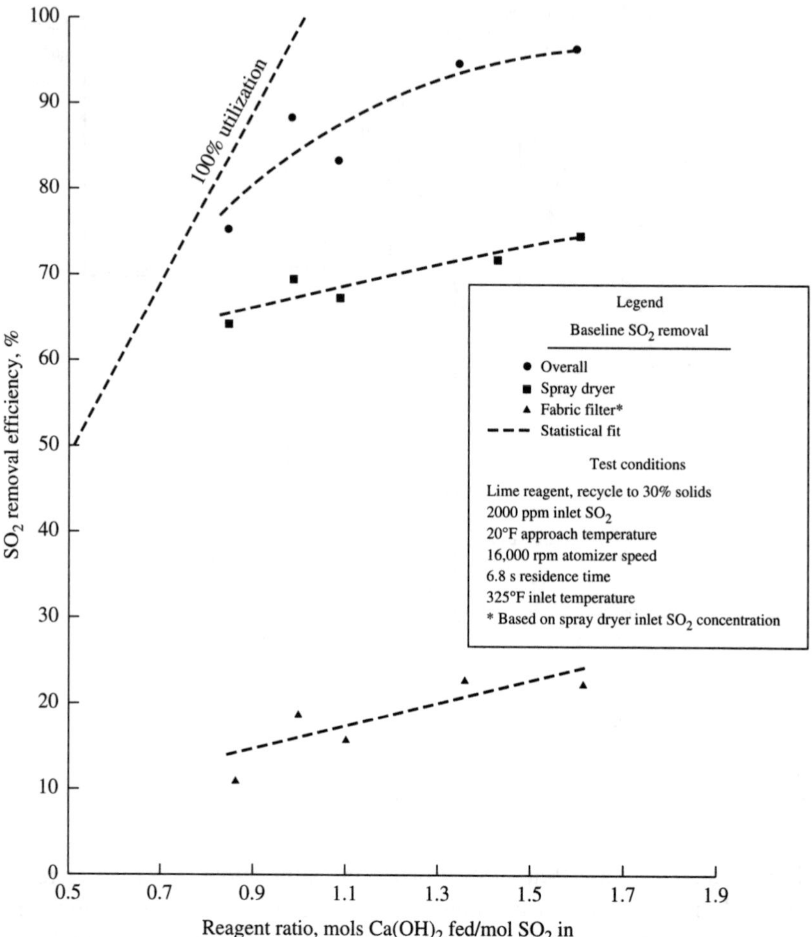

FIGURE 11.9
Test results for a pilot plant spray dryer processing 4200 scfm of flue gas with 2000 ppm SO_2 using $Ca(OH)_2$ [19]. The 100% utilization curve corresponds to all the $Ca(OH)_2$ being converted to $CaSO_4$. (Copyright ©1991. Electric Power Research Institute. EPRI GS-7449. *Evaluation of a 2.5-MW Spray Dryer/Fabric Filter SO_2 Removal System.* Reprinted with permission.)

(*c*) How much water is admitted per pound of inlet gas?
(*d*) How much solids are fed, and what fraction of those are recycled?
(*e*) In the mix of gas and particles in the spray dryer, what is the average spacing between drops as they turn into particles?

Solution

(*a*) Reading Fig. 11.9, we find an efficiency of 68% in the spray dryer and 17% in the filter. (These sum to 85%, whereas the overall curve reads 87%, showing a modest inconsistency between the curves.)

(*b*) For the filter, the incoming gas has 100% − 68% = 32% of the original SO_2, so that its SO_2 collection efficiency is

$$\eta = \frac{17\%}{32\%} = 53\%$$

Viewing these as penetrations, we calculate

$$P_{\text{overall}} = P_{\text{dryer}} \cdot P_{\text{filter}} = 0.32 \cdot 0.47 = 0.15$$

(The collection efficiency of the filter for particles is probably > 99%.)

(*c*) The inlet temperature is 325°F and the approach to adiabatic saturation is 20°F. The inlet gas moisture content is not stated, but it is assumed to be 8 mol % ≈ 0.054 lb/lb. By the procedure shown in Fig. 7.11, one finds that the corresponding adiabatic saturation temperature is about 128°F. Thus the outlet temperature is 148°F. By calculations similar to those in Example 7.12 one finds that the outlet moisture content is about 0.101 lb/lb, so that the moisture added is (0.101 − 0.054) = 0.047 lb/lb of air. (Here the properties of air are used; for flue gases the properties are slightly different.)

(*d*) The slurry is 30 wt % solids, so the solid feed rate is

$$\frac{\dot{m}_{\text{solid}}}{\dot{m}_{\text{gas}}} = \frac{0.3(\dot{m}_{\text{solid}} + \dot{m}_{\text{water}})}{\dot{m}_{\text{gas}}} = 0.3\frac{\dot{m}_{\text{solid}}}{\dot{m}_{\text{gas}}} + 0.3\frac{\dot{m}_{\text{water}}}{\dot{m}_{\text{gas}}}$$

$$\frac{\dot{m}_{\text{solid}}}{\dot{m}_{\text{gas}}}(1 - 0.3) = 0.3\frac{\dot{m}_{\text{water}}}{\dot{m}_{\text{gas}}}$$

$$\frac{\dot{m}_{\text{solid}}}{\dot{m}_{\text{gas}}} = \frac{0.3}{0.7}\frac{\dot{m}_{\text{water}}}{\dot{m}_{\text{gas}}} = \frac{0.3}{0.7}0.047\ \frac{\text{lb}}{\text{lb}} = 0.020\ \frac{\text{lb solid}}{\text{lb gas}}$$

The lime feed rate is

$$\frac{\dot{m}_{\text{Ca(OH)}_2}}{\dot{m}_{\text{gas}}} = 1.1 \cdot \frac{74}{64}\frac{\dot{m}_{\text{SO}_2}}{\dot{m}_{\text{gas}}} = 1.27 \cdot 0.002\frac{64}{29} = 0.0056\ \frac{\text{lb Ca(OH)}_2}{\text{lb gas}}$$

Thus (0.0056/0.020) = 28% of the solids is fresh $Ca(OH)_2$ and the remaining 72% is recycled solids from the baghouse or from the bottom of the spray dryer.

(*e*) For 1.0 lb of gas there are 0.047 lb of water and 0.020 lb of solids, or 0.067 lb of slurry. Estimating the specific gravity of the slurry as 1.2, we estimate the volume of slurry per pound of gas as

$$V_{\text{slurry}} = \frac{m}{\rho} = \frac{0.067\ \text{lb}}{1.2 \cdot 62.4\ \text{lb/ft}^3} = 8.9 \times 10^{-4}\ \text{ft}^3 = 2.5 \times 10^{-5}\ \text{m}^3$$

and for an assumed average droplet size of 20 μ, we calculate the number of drops per lb of gas as

$$N = \frac{V}{(\pi/6)D^3} = \frac{2.5 \times 10^{-5}\ \text{m}^3}{(\pi/6)(20 \times 10^{-6}\ \text{m})^3} = 6.1 \times 10^9$$

and the volume of 1.0 lb of gas at the average temperature of the dryer as

$$V_{gas} = \frac{m}{\rho} = \frac{1 \text{ lb}}{(0.075 \text{ lb/ft}^3)(528°\text{R}/697°\text{R})} = 17.6 \text{ ft}^3 = 0.50 \text{ m}^3$$

Dividing this value by the corresponding number of particles, we find that each particle would occupy a volume of 8.2×10^{-11} m^3, corresponding to a cube with an edge length of 0.4 mm. ■

Comparing this device to the wet limestone scrubber, we see that the drops are much smaller (20 μ/3 mm = 1/150). The time the gas spends in the scrubbing environment is roughly twice as large. However, once the particles become dry, their reactive capacity is greatly reduced compared to the drops in wet scrubbers. In addition, the co-flow pattern is much less efficient (see Problem 11.34).

The amount of water one can introduce in these devices is limited by the amount that the hot gas can evaporate. If more water is introduced than is needed to cool the gas to the adiabatic saturation temperature (see Sec. 7.11), then not all the drops will evaporate, the resulting particles will be wet and sticky (plugging the filter), and the dryer and downstream equipment will suffer severe corrosion. The legend for Fig. 11.9 lists a test condition of "20°F approach temperature," indicating that the amount of water fed in the slurry was limited to keep the gas temperature 20°F above its adiabatic saturation temperature. Test data show that the collection efficiency improves as one approaches saturation, presumably because much of the reaction takes place before the droplet is completely converted to a solid and a close temperature approach keeps the drops wet longer [20]. In addition, at high relative humidities the particles of $Ca(OH)_2$ will adsorb one or two molecular layers of water, thus greatly increasing their reactivity compared with that of totally dry $Ca(OH)_2$ at the same temperature. Water adsorption increases the collection efficiency after the droplets evaporate, both inside the spray dryer and in the baghouse. Operators carefully monitor their approach to the adiabatic saturation temperature: A close approach gives the best removal efficiency; too close an approach produces a sticky cake and corrosion.

The high solids recycle rate shown is needed to get good reagent utilization. With once-through solids use the utilization is poor. In some cases the recycle material is first ground to break the particles open and provide better access to the unreacted materials in the centers of the particles.

11.5.2.4 Regenerative systems. Table 11.4 shows an entirely different category of systems. In these some kind of absorbent or adsorbent is used to capture SO_2 from the flue gas. Then in some separate device or set of devices the adsorbent or absorbent is regenerated to produce a flow of relatively pure SO_2 or H_2SO_4. These systems were under intense study and development when it appeared that the problems with wet limestone scrubbers were insoluble. As those problems were solved, interest in regenerative systems waned. Recently, work has begun on regenerative processes that will simultaneously capture *both* SO_2 and NO_x. These systems have not yet advanced to commercial scale, but they may have a major role in future air pollution control [13].

11.5.2.5 Tomorrow's limestone control devices. The forced-oxidation limestone scrubber is a great technological accomplishment. It does a difficult task with high efficiency and reliability at a high but not impossible cost. However, industry would like a simpler, cheaper system. The manufacturers of forced-oxidation limestone scrubbers have shown in pilot plants that one can operate them at gas velocities up to 18 ft/s, if one can make the entrainment separators work well enough to capture the small drops that are carried along with the gas stream at that velocity. With this higher velocity one can use a much smaller scrubber with a large cost saving.

Industry continues to try to develop dry limestone-based processes. As discussed above the low reactivity of limestone makes these difficult. However, if one recycles most of the captured particles through the boiler or through an intermediate gas-solid contact vessel and humidifies the gas almost to adiabatic saturation, then one can get satisfactory SO_2 capture and satisfactory reagent utilization with these devices [21]. Whether they will be more economical than the forced-oxidation limestone scrubber remains to be seen.

11.6 ALTERNATIVES TO "BURN AND THEN SCRUB"

When the electric power industry first faced regulations requiring it to reduce the emissions of SO_2 from power plants, it decided for the most part to leave the power plant alone and to scrub the gas leaving the power plant. This approach is still the most common, using either wet limestone scrubbers or lime spray dryers. But the industry never entirely abandoned the investigation of alternative approaches. With strong pressure from the Clean Air Amendments of 1990 to reduce emissions of acid rain precursors, the electric power industry has renewed interest in these other possibilities [13].

11.6.1 Change to a Lower Sulfur Content Fuel

If the management of a power plant can replace a high-sulfur coal with a low-sulfur coal, it reduces the SO_2 emissions quickly, simply, and without having to install expensive SO_2 control devices or to deal with their solid effluent. (Switching coals can cause some problems in the plant, which was presumably designed for the coal originally used, but such problems are generally manageable.) Many power plants that burned high-sulfur eastern coals switched to lower-sulfur coals from the Rocky Mountain states (see Appendix C). This decision was a boon to the economies of Wyoming and Montana and a blow to the economies of the midwestern and eastern coal-producing states. This approach has been vigorously attacked, mostly on the grounds of job losses, by the midwestern and eastern coal miners and their elected representatives; it is a continuing political struggle.

11.6.2 Remove Sulfur from the Fuel

Another alternative is to remove the sulfur from the fuel before it is burned.

11.6.2.1 Coal cleaning. Pyritic sulfur can be removed by grinding the coal to a small enough size that the pyrites are mostly present as free pyrite particles. Gravity methods are then used to separate the low-density coal (s.g. = 1.1 to 1.3) from the high-density pyrites (s.g. = 5.0). This approach is particularly suited for coals in which a substantial fraction of the sulfur is present as pyrites. Unfortunately, pyrite particles are generally quite small, so that very fine grinding is needed to separate them from the rest of the coal.

11.6.2.2 Solvent-refined coal. It is also possible to dissolve coal in strong enough solvents and then to treat the solution by the same kind of catalytic hydrogenation processes that are used to remove sulfur from petroleum products. The mineral (ash-forming) materials do not dissolve, so they are rejected by filtration or settling. When the solvent is then removed for reuse, the remaining product is a very clean-burning combustible solid, free of ash and sulfur, called *solvent-refined coal.* Considerable development work on this process showed that it can be done, but so far not at a price comparable to "burn and then scrub."

11.6.3 Modify the Combustion Process

The standard way of burning large amounts of coal (pulverized-coal furnace) is to grind the coal to about 50- to 150-μ size and blow it with hot air into a large combustion chamber. There the small coal particles decompose and burn in the one to four seconds that they spend in the furnace, transferring most of the heat generated to the walls of the furnace as radiant heat. The furnace walls are made of steel tubes in which fluid (most often water turning to steam) is heated. The hot gases leaving the furnace then pass over banks of tubes and transfer much of their remaining sensible heat to the fluid being heated.

Fluidized bed combustion is an alternative way to burn coal that is currently in the demonstration plant stage [22]. In it, coal is burned in gravel-sized pieces by injecting them into a hot fluidized bed of limestone particles instead of as a finely dispersed powder in air. A fluidized bed is a dense bed of solid particles suspended in air; such beds are widely used in chemical engineering, e.g., in fluidized bed catalytic cracking. The coal spends much longer in the bed than it would in a pulverized coal furnace, because more time is needed to get complete combustion of the much larger particles.

In such a fluidized bed combustor SO_2 is formed in the presence of a large number of limestone particles and has a high probability of reacting with one of them in the combustion bed. Here the temperatures are much higher than in the dry processes discussed in Sec. 11.5.2.2, and most of the limestone has been converted to CaO, so that the reaction of SO_2 with CaO is rapid enough to provide adequate SO_2

control. The limestone in the bed is steadily replaced, and the material withdrawn has largely been converted to $CaSO_4$. Here again, a dry powder waste is produced instead of a wet scrubber sludge.

The fluidized bed has tubes full of water and steam projecting into the bed. The heat transfer between the hot bed in which the coal is burned and the tubes is much better than that between the flames and the walls of an ordinary coal-fired boiler. For this reason fluidized bed combustors are smaller and operate at lower temperatures than ordinary coal-fired boilers. This saves on some costs and greatly reduces the formation of nitrogen oxides (see Chapter 12). These boilers, however, have other problems, so that they are not yet a clear winner over conventional boilers.

A second combustion modification alternative, also in the demonstration plant stage [23], is to convert the coal to a synthetic fuel gas and then burn that in combination gas-turbine steam-turbine power plants, Fig. 11.10. This seems complex and costly, but it has the advantage that in the synthetic fuel gas the sulfur is present as H_2S; since no other acid gas is present, the sulfur can be easily removed from the gas by the methods described in Sec. 11.3. The second, and more important, advantage is that modern gas-turbine steam-turbine plants have a much higher thermal efficiency than typical coal-fired steam plants (perhaps 45 percent vs. 33 percent).

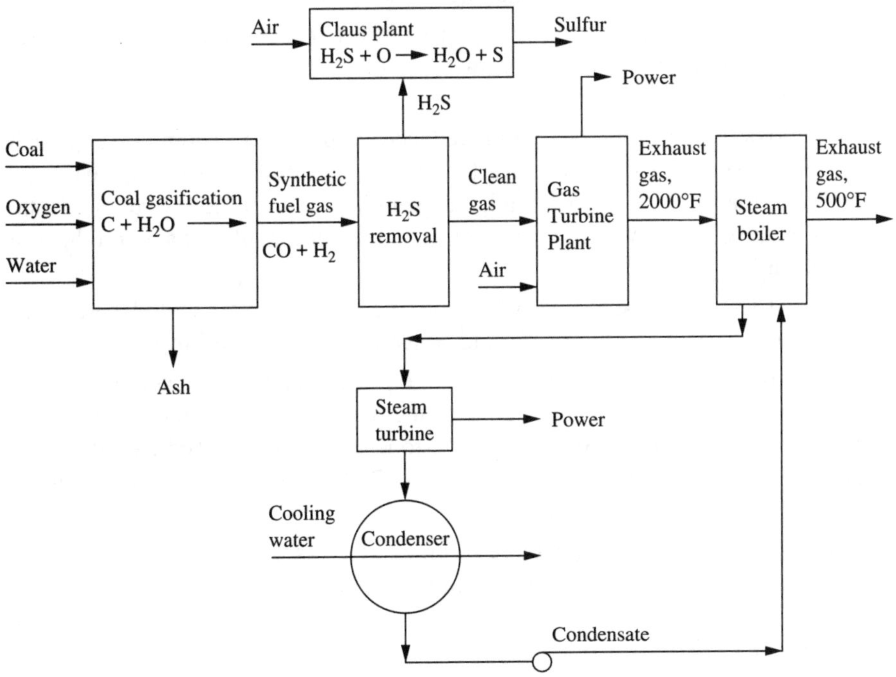

FIGURE 11.10
Schematic flow diagram of a synthetic fuel gas, gas-turbine steam-turbine power plant (also called a *combined cycle* plant). The net power produced is the sum of that from the gas turbine and from the steam turbine.

If the problems with this technology can be solved, it may offer a more efficient and economical way of converting coal to electricity than the systems currently used even though it is much more complex.

11.6.4 Don't Burn at All

The majority of the SO_2 derived from human activities comes from coal and oil combustion in electric power plants. If we can produce electricity in some other way or reduce our use of electricity, we will consequently reduce our emissions of SO_2. For this reason more efficient electric devices (lights, refrigerators, motors) are, in effect, SO_2 control devices. So also are nuclear, wind, solar, tidal, geothermal, and hydroelectric power plants. There is currently a serious effort by the U.S. EPA and by the electric utility industry to improve the efficiency of electricity usage, and to encourage production of electricity from alternative energy sources for a variety of reasons, including reduction of SO_2 emissions.

11.7 SUMMARY

1. SO_2 emissions from human activities are mostly due to the combustion of sulfur-containing fossil fuels and the smelting of metal sulfide ores.
2. The overall control strategy for SO_2 emissions is to convert the sulfur to $CaSO_4 \cdot 2H_2O$ and return it to the ground in some kind of landfill, or use it to make wallboard.
3. For liquid or gaseous fuels containing reduced sulfur, the most common approach is to use catalytic processes to convert the contained sulfur to H_2S, remove that by scrubbing the gas with a weakly alkaline solution, convert the H_2S to elemental sulfur by the Claus process, and either sell that sulfur for sulfuric acid production or place it in a landfill.
4. For metal sulfide ore smelting, which produces waste gases with 4 percent or more SO_2, the common approach is to convert that SO_2 to sulfuric acid.
5. For coal (or high-sulfur oil) used in a large power plant, the most common approach is to burn the coal and then treat the plant's exhaust gas (typically containing about 0.1 percent SO_2) with limestone or lime in a forced-oxidation wet scrubber or a spray dryer, to convert SO_2 to $CaSO_4 \cdot 2H_2O$, which will then go to a landfill or a wallboard plant.
6. Other alternatives are being explored, some in large-scale demonstrations. They may replace those just listed in the future.

PROBLEMS

See Common Units and Values for Problems and Examples, inside the back cover.

11.1. Equation (11.3) shows that the products are metallic copper and iron oxide. Why does this reaction not end either with both metals in the metallic form or with both in the oxide form?

11.2. The U.S. average gasoline is 300 ppm by weight S. There is currently a regulatory effort to reduce that to perhaps 40 ppm. U.S. average gasoline consumption is about 350 million gal/day.

(*a*) What fraction of the national SO_2 emissions (see Table 11.2) is due to gasoline?

(*b*) Is this proposed reduction mostly to reduce the SO_2 content of the atmosphere? or is it for some other purpose? Suggest a possible purpose.

11.3. As shown in Problem 10.28, the Henry's law constant for O_2 in water at 20°C is 40 100 atm. In Example 11.1 we see that the Henry's law constant for H_2S in water at 20°C is 483 atm.

(*a*) Does this mean that H_2S is more soluble or less soluble in water than O_2? By what ratio?

(*b*) Why is there this much difference is solubility?

11.4. Compute Q_L/Q_G, the ratio of volumetric flow rates, in Examples 11.1 and 11.2. Industrially it is observed that in practical systems this ratio is seldom more than 10%. Do the values computed here agree with that observation?

11.5. Repeat Example 10.15 for the H_2S absorption problem in Example 11.2. Observe that at this pressure the density of methane is $\approx$ 1.2 times the perfect gas density. The liquid density is practically that of water.

11.6. Estimate the number of transfer units required in Example 11.2,

(*a*) By direct numerical integration of

$$N = \int_{\text{bottom}}^{\text{top}} \frac{dy_i}{(y_i - y_i^*)}$$

using Eqs. (10.25) and (11.6).

(*b*) By computing the equivalent value of H for the top of the column and using that value in Eq. (10.25). Compare this result to that in part (*a*).

(*c*) Repeat part (*b*) using the equivalent value of H for the bottom of the column, and show that this leads to an impossible result. The reason is that for the higher value of H at the bottom of the column the calculated $(y - y^*)$ at the top becomes negative. This column could not perform that separation with a solvent with that high an H. Reconsider your integration in (*a*) to see that most of the N is needed at the top part of the column, where H is small.

11.7. Using the values from Problems 11.5 and 11.6, estimate the required column height. Use the same value of Ka as in Example 10.16.

11.8. As shown in the text, the U.S. EPA standard for H_2S in gas streams to be burned for heat in oil refineries is $H_2S \leq 230$ mg/dscm (dry standard cubic meter). How many parts per million is this?

11.9. Figure 11.11 on page 432 shows the estimated capital and operating costs for sulfuric acid plants processing smelter off-gases [12].

(*a*) From these figures deduce the underlying mathematical relation used to make the figures.

(*b*) Based on these figures, compute the break-even selling price of H_2SO_4 for a plant with a feed stream with 10 000 lb/h of SO_2 in the feed and feed strengths of 2%, 4%, and 8% SO_2. Assume that the plant works 80% of the time, and that the annual charge on invested capital is 20%.

11.10. The feed gas to a sulfuric acid plant has 7.8% SO_2, 10.8% O_2, and 81.4% N_2. It is brought to chemical equilibrium over a catalyst at 500°C, at which

$$K = \frac{y_{SO_3}}{(y_{SO_2})(y_{O_2})^{0.5}} = 85$$

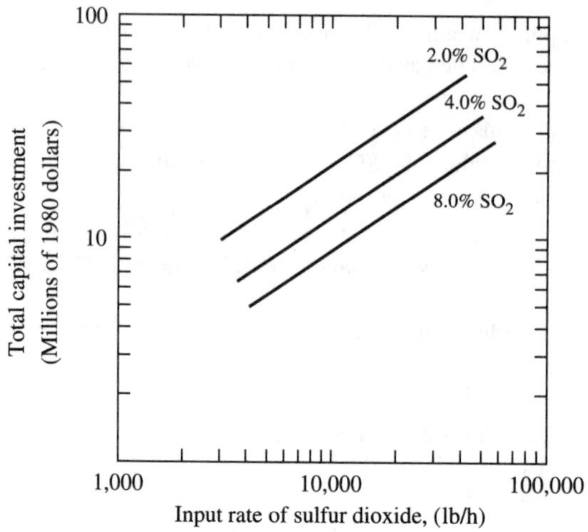

Note: Gas cooling and conditioning included.

(*a*)

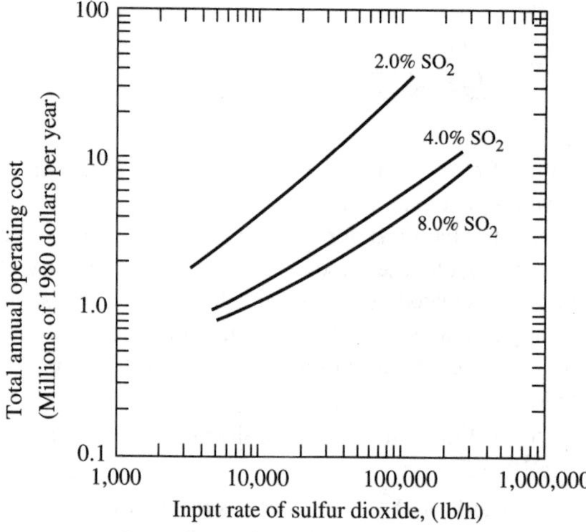

Note: Gas cooling and conditioning included.

(*b*)

FIGURE 11.11
(*a*) Total capital investment, and (*b*) total annual operating costs [12] for a plant to convert SO_2 to sulfuric acid. These are 1980 values; 1999 values are roughly twice these values. (Reprinted with permission from "Preliminary Economic Analysis of SO_2 Abatement Technologies" by J. C. Agarwal and M. J. Loreth, Figure 1, p. 71 and Fig. 2, p. 72 (pp. 67–89), from the book *Sulfur Dioxide Control in Pyrometallurgy,* edited by T. D. Chatwin and N. Kikumoto, TMS, Warrendale, PA, 1981.

What fraction of the SO_2 is converted to SO_3? (This problem is worked as an example calculation in Ref. 24. The SO_2 and O_2 do not sum to 21% because the feed gas derived from a pyrites roaster.)

11.11. In Fig. 11.2, why are two different equilibrium curves shown for differing inlet SO_2 concentrations? Why is the equilibrium conversion shown higher for the lower inlet SO_2 concentration?

11.12. Assume that a single absorption sulfuric acid plant recovers 98 percent of its SO_2 as acid, and a double absorption plant recovers 99.7 percent.

(*a*) What fraction of the SO_2 passing from the first absorber must be captured by the final conversion and absorption?

(*b*) If the final catalyst bed operates at 420°C, at which the equilibrium constant $K \approx 300$, what fraction of the equilibrium conversion is this? Use the inlet composition in Problem 11.10.

11.13. In flash smelting of copper ores, where oxygen is used instead of air, it is possible to produce an exhaust gas that is 40 percent SO_2. What economic advantages might this have?

11.14. Estimate the slope of the adiabatic conversion curves on Fig. 11.2 from the following data: The exothermic heat of reaction in Eq. (11.12) is 23 kcal/mol. The heat capacity of the gases is roughly that of air, or 7 cal/mol K. How does this calculated value compare with the value read from Fig. 11.2?

11.15. The prices of chemicals change with time, but in 1999 typical prices were \$45/ton for sulfur on the U.S. Gulf Coast and \$75/ton for sulfuric acid in the eastern United States. The railroad cost of shipping bulk cargoes in the United States was roughly \$0.03 per ton mile.

(*a*) If you buy sulfur and make it into sulfuric acid to sell, at these prices how much per ton of acid can you afford to pay for the capital and operating costs of the sulfuric acid plant in order to break even?

(*b*) If you must ship sulfur or sulfuric acid 500 miles to a customer, what is the shipping cost per ton of contained sulfur?

(*c*) What do these answers tell you about the economics of shipping sulfur or sulfuric acid?

11.16. Consider the typical limestone scrubber shown in Figs. 11.5 and 11.6 and Table 11.3.

(*a*) How much limestone is fed per ton of coal?

(*b*) How many pounds of solids (dry basis) leave the system per ton of coal burned in the gypsum product stream? Of these, what weight fraction is $CaSO_4 \cdot 2H_2O$? $CaSO_3 \cdot 0.5H_2O$? $CaCO_3$?

(*c*) How much water leaves the system with the gypsum solids per ton of coal?

(*d*) How much water is evaporated in the scrubber per ton of coal?

(*e*) Typical coals contain 0.01 to 0.1 percent chlorine. All of this is reacted to HCl in the combustion and is collected in the scrubber. It forms no solids, and none leaves with the gas, so the only way chlorine leaves the system is dissolved in the water to waste treatment and in the water contained in the gypsum product. Stainless steels are sensitive to chlorine, so our specification is that the circulating water shall contain no more than 10 000 ppm = 1% by weight of Cl (as chloride ion). What is the required flow of water to waste treatment to meet this specification, if the Cl content of the coal is 0.1%?

11.17. Figures 11.5 and 11.6 and Table 11.3 correspond roughly to the 10 scrubber modules installed in two power plants in Korea [25]. Each of those modules treats the gas equivalent to 500 MW of power production. Assuming that the power plant burns the typical Pittsburgh seam coal, with a thermodynamic efficiency of 35% (see Example 7.10),

(*a*) Estimate the diameter of each of the scrubber modules.

(*b*) Estimate the flow rate of the liquid to each module.

(*c*) Estimate how much gypsum (dry basis) is produced per year from each module.

11.18. In the previous problem estimate the air flow to the bottom of the scrubber module per ton of coal fed to the boiler.

11.19. The limestone scrubber in Fig. 11.6 also acts as a particulate scrubber and collects some of the particles in the gas stream coming to it. Estimate its collection efficiency for 1-μ diameter particles using the equations in Sec. 9.2.4.

11.20. Compare the limestone wet scrubber in Fig. 11.6 to the venturi scrubbers used for fine particulate control by estimating for each (*a*) the mean relative velocity between liquid drops and gas, (*b*) the size of the liquid drops, and (*c*) the ratio of liquid to gas volumetric flow rates. Information on venturi scrubbers for particulate control is in Sec. 9.2.4.

11.21. For the scrubber shown in Fig. 11.6, if all the drops are 3 mm in diameter and are uniformly dispersed in the body of the scrubber, what is the distance between adjacent drops?

11.22. Ordinary coal-fired electric power plants produce about 3.3 acfm (0.00156 m^3/s) of stack gas (at 400°F and 1 atm) per kilowatt of net power produced. Limestone scrubbers have pressure drops of about 10 inches of water. What fraction of the power produced by the plant must be used to overcome the pressure loss in the scrubber? Assume the following:

$$\text{Power} = (\text{flow rate})(\text{pressure drop/blower efficiency})$$

$$\text{Blower efficiency} = 0.8$$

11.23. Estimate the power required to recirculate the slurry in Figs. 11.5 and 11.6 as a fraction of the net power produced by the power plant. Use the data from Table 11.3, plus the assumptions that the pump efficiency is 0.80 and the pressure drop through the piping system, which must be overcome by the pump, is 20 psi.

11.24. The equilibria shown in all of the equations in this chapter are much simpler than the true chemistry. Figure 11.12 shows a more nearly realistic picture, although it is still a simplification. In this figure, where does the $SO_4^=$ come from? Is it desirable?

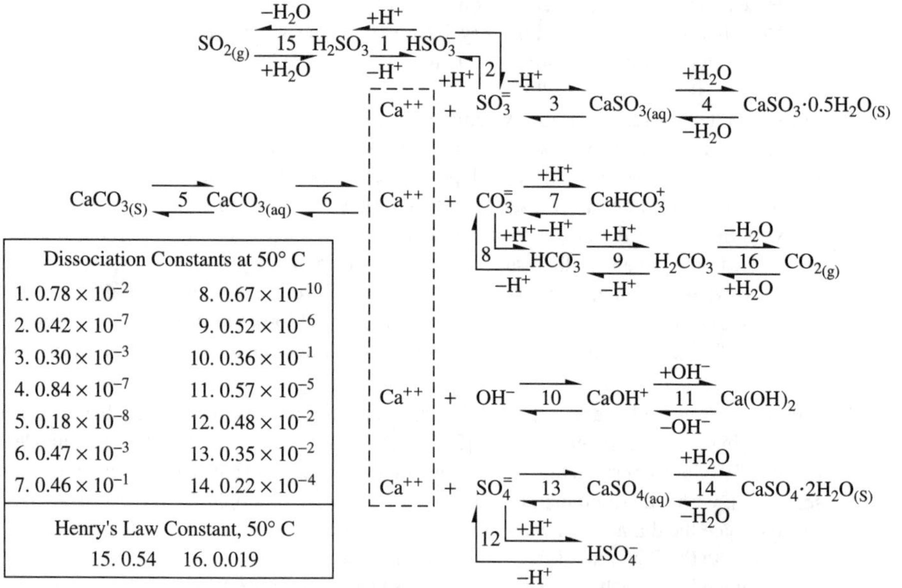

Dissociation Constants at 50° C	
1. 0.78×10^{-2}	8. 0.67×10^{-10}
2. 0.42×10^{-7}	9. 0.52×10^{-6}
3. 0.30×10^{-3}	10. 0.36×10^{-1}
4. 0.84×10^{-7}	11. 0.57×10^{-5}
5. 0.18×10^{-8}	12. 0.48×10^{-2}
6. 0.47×10^{-3}	13. 0.35×10^{-2}
7. 0.46×10^{-1}	14. 0.22×10^{-4}
Henry's Law Constant, 50° C	
15. 0.54	16. 0.019

FIGURE 11.12
Greatly simplified representation of the principal equilibria in a limestone scrubber. The dissociation and Henry's law constants are similar but not identical to those in [26]; the original article does not cite its sources. The Henry's law constants are the reciprocals of those normally used in chemical engineering. (Reproduced by permission of the Air and Waste Management Association from Ref. 15.)

11.25. In limestone scrubber calculations it is often assumed that the vapor pressure of SO_2 in equilibrium with the absorbing slurry, y^*, is so small that it is effectively zero.

(*a*) Show that this is equivalent to Henry's law with $H \approx 0$.

(*b*) Show that if one substitutes $H \approx 0$ into the integrated form of Eq. (10.25) one finds a very simple relation between N and the penetration.

(*c*) Based on the simple formula in (*b*) estimate the number of transfer units needed for 90% collection in any type of control device including a scrubber. Compare that to the N computed in Example 10.16, which has 90% collection efficiency but assumes that the vapor pressure of the absorbed material is zero.

11.26. Henry's law works well for simple absorption of gases, e.g., Example 10.16. It is much less satisfactory for solution of acid gases, which dissociate in aqueous solution. For SO_2 in water the reactions seem to be those in Eq. (11.17). Whether the free acid, H_2SO_3, exists or whether instead the dissolved $SO_2 + H_2O$ goes directly to the ions is not clear experimentally. For this treatment [26], we assume that the free acid does not exist, so that the equilibria that exist are

$$SO_2(\text{gas}) \Leftrightarrow SO_2(\text{dissolved}) \tag{11.21}$$

$$SO_2(\text{dissolved}) + H_2O \Leftrightarrow H^+ + HSO_3^- \tag{11.22}$$

$$HSO_3^- \Leftrightarrow H^+ + SO_3^{2-} \tag{11.23}$$

for which the individual equilibrium relationships for dilute solutions are

$$H_{SO_2} = Py_{SO_2}/[SO_2(\text{dissolved})] \tag{11.24}$$

$$K_1 = [HSO_3^-]\cdot[H^+]/[SO_2(\text{dissolved})] \tag{11.25}$$

$$K_2 = [SO_3^{2-}]\cdot[H^+]/[HSO_3^-] \tag{11.26}$$

In aqueous chemical equilibria the [ ] symbol almost always stands for concentration in mol/L. Here the Henry's law constant, H_{SO_2}, has dimensions of atm/(mol/L). K_1 and K_2 are the first and second ionization constants, with dimensions (mols/L). Reference 26 shows equations for these equilibrium constants (and many of the others that are important for limestone scrubbers) as a function of temperature. The following table shows the calculated values at 68°F = 20°C, and at 125°F ≈ 52°C.

	T = 68°F = 20°C	T = 125°F ≈ 52°C
H_{SO_2}, atm/(mol/L)	0.679	1.870
K_1, mol/L	0.0156	0.00678
K_2, mol/L	$6.86 \cdot 10^{-8}$	$5.25 \cdot 10^{-8}$

(*a*) Henry's law, Eq. (10.19), is most often written using the mol fraction, x_1, of the dissolved material as the liquid concentration value. In the above table, and in much of the scrubber literature, Henry's law is written with molarity (mol/L) or mass ratio or concentration (mol/kg or lbmol/ft^3) as the liquid-phase concentration value. These concentration values can all be converted one to the other, so the H for one can be converted to the H for any of the others. Estimate the value of the Henry's law constant at 20°C in Eq. (10.19), which takes mol fraction as the concentration variable (and has the dimension atm) from the value in the above table.

(*b*) Estimate the values of the concentrations at 20°C of [SO_2 (dissolved)], [H^+], [HSO_3^-] and [SO_3^{2-}] in a solution that is in equilibrium with a gas at 1 atm, with $y_{SO_2} = 0.001$. The simplest procedure is to solve Eq. (11.24) for [SO_2 (dissolved)], then assume that [H^+] and [HSO_3^-] are equal [i.e., that all the [H^+] comes from Reaction (11.22)]. This allows a direct solution of Eqs. (11.25) and (11.26). Finally, check the quality of that assumption.

(*c*) Estimate the pH of the solution in (*b*).

(*d*) The Henry's law constant shown in Eq. (11.24) is based on the dissolved SO_2 only, and not its ionization products. If one wrote it to take into account the ionized products, i.e.,

$$H_{SO_2} = P y_{SO_2} / \{[SO_2(\text{dissolved})] + [HSO_3^-] + [SO_3^{2-}]\} \tag{11.27}$$

for the situation in (*b*), what would the value of H_{SO_2} be, both in atm and in atm/(mol/L)?

11.27. *Perry's Chemical Engineers' Handbook* [4: pp. 2–128] presents a table of the vapor pressures of SO_2 over aqueous solutions. At 20°C for dissolved SO_2 of 0.01, 0.1, and 1 g/100 g of water, the reported SO_2 vapor pressures are 0.07, 3.03, and 58.4 mm Hg, respectively.

(*a*) Can these three values be represented by simple Henry's law, which requires that in dilute solutions the vapor pressure is proportional to the concentration?

(*b*) Estimate the corresponding pressures using the equilibrium data in Problem 11.26. Do they agree with the values from *Perry's*?

(*c*) If we wish to fit the water-SO_2 equilibrium into the mold of simple Henry's law with H in atm and concentration in mol fraction, what values will we use for the Henry's law constants at the above three liquid concentrations?

11.28. Using the values from Example 11.6, estimate

(*a*) How many mols/L of SO_2 are dissolved in the scrubbing slurry in one pass through the scrubber?

(*b*) If the slurry entering the scrubber had pH = 5, what will be the pH of the scrubbing slurry at the bottom of the scrubber? Here assume that there is no reaction with $CaCO_3$ in the 3 s that the drops fall through the scrubber.

11.29. Problem 11.28 shows that if there is no chemical reaction, we would expect the pH of the drops in a common limestone scrubber to fall significantly as the drop absorbs SO_2 in its descent through the scrubber. This reduces its absorptive capacity. Estimate the values of the concentrations at 52°C of [SO_2 (dissolved)], [H^+], [HSO_3^-], and [SO_3^{2-}] in a solution that is in equilibrium with a gas at 1 atm, with $y_{SO_2} = 0.001$;

(*a*) If the pH is maintained at 4 (i.e., [H^+] = 10^{-4}).

(*b*) If the pH is maintained at 5.

(*c*) If the pH is maintained at 6.

(*d*) Prepare a small table showing these values, and also the values from Problem 11.26, in which the exit pH is allowed to increase.

In limestone scrubbers we try to maintain the pH of the solution practically constant by the dissolution of limestone and the removal of HSO_3^- to form either $CaSO_3 \cdot 0.5H_2O$ or $CaSO_4 \cdot 2H_2O$. Unfortunately, these reactions are slow compared to the 3 s that a typical drop spends on one trip through the scrubber, so the pH does decrease, with a consequent decrease in collection efficiency. One solution to this problem is to include a water-soluble weak acid in the scrubbing solution. The most common is a mixture of adipic, glutaric, and succinic acids, called *di-basic acid* (DBA). These acids absorb H^+ ions when the pH falls, thus holding the pH closer to constant and improving the collection efficiency. They can do this much faster than the dissolution of limestone or the removal of solids because reactions in the liquid are faster than those involving solids.

11.30. Prices of chemicals fluctuate over time, but in 1999 the tonnage lot prices of various alkalis were NaOH (sodium hydroxide, or caustic soda, 76 percent, balance water), \$370/ton; Na_2CO_3 (soda ash), \$105/ton; $Ca(OH)_2$, \$70/ton; CaO, \$57/ton; $CaCO_3$, \$10/ton.
(*a*) Put these on a comparable basis by computing the price per lbmol of equivalent Ca.
(*b*) What do these calculations tell you about which chemicals are likely to be most economical for use in sulfur oxide control devices?
(*c*) Does shipping cost play a role in this question? Assume that the chemicals must be shipped 500 miles and that shipping costs are \$0.03/ton mile by rail or \$0.05/ton mile by truck.

11.31. For the typical limestone wet scrubber shown in Figs. 11.5 and 11.6 and described in Problem 11.16, estimate the cost of limestone in \$/ton of coal burned. Use the limestone price in Problem 11.30.

11.32. One obvious drawback with wet scrubbers is that they lower the temperature of the exhaust gas. This takes away plume buoyancy and brings the remaining pollutants to ground more easily (see Chapter 6). For a typical power plant the temperature coming into the scrubber will be about 400°F, and the exit temperature will be about 125°F. Some plants reheat the stack gas to about 175°F. List the ways that might be done.

11.33. In the spray dryer in Fig. 11.8, if we want the particles not to settle to the floor of the dryer, what gas velocity will we need in the outlet pipe to bring them along? Assume that the particles have an average diameter of 10 μ.

11.34. Figure 10.16 shows the change of y and y^* with distance for the counterflow arrangement in Fig. 10.15. Figure 11.8 shows that for solid addition systems the flow is co-flow rather than counterflow.
(*a*) Sketch the equivalent of Figure 10.16 for a co-flow system.
(*b*) Explain, based on this figure and Fig. 10.16, why counterflow is more efficient than co-flow.
(*c*) Explain why, in spite of this efficiency loss, co-flow is selected for this system and for the spray dryer system.

11.35. A typical 100-W incandescent electric bulb can be replaced with a 20-W fluorescent bulb, with equal light output. In an office building or retail store such a bulb burns about 2000 h/yr. If we replace one 100-W bulb with its 20-W fluorescent equivalent, how much SO_2 per year will we remove from the atmosphere?

Assume that all the electricity comes from coal combustion. Use the "typical coal" and burning conditions in Example 7.10 (see Appendix C) for a power plant with 35 percent thermal efficiency ("heat rate" (1/0.35 $\times$ 3413 Btu/kwh) = 9751 Btu/kWh). The power plant has a wet scrubber that removes 90 percent of the SO_2 from the stack gas.

REFERENCES

1. Urone, P.: "The Primary Air Pollutants—Gaseous: Their Occurrence, Sources and Effects," in A. C. Stern (ed.), *Air Pollution,* 3d ed., Academic Press, New York, p. 38, 1976.
2. Lide, D. R., ed.: *CRC Handbook of Chemistry and Physics,* 71st ed., CRC Press, Boca Raton, FL, p. 14-7, 1990.
3. *National Air Quality and Emissions Trends Report, 1997,* EPA 454/R-98-016, 1997.
4. Liley, P. E., G. H. Thompson, D. G. Friend, T. E. Daubert, and E. Buck: "Physical and Chemical Data," in D. W. Green and J. O. Maloney (eds.), *Perry's Chemical Engineers' Handbook,* 7th ed., McGraw-Hill, New York, pp. 2–127, 1997.
5. Kohl, A., and R. Nielsen: *Gas Purification,* 5th ed., Gulf Publishing, Houston, TX, p. 71, 1997.
6. Edwards, W. M.: "Mass Transfer and Gas Absorption," in D. W. Green and J. O. Maloney (eds.), *Perry's Chemical Engineers' Handbook*, 6th ed., McGraw-Hill, New York, Chapter 14, 1984.

7. Treybal, R.: *Mass Transfer Operations*, 3d ed., McGraw-Hill, New York, 1980.
8. Donovan, J. R., and J. M. Salamone: "Sulfuric Acid and Sulfur Trioxide," in H. F. Mark, D. F. Othmer, C. G. Overburger, and G. T. Seaborg (eds.), *Kirk-Othmer Encyclopedia of Chemical Technology*, Vol. 22, 3d ed., John Wiley & Sons, New York, p. 213, 1983.
9. Muller, T. K.: "Sulfuric Acid," in A. J. Buonicore and W. T. Davis (eds.), *Air Pollution Engineering Manual*, Van Nostrand Reinhold, New York, pp. 469–476, 1992.
10. Duecker, W. W., and J. R. West: "The Manufacture of Sulfuric Acid," *American Chemical Society Monograph 144*, Reinhold, New York, 1959.
11. Sulfuric acid is a permitted food additive in the United States; it has no taste in very dilute form and is used to adjust the acidity of foods. This was done by some French wine makers a few years ago, which is not permitted by French wine laws. The scandal was that this action was discovered when a disgruntled winery employee "blew the whistle," but not by the discriminating palates of the great French wine tasters!
12. Agarwal, J. C., and M. J. Loreth: "Preliminary Economic Analysis of SO_2 Abatement Technologies," in T. D. Chatwin and N. Kikumoto (eds.), *Sulfur Dioxide Control in Pyrometallurgy*, The Metallurgical Society of AIME, Warrendale, PA, pp. 71–72, 1981.
13. Dalton, S. M., B. Toole-O'Neil, B. K. Gullett, and C. J. Drummond, "Summary of the 1991 EPRI/EPA/DOE SO_2 Control Symposium," *J. Air Waste Manage. Assoc.*, Vol. 42, pp. 1110–1117, 1992.
14. Rees, R. L.: "The Removal of Oxides of Sulfur from Flue Gases," *J. Institute of Fuel*, Vol. 25, pp. 350–357, March 1953.
15. Slack, A. V., H. L. Falkenberry, and R. E. Harrington: "Sulfur Oxide Removal from Waste Gases: Lime-Limestone Scrubbing Technology," *J. Air Pollut. Control Assoc.*, Vol. 22, pp. 159–166, 1972.
16. Henzel, D. S., B. A. Laseke, E. O. Smith, and D. O. Swenson: "Limestone FGD Scrubbers: User's Handbook," U.S. Environmental Protection Agency Report No. EPA-600/8/81-017, U.S. Government Printing Office, Washington, DC, 1981.
17. Noblett, J. G., and J. M. Burke: "FGD Chemistry and Analytical Methods Handbook: Volume 1, Process Chemistry—Sampling, Measurement, Laboratory and Process Performance Guidelines (Revision 1)," EPRI CS-3612, Vol. 1, Revision 1, Electric Power Research Institute, Palo Alto, California, 1990.
18. Masters, K.: *Spray Drying Handbook*, John Wiley & Sons, New York, p. 481, 1985.
19. Blythe, G. M., L. R. Lepovitz, and C. M. Thompson: "Evaluation of a 2.5-MW Spray Dryer/Fabric Filter SO_2 Removal System," EPRI GS-7449, p. 5-5, Electric Power Research Institute, Palo Alto, California, 1991.
20. Klingspor, J. S.: "Improved Spray Dry Scrubbing through Grinding of FGD Recycle Material," *J. Air Pollut. Control Assoc.*, Vol. 37, pp. 801–806, 1987.
21. Madden, D. A., and M. J. Holmes: "B&W's E-LIDS™ Process—Advanced SO_2, Particulate and Air Toxics Control for the Year 2000," *EPRI-DOE-EPA Combined Utility Air Pollutant Control Symposium, The Mega Symposium,* G. Offen, L. Ruth, and D. Lachapelle (eds.), Washington, DC, August 1997.
22. Radovanovic, M., ed.: *Fluidized Bed Combustion*, Hemisphere, Washington, DC, 1986.
23. "Cool Water Gasification Program: Final Report," EPRI GS-6808, Electric Power Research Institute, Palo Alto, California, 1990.
24. Hougen, O. A., K. M. Watson, and R. A. Ragatz: *Chemical Process Principles, Part II, Thermodynamics*, 2d ed., John Wiley & Sons, New York, p. 1019, 1959.
25. Telesz, B. W.: "5000 MW FGD Project for KEPCO," presented to the 12th U.S.–Korea Joint Workshop on Energy & Environment, Taejon, October 1997.
26. Pasiuk-Bronikowska, W., and K. J. Rudzinski: "Absorption of SO_2 into Aqueous Systems," *Chem. Eng. Sci.*, Vol. 46, pp. 2281–2291, 1991. Observe that this paper uses a Henry's law definition that is the reciprocal of the common one.

CHAPTER 12

CONTROL OF NITROGEN OXIDES

12.1 AN OVERVIEW OF THE NITROGEN OXIDES PROBLEM

Most of the world's nitrogen is in the atmosphere as an inert gas. In crustal rocks it is the 34th most abundant element with an abundance of only $\approx$ 20 ppm [1]. Although nitrogen forms eight different oxides [2], our principal air pollution interest is in the two most common oxides, nitric oxide (NO) and nitrogen dioxide (NO_2). In addition, we are beginning to be concerned with nitrous oxide (N_2O). It is not a common air pollutant, but it may be a significant contributor to global warming and to the possible destruction of the ozone layer (see Chapter 14). See Sec. 11.1 for an elementary review of the oxidation-reduction chemistry of nitrogen and sulfur.

12.1.1 Comparison with Sulfur Oxides

Figure 12.1 on page 440 shows part of how nitrogen moves in the environment, as a result of human activities. Nitrogen oxides are often lumped with sulfur oxides as air pollution control problems because of the similarities between the two:

1. Nitrogen oxides and sulfur oxides react with water and oxygen in the atmosphere to form nitric and sulfuric acids, respectively. These two acids are the principal contributors to acid rain. Because the acid rain process removes both nitrogen and sulfur oxides from the atmosphere, neither is believed to be increasing in concentration in the global atmosphere.

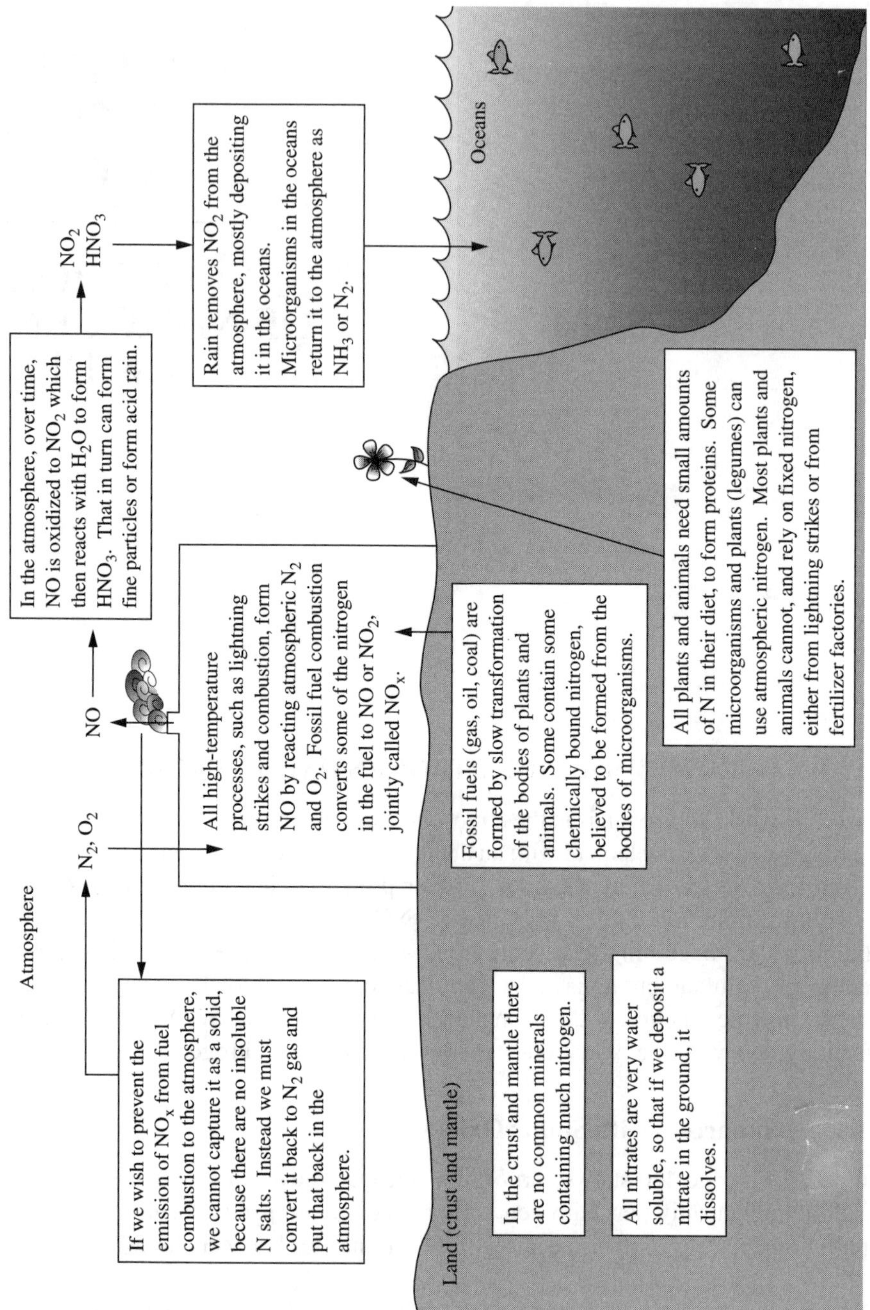

FIGURE 12.1
Principal flows of N in the environment, as influenced by humans. The atmosphere is about 78% N_2, which is mostly inert and remains in the atmosphere. This figure shows only how the oxidized and reduced compounds of nitrogen are formed, move through the environment, and are converted back to relatively inert N_2.

2. Both undergo atmospheric transformations leading to or contributing to the formation of PM_{10} and $PM_{2.5}$ in urban areas.
3. Both are released into the atmosphere in large quantities, and both are regulated pollutants for which we have NAAQS. In high concentrations sulfur oxides and nitrogen dioxide are severe respiratory irritants [3].
4. Both are released to the atmosphere by large combustion sources, particularly coal combustion sources. Table 12.1 [4] shows in more detail than Table 1.1 the emission sources of NO_x in the United States in 1997.

We see that vehicles contribute almost half of the total, and fuel combustion about a third. Natural gas appears in this table, but not in the SO_2 table, as explained later. The largest contributors to the "All other sources" are nontransportation internal combustion engines and residential combustion.

TABLE 12.1
National emissions estimates for nitrogen oxides (expressed as NO_2) for 1997 (see Table 1.1)

Source category	Emissions, thousands of tons/yr	Percent of total	
Coal combustion, utilities	5588	23.70	
industrial	614	2.60	
commercial	40	0.17	
Subtotal			26.48
Oil combustion, utilities	131	0.56	
industrial	240	1.02	
commercial	107	0.45	
Subtotal			2.03
Gas combustion, utilities	286	1.21	
industrial	1385	5.87	
commercial	241	1.02	
Subtotal			8.11
Industrial processes			
Chemical and allied	167	0.71	
Metals processing	102	0.43	
Petroleum and related	115	0.49	
Other industrial	421	1.79	
Subtotal			3.41
Vehicle emissions, on-road	7035	29.84	
off-road	4555	19.32	
Subtotal			49.16
All other sources	2549	10.81	10.81
Total	23 576	100.00	

Source: Ref. 4.

While the similarities listed above are significant, there are also major differences between nitrogen oxides and sulfur oxides. The most important of these are:

1. Motor vehicles are the major emitter of nitrogen oxides, but a very minor source of sulfur oxides. If motor vehicles had zero-sulfur fuels, they would emit no sulfur oxides. If they had zero-nitrogen fuels, which they practically do, they would still be major contributors to the nitrogen oxides problem (see Chapter 13).
2. Sulfur oxides are formed from the sulfur *contaminants* in fuels or the unwanted sulfur in sulfide ores. Removing all sulfur from the fuels would completely eliminate sulfur emissions from fuel combustion. Although some of the nitrogen oxides emitted to the atmosphere are due to nitrogen contaminants in fuels, most are not. Instead they are formed by the reaction of atmospheric nitrogen with oxygen in high-temperature flames. Removing all of the nitrogen from fuels would reduce the nationwide emissions of nitrogen oxides from fuel combustion (including autos) by only about 10 to 20 percent. (Sulfur oxides are made from things we take from the ground; nitrogen oxides are mostly made from components of the air.)
3. The formation of nitrogen oxides in flames can be greatly reduced by manipulating the time, temperature, and oxygen content of the flames. No such reductions are possible with sulfur oxides.
4. The ultimate fate of sulfur oxides removed in pollution control or fuel-cleaning processes is to be turned into $CaSO_4 \cdot 2H_2O$, which is an innocuous, low-solubility solid, and to be placed in landfills. There is no correspondingly cheap, innocuous, and insoluble salt of nitric acid, so landfilling is not a suitable fate for the nitrogen oxides we collect in pollution control devices or remove from fuels. Solids containing substantial amounts of nitrogen are only found in the driest of environments, e.g., the Atacama Desert of Chile, where the droppings of centuries of birds have formed economically useful nitrate deposits. The ultimate fate of nitrogen oxides that we wish to keep out of the atmosphere is to be converted to gaseous nitrogen and oxygen, N_2 and O_2, and be returned to the atmosphere.
5. It is relatively easy to remove SO_2 from combustion gases by dissolving SO_2 in water and reacting it with alkali. Aqueous SO_2 quickly forms sulfurous acid, which reacts with the alkali and then is oxidized to sulfate. Collecting nitrogen oxides is not nearly as easy this way because NO, the principal nitrogen oxide present in combustion gas streams, has a very low solubility in water. Unlike sulfur oxides, which quickly react with water to form acids, NO must undergo a two-step process to form an acid:

$$NO + 0.5O_2 \rightleftarrows NO_2 \tag{12.1}$$

$$3NO_2 + H_2O \rightarrow 2HNO_3 + NO \tag{12.2}$$

The first reaction is relatively slow. It is fast enough in the atmosphere to lead to the formation of acid precipitation in the several hours or days that the polluted air travels before encountering precipitation, but slow enough that it does not remove significant quantities of NO in the few seconds that a contaminated gas spends in a wet limestone scrubber used for SO_2 control. (Evidently some of the NO_2 in stack gases is removed in such scrubbers, but that is normally only a small fraction of the total nitrogen oxides.)

12.1.2 Reactions in the Atmosphere

NO is a colorless gas that has some harmful effects on health, but these effects are substantially less than those of an equivalent amount of NO_2. In the atmosphere and in industrial devices NO reacts with O_2 to form NO_2, a brown gas that is a serious respiratory irritant. Its color is strong enough that it is often possible to see a distinct brown color emerging from a power plant stack or from the vent of any process using nitric acid, which releases NO_2. NO and NO_2 are often treated together as one problem or as a quasi species, and written NO_x. Most regulations for NO_x emissions base all numerical values on the assumption that all of the NO is converted to NO_2. One sees this written as "NO_x expressed as NO_2."

There is an NAAQS for NO_2 to protect human health (see Chapter 2), which was sometimes exceeded in the 1980s, but now is never exceeded in the United States. Our principal concern with NO_x is that nitrogen oxides contribute to the formation of ozone, O_3, which is a strong respiratory irritant and one of the principal constituents of urban summer eye- and nose-irritating smog. The NAAQS for ozone is regularly exceeded in many U.S. cities. The overall reaction (see Appendix D) is

$$NO + HC + O_2 + \text{sunlight} \rightarrow NO_2 + O_3 \tag{1.2}$$

Figure 1.2 shows the typical pattern. NO, emitted during the morning commuter rush, is oxidized in the atmosphere to NO_2 over a period of several hours. The NO_2 thus formed then reacts as described in Appendix D to form O_3. The O_3 peak occurs after the NO_2 peak. We are also concerned with the slower reaction that produces nitric acid and/or its salts. The latter reaction leads to acid deposition and/or PM_{10} and $PM_{2.5}$.

Most nitrogen oxides derived from human activities are formed in flames. In Sec. 7.10 we discussed many important ideas about combustion. (Readers should review that section if they have not read it recently.) In VOC incineration, Sec. 10.5.1, the goal is to provide as high a temperature and as long a residence time (or "contact time" or "dwell time") and as much turbulence as possible. In this chapter we will see that, to prevent the formation of nitrogen oxides in flames, we generally want just the opposite, as low a temperature, as short a residence time, and (in some cases) as little turbulence as possible.

12.1.3 NO and NO_2 Equilibrium

The most important reactions for producing NO and NO_2 in flames are

$$N_2 + O_2 \rightleftarrows 2\,NO \tag{12.3}$$

TABLE 12.2
Equilibrium constants for the formation of NO by Eq. (12.3) and NO_2 by Eq. (12.1)

Temperature,		K_p of formation	
K	°F	NO	NO_2
300	80	7×10^{-31}	1.4×10^{6}
500	440	2.7×10^{-18}	130
1000	1340	7.5×10^{-9}	0.11
1500	2240	1.07×10^{-5}	0.011
2000	3140	0.00040	0.0035
2500	4040	0.0035	0.0018

Source: Ref. 6.

and Eq. (12.1). Both of these are reversible reactions that do not go to completion. Here we examine the effect of temperature and chemical composition on the equilibrium in these reactions. For any chemical reaction at equilibrium [5], the Gibbs free energy is at a minimum for the reaction's temperature and pressure. From that condition it follows that

$$\ln K = -\frac{\Delta G^0}{RT} \tag{12.4}$$

where ΔG^0 = standard Gibbs free energy change
K = equilibrium constant, defined below
R = the universal gas constant
T = absolute temperature (K or °R)

For all reactions involving simple gases, values of G^0 are published [6], so that one may compute the values of ΔG^0 and K at any temperature. Using the published G^0 values for the reactions in Eqs. (12.1) and (12.3) we may construct Table 12.2 (see Problem 12.2). Here we show K_p, the equilibrium constant based on taking the standard states as perfect gases at 1 atm pressure, and expressing gas concentrations as partial pressures. Other choices of standard state are possible, leading to different numerical values of K. All such choices lead to the same computed equilibrium concentrations if the user is careful about units. Because the number of mols does not change in Eq. 12.3, its K_p is dimensionless. However, in Eq. (12.1) the number of mols decreases by $\frac{1}{2}$, so that K_p has the dimension $(\text{atm})^{-0.5}$.

Example 12.1. Calculate the equilibrium concentrations of NO and NO_2 for air that is held at 2000 K = 3140°F long enough to reach chemical equilibrium. Assume that the only reactions of interest are Eqs. (12.1) and (12.3).

Here the definitions of the two equilibrium constants are as follows:

$$K_{12.3} = \frac{[\text{NO}]^2}{[\text{O}_2][\text{N}_2]} \tag{12.5}$$

$$K_{12.1} = \frac{[NO_2]}{[NO][O_2]^{1/2}} \tag{12.6}$$

Here [NO] stands for the *activity* of NO in the reaction mix at equilibrium. For perfect gases the activity is *identical* to the concentration, which can be expressed in a variety of units. For the low pressures and high temperatures of interest in this chapter, real gas behavior is close enough to perfect gas behavior that we will no longer use activities in any of our descriptions, equations, examples, and problems, but instead will use concentrations (in various units). We understand that, for the most precise work, the nonperfect gas behavior must be taken into account. The concentration values must be multiplied by a dimensionless *activity coefficient* that converts the concentration values we are using to the activity values that truly appear in these equations. (For a perfect gas the activity coefficient = 1.00.)

If we use K_p, then the concentrations must be expressed as *partial pressures,* equal to the (pressure · mol fraction). If the reaction equilibrium is at 1 atm, then the concentrations are numerically equal to mol fractions. Unfortunately, the symbol [X] (where X can be O_2, N_2, etc.), normally stands for a mol fraction in equilibrium calculations and for an absolute concentration, e.g., mol/cm^3, in kinetic calculations. It is used both ways in the literature and in this chapter; one must always check to be sure which is used. This chapter tries to make clear in each example which is intended.

Solving Eq. (12.3) for the equilibrium concentration of NO, we find

$$[NO] = (K_{12.3}[N_2][O_2])^{1/2}$$

Substituting values, including the value of $K_{12.3}$ from Table 12.2, into this equation, we see

$$[NO] = (0.0004\,[0.78][0.21])^{1/2} = 0.0081 = 8100 \text{ ppm}$$

In the most careful work we must account for the changes in mol fractions of nitrogen and oxygen; in this case doing so would change the answer by 1 percent (see Problem 12.3). If the oxygen concentration were lower, as in a combustion gas, the error introduced by ignoring the change in oxygen concentration would be greater.

Solving Eq. (12.6) for $[NO_2]$, we find

$$[NO_2] = K_{12.1}[NO][O_2]^{1/2}$$

Substituting values, including the value of $K_{12.1}$ from Table 12.2, into this equation, we see

$$[NO_2] = 0.0035\,[0.0081][0.21]^{1/2} = 1.3 \times 10^{-5} = 13 \text{ ppm}$$ ■

Following the same procedure as in Example (12.1), we may make up Table 12.3 on page 446, which shows the calculated equilibrium concentrations of NO and NO_2 at various temperatures, both for a starting gas that is 78 percent nitrogen, 21 percent oxygen and for a starting gas that is 78 percent nitrogen, 4 percent oxygen. The latter is more representative of combustion gases in which most of the oxygen has been consumed by the combustion.

TABLE 12.3
Calculated equilibrium concentrations of NO and NO_2

Temperature,		Starting with 78% N_2, 21% O_2		Starting with 78% N_2, 4% O_2	
K	°F	ppm NO	ppm NO_2	ppm NO	ppm NO_2
300	80	3.4×10^{-10}	2×10^{-4}	1.4×10^{-10}	4×10^{-5}
500	440	7×10^{-4}	0.04	3×10^{-4}	7.6×10^{-3}
1000	1340	35	1.9	15	0.35
1500	2240	1320	6.8	580	1.3
2000	3140	8100	13.2	3530	2.5
2500	4040	24 000	20	10 500	4.0

The values from Table 12.3 are shown in Figs. 12.2 and 12.3 on pages 447 and 448. From the table, and the figures we see the following:

1. If the atmosphere were at equilibrium (at a temperature near 300 K = 80°F), it would have less than a part per billion of NO or NO_2. The concentrations of NO and NO_2 observed in cities in the United States often exceed these equilibrium values (see Figure 1.2), so that equilibrium alone is not a satisfactory guide to the presence of NO and NO_2 in the atmosphere.
2. The equilibrium concentration of NO increases dramatically with increasing temperature. The rapid increase begins at about 2000–2500°F (1367–1644 K).
3. At low temperatures the equilibrium concentration of NO_2 is much higher than that of NO, whereas at high temperatures the reverse is true.
4. We are likely to get to the temperatures where NO and NO_2 are formed from atmospheric N_2 and O_2 only in flames and in lightning strikes. Lightning strikes are a major global source of NO_x, but combustion in our vehicles and factories is the main source of NO_x in heavily populated areas.

12.1.4 Thermal, Prompt, and Fuel NO_x

Combustion scientists classify the nitrogen oxides found in combustion gases as *thermal, prompt,* and *fuel nitrogen oxides* [7]. Thermal nitrogen oxides, which are generally the most significant, are formed by the simple heating of oxygen and nitrogen, either in a flame or by some other external heating, e.g., a lightning bolt. Prompt refers to the nitrogen oxides that form very quickly as a result of the interaction of nitrogen and oxygen with some of the active hydrocarbon species derived from the fuel in the fuel-rich parts of flames. They are not observed in flames of fuels with no carbon, e.g., H_2. They cannot be formed by simply heating oxygen and nitrogen; the participation of some active hydrocarbon species from the fuel is required. Fuel nitrogen oxide is formed by conversion of some of the nitrogen originally present in the fuel to NO_x. (Coal and some high-boiling petroleum fuels contain significant amounts of organic nitrogen [see Appendix C]; low-boiling petroleum fuels and

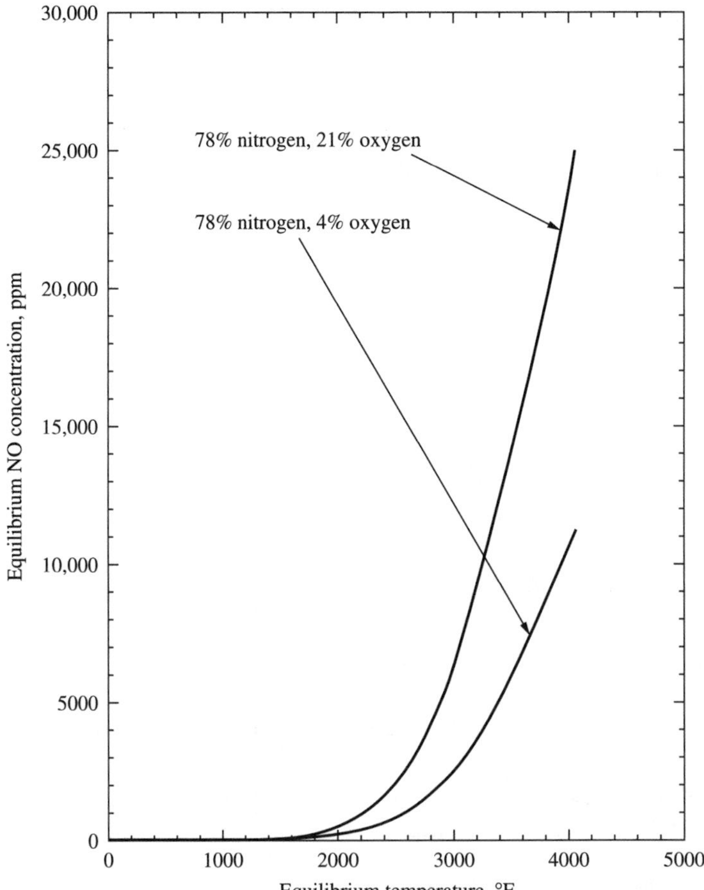

FIGURE 12.2
Calculated equilibrium concentrations of NO for two different oxygen contents of the reacting gas. See Example 12.1.

natural gas contain practically none.) All three kinds of NO_x are present in many combustion gases.

Figure 12.4 on page 449 shows estimates of the contribution of the thermal, fuel, and prompt mechanisms to the NO_x emissions from coal combustion. Below about 1300°C = 2372°F, the thermal NO mechanism is negligible compared with the other two, while at the highest temperatures it is the most important. If we had based our estimates on the thermal mechanism alone, we would predict approximately zero NO_x would be produced at temperatures below 1300°C. But, as the figure shows, the observed emissions at that temperature can be substantial.

Each of the three mechanisms for producing nitrogen oxides has its own separate kinetic pathways and corresponding rate equations. All three are governed jointly by the overall equilibrium relations already discussed and the reaction

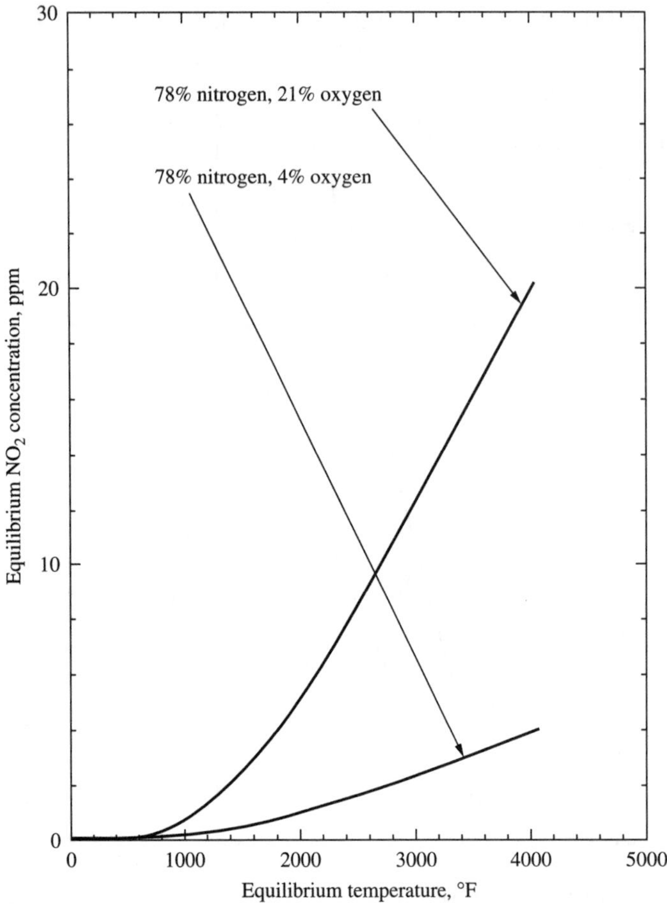

FIGURE 12.3
Calculated equilibrium concentrations of NO_2 for two different oxygen contents of the reacting gas. See Example 12.1. (Observe the large difference in scale between Figs. 12.2 and 12.3.)

kinetics discussed later in this chapter. The products from these three mechanisms interact, so that treating them separately as we do here is a serious simplification.

12.2 THERMAL NO

At the highest temperatures, the thermal mechanism is the most important of the three ways of making NO. (See Fig. 12.4).

12.2.1 The Zeldovich Kinetics of Thermal NO Formation

If we heat some air to 2000 K = 3140°F and hold it long enough for equilibrium to be reached, then we know from Example 12.1, Table 12.3, and Fig. 12.2 that it will contain 8100 ppm of NO. Here we inquire how rapidly the mixture approaches that equilibrium value.

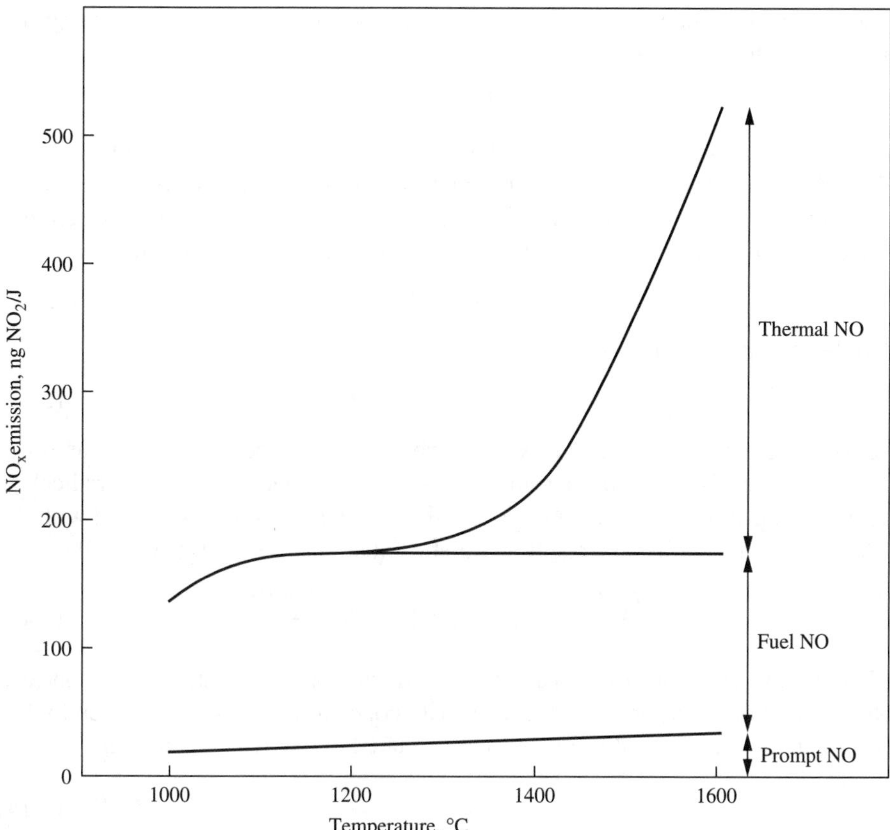

FIGURE 12.4
Estimated contributions of the three NO mechanisms to total NO_x formation in coal combustion. (Courtesy of the Air and Waste Management Association, from Ref. 7.)

The reactions shown in Eqs. (12.1) and (12.3) do not proceed as written in those equations. Rather, they proceed by means of intermediate steps involving highly energetic particles called *free radicals.* (The same is true of almost all high-temperature reactions. See Sec. 7.10.5.) The free radicals most often involved in combustion reactions are O, N, OH, H, and hydrocarbons that have lost one or more hydrogens, e.g., CH_3 or CH_2. These materials are very reactive and energetic and can exist in significant concentrations only at high temperatures. In principle they can be formed by equilibrium reactions like the following:

$$N_2 \rightleftarrows 2N \tag{12.7}$$

$$O_2 \rightleftarrows 2O \tag{12.8}$$

$$H_2O \rightleftarrows H + OH \tag{12.9}$$

However, the reactions are conventionally written with an M on both sides of the equation to indicate that another molecule, which is not chemically changed in the reaction, must collide with the N_2 or O_2 molecule to supply or remove energy for the reaction to occur, e.g.,

$$N_2 + M \rightleftarrows 2N + M \tag{12.10}$$

The concentration of M, which can be any other gas molecule (e.g., O_2 or N_2 or H_2O) does not influence the equilibrium but does influence the rate of the reactions.

The most widely quoted mechanism for the thermal NO formation reaction is that of Zeldovich [8]. It assumes that O radicals attack N_2 molecules by

$$O + N_2 \rightleftarrows NO + N \tag{12.11}$$

and that N radicals can form NO by

$$N + O_2 \rightleftarrows NO + O \tag{12.12}$$

Various degrees of simplification of this mechanism can be made. If one assumes that the O radicals are in equilibrium with O_2, that the concentration of N radicals is not changing significantly with time, and that one term is small compared with the others, one can simplify the resulting kinetic equations (see Problem 12.10) to

$$\frac{d[NO]}{dt} = k_f[N_2][O_2]^{1/2} - k_b\frac{[NO]^2}{[O_2]^{1/2}} \tag{12.13}$$

where k_f and k_b are the forward and backward reaction rate constants. Then one can observe that the equilibrium value of the NO concentration, $[NO]_e$, is given by Eq. (12.5), and that the equilibrium constant is related to the two rate constants by

$$K_{12.3} = \frac{k_f}{k_b} \tag{12.14}$$

One may show this by setting the rate of change of concentration in Eq. (12.13) to zero and comparing the result to Eq. (12.5). Making this substitution and rearranging in Eq. (12.13), one finds

$$\frac{d[NO]}{[NO]_e^2 - [NO]^2} = \frac{k_b}{[O_2]^{1/2}}\,dt \tag{12.15}$$

which may be integrated and rearranged (with an assumption of zero NO at time zero) to

$$\frac{[NO]}{[NO]_e} = \frac{1 - \exp(-\alpha t)}{1 + \exp(-\alpha t)} \tag{12.16}$$

where

$$\alpha = \frac{2[NO]_e \cdot k_b}{[O_2]^{1/2}} \tag{12.17}$$

Example 12.2. Estimate the concentration of NO in a sample containing 78 percent N_2 and 4 percent O_2 that is held for one second at 2000 K = 3140°F, according to the Zeldovich thermal mechanism.

From Table 12.3 we know the equilibrium concentration, $[NO]_e$, is 3530 ppm. Seinfeld [9] suggests a value for k_b of $4.1 \times 10^{13} \exp(-91\,600/RT)$, for T in K, R in cal/mol · K, t in seconds, and concentrations in (mol/cm^3). At 1.0 atm and 2000 K the molar density of any perfect gas is 6.1×10^{-6} mol/cm^3, so that

$$[NO]_e = 6.1 \times 10^{-6} \cdot 0.003530 = 2.15 \times 10^{-8} \frac{\text{mol}}{\text{cm}^3}$$

$$[O_2] = 6.1 \times 10^{-6} \cdot 0.04 = 2.44 \times 10^{-7} \frac{\text{mol}}{\text{cm}^3}$$

$$k_b = 4.1 \times 10^{13} \exp\left(\frac{-91\,600}{1.987 \cdot 2000}\right) = 4003 \left(\frac{\text{mol}}{\text{cm}^3}\right)^{-1/2} \left(\frac{1}{\text{s}}\right)$$

$$\alpha = \frac{2 \cdot 2.15 \times 10^{-8} \cdot 4003}{(2.44 \times 10^{-7})^{1/2}} = \frac{0.349}{\text{s}}$$

and

$$\frac{[NO]}{[NO]_e} = \frac{1 - \exp[(-0.349/\text{s})(1\ \text{s})]}{1 + \exp[(-0.349/\text{s})(1\ \text{s})]} = 0.173$$

From this expression it follows that $[NO] = 3530\ \text{ppm} \cdot 0.173 = 610\ \text{ppm}$. ■

This is a *simplified* version of the Zeldovich *simplification* of the kinetics of the thermal NO formation reaction. Using this simple relation, one may readily make up plots like Fig. 12.5 on page 452, which show the expected time-temperature relation for one specific starting gas composition. From that plot we see that at 1500 K = 2240°F not only is the calculated equilibrium concentration quite low but also the calculated rate of reaching it is slow enough that after 30 seconds the mixture is far from equilibrium. In contrast, at 2300 K = 3680°F the equilibrium concentration is much higher, and the calculated reaction rate is fast enough that equilibrium is reached in about 0.3 s. Thus to make a good estimate of the NO concentration of a combustion gas due to the thermal mechanism, we need to know its initial nitrogen-oxygen ratio and its temperature-time history.

More complex versions of the Zeldovich mechanism add another equation to the reaction list,

$$N + OH \rightleftarrows NO + H \tag{12.18}$$

with a resulting increase in mathematical complexity. A derivation and example corresponding to the preceding one, including Eq. (12.18), is given by Benítez [10]. The result cannot be reduced to any simple plot like Fig. 12.5. (See Problem 12.11.)

The Zeldovich mechanism shown here makes reasonably accurate predictions of the rate of formation of NO from N_2 and O_2 in the highest temperatures observed in flames, and in other high-temperature situations. It predicts much lower NO concentrations than those observed in low-temperature flames, like those in a kitchen stove, or in the burning of nitrogen-bearing fuels, like some coals (Fig. 12.4).

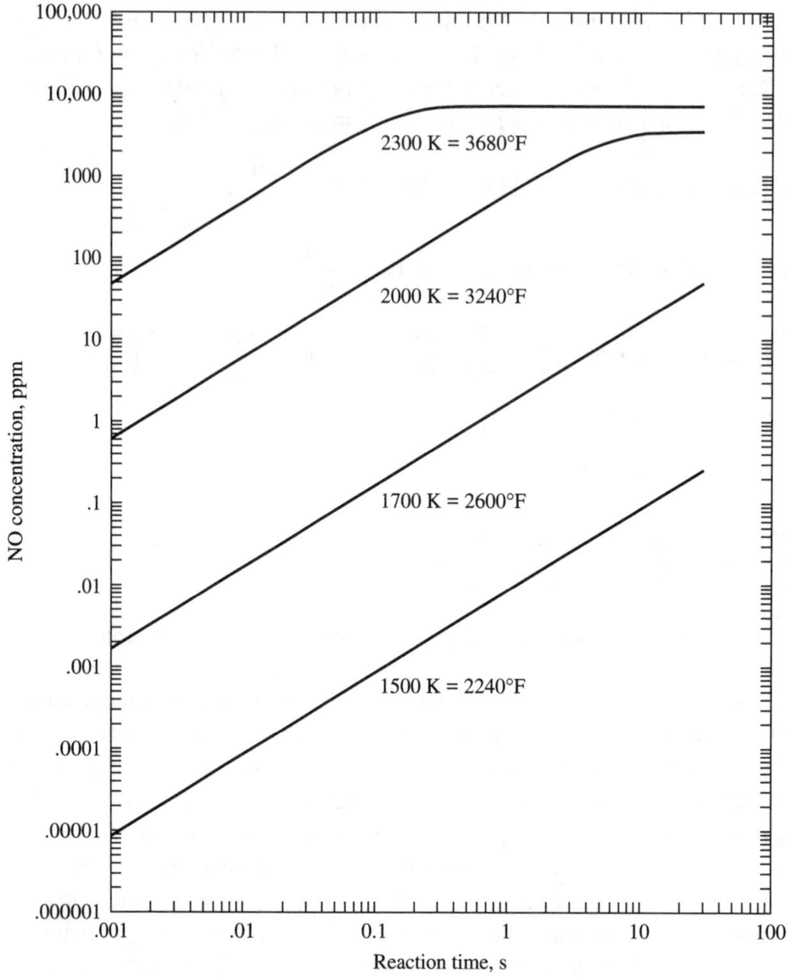

FIGURE 12.5
Calculated concentrations of thermal NO in a gas containing 78% N_2, 4% O_2 as a function of time and temperature according to the simplified form of the Zeldovich mechanism. See Example 12.2. This calculation does not include prompt or fuel NO.

Before we discuss the prompt and fuel NO mechanisms that predominate at low temperatures, we explore the temperatures and times in flames, to help understand what kinds of flames produce significant thermal NO and what kinds do not. From Table 7.2 we can see that the required temperatures for significant thermal NO occur in auto engines, and in large furnaces with preheating. For large furnaces without preheating the temperatures are barely high enough. The combustion time is short in auto engines but, as shown in Chapter 13, the temperature is high enough that even for these short times, there is substantial thermal NO.

12.2.2 Heating and Cooling Times

How much thermal NO is formed in a flame as well as how much is then converted back to N_2 and O_2 as the gases cool is a strong function of how fast the gases heat and cool in the flames. The heating and cooling rates in flames are quite impressive. Figure 12.6 on page 454 shows some possible heating and cooling patterns for flames.

The square wave of Fig. 12.6*a* leads to the easiest calculations, as shown in Example 12.2. That example implicitly assumed that cooling after the flame is instantaneous, so that none of the NO formed in the flame is converted back to N_2 and O_2. That assumption is clearly a simplification; nothing in this world is truly instantaneous.

Example 12.3. Estimate the heating and cooling rates in the flame of an ordinary gas stove. From Table 7.2 we know that the peak temperature is about 2400°F and the combustion time about 0.005 s. If we assume that the temperature-time pattern is the triangle wave shown in Figure 12.6*b*, we can say that the heating and cooling rates should be about the same, and

$$\frac{dT}{dt} \approx \frac{\Delta T}{\Delta t} = \frac{2400°\text{F} - 68°\text{F}}{0.0025\ \text{s}} = 9.3 \times 10^5\ \frac{°\text{F}}{\text{s}} = 5.2 \times 10^5\ \frac{°\text{C}}{\text{s}} \qquad \blacksquare$$

This number is startlingly large, almost a million degrees Fahrenheit per second! It is a reasonable estimate of the heating time in a gas stove. The cooling is initially of comparable speed, but the final stages of cooling are presumably slower as the hot combustion gases mix with the kitchen air and heat the pot on the stove. The assumption of a symmetrical triangle temperature-time pattern is closer to reality than the square wave and still leads to relatively simple calculations (Example 12.4). But it is also a strong simplification of the real temperature-time pattern in a flame. If we take the viewpoint of a person riding with a parcel of gas through the flame (the Lagrangian view), we can say that

$$\begin{aligned} V\rho C_P \frac{dT}{dt} &= \text{heat added by combustion} - \text{heat lost to surroundings} \\ &= r\ \Delta h_{\text{combustion}} - UA(T_{\text{flame}} - T_{\text{surroundings}}) \end{aligned} \tag{12.19}$$

where r is the rate of combustion and UA the product of the overall heat transfer coefficient and the applicable surface area. For the rising part of T-t pattern on Fig. 12.6 to be straight, the reaction rate r would have to remain constant from the start of combustion until all the fuel was burned and be much, much larger than the heat loss term. The observed behavior is that the rate starts slowly, increases as the temperature rises, and then slows to a stop as the last of the fuel is consumed. Thus the rising part of the T-t plot must be S-shaped. Heat losses make the top more rounded than one would compute from reaction rate alone. During the cooling period, if UA were a constant we would expect the temperature decline to be exponential. Figure 12.6*c* shows a more plausible T-t pattern for a flame.

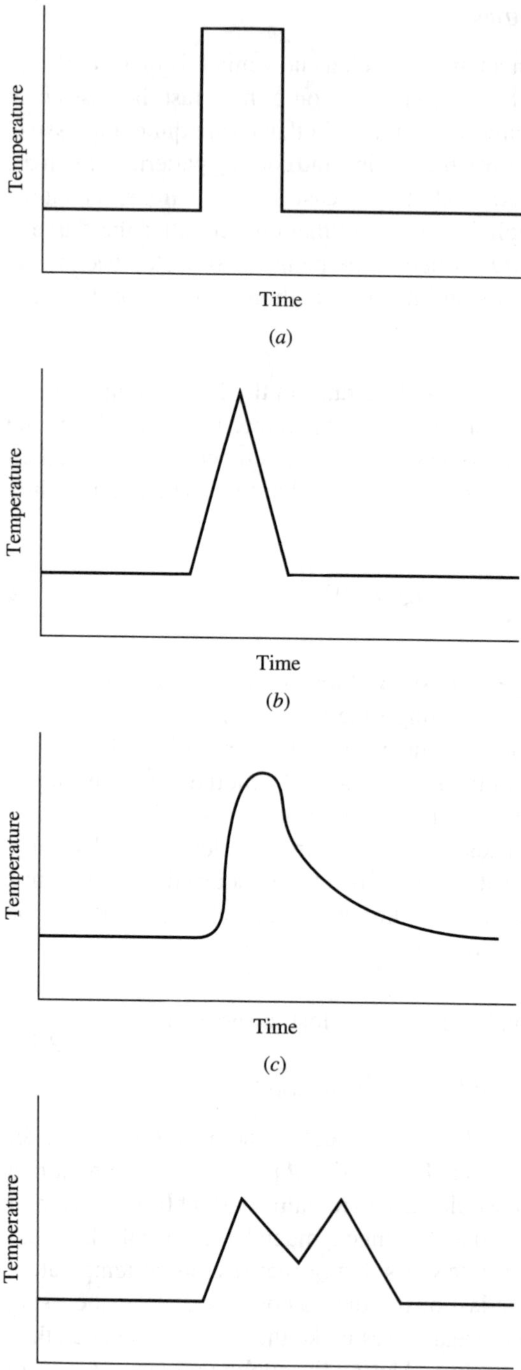

FIGURE 12.6
Four possible temperature-time patterns for flames: (*a*) square wave, (*b*) triangle wave, (*c*) plausible flame pattern, (*d*) staged combustion.

Near the peak of the T-t curve (where the major NO-forming reactions occur), dT/dt is approaching zero, which means that the two terms on the right of Eq. (12.19) must be of comparable magnitude. That fact indicates that after the peak is reached, the decline must be steeper than the rise was because the rise was proportional to the difference between these two terms but the decline is proportional to the cooling term alone. Thus, for most flames near their peak temperature, we would expect the T-t behavior to be that in Fig. 12.6c, with a more rapid cooling after burnup of all the fuel than heating during the last stages of burnup.

Example 12.4. Repeat Example 12.2 for the strong simplifying assumptions that (1) the gas heats from 293 to 2000 K in 0.5 s with a linear temperature increase, then cools from 2000 to 293 K in 0.5 s with a linear temperature decrease and (2) the initial NO concentration is zero.

Here we use the simple Zeldovich mechanism and Eq. (12.15), which we know to be only approximate. We can see from the problem statement that

$$T = 293\text{ K} + \frac{1707\text{ K}}{0.5\text{ s}}t = 293\text{ K} + 3414\frac{\text{K}}{\text{s}}\cdot t \qquad \text{for} \qquad 0 < t < 0.5\text{ s}$$

$$T = 2000\text{ K} - \frac{1707\text{ K}}{0.5\text{ s}}(t - 0.5\text{ s})$$

$$= 2000\text{ K} - 3414\frac{\text{K}}{\text{s}}\cdot(t - 0.5\text{ s}) \qquad \text{for} \qquad 0.5\text{ s} < t < 1\text{ s}$$

If we consider the time interval $0.0 \leq t \leq 0.01$ s, we see that the initial temperature is 293 K and the final temperature is $293 + 34 = 327$ K. For this time interval the average temperature is 310 K. Using that value, we may estimate the equilibrium constant (see Problem 12.8) as

$$K_p = 21.9\exp\left(-\frac{43\,400}{1.987\cdot 310}\right) = 5.5\times 10^{-30}$$

For the most careful work we should take into account the change in oxygen content as the heating occurs. However, for simplicity we will proceed as if the oxygen content were the 4 percent in that example, independent of time. Substituting this value of K_p in Eq. (12.5) and solving, we find

$$[\text{NO}]_e = (5.5\times 10^{-30}[0.78][0.04])^{0.5} = 4\times 10^{-16}$$

This is the $[\text{NO}]_e$ in partial pressure units. Equation (12.15) requires that all concentrations be in concentration units. At this temperature and one atmosphere pressure, the molar density is $(4.46\times 10^{-5})(273/310) = 3.93\times 10^{-5}$ mol/cm^3, so that in concentration units $[\text{NO}]_e = 4\times 10^{-16}(3.93\times 10^{-5}) = 1.6\times 10^{-20}$ mol/cm^3. At this temperature, using the equation from Example 12.2, we have

$$k_b = 4.1\times 10^{13}\exp\left(-\frac{91\,600}{1.987\cdot 310}\right) = 1.1\times 10^{-51}$$

and

$$\frac{d[\text{NO}]}{dt} = \frac{k_b([\text{NO}]_e^2 - [\text{NO}]^2)}{[\text{O}]_2^{1/2}} = \frac{1.1 \times 10^{-51}([1.6 \times 10^{-20}]^2 - [0]^2)}{[0.04 \cdot 3.93 \times 10^{-5}]^{1/2}}$$

$$= 2.2 \times 10^{-88} \frac{\text{mol/cm}^3}{\text{s}}$$

Multiplying this by $\Delta t = 0.01$ s, we conclude that during the first 0.01 s the concentration of NO went from zero to 2×10^{-90} mol/cm^3. This value is negligible, but we now proceed for the remaining 99 time steps, each of the same size, using a spreadsheet, and find the results shown in Fig. 12.7. The final calculated NO concentration is 20 ppm. ■

From this plot we see that the calculated equilibrium concentration goes from 5.5×10^{-24} ppm (≈ 0) to the 3530 ppm in Table 12.3, as the temperature rises, and then back to zero as the temperature falls. The reaction rate is practically zero, and hence the amount of NO actually formed is practically zero until about 0.44 s ($T \approx 1780$ K), after which the rate of NO formation becomes significant. After the flame temperature peaks, the NO concentration continues rising slowly,

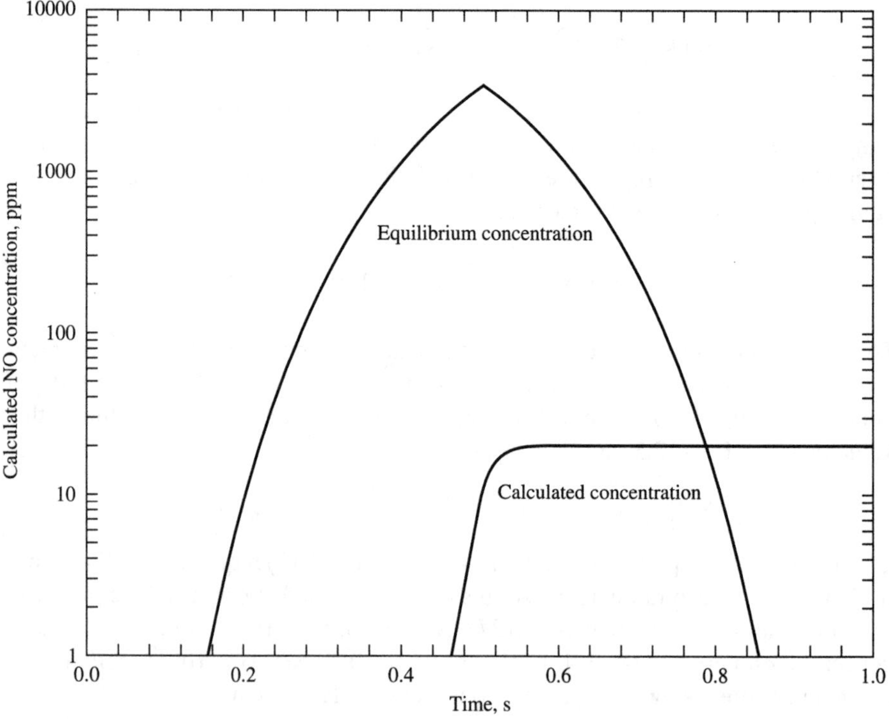

FIGURE 12.7
Calculated NO concentration, and equilibrium concentration from Example 12.5, based on the simple form of the Zeldovich thermal NO mechanism.

because the actual concentration is less than the equilibrium concentration even though the equilibrium concentration is falling rapidly with the decreasing temperature. By the time the two curves cross, the temperature is low enough that the reaction rate is practically zero, so that the NO concentration remains practically constant during the further cooling. (One speaks of the concentration being "frozen" at its higher temperature value.)

This type of calculation is obviously more time-consuming than the simpler version in Example 12.2. It is closer to reality than that calculation and would give the best available estimate if we had used the real *T-t* curve and the current best-estimate kinetics, which include the prompt NO and fuel NO described next. Such calculations are regularly made on large computers.

This example did not specify what fuel was used; the computed result is independent of that. This explains why natural gas combustion in high-temperature furnaces is a significant source of NO_x (see Table 12.1), even though it is the classic "clean fuel."

12.3 PROMPT NO

During the first part of combustion, the carbon-bearing radicals from the fuel react with nitrogen by

$$CH + N_2 \rightleftarrows HCN + N \tag{12.20}$$

and several similar reactions involving the CH_2 and C radicals. The N thus produced attacks O_2 by Eq. (12.12) to increase the amount of NO formed; and the HCN partly reacts with O_2, producing NO, partly with NO, producing N_2. There does not appear to be any simple theoretical treatment like the Zeldovich mechanism that allows a paper-and-pencil calculation or a simple plot of the amount of prompt NO formed in a flame. But it appears certain that the amounts of NO found in low-temperature flames (e.g., in a gas stove), which are much larger than one would predict from the Zeldovich thermal mechanism, are actually mostly prompt NO [11].

From Fig. 12.4 one would estimate that the production of prompt NO is only weakly dependent on temperature and averages about 30 ng/J (= g/GJ); see Example 12.5.

12.4 FUEL NO

Most gaseous and liquid fuels contain little nitrogen, and the contribution of that nitrogen to the total NO in the combustion products is minimal. However, some coals as well as the hydrocarbons derived from oil shale can contain up to 2 percent N (see Appendix C). Most of the fuel nitrogen is converted in the flame to HCN [12], which then converts to NH or NH_2. The NH and NH_2 can react with oxygen to produce NO + H_2O, or they can react with NO to produce N_2 and H_2O. Thus the fraction of the fuel nitrogen that leaves the flame as NO is dependent on the NO/O_2 ratio in the flame zone. Keeping the oxygen content of the gases in the

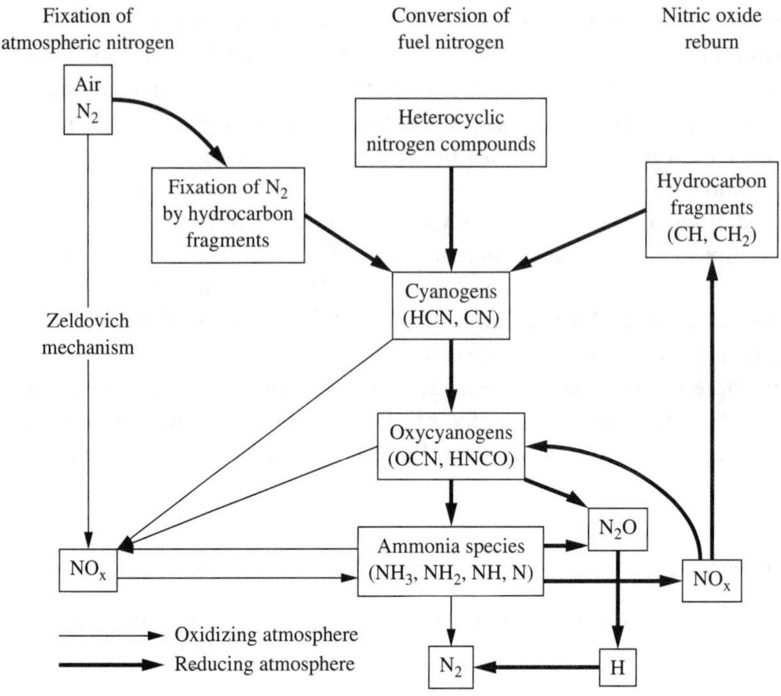

FIGURE 12.8
Nitrogen oxide formation pathways in combustion. The thermal mechanism is shown at the left, the prompt begins at the upper left and passes through the center to the lower left. The fuel nitrogen conversion begins at the top center, proceeds to the middle, and then can go to the lower left or the lower middle, depending on whether conditions in the middle are oxidizing or reducing. The plot also shows nitric oxide reburn, in which a combustion gas stream rich in NO_x is reacted at modest temperature with an HC flame, taking the products through the middle of the figure to the lower left or lower center. *Source:* Ref. 13. (Reproduced courtesy of McGraw-Hill.)

high-temperature part of the flame low significantly lowers the fraction of the fuel nitrogen converted to NO.

The fraction of the fuel nitrogen that appears as NO_x in the exhaust gas is estimated to be typically 20 percent to 50 percent, depending on furnace conditions and, to some extent, the chemical nature of the N in the fuel. The thermal, prompt, and fuel NO products interact, as do the radicals involved. Thus, although we analyze and study the mechanisms separately, flames treat them as one combined mechanism.

Figure 12.8 shows one author's schematic view of the interactions of the three mechanisms of NO formation. This is a simplification of the many reactions actually involved, but shows the three main pathways and how they can interact.

12.5 NONCOMBUSTION SOURCES OF NITROGEN OXIDES

The production and utilization of nitric acid lead to emissions of NO and NO_2, as do some other industrial and agricultural processes, e.g., silage production [14]. The concentrations of NO_x in the exhaust gases from these processes can be significantly larger than those from combustion sources, but the total volume of gases emitted is much smaller than that from combustion sources, and most such industrial sources are under fairly strict control, so that their contribution to the overall NO_x problem is generally small. Table 12.1 shows that such sources account for about 3 percent of our national NO_x emissions.

12.6 CONTROL OF NITROGEN OXIDE EMISSIONS

There are two possible approaches to controlling NO_x in combustion gases:

1. Modify the combustion processes to prevent the formation of NO_x.
2. Treat the combustion gas chemically, after the flame, to convert the NO_x to N_2.

Both approaches are used, separately or together.

For smaller industrial sources, like nitric acid plants, other control techniques are used [14], e.g., scrubbing with solutions of NaOH and $KMnO_4$. As discussed in Sec. 11.3, it is difficult to remove NO by scrubbing with an aqueous solution because NO is only slightly soluble in water. To scrub it successfully one must use a solution in which the dissolved NO is quickly converted to something else, allowing more NO to be absorbed. The strongly oxidizing $KMnO_4$ quickly converts NO to NO_2, allowing it to be captured by the alkaline NaOH. This process requires an expensive electrochemical regeneration of the $KMnO_4$; it is not suitable for large-scale combustion sources.

12.6.1 Nitrogen Oxide Control by Combustion Modification

This approach has been most widely applied. As the previous text shows, the formation of thermal NO is increased by increasing any of the following three variables: peak temperature, time at high temperature, oxygen content at the high temperature. Although the nitrogen content appears in the equations for the formation of NO, there seems to be no practical way to lower the nitrogen content of the high-temperature gas. (One could do that by carrying out all combustion reactions with oxygen instead of air. Pure oxygen costs; air is free.)

Some combustion modification schemes involve mixing part of the combustion air with the fuel, burning as much of the fuel as that amount of air will burn, transferring some of the heat from the flames to whatever is being heated, then adding the remaining air and finishing the combustion. This scheme is often called *two-stage combustion* or *reburning;* its temperature-time pattern is shown in Fig. 12.6*d*. In the first stage the maximum temperature is lowered because not all the fuel is burned, and the maximum temperature is reached when all the oxygen has been used up, so that there is not enough oxygen to form NO. In the second stage enough of the heat released in the first stage has been removed that the maximum temperature reached—in the presence of excess oxygen—is low enough that the formation of

NO is small. In some schemes a small amount of additional fuel is added for the second stage, often a low-nitrogen-content fuel like methane if the primary fuel has high nitrogen. (This is called *reburning* and is shown on Fig. 12.8).

The advantage of this approach is that it is cheap. The disadvantages are that it requires a larger firebox for the same combustion rate (or requires the firing rate of a furnace to be reduced if it is applied to an existing furnace) and that it is difficult to get complete burning of the fuel in the second stage, so that the amount of unburned fuel and/or carbon monoxide in the exhaust gas is increased.

Figure 12.9 shows an alternative approach to designing a low-NO_x burner. It is intended for use with a gaseous fuel, where the majority of the total NO_x is thermal NO_x, but can also burn oil. In normal operation the incoming gas is thoroughly mixed with air and recycled combustion products (Flue Gas Recirculation, FGR), with about 15% excess air. This thorough mixing, together with excess air, prevents there ever being any fuel-rich part of the flame, in which prompt NO can form. The gas-air-FGR mixture flows outward in a swirling hollow cone around an internal recirculation zone, which is much like the low-pressure wake behind a building, shown in Fig. 6.11. That recirculation zone stabilizes the flame. Figure 12.10 shows some test results for this burner, burning natural gas with 15% excess air and varying degrees of air preheat and FGR. With no FGR the NO_x (mostly NO) is 85 ppm with ambient air input and 210 ppm with the air preheated to 500°F. The dilution with

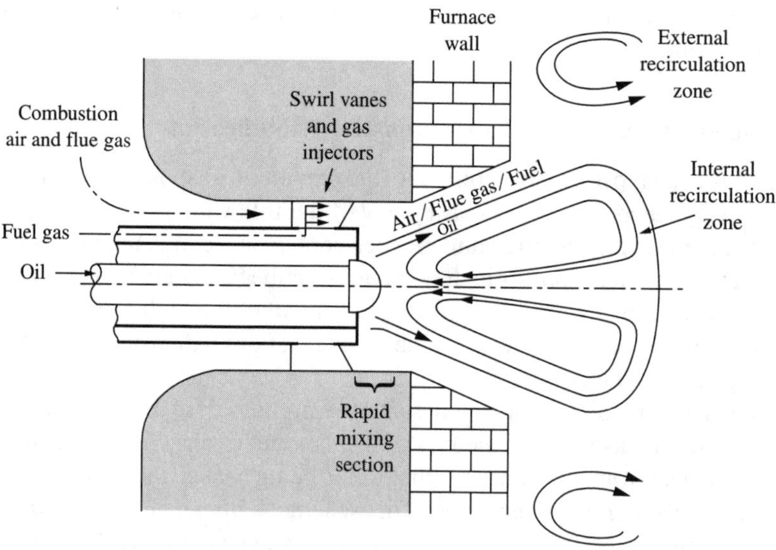

FIGURE 12.9

One type of low-NO_x burner. Gaseous fuel is introduced through the injectors, mixed with excess air and recirculated flue gas in the swirl vanes, and advanced into the combustion zone, where all components are now well mixed. This prevents prompt NO formation. FGR lowers flame temperatures, thus reducing or preventing thermal NO formation. When oil is used, thermal NO is also reduced or prevented, but not fuel or prompt NO [15]. (Courtesy of Radian International.)

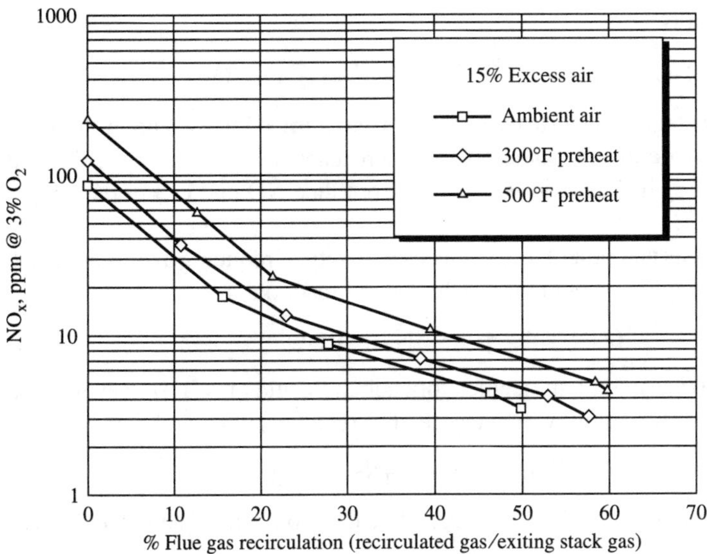

FIGURE 12.10
Experimental results for the burner shown in Fig. 12.9. With no FGR the exhaust gas has 85 to 210 ppm NO_x, almost all as thermal NO, depending on the degree of preheat. Adding FGR lowers the temperature everywhere in the flame, with a resultant reduction in thermal NO [15]. (Courtesy of Radian International.)

recycled flue gas simply lowers the temperature at every part of the flame, because the heat released must be divided among more molecules of N_2 and combustion products. As the figure shows, it is possible to reduce the NO_x to 10 ppm, which is indeed ultra low-NO_x, if one is willing to bear the extra cost of pumping 20 to 40% of the flue gas back into the burner. However, compared to other methods of achieving 10 ppm—i.e., catalytic treatment of the flue gas—the cost of FGR is very reasonable.

This burner is a moderately low-NO_x burner when it is burning fuel oil. Ultra-low NO_x levels are not achieved since most fuel oils contain bound nitrogen that is oxidized in the flames to NO_x. This fuel NO_x is not significantly impacted by the flame temperature-reducing effects of FGR. Therefore, the NO_x levels achieved by this burner on fuel oil are dependent on the nitrogen content of the oil. Also, because there will be a fuel-rich region at the surface of the evaporating oil droplets, there will be prompt NO formation. FGR helps reduce thermal NO_x formation on oil as on gas, but the fuel NO_x and prompt NO components keep the emissions well above the 10 ppm target for these burners.

12.6.2 Nitrogen Oxide Control by Postflame Treatment

Most postflame treatment processes add a reducing agent to the combustion gas stream to take oxygen away from NO. Almost any gaseous reducing agent can be used, e.g., CO, CH_4, other HCs, NH_3, various derivatives of NH_3. In modern auto engines the reaction is

$$2NO + 2CO \xrightarrow{\text{platinum-rhodium catalyst}} N_2 + 2CO_2 \tag{12.21}$$

This is particularly elegant because two pollutants are turned into two nonpollutants. The exhaust gas also contains some HCs, which react with NO as well. Careful control of the engine air-fuel ratio is needed to produce NO and CO in the proper ratio (Chapter 13).

For power plants and other large furnaces a variety of reducing agents are used, of which the most popular is ammonia. The desired reaction is

$$6NO + 4NH_3 \rightarrow 5N_2 + 6H_2O \tag{12.22}$$

which requires 0.667 ammonia molecules for each molecule of NO. However, there is always some oxygen present, which can lead to reactions like

$$4NO + 4NH_3 + O_2 \rightarrow 4N_2 + 6H_2O \tag{12.23}$$

in which one ammonia molecule is required for each molecule of NO. The NO_2 is reduced by

$$2NO_2 + 4NH_3 + O_2 \rightarrow 3N_2 + 6H_2O \tag{12.24}$$

Reactions (12.22)–(12.24) can be carried out over a variety of catalysts [16], of which zeolites and supported TiO_2 are the most popular, at about 700°F, or simply in the gas stream in the part of the furnace where the temperature is between 1600° and 1800°F. Below 1600°F the reaction rate is too slow. Above 1800°F the dominant reaction becomes

$$NH_3 + O_2 \rightarrow NO + \tfrac{3}{2}H_2O \tag{12.25}$$

which increases rather than decreases the NO content of the gas. Both catalytic and noncatalytic processes have been tried on a large scale and give good reductions in NO, but not cheaply. The catalytic processes are called Selective Catalytic Reduction, or SCR, and the higher-temperature ones, without catalysts, Selective Noncatalytic Reduction, SNCR.

There are experimental processes under development that adsorb both SO_2 and NO onto an activated carbon adsorbent [17]. They are then stripped off one at a time and treated, making S and N_2 the final products. So far these seem expensive; perhaps further work will make them more economical than the processes currently in use.

12.7 UNITS AND STANDARDS IN NO_x CONTROL

Emissions of NO_x are regularly reported and regulated in terms of ppm of NO_x, in mg/m^3, in lb/10^6 Btu, in g/GJ or μg/kcal. The lb/10^6 Btu, μg/kcal, or g/GJ standards are relatively straightforward, if we agree on which measure of heating value is to be used (higher or lower). The concentration descriptions, ppm and mg/m^3, depend on dilution. If a combustor uses a large amount of excess air, then the NO_x concentration in its exhaust gas will be lowered relative to the concentration for a facility with the same mass emission using less excess air. For that reason emission standards normally state that the concentrations shall be adjusted to a standard percentage of O_2 in the exhaust gas. The most common choices are 3 percent, 6 percent, and 7 percent O_2 [18].

Example 12.5. Wadden and Scheff present an extensive tabulation of air pollutant emission measurements from natural gas kitchen stoves [19]. The emission factor values for burners and ovens (but not pilot lights) can be summarized as follows:

$$\text{Emission factor for NO} \approx 89 \pm 37\ \mu\text{g/kcal}$$

$$\text{Emission factor for NO}_2 \approx 61 \pm 18\ \mu\text{g/kcal}$$

What are the emission rates (as NO_2) in g/GJ, lb/10^6 Btu, mg/m^3, and ppm? Base the ppm and mg/m^3 calculations on 6 percent O_2. Use the average emission factor values, ignoring the large variation in test results.

One gram of NO produces (46/30) = 1.533 grams of NO_2, so that the emissions "as NO_2" were $1.533 \cdot 89 + 61 = 197$ μg/kcal. (The NO_2 was 33 wt % of the total; most gases from large combustion sources are about 10 percent NO_2.)

The emission factor in g/GJ is

$$\text{EF} = 197\,\frac{\mu\text{g}}{\text{kcal}} \cdot \frac{0.239\ \text{kcal}}{\text{kJ}} \cdot \frac{10^6\ \text{kJ}}{\text{GJ}} \cdot \frac{\text{g}}{10^6\ \mu\text{g}} = 47.2\,\frac{\text{g}}{\text{GJ}}$$

and in English units it is

$$\text{EF} = 47.2\,\frac{\text{g}}{\text{GJ}} \cdot \frac{\text{lb}}{454\ \text{g}} \cdot \frac{1.055\ \text{GJ}}{10^6\ \text{Btu}} = 0.110\,\frac{\text{lb}}{10^6\ \text{Btu}}$$

From Example 7.9 we know that for burning methane (the principal component of natural gas) the oxygen content of the combustion gases for 1 mol of methane will be

$$n_{O_2} = E n_{\text{stoich}}$$

$$y_{O_2} = \frac{E n_{\text{stoich}}}{x + (y/2) + n_{\text{stoich}}\{[(1+E)/0.21](1+X) - 1\}}$$

which may be rearranged to

$$E = \frac{y_{O_2}\{x + (y/2) + n_{\text{stoich}}[(1+X)/(0.21) - 1]\}}{n_{\text{stoich}}\{1 - y_{O_2}[(1+X)/(0.21)]\}} \tag{12.26}$$

from which we see that for $y_{O_2} = 6\% = 0.06$, $X = 0.0116$, and for methane, which has $x = 1$, $y = 4$, and $n_{\text{stoich}} = 2$,

$$E = \frac{0.06\{1 + 4/2 + 2[(1+0.0116)/(0.21) - 1]\}}{2\{1 - 0.06[(1+0.0116)/(0.21)]\}} = 0.448 = 44.8\%$$

and

$$n_{\text{total out}} = 1 + \frac{4}{2} + 2\left[\left(\frac{1+0.448}{0.21}\right)(1+0.0116) - 1\right] = 14.95\,\frac{\text{mol}}{\text{mol CH}_4}$$

From Table 7.1 we find that the heating value of methane is 21 502 Btu/lb = 50.0 kJ/g, so that 1 mol (16 g) of fuel corresponds to 0.8 MJ of heat, which corresponds to an emission of (0.0008 GJ) (47.2 g/GJ) = 0.0378 g = 0.000821 mol (as NO_2); and the concentration is 0.000821 mol/14.95 mols = 54.9 ppm.

Finally, we compute the concentration in mg/m^3, using the standard density of air, as

$$c = 54.9 \times 10^{-6} \cdot \frac{46}{29} \cdot 1200 \frac{\text{g}}{\text{m}^3} = 0.104 \frac{\text{g}}{\text{m}^3} = 104 \frac{\text{mg}}{\text{m}^3} \quad \blacksquare$$

From this long calculation we see that the same emission factor can be expressed in numerous ways. All of them appear in the literature and in regulations of various countries. Several values of $[O_2]$ are regularly used as "standard." The emission factors for combustion of natural gas in home appliances in Example 12.5 are roughly one-fourth of the factors usually seen in gas-fired power plants and about a tenth of the factors usually seen in coal-fired power plants.

12.8 SUMMARY

1. At high temperatures thermal NO forms fairly quickly, but the times available for its formation are short enough that equilibrium may not be reached.
2. In most combustion situations the hot combustion gases are cooled so rapidly that the NO formed in the flames does not have time to revert to N_2 and O_2, as it would at equilibrium, so that the concentration is "frozen" at the value it had at a higher temperature, often close to the value it had at the peak temperature.
3. Prompt NO is formed in hydrocarbon-air flames at a rate that is only weakly dependent on temperature. At the low combustion temperatures of kitchen appliances prompt NO is the major source of NO_x.
4. Between 10 percent and 50 percent of the contained N in fuels (mostly coal) is emitted to the atmosphere as NO_x when the fuel is burned.
5. The rate of conversion of NO to NO_2 in flames is fairly slow, so that most of the NO_x in flames is NO. The proportion is typically 90 percent NO, 10 percent NO_2.
6. As discussed in Appendix D, NO reacts rapidly in the atmosphere with any available O_3 to produce NO_2. In the presence of sunlight, NO, NO_2, and O_3 come to equilibrium with each other. Hydrocarbons interfere with this equilibrium, forcing it in the direction of NO_2 and O_3 (see Fig. 1.2). The NO_2 reacts with moisture in the atmosphere to produce nitric acid, which contributes to the PM_{10}, $PM_{2.5}$, and acid rain problems.
7. NO formation in flames can be reduced by modifying the combustion process to minimize the peak temperature, the oxygen content at the peak temperature, or the time at the peak temperature.
8. The NO in combustion gases can be reconverted to N_2 by reaction with a reducing agent. The reducing agent is normally CO in internal combustion engines and NH_3

(or one of its chemical relatives) in power plants. The reduction can be thermal or catalytic.

PROBLEMS

See Common Units and Values for Problems and Examples, inside the back cover.

12.1. The Federal new source performance standards [20] for coal-fired power plants require that the emissions of nitrogen oxides be less than 0.6 lb of nitrogen oxides per million Btu of coal burned. These are to be computed based on all the nitrogen oxides being in the form of NO_2. If a plant emits 0.4 lb of NO and 0.1 lb of NO_2 per million Btu, what is its nitrogen oxide emission, reported as NO_2?

12.2. (*a*) Show the construction of the 2000 K row of Table 12.2. The reported values for the free energy changes of formation from the elements of NO and NO_2 at 2000 K are 15.548 and 38.002 kcal/mol, respectively [6]. Two mols of NO are formed in the reaction as written.

(*b*) What is the effect of increasing the total pressure on K_p?

(*c*) Some authors write Eq. (12.3) as

$$0.5N_2 + 0.5O_2 \rightleftarrows NO \qquad \text{(A)}$$

What is the value of K_p for Eq. (A) at 2000 K? Repeat the calculation of [NO] in Example 12.1 using that value of K_p. Does the answer change? Should it? Is there a simple mathematical relation between $K_{12.3}$ and K_A?

(*d*) The K_p for the formation of NO_2 in Table 12.2 is based on Eq. (12.1). In principle one could also form it by the reaction

$$0.5N_2 + O_2 \rightleftarrows NO_2 \qquad \text{(B)}$$

although that reaction apparently plays no significant role in NO_2 formation. Compute the value of K_p at 2000 K for Eq. (B). Compute the equilibrium values of NO_2 at 2000 K in Table 12.3, using that value of $K_{p\,\text{reaction B}}$ and Eq. (B). Do they agree with the values in the table? Should they?

(*e*) Is there a simple mathematical relationship between $K_{12.1}$, $K_{12.3}$, and $K_{p\,\text{reaction B}}$?

12.3. To show the effect of ignoring the changes in nitrogen and oxygen concentrations on the equilibrium concentration of NO in Example 12.1,

(*a*) Recompute that value, taking those changes into account. Express [NO] as $2x$, and $[N_2]$ as $(0.78 - x)$, etc. This computation leads to a quadratic equation in x, which is easily solved.

(*b*) Also show the change for the 4 percent oxygen values at that temperature in Table 12.3.

12.4. In the preceding problem you showed that, for the equilibrium formation of NO at 2000 K from air, ignoring the reduction in available N_2 and O_2 makes little change in the calculated equilibrium concentration. However, you also showed that as the starting O_2 concentration becomes smaller, the difference becomes more important. Estimate the value of the initial oxygen content at which the difference between the equilibrium NO concentration calculated as in Example 12.1 (ignoring changes in N_2 and O_2) and that calculated in the Problem 12.3 (taking those changes into account) is equal to 10 percent.

12.5. Although the atmosphere contains 78 percent nitrogen, combustion gases normally contain less as a result of dilution with H_2O and CO.

(*a*) Write an equation for the nitrogen content of combustion gases as a function of the percent excess air and of the molecular H/C ratio of the fuel, assuming that all the C is oxidized to CO_2. This is easy if you start with the results of Example 7.9. The molecular H/C ratio can vary from infinity (pure hydrogen) to zero (pure carbon). It is 4 for natural gas, about 1.5 to 2 for most liquid fuels, and about 1 for most coals.

(*b*) Based on the equation derived in (*a*), estimate the lowest nitrogen content that is likely to occur in any combustion gas that uses air (as opposed to an oxygen-enriched gas) as its oxygen source. Assume $X = 0.0116$.

(*c*) Estimate the difference in equilibrium NO and NO_2 concentrations on the right side of Table 12.3 that would result from taking this difference into account.

12.6. It is often stated that if we had the universal catalyst and exposed it to the atmosphere, the nitrogen and oxygen would combine and dissolve in the world's oceans, leaving behind little oxygen. Is this true? (This problem is discussed in Ref. 2.)

Use the equilibrium constants for 300 K (= 80°F) in Table 12.3. The vapor pressure of NO_2 over dilute solutions of nitric acid is given by

$$p \approx 45\text{mm Hg (wt fraction HNO}_3)^{7.8}$$

The mass of water in the oceans is approximately 100 times the mass of the atmosphere.

12.7. The equilibrium constants for Eqs. (12.7)–(12.9) are shown in Table 12.4 [6].

(*a*) Based on Table 12.4, estimate how many N and O free radicals would be expected at equilibrium in the whole atmosphere of the earth, assuming that the atmosphere is all at 300 K. The mass of the earth's atmosphere is roughly 5×10^{18} kg. This calculation is misleading because we do not have equilibrium; it is perturbed by chemical reactions with species other than nitrogen and oxygen and by dissociation by sunlight. But the calculation does give some idea of how unstable these radicals are at 300 K.

(*b*) Estimate the concentration of N, O, and OH radicals in atmospheric air at 2000 K = 3140°F.

12.8. The K_p values in Table 12.2, based on Ref. 6, are widely regarded as the most reliable that are currently available. Equation (12.14) shows that one should be able to compute the value of K_p for NO from the forward and reverse rate constants in Eq. (12.12). The reported values are $k_f = 9 \times 10^{14} \exp(-135\,000/RT)$ and $k_b = 4.1 \times 10^{13} \exp(-91\,600/RT)$.

TABLE 12.4
Equilibrium constants for the formation of N, O, and OH from N_2, O_2, and H_2O

Temperature		K_p for formation of		
K	°F	N	O	OH
300	80	10^{-158}	10^{-81}	10^{-81}
500	440	10^{-93}	10^{-45}	10^{-46}
1000	1340	10^{-43}	10^{-20}	10^{-19}
1500	2240	10^{-26}	10^{-11}	10^{-11}
2000	3140	10^{-18}	4×10^{-7}	3×10^{-7}
2500	4040	10^{-13}	2.1×10^{-4}	1.3×10^{-4}

Using those values, calculate the K_p values for NO corresponding to the temperatures in Table 12.3. Present your results as the ratio of $(K_{p\text{–Table 12.2}})/(K_{p\text{–This problem}})$ for the temperatures shown in Table 12.2.

These values of k_f and k_b are not for [NO] in pressure units. However, because there is no change in the number of mols in this reaction, the units cancel in Eq. (12.14), and the values may be used without concern for units. The same is not true for reactions in which the number of mols changes.

12.9. Show the algebra leading from Eq. (12.13) to Eq. (12.15) and then the integration and subsequent algebra leading from Eq. (12.15) to Eq. (12.16).

12.10. Show the simplifications leading to Eq. (12.13). Here, to save writing, we rename Eq. (12.11) as Eq. (A), and Eq. (12.12) as Eq. (B). Both of these are equilibrium reactions, which proceed in both directions, so taking the forward and backward rates of both into account, we have

$$\frac{d[\text{NO}]}{dt} = k_{\text{A},f}[\text{N}_2][\text{O}] + k_{\text{B},f}[\text{N}][\text{O}_2] - k_{\text{A},b}[\text{NO}][\text{N}] - k_{\text{B},b}[\text{NO}][\text{O}] \tag{C}$$

and similarly,

$$\frac{d[\text{N}]}{dt} = k_{\text{A},f}[\text{N}_2][\text{O}] - k_{\text{B},f}[\text{N}][\text{O}_2] - k_{\text{A},b}[\text{NO}][\text{N}] + k_{\text{B},b}[\text{NO}][\text{O}] \tag{D}$$

The concentration of N radicals is assumed always to be small compared to the other species and not to change significantly with time, so that $(d[\text{N}]/dt)$ is set equal to zero in Eq. (C), which is then solved for the steady-state N concentration:

$$[\text{N}]_{\text{steady state}} = \frac{k_{\text{A},f}[\text{N}_2][\text{O}] + k_{\text{B},b}[\text{NO}][\text{O}]}{k_{\text{B},f}[\text{O}_2] + k_{\text{A},b}[\text{NO}]} \tag{E}$$

Then this steady-state value from Eq. (E) is substituted in Eq. (C), like terms are canceled, and it is rearranged to

$$\frac{d[\text{NO}]}{dt} = 2\cdot[\text{O}]\frac{\{k_{\text{A},f}[\text{N}_2] - (k_{\text{B},b}k_{\text{A},b}[\text{NO}]^2/k_{\text{B},f}[\text{O}_2])\}}{1 + (k_{\text{A},b}[\text{NO}]/k_{\text{B},f}[\text{O}_2])} \tag{F}$$

We then assume that

$$\frac{k_{\text{A},b}[\text{NO}]}{k_{\text{B},f}[\text{O}_2]} \ll 1 \tag{G}$$

which changes the denominator to a 1. We next assume that O radicals are in thermodynamic equilibrium with O_2 molecules, so that

$$[\text{O}] = (K_{\text{Eq. 12.8}}\,[\text{O}_2])^{1/2} \tag{H}$$

Equations (G) and (H) are substituted into Eq. (F), and the variables renamed to find Eq. (12.13). The solution without the simplification of Eq. (G) is shown by Wark and Warner [21].

12.11. (*a*) Repeat Example 12.2 for 2250 K, 0.005 s, and a pressure of 10 atm. This corresponds, roughly, to the time, temperature, and pressure in a gas turbine engine combustor. At 2250 K, $K_p \approx 0.00133$.

(*b*) Benítez [10] shows a sample calculation for part (*a*), using the version of the Zeldovich mechanism that includes Eq. (12.18), finding a concentration of 608 ppm. Compare your result from part (*a*) with that value. Is the difference in values reasonable? Explain.

12.12. In Example 12.2, how long would the gas have to be held at 2000 K for the calculated thermal NO concentration to be 1500 ppm?

12.13. Repeat the preceding problem for a gas that has 21 percent O_2 instead of the 4 percent in that problem.

12.14. Repeat Example 12.2 for a gas that has 21 percent O_2 instead of the 4 percent in that example.

12.15. Example 12.3 shows that for an ordinary gas flame the cooling rate is about 0.9 million °F/s.

(*a*) For the time after the reaction has stopped, estimate the heat removal rate (cal/s or equivalent) using Eq. (12.19). Assume that $V = 1\ \text{cm}^3$.

(*b*) Could this amount of heat be removed by simple conduction? Assume the surface area is 3 cm^2, that the thermal conductivity of the combustion gases at that temperature is 0.05 Btu/h · °F · ft. Make plausible assumptions for the temperature gradient.

(*c*) Could this amount of heat be removed by radiation? Assume an emissivity of 0.1 and a flame temperature of 2400°F.

(*d*) How would the ratio of conductive heat removal rate estimated in (*b*) to that by radiation estimated in (*c*) change if the flame temperature increased to 3000°F?

12.16. Using a spreadsheet program, show the remaining 99 steps of Example 12.4. The suggested columns on the spreadsheet are t, T at that time (K), average T over the time interval, K_p at that temperature, equilibrium NO mol fraction, molar volume at that temperature, $[\text{NO}]_e$ in mol/cm^3 units, k_b at that temperature, [NO] in mol/cm^3 units, $d[\text{NO}]/dt$, fractional change of [NO] in that time interval due to thermal expansion, [NO] ppm, $[\text{NO}]_e$ ppm. As you calculate down the table, each value of [NO] will be the sum of the value in the preceding line plus that formed by reaction in that time interval, all multiplied by the expansion or contraction due to change in temperature.

12.17. (*a*) Sketch the flow diagram for the whole furnace that uses the low NO_x burner in Fig. 12.9.

(*b*) Does the FGR reduce the furnace efficiency (fraction of heating value of the fuel which is transferred to whatever is being heated)? Explain why.

12.18. Are the NO_x contents of the exhaust gases at zero FGR in Fig. 12.10 consistent with Fig. 12.5? Here ignore the difference between 4 percent and 3 percent O_3 and assume that the high temperature part of the flame lasts for 0.01 s. Assume that ambient temperature is 68°F, so that preheating to 300°F increases the average flame temperature by 232°F, and to 500°F increases it by 432°F.

12.19. Estimate how much 20% FGR reduces flame temperatures (see Fig. 12.10). This requires an estimate of the rate of heat loss from the flame. Assume that the actual flame temperature is 80% of the calculated adiabatic flame temperature, for any degree of FGR, and that the fuel gas is pure methane.

12.20. If we decide to remove NO_x from combustion gases with NH_3, and assume that the operative reactions will be Eqs. (12.22) and (12.24), that the NO_x is 90 mol % NO, balance NO_2 and that we will treat all the NO_x from coal combustion in Table 12.1 this way, what fraction of the total U.S. production of NH_3 ($\approx 20 \times 10^6$ tons/yr) will be needed?

12.21. The emission factor for nitrogen oxides from coal combustion in large industrial boilers (general type) is 18 lb/ton of coal burned [22]. Compute the corresponding exit gas NO_x concentration in ppm. Assume that there is 6 percent O_2 in the exit gas, and that the coal has the analysis shown in Example 7.10.

12.22. Show the derivation of Eq. (12.26) by rearrangement of the two equations preceding it.

12.23. Example 12.5 shows the concentrations in ppm and mg/m^3 for a standard oxygen content of 6 percent.

(*a*) Some countries use a standard oxygen content of 4 percent. Show the corresponding concentrations for that oxygen content.

(*b*) Some regulations require the values in ppm and mg/m^3 to be expressed on a dry basis, which would be found if all water in the gas were condensed out before the measurement was made. Compute the NO_x concentrations (in ppm) in Example 12.5 on a dry basis.

12.24. The standard way of reporting NO_x emissions from gas turbine and jet engines is in (g of NO_x as NO_2)/(kg of fuel burned). Compute the emission rate in Example 12.5 in these terms.

12.25. Using Fig. 12.4, estimate the fraction of the fuel N is emitted as NO_x. Assume that the coal is the "typical coal" shown in Example 7.6.

12.26. Example 12.5 presents a summary of measured emission factors from a gas stove. This is believed to be all prompt NO, some of which is converted to NO_2. Figure 12.4 shows the estimated emissions of prompt NO_x from coal combustion. Is the prompt emission shown in Fig. 12.4 more or less than the total emission shown in Example 12.5? Assume that the flame temperature of the stove is the same as shown in Table 7.2.

12.27. In Problem 11.35 you estimated how much SO_2 per year is kept out of the atmosphere by replacing a 100-W incandescent light bulb with an illumination-equivalent 20-W fluorescent bulb. Using the data from that problem, estimate how much NO_x (expressed as NO_2) is kept out of the atmosphere by the same bulb replacement. Assume that the electric power comes from a plant whose NO_x emissions are exactly equal to the New Source Performance Standard for NO_x. (See Table 3.1. Use the value for "most coals.")

REFERENCES

1. Lide, D. R. ed.: *CRC Handbook of Chemistry and Physics,* 71st ed., CRC Press, Boca Raton, FL, p. 14–7, 1990.
2. "Air Quality Criteria for Nitrogen Oxides," U.S. Environmental Protection Agency Report No. EPA-600/8-82-06, U.S. Government Printing Office, Washington, DC, 1982.
3. *Nitrogen Oxides,* Report of the Committee on Medical and Biological Effects of Environmental Pollutants, National Academy of Sciences, Washington, DC, 1977.
4. *National Air Quality and Emissions Trends Report, 1997,* EPA 454/R-98-016.
5. Smith, J. M., H. C. Van Ness, and M. M. Abbot: *Introduction to Chemical Engineering Thermodynamics,* 5th ed., McGraw-Hill, New York, Chapter 15, 1996.
6. "JANAF Thermochemical Tables," U.S. Government, Clearinghouse number PB 168 370, 1964.
7. Hupa, M., F. Backman, and S. Bostroem: "Nitrogen Oxide Emissions of Boilers in Finland," *J. Air Pollut. Control Assoc.,* Vol. 39, pp. 1496–1501, 1989.
8. Zeldovich, Y. B.: "The Oxidation of Nitrogen in Combustion Explosions," *Acta Physicochimica USSR,* Vol. 21, p. 577, 1946.
9. Seinfeld, J. H.: *Air Pollution, Physical and Chemical Fundamentals,* McGraw-Hill, New York, p. 373, 1975.
10. Benítez, J.: *Process Engineering and Design for Air Pollution Control,* Prentice Hall, Englewood Cliffs, New Jersey, p. 257, 1993.
11. Hayhurst, A. N., and I. M. Vince: "Nitric Oxide Formation from N_2 in Flames: The Importance of 'Prompt' NO," *Prog. Energy Combust. Sci.,* Vol. 6, pp. 35–51, 1980.
12. Miller, J. A., M. C. Branch, W. J. McLean, D. W. Chandler, M. D. Smooke, and R. J. Lee: "The Conversion of HCN to NO and N_2 in H_2-O_2-HCN-Ar Flames at Low Pressure," *Twentieth Symposium (International) on Combustion,* The Combustion Institute, pp. 673–684, 1984.
13. Loftus, P. J., et al.: "Energy Resources, Conversion and Utilization," in *Perry's Chemical Engineers' Handbook,* 7th ed., R. H. Perry and D. W. Green (eds.), McGraw-Hill, New York, Chapter 27, 1997.
14. "Control Techniques for Nitrogen Oxides Emissions from Stationary Sources—Revised Second Edition," U.S. Environmental Protection Agency Report No. EPA-450/3-83-002, U.S. Government Printing Office, Washington, DC, 1983.

15. Christman, R. C., S. J. Bortz, D. E. Shore, and M. Brecker: "The Radian Rapid Mix Burner™ for Ultra-Low NO_x Emissions," *1995 EPRI-EPA Joint Symposium on Stationary Combustion NO_x Control,* Kansas City, Missouri, May 1995.
16. Ozkan, U. S., S. K. Agarwal, and G. Maracelin, eds.: *Reduction of Nitrogen Oxide Emissions,* ACS Symposium Series 587, American Chemical Society, Washington, DC, 1995.
17. Dalton, S. M., B. Toole-O'Neil, B. K. Gullett, and C. J. Drummond: "Summary of 1991 EPRI/EPA/DOE SO_2 Control Symposium," *J. Air Waste Manage. Assoc.,* Vol. 42, pp. 1110–1117, 1992.
18. Hjalmarsson, A.-K. "NO_x Control Technologies for Coal Combustion," IEA Coal Research, London, *IEACR/24,* 1990.
19. Wadden, R. A., and P. E. Scheff: *Indoor Air Pollution: Characterization, Prediction and Control,* John Wiley & Sons, New York, p. 53, 1983.
20. "Standards of Performance for New Stationary Sources," U.S. Federal Government, *40 CFR 60,* 1991.
21. Wark, K., and C. F. Warner: *Air Pollution, Its Origin and Control,* 2d ed., Harper & Row, New York, p. 382, 1981.
22. "Compilation of Air Pollutant Emission Factors," U.S. EPA, Office of Air Quality Planning and Standards, Research Triangle Park, NC 27711, *AP-42,* 1972.

CHAPTER 13

THE MOTOR VEHICLE PROBLEM

The motor vehicle has been available in large numbers only in this century. The first gasoline-powered automobiles appeared in 1886 [1]; by 1900 world production was only about 20 000 vehicles per year, compared to about 30 million in 1999. The personal motor car has given its owners personal mobility and freedom that would have been incomprehensible two centuries ago. The author loves his car and assumes that the readers of this book love theirs too. Alas, although any one car consumes little fuel and emits small amounts of pollutants, together the roughly 500 million of them in the world consume large amounts of fuel and emit large amounts of pollutants. The motor vehicle industry, broadly defined, constitutes more than 10 percent of the total industry of industrialized countries; the health of their economies rises and falls with the health of their motor vehicle industry. More than one war has been fought over supplies of oil for them.

13.1 AN OVERVIEW OF THE PROBLEM OF AIR POLLUTION FROM MOTOR VEHICLES

13.1.1 Emissions

There are about 123 million autos in use in the United States and about 70 million trucks [2]. Their contribution to our CO, HC, and NO_x problems is shown in Table 13.1 on page 472. This table shows that motor vehicles are the source of three-fourths of our national emissions of CO, and 40 to 50 percent of our emissions of HC and NO_x. Motor vehicles also emit particles and SO_2 but their percent contribution to

TABLE 13.1
Contribution of motor vehicles to national emissions

	CO	NO_x	HC
Total U.S. emissions from human sources, millions of tons/yr, 1997	87.5	23.6	19.2
% emitted by on-road vehicles	57.4	29.8	27.7
% emitted by off-road vehicles	19.2	19.3	12.6
Total % due to transportation sources	76.6	49.1	40.3

Source: Ref. 3.

those problems is much less than the values shown in this table. Here "off-road vehicles" include aircraft, railroads, boats, construction equipment, and farm equipment. Autos and light trucks contribute much more to our emissions than do these other sources. Autos do a higher percentage of their travel in highly populated urban areas than trucks, airplanes, or boats. For these reasons our air pollution concern is mostly with autos.

Before 1970 motor vehicles were the major source of atmospheric lead particles. Since then, the removal of tetraethyl lead from gasoline has made them a much smaller contributor (see Table 1.1 and Chapter 15.)

13.1.2 The Regulatory History of Motor Vehicle Air Pollution Control

Motor vehicles did not attract much attention as air pollution sources until about 1950. Before that time there were very large, uncontrolled air pollution emissions from industry, and the emissions from coal combustion were the most important contributors to air pollution in most U.S. cities. As these sources were controlled, and as natural gas replaced coal as the principal urban heating fuel in the United States, a new type of air pollution was discovered in Los Angeles. There, the principal home and industrial heating fuel was natural gas, and there were few "smokestack" industries. However, a type of eye- and nose-irritating air pollutant, later named *smog* (a poor name, because Los Angeles smog is not connected with smoke or fog, but a name that has persisted), occurred there, mostly in the summer. (The same type of pollutant is now observed in the summer in most major U.S. cities.) Professor A. J. Haagen-Smit demonstrated that the eye-irritating materials were largely formed from emissions from autos [4]. Initially the auto manufacturers denied that autos were to blame, but eventually the scientific evidence became too great to be denied.

California began to regulate emissions from autos in 1963. In the Clean Air Act of 1970 Congress began federal regulation of autos, requiring stricter rules for any states that already had state rules (only California), but also requiring fairly strict rules for the rest of the country. In the early 1970s there was major political conflict between the auto industry and the U.S. EPA because the automotive emission regulations issued by EPA were intentionally *technology forcing*, i.e., they could not be met using existing technology. The auto manufacturers succeeded in developing the new technology and met the emission regulations within the statutory deadlines (which were extended as allowed by the law).

TABLE 13.2
Selected history of U.S. automobile air pollutant emission regulations

	Permitted emissions in g/mile				
	Tailpipe emissions			Other HC emissions	
Year	CO	NO_x	HC	Crankcase	Evaporative
Precontrol, e.g., 1960	87	3.6	8.8	3	4
1970	23	–	2.2	0	4
1971	–	–	–	0	0.8
1972	39	–	3.4	0	0.27
1973	–	3	–	–	–
1975	15	3.1	1.5	–	–
1978	–	2.0	–	0	0.8
1980	7	–	0.41	–	–
1981	3.4	1.0	–	0	0.27
1993	–	0.4	0.25	–	–
2003	1.7	0.2	0.0125	–	–

Notes:

1. This table is based on Refs. 5 and 6; it omits many details shown in those sources.

2. There are separate, more stringent standards for California, not shown here. Other states may adopt these in the near future.

3. The testing methods have changed over time; three different sets of test methods are represented here. The apparent increase in permitted CO and HC tailpipe emissions in 1972 and in evaporative emissions in 1978 are due to changes in test procedure. The actual permitted emissions probably declined.

4. Evaporative emissions are in grams per test. The trip length in the current tailpipe emission test is 7.5 miles. To get a comparable basis, one divides the evaporate emissions of 2 g/test by 7.5 miles/test.

5. A dash, –, in this table means no change from the previous standard.

6. California also has a formaldehyde standard.

7. This table is for autos. There are comparable tables for light trucks, heavy trucks, etc.

The history of these regulations is shown in Table 13.2. Over time, the permitted emissions have been substantially reduced. A car that meets the 1993 standards emits about 3 percent as much HC (tailpipe plus crankcase plus evaporative emissions), 4 percent as much CO, and 11 percent as much NO_x as a 1960 car. In spite of this significant technical achievement, we have not met our air quality goals in most major U.S. cities and, as the last row shows, we are about to introduce even more stringent emission regulations. In many other parts of the world automobiles are being produced that have emissions comparable to the 1960 values in Table 13.2.

13.2 THE INTERNAL COMBUSTION (IC) ENGINE

External combustion engines were developed before *internal combustion*, or IC, *engines*. James Watt's 1776 steam engine was the first general-purpose heat engine that converted heat from combustion to a steady flow of power to a rotating shaft. For 100 years steam engines, with combustion in a boiler external to the power-producing part of the engine, were the only combustion engines. (The early steam engines were built with crude machine tools. These steam engines launched a giant technological expansion, which, among other things, led to the development of much better machine tools. The improved machine tools made it possible to build the first IC engines, which the crude machine tools used to build the first steam engines could not have built.) The first commercially successful IC engines (combustion inside the power-producing parts) were those of Otto and Langen [1] about 1876. For a given power output these engines were substantially smaller and lighter than external combustion engines and had a higher thermal efficiency (lower fuel consumption). Those features made them the natural choice for motor vehicles. The steam engine held on in railroad locomotives until the 1950s, when the major cost savings brought by diesel engines led to its replacement [7]. Automobiles, trucks, and airplanes have always been powered almost exclusively by IC engines.

External combustion engines are now used almost exclusively for steam-electric power plants burning coal, residual oil, or sometimes natural gas. Coal is a much cheaper and more abundant fuel than oil or gas, but it is much more difficult to handle and burn because of its ash content. As the coal burns it leaves behind its ash as a hard, abrasive solid. External combustion engines handle that coal ash in a boiler that is separate from the fast-moving, close-tolerance, power-producing parts of the engine. Efforts have been made to develop coal-burning IC engines, particularly of the gas-turbine variety, but the difficulties associated with the abrasive coal ash have so far prevented the development of an economical coal-fired IC engine.

13.2.1 The Four-Stroke IC Gasoline Engine

The *four-stroke IC gasoline engine* has been the power source for 99+% of the autos and small trucks ever built. (Other types of IC engines are discussed in Sec. 13.7.) It has withstood the challenges of all other types of engines because it is relatively light and small, durable, and moderately easy and relatively inexpensive to manufacture; has fairly good fuel efficiency; responds quickly and smoothly to changes in throttle setting; requires very little maintenance and tolerates substantial abuse; and can operate efficiently over fairly wide speed and load ranges. Other engine types can beat it at one or more of those attributes, but so far none has been able to beat it at enough of them to displace it. This chapter will deal mostly with this type of engine, with occasional references to other types.

Figure 13.1 shows, in very simplified form, a cross-sectional view of a typical auto engine. It shows only one piston and cylinder; most auto engines have four such pistons and cylinders, though some have six or eight. In operation, the crankshaft rotates, causing the piston to move up and down, driven by the crank, connecting rod, and wrist pin. To begin a cycle, with the piston at the top (*top dead center,* TDC) during the first stroke the piston moves downward while the intake valve is open, so that an air-fuel mixture is sucked into the combustion chamber (the space within the cylinder, above the piston). When the piston is at the bottom (*bottom dead center,* BDC), the intake valve closes, ending the intake stroke. As the piston rises again to the top during the compression stroke, both valves are closed, so that the air-fuel mixture is compressed. Near the top of that stroke the spark plug fires, igniting the

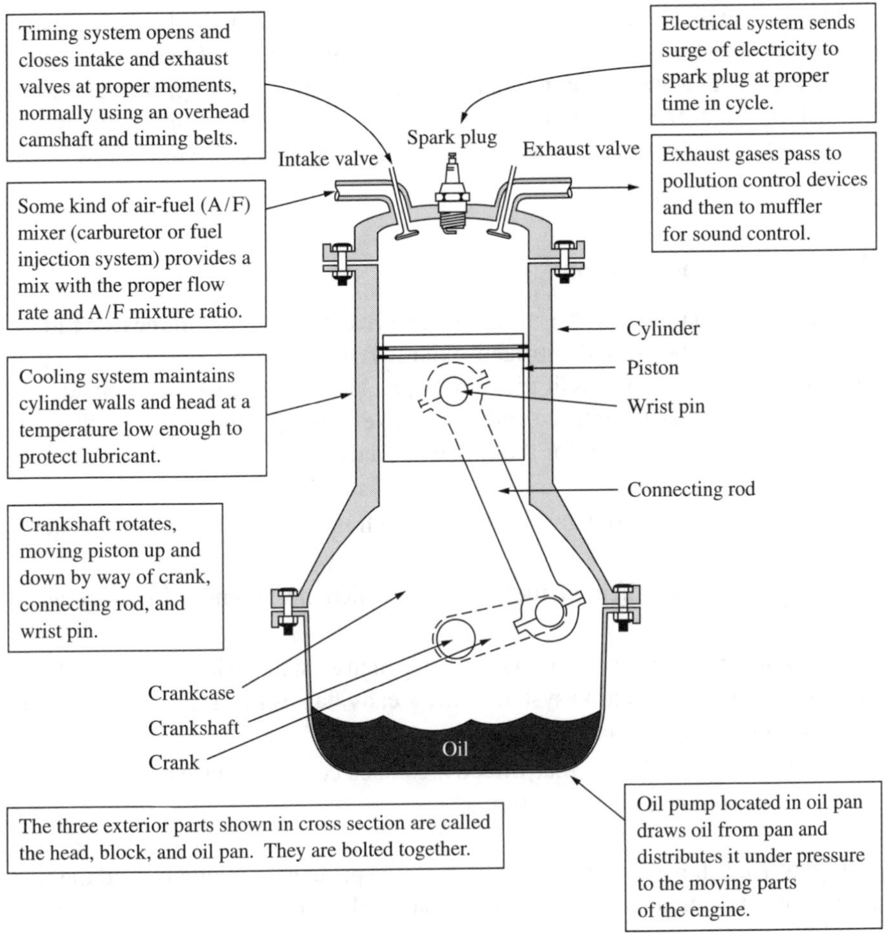

FIGURE 13.1
Very simplified schematic of one piston and cylinder of a four-stroke gasoline IC engine.

air-fuel mixture. In its next downward travel, the power stroke, the piston is driven by the high-pressure combustion gases, which do the actual work of the engine. At the bottom of the piston travel, the exhaust valve opens, and on its next upward travel the piston pushes the burned gases out into the exhaust system. The cycle is named for its *four strokes*—intake, compression, power, and exhaust. The spark plug fires every second upward travel of the piston. Power is produced only during the power stroke. Each of the other three strokes consumes power. The engine must have enough inertia so that the power produced in the power stroke will carry it through to the next power stroke. For a one-cylinder engine, this normally requires a large flywheel. For multicylinder engines the firing times of the cylinders do not coincide, so a much smaller flywheel is suitable.

This simple picture does not discuss the mechanical design of the various parts or the mechanisms for providing the proper air-fuel mixture, for producing the spark or timing it to occur at the proper moment, or for lubricating, cooling, and balancing. Those will be mentioned briefly later, only as needed to illustrate the discussion of the air pollution problems of these engines. Those topics are discussed elsewhere [8, 9].

13.2.2 Pollutant Formation

The principal pollutants emitted from simple gasoline-powered IC engines are carbon monoxide, hydrocarbons, and nitrogen oxides. All these are formed in all other combustion processes, e.g., fossil fuel power plants, kitchen stoves, campfires, and charcoal barbecues (see Chapters 7 and 12). Auto engines produce more of them per unit of fuel burned principally for the following reasons:

1. Auto engines are often oxygen deficient, which most other combustion systems are not.
2. Auto engines preheat their air-fuel mixtures, which most combustion systems do not.
3. Auto engines have unsteady combustion, in which each flame lasts about 0.0025 s. Almost all other combustion systems have steady flames that stand still while the materials burned pass through them.
4. Auto engines have flames that directly contact cooled surfaces, which is not common in other combustion systems.

Figure 13.2 shows the emissions of the three principal pollutants (and the fuel consumption) as a function of normalized air-fuel ratio (discussed below), for a typical IC engine, running steadily. We will return to this figure many times in this chapter; it is worth the student's time to study it.

13.2.2.1 Air-fuel ratio (A/F), normalized (A/F) ratio. Figure 13.2 makes clear that emissions and efficiency depend strongly on the normalized air-fuel ratio, λ. Unfortunately, there is little agreement on how to present this, and various terms are used. Although stationary combustion sources normally run at practically constant air-fuel ratios, and with substantial excess air, auto engines run with rapidly changing air-fuel ratios, and often with less than stoichiometric air.

From Sec. 7.10 we know that, for any hydrocarbon fuel with formula C_xH_y

$$n_{\text{stoich oxygen}} = x + \frac{y}{4} \tag{7.14}$$

All gasolines are mixtures of many components, but they can be characterized as having an approximate average formula C_xH_y, where for a typical gasoline x is about 8 and y is about 17. Gasoline manufacturers change these values from one location

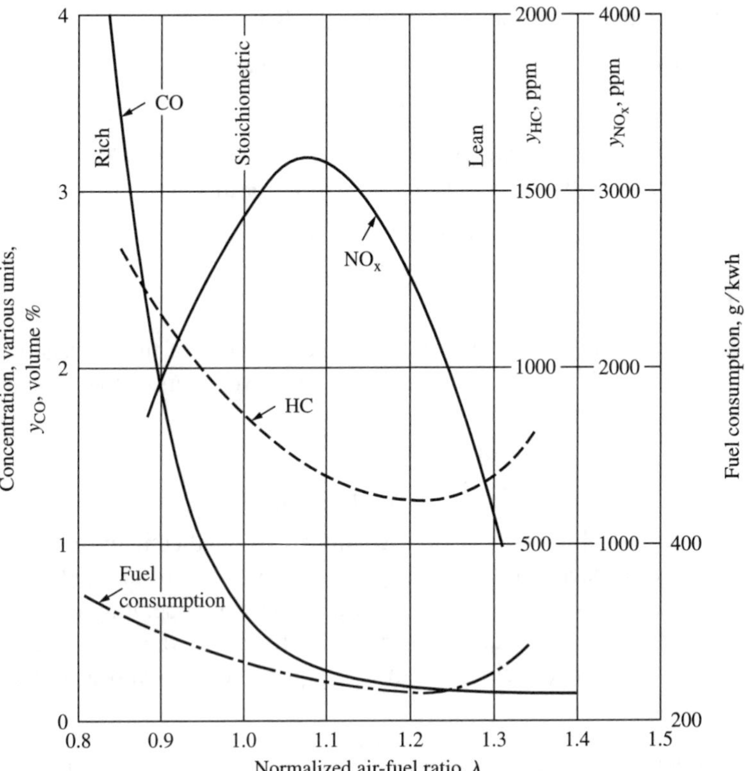

FIGURE 13.2
Emissions and fuel consumption for a typical IC engine in steady operation. Observe that the pollutant scales are different. The scale for y_{NO_2} is 1/10 that for y_{CO} and twice that for y_{HC}. These could be put on a common scale, but one rarely sees this figure that way. Also the volume % could be shown as ppm, or the reverse, but these are the concentration units most often seen. Here y is the mol fraction, not to be confused with the y in Eq. (7.14). The horizontal scale is the normalized air-fuel ratio, discussed shortly. *Source:* Ref. 10. (Courtesy of Springer-Verlag.)

to another and with season of the year (smaller values in winter and in cold climates than in summer and warm climates).

For complete combustion (stoichiometric air-fuel ratio) of this fuel, the equation is

$$C_xH_y + \left(x + \frac{y}{4}\right) O_2 \rightarrow xCO_2 + \left(\frac{y}{2}\right) H_2O \tag{13.1}$$

Example 13.1. Calculate the stoichiometric A/F for a gasoline whose average formula is C_8H_{17}.

The normalized A/F is shown as a dimensionless ratio, centered on 1.00 in Fig. 13.2, but in the "practical" IC engine literature it is normally shown in weight terms (normally lb/lb in the United States). In general it is written as

$$\left(\frac{\text{A}}{\text{F}}\right)_{\text{stoichiometric}} = \frac{[x + (y/4)](32 + 3.76 \cdot 28)}{12x + 1 \cdot y} \tag{13.2}$$

where 32 and 28 are the molecular weights of O_2 and N_2 ($3.76 = 0.79/0.21$) and 12 and 1 are the atomic weights of C and H, respectively. For this example we have

$$\left(\frac{\text{A}}{\text{F}}\right)_{\text{stoichiometric}} = \frac{(8 + 17/4)(32 + 3.76 \cdot 28)}{12 \cdot 8 + 17} = 14.88$$ ■

In the older IC literature and in "practical" IC books, most of the discussion is in terms of A/F expressed in lb/lb as shown above. In newer or more theoretical books it is most often in terms of λ.

$$\lambda = \begin{pmatrix}\text{normalized} \\ \text{A/F ratio}\end{pmatrix} = \frac{\text{A/F}}{(\text{A/F})_{\text{stoichiometric}}} \tag{13.3}$$

A/F has the advantage of dealing with the physical quantities handled in the air-fuel system. But it has the drawback that $(\text{A/F})_{\text{stoichiometric}}$ is a property of the fuel, i.e., it depends on the ratio y/x (see Problem 13.3). A given engine, operating on two different gasolines, has practically equivalent behavior for equal values of λ, but not for equal values of A/F. The reciprocals of these quantities also appear in the literature:

$$\text{Fuel-air ratio} = \frac{1}{\text{Air-fuel ratio}} = \frac{1}{\text{A/F}} = \text{F/A} \tag{13.4}$$

and the *equivalence ratio*,

$$\phi = \frac{1}{\lambda} = \begin{pmatrix}\text{equivalence} \\ \text{ratio}\end{pmatrix} = \frac{(\text{A/F})_{\text{stoichiometric}}}{\text{A/F}} \tag{13.5}$$

Example 13.2. An auto engine with the gasoline in Example 13.1 is operating with an A/F of 14.316 lb air/lb fuel. What are λ and ϕ?

$$\lambda = \frac{14.316}{14.88} = 0.95, \qquad \phi = \frac{1}{\lambda} = \frac{1}{0.95} = 1.026$$ ■

If we had chosen ϕ as our measure of the A/F or F/A relation, then Fig. 13.2 would be replaced by its mirror image. One sees both the form of Fig. 13.2 and its mirror image (using ϕ) about equally as often in the literature. For the rest of this book we will use only λ, and the form of Fig. 13.2, but the reader should be aware that the other form is widely used. In Fig. 13.2 we see along the left axis the word "Rich," which stands for "Fuel-Rich" (or air-poor). For $\lambda < 1$ there is not enough air to burn up all the fuel. Along the right axis we see the word "Lean," indicating that the fuel supplied is less than that which would react with the available air. The terms Rich and Lean are very common in discussions of IC engines.

From Fig. 13.2 we see that when an IC engine operates in the lean mode ($\lambda > 1$, positive excess air, $E > 0$ (excess air is defined and discussed in Sec. 7.10.9)) then the fuel consumption is a minimum (highest gasoline mileage) and the emissions of CO and HC are at their lowest. To the extent feasible, IC engines operate in this mode. However, as shown in Table 13.3, IC engines also operate in other modes. If automobiles only operated warmed-up, at constant speed on level roads, they would always operate with lean A/F ratios. However, we regularly ask our engines to start quickly when they are very cold or very hot, to run at a low speed at idle, to provide sudden bursts of power for starting or passing, to provide more power for hill climbing than for level-road driving, to run at greatest efficiency at fairly low loads, to respond very quickly to changes in throttle setting, and to do all of this quietly and smoothly. No one has devised an engine that can match that list of requirements all at a constant A/F ratio.

From Table 13.3 we see that, based on the flammability limits shown in Table 7.1 for typical constituents of gasoline, one would assume that an IC engine would operate satisfactorily for any normalized A/F ratio between 0.5 and 3.5; however, based on experience with actual engines the operable range is from about 0.8 to 1.2. The smaller operable range is mostly due to the large heat losses from the small amount of combustible mixture (Problem 13.7) in the cylinder to the surrounding cooled cylinder walls and head. The values in Table 7.1 are based on tests in devices with negligible heat loss.

For steady operation at most driving speeds, the best fuel economy (highest value of mechanical work produced per unit of fuel consumed, or most miles per

TABLE 13.3
A/F and normalized A/F ratios

	Rich	Stoichiometric	Lean
A/F, lb/lb	< About 14.9	About 14.9	> About 14.9
Normalized A/F, λ	< 1.00	1.00	> 1.00
Equivalence ratio, ϕ	> 1.00	1.00	< 1.00
Possible range of λ values, based on Table 7.1		0.5 to 3.5	
Actual range of λ values, based on engine operation		0.8 to 1.3	
Normal uses	Starting, idling, maximum power, e.g., passing	Used with 3-way catalysts	Steady driving at light load, e.g., freeway driving
Power	Highest	Average	Poor
Fuel economy	Worst	Average	Best
CO emissions	High	Medium	Low
HC emissions	High	Medium	Low
NO_x emissions	Low	Medium	High

gallon) occurs at a λ of about 1.2. This was the normal ratio for steady driving before the introduction of three-way catalysts, discussed in Sec. 13.4.5. For acceleration or hill climbing the requirement is not best fuel economy but maximum power output, which is found at $\lambda \approx 0.95$. Decreasing λ (operating on the rich side of stoichiometric) has the effect of burning more fuel per revolution; the efficiency of use of the fuel declines, but down to these values of λ the effect of increasing the amount of fuel burned outweighs that. Most gasoline IC engines will not idle successfully (operate smoothly at no load, at about 500 revolutions per minute [RPM]) at a λ greater than 1.0. Most engines idle more smoothly at values between 0.90 and 0.95. (At low speeds there is more time for heat losses to put the flame out.)

Cold starting poses a special problem for IC engines. When the engine is warmed up, enough heat is transferred to the incoming air-fuel mixture from the exhaust system so that the droplets of liquid fuel are almost completely vaporized in the compressed mixture in the combustion chamber before the spark is applied. However, when the engine is cold this exhaust heat is not available, and the temperature in the compressed mixture is so low that much of the liquid fuel is not vaporized. Only the most volatile parts of the fuel will be vaporized under this condition. To make λ, based on the vaporized part of the fuel, be low enough for the engine to start, one must put more total fuel into the air-fuel mixture. If only half of the fuel vaporizes, one must put in twice as much fuel. In carburetor autos, this excess fuel is added by a *choke*. This was operated by hand on older cars and is now operated by thermostatic or electronic sensors (and called an *automatic choke*) that apply it when the incoming air is cold and the engine temperature is cold, and then shut it off when the engine temperature rises. Fuel injection engines regulate the amount of fuel injected, taking the same variables into account. Gasolines are tailored to the temperature at which they are likely to be used. For winter or cold climates the content of low-boiling materials (butanes and pentanes) is increased, compared to gasoline for summer or warm climates.

Thus we see that, although we would prefer never to operate on the left side of Fig. 13.2, we must do so under some circumstances. Thus we are certain to produce CO and HC.

13.2.2.2 Carbon Monoxide (CO). Carbon monoxide is present in the combustion products from any carbon-bearing fuel, gasoline, natural gas, coal, wood, charcoal, forest fires, garbage. As Fig. 13.2 shows, the amount of CO depends strongly on the normalized A/F ratio. For lean combustion (ample excess air) there is little CO; for rich combustion (less than stoichiometric air) there is ample CO.

Example 13.3. For combustion with $\lambda = 0.95$, estimate the CO concentration in the exhaust gas. In this rich combustion situation there is an oxygen deficit; not enough oxygen is present to burn up all the HC and also to oxidize all the CO to CO_2. For this example we will assume that all of the HC is burned up, and thus that all of the oxygen deficit results in the incomplete combustion of CO.

Here we rewrite Eq. (13.1) as

$$C_xH_y + \lambda\left(x + \frac{y}{4}\right) O_2 \rightarrow zCO + (x - z)CO_2 + \frac{y}{2}H_2O \tag{13.6}$$

where z is the number of mols of CO formed per mol of gasoline. The corresponding balance on O atoms is

$$2\lambda\left(x + \frac{y}{4}\right) \rightarrow z + 2(x - z) + \frac{y}{2} = -z + 2\left(x + \frac{y}{4}\right) \tag{13.7}$$

$$z = 2(1 - \lambda) \cdot \left(x + \frac{y}{4}\right) = 2(1 - 0.95) \cdot \left(8 + \frac{17}{4}\right) = 1.225$$

$$\begin{aligned} y_{CO} &= \frac{z}{z + (x - z) + (y/2) + 3.76\lambda[x + (y/4)]} \\ &= \frac{1.225}{1.225 + (8 - 1.225) + (17/2) + 3.76 \cdot 0.95 \cdot [8 + (17/4)]} \\ &= 0.020 = 2\% \end{aligned}$$

■

Comparing this value to Fig. 13.2, we read that at $\lambda = 0.95$, $y_{CO} \approx 1.1\%$, which shows that our assumption that all the HC was burned up is incorrect. From Fig. 13.2 we see that $y_{HC} \approx 1000$ ppm. The oxygen deficit, $(1 - \lambda)$, is shared about equally between CO, which is not oxidized to CO_2, and HC, which is not oxidized (see Problem 13.6).

13.2.2.3 Hydrocarbons (HC). Figure 13.2 shows that at all values of λ one measures significant concentrations of unburned hydrocarbons, and that the dependence of unburned hydrocarbons on λ is much less than is the dependence of the y_{CO} on λ. Most of the unburned hydrocarbons are the result of *flame quenching*. IC engines must have some kind of lubrication where the piston slides up and down in the cylinder. In auto engines this is provided by the motor oil, which is pumped from a sump at the bottom of the crankcase through holes drilled or cast in the block, bearings, crankshaft, connecting rods, wrist pins, and cylinders to holes on the side of the piston. The piston rings, which are the actual sliding surface between piston and cylinder, ride on this oil film. Running the engine without oil causes the pistons to seize and destroys the engine in minutes.

Normal hydrocarbon lubricants cannot stand temperatures much higher than about 250–300°F (121–149°C) for long periods. The principal purpose of the cooling system of an auto engine is to keep the temperature of the lubricant film between the piston rings and the cylinder wall at or below that temperature. (Heavily loaded engines, in trucks or autos that pull trailers, have separate radiators to cool the oil!) If the temperature becomes significantly higher than that, the lubricants decompose, leaving behind solid carbon residues that cause the engine to seize; an engine operated without its cooling system is destroyed in a few minutes. Research engines have been built that use solid lubricants (MoS_2) that can stand very high temperatures.

These engines have no cooling system and operate at temperatures comparable to the melting point of steel. They have excellent fuel economies but very difficult materials-engineering problems. (Fame and fortune await the student who can invent an *inexpensive* lubricant that will operate at substantially higher temperatures than 250 to 300°F, thus allowing the auto manufacturers to decrease the size, weight, cost, and power requirements of their cooling systems!)

The cooling of the cylinder walls and head makes them cold enough that in a narrow *quench zone* adjacent to them the flame goes out, and the hydrocarbons in that part of the air-fuel mixture are not burned up.

Example 13.4. Estimate the hydrocarbon concentration to be expected in the exhaust gas from an engine with a piston diameter of 6 cm, a stroke of 5 cm, and a quench zone thickness of 0.2 mm at $\lambda = 1$.

Here we assume that all of the surface of the cylinder and the head has a quench zone. The top of the piston is not cooled and does not apparently play a significant role in flame quenching. The ratio of the volume of the quench zone to the volume of the combustion chamber with diameter D, piston travel L, and quench zone thickness t (assuming a flat head) is

$$\begin{pmatrix}\text{Ratio of quench}\\ \text{volume to total}\\ \text{volume}\end{pmatrix} = \frac{tA}{V} = \frac{t(\frac{\pi}{4}D^2 + \pi DL)}{\frac{\pi}{4}D^2 L} = t\left(\frac{1}{L} + \frac{4}{D}\right) \tag{13.8}$$

$$= 0.02\text{ cm}\left(\frac{1}{5\text{ cm}} + \frac{4}{6\text{ cm}}\right) = 0.017 = 1.7\%$$

Thus we would expect 1.7 percent of the total hydrocarbons in the fuel to appear in the exhaust. From Eq. (7.17), for $x = 8$, $y = 17$, and $\lambda = 1$, we find that for one mol of fuel we have

$$\begin{pmatrix}\text{Total mol of}\\ \text{combustion}\\ \text{products}\end{pmatrix} = n_{\text{total}} = 3.76\left(x + \frac{y}{4}\right) + x + \left(\frac{y}{2}\right)$$

$$= 3.76\left(8 + \frac{17}{4}\right) + 8 + \left(\frac{17}{2}\right) = 32.5\ \frac{\text{mol combustion products}}{\text{mol of fuel}}$$

and

$$\begin{pmatrix}\text{Mol fraction}\\ \text{of unburned}\\ \text{fuel in exhaust}\end{pmatrix} = y_{\text{unburned}} = \frac{0.017\ \dfrac{\text{mol unburned}}{\text{mol fuel}}}{32.3\ \dfrac{\text{mol comb. prods.}}{\text{mol fuel}}}$$

$$= 5.23 \cdot 10^{-4} = 523\text{ ppm}$$ ■

This calculation is a vast simplification of what actually goes on in an engine, but

1. The quench thickness used in this example is in the range of those values actually measured in special research engines.
2. The calculated hydrocarbon concentration is in the range of those normally measured (see Fig. 13.2).
3. The calculation shows that we would expect a higher hydrocarbon concentration in the exhaust from a small engine than a large one, which is also observed.

This example assumes that the hydrocarbons in the exhaust have the same chemical composition as those in the fuel. Table 13.4 shows the typical composition of hydrocarbons in untreated auto exhaust. The methane, ethane, acetylene, propylene, formaldehyde, and other aldehydes were not present in the fuel and must have been formed by incomplete combustion, mostly in the quench zone. The benzene, toluene, and xylenes were present in the fuel. They are the gasoline components with the slowest burning velocities, and hence the highest probability of passing, unburned, into the exhaust.

This change of HC composition from fuel to exhaust makes stating the concentration of hydrocarbons in the exhaust complex. The U.S. auto emission regulations are based on grams of nonmethane hydrocarbons (NMHC) per mile of vehicle travel. (Methane is always present in exhaust gas. Because it is quite unreactive in producing photochemical smog, it is generally not counted for air pollution purposes as an exhaust hydrocarbon.) Normally the hydrocarbon composition is measured by chromatography, and the weights of the various components are totaled. Sometimes one sees NMHC reported "as hexane" or "as C" (see Problem 13.3). One may show that the average molecular weight of the hydrocarbon mix in Table 13.4 is ≈ 45 g/mol, whereas that of C_8H_{17} is 113. This change complicates reporting of HC concentrations, which are normally stated in ppm by mole. Reporting and regulating emissions by mass rather than moles solves this problem.

TABLE 13.4
Major "unburnt" hydrocarbons, ppm

Methane	170
Ethane	160
Acetylene	120
Formaldehyde	100
Toluene	55
Aldehydes excluding formaldehyde	53
Xylenes	50
Propylene	49
C_4 alkenes	36
C_5 alkenes	35
Benzene	22

Source: Ref. 11.

Studies suggest that in addition to the quenching on the combustion chamber walls there is also flame quenching in the crevice between the piston and cylinder (above the top piston ring), in the crevice formed by the head gasket, and around the spark plug [12]. In addition some of the fuel apparently absorbs into the lubricant film on the wall during the compression stroke and then desorbs during the expansion stroke, thus contributing to the unburned HC in the exhaust gas. Thus there are more quench areas than just the cooled walls and head, but they all have the same effect—some of the fuel is not burned and some of it is only partly burned.

This calculation suggests that the percent hydrocarbon in the vehicle exhaust should be independent of air-fuel ratio. The observation is more complex. Figure 13.2 shows that at low values of λ the measured HC concentration rises, for the same reason that the measured CO concentration rises: the HC and CO are competing for the available O_2 and there is not enough to satisfy their needs. So the HC values are the sum of those due to flame quenching, more or less independent of λ, and those due to oxygen deficiency. The increase of HC with increasing λ at high values of λ is caused by *misfire*. At very lean conditions, $\lambda > 1.2$, sometimes the air-fuel mixture will fail to ignite, thus increasing the HC emissions (but not those of CO!).

We can also see from Fig. 13.2 that the CO concentration does not become zero at $\lambda > 1.0$ but rather continues at some low value, even when there is plenty of excess air in the exhaust gas. At the high temperatures of the flame the reaction that actually consumes CO [13],

$$CO + OH \rightleftarrows CO_2 + H \tag{13.9}$$

is an equilibrium reaction that does not go to completion. As the temperature is lowered toward the exhaust temperature, the equilibrium shifts strongly to the right, but the reaction becomes very slow below a temperature of 2200 K (3500°F) so that some of the CO found in the exhaust is due to the noncompletion of this reaction even though adequate oxygen is present.

Returning to Fig. 13.2, we can say that the CO consists of "oxygen deficit" CO and "incomplete reaction" CO, with the former being dominant at low values of λ and the latter being a substantially smaller value, practically independent of λ. The HC consists of "oxygen deficit" HC and "quench zone" HC, with the "quench zone" amount being about the same at all values of λ, and the "oxygen deficit" amount rising rapidly with decreasing values of λ.

13.2.2.4 NO_x. Chapter 12 showed how NO is formed from N_2 and O_2 in high-temperature flames. The flames in an auto engine certainly meet that description. Peak temperatures are of the order of 2700 K (4400°F). Figure 13.3 shows the calculated temperature and NO concentration history for a typical single combustion in an IC engine.

Example 13.5. For the example shown in Fig. 13.3 estimate the temperature before the beginning of combustion, the temperature in the burned gas just after combustion begins, and the temperature in the burned gas at the end of combustion.

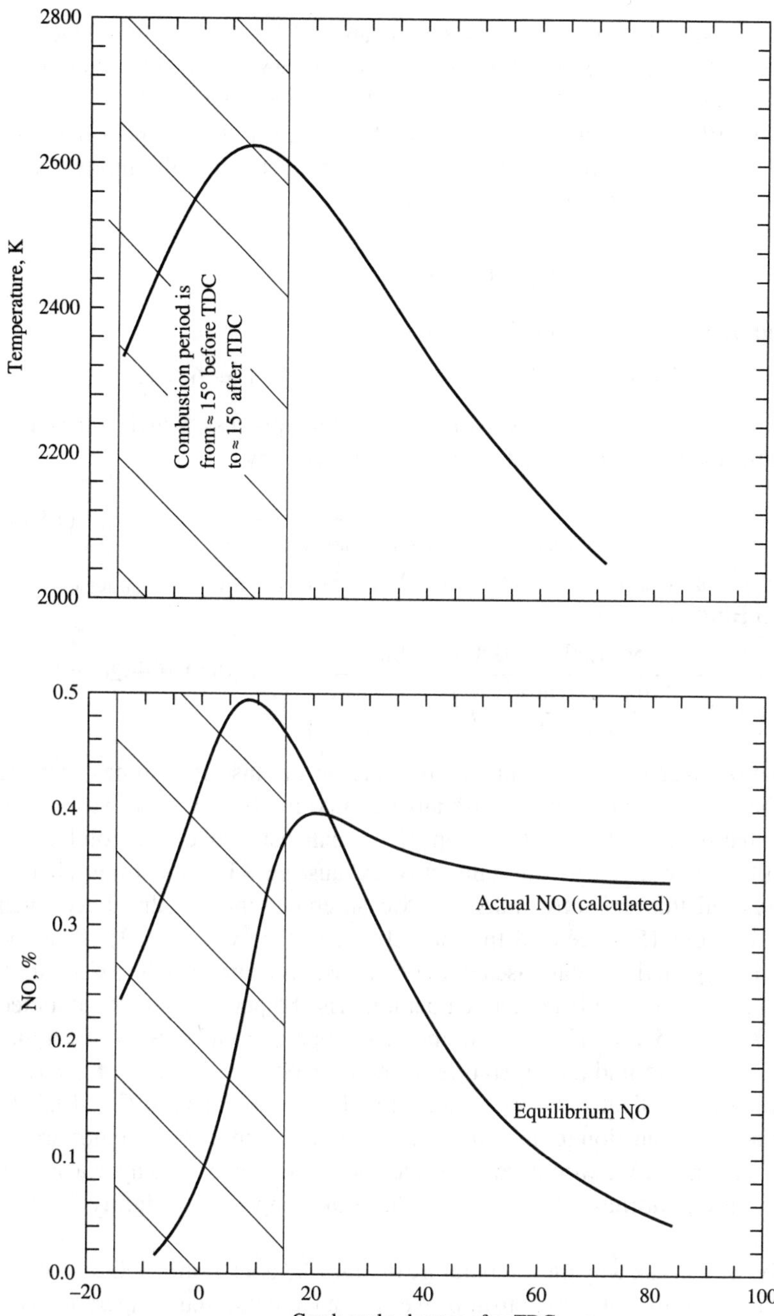

FIGURE 13.3
Calculated variation of the temperature and the NO concentration in the burned gases during combustion and initial expansion. These are average concentrations over the whole mass of burned gas for a small gasoline IC engine with a compression ratio of 7, running at 2000 RPM, and $\lambda = 1.0$. The combustion period is from 15° before TDC to 15° after TDC. Observe the similarity to Fig. 12.7. (Adapted from W. J. D. Annand, "Gasoline Engines," in *Internal Combustion Engines,* ed. C. Arcoumanis, Academic Press, p. 77, 1988.)

The figure shows that combustion begins about 15 degrees before TDC and continues until about 15 degrees after TDC. At 15 degrees before TDC the piston has traveled about 98 percent of its travel from BDC (see Problem 13.21), so we can safely ignore further piston travel and assume that the gas in the cylinder has been compressed by the stated compression ratio of 7. From standard thermodynamics texts for reversible adiabatic compression, we find

$$T_2 = T_1 \left(\frac{V_1}{V_2} \right)^{R/(C_P - R)} \tag{13.10}$$

Taking common values of T_1 and C_P, we write

$$T_2 = 528°\text{R} \cdot (7)^{R/(3.5\,R-R)} = 1154°\text{R} = 694°\text{F} = 641\text{ K}$$

The temperature rise during the first part of the burning process, which is assumed to be adiabatic and to occur at constant pressure, is given by

$$\Delta T = \frac{m_{\text{fuel}}\,\Delta h_{\text{combustion}}}{m_{\text{combustion products}}\,C_{P,\text{ combustion products}}} \tag{13.11}$$

For a typical gasoline $\Delta h_{\text{combustion}}$ is about 19 020 Btu/lbm and $C_{P,\text{ combustion products}}$ is about 0.33 Btu/lbm · °F, so that

$$\Delta T = \frac{1\text{ lbm fuel} \cdot 19\,020\text{ Btu/lbm}}{\left(15.88\,\dfrac{\text{lbm comb. prods.}}{\text{lbm fuel}}\right)\left(0.33\,\dfrac{\text{Btu}}{\text{lbm}\cdot°\text{F}}\right)} = 3629°\text{F} = 2016\text{ K}$$

This calculation overstates the temperature increase because it assumes that the whole combustion chamber is filled with air-fuel mixture. In Fig. 13.1, we see that when the exhaust valve closes at the top of the exhaust stroke, the combustion chamber will contain a substantial amount of exhaust gas. The incoming air-fuel mixture mixes with this *residual exhaust gas*. Measurements indicate that the residual exhaust gas is about 15 percent of the total charge to the cylinder [14]. Thus the heat released per pound of total gases in the combustion chamber will be about 85 percent of that just computed, and the temperature rise 85 percent of that computed above, or $\Delta T = 3085°\text{F} = 1714\text{ K}$. Adding this temperature increase to the initial temperature, we would find a temperature of $694 + 3085 = 3779°\text{F} = 4239°\text{R} = 2355\text{ K}$, which is practically the value shown at 15° before TDC in Fig. 13.3. In addition, at peak combustion temperatures the combustion reactions are not driven completely to the right. We would expect some unreacted oxygen and hydrocarbons to be present at equilibrium, thus reducing the peak temperature slightly in most auto engines.

Finally, we can ask why the temperature in Fig. 13.3 goes from 2350 K to about 2640 K during the remaining combustion. We computed the initial temperature rise by assuming the first part of the combustion occurred at constant pressure. This assumption is plausible because, as the first part (near the spark plug) burns, it expands and pushes away the remaining gas. But later, when the remaining gas burns, it also expands and hence compresses the gas that burned first so that the gas that burned first undergoes combustion at practically constant pressure, followed by an adiabatic compression. At the end of the combustion process the gas temperature in the combustion chamber will not be at a uniform temperature; the first part to burn will be hottest (Problem 13.9). But if we wish to compute the average temperature we can proceed by assuming that the whole combustion process occurs at constant volume, for which we can write

$$\Delta T = \frac{m_{\text{fuel}}\,\Delta u_{\text{combustion}}}{m_{\text{combustion products}} C_{V,\text{ combustion products}}} \tag{13.12}$$

The heat of combustion used in this example and the commonly tabulated heats of combustion (see Table 7.1) are not $\Delta u_{\text{combustion}}$ but $\Delta h_{\text{combustion}}$, normally at 1 atmosphere and 25°C. However (see Problem 13.11), one makes a negligible error in most cases by treating these as the same. $C_{V,\text{ combustion products}}$ is typically about 0.26 Btu/lbm · °F, so that for the overall combustion at constant volume we would compute

$$\Delta T = \frac{1\text{ lbm fuel}\cdot 19\,020\text{ Btu/lbm}}{\left(15.88\,\dfrac{\text{lbm comb. prods.}}{\text{lbm fuel}}\right)\left(0.26\,\dfrac{\text{Btu}}{\text{lbm}\cdot{}^\circ\text{F}}\right)} = 4607^\circ\text{F} = 2559\text{ K}$$

which shows that

$$\frac{\Delta T_{\text{constant volume}}}{\Delta T_{\text{constant pressure}}} \approx \frac{C_P}{C_V} = \frac{0.33}{0.26} = 1.27 \tag{13.13}$$

The values from Fig. 13.3 are

$$\frac{\Delta T_{\text{peak}}}{\Delta T_{\text{inital combustion}}} = \frac{2640 - 641}{2350 - 641} = 1.17$$

The differences between these values are due to heat losses from the combustion gases and to the work of expansion done by the gases on the piston. ■

After this long digression about the temperatures in Fig. 13.3, we can now consider the NO values. The equilibrium NO curve in Fig. 13.3 corresponds to the thermal NO values calculated by the Zeldovich mechanism, shown in Chapter 12 (Problem 13.14). The actual curve suggests that the rate of NO formation is negligible until the temperature reaches about 2400 K. The concentration rises rapidly toward the equilibrium value, crossing it at about 22 degrees after TDC. From then on the concentration is higher than the equilibrium concentration in the rapidly cooling gas, so the concentration falls; but when the gas temperature reaches about 2300 K the reaction rate becomes negligible, and the NO concentration is "frozen" at a value above the equilibrium value. This result is similar to that in Example 12.4. The combustion period shown corresponds to about 30° of crank angle, or $\frac{1}{12}$ of a revolution, and the time of combustion to about 2.5 ms.

There is very little nitrogen in gasoline, so the amount of fuel NO (see Chapter 12) is generally negligible. Most of the exhaust NO is thermal NO, produced in the high-temperature part of the combustion process. (Some prompt NO is also present; see Chapter 12.) The calculated NO concentration in the exhaust gas is quite close to the average experimental value shown in Fig. 13.2.

13.3 CRANKCASE AND EVAPORATIVE EMISSIONS

Crankcase emissions are mostly of historic interest, because they have been almost completely eliminated from modern U.S. autos (but not in autos in developing countries). Evaporative emissions have been substantially reduced from their 1960 values, but as tailpipe emissions are further reduced, evaporative emissions will form an increasing part of total emissions and will require additional control.

Table 13.2 shows that in 1960 crankcase emissions accounted for about 25 percent of all hydrocarbon emissions from cars. In Fig. 13.1 we see that the piston moves up and down in the cylinder. It is sealed against the sides of the cylinder by piston rings, which are seated in grooves in the piston and which are held out by spring force against the wall of the cylinder. As discussed in Chapter 10, in any such moving seal if we make the seal too tight, the frictional resistance will be too great, while if the seal is too loose, the leakage will be too great. The proper setting is one that allows some small amount of leakage.

This leakage results mostly from the movement of gases from the high-pressure combustion chamber into the low-pressure crankcase, called *blowby*. If the crankcase were totally closed, the blowby would raise its internal pressure to unsafe levels. The blowby gas is similar to exhaust gas but has a higher percentage of partly combusted products, e.g., oxygenated acids. If this gas is allowed to remain in contact with the oil in the oil sump, these acids—plus the water of combustion plus other contaminants in it—will dissolve in the oil, forming products that are corrosive to the engine. For that reason the blowby gas must be promptly removed from the crankcase.

In pre-1960 cars this problem was solved by having a simple vent from the crankcase to the atmosphere. The vent was shaped so that when the car was moving at highway speeds the air flow produced a slight vacuum on it, sucking gases out of the crankcase. The oil filler cap had vents (with dust filters) allowing this vacuum to pull fresh air through the filler cap into the crankcase and thus ventilate it. As the hot exhaust gases flowed through the crankcase they vaporized or entrained some of the hot oil in the crankcase, thus producing (as shown in Table 13.2) about 25 percent of the total hydrocarbon emissions from the engine.

The solution to this emission problem has been *positive crankcase ventilation*. The crankcase vent, instead of passing to the atmosphere, passes into the carburetor or the air-fuel intake manifold so that the hydrocarbon-laden gases are sucked into the engine and burned. This design change results in a slight increase in fuel economy; instead of throwing away those fumes and droplets, we burn them. The flow must be regulated, because at idle the extra air passing through the engine would upset the A/F ratio and cause rough idle. The positive crankcase ventilation (PCV) valve in the vent line senses the vacuum in the manifold and reduces the flow when the vacuum is high (at idle), increasing it when the vacuum drops (normal or high-speed operation), thus allowing the vapor from the crankcase to be captured. The vented oil filler cap has been replaced by a sealed one, or one that draws air in from the air cleaner. Crankcase emissions are practically zero for post-1968 U.S. vehicles. The design details of this system are described in Ref. 15.

Evaporative losses are of two kinds: evaporation from the fuel tank and evaporation from the carburetor or fuel injection system. The first type was discussed in Chapter 10; it is similar to the breathing losses on any vented tank containing a VOC. Losses from fuel tanks are more significant than those from stationary tanks because the exhaust pipe on most autos passes close to the fuel tank, heating it so that the temperature swing, cold to hot, will be much greater than for a stationary VOC storage tank. Typically this swing is from the overnight storage temperature to about 120°F (49°C) on a hot day.

Almost all new autos sold in the United States use fuel injection, and do not have carburetors. Older U.S. cars, and those sold in some developing countries, have carburetors of the type described here. Fuel injection engines store their fuel in completely closed systems and do not have the carburetor emissions described here. The carburetor evaporative losses occur during the *hot soak* period that occurs when an auto is stopped. While the auto is moving down the road, air flow under the hood limits the temperature at the carburetor to about 120°F (49°C) if the outside air is at 60°F (16°C). When the car stops and the engine is turned off, that cooling air flow stops, and the stored heat in the engine heats the carburetor to 160–180°F (71–82°C). The carburetor has a small reservoir that contains about 70 cc of liquid fuel. This is kept in the carburetor so that the carburetor can respond quickly when the driver starts to accelerate, without having to wait for the fuel pump to bring up the necessary sudden burst of fuel from the fuel tank. The constant level of this liquid also acts as part of the fuel-metering system. This level is controlled by a float valve much like that in toilet tanks. When the carburetor temperature rises during hot soak, about 30 percent of the fuel in this reservoir evaporates. (In hot weather even more evaporates. When one starts a carburetor auto on a hot day after a hot soak, one must wait for the fuel pump to fill this reservoir; for this reason most carburetor autos are slow to start after a hot soak on a hot day.) In pre-1960 cars this evaporated fuel, about 19 percent of the total HC emissions, passed to the atmosphere (Table 13.2). That is only approximate because it is a value per hot soak, and its value per mile is its value per hot soak divided by the average trip length. Details on evaporative emissions are presented in Ref. 16.

Most autos now deal with both the carburetor hot soak emissions and the breathing losses from the fuel tank with the system shown in Fig. 13.4 on page 490. In it we see that the carburetor vent, instead of being open to the atmosphere, passes through the carburetor vent line to the *charcoal canister*. This contains an adsorbent, normally activated charcoal. The vents from the fuel tank also pass through this canister. In it, the charcoal adsorbs the gasoline vapor, allowing cleaned air to pass out to the atmosphere (adsorption of HC vapors is discussed in Chapter 10). When

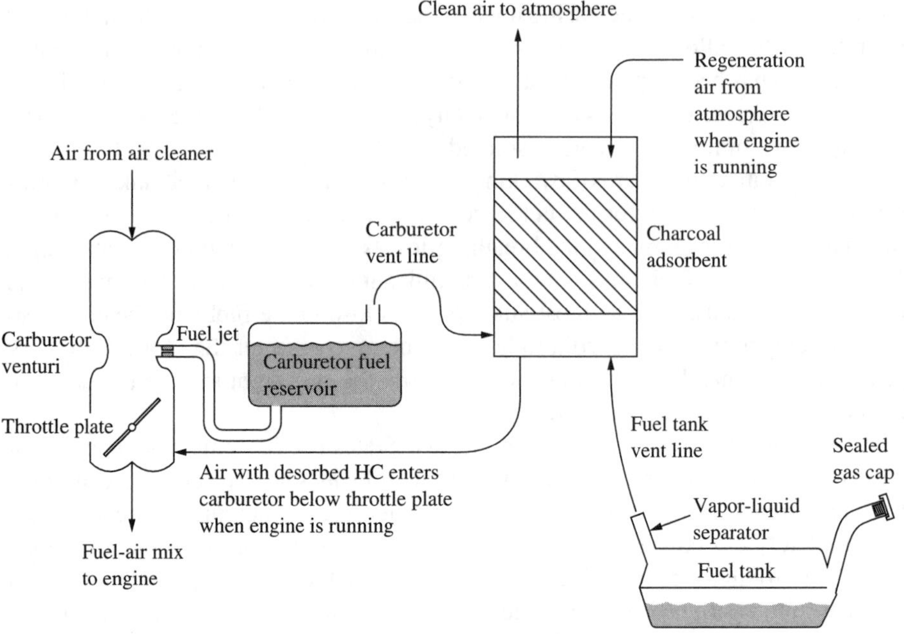

FIGURE 13.4
Control of evaporative emission. The HC-containing vapors from the carburetor and the fuel tank pass through a charcoal canister that removes the HC before those vapors are vented to the air. When the engine is running at other than idle or very low speeds, air is sucked through the canister in the reverse direction, removing the HC from the charcoal, preparing it for its next service. This regeneration air is returned to the air entry of the engine, where the HC it contains is burned. Suitable valves maintain the flow in the proper direction. The canister has a low-enough flow resistance that the pressure in the carburetor float chamber is close enough to atmospheric for proper operation of the carburetor. Fuel injection autos have no carburetor; they use this system to control vapors from the fuel tank.

the engine is running at sufficient speed, air is drawn through the charcoal canister in the opposite direction, stripping the adsorbed fuel off the charcoal and returning it to the engine air inlet, thus regenerating the adsorbent making it ready for its next adsorption task. Suitable valves are used to guarantee that the flows are always in the correct direction and at the right times.

All autos sold in the United States starting with model year 2000 (and some before) have a combination of filling tube, gas tank geometry, and charcoal canister suitable to capture these emissions during refueling. Earlier autos did not capture this emission and vented the displaced vapors back through the filling neck. Stage 2 vapor recovery (Fig. 10.6) captured those emissions, while stage 1 (Fig. 10.5) allowed them to escape to the atmosphere. As the older cars pass out of use, the existing stage 2 devices will be removed, probably about 2010.

There still remain miscellaneous HC evaporative losses in the fuel tank and the fuel handling system. As the tailpipe emissions described next are reduced to lower and lower values, these remaining evaporative losses become a more significant part of the overall problem (see Table 13.2).

13.4 TAILPIPE EMISSIONS

Table 13.2 shows that all the CO and NO_x emissions and about half the HC emissions are tailpipe emissions, i.e., those in the exhaust gas. The possibilities for dealing with these are discussed next. These possibilities are not mutually exclusive; most current autos use some combination of these alternatives. The cheapest and simplest approach is to change what goes on in the combustion process, to minimize the formation of pollutants. The first four alternatives discussed below do that. The remaining four require some change outside the combustion process. (In the literature the pollutants in the exhaust gas passing from the engine to an external control device, most often a catalyst, are called the *engine out* emissions. The tailpipe emissions are that fraction of the engine out emissions that are not destroyed in the external control device.)

13.4.1 Lean Operation

From Fig. 13.2 it is clear that lean combustion greatly reduces CO and HC emissions compared to rich combustion. Unfortunately, most mechanics have always known that an engine tuned for rich combustion starts and runs more smoothly than one tuned for lean combustion (but with poorer fuel economy). Before the 1970s mechanics regularly changed the factory carburetor settings to make the combustion richer. Their customers liked the car's smooth performance, but these changes greatly increased emissions. One of the early air pollution control steps was to modify the carburetors to make it much harder (and illegal) to change the factory settings. Older autos and autos with worn engines are set to run rich, to overcome the consequences of engine wear. Developing countries have many such autos and suffer the high CO and HC emissions that accompany rich combustion.

Many pre–catalytic converter auto engines operate in the lean mode as much as possible, for reasons of both pollution reduction and fuel economy. We will return to this topic later.

13.4.2 Exhaust Gas Recirculation (EGR)

As shown in Fig. 13.2, the peak NO emissions occur at an A/F ratio slightly leaner than stoichiometric ($\lambda \approx 1.07$). Starting in the 1970s and continuing in many current engines, that production of NO_x was reduced by Exhaust Gas Recirculation (EGR), in which the incoming combustion air is diluted with up to 20 percent exhaust gas. EGR reduces the peak flame temperatures and the O_2 content of the burned gas, by simple dilution with a gas that is mostly nitrogen, carbon dioxide, and water. Both of these effects lower the NO_x formation. EGR reduces the power output of the engine (or requires a larger engine for the same power output); it also requires extremely good control of the carburetion process and of air-fuel mixture distribution. At extremely lean operation EGR can lead to misfire and thus increased HC emissions. Most EGR systems turn off the EGR at low engine speeds and during warmup, when it can lead to very poor engine performance; besides, EGR is not needed under these circumstances because the peak temperatures are low compared to those at part or full power operation. They also turn EGR off when peak power is demanded, because EGR reduces power output and because at peak power output the fuel mixture is always rich, so that there is not enough oxygen for much NO_x formation.

Figure 13.5 shows some example data on the effect of EGR. For this particular engine and operating condition, adding 20 percent EGR reduced the NO_x emissions, expressed as g/HP·h, from 11.4 to 4.3, i.e., to 37 percent of the value with no EGR. The fuel consumption, expressed in g/HP·h, increased by only 1 percent. The exhaust temperature fell from 1250° to 1200°F and the HC emission rose by 75 percent, presumably due to occasional misfire (failure of the mixture in the combustion chamber to ignite). Compare the NO_x reduction shown here with that for a stationary burner in Fig. 12.10; the results are similar.

13.4.3 Reduce Flame Quenching

In the lean combustion mode, most of the HC and CO are formed by wall quenching of the flame. Various techniques have been used to minimize this. The most obvious is to make the combustion chamber more nearly spherical, thus reducing the surface per unit volume. Reducing the sizes of the crevices associated with the head gasket, spark plug gasket, and piston rings also contributes to reduced flame quenching.

Raising the temperature of the cylinder wall and head lowers the thickness of the quench zone. This was one of the reasons for switching the auto coolant from water in summer and ethylene glycol in winter to a year-round mixture of water and ethylene glycol. With that mixture's higher boiling point, higher coolant temperatures were possible without excessive coolant pressures. This not only reduced HC emissions but also allowed the auto manufacturers to reduce the size of all the cooling system components.

If one could place the fuel in the middle of the combustion chamber instead of having it mixed uniformly through it, one could greatly reduce flame quenching at the walls of the combustion chamber. Diesel engines do place the fuel in the middle and they have much lower CO and HC emissions than conventional auto engines; the reasons are discussed in Sec. 13.7. Several modified combustion chambers have tried to achieve the same result in auto engines, with some success, e.g., the Honda two-chamber system or various manufacturers' induced swirl systems.

13.4.4 Speed the Warmup

Much if not most of the CO and HC emissions of a typical driving cycle occur in the first minute or two while the engine is cold. As we have already discussed, to start a cold engine one must operate it with a very rich air-fuel ratio ($\lambda < 1$), normally

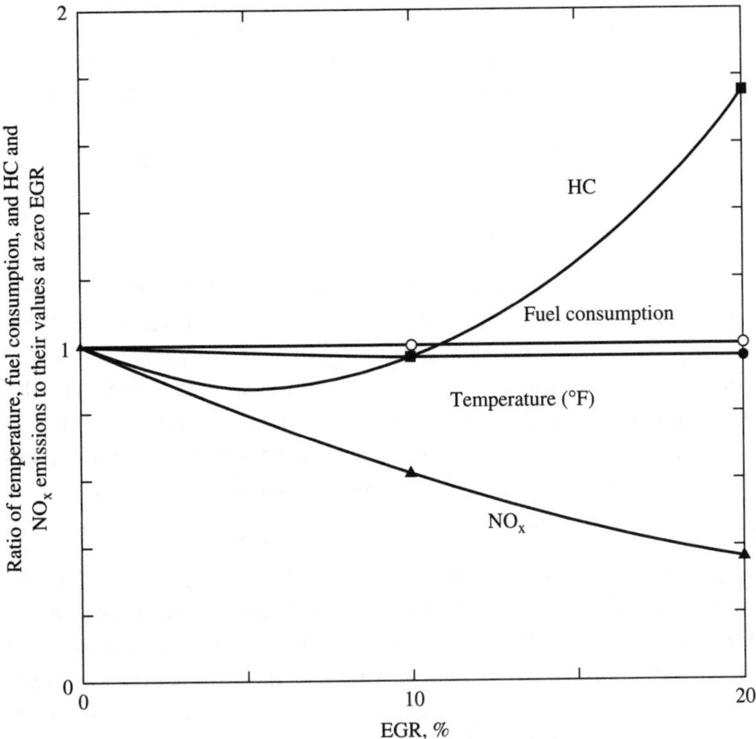

FIGURE 13.5
Some effects of EGR. These data are for a 6.56 liter, 9.3 compression ratio V-8 engine at 1700 rpm, 32 kW output, at 16.0 A/F ($\lambda \approx 1.07$) and spark 10° retard from the maximum power setting. The results are shown as ratios compared to the value with no EGR. For example, the NO_x emissions went from 11.4 to 4.3 g/HP · h as EGR was increased from 0 to 20 percent. The other zero EGR values are HC, 0.9 g/HP· h; fuel consumption, 468 g/HP · h; and exhaust temperature, 1250°F. The rise in HC emission with EGR at 20 percent EGR is presumably due to occasional misfire, which dumps unburned HC into the exhaust manifold. The curves pass through the three measured values at 0, 10, and 20 percent EGR [17].

by using an automatic choke. This rich condition, combined with the cold walls of the combustion chamber, leads to very high CO and HC emissions. As we shall see in Sec. 13.4.5, the catalyst designed to treat these emissions does not "light off" until it has been warmed by hot exhaust gas, so these emissions pass untreated to the atmosphere. All modern IC engines take various steps to speed the warmup of the engines, to minimize emissions during this period.

Most engines measure the temperature of both the underhood air (where the engine air intake snorkel is located) and the engine coolant. When these are cold the engine uses valves to draw air through a shroud over the exhaust manifold into the engine air intake to speed the warmup. When the engine is warm these valves close, allowing the engine to draw in cooler air from under the hood, thus improving engine efficiency and power. The heated air flow during warmup causes the automatic choke to turn off sooner, thus ending the high-emission startup period sooner. Some engines use electric heaters to warm the air-fuel mix during this cold start period. By warming that mix, they increase the fraction vaporized and thus allow the engine to start with less fuel enrichment. Even with these special measures, the cold start period is still responsible for a disproportionate part of the CO and HC emissions. The NAAQS for CO is rarely exceeded in U.S. cities in the summer; it is often exceeded in winter.

13.4.5 Catalytically Treat the Combustion Products

Most auto manufacturers have concluded that they cannot meet current and future emission standards by engine modifications alone. Their efforts to do that in the 1970s resulted in some very poorly performing automobiles. Instead, they have concluded that the most satisfactory solution is to modify the engine so that it produces the right mix of pollutants and then treat that mix catalytically to meet the emission standards. The first attempts used two catalysts, but since then auto manufacturers have developed the "three-way catalyst" that promotes the following reaction:

$$\mathrm{NO} + \mathrm{CO} + \mathrm{HC} \xrightarrow{\text{Platinum-palladium-rhodium catalyst}} \mathrm{N_2} + \mathrm{CO_2} + \mathrm{H_2O} \tag{13.14}$$

This reaction requires very close control of the ratio of oxidizing agent (NO) to reducing agents (CO + HC). Figure 13.6 shows that with very close control of the A/F ratio, conditions can be found that lead to about 95 percent destruction of all three pollutants. The key to doing this successfully was the development of the doped zirconium dioxide oxygen sensor, shown in Fig. 13.7 on page 496. This consists of a piece of doped ZrO_2 that is coated on both sides with a porous platinum film and inserted into the exhaust manifold. The sensor is an electrolytic cell, with the ZrO_2 acting as a solid electrolyte; its output voltage is a strong function of the oxygen content of the exhaust gas. Using the measured value of the exhaust gas oxygen content, the engine computer can control the A/F ratio to stay within the ± 0.05 A/F tolerance (λ tolerance of ± 0.003) needed to stay at the top of the curves in Fig. 13.6.

Catalysts were discussed in Chapter 7. The typical automobile exhaust gas catalyst is a (5:1) mix of platinum or palladium with rhodium, supported on an Al_2O_3 layer that is deposited on a cheaper ceramic base (see Figs. 7.13 and 7.14). The requirements for an auto exhaust catalyst are

1. Produce at least 90 percent destruction of CO, HC, and NO_x according to Eq. (13.14) in as small, lightweight, and inexpensive a package as possible. If possible, the catalyst should be formed in a pancake shape for easy installation under an auto.
2. Start to destroy CO, HC, and NO_x at as low a temperature as possible. Typical catalysts do not begin to promote the reaction until they are heated by the exhaust gas to their "light off" temperature of about 350°C (662°F). Thus they are inactive during the period of highest emissions, that is, during cold start.
3. Not have excessive heat flow to the surroundings or excessive surface temperatures, to prevent excessive heat flow to the passenger compartment of the car or the starting of grass fires. (Many cars have heat shields to solve this problem.)

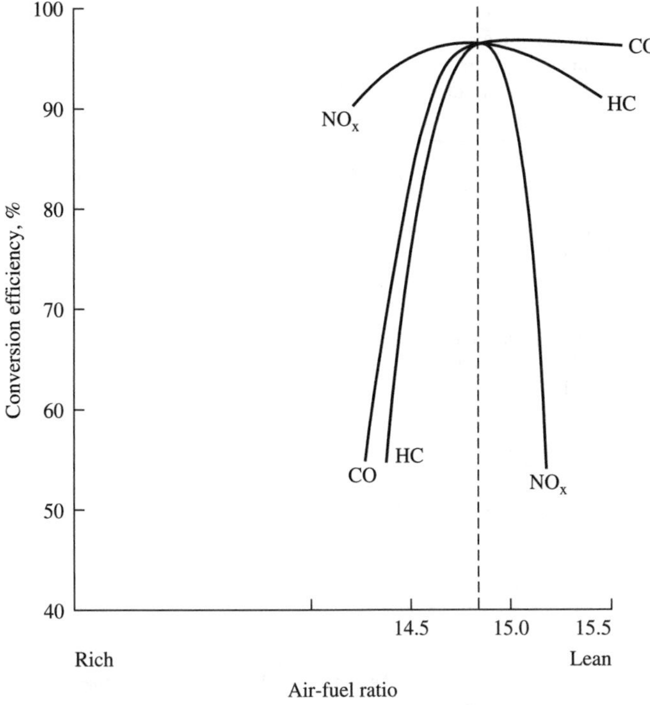

FIGURE 13.6
Conversion efficiency of a three-way catalyst system as a function of A/F ratio. From J. B. Heywood and R. J. Tabacyzniski, "Current Developments in Spark-Ignition Engines," in *A History of the Automotive Internal Combustion Engine,* SAE Publication SP-409, 1976.

4. Perform satisfactorily for at least 50,000 miles or five years (required by EPA regulations) in a very difficult environment (heat; cold; vibrations; varying input flow temperature, pressure, and chemical composition).

5. Have minimum pressure drop.

6. Cause the small amount of S in the gasoline to exhaust as SO_2 rather than the much smellier H_2S, but not oxidize the SO_2 to the more toxic SO_3.

A typical modern auto catalyst does these things. It has a volume of about one liter and a noble metal content of about 1.5 g. The most common catalyst support is that shown in Fig. 7.14, packaged as shown in Fig. 13.8 on page 496. Some auto manufacturers attack the slow light off problem by placing the catalyst as close to the exhaust manifold as possible, instead of placing it under the auto, and some use two catalysts, a small one designed to withstand the high exhaust temperature in the warmed-up state, and placed very close to the engine to deal with the cold-start emissions, and a larger one, under the auto, to handle the warmed-up emissions.

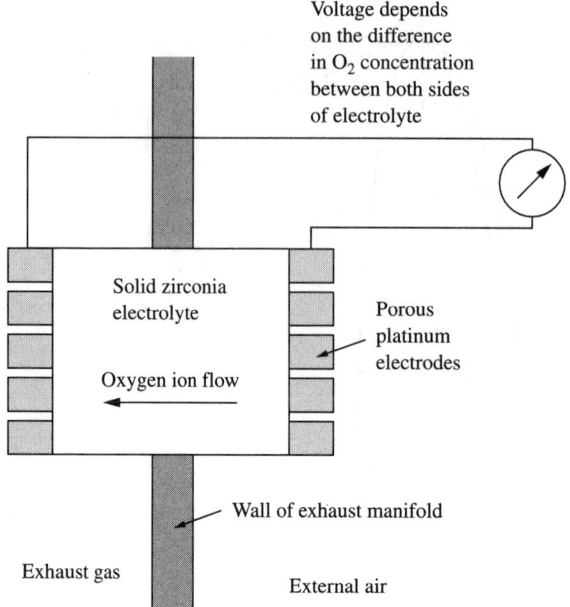

FIGURE 13.7
Schematic of the exhaust-gas oxygen sensor, used to control the NO/(CO + HC) ratio going to the "three-way catalyst." The voltage between the two porous platinum electrodes depends on the difference in O_2 content between the outside air (practically constant) and the exhaust gas. This schematic does not show the mechanical arrangements that allow the sensor to be screwed into a hole in the exhaust manifold and protect the delicate electronic parts from mechanical abuse. (From Ref. [10]. Courtesy of Springer Verlag.)

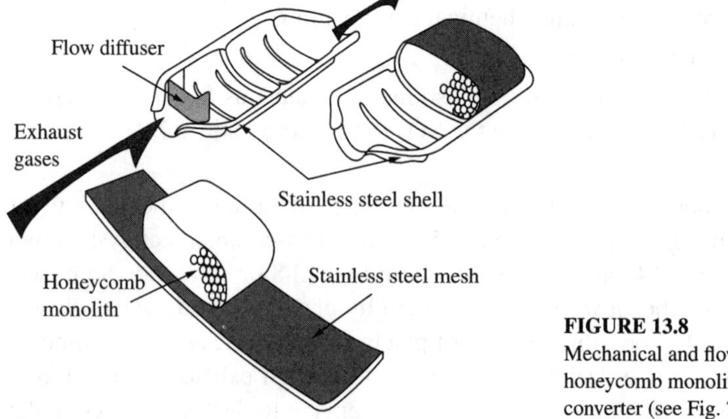

FIGURE 13.8
Mechanical and flow arrangements of a honeycomb monolith–type catalytic converter (see Fig. 7.14).

13.4.6 Change the Fuel

The requirements for the fuel for an IC spark ignition engine are

1. High heating value (Btu/lbm) so that the vehicle will have adequate range between refuelings, without carrying an excessive fuel weight
2. A high fuel density so that the fuel storage container will be of acceptable size
3. Easily handled, normally as a liquid at ambient temperatures
4. Good antiknock properties (discussed next)
5. Ability to vaporize in air-fuel system (adequate volatility)
6. Other miscellaneous properties, like good storage stability, limited toxicity, etc.

Antiknock properties, which we have not previously discussed, are a very important part of this list. Returning to Fig. 13.1, we see that the combustible mixture is ignited by the spark plug, which is normally located at or near the middle of the top of the combustion chamber. The flame starts there and spreads out through the chamber. The burned gases have a much higher volume than they had as unburned gases, so they expand and compress the unburned gas. As a result the temperature of the unburned gases rises before the flame reaches them. (The propagation speed of the pressure increase is much faster than the speed of the flame front.) If the unburned gas is heated to its autoignition temperature before the flame front reaches it, it will spontaneously ignite, producing a loud knock. This knock is annoying to the driver and destructive of the engine. Motor manufacturers and fuel suppliers have worked hard to prevent it.

It was apparent early in the history of the automobile that raising the compression ratio—that is, (the volume contained at BDC)/(the volume contained at TDC)—increased the efficiency of the engines but also increased their tendency to knock. Early engines had compression ratios of about 4, most current auto engines have compression ratios of 8 to 10, and diesel engines have compression ratios of about 16 to 20. In diesel engines the fuel and air are not premixed; fuel is sprayed into the chamber near TDC. In that case the hot gases ignite the fuel without a spark plug. Diesel engines would have a terrible knock problem if the fuel and air were premixed, but they have no problem if the fuel is added slowly and if it burns as it is added. Thus the very property of the fuel—ease of autoignition—that makes it a good diesel fuel makes it a bad fuel in auto engines because of knock. Knock sets the limits on allowable compression ratios in gasoline engines.

Different families of hydrocarbons have different resistances to knock. Straight-chain hydrocarbons are the worst, highly branched paraffins the best, olefins better than corresponding paraffins, and aromatics almost as good as highly branched paraffins. (The order of quality is just the reverse for diesel fuels.) Fuels are rated by their *octane number,* which is equivalent to the percent isooctane (2,2,3-trimethylpentane) in a blend of isooctane and n-heptane that has the same antiknock properties. (This is equivalent to assigning an octane number of 100 to isooctane and of zero to n-heptane, and assuming linearity of octane number on blending of these two.)

Octane numbers are tested in special laboratory one-cylinder engines with variable compression ratios. The engine is calibrated with carefully prepared blends of isooctane and n-heptane and then run on the fuels to be tested. The compression ratio is increased until knock occurs. The reported octane number is that of the test blend that knocks at the same compression ratio. The octane number posted on gasoline pumps is listed as (R + M)/2, indicating that it is the average of the Research Octane Number and the Motor Octane Number. These two values refer to two different sets of operating conditions in the same basic laboratory engines [18]. Typical motor fuels in the United States have octane numbers in the range of 85–90. Fuels for piston aircraft engines have octane numbers over 100; these are expensive, but they allow a higher compression ratio and thus better fuel economy and longer range. A major part of oil refining consists of modifying the chemical structure of the components of gasoline to increase the octane number by changing low-octane-number hydrocarbons into higher-octane-number hydrocarbons, mostly by catalytic reforming, isomerization, alkylation, and catalytic cracking. One can also improve the octane number with additives. The most successful has been tetraethyl lead (TEL). When it is added to gasoline at the ratio of up to 0.1 percent by weight, it can increase the octane number by up to 6. Unfortunately, although lead leaves the combustion chamber as a gas it forms fine particles on cooling. These deactivate automotive emission catalysts and increase the lead content of the atmosphere. TEL is still used in developing countries, but has practically disappeared from U.S. gasolines. The continuing search for other octane-number-improving additives that will be as effective as TEL without its disadvantages is intense; the financial reward for finding one would be immense.

Several other fuels have been used for many years in slightly modified automobile engines, for reasons of cost and availability. Natural gas is regularly used in places like natural gas pumping stations. It meets all the requirements just listed except that of easy storage on the vehicle. That is not a problem for stationary engines. Other gaseous fuels, e.g., CO or H_2, can be used in slightly modified automobile engines if a suitable storage system can be worked out. So far the only gaseous fuel for which that has been done commercially is natural gas, which is stored in high-pressure (2500 psia) containers in the vehicle. These containers are larger than gasoline tanks that give comparable ranges, and they cannot be fabricated in a pancake shape to fit easily under the vehicle, like gasoline tanks. The service stations that fill them are more complex than gasoline stations, and the fueling time is longer. However, for fleet vehicles that all take on fuel at one station and that never go to areas without natural gas fueling stations, they are quite practical and perhaps economical. (Natural gas is normally cheaper than gasoline on a \$/Btu basis, and in many states it enjoys a lower tax rate than gasoline.)

Commercial propane, sometimes called LPG or "liquefied petroleum gas," is a mixture that is typically 90+ percent propane. Its vapor pressure at 100°F is about 200 psig; it is normally stored and transported in tanks at that pressure. It is a perfectly satisfactory fuel for gasoline engines, many of which are equipped with "dual fuel" capabilities, enabling them to switch from propane to gasoline depending on fuel availability and price.

The two lowest molecular weight alcohols—methanol, CH_3OH, and ethanol, C_2H_5OH—can be used in slightly modified gasoline engines. For instance, Brazil has very little domestic petroleum; they have developed an ethanol-based auto fuel system, using ethanol produced by fermentation of sugar and other domestically grown crops, thus reducing their balance of trade problem.

All of the fuels discussed in this section produce lower air pollutant emissions for the following reasons:

1. All of these fuels have lower boiling points than the highest molecular weight components of gasoline, so that they are more easily converted to the all-gas state. That leads to better and more complete mixing with air than can be done with gasoline. This promotes complete combustion. Under some circumstances there is liquid gasoline in the combustion chamber of gasoline engines; it passes to the cylinder walls and is not burned in the quench zone. Liquid formation on the cylinder walls does not happen at all with natural gas and propane, and much less with the alcohols than with gasoline. For this reason all of these fuels produce lower HC emissions than gasoline.
2. All of these fuels have simpler molecules than gasoline and thus take fewer chemical steps to be totally combusted to CO_2 and water. Combustion is more likely to be complete with them than with gasoline, leading to lower HC and CO emissions. (Likewise, fewer products of incomplete combustion and/or polymerization form, so that with these fuels the engine is cleaner, engine life is longer, and motor oil changes are less frequent.)
3. These fuels generally contain practically zero N and S, whereas gasoline contains a small amount. That N and S make the catalysts operate less effectively, so these fuels are easier for catalysts to treat.

For all of these reasons, natural gas, propane, methanol, and ethanol are being promoted as "clean fuels." In addition, all of them have high octane numbers, so that the engines that use them can have higher compression ratios and better fuel economy than ordinary gasoline engines.

The observation that oxygen-bearing fuels lead to lower HC and CO emissions (mostly lower CO emissions) has led to the 1990 Clean Air Amendments [19] requiring regions with severe winter CO problems to use only gasoline that contains at least 2.7 weight percent oxygen during the winter months. It appears that this requirement will be mostly met by blending into the gasoline methanol or ethanol or methyl *tert*-butyl ether (MTBE), which is made from isobutene and methanol. All three of these improve the octane number of the fuel they are blended into. These all lower emissions by tricking the engine's control system to operate in a leaner mode than it would with ordinary gasoline. During cold start and maximum power, the engines' control systems ignore the reading of the exhaust gas oxygen sensor (which reads ≈ 0 in these circumstances) and mix fuel and air by weight. The oxygenated fuels have lower values of $(A/F)_{stoichiometric}$, so that at an equal (A/F) they have a larger λ than a conventional gasoline. (See Problem 13.4.) In the warmed-up, steady driving situation the engine control system adjusts the (A/F) to have the correct O_2 content for the three-way catalyst; the oxygen in the fuel has negligible effect in this circumstance.

Currently EPA is enforcing *Reformulated Gasoline* regulations. These not only mandate the inclusion of oxygen but also limit the content of several toxic materials like benzene, aromatics, and butadiene, and mandate volatility changes to minimize evaporative emissions [20].

13.4.7 Computer Control

Before about 1980 all automobiles controlled A/F, spark timing, and EGR rate with mechanical or pneumatic devices that sensed engine speed, throttle setting, manifold vacuum, and various temperatures. Starting in the 1980s, auto manufacturers switched to computer control, using the same kinds of microchips that are used in personal computers. With them, the controls can react to changes faster than the mechanical or pneumatic devices could and can use more complicated control algorithms. For example, the spark normally does not fire at the top of the piston travel (TDC) but rather some time before that ("spark advance"). The number of degrees of spark advance required is a function of engine speed and load. However, increasing the spark advance, which improves performance, also increases the probability of knock. In precomputer cars the spark advance was controlled mechanically, based on measurements of engine speed and intake manifold vacuum. Now some computer-controlled cars include a knock-meter (a microphone that senses knock, attached to the engine) allowing them to operate very close to the knock limit of spark advance and thus to improve fuel economy. This type of improved engine control opens the door both to improved fuel economy and to improved emission control.

13.4.8 Lean Burn

One serious drawback with the three-way catalyst is that it requires operating the engine at or near $\lambda = 1$. This reduces overall fuel economy by about 10 percent. In the early 1970s some auto manufacturers tried to solve their emissions problems by *lean burn,* an engine design based on operating almost always at $\lambda = 1.05$–1.1. This gave good fuel economy, but normally it put the engines near the NO_x peak on Fig. 13.2, which ultimately drove this kind of engine off the market. With the advent of computer controls it became possible to operate engines successfully closer to the limits of stable combustion than with the previous mechanical controls, so there was renewed interest in lean burn. The goal is to operate lean enough to get to the right side of the NO_x hump on Fig. 13.2 and thus reduce NO_x production enough to meet the emission standards (with EGR and a catalyst), while gaining the fuel economy advantage of operating at a high value of λ. In the 1990s considerable effort was expended to develop such engines, but researchers could never get the NO_x emissions low enough for U.S. standards. There is still some interest in them in Europe, where the NO_x standards are less stringent, and where the high fuel price (mostly taxes) makes the $\approx$ 10% fuel economy benefit seem worth pursuing.

13.5 TAMPERING AND EMISSION TESTING

Auto manufacturers go through rigorous certification testing of various engine and power train combinations at the EPA test facility, before the engine and power train combinations go into mass production. Manufacturers are required to show that in normal use with regularly scheduled maintenance, their engines and power trains will meet the emissions standards in Table 13.2 for five years or 50,000 miles.

Not all auto users maintain their vehicles properly. With many of the 1970 vehicles, removing the emission control devices improved fuel economy and "drive-ability." Although the law forbade such removal (called *tampering* in EPA literature), much of it occurred. Measured pollutant concentrations near highways or in tunnels showed that the average emissions from autos were substantially larger than those from properly maintained vehicles that had not been tampered with. This difference led to auto emission testing, in which states required autos to be regularly inspected to see that they had not been tampered with, and that their emissions were comparable to new car emissions (for their make and model year). Ample test data show that most autos have emissions close to the factory specifications, but a few have emissions many times factory specifications. This includes not only old junkers but also some fairly new cars. These high-emission vehicles contribute a disproportionate share of the total emissions. The main goal of emissions testing is to identify and fix these vehicles.

The average lifetime of an auto is about 10 years. Thus average vehicle emissions will always be greater than the emissions from the newest vehicles that are made to meet the current standards.

13.6 STORAGE AND TRANSFER EMISSIONS

A systems approach to the motor vehicle problem shows that we need to consider what happens from the time the oil comes out of the ground until it has all been burned. There are emissions in oil production, oil transport, oil refining, gasoline storage, gasoline transfer to the service station, gasoline transfer to the customer's auto, and then in the use by the auto. Some of these are discussed in Chapter 10, some in this chapter. As we control any one of these, the remaining ones become a larger part of the remaining problem.

13.7 ALTERNATIVE POWER PLANTS

Other types of IC engine have been used or proposed for use in automobiles. The most prominent of these are discussed in this section.

13.7.1 Diesel Engines

Table 13.5 shows the differences between conventional gasoline and diesel engines. In a diesel engine ignition is caused not by a spark but by spontaneous ignition of the fuel in contact with the hot air in the combustion chamber. That air is hotter than the corresponding air in the combustion chamber of an auto engine because of the diesel's much higher compression ratio (16–20 vs. 8–10 for a gasoline engine). When a gasoline engine is running at its maximum temperature, it will sometimes continue to run even after the ignition is turned off as a result of this same kind of ignition. Most auto engines have "anti-dieseling" devices in their fuel systems to prevent this.

In a conventional auto the power and speed of the engine are controlled by the position of the throttle plate in the carburetor (or air intake device of fuel-injected engines), which is directly linked to the accelerator on the floor. At idle (closed throttle plate) a small amount of air per revolution enters the engine and at wide-open throttle a much larger amount enters (the ratio for a typical engine between

TABLE 13.5
Comparison of common gasoline and diesel engines

	Conventional automobile engine	Conventional diesel engine
Cause of ignition	Electric spark	Spontaneous ignition of fuel by mixing with hot air
Compression ratio	8–10	16–20
Mass of air admitted to engine per revolution	Varies, low at idle, high at wide-open throttle	Practically constant
Air-fuel ratio	Practically constant, near stoichiometric	Highly variable, very lean except at maximum load
Means of controlling power output and speed	Restricting the air flow to the engine with throttle plate in carburetor or air inlet device	Varying the air-fuel ratio by varying the amount of fuel per injection
Placement of air-fuel mixture	In the whole combustion chamber by a carburetor or by injection into the inlet valve before it opens	In the center of the combustion chamber, away from the walls, by injection into the center of the combustion chamber
Untreated emissions of CO and HC	High	Low
Untreated emissions of NO_x	Low	Medium
Other problems	—	Particulates (black soot), odor, noise, harder to start than auto engine
Fuel economy	Good	Best of any type of combustion engine
Cost and weight for a given power output	—	Higher than a typical gasoline engine

maximum and minimum air flow is about 80:1). Diesel engines have no throttle; they put practically the same amount of air per revolution into the engine at idle and at full power. They control engine speed and power by varying the amount of fuel in each injection; at idle they use very little fuel, at full power much more.

Except at full power, diesel engines run very lean overall. In the center of the combustion chamber, where the fuel is sprayed, combustion is rich, but the flame is small and does not reach the cooled combustion chamber walls. At full power the flame is much larger, and the overall mixture is often rich enough to produce the black smoke seen when diesel vehicles operate at full power (e.g., starting from a dead stop or climbing hills).

Diesels also produce a kind of two-stage combustion, in which most of the combustion takes place in the fuel-rich center of the combustion chamber and then the final combustion takes place in the fuel-lean periphery. This leads to lower emissions of NO_x compared to one-stage combustion (see Chapter 12).

Diesel engines have a higher thermal efficiency than any other type of thermal power plant (with the possible exception of combined cycle gas-steam turbine power plants). They are used in many trucks, almost all railroad engines, many ships, and some small electric power plants. Some autos also use diesel engines. They are more expensive, heavier, harder to start, noisier, and smellier than comparable gasoline engines. Many believe that for countries like the United States, their superior fuel economy and potentially very low emission rates will make them the engine of the future for autos. The diesel optimists believe that problems of noise, cost, and soot formation are solvable.

13.7.2 Gasoline-Powered Two-Stroke Engines

These engines discharge exhaust gas and bring in inlet gas simultaneously. Their spark plugs fire every revolution. They do not have mechanical valves, but use ports, passages, and crankcase compression to move the inlet gas into the cylinder. They are used in some lawnmowers, chain saws, many portable power tools, some motorcycles, and many outboard boat motors. They are simpler, cheaper, smaller, and lighter for a given power output than typical automobile engines; but they are less fuel efficient and have much higher exhaust emissions. Early in the 1990s there was revived interest in them because some believed that with computer control and fuel injection, it would be possible to improve their fuel efficiency and control their exhaust emissions, while retaining their size, simplicity, cost, and weight advantages over four-stroke engines. At the end of the 1990s this interest has faded; solving the efficiency and emissions problems proved to be difficult.

13.7.3 Gas Turbine Engines

These engines do their compression and expansion with rapidly rotating (e.g., 20,000 RPM) compressors and turbines instead of the piston and cylinder arrangement in most IC engines. Their burners operate in steady flow instead of intermittently, as in other IC engines, and they use large amounts of excess air. They are used in jet aircraft, helicopters, small electric power installations, and a few trucks. Turbines have less weight for a given power output than a typical automobile engine, so they are the standard power plant in helicopters and, in modified form, jet aircraft. They have satisfactory fuel economy at full load but very poor fuel economy at part load; and they respond slowly to changes in throttle setting. These drawbacks have defeated all efforts to build a cost-competitive automotive gas turbine engine.

13.7.4 Electric Vehicles

Perhaps the best solution is to abandon the IC engine and use some other power plant. The problem is, what other power plant? The most common suggestion is the battery-powered auto. It produces negligible local air pollutant emissions (the emissions occur at the power plant that charges the batteries, which may be in a remote area, not at the vehicle in the downtown of a city). An electric auto is also quieter than the IC-powered auto. Unfortunately no one knows how to build one with a range and ease of refueling that are comparable to the IC engine. Remember that the IC engine takes its oxidizer from the air, typically 14 pounds of air per pound of fuel. The battery car carries both fuel and oxidizer with it. The extra weight is a major handicap for cars using traditional lead-acid batteries. Other types of batteries that use air (e.g., the zinc-air battery) have a lower weight per unit of energy than the lead-acid battery, but no known battery system comes close to the combination of gasoline tank and IC engine in (power · time)/weight. In addition, a typical gasoline auto can be refueled in a minute; recharging the batteries is much slower.

13.7.5 Hybrid Vehicles

One solution to the poor range of electric vehicles is to charge the batteries as you drive. This is done by *hybrid vehicles,* see Fig. 13.9.

One would not think the arrangement in Fig. 13.9 would have any advantages to offset its greater complexity compared to the common auto. But it does. Much of the cost, fuel consumption, and emissions of an ordinary auto are associated with

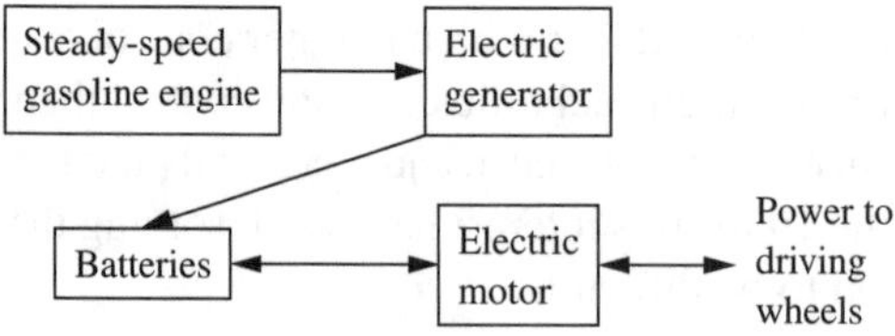

FIGURE 13.9
Simplified schematic of a hybrid vehicle. Batteries and an electric motor drive the auto. A small, steady-speed gasoline engine drives a generator that continually recharges the batteries. The arrows at the right are two-headed because the car is braked by having it drive the electric motor in reverse and thus charge the batteries. This is the "series" type hybrid vehicle. In the "parallel" type some power flows directly from the gasoline engine to the driving wheels.

the facts that the engine must be large enough to meet the maximum power demands (passing, hill climbing) and that the engine speed changes often and rapidly. In steady level highway driving only perhaps 10 percent of the engine's maximum power is used. A hybrid vehicle has a much smaller gasoline engine than the same-sized conventional auto, and it runs steadily whenever the auto is in use. That makes it much more efficient and makes its emissions much less than the larger, variable speed engine of a conventional auto. The battery system absorbs power in normal driving and provides bursts of power for passing or hill climbing. The size of battery required is much less than for an all-electric vehicle. This figure shows only one version of this vehicle; there are several others.

There are numerous buses and private autos now using the hybrid system in Europe and Japan, and plans to introduce such autos in the United States have been announced. The manufacturers claim that they have about twice the fuel economy and one-tenth the pollutant emissions of conventional vehicles of comparable size and performance [21].

13.7.6 Other Options

External combustion engines like steam engines, Stirling engines, etc. are regularly proposed for automotive use. Because they use steady combustors, do not compress the combustion gas, and do not need to lubricate and cool the walls of the combustion chamber, they can be operated with very low emissions (comparable to those of household water heaters and furnaces). However, their weight, cost, size, and poor fuel economy have prevented them from competing successfully with IC engines.

Fuel cells react fuel electrochemically with oxygen without burning it. They have a higher thermal efficiency than any fuel-burning engine. To date, they are only practical for hydrogen as a fuel, which is satisfactory for space travel, but not for autos. Serious efforts are underway to develop an automotive fuel cell powerplant.

California regulations require that 10 percent of all motor vehicles sold by any manufacturer in Los Angeles must be zero-emission vehicles by 2003. Most auto manufacturers plan to meet this requirement with battery-powered cars, with enough driving range between recharges for ordinary commuting but not for long trips. The world is waiting for the invention of a suitable substitute for the gasoline-powered IC engine.

13.8 REDUCING OUR DEPENDENCE ON MOTOR VEHICLES

The total daily emissions from automobiles in one metropolitan area are

$$\text{Daily emissions} = \left(\frac{\text{vehicle miles driven}}{\text{day}}\right)\left(\frac{\text{emissions}}{\text{vehicle mile}}\right) \tag{13.15}$$

The previous parts of this chapter have concerned the second term on the right. At some point it becomes easier and more cost-effective to work on the first term. Los Angeles, which has the most severe auto-related air pollution problems in the United

States, has forced employers to reduce this factor for their employees by mandatory ride pooling, van pooling, and whatever other measures they can take. Most of us are perfectly willing for other people to give up some comfort and convenience by sharing rides and using public transit (or walking or using bicycles or rollerblades). Deciding to give up our own personal comfort and convenience is less popular. In the future we may have to.

13.9 SUMMARY

1. Transportation sources contribute three-fourths of the CO and more than a third of the HC and NO_x emitted in the United States. Although any one vehicle emits little, the 193 million of them in the United States together emit enough to cause serious air pollution problems.
2. Almost all autos and light trucks use four-stroke gasoline engines. These produce HC and CO emissions by sometimes operating in an oxygen-deficient mode, by incomplete combustion caused by flame quenching at the walls of the combustion chamber, and by incomplete combustion due to kinetic limitations in the cooling gas. They produce NO mostly by the thermal mechanism.
3. Most modern U.S. autos use three-way catalysts with computer controls to meet their emission standards and fuel economy and performance requirements.
4. Modern U.S. autos have zero crankcase emissions. Their evaporative emissions are much less than for 1960 autos, but not quite zero.
5. Future emissions requirements may allow some other power plant to challenge the four-stroke gasoline engine. It has defeated all such challenges for the past 100 years.

PROBLEMS

See Common Values for Problems and Examples, inside the back cover.

13.1. A typical gasoline-powered lawnmower emits 59 g/h of hydrocarbons [22]. If the average vehicle speed of an auto (including both city and highway driving) is 25 mi/h, what is the ratio of (emissions per hour of a lawnmower)/(emission per hour of a 1999 auto)?

13.2. Estimate how much CO, HC, and NO_x are removed from the atmosphere per year by removing a 1960 auto from the road and junking it, assuming that it is replaced by a 1993 model. Assume that all autos travel 10 000 mi/yr.

13.3. Show the equation for the stoichiometric A/F of an HC in terms of the ratio y/x. Real gasolines always contain some O, N, and S, so this treatment, which assumes only C and H are present, is an approximation, generally a good one. Of the common components in gasoline, the lowest y/x ratio is 1.0, for benzene, and the highest is 2.5, for butane. In Examples 13.1–13.4 the ratio $y/x = 17/8 = 2.125$.

13.4. In areas with severe CO problems during the winter months, all the gasoline sold must contain at least 2.7 percent by weight oxygen (see Sec. 13.4.6). The gasoline companies are meeting this requirement by blending into their ordinary summer gasoline enough methanol (CH_3OH), ethanol (C_2H_5OH), or methyl *tert*-butyl ether, MTBE ($CH_3OC_4H_9$), to meet the required 2.7 percent by weight oxygen.

(*a*) If we assume that their ordinary summer gasoline is the equivalent of C_8H_{17}, and that they plan to meet this oxygen requirement by blending enough MTBE with their ordinary summer gasoline, what weight percent MTBE must there be in the final blend? Assume that the densities of both ordinary gasoline and MTBE are the same, about 0.75 g/cm^3.

(*b*) What is the stoichiometric A/F ratio (lb/lb) for this mixture?

13.5. (*a*) Repeat the previous problem for ethanol as the oxygen-bearing additive.

(*b*) Ethanol has a high affinity for water, which MTBE does not. Suggest what special handling procedures should be used to deal with this problem when ethanol is used as a gasoline additive.

13.6. In Example 13.3 we assumed that all the HC was combusted, and thus all the oxygen deficit appeared as CO. The calculated CO concentration was $(2\%/1.1\%) = 1.8$ times ≈ 2 times what one observes in Fig. 13.2. This suggests that the oxygen deficit is shared roughly equally by unconverted CO and unburned HC. To test this idea:

(*a*) Read the HC concentration for $\lambda = 0.95$ from Fig. 13.2.

(*b*) Write the equation equivalent to Eq. (13.6) taking into account the possibility that some HC is not combusted. Here show on the right a term $\alpha C_x H_y$, where α represents the mols of unburned HC.

(*c*) Then write the oxygen balance equivalent to Eq. (13.7), finding that it is one equation with two unknowns, z and α.

(*d*) Solve it for α, on the assumptions that z is about half the value calculated in Example 13.3 and that the unburned HC has the same composition as the fuel, C_8H_{17}.

(*e*) Compare that result to the one found in part (*a*).

13.7. How much gasoline is inserted into the combustion chamber of an auto engine for each combustion? Assume 2000 RPM, 60 mi/h, 25 mi/gal, 4-cylinder engine.

13.8. The temperature calculations in Example 13.5 ignore the energy input necessary to vaporize the fuel. Estimate the decrease in mixture temperature if liquid gasoline and air are mixed at 68°F, at $\lambda = 1$; assume all the gasoline vaporizes. This problem is explored for a variety of fuels in Ref. 23. The latent heat of vaporization of gasoline is approximately 140 Btu/lb.

13.9. Why is the first part of the gas to burn the hottest? It undergoes combustion heating followed by compression, whereas the last part to burn undergoes compression followed by heating. The temperature increase due to heating is given by Eq. (13.12). That due to compression is given approximately by the equation for isentropic compression for an ideal gas,

$$\frac{T_2}{T_1} = \left(\frac{P_2}{P_1}\right)^{R/C_P}$$

Applying these equations, show that heating followed by compression and compression followed by heating lead to different final temperatures. Heat transfer complicates the picture, but the difference illustrated here is the main reason for the result. Experimental results show that this difference can be up to 400 K [24]. Is that plausible in terms of the calculations in this problem?

13.10. The heat of combustion used in Example 13.5 is taken from standard tables, which are almost always the values at 25°C = 298 K. In that example the combustion begins at 641 K, so we should use the heat of combustion at that temperature. Estimate how large an error is made by using the standard value. The method of computing this change is shown in all chemical engineering textbooks on heat and material balances.

13.11. (*a*) Show the relationship between $\Delta u_{\text{combustion}}$ and $\Delta h_{\text{combustion}}$ for reactions involving only perfect gases.

(*b*) For a combustion reaction in which the number of mols does not change, show that these are identical.

(*c*) For the reaction shown in Eq. (13.1) compute the change in number of mols, and then the difference between $\Delta u_{\text{combustion}}$ and $\Delta h_{\text{combustion}}$ based on your results from part (*a*).

13.12. (*a*) Using the methods in Example 13.5, estimate the gas temperature of the exhaust gas (equivalent to 90° crank angle after TDC).

(*b*) The observed temperatures are generally lower than this. They are much lower at idle than at full power even though both of those conditions have roughly the same A/F ratio. Why?

13.13. The current Clean Air Act [25] refers to HC as NMOG. What is that likely an acronym for?

13.14. Do the NO equilibrium values in Figure 13.3 really agree with the data presented in Chapter 12? There we always knew the O_2 concentration because we specified it; here we do not. If we had complete combustion, it would be zero! As best the author can read that graph, the matched readings are as follows: (NO mol %: T K) 0.493: 2640; 0.4: 2560; 0.3: 2470; 0.2: 2340. Based on these readings, estimate the O_2 concentration that has been assumed in making up that plot. Is it a constant? Is it plausible? For this problem the equilibrium constant for Eq. (12.5) may be represented by $K_{12.5} = 21.9 \exp(-21\ 842/T)$, with T expressed in K.

13.15. Figure 13.3 shows the calculated NO concentration rising to a peak and then declining before leveling off in its "frozen" condition. Figure 12.7, based on the results of a similar calculation, shows no such decline from the peak concentration. Explain this.

13.16. The charcoal canister in a typical 1993 auto contains 700 to 800 g of charcoal and can hold roughly 0.3 g HC/g charcoal.

(*a*) If a typical gasoline tank filling is 12 gallons and the vapor in the tank before filling is in equilibrium with liquid gasoline at 100°F, how much HC will be in the displaced vapor? Assume that the gasoline has the same vapor pressure as Eq. (10.15).

(*b*) Must the charcoal canister be enlarged if it must accommodate this amount of HC? By how much?

13.17. Assume that the typical charcoal canister (see Problem 13.16) is a cylinder with height = 1.5 diameter and that the charcoal pieces are spheres with diameter $\frac{1}{8}$ inch. The bulk density of the charcoal is 30 lb/ft^3 and the external porosity $\varepsilon \approx 0.3$.

(*a*) Estimate their pressure drop–volumetric flow rate relationship. See any fluid mechanics book for data on flow through porous media.

(*b*) Estimate the pressure drop if a charcoal canister is used to capture the vapors from gas tank fueling, which occurs at roughly 10 gal/min.

13.18. Most of the gasoline moving from the gasoline tank to the engine, and through the other parts of the fuel system, is contained in metal tubing, which is impervious to gasoline. But because of the vibrations of the engine relative to the auto body, some of the connections must be made of a flexible hose, which is a very high-tech version of an ordinary flexible garden hose. The "ordinary" hose has a gasoline permeation rate of 100 g/(m$^2\cdot$day), whereas a technically advanced hose has a permeation rate of less than 10 g/(m$^2\cdot$day) [26].

(*a*) If we replace the ordinary hose with the technically advanced hose, by how much will the HC emissions of an ordinary auto be reduced? Assume that the total length of such hose in an auto is 1 m and that its diameter is 2 cm.

(*b*) If all the autos in the United States have the "ordinary" hose, what fraction of the total HC emissions in the United States (Table 13.1) is due to permeation through the hoses?

13.19. Explain the dome-shaped curve for NO_x in Fig. 13.2. *Hint:* sketch plots of T and of mol fraction oxygen in the combustion gases, both as a function of λ. Then apply the ideas about thermal NO formation from Chapter 12.

13.20. In Fig. 13.3 the peak temperature is reached before the end of the combustion period. Explain why.

13.21. Show how the 98 percent value in Example 13.5 is calculated. Assume sinusoidal movement of the piston. Comment on why the true movement is not exactly sinusoidal. How much difference is that likely to make in this problem?

13.22. How much does the platinum or palladium and rhodium in a typical automotive catalyst cost? Typically there are about 1.5 grams of precious metal. Precious metal prices fluctuate, but in 1999 they are in the order of \$400/troy ounce. (See the front papers for conversion factors.) Is it likely to be economic to salvage and reprocess catalytic converters? Is it done?

13.23. Why do autos need exhaust mufflers? *Hint:* calculate the combustion chamber pressure at the beginning of the exhaust stroke for the conditions in Example 13.5. Assume that all the gases behave as ideal gases with a constant $C_P = 3.5\ R$.

13.24. Estimate the rate of cooling (K/s) during the power stroke in Example 13.5.

REFERENCES

1. Monaghan, M. L.: "Introduction," in C. Arcoumanis (ed.), *Internal Combustion Engines,* Academic Press, London, p. 6, 1988.
2. *Statistical Abstract of the United States 1998,* 118th ed., U.S. Department of Commerce, Bureau of the Census, Washington, DC, 1998.
3. *National Air Quality and Emissions Trends Report, 1997,* EPA 454/R-98-016.
4. Chambers, L. A.: "Classification and Extent of Air Pollution Problems," in A. C. Stern (ed.), *Air Pollution,* Vol. I, 3d ed., Academic Press, New York, p. 11, 1976.
5. Clean Air Amendments of 1990, PL 101-549, Sections 203, "Emission Standards for Conventional Motor Vehicles," and 243, "Standards for Light-Duty Clean Fuel Vehicles."
6. "Supplement A to Compilation of Air Pollutant Emission Factors, Volume II: Mobile Sources," U.S. Environmental Protection Agency Report AP-42 Supplement A, A-2, U.S. EPA, Ann Arbor, MI, January 1991.
7. Klein, M., "The Diesel Revolution," *American Heritage of Invention and Technology,* Vol. 6, Issue 3, pp. 16–22, 1991.
8. Arcoumanis, C., ed.: *Internal Combustion Engines,* Academic Press, London, 1988.
9. Taylor, C. F.: *The Internal-Combustion Engine in Theory and Practice,* Vol. I, rev. ed., The MIT Press, Cambridge, MA, p. 56, 1985.
10. Schaefer, F., and R. van Basshuysen: *Reduced Emissions and Fuel Consumption in Automobile Engines,* Springer-Verlag, New York, and SAE International, 1995.
11. Campbell, I. M.: *Energy and the Atmosphere,* 2d ed., John Wiley & Sons, New York, p. 70, 1986.
12. Ramos, J. I.: *Internal Combustion Engine Modeling,* Hemisphere, New York, p. 290, 1989.
13. Annand, W. J. D.: "Gasoline Engines," in C. Arcoumanis (ed.), *Internal Combustion Engines,* Academic Press, London, p. 69, 1988.
14. Taylor, C. F.: op. cit., Vol. I, rev. ed., p. 512.
15. Patterson, D. J., and N. A. Henein: *Emissions from Combustion Engines and Their Control,* Ann Arbor Science, Ann Arbor, MI, p. 197, 1972.
16. Ibid., p. 181.
17. Morgan, C. R., and S. S. Hetrick: "The Effects of Engine Variables and Exhaust Gas Recirculation on Emissions, Fuel Economy and Knock—Part II," *Trans. Soc. Automotive Engineers,* Vol. 85, pp. 893–900, 1976.

18. Thomas, A.: "Automotive Fuels," in C. Arcoumanis (ed.), *Internal Combustion Engines,* Academic Press, London, p. 228, 1988.
19. Clean Air Amendments of 1990, PL 101-549, Section 219, "Reformulated Gasoline and Oxygenated Gasoline."
20. "Reformulated Gasoline," *40CFR80.40.*
21. Toyota press release, "Toyota to Introduce Prius in Overseas Markets," (www.toyota.co.jp) July 14, 1998.
22. *EPRI Journ.*, Vol. 21, Issue 2, p. 18, March/April 1996.
23. Taylor, C. F.: op. cit., p. 158.
24. Lavoie, G. A., J. B. Heywood, and J. C. Keck: "Experimental and Theoretical Study of Nitric Oxide Formation in Internal Combustion Engines," MIT Fluid Mechanics Laboratory, Publication No. 69-10, Cambridge, MA, 1969.
25. Clean Air Amendments of 1990, PL 101-549, Section 202(a)(6), "Control of Vehicle Refueling Emissions."
26. Anon., DuPont F2000 Automotive Fuel Line Hose, DuPont Automotive Publication H-63433.

CHAPTER 14

AIR POLLUTANTS AND GLOBAL CLIMATE

Humans and other living things can make major local and global changes in the earth. For example, as far as we know, all the free oxygen in the atmosphere was put there by green plants using Eq. (14.1),

$$n\text{CO}_2 + n\text{H}_2\text{O} + \text{light} \rightarrow (\text{CH}_2\text{O})_n + n\text{O}_2 \tag{14.1}$$

All of the fossil fuels (coal, oil, natural gas) in the world were produced by living things. Human beings have produced deserts by grazing livestock in numbers greater than the normal rainfall would support: the livestock simply ate away the native plants. Much of the world's farmlands were originally forests that our remote ancestors cleared. Humans have produced areas of nearly sterile soil near industrial plants in the days before air pollution control; some of those areas, devastated by exposure to high concentrations of sulfur oxides, have not yet been recolonized by plants, decades after we stopped polluting them. So it is clearly within our powers to change the surface of the earth and presumably to modify the regional and global climate.

In this chapter we consider three air pollution problems in which humans may be making large-scale changes in our planet. These pose a severe political challenge. Air pollution laws in the United States and most other countries (see Chapter 3) are based on the assumption that air pollution is a local matter. The smoky or stinky or potentially toxic factory is a nuisance or a hazard to its neighbors, who can go to the local government and ask it to clean up or close the factory. The local government has to balance the interests of the offended public with those of the factory owners and workers. These officials face all three groups at the next election. If one believes in the democratic process, then one should believe that the problem will be solved correctly. Local solutions are not available for global problems or for problems of pollutants like acid rain that cross international boundaries. No international elections can be held to settle such problems. Our political systems are responding slowly to this challenge.

In addition, we are concerned by the long lifetimes of some of the potentially climate-modifying chemicals we emit. If we decide to reduce our emissions of them, the amount already emitted and stored in the atmosphere may still cause serious problems. Some parts of the global climate *overshoot*; they continue to change in the direction they are changing even after the cause of the change has been reduced or withdrawn. Existing political and regulatory systems are not good at dealing with long-term consequences of current actions nor with systems with overshoot.

14.1 GLOBAL WARMING

Humans are putting gaseous materials into the atmosphere that may cause the earth's average temperature to rise. This is called *global warming*, or *the greenhouse effect.*

Example 14.1. Estimate the average temperature that the earth would have if it had no atmosphere.

The total radiant energy flux from the sun, just outside the earth's atmosphere, is 1.353 kW/m^2 (429 Btu/h $\cdot$ ft^2). The diameter of the earth is 12.75×10^6 m so that, if all the incoming solar energy were absorbed by the earth, the total heat flow in from the sun would be

$$\text{Total heat flow in from the sun} = \frac{\pi}{4} D^2 \cdot \text{flux} = \frac{\pi}{4} (12.75 \times 10^6 \text{ m})^2 \cdot 1.353 \frac{\text{kW}}{\text{m}^2}$$

$$= 1.73 \times 10^{14} \text{ kW} = 1.64 \times 10^{14} \frac{\text{Btu}}{\text{s}}$$

The total heat radiated to outer space would be this amount plus the amount produced on earth by nuclear decay and tidal friction with the moon, which together are less than 0.1 percent of the solar energy inflow and can be safely ignored. The outward radiation (assuming a zero temperature for outer space and blackbody radiation), using the surface area of the earth rather than the projected area, is

$$\text{Total heat flow out} = \pi D^2 \sigma T^4 = \pi (12.75 \times 10^6 \text{ m})^2 \cdot 5.672 \times 10^{-11} \frac{\text{kW}}{\text{m}^2\text{K}^4} T^4$$

where σ = the Stefan-Boltzmann constant = 5.672×10^{-11} kW/(m$^2 \cdot$ K^4). Setting these equal and solving for T, we find 278 K = 5°C = 41°F. ■

This is approximately 10°C, or 18°F, below the observed average surface temperature of the earth, which is about 15°C or 59°F. Thus, the net effect of having an atmosphere is to raise the average temperature of the earth about 10°C (= 18°F) above the value it would have with no atmosphere, if the earth absorbed all incoming sunlight. This effect is even more impressive when we consider that not all the incoming solar radiation is absorbed. The earth reflects roughly 30 percent of all the incoming solar radiation back to outer space from the tops of clouds, icy surfaces, oceans, etc. (Technically, the earth's *albedo* is about 0.3.) The moon, which has no atmosphere and hence no clouds, surface water, or ice sheets, reflects about 12 percent of its incoming solar radiation. (If it absorbed it all and reflected none, we would see it as a totally black circle against the starry night sky! If it reflected the same percentage as the earth, it would be 2.5 times as bright as it is.) If the atmosphere let the same amount of sunlight in as it actually does but did not prevent the outward flow of radiant heat, then we should multiply the incoming solar radiation in Example 14.1 by 0.7, finding an average surface temperature of 254 K = −19°C = −2°F, and a frozen world.

The fact that the observed average world temperature is higher than the value in Example 14.1 demonstrates that the atmosphere must block a higher proportion of the outgoing radiation than it does of the incoming radiation.

Example 14.2. What fraction of the outgoing radiation from the earth is blocked by the atmosphere?

As just discussed, we assume that 30 percent of the incoming solar radiation is reflected away, and use an average surface temperature over the whole planet of approximately 15°C = 59°F = 288.15 K. Then setting incoming approximately equal to outgoing and solving for the fraction emitted, we have

$$\text{Fraction emitted} \approx \frac{0.7(\text{total solar input})}{\pi D^2 \sigma T^4}$$

$$= \frac{0.7(1.73 \times 10^{14})\ \text{kW}}{\pi (12.75 \times 10^6\ \text{m})^2 \cdot 5.672 \times 10^{-11}\ \dfrac{\text{kW}}{\text{m}^2\text{K}^4}(288.15\ \text{K})^4}$$

$$= 0.606$$

■

We see that for the earth's surface temperature to average about 15°C = 59°F, the atmospheric outward transmission of radiant energy must be (0.606/0.7), or 86 percent of the inward transmission of solar energy. We also see that if something changes this ratio, then the earth will balance these energy flows by changing the average surface temperature. The possibility that humans may be doing one or more things to change that ratio is the cause of our concern with global warming.

Clouds block radiation, both inbound and outbound. (Cloudy days are cool and cloudy nights are warm relative to clear days and nights at the same season.) They are more or less equal in their resistance to incoming and outgoing radiation. The same is not true for clear air, which contains CO_2, H_2O, CH_4, and some other gases that can absorb radiant energy. If the wavelengths of the incoming and outgoing radiant energies were the same, then these gases would block equal amounts in both directions. But the wavelengths are quite different.

Figure 14.1 on page 514 shows the absorptive properties of the clear atmosphere (without clouds, dust, birds, insects) and some properties of the incoming solar radiation and the outgoing thermal radiation from the earth. The upper part

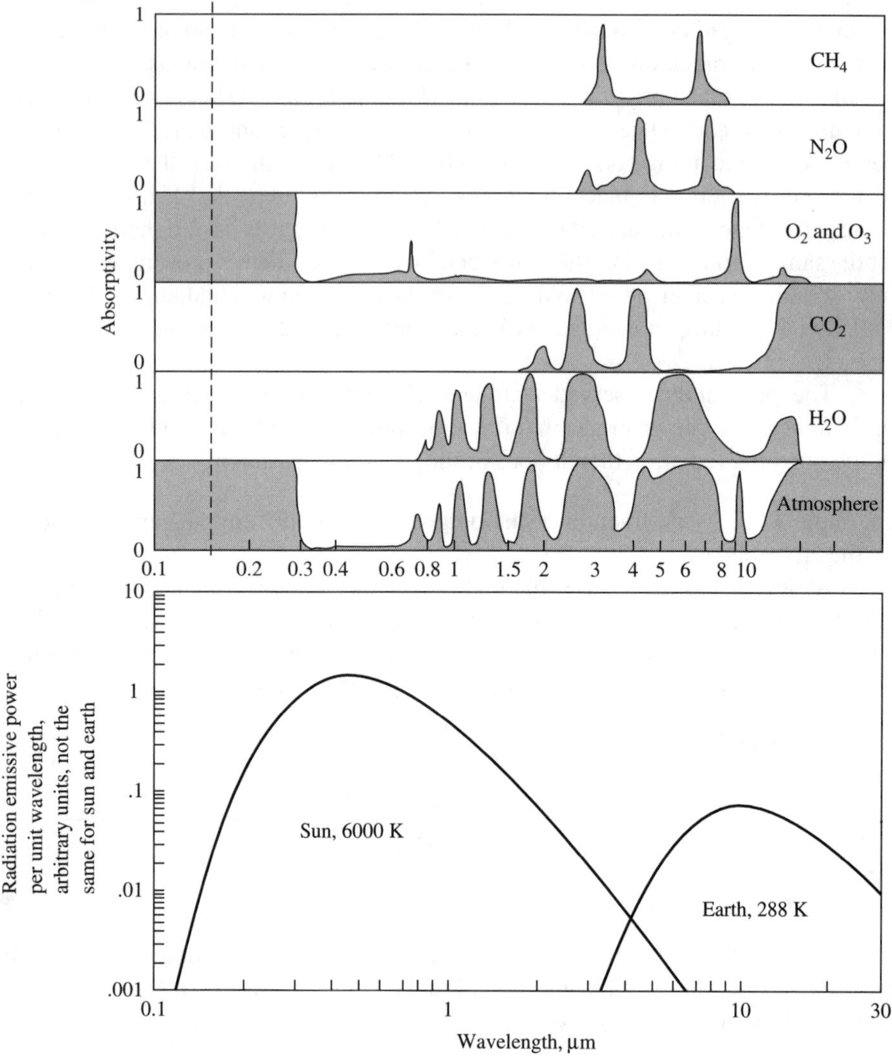

FIGURE 14.1
Absorptive properties of the atmosphere as a function of wavelength, and approximate emission spectra (see Problem 14.1) of sun and earth. The absorptive properties are after Oke [1]. At any one wavelength, the absorptive properties are the same for incoming and outgoing radiation.

of this figure shows the contribution to the absorptive properties of the atmosphere of CH_4, N_2O, O_2 and O_3, CO_2, and H_2O. The section labeled "Atmosphere" is the sum of the five sections above it. As a simple illustration of this part of the figure, if we sketched on this section the absorption curve for a sheet of blue-colored glass, which only allows blue light to pass through, the glass would show an absorptivity of 1 for all wavelengths except that of blue light ($\approx 0.5\ \mu$) and a "window" of low absorptivity for that wavelength. Applying the same logic to Fig. 14.1, we see that O_2 and O_3 block all light with wavelengths less than about 0.28 μ and that CO_2 blocks all light with wavelengths more than about 15 μ. The combined absorption spectrum for the whole atmosphere is largely transparent in the wavelength range from 0.3 μ to 0.7 μ, and has various "windows" at other wavelengths, of which the most important is the two-part window between about 8 μ and 12 μ.

The interaction of a photon with a gas molecule is quite different from that with a cloud droplet or with a fine particle (discussed later). A gas molecule will absorb a light photon if the gas molecule can make an internal rearrangement that requires the same amount of energy as that carried by the photon. One may think of this as a "tuning" or "resonance" phenomenon. (Your radio responds only to the discrete frequency to which it is tuned.) For wavelengths shorter than about 0.28 μ, the internal transitions involve shifts of electrons in their orbitals around the nuclei of one or more of the atoms that make up the molecule, but not any change in the relation of one atom to another within the molecule. For the wavelengths longer than about 1 μ the changes are not within the individual atoms but are those associated with the vibrations of the various atoms in the molecule, relative to each other. In the 0.28 to 1 μ window, the photons have too little energy to cause shifts of electron orbitals, and too much energy to be in tune with intramolecular vibrations (for the molecules in the air). The H_2O absorption peaks shown on Fig. 14.1 are caused by the various intramolecular vibration modes of the water molecule (three fundamental vibrations, plus overtones).

The lower part of the figure shows the distribution of energy in sunlight and in the infrared radiation from the earth. These are idealized values for blackbody radiators at 6000 and 288 K, which correspond roughly to the average surface temperatures of sun and earth. The real spectra are more complex, but these simplified spectra are close to correct. The quantity plotted is the fraction of the total emitted energy per micron of wavelength, which has a higher maximum (140 percent per micron) for the sun than for the earth (7 percent per micron) because the sun's spectrum is narrower. (Observe the logarithmic scale for wavelength.)

This lower section shows that radiant energy is distributed over a range of wavelengths, which is narrower for hotter bodies. Wein's law for blackbody radiation is

$$\begin{pmatrix}\text{Wavelength of} \\ \text{maximum emission}\end{pmatrix} = \lambda_{\text{max}} = \frac{2.987 \times 10^3\ \mu\text{m} \cdot \text{K}}{T} \tag{14.2}$$

which shows that for the temperature of the sun's surface, about 6000 K, the peak intensity is at 0.50 μ, corresponding to visible light. For the earth's surface temperature of about 288 K the peak intensity is 10.3 μ, which is in the infrared region, not visible to our eyes. (We do not see the earth glowing in the dark, but if we had infrared-sensing eyes we would.) The sun's energy comes to the earth mostly as visible light; the earth sends energy out mostly as infrared radiation.

Comparing the lower and upper parts of the diagram, we see that sunlight comes to the surface practically unimpeded except for cloudy areas, whereas the peak radiation from the earth is close to the 8- to 12-μ window, which is not as wide nor as completely open as the window for solar energy. This is the main reason that the atmosphere is less transparent for outgoing infrared energy than it is for incoming solar energy.

Figure 14.1 is made up for the current concentrations in the atmosphere of the gases shown. It shows that CO_2, CH_4, N_2O, and H_2O all have some absorption in the 8- to 12-μ window. (The same is also true for *chlorofluorocarbons*, or CFCs, not shown on this figure but discussed later in this chapter.) If we were to increase the concentration of any or all of these gases in the atmosphere, the 8- to 12-μ window on Fig. 14.1 would become less transparent, thus making the atmosphere's ratio of outgoing to incoming transparency decline. In turn, the average temperature of the earth would rise, thus producing the so-called greenhouse effect. (This is technically a poor name; greenhouses work mostly by cutting off wind and air circulation while letting in sunlight [2]. However, this name is in common usage and will be used here. The group consisting of CO_2, CH_4, N_2O, and CFCs are collectively called *greenhouse gases*.)

Human activities are increasing the concentrations of greenhouse gases in the atmosphere. Of these gases, the strongest contributor to reducing the transparency of the 8- to 12-μ window is water vapor. However, humans do not directly influence its concentration in the atmosphere, and it is not normally a part of the discussion of the greenhouse effect. Figure 14.2 shows a very simplified view of the interactions

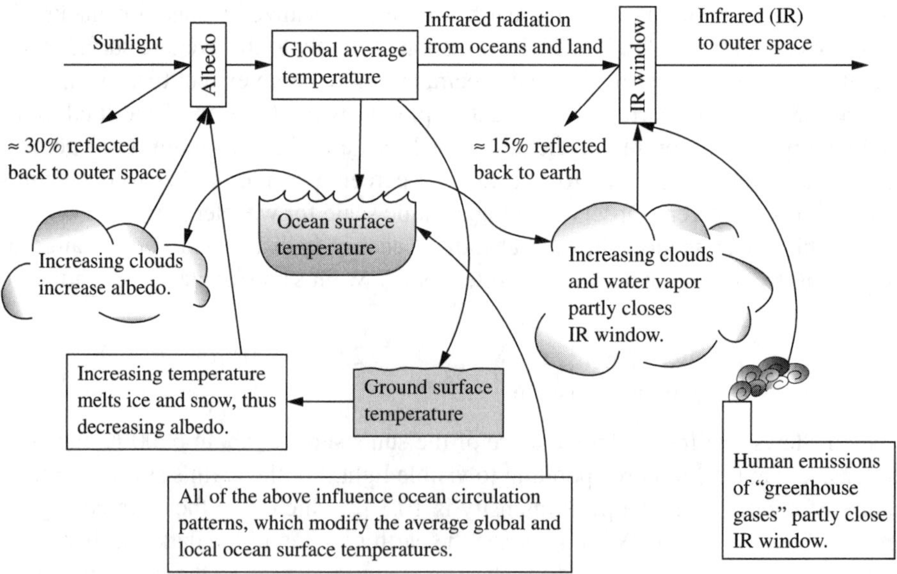

FIGURE 14.2
Simplified view of the interactions and feedback loops involved in the global warming problem. These are believed to be the major effects; there are certainly others.

and feedback loops involved in the global temperature. The previous examples have only considered the top row of the figure—sunlight in, infrared (IR) out. Increasing the global temperature by adding greenhouse gases will have positive and negative effects on the albedo by increasing cloudiness and reducing snow and ice cover, and will further close the IR window by increasing the average water vapor content of the atmosphere and the average cloudiness. Best current estimates suggest that the overall effect of all clouds is to slightly cool the earth. Perhaps the most uncertain and frightening consequences are those of changing ocean circulation patterns, like the Gulf Stream, which keeps Europe from freezing, and the El Niño currents in the central Pacific Ocean, which have very widespread weather consequences.

The previous discussion and this model are all strong simplifications of the true complexity of the earth's energy balance. They ignore the mixing and energy transport within the atmosphere and the adsorption of IR by greenhouse gases and then its partial reradiation back to the surface and its partial radiation to outer space. More detailed accounts, which do not hide this complexity, are available [3, 4].

Figure 14.3 shows the calculated relative contributions of the various greenhouse gases to the reduction of transparency of the atmosphere in the 8- to 12-μ window for the period 1980 to 1990 [5]. We see that CO_2 contributed more than half, followed by the CFCs, methane, and N_2O.

If the greenhouse effect causes the earth's mean temperature to rise even slightly, climatic changes will result. The form these changes will take is unknown. Large-scale computer modeling of the atmosphere *suggests* what will happen, but the calculations are still considered somewhat speculative. If rising temperatures were to melt the ice cap in Antarctica, the world sea level would rise several hundred feet, flooding most of the coastal cities and agricultural areas of the world. Temperature increases much smaller than those needed to melt the ice caps would cause the deserts and the temperate zones to extend farther from the equator. Agricultural

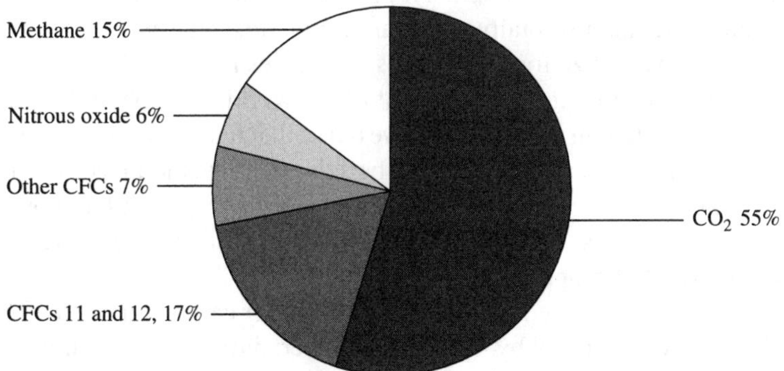

FIGURE 14.3
The contribution of human-caused greenhouse gases to the change in transparency of the 8- to 12-μ window of the atmosphere for the period 1980–1990. Tropospheric ozone may also play a role but its magnitude is uncertain [5].

areas that are currently highly productive would become dryer and hotter, while sub-Arctic regions would become warmer and wetter. Good current discussions are given in Refs. 5–7.

The best current estimates are that, for a "business as usual" projection of future emissions of all greenhouse gases, the global mean temperature will increase by 0.2° to 0.5°C (best estimate 0.3°C) per decade for the next century [5]. This change in global temperature is more rapid than any other that has occurred in the past 10 000 years (since the end of the last ice age, which was apparently quite sudden compared to the speed of other geologic events). The corresponding projection of world sea level is for a rise of 3 to 10 cm/decade (best estimate 6 cm/decade) over the same period.

The greenhouse problem has a strong overshoot. The oceans have several hundred times the heat storage capacity of the atmosphere. As greenhouse warming raises the temperature of the atmosphere, at first the cooler oceans will remove heat, slowing the rate of atmospheric temperature increase. However, as the surface layers of the ocean become warmer, that cooling effect will decline, thus producing a temperature overshoot in the atmosphere. This overshoot is likely to have a time scale of several hundred years at current emission rates [5]. Thus if we were to take steps that guaranteed that the composition of the atmosphere remained at its current state, there would still be a significant atmospheric temperature increase in the next few decades due to this overshoot.

14.1.1 Carbon Dioxide

Carbon dioxide (CO_2) is a colorless, tasteless gas that provides the "carbonation" in soft drinks and sparkling wines. It has been part of the earth's atmosphere as long as the earth has had an atmosphere. The current carbon dioxide concentration in the world atmosphere is approximately 360 ppm. At that concentration it has no known harmful effects to humans, and it is totally necessary for photosynthesis, Eq. (14.1). Our bodies produce it as we utilize foods; it leaves our bodies in exhaled breath.

Geologic records show that the CO_2 content of the world atmosphere before about A.D. 1750 was 280 ± 10 ppm and did not move out of that range for hundreds or thousands of years. About 1750 humans began to burn increasing amounts of fossil fuels, and the CO_2 content of the global atmosphere has risen. Figure 14.4 shows CO_2 concentrations from the past 30 years. During that period, the annual increase in CO_2 concentration was $\approx$ 1.5 ppm/yr.

Figure 14.5 shows the estimated reservoirs and flows for carbon on earth. (To convert from carbon to CO_2 multiply by 44/12). Our uncertainty in the magnitudes of the natural flows is probably greater than the magnitude of the man-made flows, but the natural flows are apparently in balance so that the increase in the atmosphere shown in Figure 14.4 is apparently due to man-made emissions.

On a geological time scale, such changes in atmospheric CO_2 content are unimportant. Over geologic time the atmosphere's CO_2 content has changed, and

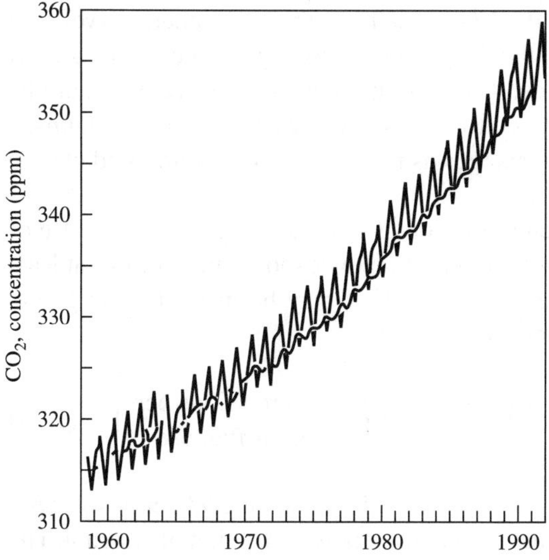

FIGURE 14.4
Recent history of the CO_2 content of the global atmosphere. The oscillating curve is from the top of Mauna Kea (13 796 ft), far from any human-caused source of CO_2 emissions. The annual oscillation of $\approx$ 6 ppm is caused by the large uptake of CO_2 by vegetation in the late spring and early summer, which produces an annual minimum. The much more regular curve is from Antarctica, which shows less seasonal variation, and generally lags on the Mauna Kea values by about 2 ppm, because most of the combustion of fossil fuels occurs in the Northern Hemisphere. (From [4], courtesy of The Cambridge University Press.)

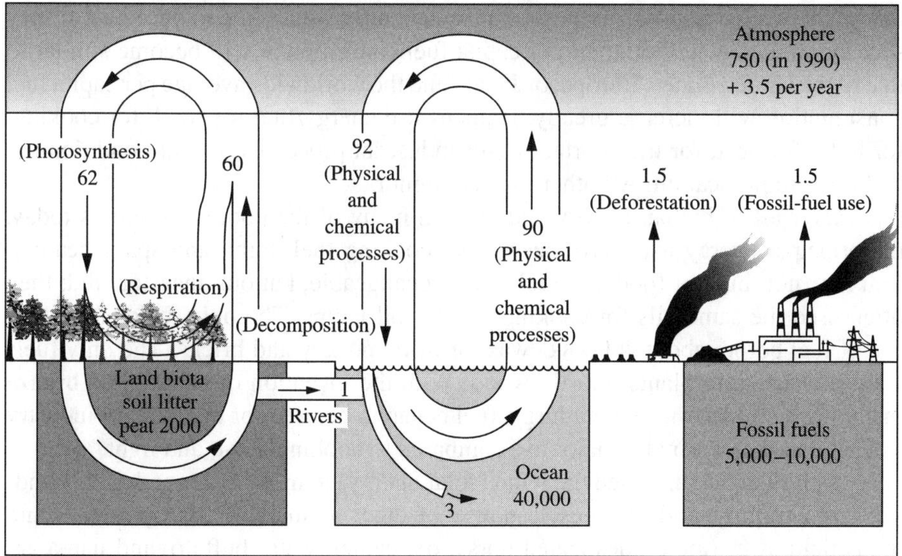

FIGURE 14.5
Estimated global flows and reservoirs of carbon (in various chemical combinations). The flow units are 10^9 metric tons (Gt)/yr; the reservoir units are Gt. We see that the oceans contain $\approx$ (40 000/750) $\approx$ 53 times as much as the atmosphere, and 4 to 8 times as much as all the fossil fuels. (From [4], courtesy of The Cambridge University Press.)

will change again. To some extent, the CO_2 content of the atmosphere serves as a global temperature regulator; when the CO_2 content rises, geologic forces are set into motion that cause it to be reduced [8]. But on a human time scale it could be a disaster. Our remote ancestors adapted to the last ice ages, but since then humans have never had to deal with climate changes as rapid as those that are predicted to occur as a result of the greenhouse effect.

If, as seems likely, increased emissions of CO_2 are the largest single cause of global warming, then our only control option is to reduce those emissions or at least reduce the rate of increase of those emissions. That may be difficult. The global annual fuel combustion CO_2 emissions are

$$\begin{pmatrix}\text{Global fuel}\\ \text{combustion}\\ \text{CO}_2\text{ emission}\end{pmatrix} = \begin{pmatrix}\text{global}\\ \text{population}\end{pmatrix}\begin{pmatrix}\text{per capita}\\ \text{fuel use}\end{pmatrix}\begin{pmatrix}\text{CO}_2\text{ emissions per}\\ \text{unit of fuel use}\end{pmatrix} \tag{14.3}$$

The first term, the global population, is growing at about 1.4 percent per year (population doubles every 50 years), and that growth rate shows little sign of slowing. The second term is highly variable from country to country. It is highest in the United States (where fuel has traditionally been cheap); next highest in Europe and Japan, where the standards of living are comparable to that in the United States, but governments have intentionally taxed fuels to keep the price relatively high to encourage fuel economy; and lowest in the Third World countries, which have much lower material standards of living than ours.

Those parts of the human population that do not have a standard of living like that in the United States, Europe, or Japan generally would like to have one. If they succeed in that goal, then their per capita fuel consumption will become similar to that of the United States, Europe, or Japan, and the worldwide average per capita fuel consumption will increase greatly. Humans use energy/fuel for food, for cooking, for light, for heat, for transportation, for industrial processes, for air conditioning, and for communication. All other uses are minor.

Over all of human history and even in many of the poorest countries today, the principal energy uses have been food, cooking fuel, light, and space heating. You may not think of food and fuel as interchangeable, but our ancestors did; they often used the same oils for cooking and for oil lamps. The only major industrial use of fuel before about 4000 B.C. was for firing pottery and bricks. The only fuels were derived from plants, mostly wood. With the invention of copper and bronze metallurgy, and later iron metallurgy, fuel began to be used for that kind of industry as well, but that was not a major use compared to cooking, heat, and light.

Although humans used the wind (solar energy) to move boats and drive windmills in antiquity, and the muscle power of other animals for transportation and industrial power (the domesticated horse, ox, camel, water buffalo, and llama are devices for converting plant materials to mechanical power), the first use of fossil fuels for driving machines was in 1776 with Watt's first general-purpose steam engine. The first use for transportation was in 1825 when Trethivic put a small, high-speed steam engine on a small railroad car of a horse-drawn railroad. The result was the complete revolution in our ability to produce things and to move about. In the period from the domestication of the horse about 2000 B.C. to A.D. 1825, the fastest that humans could travel was the speed of a horse, about 30 miles an hour over a one-mile course, about 3–4 miles an hour going all day. From 1825 to 1900 that speed increased to 100 miles an hour all day, in fast trains. In the period from 1900 to 1990 that increased to 18 000 miles an hour for astronauts, and 600 miles an hour in commercial airlines. The change in personal mobility caused by the auto has been even more spectacular. We now think little of going 300 miles each way for a weekend outing; it takes 5 or 6 hours each way. Our great-grandparents would have needed 8 to 10 days each way for such a trip.

Humans first began to make consistent efforts to store food under refrigeration about 1860, and the effort became large-scale about 1920. As a result, we have a much more varied, tasty, healthful, and bacteriologically safe diet than our grandparents did. Air conditioning first appeared about 1900. It became available on a large scale about 1950. Since then people in the United States have spent enormous amounts of fuel cooling their homes, cars, and offices in the summer. We first had significant use of home clothes dryers about 1950. All of these appliances have made our lives more comfortable, and increased our use of fossil fuels.

To compare energy uses, we need a proper standard energy unit. The most intuitive unit is the minimal energy intake, as food, that a normal human needs, about 4 million BTU per year ($\approx$ 2750 kcal/day; the "calorie" in diet books is the kcal). That is the minimum amount of energy as fuel—food—that a "standard" human needs to live an active life for a year. Using it, we can make Table 14.1.

In the United States we use a total of about 79 times as much fuel as the minimum needed to feed ourselves. About a quarter of it goes for transportation (autos, trucks, trains, airplanes, ships), an eighth to heat houses, almost a quarter for all industry, and about a third for electricity. Of that electricity, two-thirds goes for light, heat, and air conditioning, the rest for industrial uses.

TABLE 14.1
Average U.S. per capita annual fuel usage, 1988

Food	1
Transportation	22
Residential and commercial space heat	10
Industry	18
Electricity (64% residential and commercial, 36% industrial)	28
Total	79

Note: Here, 1 = the basic food energy needed for a human = about 4 million BTU/yr
Source: Ref. 9

If we had made a similar table for the average person in the United States in 1850, or the average person in the Third World today, we would have seen that they used or use perhaps three to five times the energy they needed as food, mostly in the form of fodder for their animals and firewood for cooking and heating their homes. We use probably 15 to 30 times as much fuel per person as they do and live a much more physically comfortable life. If they are to live as we do, then world fuel consumption will grow dramatically.

The third term in Eq. (14.3) depends on the hydrogen/carbon ratio of the fuel burned. For equal amounts of energy released, the relative CO_2 release rates are approximately coal, 1.0; oil, 0.8; and natural gas, 0.6. Switching from coal to oil or from coal or oil to natural gas lowers the emission rate of CO_2. Our long-term supply of coal is much larger than that of oil or natural gas. It is possible to capture CO_2 from combustion exhaust gas and prevent its release, but only by using chemicals like CaO, whose production leads to the release of more CO_2. On a geological scale the process for removing CO_2 from the atmosphere is

$$CO_2 + Ca^{2+} + 2OH^- \rightarrow CaCO_3 + H_2O \tag{14.4}$$

which occurs in all of the the world's oceans, depositing solid $CaCO_3$ (limestone, calcite, or some other variants) on the ocean floor. It has been proposed to disperse the CO_2 from large power plants into the deep oceans in order to speed this reaction. So far that idea has not gotten beyond paper and pencil study. Its effects on the oceans are not known.

The only methods we now know to slow or stop the buildup of CO_2 in the atmosphere are to reduce the use of fossil fuels (gas, oil, coal, peat, lignite) and to stop the deforestation of the tropical rain forests. Solar, wind, hydroelectric, geothermal, and nuclear energy release no CO_2. Environmental activists suggest that we switch to solar and wind energy as one of the ways to limit CO_2 emissions. Others think we should switch to nuclear energy to solve our CO_2 emissions problem. If we decide that the global warming problem is as serious as some believe, then both of these suggestions will have to be taken very seriously. (Currently, instead of reducing the rate of fossil fuel usage, the human race is increasing it; our short-term goal is not to reduce the rate of fossil fuel usage, but to *reduce the rate of increase* in fossil fuel usage!)

The deforestation of the tropical rain forests is driven by population growth in the countries that have such forests. People without land to farm seek it by cutting down the forest. As the population grows, the demand for new agricultural and grazing land grows. The only way to stop this pressure on the remaining forests is to stop or slow the population growth in those countries.

14.1.2 Other Greenhouse Gases, Aerosols

Figure 14.3 shows that CFCs are apparently next in greenhouse effect after CO_2. They are discussed in Sec. 14.2. Next in importance is methane, the principal component of natural gas, which is formed in many anaerobic biological processes. It is the principal component of "swamp gas," is produced by bacterial decay of woody matter, and is a major component of the waste gases produced by landfills and sewage treatment plants. It is also emitted by almost all animals; our domestic dairy and meat cattle and pigs are a significant worldwide source [10].

In preindustrial times the world atmosphere contained ~ 0.7 ppm of methane. Over the past century that has increased to ~ 1.7 ppm, and it is increasing by about 0.01 ppm/yr [3]. A methane molecule is roughly 20 times as strong an infrared absorber as a CO_2 molecule (in the 8- to 12- μ window), so that even at this low concentration methane can play a significant role. The principal emissions of methane attributable to human activities are incomplete combustion (it is the most prominent hydrocarbon in automobile exhaust gases) and agricultural activities (rice paddies and animal husbandry). It is also emitted by coal mining and in the production and distribution of natural gas.

The remaining important greenhouse gas is nitrous oxide, N_2O, which formerly was often used as a dental anesthetic ("laughing gas"). It is not believed to have any harmful effects as an air pollutant except in its role as a greenhouse gas. One N_2O molecule is roughly 200 times as effective as one CO_2 molecule in reducing the transmission in the 8- to 12-μ window. The sources and sinks for N_2O are not as well known as those for the other greenhouse gases. Major human sources are not known. There is some concern that the NO_x control technologies that reduce NO with NH_3 and its near chemical relatives may produce significant amounts of N_2O.

Table 14.2, based on data from Ref. 5, summarizes current information on greenhouse gases.

The earth's average temperature can also be altered by an increase in the content of fine particles of the atmosphere. This is probably less threatening than the problem of greenhouse gases, because of the short time that fine particles spend in the atmosphere compared with the much longer time carbon dioxide and other greenhouse gases spend. However, the effects of these particles can be significant,

TABLE 14.2
Greenhouse gases

Name	Formula	Concentration in world atmosphere, 1994	Annual growth rate of concentration	Effectiveness as a greenhouse gas, per molecule, relative to CO_2
Carbon dioxide	CO_2	360 ppm	0.4%	1
Methane	CH_4	1.7 ppm	0.6%	20
CFC-11	CCl_3F	0.28 ppb	0	12 000
CFC-12	CCl_2F_2	0.48 ppb	0	16 000
Nitrous oxide	N_2O	0.31 ppb	0.25%	200

Source: Ref. 5.

as discussed in Ref. 11. For atmospheric particles to have effects lasting more than a few days, they must be injected into the stratosphere (above about 36 000 ft) because, as discussed in Chapter 5, there is little mixing between the stratosphere and the troposphere, so that particles in the stratosphere have lifetimes measured in years. In the troposphere (below that) there is fairly good atmospheric mixing, and particles are removed in days or weeks. Few human activities place many particles in the stratosphere.

Such particles can be injected into the stratosphere in large quantities by major volcanic eruptions. There they cause a lowering of the global temperature, generally for only a year or two after the eruption [12]. They lower global temperature because they are generally close in size to the wavelength of light (0.3 to 0.6 μ) and hence effective in scattering light and reducing the amount of incoming sunlight. However, these particles are much smaller than the wavelength of outgoing infrared radiation, and hence less effective in scattering it. It is widely but not universally believed that one of the causes of the extinction of the dinosaurs was the injection of a large mass of such particles into the atmosphere by a meteorite collision with the earth, which caused a severe and protracted cold spell [13, 14]. It is also widely but not universally believed that a major nuclear war would cause a "nuclear winter" by placing large amounts of particles in the high atmosphere [15].

14.2 STRATOSPHERIC OZONE DEPLETION AND CHLOROFLUOROCARBONS

The second global problem concerns the possible destruction of the stratospheric ozone layer. At ground level, ozone, O_3, is a strong eye and respiratory irritant and a major component of photochemical smog (see Fig. 1.2). It may also act as a greenhouse gas. In the stratosphere, 10 to 20 km above the earth's surface, is a layer of low-density air containing 300 to 500 ppb of ozone. Figure 14.1 shows the combined effects of O_2 and O_3; in the wavelength range below 0.28 μ, the effect is mostly due to ozone. That ozone prevents the components of sunlight with wavelengths less than about 0.28 μ from reaching the earth's surface. Ozone is the only component of the atmosphere that absorbs significantly at that wavelength (far ultraviolet). Figure 14.1 also shows that the calculated solar intensity is significant down to about 0.2 μ. (The true solar spectrum is more complex than the simple blackbody radiative flux shown here.) If that ozone layer were removed, we would expect large amounts of ultraviolet light in the wavelength range 0.2 to 0.28 μ to reach the surface of the earth.

The energy of a photon of light is proportional to (1/wavelength), so that shorter wavelength photons are more chemically active than longer wavelength photons. If high-energy photons reach the earth's surface, they will cause chemical reactions in the surfaces they contact, including human skin; they are expected to cause increased rates of skin cancer in humans. People can put on sunhats and sunscreen, but plants and animals cannot. (Sunscreens contain chemicals that are largely opaque to ultraviolet light; they convert chemically active ultraviolet light into chemically inactive heat.) It is not clear how plants and animals would respond to the appearance of this shorter wavelength (more chemically active) radiation. Thus ozone is a harmful pollutant at ground level, but a beneficial ultraviolet shield in the stratosphere.

Destruction of the ozone layer is a complex technical question that has produced a substantial literature [16]. That destruction is mostly caused by elemental chlorine atoms; the mechanism apparently involves two reactions,

$$Cl + O_3 \rightarrow ClO + O_2 \tag{14.5}$$

$$ClO + O_3 \rightarrow Cl + 2O_2 \tag{14.6}$$

Other reactions are going on in the stratosphere that modify and compete with these two, but if we ignore other reactions, add these two reactions, and cancel like terms, we see that the overall reaction is

$$2O_3 \rightarrow 3O_2 \tag{14.7}$$

with no net consumption of Cl atoms. Thus one Cl atom can convert many ozone molecules to ordinary oxygen molecules. One sees estimates from 10^4 to 10^6 O_3 molecules destroyed by one Cl atom. (This mechanism is often referred to as catalytic destruction of ozone, because the chlorine atom acts as a nonconsumed catalyst for the reaction.)

Most of the chlorine in the world is in the form of chemically stable NaCl (table salt) either dissolved in the oceans or in underground salt deposits formed by the evaporation of ancient oceans. Elemental chlorine, a very reactive chemical, has a short lifetime in the lower atmosphere and has few natural ways to get from the lower atmosphere up to the ozone layer. The only naturally occurring chemical that can transport much chlorine high enough into the atmosphere to damage the ozone layer is methyl chloride, CH_3Cl, which is produced in large quantities by biological processes in the shallow oceans. Most of it is destroyed in the troposphere, but an estimated 3 percent of worldwide methyl chloride emissions reaches the stratospheric ozone layer [17]. Chemically active ultraviolet light in the 0.2- to 0.28-μ range, which enters the ozone layer but does not penetrate below it, is strong enough to split up methyl chloride (and the other chlorine compounds, discussed shortly), releasing Cl atoms, which initiate Reaction (14.5). Before we had synthetic halogen compounds, methyl chloride was probably the principal natural destroyer of the ozone layer; its destruction of the ozone was in balance with natural production mechanisms, leading to a steady-state ozone layer.

Starting about 1900, humans began releasing into the atmosphere synthetic chlorine-containing compounds in significant amounts. Those like methyl chloride that have hydrogen atoms can be attacked in the atmosphere by the OH radical; for this reason most of them do not survive to reach the stratosphere. Carbon tetrachloride, CCl_4, has no hydrogen; most of it is believed to reach the stratosphere and to participate in the destruction of the ozone layer. Its world production has been more or less constant over the past 60 years.

CFCs (compounds containing chlorine, fluorine, and carbon, commonly called *Freons*) were first developed by General Motors for use in household refrigerators.

One of their virtues is their chemical inertness; they are nontoxic, nonflammable, invisible, tasteless, odorless, non–almost everything else. They replaced toxic sulfur dioxide and ammonia in household refrigerators. *This replacement saved many lives.* Later, their inertness led to the widespread use of CFCs as propellants in spray cans and as a blowing agent in the production of plastic foams. Then in the 1950s we began to use air conditioners in autos. The CFCs used as refrigerants in these are much more likely to leak to the atmosphere than the CFCs in refrigerators and home air conditioners because the shaft-sealing problem on belt-driven auto air conditioners is more difficult than that on electric-driven refrigerators and home air conditioners; auto air conditioners became a major source of CFC emissions.

There are many different CFCs; the two most widely used are CFC 12, CF_2Cl_2, and CFC 11, $CFCl_3$ (the first digit in the name is the number of carbon atoms, the second is the number of F atoms). CFCs have no H, so they cannot be attacked by atmospheric OH. As far as we know, the only process for removing CFCs from the atmosphere is their slow transport to the top of the ozone layer, where they are attacked by shortwave ultraviolet light and thus destroyed, releasing their Cl to participate in the reactions with ozone.

Trichloroethane, CH_3CCl_3, is not a CFC because it has hydrogens and thus can be attacked by atmospheric OH. But that attack is relatively slow, so that an estimated 9 percent of this material that is emitted to the atmosphere makes its way to the stratosphere and participates in ozone destruction. Trichloroethane is a widely used cleaning solvent; large amounts of it are emitted. Table 14.3 shows the concentrations, lifetimes, and expected contribution to delivery of elemental chlorine to the stratosphere for these chemicals.

Some other gases can attack the ozone layer, e.g., NO from stratospheric airplanes and relatively inert N_2O, if we release much of it at ground level. NO, released

TABLE 14.3
Chlorine-containing compounds believed to attack the ozone layer

Name	Formula	Global atmospheric concentration, ppb	Estimated atmospheric lifetime, years	% of emissions that reach the stratosphere	Annual destruction in the stratosphere, (10^7 kg/year)
Methyl chloride	CH_3Cl	0.62	2 to 3	≤ 3	6.1
CFC 12	CF_2Cl_2	0.48	> 80	100	3.9
CFC 11	$CFCl_3$	0.28	≈ 83	100	2.7
Carbon tetrachloride	CCl_4	0.12	50	≤ 100	1.2
Trichloro-ethane	CH_3CCl_3	0.12	≈ 9	9	3.8

Source: Ref. 17.

by high-flying aircraft, can contribute to ozone depletion by the reaction

$$NO + O_3 \rightarrow NO_2 + O_2 \tag{14.8}$$

which is swift and practically irreversible. But it is not a catalytic reaction like the chlorine reaction; one NO molecule only destroys one O_3 molecule, so much more NO is needed per unit of damage than Cl.

The only method we know to protect the ozone layer is to limit the emission of those materials that can harm it. (No one knows of another material we could send into the air to protect it.) The threat seems severe enough that international conferences have been held and declarations and treaties adopted that commit the nations that produce the CFCs to restrict and eventually eliminate their use [18]. For some of their applications satisfactory replacements are available. For others they are being sought. Many of the proposed substitutes are chlorohydrofluorocarbons, which contain at least one H atom, and hence are susceptible to OH attack in the atmosphere. Table 14.3 shows that trichloroethane and methyl chloride (chlorohydrocarbons, chemically similar to chlorohydrofluorocarbons) are not completely removed in the troposphere, but they contribute much less, per molecule, to damage to the ozone layer than the CFCs, which are apparently not attacked at all below the stratosphere.

Section 14.1 shows that CFCs are also potent greenhouse gases. Most efforts to reduce the emission of CFCs are based on their role in attacking the ozone layer. Those efforts also help prevent global warming.

14.3 ACID RAIN

The average acidity of rainfall in Scandinavia, the northeastern United States and Canada, and parts of Europe has increased over the past 40 years. There seems no question that this change is primarily due to the increased emissions of sulfur oxides and nitrogen oxides that have accompanied the greatly increased economic activity (and hence increased combustion of fuels) in or upwind of these regions. As discussed in Chapter 11, these sulfur oxides and nitrogen oxides are oxidized over several hours to several days to sulfuric and nitric acids. They then are generally captured by raindrops and returned to the surface as acid precipitation. The common name is "acid rain," but the complete description includes acidic rain, acidic snow or hail, acids adsorbed on falling dust particles, etc.

The normal technical measure of acidity is pH, defined as

$$\text{pH} = -\log_{10} (\text{activity of H}^{+}\text{ions, expressed in mol/liter}) \tag{14.9}$$

As discussed in Sec. 12.1.3, for ideal gases the activity is *identical* to the concentration. For dilute solutions of acids and bases in water the activity of H^+ ions is very close to, but not identical to, the concentration of those ions. In this book we will use Eq. (14.9) *as if* concentration and activity were identical for H^+ ions, and replace "activity" by "concentration" in Eq. (14.9). Although chemists are familiar with this definition and concept, many others are not. See Table 14.4 on page 528.

TABLE 14.4
pH of various substances

By definition, pH $= -\log_{10}$ (activity of H^+ ions, expressed in mol/L). As discussed in Sec. 14.3, for the dilute solutions of practical interest, activity and concentration are almost the same, and we normally apply this equation using concentration in the place of activity. The possible pH range is from about 0 to about 14. Because of the log scale, a change of 1 in pH corresponds to a change of 10 in H^+ concentration.

Low pH values (< 7) indicate acidity (high H^+ concentration)
pH $= 7$ is neutral, neither alkaline nor acid
High pH values (> 7) indicate alkalinity (low H^+ concentration)

Most foods are acid. Soils in humid climates are acid; in dry climates they are alkaline. The pH of surface waters and sea water varies due to the variation in dissolved CO_2 and other materials; sea water is slightly alkaline.

pH	Substance
0.1	HCl (1 N)
0.3	H_2SO_4 (1 N)
1 to 3	Human stomach contents
2 to 4	Soft drinks
2.2 to 2.4	Lemons
2.4	Acetic acid (1 N)
2.4 to 3.4	Vinegar
2.8 to 3.8	Wines
2.9 to 3.3	Apples
2.9 to 3.3	Cider
3	Mine drainage waters
3.2 to 3.6	Pickles
4 to 5	Beers
5.6	Unpolluted rainfall
6.1 to 6.3	Salmon
6.3 to 6.6	Cow's milk
6.5 to 7.5	Human saliva
6.8 to 8.0	Hominy
7	Very pure water
7.3 to 7.5	Human blood
8	Sea water
8 to 9	Soaps, shampoos
9.4	Calcium carbonate (saturated)
11.6	NH_3 (1 N)
12.4	Lime (saturated)
14	NaOH (1 N)

Rain falling through a perfectly unpolluted atmosphere will arrive at the earth with a pH of about 5.6 because of the carbon dioxide in the atmosphere, which reacts with rainwater by these reactions:

$$CO_2 + H_2O \rightleftarrows \underset{\text{carbonic acid}}{H_2CO_3} \rightleftarrows H^+ + HCO_3^- \tag{14.10}$$

Carbonic acid is a weak acid, and Reaction (14.10) is reversible, with the acid concentration in the rain depending on the concentration of carbon dioxide in the air. Generally, any rain with a pH less than the 5.6 (i.e., a H^+ concentration more than $10^{-5.6}$ mol/L) is considered acidic, but damage to plants and animals (or fish) does not begin to become apparent until a pH of about 4.5 or less is reached.

Although damage to human health has not been shown, acid damage to the ecology of mountain lakes and forests is apparent. The transport distances between emission and precipitation are generally hundreds of miles, so that local control seems impossible. Acid rain has a substantial literature [19–21].

Example 14.3. Table 1.1 shows that the total annual U.S. emissions of SO_2 in 1997 were 20.4 million tons. If we assume that 25 percent of that was in the Midwest–Ohio Valley area, and that 50 percent of that came to ground as acid precipitation in a 1000 km by 1000 km area in the northeastern United States and southeastern Canada, and that the average precipitation over that area is 1 m/yr, by how much would this sulfur dioxide (if all converted to H_2SO_4) change the pH of the rainwater?

The annual U.S. emissions are

$$\begin{pmatrix}\text{Annual U.S.}\\ \text{SO}_2\text{ emissions}\end{pmatrix} = 20.4 \times 10^6 \text{ ton} = 18.5 \times 10^{12} \text{ g} \cdot \frac{\text{mol}}{64 \text{ g}} = 2.9 \times 10^{11} \text{ mol}$$

of which an estimated $25\% \times 50\% = 12.5\%$ falls on the affected area. The estimated annual precipitation for that area would be

$$\begin{pmatrix}\text{Annual}\\ \text{precipitation}\end{pmatrix} = \text{area} \cdot \text{depth} = (10^6 \text{ m})(10^6 \text{ m})(1 \text{ m}) = 10^{12} \text{ m}^3 = 10^{15} \text{ L}$$

Each mol of SO_2 produces two mols of H^+, so the increase of H^+ above the naturally occurring value is

$$(\text{Increase in H}^+) = \frac{2 \cdot 0.125 \cdot 2.9 \times 10^{11} \text{ mol}}{10^{15} \text{ L}} = 7.23 \times 10^{-5} \frac{\text{mol}}{\text{L}}$$

The original (natural) rainfall is assumed to have a pH of 5.6, or a H^+ concentration of

$$\text{H}^+ = 10^{-\text{pH}} = 10^{-5.6} = 2.51 \times 10^{-6} \frac{\text{mol}}{\text{L}}$$

Adding these two values, we find an H^+ concentration of 7.48×10^{-5} mol/L, or

$$\text{pH} = -\log(\text{H}^+) = -\log(7.48 \times 10^{-5}) = 4.13$$ ■

This approximate calculation shows that the amount of SO_2 emitted in the United States is enough to acidify the precipitation over a large area for certain meteorological conditions. It also shows that because of the log in the definition of pH, we could have ignored the mild acidity of the unpolluted rainfall with negligible error.

How harmful acid precipitation is to a given area is strongly dependent on the buffering capacity of the soil. If the local soil contains significant amounts of limestone, $CaCO_3$, then the acid will react by

$$\text{CaCO}_3 + \text{H}^+ \rightarrow \text{Ca}^{2+} + \text{HCO}_3^- \tag{14.11}$$

thus removing the H^+ ion. If the soil contains little limestone, then the observed pH values of the surface waters will be lower and the biologic effects more pronounced. The acidity need not act alone on plants and fish. It is also believed that in some areas the increased rainfall acidity has speeded the dissolution of metals from the soil, e.g., aluminum, thus raising the content of those metals in the water. These dissolved metals may be the true agents of destruction for fish or plants.

14.4 THE REGULATORY SITUATION

The Clean Air Act [22] requires a steady reduction in the emission of acid rain precursors, which mostly means power plants in the central United States. That reduction has begun, producing a measurable reduction in the rate of acid deposition in the northeastern United States [23]. The act calls for continuing reductions, so it is probable that there will be continued reductions in that deposition. There is still some argument between the United States and Canada about whose emissions cause whose acid deposition, but that quarrel has become less intense as emission controls on both sides of the border have made the problem less severe.

The Montreal Protocol and its Amendments and Adjustments have caused enough reduction in emissions that the total abundance of ozone-depleting chemicals in the atmosphere peaked in 1994, and is now slowly declining [24]. The U.S. regulations under that protocol are in Ref. 22.

Global warming has been a much more contentious issue. Conferences in Kyoto, Japan, and Rio de Janiero, Brazil, formulated some guidelines but did not end the basic conflict. The only known way to reduce CO_2 emissions (or to slow the growth of those emissions) is to reduce (or slow the rate of increase) of fossil fuel combustion and/or rain forest destruction. The industrial countries generally take the view that all countries should make roughly proportional reductions in their current emissions. The developing countries take the view that all countries should be entitled to equal emissions per capita. They say that if the industrial countries' view is adopted, they will be permanently stuck at a lower level of economic development than the industrial countries. Some industrial countries have committed to reduce the CO_2 emissions; how this will be accomplished is not yet clear.

14.5 HOW SURE ARE WE?

Most scientists believe that the global warming, ozone depletion, and acid rain problems are real. There is much less agreement as to the magnitude of the problems and/or the need for prompt (and expensive) action [25–28]. The evidence that global warming has been occurring is not strong enough to be distinguished from the annual random variation of climate with *complete certainty*. The evidence for destruction of the ozone layer over Antarctica seems strong, suggesting that the stratospheric ozone problem is severe and may cause significant effects in a time scale of years rather than decades or centuries. The acid rain effects in the worst affected areas seem certain.

On the basis of the preceding estimates of the state of the problems, it seems prudent to restrict the emissions of ozone-depleting gases, because replacements can be found without cataclysmic costs. Large expenditures are being made to reduce emissions of sulfur oxides, which will minimize acid precipitation, and have some other benefits such as lowering the fine particle content of the affected atmospheres. Similar actions for nitrogen oxides have not been taken in the United States, but they are mandated for the future by the acid rain provisions of the Clean Air Act. Whether we should take active steps to reduce emissions of CO_2 and other greenhouse gases is less certain and the subject of intense political debate in the United States. The complexity of world climate models (of which Fig. 14.2 is a very simplified schematic) lends support to those who doubt that we know enough to justify costly and difficult actions. Some steps, like improving the efficiency with which we use all fuels and supporting the development of solar, biomass, and wind energies, are certainly worthwhile. Other more drastic steps may or may not be justified based on current knowledge.

There is little record in human history of situations in which the whole human race jointly restricted its activities for the global common good. The International Whaling Commission restrictions on whaling, the international ban on the use of poison gases, and the nuclear nonproliferation treaty may be the only examples. (All three of these treaties have been violated by some of their signatories.) None of these reaches into the daily lives of most citizens (except those who consider whale meat a delicacy). The CFC restrictions seem to be being obeyed, and the U.S. acid rain reduction provisions are being carried out. Ordinary citizens will see the effects of these as a very small change in their electric bills and the cost of air conditioners (particularly automotive air conditioners) and refrigerators. If we decide to significantly limit the emissions of CO_2, that will have economic consequences much more serious than those of any of these examples. If the projections of the IPCC [5] are correct, then failing to do so will have negative economic consequences far greater than those of doing so.

14.6 SUMMARY

1. Global warming, destruction of the ozone layer, and acid rain are air pollution problems that cannot be dealt with by the mostly local regulations used for other air pollutants. If action is to be taken, worldwide action is needed.
2. Our uncertainty of the relation between emissions and effects is greater with these problems than with most other air pollution problems.
3. The current U.S. course of action is to attempt to reduce global emissions of ozone-destroying materials, to limit the emission of acid rain precursors, and, as of yet, to take no serious action other than study and debate on emissions of greenhouse gases.

PROBLEMS

See Common Values for Problems and Examples, inside the back cover.

14.1. Show the calculations leading to the sun and earth spectra on Fig. 14.1. The quantity plotted is a fraction of total emission per micron of wavelength. The emission spectrum is given by Planck's law

$$E_{b\lambda}(T) = \frac{C_1}{\lambda^5\left[\left(\exp\dfrac{C_2}{\lambda T}\right) - 1\right]}$$

where $E_{b\lambda}$ = monochromatic emissive power of a blackbody at temperature T, in W/m^3 or equivalent

λ = wavelength, in m or equivalent

T = absolute temperature

$C_1 = 374.15 \times 10^{-18}\ W \cdot m^2$

$C_2 = 14.388 \times 10^{-3}\ m \cdot K$.

14.2. Based on Fig. 14.5, is the net mass of CO_2 flowing into the world's oceans more or less than the net mass flowing out of the oceans? Are we concerned about that? Should we be?

14.3. Based on Fig 14.5, estimate the average length of time that a CO_2 molecule spends in the atmosphere.

14.4. One way to reduce the emissions of CO_2 from coal combustion would be to scrub the exhaust gas in the way now used to remove SO_2 (see Chapter 11) but using a solution that is alkaline enough to remove both SO_2 and CO_2.

(*a*) What reagents might be used for this kind of scrubbing on a massive scale?

(*b*) The electric power industry has estimated the costs of doing this [29] and found them very high. They suggest that it would be cheaper per ton of CO_2 removed from the atmosphere for the electric industry to go into the tree farming business, growing trees or some other biomass to take CO_2 out of the air. Discuss the advantages and disadvantages of this proposal.

14.5. Visible light, as our eyes perceive it, is electromagnetic radiation in the wavelength range of 0.3 to 0.7 micron. That corresponds to the peak intensity of solar radiation. What are the plausible explanations for this practically exact correspondence?

14.6. The average depth of the world's oceans is 3.8 km. Most of the deep ocean is at about 4°C, where the coefficient of thermal expansion of sea water is zero. However, the top kilometer of the oceans is at an average temperature of approximately 10°C, and this water has an average coefficient of thermal expansion of 0.00012/°C. Estimate how much the sea level would rise if the temperature of the top kilometer of the ocean increased by 1°C.

14.7. In my May 1996 electric bill from Utah Power and Light, they tell me that they are combating the CO_2 buildup in the atmosphere by buying rain forest in Belize and managing it for CO_2 removal rather than seeing it cleared and used for agriculture. They indicate that they will own or manage 120 000 acres, and that so doing ". . . could reduce greenhouse gas emissions by 5.2 million tons over 40 years."

(*a*) Estimate the emission reduction in tons of CO_2 per acre per year.

(*b*) Assuming that this is done by making wood, how many tons/(acre · year) of wood (dry basis) do they expect to be produced? Here assume that wood is 50 wt% carbon (dry basis).

(*c*) Assuming that there are 400 trees/acre, estimate how many pounds of wood growth (dry basis) per tree per year they are assuming.

(*d*) UP and L's Huntington Power Plant (one of several they have) generates 840 MW of electricity. Its average use over a year is about 70% of the maximum possible (equivalent to operating at full capacity for 70% of the hours of the year). Its thermal efficiency is $\approx$ 35%. If they use the "Typical Pittsburgh Seam Coal," in the examples in this book, how many tons of CO_2 will that plant emit in 40 years?

14.8. If we put enough "greenhouse gases" into the atmosphere so that the fraction of the outgoing terrestrial radiation that escapes to outer space (see Example 14.2) is reduced to 90% of its current value how much would we expect the average temperature of the earth's surface to increase? For this problem ignore all the feedback loops on Fig. 14.2

14.9. The ideal refrigerant to be used in a compression-expansion refrigerator (the most common kind) would have the following properties: a moderate vapor pressure at all temperatures in the cycle, a high vapor density at the lowest pressure in the cycle, a freezing point below the lowest temperature in the cycle, a low liquid specific heat, a high vapor specific heat, a high latent heat of vaporization, and good liquid film condensing coefficients (see any book on heat transfer).

(*a*) Compare these values for Freon 12, the most widely used refrigerant, with those of possible replacements for it.

(*b*) What are the possible replacements for use in household refrigerators where the refrigerant must be nonflammable and nontoxic?

14.10. Several years ago the French wine industry was rocked by the scandalous discovery that some French wine makers had used sulfuric acid to adjust the acidity of their wines. Sulfuric acid is a permitted food additive in the United States, and probably in most other countries; it is used to adjust the acidity of many foods because it has no taste. But the rules about additives for wine making are more strict than the rules for general food additives and this usage was forbidden by French wine law.

(*a*) If we assume that the wine makers wanted to change the pH of their wines from 3.8 to 2.8, how many grams of sulfuric acid, H_2SO_4, would they have had to add per liter of wine? Assume complete ionization of the acid, i.e.,

$$H_2SO_4 \rightarrow 2H^+ + SO_4^{2-}$$

(*b*) The true situation is more complex. We actually have this situation:

$$H_2SO_4 \rightarrow H^+ + HSO_4^-$$

$$HSO_4^- \rightleftarrows H^+ + SO_4^{2-}$$

The first of these is practically complete at room temperature,

$$K_1 = \frac{[H^+][HSO_4^-]}{[H_2SO_4]} \approx 100 \frac{\text{mol}}{\text{L}}$$

but the second is not,

$$K_2 = \frac{[H^+][SO_4^{2-}]}{[HSO_4^-]} = 0.012 \frac{\text{mol}}{\text{L}}$$

Repeat part (*a*), taking into account the fact that the second ionization is not complete.

14.11. Why does Table 14.4 show that 1 N H_2SO_4 is a weaker acid (has a higher pH) than 1 N HCl?

14.12. One defense against acid rain is the liming of lakes. Ground limestone is added to offset the effects of acid rain. If we assume that a lake drains 10 km^2 and that the annual precipitation in the entire drainage basin is 1 m (1 m^3/m^2 of surface), how much limestone, $CaCO_3$, should be added to the lake to change the acid rainfall (pH 4.5) to the equivalent of normal rainfall (pH 5.6)?

14.13. Repeat Example 14.3, leaving out of account the H^+ ions in the naturally occurring rainwater. How much does this change the calculated pH of the acidified rainwater?

REFERENCES

1. Oke, T. R.: *Boundary Layer Climates,* 2d ed., Methuen, London, p. 14, 1987.
2. Fleagle, R. G., and J. A. Businger: *An Introduction to Atmospheric Physics,* 2d ed., Academic Press, New York, p. 233, 1980.
3. Houghton, J. T., et al., eds.: *Climate Change 1995: The Science of Climate Change,* IPCC, Cambridge University Press, Cambridge, 1996.
4. Houghton, J.: *Global Warming, The Complete Briefing,* 2d ed., Cambridge University Press, Cambridge, 1997.
5. Houghton, J. T., G. J. Jenkins, and J. J. Ephramus, eds.: *Climate Change: The IPCC Scientific Assessment,* Cambridge University Press, Cambridge, 1990.
6. Smith, J. B., and D. A. Tirpak: *The Potential Effects of Global Climate Change on the United States,* Hemisphere, New York, 1990.
7. Houghton, R. A., and G. M. Woodwell: "Global Climatic Change," *Sci. Amer.*, Vol. 260, Issue 4, pp. 36–44, 1989.
8. Kasting, J. F., O. B. Toon, and J. B. Pollack: "How Climate Evolved on the Terrestrial Planets," *Sci. Amer.,* Vol. 258, Issue 2, pp. 90–97, 1988.
9. de Nevers, N.: "Global Energy Use and Global Warming," in G. C. Bryner (ed.), *Global Warming and the Challenge of International Cooperation: An Interdisciplinary Assessment,* David M. Kennedy Center for International Studies, Brigham Young University, Provo, UT, 1992.
10. Blake, D. E.: "Trace Gases in the Atmosphere: Temporal and Spatial Trends," in B. G. Levi, D. Hafenmeister, and R. Scribner (eds.), *Global Warming: Physics and Facts AIP Conference Proceedings 247,* American Institute of Physics, New York, 1992.
11. Charlson, R. J., and T. M. L. Wigley: "Sulfate Aerosols and Climatic Change," *Sci. Amer.*, Vol. 270, Issue 2, pp. 48–57, February 1994.
12. Harington, C. R., ed.: *The Year Without a Summer? World Climate in 1816,* Canadian Museum of Nature, Ottawa, 1992.
13. Courtillot, V. E., "What Caused Mass Extinction: A Volcanic Eruption," *Sci. Amer.,* Vol. 263, Issue 4, pp. 85–92, 1990.
14. Alvarez, W., and F. Asaro: "What Caused Mass Extinction: An Extraterrestrial Impact," *Sci. Amer.,* Vol. 263, Issue 4, pp. 76–84, 1990.
15. Turco, R. P., O. B. Toon, and T. P. Ackerman: "The Climatic Effects of Nuclear War," *Sci. Amer.,* Vol. 252, Issue 2, pp. 33–43, 1984.
16. National Research Council, "Protection against Depletion of Stratospheric Ozone by Chlorofluorocarbons," National Academy of Sciences, Washington, DC, 1979.
17. Campbell, I. M.: *Energy and the Atmosphere: A Physical-Chemical Approach,* 2d ed., John Wiley & Sons, Chichester, England, p. 194, 1986.
18. United Nations Environmental Program (UNEP): "Montreal Protocol on Substances That Deplete the Ozone Layer," UNEP Conference Services Number 87-6106.
19. Mohnen, V. A.: "The Challenge of Acid Rain," *Sci. Amer.,* Vol. 259, Issue 2, pp. 30–38, 1988.
20. Bresser, A. H. M., and W. Salomons, eds.: *Acidic Precipitation,* Volumes 1 to 5, Springer-Verlag, New York, 1990.
21. Regens, J. L., and R. W. Rycroft: *The Acid Rain Controversy,* University of Pittsburgh Press, Pittsburgh, PA, 1988.

22. Clean Air Act Amendments of 1990, PL 101-549, Title IV, Acid Deposition and Title VI, Stratospheric Ozone Protection.
23. *National Air Quality and Emissions Trends Report, 1997,* EPA 454/R-98-016, Chapter 7, "Acid Deposition."
24. *Scientific Assessment of Ozone Depletion, 1998,* World Meteorological Organization/United Nations Environmental Programme, Preprint released June 1998.
25. Ray, D. L., and L. R. Guzzo: *Trashing the Planet,* Harper Collins, New York, 1990.
26. Seitz, F.: *Global Warming and the Ozone Hole Controversies: A Challenge to Scientific Judgment,* George C. Marshall Institute, Washington, DC, 1994.
27. Parsons, M. L.: *Global Warming: The Truth behind the Myth,* Plenum Press, New York, 1995.
28. Moore, T. G.: *Global Warming, A Boon to Humans and Other Animals,* Hoover Institute of War, Revolution and Peace, Stanford University, Palo Alto, CA, 1995.
29. Alpert, S.: "Biological Approaches to Reducing Atmospheric CO_2," *EPRI Jour.,* Vol. 18, Issue 3, pp. 44–47, 1993.

CHAPTER 15

OTHER TOPICS

This chapter considers five additional topics, each of which could form a complete chapter or a complete book.

We devoted full chapters to particulates, VOCs, sulfur oxides, and nitrogen oxides. Although CO and lead are also major pollutants that the law treats the same way as particulates, VOCs, sulfur oxides, and nitrogen oxides, their most important sources, are not subject to local control, and hence CO and lead are less likely to be part of the work of an air pollution control engineer. There are other differences as well, discussed below.

15.1 CARBON MONOXIDE (CO)

Three-fourths of the CO that enters the air from human activities comes from vehicles with IC engines (Chapter 13). The highest ambient CO concentrations are measured in the downtown of major cities, where almost all of the CO comes from motor vehicles. The most effective control of CO is to reduce emissions from various kinds of motor vehicles. CO is also produced in much lesser amounts by almost any combustion process, e.g., gas stoves, forest fires, industrial processes. For these sources the control measures are mostly requirements that the combustion use ample excess air and good fuel-air mixing, as described in Sec. 7.10. Ultimately we deal with CO by oxidizing it to biologically harmless CO_2.

Example 15.1. The average U.S. family house uses about 160×10^6 Btu/yr of natural gas, mostly for heating in the winter months, but also for water heating, clothes drying, and cooking all year. The CO emission factor for residential natural gas usage is 0.020 lb/10^6 Btu [1]. The same average family drives its auto(s) an average of 10 000 mi/yr. The CO emission factor for new autos is 3.4 g/mi (Chapter 13). What is the ratio of CO emitted by their auto(s) to CO emitted by all the natural gas appliances in their house?

The automobile emissions are

$$\dot{m} = \left(3.4\ \frac{\text{g}}{\text{mi}}\right)\left(10\ 000\ \frac{\text{mi}}{\text{yr}}\right)\left(\frac{\text{lb}}{454\ \text{g}}\right) = 74.9\ \frac{\text{lb}}{\text{yr}}$$

The residential gas usage emissions are

$$\dot{m} = \left(160 \times 10^6\ \frac{\text{Btu}}{\text{yr}}\right)\left(0.020\ \frac{\text{lb}}{10^6\ \text{Btu}}\right) = 3.2\ \frac{\text{lb}}{\text{yr}}$$

And the ratio is (74.9/3.2) = 23.4 ■

CO is different from most other air pollutants in its acute health effects. The United States averages a few hundred fatalities annually due to exposure to high concentrations of CO, mostly inside buildings with improperly vented heating systems, in idling parked cars with faulty exhaust systems, and also in several industrial settings. CO poisoning is the major cause of death in residential fires (about 4000 deaths per year) and coal mine fires (about 10 per year). If one does not die of such exposure then there probably will be no permanent health damage; the effects are practically reversible. In contrast, most other air pollutants rarely cause fatalities due to short-term (acute) exposures, and their effects are much less likely to be reversible.

CO causes its harm by binding with the hemoglobin in our blood, forming carboxyhemoglobin (COHb). CO attaches to hemoglobin roughly 220 times more strongly than does oxygen, so that small amounts of CO in the air we breathe can cause significant amounts of our hemoglobin to be tied up as COHb. The hemoglobin thus tied up cannot serve its normal function, to transport oxygen in the blood (as oxyhemoglobin, O_2Hb). Thus as the blood's ability to transport oxygen declines, various parts of the body suffer oxygen deprivation. Table 15.1 on page 538 shows the effects corresponding to various fractions of the blood hemoglobin thus immobilized. Seventy or more percent COHb is normally fatal.

The subject of CO effects on health has been thoroughly studied, but we are not certain that we know all the subtle effects [2, 3]. The reduced oxygen supply to the fetus due to CO from smoking may be the way in which a mother's smoking causes low birth weight and increased infant mortality.

Because the uptake of CO by the hemoglobin is practically reversible and because it depends on the time and concentration to which one is exposed, computation of the blood COHb concentration is fairly simple and reliable. Figure 15.1 on page 539 shows a comparison of the measured blood COHb of human subjects with the predictions of the model of Coburn, Forster, and Kane [4]. The agreement is very good. For more on this model, see Problems 15.1–15.4.

The NAAQS for CO (Table 2.3) are based on the combination of time and concentration that leads to a COHb of 3 percent in nonsmokers, the lowest concentration at which demonstrable effects occur (see Table 15.1). Over the period

TABLE 15.1
Effects of CO

% of blood hemoglobin converted to COHb	Effects
0.3–0.7	Physiologic norm for nonsmokers
2.5–3.0	Cardiac function decrements in impaired individuals; blood flow alterations; and, after extended exposure, changes in red blood cell concentration
4.0–6.0	Visual impairments, vigilance decrements, reduced maximal work capacity
3.0–8.0	Routine values in smokers. Smokers develop more red blood cells than nonsmokers to compensate for this, as do people who live at high elevations, to compensate for the lower atmospheric pressure
10.0–20.0	Slight headache, lassitude, breathlessness from exertion, dilation of blood cells in the skin, abnormal vision, potential damage to fetuses
20.0–30.0	Severe headaches, nausea, abnormal manual dexterity
30.0–40.0	Weak muscles, nausea, vomiting, dimness of vision, severe headaches, irritability, and impaired judgment
50.0–60.0	Fainting, convulsions, coma
60.0–70.0	Coma, depressed cardiac activity and respiration, sometimes fatal
>70.0	Fatal

1988–1997 estimated nationwide CO emissions decreased 38 percent [5], mostly as a result of the steady replacement of older autos with newer, more strictly controlled autos. Nonetheless, 20 air quality control regions in the United States are classified as nonattainment for failure to meet the CO standard in 1997. Most of the sites that exceeded the standard exceeded the 8 hour, 9 ppm standard, not the 1 hour, 35 ppm standard. The exceedences mostly occur in winter during prolonged inversions with low wind speeds.

15.2 LEAD

Lead has been recognized as a health hazard for at least 2000 years [6]. The effects of lead entering the body with food and water and air seem to be cumulative and independent of the mode of entering the body. The NAAQS for lead (see Table 2.3) was established to protect children who had high exposure to dietary lead; the goal is to hold their blood lead content below 30 μg/dL = 300 μg/L $\approx$ 0.3 ppm.

Before the adoption of the lead standard about 85 percent of the lead entering the atmosphere came from tetraethyl lead $(C_2H_5)_4Pb$, which was used as an octane improver (see Chapter 13) in most of the motor gasoline in the world. Under the pressure of EPA regulations, the gasoline producers in the United States have removed practically all the lead from the gasoline sold in the United States. As a result, the 1997 lead emissions in the United States were about 1.7 percent of those

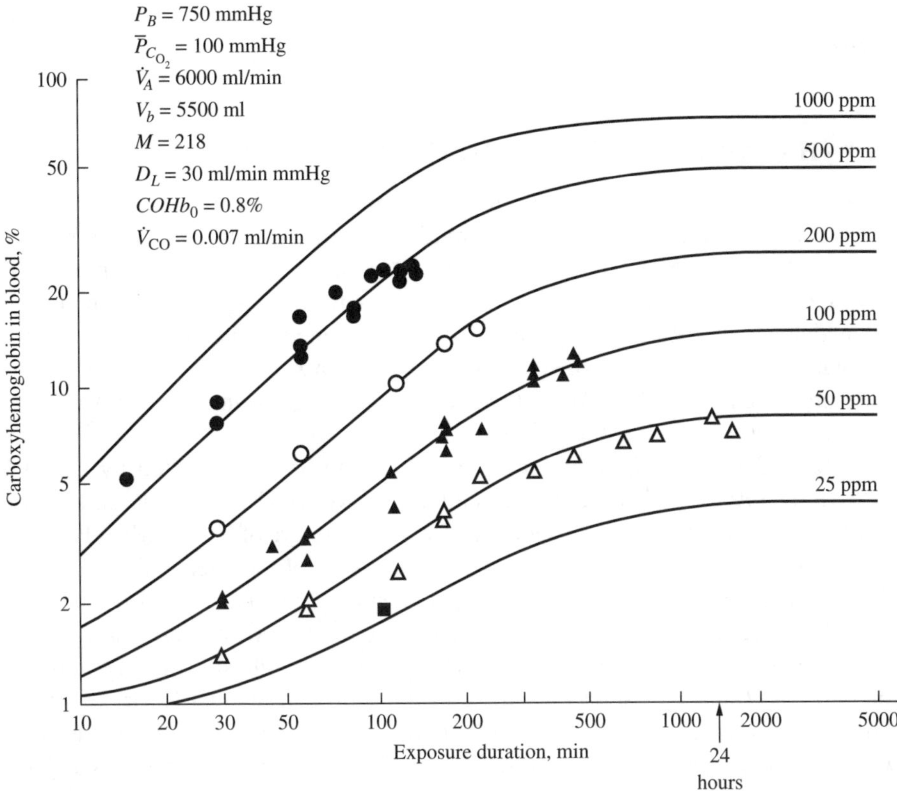

FIGURE 15.1
Fraction of blood hemoglobin tied up as COHb as a function of time of exposure and CO concentration in the air breathed. The curves are from the model of Coburn, Forster, and Kane [4]; the data represent experimental values. See Problem 15.1 for a description of the values shown on the figure. (Copied from Ref. 7, reproduced courtesy of Prentice Hall.)

in 1970. The remaining sources are the remaining small amount of leaded gasoline, miscellaneous industrial processes, and waste incineration.

Most nonautomotive lead emission sources vaporize the lead (normal boiling point 1740°C = 3164°F) or vaporize some lead salt like lead chloride (normal boiling point 954°C = 1749°F) in some kind of combustion process. Then as the combustion gas cools, the lead condenses, forming fine particles (see Chapter 8). These particles are small enough that they are not captured in low-quality particulate control devices. High-quality devices capture them (Chapter 9). Thus the appropriate emission control for lead from nonautomotive sources is to cool the exhaust gas enough that the lead vapor all condenses as solid lead particles (or particles of lead compounds) and then to follow with high-quality particulate capture. The captured solids normally are placed in some kind of landfill, where the possibility of the lead dissolving and entering the ground water is a serious problem.

As of 1997 the EPA did not classify any AQCR as failing to meet the lead AAQS [5].

15.3 HAZARDOUS AIR POLLUTANTS, HAP, (AIR TOXICS)

Chapter 3 shows that U.S. air pollution law deals with pollutants that are believed to have no threshold (NESHAPS) differently from the way it deals with threshold value pollutants (NAAQS). The no-threshold pollutants are regulated mostly by emission standards, which restrict the emissions from all sources as much as is deemed practical. EPA refers to emission standards when applied to this category of pollutants as *technology-based* standards. In the Clean Air Act (1990 Amendments) these are called *Hazardous Air Pollutants*, or HAP, in Section 301, and *Air Toxics* in Section 206. The two names are equivalent.

Table 2.1 shows the eight substances that had been designated by the EPA as HAP by 1998. The list includes three metals, two VOCs, one mineral, and two others, of which coke oven emissions are largely a mixture of VOCs. Most HAP are known or suspected carcinogens. In the 1970 version of the Clean Air Act, it was clearly intended that the EPA would examine substances that should probably be regulated as HAP, one at a time, and add them to the list (with appropriate control regulations) at a steady pace. Twenty years later only the eight shown had completed this process. Table 15.2 shows a sample of the Emission Standards for Hazardous Air Pollutants issued under the provisions of the 1970 Clean Air Act. From that table we see that the regulations are a combination of simple weight emissions standards and ambient air quality standards. In addition there are some prohibitive and work practice standards, mostly relating to asbestos.

TABLE 15.2
National Emission Standards for Hazardous Air Pollutants

This list is an excerpt from the 1998 version of *40CFR61*. Standards are listed there for 18 categories of HAP emitters. This excerpt shows the kind of regulations that are contained in that much larger compilation, which also includes test procedures and reporting details.

1. Emissions of radon-222 to the ambient air from an underground uranium mine shall not exceed those amounts that would cause any member of the public to receive in any year an effective dose equivalent of 10 mrem/yr.
2. Emissions of beryllium from a stationary source shall not exceed 10 g/24-h period, or the source may show that its emissions do not cause an ambient concentration greater than 0.01 $\mu g/m^3$ averaged over a 30-day period.
3. Emissions of mercury from mercury ore processing facilities and mercury chlor-alkali plants shall not exceed 2.3 kg/24 h. Emissions of mercury from sewage sludge incineration plants shall not exceed 3.2 kg/24 h.
4. The concentration of vinyl chloride in the exhaust stream from any vinyl chloride formation and purification equipment shall not exceed 10 ppm (averaged over 3 hours). For each loading or unloading operation the quantity of vinyl chloride transferred to the atmosphere shall be less than 0.0038 m^3 (0.13 ft^3) at stp. There are also detailed rules for release of vinyl chloride from relief valves, strippers, pump seals, etc.

In the 1990 amendments to the Clean Air Act, Congress took a different approach by listing 189 materials as hazardous air pollutants, including the eight previously designated. Congress requires the EPA to issue control regulations for any source that emits 10 tons/yr of any of these 189 materials or 25 tons/yr of any combination of them, as well as requiring regulations for smaller sources under some circumstances. For example, the EPA published the following information on ethylene oxide sterilizers [8]:

> Ethylene oxide is widely used as a sterilant/fumigant in the production of medical equipment and in sterilization and fumigation operations. Current estimates indicate that there are about 190 facilities in the U.S. performing ethylene oxide commercial sterilization. Commercial sterilization is performed by medical equipment suppliers, pharmaceutical manufacturers, spice manufacturers, contract sterilizers, libraries, museums and archives, and laboratories. Emissions of ethylene oxide are estimated at 1.1 million kg/yr (2.4 million lb/yr) from commercial facilities.... The adverse health effects from ethylene oxide are well documented.... Many ethylene oxide sterilizers are located near population centers and may pose a threat to the surrounding public.... The Agency estimates that the maximum individual lifetime cancer risk associated with any commercial sterilizer is as high as one in 100.... currently there are no Federal regulations covering ethylene oxide sterilizer emissions except...OSHA requirements for workplace exposure levels.... there are few commercial sterilizers that exceed 9.07 Mg/yr (10 tpy) of ethylene oxide emissions....

We see that the number of sources is small (190), that the nationwide emissions from those sources are about 1200 tons/yr (compared to the millions of tons per year in Table 1.1), and that few of the 190 sources emit 10 tons/yr, which is a typical minimum value for a source to have individual control regulations and to require a state permit. But the material involved is sufficiently toxic that it may pose a serious health threat to those living near such a facility. The Hazardous Air Pollutant regulations are designed for the large number of chemicals believed to be of this type.

The list of 189 substances is in section 112 of the 1990 Clean Air Amendments. Of that list,

Sixteen are hydrocarbons, such as hexane, isooctane, and toluene.

One hundred forty-eight are organic, containing atoms other than H and C, such as methanol, vinyl chloride, and phenol.

Sixty-four are halogenated, such as carbon tetrachloride, chloroacetic acid, and epichlorohydrin.

Twenty are insecticides, herbicides, fungicides, or other compounds used to attack or deter some kind of living organism, such as DDT, 2,4-D, parathion, and pentachlorophenol.

Nine are metals or their compounds, such as mercury compounds, beryllium compounds, and selenium compounds.

Fifteen are not obvious members of any of these categories, such as radionuclides, asbestos, and phosphorus.

These numbers do not sum to 189 because some compounds fit in several categories, e.g., DDT, which is an organic compound, a halogenated compound, and an insecticide.

The control methods for the current eight substances and those likely to be used for these 189 will be similar to those discussed in previous sections of this book. Particulate matter will be controlled by the methods in Chapter 9, volatile materials by those in Chapter 10, gases capturable by liquid absorption by the methods in Chapter 10, etc.

The air toxic currently receiving the most attention (1999) is probably mercury. Metallic mercury is toxic, but its organic derivatives methyl mercury and ethyl mercury are far more toxic and have caused terrible poisonings. Our principal concern is the pathway, mercury vapor emissions → accumulation in lakes and shallow oceans → consumption by microscopic marine life, which converts elemental mercury to organic derivatives → mercury concentration as it passes up the food chain → human mercury poisoning by eating contaminated seafood. Total U.S. emissions of mercury (1995) are estimated at 158 tons/yr (compare to the millions of tons/yr of major pollutants in Table 1.1), of which 87% comes from combustion sources [9].

15.4 INDOOR AIR POLLUTION

Air pollution normally means contamination of the open (ambient) air outside our structures. Most air pollution law is directed at reducing the contamination of that air. Air contamination inside our factories and workplaces is regulated by an entirely different set of laws and a different agency of the government (OSHA) than the one that regulates outdoor air pollution (EPA). The average person spends 70 percent of her or his time in his or her residence. No one now regulates pollution inside residences and nonindustrial buildings.

15.4.1 Indoor and Outdoor Concentrations

Measurements regularly indicate that pollutant concentrations inside houses, indoor air pollution, can be higher than those outside. Figures 15.2 and 15.3, on pages 543 and 544, show some examples of comparison of indoor and outdoor pollutant concentrations.

Figure 15.2 shows average measurements of NO_2 taken outdoors, in the kitchens, and in the bedrooms of 70 houses with gas cook stoves in Topeka, Kansas, over a 12-month period. Gas cook stoves are a significant source of NO_2 (see Chapter 12), so we would expect a higher concentration indoors than out, and a higher concentration in the kitchen than in the bedroom. This set of data suggests that there is some air interchange between the kitchen and the bedroom, but they are not totally mixed. The plot also shows that although the outdoor concentration varies little from month to month, the indoor concentrations are higher in winter than summer. This

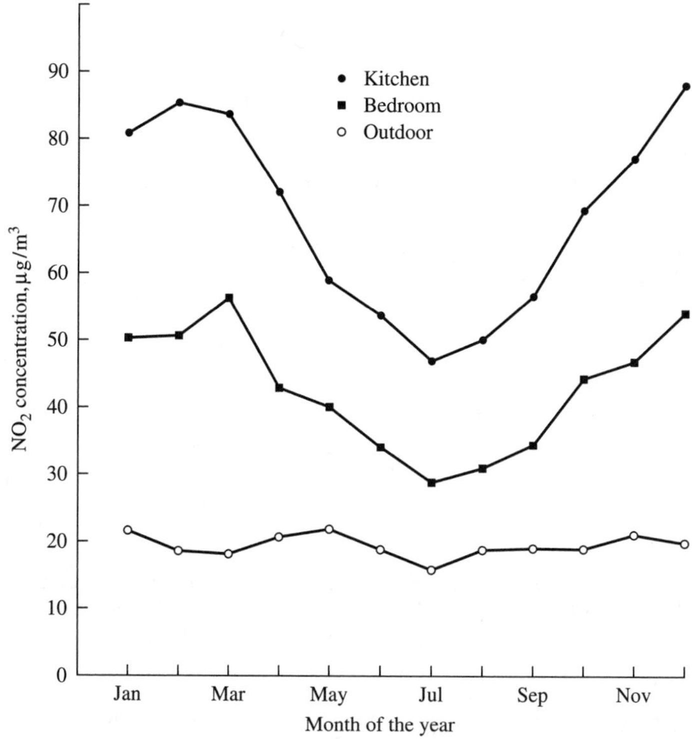

FIGURE 15.2
Comparison of monthly average indoor and outdoor concentrations of NO_2 in homes with gas cooking stoves, for each month of the year, in Topeka, Kansas [10]. The health-based NAAQS for NO_2 (see Table 2.3) is 100 $\mu g/m^3$, annual average. The outdoor value is $\approx$ 20% of the NAAQS and the kitchen annual average is $\approx$ 70% of the NAAQS.

difference is most likely due to greater air exchange between the inside and outside of the house in summer as a result of having windows open for cooling.

Figure 15.3 shows respirable particle concentrations, measured outdoors, inside houses, and personally. Volunteers wore a personal monitor, which is like a very small, battery-powered vacuum cleaner with its inlet near the wearer's breathing zone, for an entire day to obtain these personal values. Some of the 48 volunteers who carried the monitors spent their days in the home, while others wore them at work, during their commute to and from work, and in the home. The indoor and personal measurements go up and down with the outdoor values, but are generally larger. This pattern indicates that the outdoor concentration influences the indoor concentration but that particulate sources in the home add to the outdoor value. The personal values indicate that our individual activities also generate particles, so that the concentration near our breathing zones will generally be higher than that in parts of the house or workplace where there is no current human activity. Those who smoke or who are near smokers have significantly higher personal respirable particle exposure.

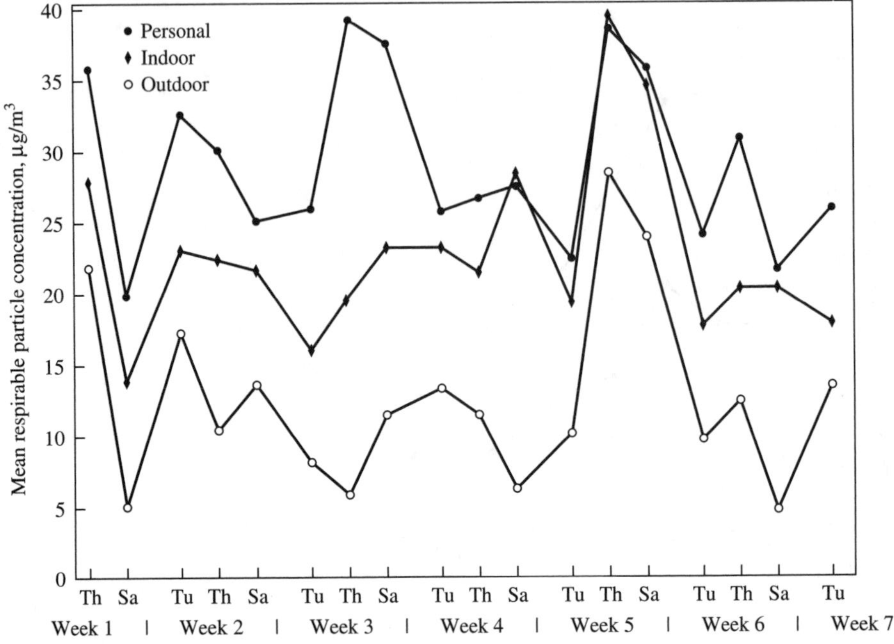

FIGURE 15.3
Comparison of daily outdoor, indoor, and personal sampler concentrations of respirable particles in Topeka, Kansas [11]. Respirable particles are $\approx PM_{2.5}$ for which the NAAQS (see Table 2.3) is 25 $\mu g/m^3$ annual average and 65 $\mu g/m^3$ maximum 24-hour average. None of the values shown here is more than about two-thirds of the 24-hour standard. The personal sampler values are above the annual average standard, and the indoor values close to it.

For ozone the opposite is observed: indoor concentrations are generally less than outdoor concentrations. Ozone is a very active chemical, which reacts with almost any solid surface. Thus the ozone coming into the house with outside air is destroyed by those surfaces. There are few indoor sources of ozone (such as copying machines and electrostatic air cleaners), so the observed indoor concentrations are lower than those outdoors [12]. For homes without indoor NO_2 sources the same is observed: indoor concentrations of NO_2 are lower than those outdoors, presumably for the same reasons [13].

15.4.2 Models

Just as we have air pollutant concentration models for ambient air (Chapter 6), we have such models for indoor air. As in Chapter 6 there are simple models and more complex models. Figure 15.4 shows the flow diagram for one regularly used model.

15.4.2.1 Simple Box Model. To construct the simplest model, which is virtually identical to the simple fixed-box model in Section 6.2, we consider a building whose internal air is always perfectly mixed, with only one flow in and one flow out, in this

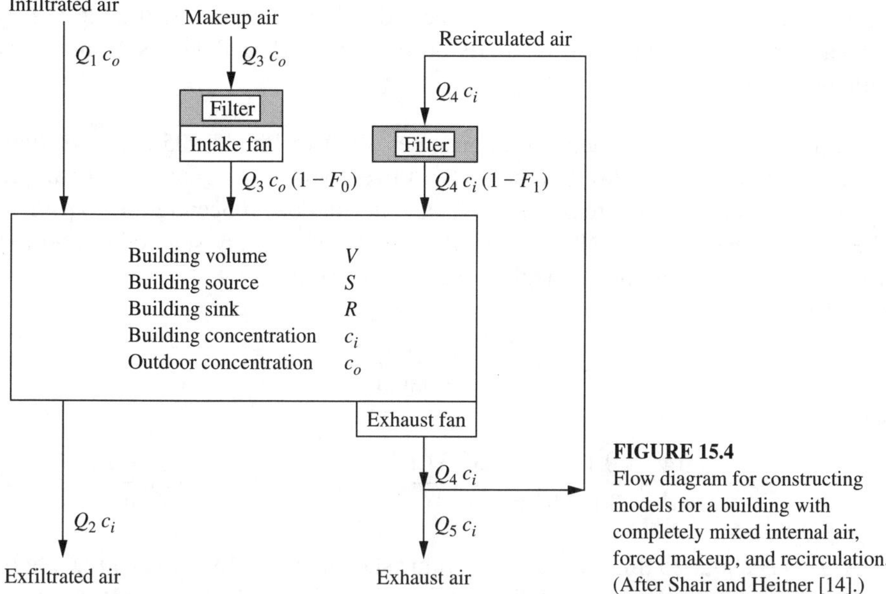

FIGURE 15.4
Flow diagram for constructing models for a building with completely mixed internal air, forced makeup, and recirculation. (After Shair and Heitner [14].)

case the *infiltration* and *exfiltration*. (These terms describe the unintentional flow of air into and out of the structure through window leakage, cracks in walls and roof, etc., driven by the wind and/or the difference in temperature between the inside and outside of the house. The observed values are discussed below.) The makeup, recirculated, and exhaust air on Fig. 15.4 are assumed to be zero for the simple box model.

A steady-state material balance for the pollutant for this set of assumptions, using the terms defined on Fig. 15.4 is

$$\begin{pmatrix}\text{Flow of}\\\text{pollutant out}\end{pmatrix} = \begin{pmatrix}\text{flow of}\\\text{pollutant in}\end{pmatrix} + \begin{pmatrix}\text{pollutant generated}\\\text{in structure}\end{pmatrix} - \begin{pmatrix}\text{pollutant removed}\\\text{in structure}\end{pmatrix} \tag{15.1}$$

$$Q_2 c_i = Q_1 c_o + S - R$$

If we ignore the change in temperature and humidity of the air passing through the structure, then Q_2 will be approximately equal to Q_1, so we may simplify Eq. (15.1) to

$$c_i = c_o + \frac{(S - R)}{Q_1} \tag{15.2}$$

Comparing Eq. (15.2) to the simple box model of Eq. (6.7), we see that c_o plays the same role as the background concentration b, and that $(S - R)/Q_1$ is the amount by which internal emissions, minus pollutant removal, increase the concentration in the structure; this quantity is comparable to qL/uH in Eq. (6.7). If the indoor removal rate is greater than the indoor emission rate, $R > S$, then the second term on the right in Eq. (15.2) will be negative and the indoor concentration will be less than the outdoor concentration, as is regularly observed with O_3 and with NO_2 in houses without gas cook stoves.

Example 15.2. A gas kitchen stove consumes 3000 kcal/h (11 905 Btu/h) of fuel. The emission factor for NO_2 for that kind of stove is about 61 μg/kcal (see Example 12.5). The air infiltration rate is 3000 ft^3/h, and the outdoor concentration 20 μg/m^3. There is no destruction of NO_2 ($R = 0$). Estimate the steady-state concentration when the stove has run long enough to reach steady-state.

Here

$$S = 3000\,\frac{\text{kcal}}{\text{h}} \cdot 61\,\frac{\mu\text{g}}{\text{kcal}} = 0.183\,\frac{\text{g}}{\text{h}}$$

and

$$c_i = 20\,\frac{\mu\text{g}}{\text{m}^3} + \frac{0.183\text{ g/h}}{3000\text{ ft}^3\text{/h}} \cdot \frac{35.31\text{ ft}^3}{\text{m}^3} = 20 + 2154 = 2174\,\frac{\mu\text{g}}{\text{m}^3} \qquad \blacksquare$$

This is about 20 times the average value shown in Fig. 15.2, indicating a high concentration at steady state with the stove running. Recall, however, that the stove does not run (full on) for more than perhaps 1/20th of the day. We will see in the next example that it would take a long time for the concentration in the kitchen to approach this steady-state value.

15.4.2.2 More Complex Models. The simple box model of Eq. (15.2) does not tell us how the concentration changes with time (e.g., after the stove is turned off) nor does it give us any way of estimating the concentration in the bedroom in Fig. 15.2. The next level of model complexity is based on all the flows in Fig. 15.4 and takes changes with time into account. The makeup air, exhaust air, and recirculated air in Fig. 15.4 are part of the ventilating system of any modern office, commercial, or industrial building. The hot-air furnace in a typical house, when it is running, acts as a recirculating air loop as shown, normally with a filter that removes large particles but not other pollutants. The filters shown in Fig. 15.4 are designed to remove particles, but some buildings also use charcoal adsorbers to remove some gases, so the treatment here is general if we consider the filter to be any air pollution control device.

Writing the general balance equation, Eq. (6.1), for the well-mixed house with all the flows shown in Fig. 15.4, we find

$$V\frac{dc_i}{dt} = [Q_1c_o + Q_3c_o(1 - F_0) + Q_4c_i(1 - F_1)] - (Q_2 + Q_4 + Q_5)c_i + S - R \tag{15.3}$$

Here the Fs represent filter efficiencies for the pollutant of interest.

Example 15.3. In Example 15.2, how long must the stove run for the NO_2 concentration in the kitchen to reach 90% of the steady-state value computed in Example

15.2? We assume that the volume of the kitchen is $V = 1000\ \text{ft}^3$. Then for the same conditions assumed in Example 15.2, we may write

$$V\frac{dc_i}{dt} = Q_1 c_o - Q_2 c_i + S$$

$$\frac{dc_i}{c_o - c_i + S/Q} = \frac{Q}{V}\,dt$$

$$\ln\frac{(c_o - c_i + S/Q)}{(c_o - c_i + S/Q)_{\text{initial}}} = -\frac{Q}{V}\,\Delta t$$

$$\Delta t = \frac{Q}{V}\ln\frac{(c_o - c_i + S/Q)_{\text{initial}}}{(c_o - c_i + S/Q)}$$

Here

$$c_o = c_{i_{\text{initial}}} = 20\ \frac{\mu\text{g}}{\text{m}^3} \text{ and } c_i = 0.9 \cdot 2174 = 1957\ \frac{\mu\text{g}}{\text{m}^3}$$

$$\frac{S}{Q} = \frac{0.183\ \text{g/h}}{3000\ \text{ft}^3/\text{h}} \cdot \frac{35.31\ \text{ft}^3}{\text{m}^3} = 2154\ \frac{\mu\text{g}}{\text{m}^3}$$

so that

$$\Delta t = \frac{1000\ \text{ft}^3}{3000\ \text{ft}^3/\text{h}}\ln\frac{20 - 20 + 2154}{(20 - 1957 + 2154)} = 0.77\ \text{h}$$ ■

We see that the time to come to 90% of steady state is comparable to the normal cooking time for a meal, so that while the simple model in Example 5.2 shows the probable maximum concentration, normal short-term stove use is unlikely to produce room concentrations that high. The model based on Fig. 15.4 and Eq. (15.3) can be modified to take into account incomplete mixing in the building, multiple rooms with differing air flow and emission rates, etc. These and several other models are reviewed in Ref. 15. See Problems 15.14 and 15.15.

15.4.3 Control of Indoor Air Quality

From Eq. (15.2) we see the following:

1. If Q_1 is very large (all the windows are open), then the indoor and outdoor concentrations will be the same.
2. The difference between the indoor and outdoor concentrations, $(c_i - c_o)$, depends on $(S - R)$ and Q_1. We can make the indoor air cleaner than the outdoor air by increasing R, which normally means using some kind of filter or adsorber to treat the makeup or recirculated air, as shown in Fig. 15.4. If we do not do that, and the naturally occurring value of R is negligible, then we cannot make the indoor air cleaner than the outdoor air.
3. If the indoor air has a higher concentration than the outdoor air, $(c_i > c_o)$, we can reduce c_i by increasing Q_1 or by reducing S.

15.4.3.1 Air Flow. Almost all modern office, commercial, and industrial buildings have forced air ventilation systems, as sketched in Fig. 15.4. Most air-conditioned buildings do not have windows that can be opened, so for them the makeup air on Fig. 15.4 is the principal air flow. The makeup air flow rate is normally selected from ASHRAE (American Society of Heating, Refrigeration and Air Conditioning Engineers, Inc.) standards [16] in the United States and from comparable standards in other countries. These standards specify the estimated occupancy for a typical facility and the required fresh air supply.

Example 15.4. Estimate the amount of outdoor air that must be supplied to a 1000 ft^2 (floor area) dining room.

From the ASHRAE standard we find that the estimated occupancy for dining rooms is 70 people/1000 ft^2, and that the minimum outdoor air flow required is 35 cfm/person for a smoking area and 7 cfm/person for nonsmoking. Thus, using the terminology in Fig. 15.4 and assuming this is a smoking dining room, we may write

$$Q_3 = (70 \text{ people}) \left(35 \, \frac{\text{cfm}}{\text{person}} \right) = 2450 \text{ cfm} = 1.16 \frac{\text{m}^3}{\text{s}}$$

If it is a nonsmoking room we need only $7/35 = 0.20$ times that value. ■

This calculation assumes that there is no removal of pollutants from the recirculating flow shown in Fig. 15.4 ($F_1 = 0$), so that all pollutant removal occurs via the exhaust or exfiltrated air. However, to save on heating and air-conditioning costs, we try to keep this fresh air flow as small as possible, because we must reject a comparable amount of heated or air-conditioned air via Stream 5. Some facilities cannot take in fresh air (submarines, space capsules), and they must design air treatment systems good enough that the recirculating system keeps the air contaminant concentrations at acceptable levels. Most buildings recirculate much of their air.

The recirculation rate is chosen to keep the air movement in the room high enough for the comfort of the occupants. The normal recommendation is that the total air flow to a room should be six to nine times the room volume each hour. Thus, in Example 15.4 if the ceiling height were 10 ft, so that the volume of the room were 10 000 ft^3, the recommended flow would be 60 000 to 90 000 cfh $=$ 1000 to 1500 cfm. For the smoking dining room the outdoor air requirement is greater than this, so no recirculated air would be used. For the nonsmoking dining room the outdoor air requirement is $(2450)(7/35) = 490$ cfm, so we would recirculate one to two times the amount of outdoor air supplied. This treatment of outdoor air and recirculation requirements is a great simplification of that presented in the air-conditioning literature.

For buildings without forced ventilation, the air exchange with the surroundings is by infiltration, the principal mode of air exchange for most houses, trailers, etc. When the outside temperature is close to our desired comfort temperature we open the windows and the infiltration rate is very high. When the outdoor temperature is colder or hotter than our comfort temperature, we close the windows to try to prevent infiltration. For a typical building without forced ventilation, with the windows closed, one may estimate the infiltration from Table 15.3.

On the basis of this table, and assuming that the kitchen in Examples 15.2 and 15.3 is a corner room with windows and doors on two sides, we would assume that it had 1.5 air changes per hour, or $Q_1 = 1500$ cfh. In those examples we used 3000 cfh, and comparison with Fig. 15.2 suggests that this is close to correct (in winter). Kitchen fans and their external closures are often leaky, so that the observed infiltration rate in kitchens is greater than Table 15.3 would suggest. The values in Table 15.3 allow only the roughest of estimates. Much more detailed infiltration estimating procedures are widely available [17]. (Example 15.2 does not consider exchange of air between the kitchen and the rest of the house, which also occurs. It has the consequence of lowering the NO_2 concentrations in the kitchen, compared to what we calculate in Example 15.2. The high concentration of NO_2 in the bedroom is presumably due to this internal exchange. See Problem 15.14.)

After the energy crises of 1972 and 1979, the price of home heating fuel in the United States rose significantly. This increase led to efforts to build more energy-efficient houses and to modify existing houses to cut down on heat leakage. As a consequence, the rate of air infiltration and exfiltration of houses was greatly reduced. Unfortunately, as Eq. (15.2) shows, that increases the concentration of indoor air pollutants. One possible solution to the problem is to increase the flow of air in and out, using air-to-air heat exchangers to recover the heating or air conditioning from the exhaust air. Those allow us to keep the ventilation rate high, without excessive heating or cooling costs. Such devices are available, but to date their economics have not been very attractive.

15.4.3.2 Indoor Emissions. The other alternative is to reduce the rate of emission of air pollutants inside structures. The major sources of pollutant emissions inside structures are smoking, combustion, solvent and household chemical usage, and emissions from the building materials of the structure itself. Combustion in furnaces

TABLE 15.3
Estimating values for air infiltration into typical residences whose windows are closed, expressed as air changes per hour (based on ASHRAE documents)

Type of room	Air changes per hour
No windows or exterior doors	0.5
Windows or exterior doors on one side	1
Windows or exterior doors on two sides	1.5
Windows or exterior doors on three sides	2
Entrance halls	2

For rooms with weather-stripped windows or with storm sash, use two-thirds of these values.

and water heaters normally does not contribute significantly to indoor air pollution because the combustion products from those are vented (carried by flues to the outside). But kitchen stoves, kerosene heaters, some space heaters, candles, etc., are not vented to the outside and are a major source of indoor air pollutants. Good summaries of indoor air pollution emission sources have been published [12, 18, 19]. Currently the Consumer Product Safety Commission is attempting to regulate those consumer products that emit the most indoor air pollutants so as to reduce their emissions (thus bringing a third federal agency into the air pollution field!) [20].

Indoor air pollution involves not only the major pollutants that are regulated in the outdoor air but also bacteria, molds, etc. that can grow on the filters of air-conditioning systems in humid climates, and various miscellaneous chemicals that are not believed to be problems in the outdoor air. As we have gone to tighter buildings, with less and less fresh air input, for energy conservation reasons, we have observed the *sick building syndrome,* in which workers in such a building report various health complaints believed to be caused by the building. This topic also has a substantial literature [21]. Radon, discussed next, is apparently not a serious problem outdoors but is a problem in buildings (and uranium mines).

15.5 THE RADON PROBLEM

Radon is a natural by-product of the decay of radioactive materials in the earth. Radon emits an alpha particle, which has little penetrating power, so that it causes health damage only if the alpha particle is emitted deep inside the lungs. Radon is a demonstrated cause of lung cancer. Underground uranium miners in the 1950s were exposed to significant amounts of radon. After their lung cancers were detected, the ventilation requirements for uranium mining were significantly increased. The combination of radon and smoking is particularly deadly!

Radon is the first decay product from radium. It seeps slowly out of the ground, all over the world. The decay sequence is

$$\text{Uranium 238} \xrightarrow{\text{4.4 billion years}} \text{Radium 226} \xrightarrow{\text{1620 years}} \text{Radon 222} \xrightarrow{\text{3.8 days}}$$
$$\text{Polonium 218} \xrightarrow{\text{3.05 minutes}} \text{Lead 214} \xrightarrow{\text{26.8 minutes}} \text{Bismuth 214} \xrightarrow{\text{19.7 minutes}}$$
$$\text{Polonium 214} \xrightarrow{\text{4.44 hours}} \text{Lead 210}$$

The numbers above the arrows show the radioactive decay half-lives [22]. Radon has a half-life of 3.8 days, which means that it does not last long in the atmosphere. For it to be a problem in our houses it must be continuously replenished by flow from the soil (or be released from being dissolved in drinking water). The *daughter products* between radon and lead 210 all have short half-lives so that once radon decays, giving off its radiation, its daughter products do so very quickly thereafter. For this reason most discussion simply lumps the daughter products with radon, combining their concentration, emissions, and effects. These daughter products are solids that rapidly attach themselves to ambient particles and thus can be transported deep within the lungs, where they can be deposited. They are the principal source of health damage from radon [23].

TABLE 15.4
Typical values of radon concentration, in pCi/L

	Range	Average
Ambient air	0.1–30	0.2
Indoor air	1–3000	1–2
Pore space in soil	20–100,000	
Dissolved in ground water	100–3 million	
EPA recommended maximum in air in a residence before remedial action should be taken	4	

Concentrations of radon are measured in pCi/L (see Problem 15.18). Typical values are shown in Table 15.4, which shows that the recommended maximum concentration above which remedial action is recommended is 4 pCi/L. This is roughly 2 percent of the permitted concentration in uranium mines.

The radon seepage rate from soil to a basement is higher in some parts of the country than others due to higher concentrations of radioactive materials in some geologic formations than others (igneous rocks generally have higher concentrations than sedimentary rocks). As we have made our houses tighter against air exchange with the surroundings, the concentration of radon in the houses due to leakage from the ground into house basements has increased. Additionally, some building materials can emit radon. Portland cement foundations made with a sand high in uranium or radium can be a serious radon emitter. In the 1950s tailings from uranium processing plants were used as sand in some building construction. Those buildings were all demolished when their high radon concentrations were discovered.

Example 15.5. The worldwide average rate of radon emission from the ground [22] is estimated at $q = 0.42\ \text{pCi/m}^2 \cdot \text{s}$. Let us assume the emission rate from under our house is equal to the worldwide average. Our house has a crawl space with a dirt floor, so all the radon being emitted from the soil may be assumed to enter the crawl space. If it is 4 feet high and has an infiltration equal to one air turnover per hour, estimate the radon concentration in the crawl space. The outdoor concentration c_o equals 1 pCi/L.

In Eq. (15.2), we set $R = 0$ because the radioactive decay rate is small compared with the ventilation rate. The volume of the crawl space is hA, where h is the height and A is the floor area; and the ventilation rate is $Q = hA/\Delta t$. The source S is qA, so that we may write

$$c_i = c_o + \frac{qA}{hA/\Delta t} = c_o + \frac{q\,\Delta t}{h}$$

$$= 1\,\frac{\text{pCi}}{\text{L}} + \frac{(0.42\ \text{pCi/m}^2\cdot\text{s})(1\ \text{h})}{(4\ \text{ft})}\left(\frac{3600\ \text{s}}{\text{h}}\right)\left(\frac{3.28\ \text{ft}}{\text{m}}\right)\left(\frac{\text{m}^3}{1000\ \text{L}}\right)$$

$$= 1 + 1.24 = 2.24\,\frac{\text{pCi}}{\text{L}}$$

■

This example shows that for the world average emission rate, the expected concentrations are less than the EPA action level of 4 pCi/L. We would expect concentrations higher than that only in geographic areas where the emission rate from the soil was significantly larger than the worldwide average. The actions to be taken in a house with excessive radon can include any combination of the following:

1. Seal the basement or crawl space as thoroughly as practical, closing all sources of air entry from the soil.
2. Provide forced ventilation of the basement or crawl space, using outdoor air, thus forcing the radon-contaminated air out of the building.
3. Install ventilation in the soil under and around the building to remove the radon before it can enter the building.
4. Use a fan to place the basement or crawl space under slight positive pressure so that any air leakage will be outward, not inward.

Radon in houses is a serious concern, for which the EPA and state and local governments are trying to devise a control program. It does not fit into the legislative framework for either ambient or workplace air pollution.

15.6 SUMMARY

1. CO is mostly an automotive problem, particularly in city centers, where the measured concentrations are highest. We control CO by oxidizing it to CO_2.
2. Lead was formerly emitted from the tetraethyl lead octane improver in gasoline. As that is being removed from gasoline (in the United States, but not in all other countries) the lead concentration in the atmosphere has fallen dramatically. The remaining lead in the atmosphere comes from various combustion sources.
3. Hazardous air pollutants are mostly VOCs. They are regulated by very specific limitations on emitters, which are mostly industrial.
4. Pollutant concentrations indoors are related to those outdoors because of the intentional or unintentional exchange of air from inside to outside of our buildings. If there are indoor sources, e.g., smoking or cooking, indoor pollutant concentrations can be higher than those outdoors. For very reactive pollutants like ozone, the indoor concentrations are mostly lower than those outdoors.
5. Radon is not a regulated pollutant, but it is found in most houses as a result of seepage from the underlying soil.

PROBLEMS

See Common Values for Problems and Examples, inside the back cover.

15.1. The Coburn model for uptake of CO by a human being [4] is

$$\frac{d(\mathrm{CO})}{dt} = \dot{V}_{\mathrm{CO}} - \frac{[\mathrm{COHb}]}{[\mathrm{O_2Hb}]}\frac{(\overline{P}_{\mathrm{CO_2}})}{(M)}\frac{1}{[1/D_L + (P_B - P_{\mathrm{H_2O}})/\dot{V}_A]} + \frac{P_{I\mathrm{CO}}}{[1/D_L + (P_B - P_{\mathrm{H_2O}})/\dot{V}_A]} \tag{15.4}$$

Here the symbols are the same as in the original reference and the medical literature, which are different from normal engineering usage; they are not shown in this book's table of nomenclature.

$\dfrac{d(\mathrm{CO})}{dt}$ is the net rate of CO uptake by the body, mL(stp)/minute

$\dot{V}_{\mathrm{CO}}$ is the rate at which the body produces CO, mL(stp)/minute

[COHb] is the concentration of carboxyhemoglobin in the blood, expressed as mL CO (stp)/mL of liquid blood

$[\mathrm{O_2Hb}]$ is the concentration of oxyhemoglobin in the blood expressed as mL O_2 (stp)/mL of liquid blood

$\overline{P}_{\mathrm{CO_2}}$ is the partial pressure of oxygen, measured in the lungs (which is less than that in the atmosphere, due to partial removal of O_2 by the blood)

M is the Haldane constant, the ratio of the blood's affinity for CO to its affinity for O_2 (220 at a blood pH of 7.4)

D_L is the [(area)(mass transfer coefficient)] for CO in the lungs, mL(stp)/ (min · mm Hg)

P_B is the atmospheric pressure, mm Hg

$P_{\mathrm{H_2O}}$ is the vapor pressure of water at body temperature = 47 mm Hg

$\dot{V}_A$ is the rate of air exchange between the atmosphere and the active (alveolar) region of the lungs, mL(stp)/min

$P_{I\mathrm{CO}}$ is the CO partial pressure in the air breathed in

(*a*) Show the logical basis for this equation. What does each term represent physically?
(*b*) Show that if one assumes that $d(\mathrm{CO}) = V_b\, d[\mathrm{COHb}]$, where V_b is the total volume of blood in the body, and that if one further assumes that the only variable quantity on the right side of Eq. (15.4) is [COHb], one can separate variables and integrate from 0 to t, finding

$$\frac{A[\mathrm{COHb}]_t - B\dot{V}_{\mathrm{CO}} - P_{I\mathrm{CO}}}{A[\mathrm{COHB}]_0 - B\dot{V}_{\mathrm{CO}} - P_{I\mathrm{CO}}} = \exp\frac{-tA}{BV_b} \tag{15.5}$$

where $A = \overline{P}_{\mathrm{CO_2}}/M[\mathrm{O_2Hb}]$

$B = (1/D_L) + (P_L/\dot{V}_A)$ and $P_L = P_B - P_{\mathrm{H_2O}}$

This is the form of the equation most commonly seen.

15.2. (*a*) Show the equation for the steady-state value of [COHb] from Eq. (15.5).
(*b*) Check to see if the values calculated from this steady-state equation correspond to the concentrations shown on Fig. 15.1, using the values of the parameters shown there. In addition you need

$$[\mathrm{O_2Hb}] \approx 0.20 \text{ mL(stp)/mL}$$

$$\%\mathrm{COHb} \approx [\mathrm{COHb}]/0.20 \text{ mL(stp)/mL}$$

You will find that the values agree quite well for the lower concentrations but not for the higher ones. The reason is that in integrating Eq. (15.4) we assumed that $[\mathrm{O_2Hb}]$

was a constant, independent of [COHb]. For low concentrations that is a satisfactory approximation, but not for high ones [24]. In making up Fig. 15.1 its authors integrated Eq. (15.4) numerically, not making that assumption.

(*c*) In the course of solving (*b*) you will find the predicted percent COHb for a person breathing air containing no CO at all. How much is that?

15.3. (*a*) Verify that the low concentration curves on Fig. 15.1 were made up from Eq. (15.5) by computing the value of the percent COHb at 100 minutes and 100 ppm, and comparing it to the value shown on that figure. Use the data on the figure plus those given in Problem 15.2. As discussed in Problem 15.2, a more complex equation was used for the higher concentrations.

(*b*) Estimate the concentration of CO being breathed that will cause COHb to be 70 percent (and presumably be fatal) in an exposure of 5 minutes, using Eq. (15.5). This calculation somewhat underestimates the required concentration. However, the OSHA lists any concentration greater than 1500 ppm as IDLH (Immediately Dangerous to Life and Health). On the basis of Fig. 15.1, do you think that is conservative?

15.4. In movies and books a favorite way to commit suicide is to sit in an idling auto in a closed garage and die of CO poisoning. Is that feasible with a properly tuned, current catalyst-equipped, computer-controlled auto?

My auto (a 1993 Subaru) at its 1997 emission test had an idling speed of 715 RPM. The measured CO content of the exhaust gas was 300 ppm. The engine displacement is 1.8 L. Like all four-stroke engines it fires every second revolution. The O_2 content of the exhaust gas was 0.9%, which may be taken for the purposes of this problem as equal to zero.

(*a*) Estimate the emission rate of CO, at idle, in g/h or equivalent.

(*b*) Estimate the consumption rate of O_2 in g/h or equivalent.

(*c*) If my garage is 20 ft · 20 ft · 8 ft, and by natural ventilation it has one air exchange per hour, estimate the steady-state concentrations of CO and O_2 of the garage.

(*d*) Comment on the feasibility of this mode of suicide. O_2 concentrations less than $\approx$ 16% are dangerous and those below $\approx$ 6% are fatal. Don't try this yourself unless you have total faith in your calculations (and my assumptions)!

15.5. Figure 2.8 shows the effect of NO_2 on plants on concentration-time coordinates. What would Eq. (15.5) look like on those coordinates? Presumably the effect is the same for any percent COHb, so the question is really what a curve of constant percent COHb looks like on that plot. Sketch a curve for 10 percent COHb on such a plot.

15.6. Estimate the temperature at which the vapor pressure of metallic lead is 10 mm Hg and 100 mm Hg, using the values in Appendix A. Are these temperatures likely to be observed in incinerators? In coal combustion?

15.7. The published emission factor [25] for lead from municipal incinerators (before the particulate control device) is 0.065 to 0.09 kg/tonne of waste burned.

(*a*) Assuming that all the lead in the waste is in the gas stream leaving the combustor (and going to the control device), estimate the lead content of municipal waste.

(*b*) List the likely sources of lead in municipal waste that goes to an incinerator.

15.8. A typical kitchen fan has a flow rate of 200 cfm. It takes air from the kitchen and blows it outside the house. Rework Examples 15.2 and 15.3 on the assumptions that the kitchen has such a fan, and that it runs all the time.

15.9. In the morning the kitchen stove in Examples 15.2 and 15.3 is used for 15 minutes to prepare breakfast. Estimate the maximum NO_2 concentration in the kitchen, using the values in those two examples. Assume that at the start of breakfast cooking $c_i = c_o$.

15.10. The kitchen stove in Example 15.3 operates for 30 minutes, is then turned off for 15 minutes, then operates again for 15 minutes. Estimate the maximum NO_2 concentration that would be observed in the kitchen.

15.11. Modern submarines stay submerged for up to 60 days at a time.

(*a*) Show the version of Eq. (15.3) that is applicable to the air in such a submarine.

(*b*) A typical person doing desk work emits about 0.63 scf/h of CO_2. If the submarine has a crew of 100, how many pounds per hour of CO_2 must be removed from the circulating atmosphere of the submarine? How should this be done?

(*c*) Wadden and Scheff give a table of odorous gases emitted by humans [26]. How are those dealt with in houses? In commercial buildings? In submarines?

15.12. (*a*) Rework Example 15.2 for CO. The average emission factor for CO for all parts of a kitchen stove is 645 μg/kcal [27]. Assume that $c_o = 1$ ppm.

(*b*) What is the ratio of the predicted maximum CO concentration for any 1-hour period to the 1-hour NAAQS? What is the comparable ratio for the average NO_2 concentration to the annual average NAAQS?

15.13. Find the simplest formula that will adequately represent the relation between the concentration in the kitchen and the bedroom in Fig. 15.2. What kind of model does this suggest?

15.14. A more realistic model for Examples 15.2 and 15.3 is sketched in Fig. 15.5. In it the house is divided into two parts, the kitchen and the rest of the house. Each part is considered to be internally well mixed. The volumes of each are shown. The infiltration and exfiltration rates for each are shown. The air exchanges between the kitchen and the rest of the house are shown, and equal. Based on this model, estimate the NO_2 24-hour average concentrations in the kitchen and the bedroom

(*a*) If the stove runs for 24 hours a day.

(*b*) If the stove runs 0.5 h in the morning, 0.5 h at noon, and 1 h at 6 PM.

15.15. Repeat Problem 15.9 for the house whose flow diagram is shown in Fig. 15.5.

15.16. A typical U.S. house has an indoor floor area of 1500 ft^2 and a ceiling height of 8 ft. When its furnace is running it releases 90 000 Btu/h. Its flow diagram and appropriate temperatures are shown in Fig. 7.8*a*. If all of the combustion and dilution air is drawn from the house, what is the ratio of the air used per hour to the volume of the house? How does this compare with the typical infiltration rate of ≈ 1 house volume per hour? The fuel is

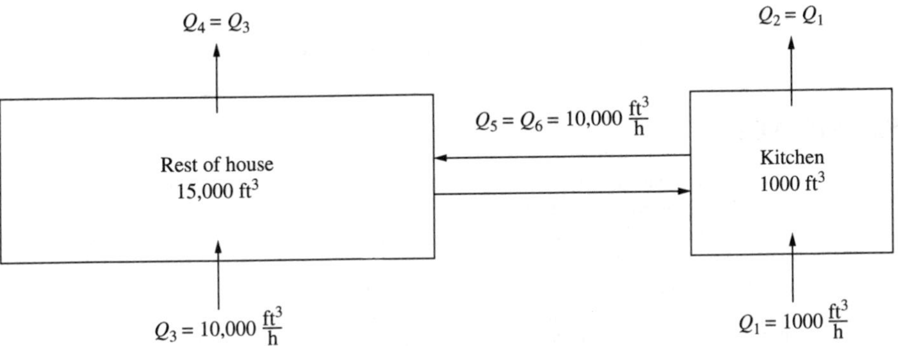

FIGURE 15.5
More realistic flow diagram for modeling concentrations of indoor pollutants emitted in the kitchen, used in Problems 15.14 and 15.15.

natural gas (≈ methane), see Table 7.1 and Example 7.9. (Most such furnaces take their air from under the house in the crawl space or from some other outdoor source, not from the living space of the house. Some do take their air from the living space.)

15.17. For the typical U.S. house in Problem 15.16, what is the winter daily fuel cost to heat the air that passes through the house by infiltration? Assume that the house is at 70°F and that the outside air is at 30°F. The heat capacity of air is roughly 0.25 Btu/lb · °F, and the cost of fuel (delivered to the house) is $\$7/10^6$ Btu. (The total heating bill is due to this infiltration loss plus the larger losses resulting from conduction through walls and ceilings.)

15.18. The concentrations of radioactive materials are regularly expressed in Ci or fractions thereof. For radon the unit used is normally the pCi $= 10^{-12}$ Ci. The curie is the approximate radiative power of 1 gram of pure radium and is defined as 3.7×10^{10} nuclear disintegrations per second. (The Bequerel is defined at 1 disintegration/s, so Ci $= 3.7 \times 10^{10}$ Bequerel.)

(*a*) How many atoms of radon produce one pCi of radiation? See Sec. 6.6 for the relation between half-life and decay rate.

(*b*) What is the mole fraction of radon in an air sample that just equals the EPA action level of 4 pCi/L?

15.19. Measurements in the basement of our house indicate a radon concentration of 6 pCi/L. The outdoor air in our area has a radon concentration of 1 pCi/L. The basement, which has a volume of 8000 ft^3, is believed to have an infiltration rate of outside air of 0.5 air change per hour. The air in the soil under the house has a radon concentration of 1000 pCi/L.

(*a*) Estimate the rate at which air from underneath the house leaks into the basement.

(*b*) If we decide to force outdoor air into the basement to increase the ventilation (i.e., make $Q_3 = Q_5 =$ some significant number), how large must we make it to reduce the radon concentration to 4 pCi/L?

(*c*) Does the half-life of radon play any significant role in these calculations?

(*d*) What other alternatives could we try to reduce the radon concentration in the basement?

15.20. In Example 15.5, how much would the answer change if we took the radioactive decay of radon into account?

15.21. The average concentration of radon in U.S. natural gas is about 23 pCi/L [22].

(*a*) Estimate the steady-state radon concentration in the kitchen in Example 15.2, when the kitchen stove is running continuously. The outdoor concentration may be estimated as 1 pCi/L.

(*b*) Should we be seriously concerned about radon in natural gas?

REFERENCES

1. "Compilation of Air Pollutant Emission Factors," U.S. Environmental Protection Agency Report No. AP-42, 4th ed., 9/91 supplement, p. 1.4-2, 1985.
2. Committee on Biological Effects of Environmental Pollutants: *Carbon Monoxide,* National Academy of Sciences, Washington, DC, 1977.
3. "Air Quality Criteria for Carbon Monoxide," U.S. Environmental Protection Agency Report No. EPA-600/8-79-022, U.S. Government Printing Office, Washington, DC, 1979.
4. Coburn, R. F., R. E. Forster, and P. B. Kane: "Considerations of the Physiological Variables That Determine the Blood Carboxyhemoglobin Concentration in Man," *J. Clinical Invest.*, Vol. 44, pp. 1899–1910, 1965.
5. *National Air Quality and Emissions Trends Report, 1997,* EPA 454/R-98-016, Chapter 2.
6. Fischbein, A.: "Environmental and Occupational Lead Exposure," in W. N. Rom (ed.), *Environmental and Occupational Medicine*, Little, Brown, Boston, MA, Chapter 38, 1983.
7. Peterson, J. E.: *Industrial Health*, Prentice Hall, Englewood Cliffs, NJ, p. 34, 1977.

8. U.S. Environmental Protection Agency: "Initial List of Categories of Sources under Section 112(c)(1) of the Clean Air Act Amendments of 1990," 57-FR31576, July 16, 1992.
9. *Mercury Study Report to Congress, Vol. 1, Executive Summary,* EPA-452/R-97-0032, December 1997.
10. Letz, R., D. P. Miller, and J. D. Spengler: "Relationships of Measured NO_2 Concentrations at Discrete Sampling Locations in Residences," in *Proceedings: National Symposium on Recent Advances in Pollutant Monitoring of Ambient Air and Stationary Sources, Raleigh, NC, May 5–6, 1983,* U.S. Environmental Protection Agency Report No. 600/9-84-001, U.S. Government Printing Office, Washington, DC, January 1984.
11. Spengler, J., and T. Tosteson: "Personal Exposure to Respirable Particles," *Environmetrics '81, Conference of the Society for Industrial and Applied Mathematics,* Alexandria, VA, 1981.
12. Wadden, R. A., and P. E. Scheff: *Indoor Air Pollution: Characterization, Prediction and Control,* John Wiley & Sons, New York, 1983.
13. Spengler, J. D., C. P. Duffy, R. Letz, T. W. Tibbits, and B. G. J. Ferris: "Nitrogen Dioxide inside and outside 137 Homes and Implications for Ambient Air Standards and Health Effects Research," *Environ. Sci. Technol.*, Vol. 17, pp. 164–168, 1983.
14. Shair, F. H., and K. L. Heitner: "Theoretical Model for Relating Indoor Pollutant Concentrations to Those Outside," *Environ. Sci. Technol.*, Vol. 8, pp. 444–451, 1974.
15. Wadden, R. A., and P. E. Scheff: op. cit, Chapter 6.
16. ASHRAE: *Standards for Ventilation Required for Minimum Acceptable Indoor Air Quality: ASHRAE 62-73R,* American Society of Heating, Refrigeration and Air-Conditioning Engineers, Inc., New York, 1980.
17. Wadden, R. A., and P. E. Scheff: op. cit, p. 107.
18. Meyer, B.: *Indoor Air Quality*, Addison-Wesley, Reading, MA, 1983.
19. Godish, T.: *Indoor Air Pollution Control,* Lewis, Chelsea, MI, 1989.
20. "The Inside Story, A Guide to Indoor Air Quality," U.S. Environmental Protection Agency Report No. EPA/400/1-88/004, U.S. Government Printing Office, Washington, DC, 1988.
21. *Sick Building Syndrome*: Citations from the NTIS Bibliographic Data Base, January 1990–March 1993, U.S. Department of Commerce, NTIS.
22. Colle, R.: "The Physics and Interaction Properties of Radon and Its Progeny," in R. Colle and P. E. J. McNall (eds.), *Radon in Buildings: NBS Special Publication 581,* U.S. Department of Commerce, National Bureau of Standards, Washington, DC, 1980.
23. Morken, D. A.: "The Biological and Health Effects of Radon: A Review" in R. Colle and P. E. J. McNall (eds.), *Radon in Buildings: NBS Special Publication 581,* U.S. Department of Commerce, National Bureau of Standards, Washington, DC, 1980.
24. Peterson, J. E., and R. D. Stewart: "Predicting the Carboxyhemoglobin Levels Resulting from Carbon Monoxide Exposures," *J. Appl. Physiol.*, Vol. 39, pp. 633–638, 1975.
25. *AP-42*, op. cit., p. 2.1-7, 9/90 supplement, 1985.
26. Wadden, R. A., and P. E. Scheff: op. cit, p. 73.
27. Wadden, R. A., and P. E. Scheff: op. cit, p. 53.

APPENDIX

USEFUL VALUES

A.1 VALUES OF THE UNIVERSAL GAS CONSTANT

$$
\begin{aligned}
R &= \frac{10.73\ (\text{lbf/in.}^2)\text{ft}^3}{\text{lbmol}\cdot{}^\circ\text{R}} = \frac{0.7302\ \text{atm}\cdot\text{ft}^3}{\text{lbmol}\cdot{}^\circ\text{R}} \\
&= \frac{8.314\ \text{m}^3\cdot\text{Pa}}{\text{mol}\cdot\text{K}} = \frac{0.08206\ \text{L}\cdot\text{atm}}{\text{mol}\cdot\text{K}} = \frac{0.08314\ \text{L}\cdot\text{bar}}{\text{mol}\cdot\text{K}} \\
&= \frac{1.987\ \text{Btu}}{\text{lbmol}\cdot{}^\circ\text{R}} = \frac{1.987\ \text{cal}}{\text{mol}\cdot\text{K}} = \frac{1.987\ \text{kcal}}{\text{kgmol}\cdot\text{K}} = \frac{8.314\ \text{J}}{\text{mol}\cdot\text{K}}
\end{aligned}
$$

Chemical engineers normally work with the Universal Gas Constant R. Several other branches of engineering use separate values of R for each gas. These are defined by $R_{\text{individual}} = R_{\text{universal}}/M$, so that, for example,

$$
R_{\text{air}} = \frac{R_{\text{universal}}}{M_{\text{air}}} = \frac{10.73\ \dfrac{(\text{lbf/in.}^2)\text{ft}^3}{\text{lbmol}\cdot{}^\circ\text{R}}}{28.96\ \text{lb/lbmol}} = 0.3705\ \frac{(\text{lbf/in.}^2)\text{ft}^3}{\text{lb}\cdot{}^\circ\text{R}} = 53.35\ \frac{\text{ft}\cdot\text{lbf}}{\text{lb}\cdot{}^\circ\text{R}}
$$

A.2 VAPOR PRESSURE EQUATIONS

Antoine equation constants, $\log_{10} p = A - \dfrac{B}{T + C}$, p in mm Hg, T in °C

Substance	Formula	Range, °C	A	B	C
Acetaldehyde	C_2H_4O	−45 to +70	6.81089	992.0	230
Acetic acid	$C_2H_4O_2$	0 to +36	7.80307	1651.1	225
		+36 to +170	7.18807	1416.7	211
Acetone	C_3H_6O	—	7.02447	1161.0	224
Ammonia	NH_3	−83 to +60	7.55466	1002.711	247.885
Benzene	C_6H_6	—	6.90565	1211.033	220.790
Carbon tetrachloride	CCl_4	—	6.93390	1242.43	230.0
Chlorobenzene	C_6H_5Cl	0 to +42	7.10690	1500.0	224.0
		+42 to +230	6.94504	1413.12	216.0
Chloroform	$CHCl_3$	−30 to +150	6.90328	1163.03	227.4
Cyclohexane	C_6H_{12}	−50 to +200	6.84498	1203.526	222.863
Ethyl acetate	$C_4H_8O_2$	−20 to +150	7.09808	1238.71	217.0
Ethyl alcohol	C_2H_6O	—	8.04494	1554.3	222.65
Ethylbenzene	C_8H_{10}	—	6.95719	1424.255	213.206
n-Heptane	C_7H_{16}	—	6.90240	1268.115	216.900
n-Hexane	C_6H_{14}	—	6.87776	1171.530	224.366
Lead	Pb	525 to 1325	7.827	9845.4	273.15
Mercury	Hg	—	7.975756	3255.61	281.988
Methyl alcohol	CH_4O	−20 to +140	7.87863	1473.11	230.0
Methyl ethyl ketone	C_4H_8O	—	6.97421	1209.6	216
n-Pentane	C_5H_{12}	—	6.85221	1064.63	232.000
Isopentane	C_5H_{12}	—	6.78967	1020.012	233.097
Styrene	C_8H_8	—	6.92409	1420.0	206
Toluene	C_7H_8	—	6.95334	1343.943	219.377
Water	H_2O	0 to 60	8.10765	1750.286	235.0
		60 to 150	7.96681	1668.21	228.0

Note: These are taken from a variety of sources. Longer lists of values are in J. A. Dean, *Lange's Handbook of Chemistry*, 12th ed., McGraw-Hill, New York, 1979, pp. 10-29 through 10-54; R. C. Reid, J. M. Prausnitz, and T. K. Sherwood, *The Properties of Liquids and Gases*, 3d ed., McGraw-Hill, New York, Appendix A, 1977; and D. R. Lide, ed. in chief, *CRC Handbook of Chemistry and Physics*, 71st ed., CRC Press, Boca Raton, FL, p. 6-70, 1990.

APPENDIX B

TABLE OF ACRONYMS

The following list of acronyms may help the reader understand some of the air pollution literature more easily.

ACFM	Actual Cubic Feet per Minute
AQCR	Air Quality Control Region
BACT	Best Available Control Technology; some specified type of control or emission rate that is judged to be "demonstrated" and thus reasonable to expect of all new sources, and perhaps of existing sources
BDC	Bottom Dead Center (auto engines)
CAA	Clean Air Act, generally the Clean Air Act of 1970. Sometimes, the Clean Air Amendments of other dates
CERCLA	Comprehensive Environmental Response, Compensation and Liability Act (also called the *superfund* act)
CFC	Chlorofluorocarbon
EIS	Environmental Impact Statement
EPA	Environmental Protection Agency, the principal federal air pollution agency
EPRI	Electric Power Research Institute
ESP	Electrostatic Precipitator
FGD	Flue Gas Desulfurizer (or desulfurization)
FGR	Flue Gas Recirculation
HAP	Hazardous Air Pollutants
HC	Hydrocarbon

IPCC	Intergovernmental Panel on Climate Change
IPP	Implementation Planning Program (see Chapter 6)
LAER	Lowest Achievable Emission Rate; somewhat more stringent than BACT, but applicable in severe situations where above-average costs may be justified
LEL	Lower Explosive Limit
MACT	Maximum Available Control Technology, similar to LAER
MTBE	Methyl Tertiary-Butyl Ether
NAAQS	National Ambient Air Quality Standard
NESHAP	National Emission Standard for Hazardous Air Pollutants
NMHC	Nonmethane Hydrocarbons
NOAA	National Oceanic and Atmospheric Administration, the custodian of weather data
NSPS	New Source Performance Standards (Officially these are Standards of Performance for New Stationary Sources, but they are almost always referred to as NSPS.)
NTIS	National Technical Information Service
NWS	National Weather Service
ODC	Ozone-Depleting Chemicals
$PM_{2.5}$	Particulate Matter, 2.5 microns in diameter or smaller
PM_{10}	Particulate Matter, 10 microns in diameter or smaller
POHC	Principal Organic Hazardous Constituent
PSD	Prevention of Significant Deterioration
RACT	Reasonably Available Control Technology
RCRA	Resource Conservation and Recovery Act (pronounced “reck-ra”)
RH	Relative Humidity
ROG	Reactive Organic Gases
RPM	Revolutions per Minute
RVP	Reid Vapor Pressure
SCFM	Standard Cubic Feet per Minute
SCR	Selective Catalytic Reduction (of NO_x)
SIP	State Implementation Plan
SNCR	Selective Noncatalytic Reduction (of NO_x)
STAPPA	State and Territorial Air Pollution Program Administrators
STP	Standard Temperature and Pressure
TDC	Top Dead Center (auto engines)
TLV	Threshold Limit Value (permissible concentration for exposure of industrial workers)
TSP	Total Suspended Particulates
TWA	Time-Weighted Average
UAM	Urban Airshed Model
UEL	Upper Explosive Limit
VOC	Volatile Organic Compound

APPENDIX C

FUELS

Most air pollutants are emitted from combustion or from high-temperature processes. Section 7.10 discusses combustion, assuming that the readers are familiar with fuels. This appendix discusses the elementary organic chemistry of fuels and provides some details on the various kinds of fuels in common usage.

C.1 WHERE FUELS COME FROM, HOW THEY BURN

Almost all the fuels we burn are ultimately based on solar energy, in the form of the photosynthesis reaction

$$nCO_2 + nH_2O + \text{sunlight} \rightarrow (CH_2O)_n + nO_2 \tag{C.1}$$

Here $(CH_2O)_n$ is the generic formula for carbohydrates. The carbohydrates with which the student is most likely familiar are ordinary sugar, $C_{12}(H_2O)_{11}$, in which one H_2O was lost when two six-membered subunits joined together, and cellulose, the structural material of all plants, trees, wood, and paper, which is practically $(CH_2O)_n$, with n being a very large number. Almost all life on this planet is based on eating carbohydrate-containing foods.

Plants and animals convert carbohydrates to fats, which is mostly an oxygen removal process. Fats are $(CH_2)_nO_m$ in which m is much smaller than n so that they are close to being $(CH_2)_n$. The stored energy per pound or gram of fats is roughly twice that of carbohydrates, because removing the oxygen reduces the mass by about a factor of 2 without reducing the fuel value much. Plant fats occur almost exclusively in plant seeds (which is why cooking oils come from plant seeds, not plant leaves or stems). Plants do this because seeds need a high stored energy/weight ratio to germinate and get their leaves up to the sunlight. Animals convert carbohydrates to fats for storage in their bodies so that they can overeat in times of food abundance and live off the stored fat in times of scarcity. The carbohydrate-to-fat conversion reduces the weight of the stored food by a factor of 2.

When we burn fuels (or burn anything) we essentially reverse Eq. (C.1), producing CO_2 and H_2O, and releasing the stored solar energy as heat. There are a few other chemical elements that will burn (i.e., react with oxygen, liberating enough heat that the reaction is spontaneous, and often form a visible flame), for example, sulfur, phosphorus, iron (at the temperature of an oxyacetylene flame), magnesium (used in pyrotechnics), and aluminum. In the case of phosphorus, magnesium, and aluminum, the combustion is really releasing stored solar energy as well, because those materials do not occur naturally on earth but are produced in industrial processes with large inputs of electricity, most of which is ultimately based on solar energy (either current solar energy in hydroelectric plants or stored solar energy in fuels).

C.2 NATURAL GAS

Natural gas is principally methane, CH_4. Many natural gas wells produce mixtures of methane with CO_2, H_2S, N_2, and the higher molecular weight hydrocarbons ethane, propane, etc. CO_2, H_2S, and N_2 are removed and disposed of because they are unwanted contaminants; the ethane, propane, and higher molecular weight hydrocarbons are removed because they command a higher price as separate components then their value as a component of natural gas. Methane is a common product of anaerobic bacterial decomposition of organic matter. It occurs in the gases from landfills and sewage treatment plants. But the majority of the world's natural gas occurs in deposits associated with or related to petroleum deposits, which (as discussed below) are believed to be derived from the decomposition of the remains of microscopic animals.

Natural gas (methane) is a very fine fuel, easy to transport on land and distribute to stationary sources, and easy to burn very cleanly in simple, safe, cheap burners. It is transported long distances in pipelines up to several feet in diameter, at pressures of typically about 60–100 atm. Inside cities the pressure is reduced to about 2 atm (gauge) for distribution in underground pipes to customers. At the customer's meter, the pressure is further reduced to about 0.01 atm gauge (4 inches of H_2O) for distribution to the appliances in the building. But natural gas is hard to store because of its large volume. Where suitable geologic structures exist (e.g., depleted oil or gas fields) it can be injected, stored, and then withdrawn as needed. Where such suitable geologic structures do not exist, the gas is sometimes stored in huge, ugly gas holders, or liquefied and stored as a liquid.

The storage problem has made natural gas an unattractive fuel for vehicles. Liquid fuels are easier to store and easier to refuel in the customer's vehicle. The natural gas industry has developed high-pressure tanks and filling systems for vehicles and found some sales to taxis, delivery trucks, and buses, which can all be fueled at a central point and whose owners will accept a 5 minute tank-refilling time. But to date natural gas has not found wide acceptance as a fuel for private autos.

Methane is odorless. Commercial natural gas is odorized with a sulfur compound (normally a mercaptan) to warn users of leaks. If we did not do this, a burner left on in the kitchen could lead to a disastrous explosion.

C.3 LIQUID PETROLEUM GAS, PROPANE, AND BUTANE

Methane is the first member of the *aliphatic* family of hydrocarbons (also called the *paraffin* family) whose general formula is C_nH_{n+2}. For methane, $n = 1$. Table C.1 shows some properties of the first few members of this family.

We see that methane cannot exist as a liquid at room temperature (68°F = 20°C) because that is above its critical temperature. All of the others in Table C.1 can exist as liquids at this temperature, although some must be held at high pressure to exist as a liquid. Ethane, propane, and *n*-butane have boiling points below room temperature, so they can only exist as liquids at this temperature if held at pressures above atmospheric.

Liquefied petroleum gas, or LPG, or "Propane" is widely used as a fuel. Commercial propane is ≈ 90+% propane, with some dissolved ethane or butane and an odorant added for safety. It is distributed as a liquid from tank trucks and stored in tanks with design pressures of 15–25 atm. In urban areas it is generally economical to serve each household with natural gas by underground pipelines. In rural areas with a more dispersed population, it is generally more economical for each residence to have a propane tank, which is refilled by the tank truck of a propane distributor. Generally residents pay about twice as much for their propane as do natural gas customers, so rural areas try to get the natural gas companies to serve them; the natural gas companies will only make the large investments in underground pipelines if there is a large enough customer base to pay off that investment.

In addition, propane is widely used as a barbecue and recreational vehicle fuel, and as the fuel for some motor vehicles. It is the standard fuel for fork-lift trucks that operate inside warehouses and factories, because engines burning it generally emit fewer air pollutants than do gasoline-powered engines. It is widely promoted as a motor vehicle fuel because emissions from burning it are less than from gasoline. Ordinary gasoline engines can be adapted to burn it at modest cost and inconvenience.

TABLE C.1
The low molecular weight aliphatic (paraffinic) hydrocarbons

n	Name	Formula	Atmospheric boiling point, °F	Critical temperature, °F
1	Methane	CH_4	−258.94	−116.92
2	Ethane	C_2H_6	−127.9	89.72
3	Propane	C_3H_8	−44.02	205.64
4	*n*-Butane	C_4H_{10}	30.86	305.36
5	*n*-Pentane	C_5H_{12}	96.56	385.28
6	*n*-Hexane	C_6H_{14}	155.42	453.32

LPG has the drawback that refilling the tank is more complex than filling a gas tank; dealers generally will not let customers refill their own tanks, for safety reasons. In warm climates where residences do not have space heating, it is uneconomic to install natural gas pipelines, so that propane, distributed in $\approx$ 5-gallon containers, is the common cooking fuel.

Butane, the next largest aliphatic hydrocarbon, has a low-enough vapor pressure at room temperature that it is widely used as a fuel in throw-away cigarette lighters. It is also used in warm climates the same way as propane is used in cold climates (where an outdoor butane tank might not supply enough fuel pressure on the coldest day of the winter, but a propane tank will).

C.4 LIQUID FUELS

Liquid fuels, mostly based on petroleum, are used in most autos, trucks, and buses in the world. Their ease of storage and transfer allows for self-service gasoline stations and simple low-pressure tanks in the vehicles, which natural gas and LPG do not. Petroleum (oil, "earth oil") is a mixture of many types of hydrocarbons (discussed below). It is believed to have been formed by the decomposition–chemical transformation of the fatty parts of the bodies of countless numbers of very small to microscopic aquatic animals, trapped in sediments and buried deeply enough for the temperature to be high enough for such chemical transformations to occur over geologic time.

The principal petroleum fuels are LPG (discussed above), gasoline, jet fuel, diesel fuel, heating oil, and residual oil. These are listed in order of increasing average molecular weight. This means that they are listed in order of decreasing vapor pressure (or increasing normal boiling point) and increasing viscosity. Heating oil and residual oils can be solids at low temperatures. These fuels are mixtures, with typical properties shown in Table C.2.

This table shows that the boiling point of LPG is practically a single number, because it is practically pure propane, but that for all the others there is a range of boiling temperatures and a range of molecular weights. The boiling point ranges and molecular weights overlap; a C_{11} molecule would be quite at home in gasoline or in kerosine and jet fuel.

TABLE C.2
Liquid petroleum products

Product	Atmospheric boiling temperature range, °C	Carbon number range
LPG (commercial propane)	-42 ± 2	2 to 4, 90+% C_3
Gasoline	30 to 210	5 to 12
Kerosine and jet fuel	150 to 250	11 to 13
Diesel fuel and light fuel oil	160 to 400	13 to 17
Heavy fuel oil	315 to 540	20 to 45
Residual fuel oil	450+	30+

Although methane, ethane, and propane each comes in only one variety, all higher molecular weight hydrocarbons come in a variety of forms. Butane (C_4H_{10}) exists as normal butane (*n*-butane) in which the four carbon atoms are arranged in a straight chain, and isobutane (*i*-butane) in which three of the carbon atoms are all joined to one central carbon atom. For any hydrocarbon with five or more carbon atoms there are more than two possible arrangements. These *branched aliphatics* (or *branched paraffins* or *branched chain aliphatics* or *isoparaffins* or *isoaliphatics*) make better gasoline and better lubricating oils than the *straight-chain paraffins,* so they command a higher price.

In addition to aliphatics, petroleum contains two other broad classes of hydrocarbon, *naphthenes* or *cycloaliphatics,* in which a straight-chain aliphatic has been formed into a loop or ring (with the loss of two hydrogen atoms), and *aromatics,* in which such rings have lost half of their hydrogens. So, for example, *n*-hexane, C_6H_{14}, can form a ring, cyclohexane, C_6H_{12}, which is a naphthene and has different properties from *n*-hexane. In turn cyclohexane can lose six hydrogen atoms to form an *aromatic* compound, benzene, C_6H_6. All three of these burn satisfactorily, but their detailed burning properties are different from one another, leading to different antiknock properties in auto engines and different difficulties in combustion. They also differ in toxicity.

The lower molecular weight liquid fuels are practically pure hydrocarbon (compounds of carbon and hydrogen). As discussed above, strong-smelling sulfur compounds are added to them in very small quantities for safety. Motor fuels contain additives to improve their antiknock properties, keep the engine parts clean, and serve some other purposes; these additives contain atoms other than carbon and hydrogen, but they are present in low concentrations. The higher the molecular weight, the more likely a liquid fuel is to contain substantial amounts of sulfur, nitrogen, oxygen, or ash. These are built into the molecules and can only be removed by breaking chemical bonds. Most diesel oil sold in the United States is chemically treated in the refining process to reduce its sulfur content for air pollution reasons. Heavy fuel oil and residual fuel oil often contain enough sulfur and ash that their combustion requires air pollution control equipment to capture that ash or sulfur.

C.5 SOLID FUELS

Almost all solid fuels are derived from the bodies of plants. Although wood waste, used rubber tires, sugar cane stems ("bagasse"), and municipal garbage are all burned for fuel (or for waste disposal), the principal solid fuel in the world is coal or its variants peat and lignite. Coal is the fuel for about 52 percent of the electricity generated in the United States. It is by far our most abundant fossil fuel. Much of the history of air pollution control is the history of controlling emissions from coal combustion. Table 1.1 shows that fuel combustion (mostly coal combustion) is the principal source of SO_2 and NO_x emissions in the United States and a significant source of the other major pollutants. Many examples in this book deal with coal combustion.

There is no such thing as a "typical coal." Rather there are several major kinds of coal. For each of these kinds we can define a "typical coal of that type." We can average among types and thus define an "average coal," but that average will not represent any particular coal very well. (The "average human being" is 50 percent male and 50 percent female; there are no "typical humans" with that description.)

Coal is derived from wood and woody materials. Table C.3 on page 568 shows the chemical changes that correspond to the transformation of wood → peat → lignite → subbituminous coal → bituminous coal → anthracite coal. Each of these steps shows a decrease in oxygen and hydrogen content with a consequent increase in the remaining percentage of carbon. Thus the maturation process of coal, which takes place over millions of years deep underground at significant temperatures and pressures, is mostly a process of removal of oxygen and hydrogen. The names on the left of the table, starting with lignite, correspond to the "ranks" of coal, which have technical definitions; the rank generally increases as the percent carbon increases.

The table also shows that the typical "as delivered" heating value is substantially less than the typical dry heating value. The difference is attributable to the moisture content of the coal, as mined. Comparing the two heating values, one would conclude that the dry weight of subbituminous coal was ($\approx 8741/13\ 606) = 64\%$ of the wet weight. This value is typical of that kind of coal.

Wood typically contains less than 0.1 percent S; coal, derived from wood, is typically 0.5 percent to 3 percent S. The sulfur in the coal did not come from the wood. Instead, it came from the reaction between the buried wood and groundwater percolating through it, containing small amounts of dissolved sulfates (mostly calcium sulfate). The buried wood is a reducing agent, capable of the overall reaction

$$\begin{gathered}\text{Sulfates} + \text{any reducing agent} + \text{bacteria} \rightarrow \\ \text{water} + \text{organically bound sulfur}\end{gathered} \tag{C.2}$$

Coals formed from woody material deposited in saltwater environments have higher sulfur contents than those formed from woody material deposited in freshwater environments, presumably because of Reaction (C.2) and the higher amount of sulfates in salt water. Similarly the ash in coal did not come from the wood, but rather was deposited (and/or chemically bound) in the coal from circulating groundwater solutions and/or from mineral matter incorporated with wood during its original burial process.

The nitrogen in coals is also not from the woody material but is thought to be derived from the bodies of nitrogen-fixing bacteria that participated in the early stages of the conversion of the woody material to coal.

Coal contains trace amounts of almost all the elements of the periodic table. As the coal burns, some of these are released in gaseous form (e.g., Hg, HCl), and some are trapped in the ash.

Sixty percent of U.S. coal is produced in surface mines (strip mines), in which the overlying rock is stripped away and then the coal excavated. The remaining 40 percent is produced in underground mines (deep mines), in which the coal seam is mined out from between the underlying and overlying rock.

TABLE C.3
Some values for fuels derived from wood

Material	Ultimate analysis, weight %, dry basis, typical values						Heating value, Btu/lb[g]		% by weight of U.S. electric generation coal, 1997[a]
	C	H	O	N	S	Ash	Dry basis[h]	Average wet basis, as delivered, U.S., 1991[a]	
Wood[b]	52.3	6.3	40.5	0.1	< 0.1	0.8	9050		
Peat[c]	57.0	5.5	31.0	1.5	0.2	4.8	9300		
Lignite (also called *brown coal*)[d]	55.0	4.4	13	1.0	1.7	24.9	9727	6372	8.4
Subbituminous coal[e]	72.5	6.1	17.2	0.7	0.4	3.1	13 006	8741	40.3
Bituminous coal[f]	75.8	5.0	7.4	1.5	1.6	8.7	13 600	11 964	51.2
Anthracite coal	82.1	2.3	2.0	0.8	0.6	12.2	13 258		0.1
U.S. average coal for electric generation, 1997[a]					1.11	9.36		10 387	100.0

Notes:

[a] The values for average U.S. coal, average heating value, and % by weight for electric generation are from "Cost and Quality of Fuels for Electric Utility Plants, 1997 Tables," available only as an Internet document at ⟨**www.eia.doe.gov**⟩.

[b] The wood values are those for Douglas fir; other kinds of wood are similar but not identical. These values are from D. A. Tillman, A. J. Rossi, and W. D. Kitto, *Wood Combustion: Principles, Processes and Economics*, Academic Press, New York, p. 43, 1981.

[c] The peat values are from B. F. Haanel, *Final Report of the Peat Committee*, A. C. Acland, Ottawa, p. 7, 1925.

[d] The lignite values are for a typical Texas lignite, courtesy of the Texas Mining and Reclamation Association.

[e] The subbituminous values are for a typical Powder River Basin coal, courtesy of the Western Research Institute.

[f] The bituminous value is for a "typical Pittsburgh seam coal." Throughout this book in examples and problems the values for this coal are used, except when it is stated to the contrary.

[g] In the United States heating values of coal are generally stated as the higher heating value, as shown here. In Europe they are generally stated as the lower heating value. See the discussion of the difference in Section 7.10.2.

[h] The dry basis heating values are computed from the equation

$$\begin{pmatrix}\text{Higher heating}\\ \text{value, Btu/lb}\end{pmatrix} = 14\,544\text{C} + 62\,028\left(\text{H} - \frac{\text{O}}{8}\right) + 4050\text{S}$$

where C, H, O, and S are the weight fractions of carbon, hydrogen, oxygen, and sulfur, respectively. This formula is reported to give values within ± 3% of experimental results.

Table C.3 shows that subbituminous coal has a lower heating value than bituminous coal, but it also has a lower sulfur and ash content. Most of the subbituminous coal mined in the United States is mined in the Powder River Basin of eastern Wyoming and Montana (making these the first and eighth largest coal-producing states). The seven largest coal mines in the United States, which produce 14 to 31 million tons per year, are all located in Wyoming or Montana. This coal occurs in very thick seams (up to 100 ft thick) close to the surface. It is easy and cheap to surface-mine on a very large scale. This coal sells at the mine for about $4 per ton, compared with $20 to $30 per ton for most of the remaining coal in the United States, which is produced from much thinner seams (typically 4 to 10 ft thick).

Because coal is a low-cost material ($20/ton = 1¢/lb), shipping costs are important. The cheapest coal transport is by barge on major waterways, which costs about 0.4 to 0.5¢ per ton mile. Large unit trains, which transport only coal, nonstop, from the mine to a single consumer, cost about 1 to 2¢ per ton mile (1¢ for very long hauls, 2¢ for shorter ones), so that, for example, a $4/ton coal, delivered 1000 miles from the mine by rail, costs the customer about $14 to $24 per ton. Truck shipment costs about 5¢ per ton mile and is seldom used for distances longer than 100 miles. Transoceanic shipments in bulk carriers cost 0.1 to 0.2¢ per ton mile.

The low sulfur content of subbituminous coal from Wyoming and Montana, combined with its low price, has made it very attractive to utilities located within about 1500 miles of the mines, if their alternative coals are high in sulfur and if they are under regulatory pressure to reduce sulfur emissions. Large amounts of this coal come to the central United States.

Lignite is similar to subbituminous coal in having a very substantial moisture content and hence a low heating value per pound as mined. Texas and North Dakota, taken together, produce 93 percent of the U.S. lignite, almost all of which goes directly to mine-mouth power plants. Lignite is also a major electricity-producing fuel in Germany (called brown coal there) and some other countries.

To simplify calculations and examples, unless it is stated to the contrary, coals in examples and problems in this book will be assumed to be a "typical Pittsburgh seam coal," which is shown as "typical bituminous" in Table C.3, and has the following "*ultimate analysis*" by weight: carbon, 75.8%; hydrogen, 5.0%; oxygen, 7.4%; nitrogen, 1.5%; sulfur, 1.6%; ash, 8.7%; heating value, 13 600 Btu/lbm (dry).

C.6 COMPARING FUEL PRICES AND EMISSIONS

In comparing fuels the appropriate basis of comparison is some fixed amount of heat released; in the United States the most common unit is 10^6 Btu. One may make fair estimates of the heating values of all fuels by assuming that they are made of carbon and hydrogen plus inert materials, that the carbon provides $\approx$ 14 100 Btu/lb, and that the hydrogen provides 51 600 Btu/lb. These are lower heating values, see Sec. 7.10.2. The values computed this way overstate the heating values by a few percent because they do not take account of the heat of formation of the compounds from the elements.

Example C.1. Estimate the heating value of propane.
Propane, C_3H_8, has molecular weight of 44, so that by weight it is

$$\frac{3 \cdot 12}{3 \cdot 12 + 8 \cdot 1} = 0.818 \text{ weight fraction C, and}$$
$$\frac{8 \cdot 1}{3 \cdot 12 + 8 \cdot 1} = 0.182 \text{ weight fraction H}$$

and its estimated heating value is

$$0.818 \cdot 14\,100 + 0.182 \cdot 51\,600 = 20\,925 \frac{\text{Btu}}{\text{lb}}$$

The published value is 19 944 Btu/lb, which reminds us that this estimating procedure is only of fair quality (5% high in this case). ■

The different common units of measure of various fuels make it difficult to compare them. Table C.4 shows the basis for comparison. The prices shown are wholesale, before tax prices. (Yes, gasoline does sell for about 50¢/gal, in 42 000 gallon lots, before state and federal taxes!) This table shows that coal is much cheaper than liquid fuels, and that natural gas, in ≈ \$20 000 lots, sells for about half the price of liquid fuels.

Because of the concern with global climate change (see Chapter 14), we are concerned with the CO_2 (or C) emission per unit of fuel burned. If the fuel contains only C and H (or if the other constituent is inert like the ash in coal) and can be represented as C_xH_y (see Chapter 7) then the carbon emission per unit of heat release (LHV) is approximately

$$\left(\frac{\text{Carbon emission}}{\text{Heat release}}\right) = \frac{12x}{12 \cdot x \cdot 14\,100 + 1 \cdot y \cdot 51\,600} \frac{\text{lb}}{\text{Btu}} \tag{C.3}$$

and the values (in lb/10^6 Btu) for natural gas, propane, gasoline, and coal are approximately 32, 39, 43, and 54, respectively. This comparison explains the opposition to the use of coal as a fuel, based on the threat of global climate change.

TABLE C.4
Comparison of fuel prices

Fuel	Common unit	Pounds/unit	Btu/lb	1998 price	\$/10^6 Btu
Natural gas	1000 scf	41.6	21 500	\$2/1000 scf	2.2
Propane	Gallon	4.2	19 900	\$0.4/gal	4.8
Gasoline	Gallon	6	19 500	\$0.5/gal	4.3
Diesel fuel	Gallon	7	18 500	\$0.4/gal	3.1
Crude oil	Bbl (42 gal)	≈ 300	≈ 18 000	\$12/bbl	2.2
Coal	Ton	2000	13 600	\$20/ton	0.73

APPENDIX D

ELEMENTARY CHEMISTRY OF OZONE PRODUCTION

A complex set of simultaneous atmospheric reactions is summarized in Eq. (1.2):

$$NO + HC + O_2 + \text{sunlight} \rightarrow NO_2 + O_3 \tag{1.2}$$

This appendix presents an intermediate-level description of those reactions. More detailed descriptions are available in Refs. 1 and 2. Here HC stands for hydrocarbons; they are also called VOC (volatile organic compound, Chapter 10), ROG (reactive organic gas), NMHC (nonmethane hydrocarbon), and NMOG (nonmethane organic gas). The latter two names reflect the observation that methane is much less reactive in the atmosphere than are higher molecular weight hydrocarbons. For the rest of this appendix we will use the term VOC.

NO, NO_2, and O_3 interact in the atmosphere by the reactions

$$NO_2 + h\nu \rightarrow O + NO \tag{D.1}$$

where $h\nu$ represents a photon of light of proper wavelength,

$$O + O_2 + M \rightarrow O_3 + M \tag{D.2}$$

where M represents any other molecule (usually N_2 or O_2), which must carry away some of the energy released in the reaction if the O_3 is to be stable, and

$$NO + O_3 \rightarrow NO_2 + O_2 \tag{D.3}$$

NO_2 is decomposed by a light photon to produce NO and the oxygen radical O. That radical reacts with O_2 to form O_3. O_3 then reacts with NO to form NO_2 and releases an O_2 molecule. If one assumes that all three of these reactions have equal rates, which must be true at steady state, then one can solve for the O_3 concentration, finding

$$[O_3] = \frac{[h\nu][NO_2]}{k_{D.3}[NO]} \tag{D.4}$$

in which $[h\nu]$ is the solar intensity in appropriate units and $k_{D.3}$ is the rate constant for Reaction (D.3) [3]. In the absence of VOC this equation leads to an equilibrium between NO, NO_2, and O_3 that is dependent on the intensity of the solar radiation but is not strongly dependent on the amount of NO_x because only the ratio of the two kinds of NO_x appears. Every NO_2 molecule that is split produces both an NO and (indirectly) an O_3 molecule that can then react with NO to reverse the reaction.

The role of VOC is to convert NO to NO_2 without using up O_3, so that there are not enough NO molecules to react with all the O_3 molecules, and O_3 accumulates. One mechanism by which VOC does this is described by these reactions:

$$OH + VOC \rightarrow RO_2 + H_2O \tag{D.5}$$

$$RO_2 + NO \rightarrow NO_2 + RO \tag{D.6}$$

$$RO + O_2 \rightarrow RCHO + HO_2 \tag{D.7}$$

$$HO_2 + NO \rightarrow NO_2 + OH \tag{D.8}$$

Adding these four reactions and canceling like terms, we see that the overall reaction is

$$VOC + 2\,NO + O_2 \rightarrow H_2O + RCHO + 2\,NO_2 \tag{D.9}$$

Here R stands for any hydrocarbon; these reactions, as normally written, are not balanced for O.

NO_2 complicates this process because it can react with an OH radical to produce nitric acid,

$$OH + NO_2 \rightarrow HNO_3 \tag{D.10}$$

thus reducing the availability of OH radicals. This equation suggests that in some situations adding NO_2 to the atmosphere will reduce the amount of O_3 formed. The relationships are shown in Fig. D.1. At the lower right in the "NO_x-limited" region, the ozone concentration depends only on the amount of NO_x available. At the upper left, in the "VOC-limited" region, increasing the amount of NO_x lowers the calculated O_3 concentration because Reaction (D.10) reduces the supply of OH and hence the efficiency of use of the limited amount of VOC. (This treatment, which is a strong simplification of the true situation, follows that in Ref. 4.)

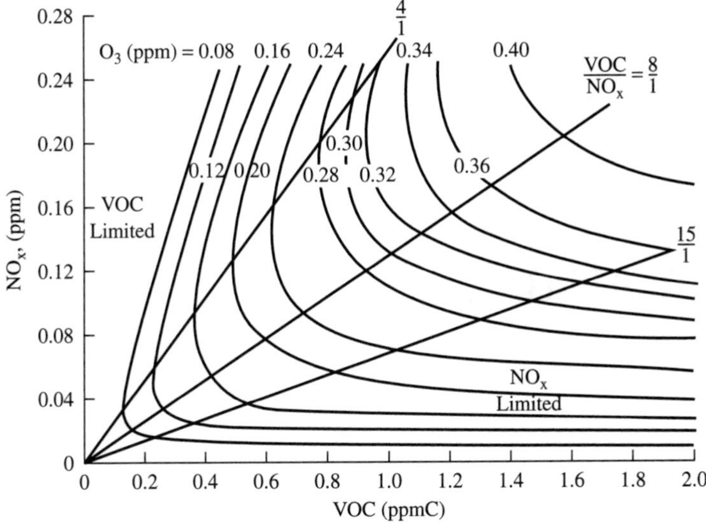

FIGURE D.1
Calculated maximum afternoon O_3 concentrations as a function of the morning NO and VOC concentrations for the same air mass [2]. The permitted concentration of O_3 (NAAQS, one hour maximum, not to be exceeded more than once per year) is 0.08 ppm. Most U.S. cities have VOC/NO_x ratios of 8 to 15 [2].

Although O_3 is the principal contributor (and only regulated component) of photochemical smog, aldehydes (RCHO) and peroxyacetylnitrate ($RC(O)OONO_2$) are also formed by reactions similar to those shown here and contribute to the eye and nasal irritation attributed to these smogs.

REFERENCES

1. Campbell, I. M.: *Energy and the Atmosphere: A Physical-Chemical Approach*, 2d ed., John Wiley & Sons, New York, 1986.
2. National Research Council: *Rethinking the Ozone Problem in Urban and Regional Air Pollution*, National Academy Press, Washington, DC, 1991.
3. Campbell, I. M.: op. cit., p. 211.
4. Blanchard, C. L.: "The Use of Photochemical Air Quality Models for Evaluating Emission Control Strategies," Electric Power Research Institute Report No. TR-101918, 1992.

APPENDIX E

ADSORBER BREAKTHROUGH TIME

Section 10.4.2 discusses adsorbers for VOC control, and it presents a simple example of the estimated breakthrough time for a fixed-bed adsorber. This appendix shows the derivation of the relations and values used in that example. This derivation follows Ref. 1, which is a simplified version of the original paper [2].

Figure E.1 shows the material balance boundaries for a section of the adsorbent bed with cross-sectional area A perpendicular to the flow and length in the flow direction of Δx. The mass of adsorbent in the bed is equal to $A\,\Delta x \rho_B$, where ρ_B = the *bulk density* of the adsorbent bed; and the mass of external gas in the bed is equal to $A\,\Delta x \rho_G F$, where ρ_G = the gas density and F = the external porosity (also called *external void fraction*) of the bed. Here we specify external porosity because the individual adsorbent particles have substantial internal porosity as well. There is also gas inside that internal pore space, which is ignored in this derivation.

The mass flow rate of gas into this section is $\dot{m} = AG(1 + Y_{A_{\text{in}}})$, where G = the mass velocity *of the inert parts* of the gas flow (= velocity · density), normally expressed in (mass/area · time). The flow of adsorbable component into the section is equal to AGY_A, where Y_A = the weight of adsorbable component (A) per unit weight of inert components, normally expressed as (mass/mass). (See Sec. 7.9.)

With these definitions we make a material balance for adsorbed component A in this section of the bed for time element Δt

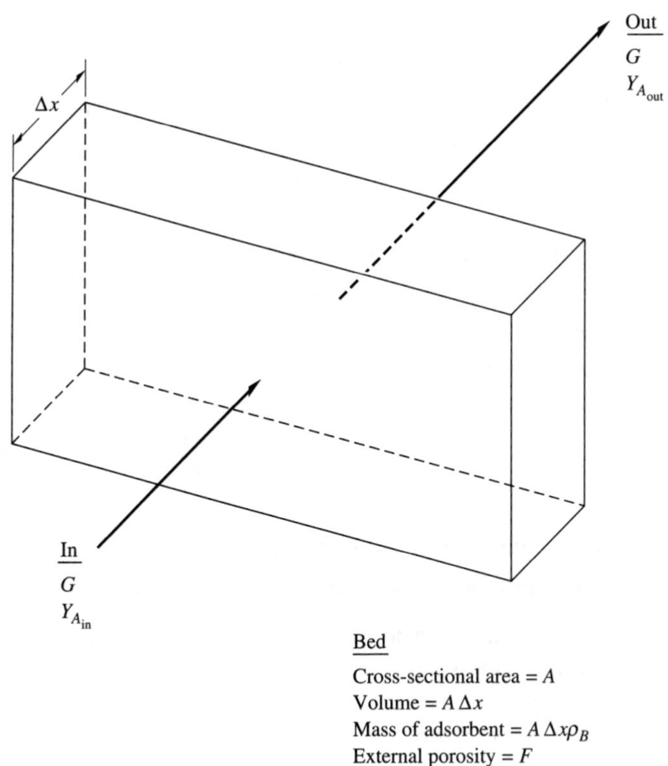

FIGURE E.1
Material balance region.

$$GA(Y_{A_{in}} - Y_{A_{out}})\Delta t = A\,\Delta x \cdot \Delta\,(\rho_B w + \rho_G F Y_A) \tag{E.1}$$

where w = the adsorbent's content of component A. The left term shows the amount of adsorbable component A that is removed from the gas in this section of the bed during time interval Δt, which is equal to the increase during this time interval of component A present in this section of the bed. Dividing both sides of Eq. (E.1) by $A\,\Delta x\,\Delta t$ and taking the limit as all the Δx and Δt approach zero, we find

$$-G\frac{dY_A}{dx} = \frac{d(\rho_B w + \rho_G F Y_A)}{dt} \tag{E.2}$$

The density of the gas is normally small enough that the $\rho_G F Y_A$ term on the right is dropped, to give

$$-G\frac{dY_A}{dx} = \rho_B\,\frac{dw}{dt} \tag{E.3}$$

The rate of transfer of component A from the gas to the solid is expressed as

$$r = k_G a_v(p_A - p_A^*) \tag{E.4}$$

where r = the mass transfer rate, normally in (mol/m^3· s)
k_G = the mass transfer coefficient, normally in (mol/m^2 · s · atm)
a_v = the external surface area of the adsorbent, normally in (m^2/m^3)
p_A = the partial pressure of component A in the gas, which for low pressures is practically equal to P, the total pressure, times y_A, the mol fraction of A
p_A^* = the partial pressure of component A in the gas that would be in equilibrium with solid adsorbent that had an A concentration of w

For pressures low enough for the perfect gas law to be applicable and for low concentrations of A, for which $P \gg p_A$,

$$Y_A = \frac{M_A y_A}{M_G(1 - y_A)} = \frac{M_A}{M_G} \cdot \frac{p_A}{P - p_A} \approx \frac{M_A}{M_G} \cdot \frac{p_A}{P} \tag{E.5}$$

where M_A = the molecular weight of the adsorbed component, A
M_G = the molecular weight of the inert part of the gas stream

If we know the equilibrium relationship between p_A^* and w, we can solve these equations together to find the unsteady-state adsorption behavior of an adsorbent bed. Example 10.11 shows how one estimates such a relation, and Fig. 10.12 shows a typical result. Using it and Eqs. (E.3)–(E.5), one can make such a calculation numerically [3]. Such computations are simple on large computers but have limited intuitive content.

The corresponding heat transfer problem is that of heating a bed of solids by passing a hot fluid through it. Such devices are widely used in industry, e.g., brickwork regenerators on blast furnaces and other similar devices [4]. One may convert the above adsorption equations to the equations for that heat transfer problem by replacing concentrations with temperatures and the mass transfer coefficient with a heat transfer coefficient, and by introducing appropriate heat capacities of gas and solid (see Problem E.1). In that case the equilibrium relationship is very simple; at equilibrium $T_{\text{gas}} = T_{\text{solid}}$. With that simplification the corresponding heat transfer problem has a well-known solution. Hougen and Marshall [2] showed that if we replace the true complex equilibrium ($p_A^* - w$) relation with a linear one,

$$p_A^* = cPw \tag{E.6}$$

where c = a dimensionless data-fitting constant, then the known solution to the heat transfer problem could be used for adsorbers. On Fig. 10.12, Eq. (E.6) would form a straight line through the origin, with slope c. That is not a good representation of the whole curve on Fig. 10.12, but if one draws it through the value of the equilibrium curve at the concentration of interest, then it is close to correct near that concentration.

To fit the adsorber problem into the mold of the heat transfer solution, Hougen and Marshall defined the quantities a and b by

$$a = \frac{a_v k_G P M_G}{G} \tag{E.7}$$

$$b = \frac{Gca}{\rho_B}\frac{M_A}{M_G} \tag{E.8}$$

With these substitutions, and the definition $N = ax$, adsorption problems can be solved on the corresponding heat transfer plot, which is shown as Fig. 10.14.

Example 10.13, continued. Let us compute the values of a, b, and N in Example 10.13, using the values in Examples 10.11 and 10.12, plus additional information from Ref. 1. Here the values of particle diameter and bulk density correspond to 4×6 mesh activated carbon in Table II of Ref. 1. That same table shows that $a_v = 310\ \text{ft}^2/\text{ft}^3$. From Fig. 10.12, for $Py_{\text{toluene}} = 0.005$ atm we read $w^* = 0.29$. Then from Eq. (E.6),

$$c = \frac{p_{\text{toluene}}}{Pw^*} = \frac{0.005\ \text{atm}}{1\ \text{atm} \cdot 0.29} = 1.72 \times 10^{-2}$$

The mass velocity is given by

$$G = \frac{\dot{m}}{A} = \frac{2.59\ \dfrac{\text{lbmol}}{\text{min}} \cdot 29\ \dfrac{\text{lb}}{\text{lbmol}}}{16\ \text{ft}^2} = 4.69\ \frac{\text{lb}}{\text{min} \cdot \text{ft}^2} = 0.382\ \frac{\text{kg}}{\text{m}^2 \cdot \text{s}}$$

The mass transfer coefficient is estimated by

$$k_G = \frac{j_d G}{P M_G (Sc_G)^{2/3}} \tag{E.9}$$

where Sc_G = Schmidt number of the gas $= \mu_G/\rho_G \mathcal{D}_G$

$\mathcal{D}_G$ = diffusivity of A in the gas

j_D = Chilton-Colburn analogy factor for mass transfer (see below)

For air at 100°F, we have $\mu_G = 1.85 \times 10^{-5}$ Pa · s and $\rho_G = 1.13\ \text{kg/m}^3$, and for toluene diffusing in air, $\mathcal{D}_G = 0.88 \times 10^{-5}\ \text{m}^2/\text{s}$ so that

$$Sc = \frac{1.85 \times 10^{-5}\ \text{Pa} \cdot \text{s}}{1.13\ \dfrac{\text{kg}}{\text{m}^3} \cdot 0.88 \times 10^{-5}\ \dfrac{\text{m}^2}{\text{s}}} = 1.86$$

Figure 9 of Ref. 1 shows a graphical correlation of experimental values of j_D as a function of the particle Reynolds number and the size of the particles of the adsorbent. That plot can be satisfactorily approximated for $0.0027\ \text{ft} \le D \le 0.0128$ ft and $10 \le \mathcal{R}_p \le 1000$ by

$$j_D = 0.24\left(\frac{D}{0.0128\ \text{ft}}\right)^{0.65} (\mathcal{R}_p)^{-0.3} \tag{E.10}$$

Here

$$\begin{pmatrix}\text{Particle}\\ \text{Reynolds}\\ \text{number}\end{pmatrix} = \mathcal{R}_p = \frac{D_{\text{part}}G}{\mu_G} = \frac{0.0128\ \text{ft} \cdot 0.382\ \dfrac{\text{kg}}{\text{m}^2 \cdot \text{s}}}{1.85 \times 10^{-5}\ \text{Pa} \cdot \text{s}} \cdot \frac{\text{m}}{3.28\ \text{ft}} = 69.2$$

for which (with $D = 0.0128$ ft) Eq. (E.10) gives $j_D = 0.07$. Thus we compute

$$k_G = \frac{0.07 \cdot 0.382\ \dfrac{\text{kg}}{\text{m}^2 \cdot \text{s}}}{1\ \text{atm} \cdot 0.029\ \dfrac{\text{kg}}{\text{mol}} \cdot (1.86)^{2/3}} = 0.612\ \frac{\text{mol}}{\text{m}^2 \cdot \text{s} \cdot \text{atm}} = 0.451\ \frac{\text{lbmol}}{\text{ft}^2 \cdot \text{hr} \cdot \text{atm}}$$

$$a = \frac{310\ \dfrac{\text{ft}^2}{\text{ft}^3} \cdot 0.612\ \dfrac{\text{mol}}{\text{m}^2 \cdot \text{s} \cdot \text{atm}} \cdot 1\ \text{atm} \cdot 0.029\ \dfrac{\text{kg}}{\text{mol}}}{0.382\ \dfrac{\text{kg}}{\text{m}^2 \cdot \text{s}}} = 14.4/\text{ft}$$

$$b = \frac{4.69\ \dfrac{\text{lb}}{\text{ft}^2 \cdot \text{min}} \cdot 1.72 \times 10^{-2} \cdot 14.4/\text{ft} \cdot 0.092\ \dfrac{\text{kg}}{\text{mol}}}{30\ \dfrac{\text{lb}}{\text{ft}^3} \cdot 0.029\ \dfrac{\text{kg}}{\text{mol}}} = 0.123/\text{min} = 7.4/\text{h} \quad \blacksquare$$

These are the values used in Example 10.13.

PROBLEMS

See Common Values for Problems and Examples, inside the back cover.

E.1. Show the heat transfer analogs of Eqs. (E.3) through (E.6).

E.2. What would happen in Example 10.13 if the mass transfer coefficient became infinite? If it became zero? (*Hint:* to avoid indeterminate forms, try mass transfer coefficients 1000 and 0.001 times as large as those in the example, and then generalize to infinity and zero.)

E.3. In Example 10.13, at what time does the outlet value $Y_{A_{\text{out}}} = 0.1 Y_{A_{\text{in}}}$?

E.4. In Example 10.13, at breakthrough (as defined in that example), what is the concentration of toluene in the gas at the point halfway from inlet to outlet of the bed?

E.5. In Example 10.13, we used the value of c corresponding to the inlet concentration. One can argue that we should have used the value of c corresponding to the outlet concentration. Calculate the appropriate value of c for the outlet concentration, and then show how using that value would change the answer to Example 10.13.

E.6. In Example 10.13, we have decided to change the bed design from a 4-ft cube to a thin bed 1 ft long in the flow direction and 8 × 8 ft in cross section. This will reduce the gas velocity by a factor of 4.

(*a*) Compare the calculated breakthrough time for this revised design to that calculated in Example 10.13. Does this result make physical sense?

(*b*) Estimate the pressure drop through this revised design (see Problem 10.25).

REFERENCES

1. Fair, J. R.: "Sorption Processes for Gas Separation," *Chem. Eng.*, Vol. 76, Issue 15, pp. 90–110, July 14, 1969.
2. Hougen, O. A., and W. R. Marshall: "Adsorption from a Fluid Stream Flowing through a Stationary Granular Bed," *Chem. Eng. Prog.,* Vol. 43, pp. 197–208, 1947.
3. Noll, K. E., V. Gounaris, and W.-S. Hou: *Adsorption Technology for Air and Water Pollution Control,* Lewis, Chelsea, MI, Chapter 9, 1992.
4. Corey, R. C.: "Heat Regeneration," in "Energy Utilization, Conversion and Resource Conservation," in D. W. Green and J. O. Maloney (eds.), *Perry's Chemical Engineers' Handbook,* 6th ed., McGraw-Hill, New York, pp. 9-89 to 9-92, 1984.

APPENDIX F

ANSWERS TO SELECTED PROBLEMS

1.1. (*a*) ≈ 0.02 psi, (*b*) 11 000 MW

2.2. (*a*) $25 \cdot 10^{-9}$ g, (*b*) $2.4 \cdot 10^4$
2.8. (*a*) 24 km, (*b*) 7.45 μg/m^3, (*c*) −1.63%, −14.3%

3.1. 0.225%
3.2. (*a*) 99.53%, (*c*) 4.08% ash

4.1. $4.3 \cdot 10^{-6}$ lb/ft^3
4.4. 0.473 g/m^3
4.8. 626.78 μg/m^3
4.11. (*a*) 0.25 μg
4.15. 62.4 lb/ton

5.1. (*a*) 16.67 ft, (*b*) 0.0075 inch
5.4. (*b*) 1 part in 365

6.2. 2.5 mg/m^3
6.6. ≈ 640 μg/m^3
6.7. 309 m
6.8. ≈ 0.55 km
6.11. (*a*) 566 μg/m^3, (*b*) 356 μg/m^3
6.12. (*a*) 661 μg/m^3, (*b*) 1.85 km
6.18. 910 m, etc.
6.20. At 1 km, 64 μg/m^3
6.27. $5 \cdot 10^{-4}$ g/m^{2}s
6.31. 861 m
6.33. ≈ 968 cm^2/s

7.1. 2.4%
7.2. 99.66%
7.4. $n = 6$
7.8. $\approx 12\%$
7.19. 6266 scfm, 10 200 acfm
7.24. (*a*) 103°F, (*b*) 147°F
7.25. −4.7%

8.2. For 1 μ, $A = 29\ 300$ ft^2 $= 0.7$ acre
8.4. 46 μ

8.8. (*a*) 16%, (*b*) 229 g/m^3, (*c*) $6.2 \cdot 10^9$
8.11. 0.0356 cm/s
8.16. (*a*) 0.012 cm/s, (*b*) 0.014 cm/s
8.18. (*a*) ∞, (*b*) $3.2 \cdot 10^{-5}$ s
8.22. (*a*) 2.3%, (*b*) 65%
8.26. 4.0 μ, 1.1 μ
8.30. 33.3%
8.32. 66.7%
8.35. (*a*) 34%, (*b*) 5.1 μ

9.10. (*a*) 3.28 μ, (*b*) 8.5%
9.15. 7.2 μ
9.17. 45%
9.23. $V_{\text{in}} = (8C_2C_3/KC_1\rho_G)^{1/3}$
9.27. $\eta = 0.78 = 78\%$
9.28. (*a*) Six, (*b*) either way, it makes no difference
9.30. $3.4 \cdot 10^3$ ft^2
9.34. 99.07%
9.45. (*a*) $3.0 \cdot 10^5$ ft^2, (*b*) 2388 bags
9.51. $N_S = 1.88$, $\eta_t \approx 0.69$
9.53. 40%
9.55. (*a*) $\approx$ 7.5 μ, (*b*) 1.003 μ
9.61. 7.1 drops/ft^3
9.67. 30%

10.1. (*b*) 0.08865 psia
10.2. (*a*) $2.06 \cdot 10^{-6}$ atm, (*b*) 273.15
10.5. (*a*) 1205 years, (*b*) 61.5 s, (*c*) 56.6 s
10.10. 47%
10.14. 0.00015 inch
10.28. $5.2 \cdot 10^{-6}$ mol fraction
10.30. 0.9%
10.36. 1.58 s
10.39. 1342°F
10.44. 0.5 s
10.46. 78%
10.51. 0.006 g/cm$^2 \cdot$ s

11.4. $Q_L/Q_G = 0.544$
11.8. 163 ppm
11.10. 95.8% conversion
11.12. (*a*) 85%, (*b*) 86%
11.15. (*a*) $60/ton, (*b*) $15/ton
11.19. Practically zero
11.22. $\approx$ 1.0%
11.31. $0.55/ton of coal
11.55. 0.37 lb SO_2/yr

12.1. 0.713 lb/10^6 Btu
12.12. 2.47 s
12.20. $\approx$ 9%
12.23. (*a*) 62.5 ppm, (*b*) 63.9 ppm
12.25. $\approx$ 9%

13.4. (*a*) 15%, (*b*) 14.4 lb/lb
13.8. ≈ -40°F
13.12. (*a*) 1486°F
13.24. $-1.2 \cdot 10^5$ K/s

14.6. 4.7 inches
14.10. (*a*) 0.070 g/L, (*b*) 0.074 g/L
14.12. 14.6 tonnes/yr

15.2. (*c*) 0.23%
15.6. 2136°F, 2581°F
15.17. about $45/month
15.18. (*a*) 17 500 atoms, (*b*) $2.8 \cdot 10^{-18}$
15.19. (*a*) 20 ft^3/hr, (*b*) 111 ft^3/min

INDEX

COMMON UNITS AND VALUES FOR PROBLEMS AND EXAMPLES

For all problems and examples in this book, unless it is stated explicitly to the contrary, assume the following:

The acceleration of gravity is $g = 32.17\ \text{ft/s}^2 = 9.81\ \text{m/s}^2$.

The surrounding atmospheric pressure is the "standard atmospheric pressure," $P_{\text{surroundings}} = P_{\text{atmospheric}} = 1$ atm $= 14.696$ (≈ 14.7) lbf/in.^{2} $= 33.89$ ft of water $=$ 10.33 m of water $=$ 29.92 in. of mercury $=$ 760 mm of mercury $=$ 760 torr $=$ 101.3 kilopascal $=$ 1.013 bar $=$ 1.033 kgf/cm^2.

If the fluid in the problem or example is water, then it is water at 1 atm pressure and 20°C = 68°F, for which

$$\rho = 62.3\ \text{lbm/ft}^3 = 998.2\ \text{kg/m}^3$$

$$\mu = 1.002\ \text{centipoise} = 1.002 \times 10^{-3}\ \text{Pa} \cdot \text{s} = 1.002 \times 10^{-3}\ \text{kg/m} \cdot \text{s}$$

$$= 6.73 \times 10^{-4}\ \text{lbm/ft} \cdot \text{s} = 2.09 \times 10^{-5}\ \text{lbf} \cdot \text{s/ft}^2$$

$$\nu = \mu/\rho = 1.004 \times 10^{-6}\ \text{m}^2/\text{s} = 1.004\ \text{centistoke} = 1.077 \times 10^{-5}\ \text{ft}^2/\text{s}$$

If the fluid in the problem or example is air, then it is air at 1 atm pressure and 20°C = 68°F = 528°R = 293.15 K, for which

$$\rho = 0.075\ \text{lbm/ft}^3 = 1.20\ \text{kg/m}^3 = 2.59 \times 10^{-3}\ \text{lbmol/ft}^3 = 41.6\ \text{mol/m}^3$$

$$\mu = 0.018\ \text{centipoise} = 1.8 \times 10^{-5}\ \text{Pa} \cdot \text{s} = 1.8 \times 10^{-5}\ \text{kg/m} \cdot \text{s}$$

$$= 1.21 \times 10^{-5}\ \text{lbm/ft} \cdot \text{s} = 3.76 \times 10^{-7}\ \text{lbf} \cdot \text{s/ft}^2$$

$$\nu = \mu/\rho = 1.613 \times 10^{-4}\ \text{ft}^2/\text{s} = 1.488 \times 10^{-5}\ \text{m}^2/\text{s}$$

$$C_P = 3.5\ R = 6.95\ \text{Btu/lbmol} \cdot {}^\circ\text{R} = 6.95\ \text{cal/mol} \cdot \text{K}$$

Any unspecified gas will be assumed to have the properties of air at 1 atm and 20°C shown above.

Any unspecified particle will be assumed to have a specific gravity of 2.00, hence a density of 2000 kg/m^3 = 124.8 lbm/ft^3.

Air is assumed to be a perfect gas with $M = 28.964 \approx 29$ g/mol, with a chemical composition, on a dry basis, expressed as mol fraction, of 78.08% N_2, 20.95% O_2, 0.93% Ar, 0.03% CO_2, all others less than 0.01%. Because argon behaves in most situations, e.g., combustion, the same way as nitrogen, the composition is normally simplified and rounded to 79% N_2, 21% O_2. Although the values are normally given on a dry basis, it is often important to know the water content of the atmosphere. At 20°C and a relative humidity of 50%, water is 0.0116 mol water/mol air = 0.0072 lb water/lb air.